A HISTORY OF TECHNOLOGY
&
INVENTION

Progress Through the Ages

VOLUME II

Contributors to This Work

Maurice AUDIN, master printer, curator of the Lyon Museum of Printing Part Two, Section Seven, Chapter 24

Maurice DAUMAS, Curator, Museum of the National Conservatory of Arts and Crafts, Introduction, Part Two, Section One, Chapter 10, and Section Two, Chapter 12

Marguerite DUBUISSON, Curator, Knitting Museum, Volume I, Part Two, Chapter 11: The Textile Industry

Daniel FAUCHER, Honorary Dean, Faculty of Letters, Toulouse, corresponding member of the Institute, Part Two, Section One, Chapter 9

André GARANGER, I.C.F., former director of the Syndicate of French Builders of Machine-tools, Section Two, Chapter 2 (collaborator)

Bertrand GILLE, professor, Faculty of Letters and Humanities, Clermont-Ferrand, Part Two

Paul GILLE, Chief Engineer of Naval Construction (E.R.), Part Two, Sections Three, Four, Five, and Six

Hubert LANDAIS, Chief Curator, National Museums, Part Two, Section 7, Chapter 23

Armand MACHABEY, doctor, University of Paris, Chief of Administrative Studies and Documentation of the Service of Measuring Instruments, Ministry of Industry, Part Two, Section Two, Chapter 14

Pierre MESNAGE, director, Advanced National School of Chronometry and Mechanics, University of Besançon, Part Two, Section Two, Chapter 13

Melvin KRANZBERG, Professor of History, Case Western Reserve University, and Editor in Chief of *Technology and Culture,* the bibliographies specially prepared for this edition with the collaboration of the Documentation Center of the History of Technology

Jacques PAYEN, palaeographer, Project Chief, Practical School of Higher Studies, Part One, Section One, Chapter 11 (collaborator); Index and tables

Jean PILISI, textile engineer, graduate of E.N.S.A.I.T. (Roubaix), editor-in-chief of *The Textile Industry,* Part Two, Section One, Chapter 11 (collaborator)

A HISTORY OF TECHNOLOGY
&
INVENTION

Progress Through the Ages

VOLUME II

The First Stages of Mechanization

EDITED AND WITH AN INTRODUCTION BY

MAURICE DAUMAS

TRANSLATED BY

EILEEN B. HENNESSY

CROWN PUBLISHERS, INC. NEW YORK

ORIGINALLY PUBLISHED AS *Histoire Generale des Techniques*,
UNDER THE DIRECTION OF MAURICE DAUMAS.

LIBRARY OF CONGRESS CATALOG CARD NUMBER: 71–93403
ISBN 0-517-507285
PRINTED IN THE UNITED STATES OF AMERICA
PUBLISHED SIMULTANEOUSLY IN CANADA BY GENERAL PUBLISHING COMPANY LIMITED
10 9 8 7 6 5 4 3 2

CONTENTS

PART ONE

THE FIFTEENTH AND SIXTEENTH CENTURIES IN THE WESTERN WORLD

By BERTRAND GILLE

PART TWO

THE MAJOR STAGES OF TRANSITION

Section One

The Exploitation of Raw Materials

Section Two

The Mechanical Arts

Section Three

Land and Water Transportation

By PAUL GILLE

Section Four

The Production of Power

By PAUL GILLE

Section Five

Military Techniques

By PAUL GILLE

Section Six

Construction and Building

By PAUL GILLE

Section Seven

Techniques of Expression

A HISTORY OF TECHNOLOGY & INVENTION

Progress Through the Ages

VOLUME II

INTRODUCTION

THE SECOND VOLUME of *Technology: A History of Progress Through the Ages* is devoted to a period that extends from the second half of the fifteenth century into the first third of the eighteenth century. The technological developments of this period are little known. While the notebooks of Leonardo da Vinci and the treatises dating from the beginning of the sixteenth century (Agricola's treatise is the one most frequently discussed) furnish a fairly complete picture of the methods at the disposal of the technicians of the Renaissance, it could seem that the progress of technology remained insignificant until that rather uncertain period in which the birth of the Industrial Revolution is usually situated.

This manner of interpreting the evolution of technology in the course of these two centuries (mid-sixteenth to mid-eighteenth) constitutes an error we have attempted to correct in this volume. This long period, which witnessed great historical revolutions of all kinds, could not be content to accept methods of production, transformation, and construction inherited from the Middle Ages and the Renaissance. Its contribution to technical progress, far from being negligible, is of a very special type for which no equivalent exists in any other period of the history of our civilization. We would be tempted to define it as a period of fundamental transition, if the term "transition" had not lost much of its significance through its application to periods of time for which no important event whatsoever can be claimed.

Such is not true during this period. Profound changes intervened in the mode of development of technology, and the overall picture of this history cannot be completely grasped without an exact idea of these transformations.

The rhythm of progress

To begin with, a modification of the rhythm of progress appeared for the first time in the history of technology. This rhythm had been constantly increasing in speed since the birth of civilization, but the historian has difficulty proving this fact because of the lack of elements for a precise chronology available to him. The great evolutionary phases that preceded the period we are discussing are not delimited by even approximately definite dates. At first the units of time were millennia, then centuries. At the end of the medieval period in the West, we begin to advance century by century, but very soon the half century, followed by the quarter century becomes the dividing line. In this way, for example, we are able to fix dates for the appearance of firearms, the mechanical clock, the treadle lathe, printing, and so on. In the second part of this volume the evolution of techniques is determined by decades, and the first inventions appear that can be dated without difficulty, for example, the inventions of precision optics and the steam engine.

The geographical evolution

Technical progress is not characterized exclusively by the acceleration of the rhythm to which it is subject; it is also characterized by the shifting of the geographical zones in which it is manifested. In the course of the three centuries covered by this volume, this latter characteristic appears as conspicuously as the former. The shifting of the geographical areas responsible for the progress of technology is not a new phenomenon. We recall that central Asia, Mesopotamia, and Egypt on the one hand, and the Far East on the other, witnessed the first stages of a creation that permitted man to dominate the environment in which he lived and to begin to modify it to improve the conditions of his existence. Southeastern Europe, Asia Minor, and gradually the area of the Mediterranean basin in turn were the scene of the development of the partially perfected elements of a technical civilization that slowly covered continental Europe, an area that played no particular innovative role during the first millennium of the Christian era.

During this millennium technology continued to be perfected in the Far East, particularly in China. In this part of the world we have already witnessed a certain recession of the wave of progress, since the great countries of central and southern Asia seem to have ceased to participate in it. Their populations clung to traditions inherited from the prehistoric ages, while technology actively participated in the flowering of the Chinese civilization.

Although these two geographical poles of technical invention were separated by vast sterile areas, there were exchanges of great importance between the Far East and the West. Our knowledge of the nature and process of these exchanges is not completely satisfactory, but we are able to say that thanks to these exchanges technology attained a comparable degree of development in both areas by about the end of the fourteenth century or the beginning of the fifteenth. This equilibrium was maintained until the beginning of the seventeenth century. In less than a century the balance was to be destroyed, in favor of the West, which itself was to witness the shift of the creative spirit from the south to the north and from east to west.

The retreat to western Europe

It is remarkable that technical progress was halted in China during the decadent period of the Ming Dynasty. The Chinese, to whom is attributed (perhaps correctly) the invention of gunpowder, were not able to make firearms; it was the Jesuit missionaries who cast the first cannons for the Chinese armies battling the Manchu attacks. Again, while they invented the principles of the production of paper pulp, in printing they did not progress beyond the stage of xylography. The same can be said for the mechanical and chemical arts, metalworking, and many other techniques that never progressed beyond the craft stage; printing, precision optics, the steam engine, and many other inventions remained in a primitive stage, when they were not totally unknown until imported by the Europeans.

Causes of cessation (if not the decline) of the development of the Chinese civilization are numerous and complex. They arise from the philosophical and religious conceptions of this race, its social structure, the demographic evolution, and the problem of the production of food, as well as from political and martial conflicts. Not only did the spirit of invention cease to manifest itself; in addition, the empire, which regained its strength under the Manchu Dynasty, proudly closed its doors to foreign penetration, thus turning away the ferments of creation that could in return have been contributed by the Westerners.

Technology thus ceased to evolve (and thereby to progress) in China, and the same was true of all the other elements of its civilization — art, literature, science, and philosophy, as well as economic and social activity, and this by virtue of the sclerosis that overtook all these phases of civilization.

In the same period the spirit of initiative and invention was concentrated in western Europe, which thus ensured a historical continuity in the progress of civilization. The privileged area shifted from the Mediterranean countries to those open to the ocean at the same time that it became concentrated in area. After a long period of initiation and establishment, during which a variety of races had to participate in the collective effort, the human race began to concentrate its technical development.

The truth is, however, that a more rational explanation must be substituted for these deterministic views. It was the shifting of the great commercial routes, as a result of the maritime discoveries, that caused this migration of the creative power.

The consequences of the discoveries

The opening of the sea routes toward the American continent and India beginning at the end of the fifteenth century shifted the centers of commercial activity. The appearance on the European market of new quantities of precious metals gave birth in Europe to a desire for the accumulation of wealth, thanks to the exploitation of hitherto unknown forms of commerce. Modern historians have designated these commercial policies by the general name of "mercantilism." These policies themselves encouraged the development of methods of production. Thus the spirit of invention in turn encouraged, and in answer to new needs the creations of technology themselves encountered, conditions favorable to their development.

The countries of western Europe did not all participate in the same manner and at the same time in this enterprise because mercantilism took different forms in each country. The Iberian Peninsula, which was the first to witness the influx of the precious metals and foodstuffs of the newly discovered countries, displayed practically no industrial qualities. The accumulation of money and gold was the immediate goal of its mercantilism, and was practiced in the most elementary fashion. The precious metals brought into the peninsula's ports by the ships were not used to create new wealth by providing an impetus for international trade and industry. Spain's resources of raw material, and especially her mineral wealth, were not negligible; for a long time they continued to supply part of European industry, as they had been doing since the end of the Roman Empire. But the extraction and transformation of ores benefited from no additional methods of exploitation, which the wealth won from the new continent would have made it possible to create for them. In the same way, the textile industry, and particularly the woolen industry, experienced no renewal of activity, while the exorbitant rights of the livestock breeders over the Spanish land ruined the country's agriculture.

The peninsula's trade flourished during the sixteenth century alone, during which period Spain and Portugal retained their mastery over the seas. The seventeenth century witnessed its decline, to the benefit of the pirates and the merchants of the Netherlands and Great Britain.

The activity of the Netherlands

The roles these two countries played in industrial growth during the seventeenth century are also quite different. The Dutch very quickly established themselves as the chief competitors of the Spanish and Portuguese on the routes to the Indies and America. The movement of international trade, concentrated at first in the port of Antwerp, and later in Amsterdam when Antwerp was reduced to ruins by the Spanish soldiers, determined the expansion of industry not only in Holland but also in Germany and the Scandinavian countries.

The Dutch ships reexported to the north the products they brought back from India and the West Indies, as well as the products manufactured in their own country. On the sea-lanes of the North Sea and the Baltic they built up an effective competition against the Hansa ships, which prior to this period distributed local products or products imported from southern Europe and the East over the continental routes. During this period the Dutch were exploring the far-northern route, and succeeded in linking Arkhangelsk with western Europe.

These vast enterprises of navigation were based on an increasingly powerful commercial organization crowned by the creation of the Bank of Amsterdam. More abundant supplies of raw materials were available to European industry, and even new materials, particularly metallurgical products from Sweden and Germany, potassium and chemical products from Silesia and Poland, wood and furs from the various regions reached by the merchants.

The mercantilism of the Dutch, shrewder and more enterprising than the Spanish variety, was not aimed at the accumulation of money and precious metals likely to become less valuable as the supply increased. It was by circulating these metals that the Dutch sought the means of increasing their wealth. Such an investment of capital could not fail to be favorable to industrial activity. Since the Netherlands possessed no special resources to place at the service of their industry, it remained an industry of transformation concerned chiefly with textiles, chemical products, and dyestuffs. The treatment and finishing of these raw materials was not a new development for the Dutch; the weaving of woolen fabrics had been a prosperous industry in Flanders during the Middle Ages, and during the sixteenth and seventeenth centuries it developed in the provinces of the Netherlands. In the seventeenth century the Dutch manufacturers were innovators in the processes of the chemical industry. However, they lacked fuel, which could be supplied only by large forests; coal was as yet little exploited. Thus the Dutch could not devote themselves to metallurgy and glassworking, as the British were doing at this period.

The British development The mercantile activity of these island dwellers had certain characteristics in common with that of the Dutch, but was based on even more favorable conditions that permitted them, at the end of the seventeenth century, to achieve uncontested mastery in international trade and to take the lead in creative activity in the industrial techniques. England, like Holland, benefited by her geographical situation, which made her ports indispensable for the trade between America and the Indies on the one hand, and continental and northern Europe on the other. In addition, her situation as an island kept her apart from the great international conflicts, although it did not protect the country from internal conflicts. It was to these international conflicts in particular that England owed her ability to acquire the maritime supremacy formerly enjoyed by Holland.

The fact that England possessed numerous exploitable resources made this commercial growth all the more favorable to industrial expansion. The expanse of her forests, which had not yet been exhausted in the seventeenth century, supplied the fuel that was indispensable for the treatment of iron, copper, antimony and zinc ores and for the making of glass, as well as materials for naval construction. When wood began to become scarce, in the first half of the eighteenth century, the earth supplied coal, and the English were the first to prepare coke for the blast furnaces; Swedish ore was easily imported, as well as pines from the Scandinavian countries, whence came the best ships' masts. In the seventeenth century England was able to reduce the size of the labor force employed in agriculture without completely sacrificing this activity. The breeding of wool-producing animals was carefully developed, and the spinning and weaving of wool became the principal base of its industry, opening the door to new activities and leading to creative experiments in other areas, particularly metalworking. In addition to wool, cotton, a vegetable fiber rather new to Europe, and directly imported from the American colonies, became a source of abundant profits during this period. In several chapters of this volume we shall see how technical invention benefited from this mercantile activity, which appeared less liberal than its Dutch counterpart. The goal of all the measures of control was to ensure for the English navigators and merchants an absolute monopoly over the trade of all the European nations with India and the West Indies. The scope of the successive Navigation Acts, the most famous of which was the Act of 1651 passed under Cromwell, has often been mentioned. Here, as in Holland, the capitalist structure was completed by the creation of the Bank of England, on which was based, with commercial hegemony, the growth of manufacturing.

The combination of these circumstances created a framework particularly favorable to the exploitation of the industrious spirit of the British middle class, which thus made major progress over the countries of western Europe and thereby over the entire world. Thanks to this progress England was able, beginning in the mid-seventeenth century, to give birth to the major inventions that were to constitute the foundation of modern mechanization. Despite the wars of the eighteenth century, the secession of the colonies in the New World, and the wars of the empire, which exhausted Britain as they did the continental nations, Great Britain was to preserve her hegemony until around the middle of the nineteenth century.

"Colbertism" and its effects

In France, "Colbertism" was the most powerful factor in industrial growth during the second half of the seventeenth century. But the great international influences, in particular those arising from the maritime conquests, were already being felt in France during the reign of Henri IV, at the end of the sixteenth century. The first efforts after the end of the religious wars were devoted to the renewal of agriculture, which benefited the production of dye plants, vegetable fibers, and, indirectly, certain chemical products. France, being rich in ores of all kinds, as well as in forests, possessed a widely dispersed metallurgical industry that was not yet capable of furnishing unaided a large quantity of high-quality steel. The glass-manufacturing industry was already developing. As in most countries, however, textile production was the most important French industry. During the second half of the sixteenth century, and especially during the seventeenth, the silk industry was imported from Italy, and the first efforts at automation in France were made in the silk industry even before they were attempted in the woolen industry. Knitting, which originally was involved almost exclusively with wool, was another activity that witnessed laborious attempts at mechanization. The Duc de Sully welcomed to France, and installed in a suburb of Rouen, the unlucky inventor of an early mechanical knitting machine. While William Lee's efforts do not appear to have met with very conclusive results, three-quarters of a century later Hindret, under the protection of Jean-Baptiste Colbert, had more definite success. During this period the invention and effective utilization of the knitting loom was one of the most typical examples of a long and laborious collective creation. Practically all the inventions that led up to the modern period presented the same characteristics and followed the same process, which was comparable to the conditions of technical invention in modern times.

The principal goal of the French mercantilism established by Colbert was the development of foreign trade by the high quality of manufactured products offered for export. Customs prohibitions, and numerous strict rules concerning working conditions in the shops, the methods utilized, and the products manufactured created a rigid framework for all areas of production. With some vicissitudes the system remained practically intact after Colbert, until the appearance of Robert Jacques Turgot. While its effect later became paralyzing, at the time of its origin it exercised a beneficent influence on the industrial equipment of France and the capacities of her factories. But for almost two centuries there appeared no great French creative mind who could have acted as a precursor and kept a step ahead of the English inventors. If French technology managed to keep up with the times, it was thanks to industrial espionage and to the enticement of English, Italian, and Flemish specialists.

The interminable conflicts with her neighbors that France had to sustain, from Richelieu to Napoleon, unquestionably paralyzed her economic activity, and thereby considerably reduced the innovative role she could have played in technical matters under other circumstances. Unlike other countries, however, she was able to keep pace with the technical level of the period.

One of the most famous political actions of Louis XIV, the revocation of the Edict of Nantes, had an undeniable repercussion on the spread of technology in certain European countries. The facts, which are well known, prove that if England, the Netherlands, and especially Germany benefited industrially from the emigration of the Protestants, it was because France possessed numerous experienced technicians and industrial workers.

This fortuitous contribution of a hard-working people was particularly favorable to Germany, where Prussia began shortly thereafter to make her appearance as a great modern state. Thus began the development of the conditions that were to permit Prussia, after the fall of the Napoleonic Empire, to take her place among the nations responsible for technical progress.

The first stages in industrial mechanization

It is within the framework of these historical events that the first stages of industrial mechanization unfolded between the beginning of the sixteenth and the beginning of the eighteenth centuries. From the point of view of the history of technology, the character of this period has not yet been satisfactorily defined, because to date this history has been studied by devoting much consideration to the economic and social aspects of the inventions and their industrial exploitation rather than to the development of technical methods as such; thus little is known about this development. Our principal goal is a brief but precise summary of our knowledge of the technical history of technology. This is not done for the purpose of proposing a definition of the history of technology that would exclude the economic, social, and political factors, all of which had a great influence on that history. The period under consideration seems to us the most propitious time to recall these influences on the evolution of technology, particularly since the period immediately following it (mid-eighteenth to mid-nineteenth centuries) seems to include in its entirety the development of what has been called the Industrial Revolution. The Industrial Revolution of the eighteenth century was brought about more by the transformation of the organization of factories, the system of commercial diffusion, and the social structures than by a rapid revolution in the methods of production.

Industrial mechanization was not created *ab nihilo* in the course of the last decades of the eighteenth century, as most of the writings on this period would lead us to believe. This confusion results from the fact that in most of these studies only the mechanization of spinning and weaving of wool and cotton is taken into consideration because textiles was the industry that experienced the most rapid growth in England. In reality, however, the renewal of economic activity after the beginning of the sixteenth century had much more widespread consequences for technology.

The opening up and exploitation of the distant sea-lanes naturally had a direct effect on naval construction and port installations. But these foreign "highways" had to be matched by an appropriate network of internal roads. At the beginning of the sixteenth century, roads open to wagon traffic were practically nonexistent in all the European countries. Not only did the major countries of western Europe bend their energies to rebuilding or creating a network of roads, they also began to dig the major canals and to improve the flow of certain rivers and streams for internal navigation.

While the construction of works of art and public works was thus experiencing an increase in activity, the growth and development of the cities was favoring the building trades. Methods of lifting and transportation were gradually perfected, and in their improved form participated in the equipping of the heavy industry of the period. Mines and metallurgy were the first industries to benefit from these adaptations, followed by the industries of transformation – casting, forging, wiredrawing, sharpening, sawing, milling, and so on – to which the first elements of mechanization gradually became available. Through various channels these elements gradually reached other areas of manufacturing: clockmaking, major and minor mechanics, knitting, weaving, as well as the preparation of fabrics and dyeing, the making of chemical products, and mirrors.

Although the production of power was not renewed by any fundamental invention – the steam engine was still in its infancy – the harnessing of waterfalls and waterways, the construction of hydraulic wheels and horse-driven treadmills, and especially the utilization of elements of transmission of movement in machines, evolved in significant fashion. Drop hammers proliferated and were adapted to increasingly varied and detailed projects; in this way, too, more diversified shops could be equipped in such a way as to at least partially replace human labor by that of the machine. However rudimentary this laborious evolution may now appear, it was nevertheless decisive in the elaboration of our modern technical civilization. The process of invention is determined by a consistent and simultaneous improvement in the general technical level, an improvement not necessarily marked by new inventions. Putting into use and exploiting new inventions are equally important stages that are sometimes long and difficult. Invention does not consist only of a few episodic factors (expression of an original idea, first attempts at embodiment), but also of the total process of development that leads (generally thanks to a collective effort) from the initial conception to producion.

The inventions of the eighteenth century in which historians have attempted to see the causes of the Industrial Revolution were simply the fruit of the general achievements accomplished in the course of the preceding centuries. They were able to have the profound repercussion they had only because all the other factors that directly or indirectly control technical progress had themselves reached a quite advanced state of development. In the words of C. J. Gignoux, speaking of the situation at the end of the seventeenth century:

"We have not yet arrived at the great Industrial Revolution of the eighteenth century, but all the preparations for its welcome are falling into place . . . while the 'peak' of the industrial revolution occurs in England (as elsewhere) between 1760 and 1815, it was preluded by a series of gradual transformations" (*Histoire du commerce,* Vol. IV).

The relations between science and technology

The economic factors mentioned in the preceding paragraphs were not the only factors to exercise an influence on the evolution of technology. Beginning at a point which it is quite difficult to define precisely, there was a continuing relationship between the development of scientific knowledge and that of technology. In earlier periods this relationship was quite vague, and we may say that the awakening of scientific curiosity was caused by the first knowledge of matter and natural phenomena acquired by the technicians. Science was of very little help to the progress of technology until around the fifteenth or sixteenth century of the Christian era. Only in the course of the three centuries that constitute the chronological limits of this volume did mutual influence really begin to develop.

A commonly held opinion claims that technological progress was solely a consequence of progress in the sciences, particularly the change in the rhythm of technological progress and production during the period in which historians place the Industrial Revolution. Insofar as these phenomena can be analyzed, they seem far more complex. Even in the modern era techniques are not only scientific applications, and during the preceding centuries the sciences learned much from technology, as indeed they still do.

The period under discussion possesses an originality of its own, and its history makes possible a more precise understanding of the nature and evolution of the relations between science and technology from the beginning of prehistoric times to the present day.

A quick glance at the centuries and even the millenia that preceded this period does seem useful. During this long period of time, science and technology began by developing independently. The first relations were established between them along the lines mentioned above. An influence of geometry and arithmetic in certain technical domains is barely discernible, and even this influence was not a direct one.

Technical literature

The points or areas of contact became more numerous around the end of the medieval period and in the two or three centuries immediately thereafter. The areas involved in this process were mechanics, chemistry and metallurgy, astronomy and surveying, the art of navigation, and the measurement of time. In some of these fields only the empirical attempts of technicians to perfect their methods were as yet in evidence, but these efforts became more effective thanks to the dissemination of printed treatises beginning in the sixteenth century. Prior to the invention of printing, many engineers preserved the evidence of some of their observations or new ideas for inventions in the form of notes and drawings. These notebooks were unknown to their contemporaries, but in the sixteenth century numerous authors wrote treatises destined to be printed. Less than one century after the appearance of printing, abundant material relative to technology was available for printing. Not only was the experience accumulated in the course of the preceding centuries valuable, but also the desire to transmit this experience more rapidly, and the need which this desire filled, were demonstrably urgent.

A more extensive and faster method of dissemination was now added to the old methods of transmission by word of mouth and example, and by apprenticeship. These books, which were addressed to practitioners, contain no scientific material; this is particularly true of those that deal with mineralogy, the exploitation of mines, and ceramics.

Treatises on chemistry, which proliferated in the course of the seventeenth century, did contain a certain amount of theory. In this period the chemists played a special role in the rapprochement of science and technology, but solely to the benefit of the former. In this regard the work of Paracelsus is typical. Paracelsus was trained both as a mining technician and as a doctor; many other chemists after him also acquired this dual education. By their commentaries on and discussions of the traditional procedures used for the treatment of ores and the preparation of metals and the then-known elements and compounds, they worked more to develop chemical theory than to perfect the processes of the chemical industry.

In other areas the relationship between science and technology was closer and more sustained. We are speaking of the body of knowledge that includes astronomy, surveying, and navigation. These were experimental sciences whose progress was completely due to the perfections made by the technicians in the instruments utilized by these sciences. In turn, the corresponding techniques themselves, which involved a great number of practitioners, owed their development solely to progress in scientific knowledge.

The construction of mathematical instruments, that is, instruments of measurement and for the observation of the stars, was by the beginning of the sixteenth century a flourishing industry. It developed first in Italy, then in England; numerous such workshops were also located in southern Germany, Holland, and France, before spreading to the other European countries. The craftsmen of this industry have been called by E. G. R. Taylor "mathematical practitioners," that is, technologists of mathematics. We know of about one hundred such men in England in the sixteenth century, more than four hundred in the seventeenth and first half of the eighteenth centuries. Although they were simply workers and technicians, almost all of them published treatises on the construction and utilization of the instruments they made: rulers, compasses, quadrants, astrolabes, circles, theodolites, and so on. Mr. Taylor has drawn up a list of 628 works of this type (including two incunabula) published between 1496 and 1715. To draw up such an impressive list for any of the other phases of science and technology would be an impossibility.

Appearance of "technology"

This literature and the activity it reveals are typical of this period, and suggest the role played in the general progress of science and technology by individuals who were no longer purely craftsmen and inventors but who were not yet scientists. These men are related (although their training was more elementary) to the modern engineers. Their training, combined with the rudiments of science they had acquired, permitted them to dominate the ensemble of problems particular to their respective areas of technology. They were able to transmit to their contemporaries a written body of knowledge rationally presented. During this period these men began to create a new form of activity conventionally designated by the term "technology," as distinct from both simple applied techniques and the science of discovery. Technology falls between science and individual techniques, and is characterized by their interpenetration. Here we use this term to designate, somewhat arbitrarily, a kind of higher form of technicity, of scholarly technicity, or, better still, the science of technicity. We are seeking to call attention to that area of activity that is common to the sciences and techniques but at the same time differs from each of them, the area within which their contacts and reciprocal collaboration is established for their greater individual benefit. For the moment we need only note the first developments in this domain. The preceding examples make it possible to understand how during the seventeenth century technology strengthened its importance in the general progress of the sciences and techniques and the formation of our modern industrial civilization.

Science, technology, and individual techniques

Other examples that will be found in this volume make possible a complete understanding of the nature of this phenomenon. There was first the somewhat mysterious appearance, at the end of the thirteenth century, of mechanical clocks, and four centuries later the invention of the measurement of time, whose history is very well known. Christiaan Huygen's adoption of the pendulum as a governor for clocks is one of the most frequently mentioned examples of the repercussion of a scientific discovery on achievement in an individual technique. The fact that this example belongs within a body of research, discoveries and inventions to which both science and technicity contributed, thanks to their comparable degree of maturity, makes it all the more significant. At the end of the seventeenth century progress in the measurement of time began to be stimulated by a problem that interested both navigators and astronomers: the measurement of longitude in order to take bearings at sea. In this period the English scientists and sailors were naturally the ones most interested in the solution of this problem, and it was therefore decided to construct the first large modern observatory of astronomy at Greenwich. This example was very quickly followed by the continental countries, and reached eastern Europe. Inside fifty years the manufacture of large observation instruments and instruments of marine astronomy became a highly specialized industry, and its techniques quickly developed in order to meet the new needs of the astronomers and navigators. Precision optics and mechanics thus found the economic and scientific bases for their early development. As for the problem of marine chronometry, it was solved by the technicians themselves, with the participation of the great eighteenth-century clockmakers in England and France.

The same unification of efforts was revealed in the same period, with definite although less spectacular results, in several other areas. Stereotomy and the study of perspective, for example, contributed new methods to construction techniques in building and public works; scientific knowledge was adapted in naval construction, hydraulics, and later mechanics and the construction of machines.

It was this latter area of industrial technology that was to open the door to the contemporary period. Thanks to its transformations and to the appearance of modern methods of metalworking, this step was taken at the beginning of the second half of the nineteenth century.

The object of the preceding discussion was essentially to draw attention to a phase and an aspect of the history of technology that until now have been too neglected, and our knowledge of which is still far from satisfactory.

Plan of work

In the first volume we adopted a chronological and geographical exposition that consisted of summarizing the contribution made by each civilization to technical progress. It is certain that each civilization before the beginning of the modern era had its own creative genius, but as far as we are concerned it is equally certain that this genius was fertilized by the heritage of earlier civilizations. The art of technology spread across the surface of the world in waves whose flux seems to have been fed from all sides by each of the civilizations that successively flourished on the earth.

The first part of this volume, however, is devoted to the analysis of the technical level reached by European civilization at the time of the Renaissance. Taking into account the stagnation of Chinese civilization, which was already imposing its crushing blanket of immobility on the Far East, we thus possess an exact account of all the possibilities at the service of the human creative spirit.

Once past this phase, the development of the history of technology can be treated only by following each of its major areas of activity. Naturally, we have encountered the irritating problem of classification, which appears destined to find only approximate solutions. The division of subjects into sections and chapters, as it appears in the table of contents, does not arise from a reasoned attempt at classification; it has been established solely with a view to providing a logical continuity. Beginning with the techniques of extraction and treatment of raw materials, and continuing with those that permitted their increasingly complex utilization, we have attempted by discussing the techniques of construction, equipment, and defense to terminate our exposition with those techniques that provided man with his highest methods of expression.

In order to ensure a continuity between the earlier method of exposition and the one that is to be followed to the end of this work, and also to fill certain gaps that may have persisted because of this change in presentation, it has seemed useful in certain cases to summarize a group of facts that antedate the period under consideration. Thus, the authors of certain chapters have been able to present a sustained development of the subjects they have undertaken to discuss.

This division by discipline of the history of technology naturally presents disadvantages. But beginning with the sixteenth or seventeenth century, the geographical division is no longer justified, for the reasons mentioned above. While a certain geographical zone, namely western Europe, became solely responsible during approximately two centuries for the continuity of the creative spirit, the full flowering of this spirit could thereafter be ensured only by its dispersion over the entire surface of the earth. The progress of technology became to an increasing degree the consequence of a collective, consistent effort. In addition, all the factors on which it depended at the time of its withdrawal into Europe, and on which it depends in the modern world, are factors that simultaneously concern the entire world rather than certain carefully delimited geographical zones, as in the time of the ancient and medieval civilizations.

The continuity of technical progress

The global division by chronological periods alone is not sufficient to describe the stages of progress. It is important to avoid blurring certain essential characteristics of technical progress, in particular its continuity and its homogeneity. There is no distortion in this history; everything happened, and happens, as if a certain equilibrium were constantly being reestablished on the one hand between the various areas of activity and on the other hand between technical creations and the economic possibilities of exploitation. Although in the framework of this book we have kept to the technical aspect of this history, we have attempted to avoid sacrificing the phenomena of tension between the various techniques that are responsible for the general progress. For this reason the chronological limits have not been — could not be — rigidly fixed. Thus certain chapters carry the reader to the end of the eighteenth century, so as to complete a phase of development in a specific area of activity, while other chapters end with the middle of the century. The history of technology, like any living creation, does not adapt itself to the caesura of the centuries, and we shall later see that it even breaks free of the framework within which classical history would like to enclose it.

MAURICE DAUMAS

PART ONE

THE FIFTEENTH AND SIXTEENTH CENTURIES IN THE WESTERN WORLD

FOREWORD

For western Europe the mid-fourteenth century was a time of troubles, a period unfavorable to technological development. The onslaught of the Hundred Years' War, the crash of the Italian banks, and the Great Plague had brought in their wake both a major economic contraction and a major demographic regression. The second half of the century was a period of stagnation, which, however, did not preclude the expansion of a few techniques. This period saw the spread of papermaking and the use of the mechanical clock, the cannon, and gunpowder. It is nevertheless true that not until the fifteenth century did a certain expansion of technological progress begin.

This expansion was all the more remarkable in that the entire civilization participated in it. The ambiguity of the term "Renaissance" has been discussed on numerous occasions; authorities in various areas of investigation have pointed out a certain spirit of originality that appeared in the fifteenth century and became increasingly apparent during the next fifty years — a spirit that constituted a wide gulf between this period and earlier and later ages. A discussion of some of the principal manifestations of this new spirit will certainly be useful: their influence on the evolution of technology will become more apparent.

The first manifestation was political and social in nature. The feudal regime was gradually disappearing, and with it an entire conception of political power. The old powers, the nobility and the church, were gradually being superseded by a middle class whose strength had long been steadily increasing. The city was becoming a dominant factor in the general life of western Europe. A realistic, middle-class order was replacing the universal hierarchy; the abstract belief in a higher unity was abandoned. Existence was understood in more direct fashion, and the eyes of humanity were turning with increasing frequency to the outside world. There is no doubt that the political rise of the middle class was the cause of a transformation in the mentality of the ruling classes. Men were turning away from the medieval spiritualism, and were beginning to become intensely interested in the real properties of objects, the nature of things, and the drama of the universe. The face of civilization was changing. This realistic, middle-class order was to make a special place for the details of existence, and therefore to technical problems; the new middle class, in fact, was very much involved in the circulation of merchandise and industrial production.

Insofar as the sciences were concerned, there occurred a remarkable change of orientation that is now very familiar. Here, again, the essential role of the middle class can be foreseen. The development of economic activity in Flanders and Italy was creating new needs, and science was to profit by a certain utilitarianism. The empirical, experimental, and mathematical tendency was accentuated. The monetary concepts of Nicole Oresme, the ideas of Pierre Dubois on the education of women, Marino Sanudo's theories on economic warfare and the efficiency of blockades proved, if proof was needed, that the human mind was opening to the outside world, a development that gave a large role to concerns that were essentially practical. One century later François Rabelais was to keep this in mind when writing about the education of Gargantua.

The teaching of mathematics had gained a toehold in the universities, first at Oxford and then at Paris. Shortly before 1366, candidates for the teaching license were obliged to take an oath that they had done at least one hundred lessons in mathematics – a formula interpreted, it is true, in a clearly limited sense. In the same period mathematics appeared in the degree program, though in vague fashion. The University of Paris thus tolerated a kind of private instruction in mathematics, a course of instruction that was to become increasingly important. This intrusion of mathematics into academic instruction actually represents a triumph of technology. By the middle of the medieval period, mathematics was considered useful only for the "mechanical arts" (an attitude shared by René Descartes). Geometry was one of the elements needed by carpenters, architects, and surveyors. By the middle of the twelfth century Gondisalvo was already insisting on the importance of geometry for the operating of "engines" and for architecture; it was by nature a technician's science. Arithmetic was the principal base of the art of commerce. With the expansion of new methods of bookkeeping, schools for teaching the use of the abacus (where the rudiments of arithmetic were taught) proliferated, especially in Italy.

Thus the efforts of certain scientists, beginning in the twelfth century, and adopted and developed by Oxford University, tended to place mathematics, and therefore technology, on the same level with the other subjects traditionally taught in the universities. Guy Beaujouan notes that business arithmetic, which was derived from Fibonacci's *Liber abaci,* became associated in the abacus schools with the dissemination of double-entry bookkeeping, which was born around 1340. The interest taken by the Oxford dons in Alhazen's optics was not unrelated to the invention of glasses and the introduction of mathematical perspective in Renaissance painting. Jean Buridan, the theoretician of *impetus,* laid the groundwork for ballistics. The science of anatomy began to develop, thanks to the first dissections, on the eve of the fourteenth century.

Even art was influenced by this new atmosphere. The result was both a certain naturalism, the fruit of greater experience, and a certain technicity, which was manifested in the material domain by a mathematical system of harmony and a characteristic insistence on all forms of vitality, in addition to experiments, not all of which were successful.

This return to reality on the part of an entire civilization completely parallels that demand for efficiency in all phases of activity, whether political or economic, that was to characterize the fourteenth and fifteenth centuries. The unity of all human activity had formerly been achieved in a divine absolute, which favored the compartimentalization of knowledge. Henceforth this search for the real could be carried out by human methods (experimentation and logical procedures). Whatever his training or profession, the human being always has the possibility of attaining universal truth. This search, which was pursued with passion, brought into play, as a substitute for broad knowledge, an attempt at inventiveness that replaced the scholastic spirit.

M. Francastel regards this replacement of a universe of essences by an experimental universe as an essential development. The idea that the world was simply an attribute of God was abandoned. This conception of nature, and of man as an actor on the stage of the world, was accompanied by that extraordinary exploration of the universe that was the great work of the men of the Renaissance. Their idea of invention implied the idea of a logical coherence of the universe, a fact that, according to Francastel, "explains why the major objective of these inventors was to devote themselves to understanding general effects and organizational schemes of the universe, and to enriching to an infinite degree the collection of individual cases. Their principal work lay in a kind of selection and isolation of the possibilities of action and intellection offered them by technology."

Another result of their viewpoint was the interpenetration of fields of knowledge, which were no longer the isolated individual stones of a building, but separate, converging methods of approaching truth. André Chastel distinguishes, in the course of the fifteenth century, the claim of certain artistic milieus that art was on the same level as the liberal arts (testament of Ghiberti), and, conversely, the interest of certain scientific circles in the applications of knowledge and experimental research. All this led to a "decompartmentalization" of knowledge, to the benefit of art and technology, the result being a sometimes close association and intimacy between scientists and artists (perspective, anatomy), and between scholars (archaeology) and artists, in which technology played the vital role of liaison.

Thus scientific and technological progress, which are closely linked, must now be studied in a new context. We shall meet new men, men of wide-ranging interests, some of whom have left valuable writings; Leonardo da Vinci is the most famous, but not the only, example. But while they understood the technological progress that was being achieved at least before their eyes, if not with their help, they were not its sole authors; perhaps their role was only to prepare and guide.

By drawing up an extremely schematic, brief diagram of the stages in this "Renaissance," we shall be better able to place the actors in this great drama in their proper roles. A major step appears to have been completed by the end of the fourteenth century, the consequences of which were to be extremely important. Three fundamental "inventions" paved the way for successive achievements: these were the discovery of the blast furnace and casting, which would make possible the expansion of the use of metal, hitherto seldom utilized; the perfecting of the movable forecarriage, which was profoundly to transform methods of transportaton; and finally the appearance of the crank-and-connecting-rod system, thanks to which mechanization made remarkable progress. Though obviously not all developments can be traced to these three technical transformations, they seem to be fairly characteristic of this first period.

The second phase was much more important; its dates can apparently be fixed with a fair amount of precision between 1450 and 1470. It saw the invention of printing, the spiral spring (1459), modern fortifications (1461), the first deliberately planned public square, at Pienza (1462), the spinning wheel with flyer (1470), and rolling mills (1470). An attentive study of the first incunabulas reveals the importance of technology in the minds of men of that age: Flavius Vegetius, printed for the first time in 1471, was to have eight editions in the fifteenth century; Vitruvius Pollio (published in 1487), two editions; Sextus Frontinus (printed in 1480), five editions; while Pliny the Elder appeared in 1469. Among later authors were the agronomist Pietro de' Crescenzi, the first edition of whose work (1471) was followed by twelve others in the fifteenth century, and Roberto Valturio (published in 1472), with four editions. Battista Alberti's treatise on architecture was published in 1485. This was both a major technical renewal and a complete change of mentality, which made it possible to place at the disposition of the educated man problems that only one century earlier had been completely foreign to him. Undoubtedly it was these successes that now made it possible for technology to become an integral part of a certain humanistic culture and, in addition, to call the attention of the governing powers to questions of continually increasing political, economic, and military importance.

This evolution, whose importance cannot too often be stressed, was aided by other favorable circumstances. Technology became a concern of the ruling powers from two points of view. On the one hand technical progress was now synonymous with military power; the engineers who have left numerous writings were almost all military engineers. In the last quarter of the fifteenth century the art of warfare was completely transformed. New firearms were developed whose production and handling, like the art of defense, required increasingly complex knowledge. Nor were industrial and more generally economic problems foreign to this state of mind of the governing classes. In the same period there appeared the first forms of mercantilist doctrines, which as late as the eighteenth century had not yet outlived their success. Technology was at the heart of this industrial mercantilism, which was to be that of a large part of western and continental Europe.

Economic progress, which was simultaneously the result of and the basis for technical progress, nevertheless required other conditions. At first, between 1450 and 1470, we find a relative political calm following upon the troubled period at the end of the Middle Ages. Between the end of the Hundred Years' War and the first French expeditions into Italy (1453–1494), the continent enjoyed almost universal peace; wealth returned, the wounds of the war were healed. This effort at reconstruction permitted, after the last convulsions of the Burgundian war, a monetary stabilization that undeniably favored this economic progress. Florence (1464), England (1470), Spain (1471), Venice (1472), and France (1475) all achieved more stable monetary systems.

There is no doubt that the great discoveries were one of the most remarkable results of these phenomena. The discovery of a new world certainly upset a certain number of scientific beliefs then held. But from the technical point of view this new acquisition was less important; the only novelties it produced were botanical and thus agricultural, and even these changes were limited to the introduction of a few additional cultivated species. The new continent perhaps benefited more than the old from the arrival of Columbus. The requirements of long ocean voyages nevertheless contributed in large measure to the development of naval techniques, both those relating to shipbuilding and those concerned with navigation as such.

Thus the great geographical discoveries, a phenomenon of immense significance in numerous areas, do not appear to have caused the immediate technical revolution some authorities have imagined. In our opinion it would be much more profitable to investigate the technical conditions that made these great discoveries possible than to attempt to determine their possible consequences. The work of the fifteenth-century technicians and the progress achieved in the second half of this century are undoubtedly of greater importance for the history of technology than the discovery of America. This is partly demonstrated by the fact that the crossing of the Atlantic was achieved to the benefit of the Spaniards, who of all the peoples of western Europe were the most backward from the technological point of view. The influence of this discovery on the other European countries was to be felt only after a longer period of time, and in very varying degrees, depending on whether or not direct relations could be established.

MAURICE DAUMAS

CHAPTER 1

ENGINEERS AND TECHNICIANS OF THE RENAISSANCE

It is in this context that the men whom we call "engineers" of the Renaissance should be studied. On the one hand they continued the slow work of earlier centuries that we have already studied, especially on those interminable construction sites whose goal was the construction either of buildings like the cathedrals or of major public works such as ports and canals. But their inventiveness also led them along the new paths in which modern technology was beginning to develop.

These men should be regarded less as "inventors," in the modern sense of the term — this is true even of Leonardo da Vinci, to whom impressive lists of technical discoveries are attributed — than as men who realized that a process of technological progress was under way, and who were able to determine its guidelines and to direct its developments, at least to a certain extent. They were admirable representatives of the technical spirit of the Renaissance who above all else eagerly sought to make of technology something more than a gradually improved routine.

To this first limitation, whose importance should not escape us, yet another must be added. These engineers were by no means concerned with technology in general, and in fact their work may even seem strikingly narrow in scope. The art of warfare in all its aspects and the major hydraulic works constituted the essence of their studies. At the end of the fifteenth century, however, problems of another kind appeared: the development of mechanization, both in its general aspects and in some of its special applications, made a strong impression on them. This is why their notebooks contained practically no notes on techniques we now regard as being of prime importance in the history of civilization; printing is the best example. Architecture, civil engineering, and military projects, in that order, were the poles of their activity.

On the other hand, these Renaissance engineers were not unaware of the intellectual and artistic movement to which we have already alluded. Some of them began as engineers, others were artists, while those of the second half of the fifteenth century almost all began as artists, but rapidly surpassed their early training and their engineering profession in an attempt ultimately to achieve a universal vision of the world. From a certain point of view, therefore, they are less interesting as pure technicians than for the interrelationships they established with the other areas of knowledge.

At the end of the fifteenth century and in the early years of the sixteenth, the unity of knowledge was once again compromised, because more thorough research demands specialization. Thus the artist confined himself to art, while the union of science and technology, though not total, was maintained at least temporarily. Galileo was perhaps one of the last technician-scientists. The technicians, however, had thereby gained a breadth of outlook which they had hitherto completely lacked. Not until a century later, however, was this wider viewpoint able to benefit fully from the scientific progress in whose birth it had played perhaps the dominant role. The technical treatises that continued this movement, which had culminated with Leonardo da Vinci, were now limited to well-defined techniques; Georgius Agricola's work on mines and metallurgy is the best example.

The work of collecting the manuscripts of these engineers, now dispersed throughout the libraries of Europe, is unfortunately quite difficult, and has been only partly completed. Nevertheless, we now have at our disposal a sufficient number of elements to permit us to attempt a valid study. These engineers were natives of southern Germany and of a few Italian cities. Apparently it was in these areas that technology began its decisive progress, for reasons that it is difficult to pinpoint. But while the Germans continued to be preoccupied particularly by military problems, which limited their desire for invention, the Italians departed more quickly from this narrow path and achieved the perfection of a Leonardo da Vinci. This development has no equivalent in any other country, not even in France, which did, however, experience a certain technological development.

The German school: Kyeser and his successors

The importance acquired by this research movement, which far surpassed a limited circle of initiates, can immediately be grasped from the first extant writings of this period. The work of this first author, whose name was Konrad Kyeser, was reproduced in a great number of manuscript copies, and even seems to have been the object of actual "editions," both in Latin and in German; it was so successful throughout the fifteenth century that its illustrations were used in the first printed editions of the works of Vegetius. But of the author himself little is known. He was born in 1366 at Eichstadt, a small city in Bavarian Franconia, halfway between Munich and Nuremberg. After 1396, political and military circumstances forced him into exile in the mountains of Bohemia, and there he composed his work, which he dedicated in 1405 to Emperor Ruprecht of the Palatinate.

Pierre Duhem tells us that we are here confronted not so much by the work of a single man as by the product of a development of several centuries. Kyeser was a man of simple mentality, who passed for the inventor of the most diabolical machines. His authority as an engineer was, however, so well known and so highly valued that his treatise continued to be the foundation of the science of machinery, as expressed in the drawings in numerous works, for more than a century. He was first and foremost a military engineer, and his *Bellifortis* is addressed particularly to army commanders; certain machines that appear to be for peaceful purposes (particularly the entire section on hydraulics) were used by troops in the field.

Kyeser's artillery is still very primitive, with the trebuchet continuing to

FIG. 1. Kyeser's elevator.
Around 1400
(Göttingen, *Cod. phil.* 63).

play an important role. His cannons are heavy and cumbersome; they are loaded onto gun carriages by means of pulley blocks. The culverins, however, appear to have mounts. Kyeser has been credited with the first "tanks" and machine guns. His escalade machines are of the same type shown in all the military manuscripts of this century.

The hydraulic techniques are perhaps more interesting. We find no piston-driven pump, but norias and Archimedean screws are represented. A water mill with overshot wheel and an ordinary windmill on a tripod present nothing out of the ordinary, but there are noteworthy features in the mechanisms used for various machines. The pole is used as a spring in pounders. Especially noteworthy is the fact that the crank-and-connecting-rod system makes its first appearance in a hand-operated mill, which in addition has a wheel that acts as a flywheel to eliminate the difficulty of the dead-center point. Kyeser's lifting machines present no novelties. The section on the manufacture of saltpeter will be discussed later.

Kyeser's work thus appears as a collection of machines and instruments, most of which had long been in existence. We are still very close to the Middle Ages, as is proved by the anthropomorphism of some of his war engines. There is no systematic research along given lines. However, his treatise does contain new ideas which were to undergo considerable development, for example the crank-and-connecting-rod system, which for purely material reasons spread very slowly, and cannon mounts. In short, this work is an anthology rather than a piece of genuine research.

The next German work is a notebook of observations rather than a formal treatise like that of Kyeser; only a single copy of it is still extant. German scholars refer to its author as "the Anonymous Author of the Hussite War," after a remark in the book. He, too, was a military engineer, a man of an inquiring mind who wrote down not so much his own ideas as those of which he had heard. Lifting devices are the most numerous in his book; the only original one appears to represent a device for bringing mining cars, traveling on rails, up to the surface. The artillery pictured is approximately the same as that depicted by Kyeser; trebuchets still appear in this manuscript, which dates from around 1430. Assault machines are equally numerous, but contain no novelties. "Prefabricated" bridges for crossing moats can already be seen in Kyeser's treatise.

There are numerous water mills; one of them may have had a horizontal wheel. Hand-operated mills have crank-and-connecting-rod systems with flywheels. The manuscript does contain a few novelties, however: one example is a machine for drilling wooden pipes, attributed to the inhabitants of Nuremberg, and a machine for digging holes in the earth, whose digging device seems very well thought out. In this manuscript a machine for polishing precious stones appears for the first time. The most amusing drawing — its practical realization seems doubtful — is that of the deep-sea diver. The reader feels that the author is not so much a man endowed with the didactic spirit as a practitioner curious about his trade and open to the most ingenious "inventions." Here he is perhaps closer to Villard de Honnecourt than to his immediate predecessors.

The original German school can be said to end here. Numerous fifteenth-century manuscripts on the military art are still extant; they are devoted especially to firearms, and in them we are able to follow the progress of artillery. From the technological point of view, however, they do not present the same interest. We shall not enter here into detailed descriptions of these works, which greatly resemble each other, but we shall later return to this subject. There is, for example, the *Hausbuch* in the possession of the family von Waldburg-Wolfegg, dated 1470. This manuscript consists of a great variety of elements, some of which belong rather to the domain of the history of art, although they are also useful to the historian of technology; in it we find the spinning wheel with flyer, and, most notably, remarkable pictures of mines that will be discussed later. It is a collection of artists' sketches rather than an engineer's notebook. A series of manuscripts kept in the libraries of southern Germany and Austria represent another very special class; these are books of combat, which depict sometimes military devices, sometimes genuine miniature battle scenes. In a manuscript of 1496 preserved in the library of Heidelberg we find another picture of a trebuchet.

The Italian school: the first generation

The superiority of Italian technology is apparent right from the beginning of the existence of the Italian school. It is fairly difficult to reconstruct the atmosphere of technical enthusiasm of this first generation. Not all its engineers left notebooks, but in those that are extant we sense a very clear feeling of technical progress that is limitless.

Filippo Brunelleschi (1377–1446) epitomizes the typical artist-technician of the Renaissance, as his very training testifies. Like most of his successors, he began as a goldsmith, sculptor, and probably a metal caster, before becoming an architect. He was more of a technician than an artist; at Santa Maria del Fiore, the cathedral of Florence, it was not the cupola that he rediscovered (it had never been forgotten) but new methods of construction. He was also a fertile inventor of machines, some of which may appear in the notebooks of other engineers.

With Fontana (1393–1455) we enter another universe, glimpses of which can be seen in the work of several technicians of the succeeding century. There are numerous enigmas in the various stages of his career. In 1418 he studied medicine and the arts at Padua. He was official doctor of the Republic of Venice, and practiced in the same capacity at Brescia. His work is above all scientific, and is concerned more specifically with natural history, but he appears to have been interested in the problems of military technology. A portion of his work, notably his collection of drawings of machines, is written in cryptographic symbols. This is still a collection of machines, like Kyeser's work a forecast of the "theaters" of the following century rather than a treatise, the work of an inquiring man rather than of a technician. The quality of its drawings, however, is clearly superior to that of the German manuscripts of the same period, and the ideas it contains are also more original, insofar as we feel, for example as regards certain problems of hydraulics, that the author attempted to reach an almost scientific level; he appears to have been familiar with certain principles of physics. (The Bibliothèque Nationale in Paris possesses a treatise on physics written in the same cryptographic writing.)

Jacopo Mariano ("il Taccola") was born at Siena in 1381 – the first appearance in our story of this city that was to produce so many famous engineers. He was the son of a simple grape grower, and he himself claimed to be an uneducated man. His career was exclusively that of a military engineer, from its inception until his death sometime before 1458. His treatise on machines is a rather heterogeneous collection; its drawings resemble those of the German manuscripts discussed above. Taccola's importance in the evolution of military technology appears to have been considerable; moreover, his reputation was

FIG. 2. Taccola's sea diver. Around 1440 (Munich, SB, lat. 197).

very great and his influence profound, as we shall see on numerous occasions. Was he the first to conceive of modern fortifications with ramparts? Was he the first to utilize mines and powder to destroy fortresses? It is difficult to say. In any case he was a prodigious amateur of machinery of all kinds, and the surname of the "Sienese Archimedes" bestowed upon him leads us to believe that his contemporaries were astonished both by his mechanical virtuosity and his scientific tastes. His writings also betray, however, that naïve self-satisfaction of which not even Leonardo himself was free. "I am holding back some of what I am capable of doing," he notes. "Don't think that I do anything without a price." He probably trained pupils at Siena.

In Battista Alberti (1404–1472) we undoubtedly have one of the first great engineers of the Renaissance. A member of a noble Florentine family exiled by political struggles, Alberti, unlike his predecessors, was a thoroughly educated man. He studied mathematics at Bologna, acquired a good education in the natural sciences, and throughout his life remained a man hungry for knowledge. His attraction to mathematics is symptomatic of certain future developments of technology. He was undoubtedly more of an architect than an engineer; however, he did write an essay on the salvaging of the Roman ships sunk in Lake Nemi, a project for which Taccola may have done a few drawings. That he influenced the training of the architects and city planners of the Renaissance is also undeniable. His treatise on architecture, which was one of the first technical treatises printed, was a standard work in the library of every engineer in the last quarter of the fifteenth century.

Other men of this early Italian Renaissance who were artists rather than engineers were nonetheless interested in technology. Filarete (Antonio di Pietro Averlino) wrote a treatise on architecture that was less original than Alberti's work (Filarete was not as well educated as Alberti, to whom he owed much). Lorenzo Ghiberti (1378–1455) was a bronze caster, and as such was interested in a certain number of technical problems, probably including that of artillery, which was such a major preoccupation of the men of this generation. Among the painters, special mention should be made of Paolo Uccello (1396–1475), who did not invent perspective but who introduced into his art geometrical tendencies from which his successors were to profit. Piero della Francesca (1406–1492) served as a bridge between the two generations. His treatise on geometric painting, a continuation of the ideas of Alberti, was written around 1460.

The importance of the work accomplished by this first generation of the Renaissance thinkers is obvious. These men, practitioners of all trades — doctors, painters, engineers, architects, metal casters — together helped to build a more complete knowledge of nature, centered on technical problems. Their practical achievements, although interesting (some of them will be studied later), should not lead us to overlook the frame of mind within which these achievements were accomplished. Their more systematic research, a very clear idea of progress, and a relationship with all the other human activities are all symptomatic of the major growth that was to occur in the second half of the fifteenth century.

The Italian school: the second generation

The social, political, and intellectual climate, moreover, was favorable to this flowering of technical thought, as could be foreseen on the eve of the appearance of printing in the Western world. Nowhere were conditions more favorable for this thought than in Italy. The decline of the city-republics and the rise of the princely families favored the transformation of military practices and the explosion of a new scientific and artistic culture.

In Rimini, the Malatesta family dominated the life of the city. Sigismondo Malatesta created what was perhaps the first of these centers of intellectual activity. He was a military leader as well as a military engineer, and while he left no written works, his achievements were undoubtedly important in the development of fortifications. The citadel of Rhodes and Ragusa may have been built under his supervision. In any event he constructed the Rocca Malatestina (1438–1446), which is still very medieval in appearance. Sigismondo, who surrounded himself with scientists, architects, and technicians, seems to have owed much to Alberti, who came to work at Rimini. Included in this group at Sigismondo's court was Roberto Valturio (born 1413), who belonged to an ancient family of Macerata that had settled in Rimini, and held various hereditary administrative posts at the Vatican before returning to Rimini in 1446. At the request of his prince, Valturio wrote a treatise on military art, which is less the work of a technician than of a man of letters, perhaps even a simple secretary (some authorities feel that he merely put into writing his employer's ideas). The treatise was completed in 1455, and although often dull it was tremendously successful, if we can judge by the number of manuscript copies and by the fact that it was the first modern technical treatise to be printed (1472). Quotations from Taccola frequently appear in its pages.

Perhaps even more open-minded than the Malatestas, the Sforza family played a similar role in Milan. Francesco Sforza, who ruled from 1450 to 1466, was the originator of the family's intellectual role. Great hydraulics projects, rather than warfare, were his major interest, notably the projects for controlling the level of the Po and for opening navigable waterways for the city of Milan. The Martesana Canal, which linked the city with Lake Como, was constructed between 1457 and 1460 by Bertola da Novate; it was followed by the canal between Pavia and the Po. Filarete was one of the men who worked at Milan. Fioravanti of Bologna may have been the inventor of the locks with movable gates built on this occasion. His son Aristotele began his career as a specialist in difficult engineering problems; he reerected the tower of the Palazza del Podestà at Bologna, moving (in 1455) a church tower that weighed more than four hundred tons. He was employed on Sforza's canal projects prior to his departure for Russia, where he worked as an architect, engineer, caster of cannons and bells, and minter.

Urbino, the city of the Montefeltro family, enjoyed an exceptional importance, especially during the period of Federico II (1422–1482) and his son Guidobaldo (1472–1508). Like the Malatestas, the Montefeltros were above all military leaders who surrounded themselves with such military engineers as Luciano, Pipo il Fiorentino, Fra Carnevale, Sirro da Castel Durante, Bacio Pontelli, and especially the great Francesco di Giorgio Martini. Federico was also an educated man, however; he had been taught at Mantua by the humanist grammarian Vittorino da Feltre. Artists and scientists were welcomed to his court along with men of letters; here Piero della Francesco wrote his treatise; here Luca Pacioli worked. Thus the fusion begun all over Italy by the preceding generation was successfully completed. The marvelous library of Urbino contained the authors of classical antiquity together with the latest Italian works.

Francesco di Giorgio (1439–1502) was perhaps the most remarkable member of the Urbino group. By birth a Sienese and a member of a humble family, he undoubtedly studied painting and sculpture at Orvieto under Angelino da Fiesole. His teacher in architecture was Lorenzo di Pietro (1412–1480), who had been first a painter, a friend of Donatello, then an architect and engineer (he constructed fortifications at Sartaneo and Urbetello). Francesco began his career as a sculptor. In 1469 he was placed in charge of the water supply, fountains, and aqueducts of Siena. In 1477, Federico di Montefeltro called him to Urbino, and entrusted to him the construction of a certain number of fortresses and probably the completion of the beautiful palace of Urbino. Between 1477 and 1480 he wrote, at his master's request, his treatise on architecture, city planning, fortification, and mechanics; it enjoyed great fame, although it was not printed until the nineteenth century. In 1485 Francesco returned to his native city to supervise military operations. Thenceforth he divided his time between these occupations and work as a consultant, invitations for which activity poured in from all sides. In 1490 he was called upon to advise on the construction of the cathedral of Milan, then on the cathedral of Pavia, on which occasion young Leonardo da Vinci made his acquaintance.

His treatise on architecture certainly owes much to the writings of Alberti and Filarete, especially to the former; he borrowed their conception of the ideal city and the layout of the major public buildings. The still unpublished section on hydraulics and mechanics is much more original, although it occasionally reveals the influence of Taccola. The devices he describes are in many cases more advanced than those of Leonardo da Vinci, and his drawings have a clarity, precision, and perspective that make Francesco di Giorgio the unquestionable precursor of the art of drawing machines. A well-known artist and a talented engineer, di Giorgio was certainly one of the men most representative of the Renaissance technicians, although he perhaps lacked that intellectual, scientific, and philosophical curiosity that were to make of Leonardo da Vinci the outstanding genius that he was. His notebooks and notes would have to be more thoroughly studied, and an accurate or at least integral edition of his treatise (which does not yet exist) would be needed, for a more complete assessment of his personality; his biographers have been attracted much more by his artistic and architectural (even military architectural) work than his mechanical ideas.

Giuliano da Francesco Giamberti ("da San Gallo") (1445–1515), a native of Florence, was slightly older than Leonardo. His father was a woodworker, cabinetmaker, carpenter, and then an architect; he worked on the fortifications of Sarzana and Pietra Santa. In 1464 Giuliano was employed on the construction of the Venezia Palace in Rome, and later (in 1470) at the Vatican. By 1478 he was a military engineer, and, on behalf of his fellow citizens, was defending Castellina, which was being besieged by Juliano della Rovere with the help of Francesco di Giorgio. In 1483 he constructed the citadel of Ostia, an intermediate stage between the medieval and the new systems of fortification. After this he traveled considerably, ultimately reaching France. Toward the end of his life he became very active militarily, directing the sieges of Pisa (1509) and Mirandola (1511). His drawings of buildings and machines reveal the powerful influence exerted on him by Francesco di Giorgio.

His brother Antonio (1455–1534) had a similar career. He began it in Rome with the Castel Sant' Angelo, which he surrounded with triangular bastions; in 1494 he worked on the citadel of Civita Castellano. After this, he constructed numerous military projects all over central Italy.

Leonardo da Vinci (1452–1519), began his career in Verrochio's studio, where he became familiar with the techniques of casting at the same time that he received a good artistic education. In 1482 he wrote his famous letter to Ludovico Sforza, in which he appears essentially as a military engineer, like his predecessors, with the same interests as Francesco di Giorgio. It is not surprising, then, to see him participating in the canalization projects in the plain of the Po and the work on the Sforza castle in Milan. He worked in Milan from 1482 to 1499 as an engineer. Here he met Francesco di Giorgio, whose treatise he had read (a manuscript copy of this work annotated in Leonardo's handwriting is still in existence); he was also a friend of the mathematicians Luca Pacioli (whose work he illustrated) and Giorgio Valla.

Leonardo's training thus appears to have been the classic training of all the engineers of his day. He knew some of these men and read their works, including that of Taccola, from which he adapted several drawings. His work as a metal caster naturally aroused in him an interest in the problem of artillery. The question of the originality of Leonardo's technical ideas will long be a topic of discussion. Undoubtedly some of his drawings were directly inspired by those of his predecessors. He appears, however, to have surpassed their somewhat limited interests, which were essentially those of architects and military engineers. Machines that were undoubtedly created in this last portion of the fifteenth century and that were adapted to a multitude of industries occupy a major place in his notebooks. Leonardo seems to have been more attracted to metalworking, which was one of the new developments of this period, than his predecessors had been. He also seems to have been more anxious to surpass the stage of individual techniques; the most interesting part of his research is that portion in which he attempts to solve several problems posed in general terms, as for example friction in various types of gears, and the question of the erosion of riverbeds. For the latter problem he may even have constructed small wooden models, in which he attempted with sand and water to reconstruct all the possible situations, after the fashion of modern hydraulics engineers. This preference for the general, these attempts at experimentation, and his insistence on the use of mathematics are undoubtedly his true claims to fame rather than the hypothetical inventions for which he has been praised and which in many cases do not correspond to any real object. His notebooks are certainly research notes rather than collections of patents.

The last years of Leonardo's life indicate, moreover, that he was a thinker rather than a doer. From 1506 to 1513 he remained in Milan, more occupied with research than with actual work. Then he went to Rome. In 1516 he answered an appeal from François I and settled in Amboise, where he died a short time later. It is a fact that unlike other great men Leonardo trained no pupils, and his notebooks remained for a long time unknown. His influence, therefore, can only have been very limited.

The technical literature of the sixteenth century

The attempts of the fifteenth-century engineers to be "universal geniuses" could not be continued indefinitely, by virtue of the very progress achieved in technology. For the sake of more thorough research it was essential to specialize, while hoping for new syntheses and even for greater effectiveness. It is perhaps this fact that was grasped by the printers who became interested in the technical movement; the first material printed in the technical field did in fact deal only with special techniques. We have mentioned the works printed at the end of the fifteenth century; except for the ancient and medieval authors, only Valturio's military treatise and Alberti's treatise on architecture were printed, in 1472 and 1485 respectively.

In this first half of the sixteenth century, it was metallurgy that benefited most from these efforts at dissemination, probably for economic reasons it would be relatively easy to determine. The *Bergbüchlein* is the oldest printed work dealing with the origin of and search for metal lodes; its first edition, dated 1505, appeared in Augsburg, and enjoyed a certain success, for following a slightly later second edition (probably printed at Leipzig), there were six German editions between 1518 and 1539. A certain Rublein von Kalbe or Calbus of Freiberg, which was one of the most important mining and metallurgical centers of Germany, published a similar work in 1527; it bears a strong resemblance to the *Bergbüchlein*. Also along the same lines, in 1530 Agricola published his first treatise, the still very incomplete *Bermannus,* and later wrote *De re metallica.*

Georgius Agricola is the best known and therefore best studied of these authors. He was born in 1494 in Saxony. By 1518 he had become rector at Zwickau; in 1522 he filled the same post at Leipzig, by which time he had acquired a solid classical culture. He studied medicine at Leipzig, and then spent two years in Italy. He returned to Germany, and in 1527 settled down to practicing medicine at Joachimstahl, where he became magistrate of the city and a counselor to the prince. He followed the intellectual movement of the day, and corresponded with Erasmus, but resisted the lure of the Reformation and remained a very fervent Catholic. Being now in the heart of the mining district, he became interested in mines and the production of metals. In 1531 he settled at Chemnitz, where he wrote a series of works on natural history. Here too he composed his *De re metallica,* a summary of the mining and metallurgical knowledge of the period and the region. The work appeared posthumously in 1556 at Basel, and had an immediate success.

Agricola had been preceded in his work by the Sienese Vanoccio Biringuccio (1480–1539), the heir of the ideas of the Italian military engineers. He worked successively for the Duke of Parma, the Duke of Ferrara, and the city of Venice before being engaged by Pope Paul III as a caster and director of artillery. He traveled extensively in the region of Como, in Styria, and in Germany. His work *De la pirotechnia libri X* appeared posthumously in 1540 at Venice, and quickly became known. It treats of the metallurgy of precious metals, the art of casting, and the making of cannons; its object is thus much more specialized than that of Agricola's treatise, whose scope still astonishes us.

The other techniques are infinitely less rich in printed treatises. As regards architecture, between Alberti's treatise, published in 1485, and Jacques Androuet Du Cerceau's *Livre d'architecture,* which appeared in 1559, there is practically nothing, except for Albrecht Dürer's treatise on fortifications, published in Nuremberg in 1527, which represents a major step toward the modern system

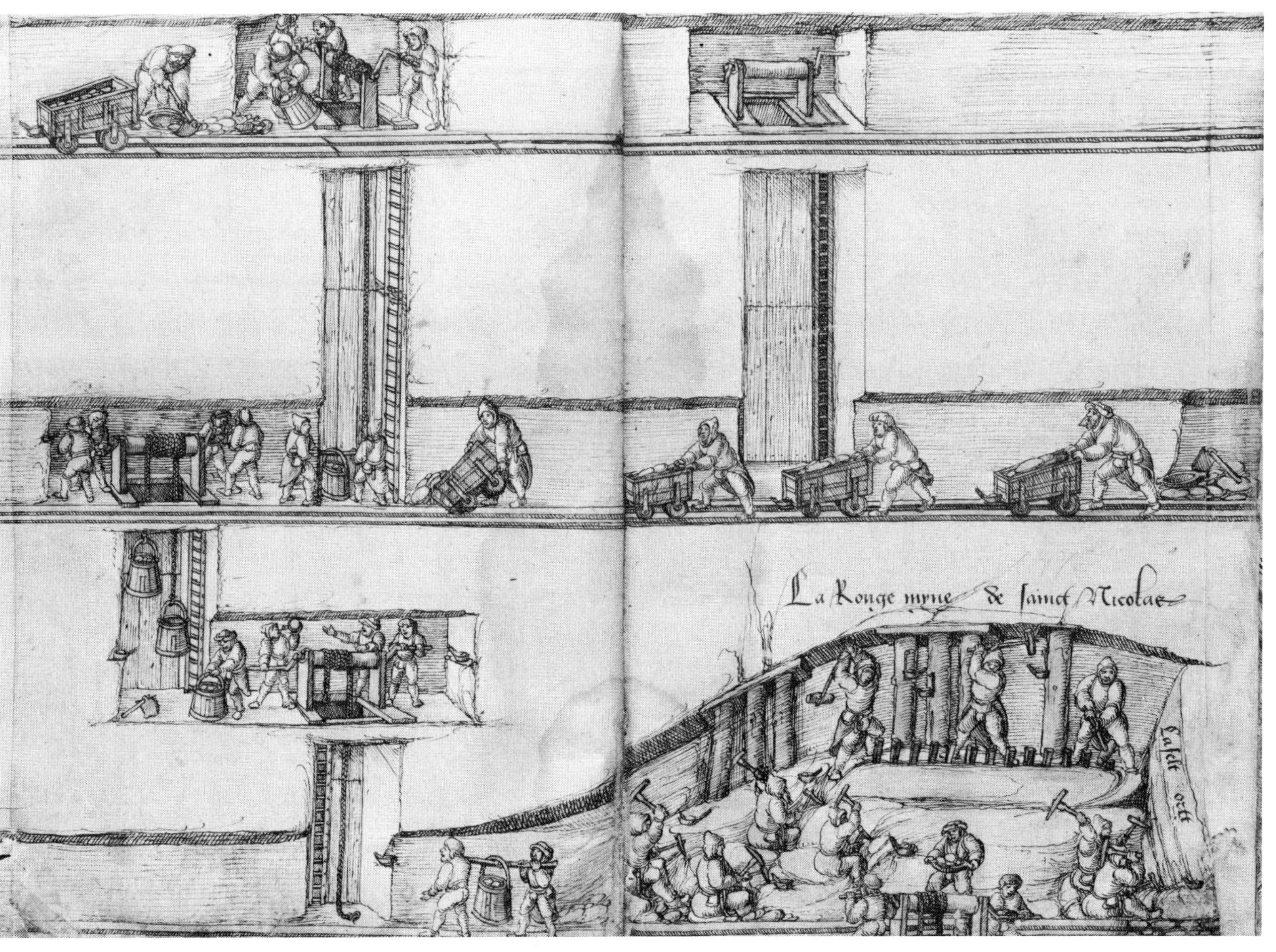

Plate 1. Extraction of silver ore at La Croix-aux-Mines in Lorraine. Engraving taken from *La Rouge Myne de Sainct Nicolas,* folios 13 v⁰ and 14, Bibliothèque de l'École des Beaux-Arts, Paris. *Photo Giraudon.*

Plate 2. Hydraulic windlass with reversing movement. Engraving taken from Agricola, *De re metallica*. Bibliothèque Nationale.
Photo Bibliothèque Nationale.

of fortifications. Works on distillation also enjoyed a certain amount of success. The German Brunschwygk published his first treatise, *Liber de arte distillandi de simplicibus,* at Strasbourg in 1500, and the second, *Liber de arte distillandi de compositis,* in 1512; the latter had five editions in the sixteenth century, as well as an English translation in 1527 and two Flemish translations (1517 and 1520).

Publications on agriculture are less numerous than we might expect; there is no comparison with the flood of works that were to appear in the last quarter of the sixteenth century and whose success was to continue into the eighteenth century. One English, one Italian, and one Spanish work form a prelude to the studies of Charles Estienne; the *Vinetum,* which appeared in 1537, and the *Arbustum* (1538) were combined to form the *Praedium rusticum* of 1554.

As is evident, this picture is relatively modest. As yet no wide public existed for works of this type; in fact, we have difficulty explaining the success of certain works, the number of editions of which can be justified only by a kind of unreasonable enthusiasm. As interesting as this source may be for a history of technology, it cannot answer all our questions. Exhaustive research on Agricola's treatise reveals that we have here an admirable description of Saxon technology, but one that is perhaps not valid for all the mining and metallurgical regions. It has already been remarked that a considerable number of these treatises were written by men who were not technicians: Agricola was a humanist and naturalist, Dürer an artist.

A genuine study of technology thus inevitably presupposes some other source. Extant engineers' notebooks of the fifteenth century are very numerous, but it would undoubtedly be possible to uncover notebooks from the following century as well; specimens of equal quality certainly exist. But as yet there has been no attempt at systematic research like that which has been begun for the fifteenth century. The sixteenth century also appears to have been much less fertile than the fifteenth. Perhaps technical enthusiasm had decreased; moreover, opinion became much less favorable to technical preoccupations; we shall later see remarkable examples of this attitude. A realistic, practical period that was one of technical progress and mercantilist policies was to be followed, from approximately 1540 to 1640, by a period much less interesting and much poorer in the activities under consideration. In addition, the progress that had just been achieved had to be absorbed before further progress could be made; in the first half of the seventeenth century technology experienced a new period of growth. The history of technology, as we have already noted, was formed by this alternation of favorable periods and periods of stagnation.

In addition, we must not isolate the engineers whose portrait we have just rapidly sketched from their historical context. Frequently it is a lack of a thorough examination of the technical milieu in which they lived that has made it possible to distort the profile of some of the authors we have mentioned. The example of Leonardo da Vinci, who is presented as a genial precursor in a multitude of areas, is perhaps the result of these incomplete studies.

CHAPTER 2

TRANSPORTATION

THE TRANSPORTATION problem developed as the outgrowth of a series of circumstances. The overland movement of merchandise had increased in considerable proportions as a result of the political evolution, which gave birth to centralized states and thus abolished in part the medieval economic compartmentalization. Traffic, both national and international, was becoming more extensive. The transportation of artillery, which was extremely difficult with the material already in existence, was another major concern of the engineers whom we have just discussed.

The greatest influence on naval construction was unquestionably the discovery of America. The shipping lanes grew longer, requiring ships that were larger and easier to handle, and the practice of navigating by the stars. Moreover, the development of the hull and the beginnings of this type of navigation were undoubtedly the cause, or one of the causes, of these discoveries.

The transformation of wagons

There was certainly no need to modify the system of harnessing, which since the discovery of the modern style of harness was perfectly suited to every requirement. But, as we have already mentioned, the heavy, four-wheeled wagons were very difficult to maneuver. The situation was improved by the appearance of the movable forecarriage, which permitted the wagon to turn easily. In addition, there was the problem of suspension, which in the second half of the sixteenth century had not yet been completely solved. More durable methods for protecting the wheels from abrasion were finally devised.

A detailed study of the movable forecarriage has yet to be made; the establishment of the facts and, in particular, the cataloguing of extant pictures has only begun. Thus hypotheses rather than precise conclusions are all that can be discussed here. Taking into account these reservations and the lack of detail of certain drawings, we can say that the wagon that appears on the seal of Francesco di Carrara and dates from the very end of the fourteenth century seems to depict the first wagon with a movable forecarriage. In any event, we find no other example either in the works of the German writers or in those of the Italian engineers of the first half of the fifteenth century. Not until the second half of the century do we find another, this time definite, picture, in the *Hausbuch* now in the possession of the von Waldburg-Wolfegg family, and dating from around 1470. Following this, the notes and drawings of Francesco di Giorgio unquestionably presuppose a knowledge of this mechanism. The origin of the movable forecarriage seems to spring directly from the gun carriage, which consisted of two single-shaft carriages, one of the carriages being attached by its shaft to a

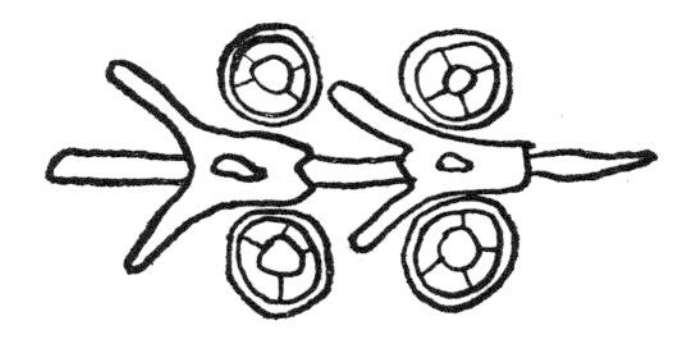

FIG. 3. Wagon with movable forecarriage, 1396. Seal of Francesco da Carrara.

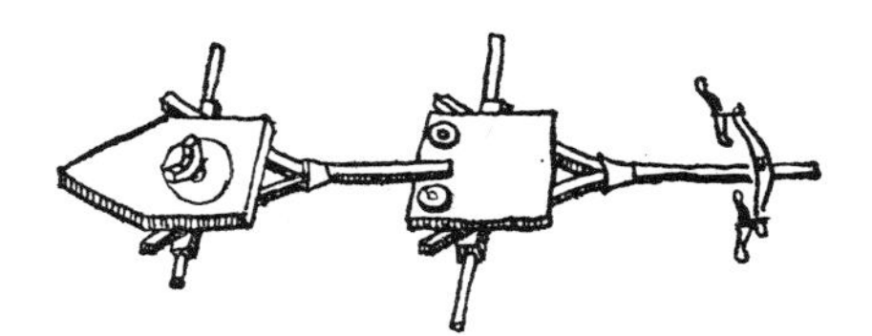

FIG. 4. Carriage support with movable forecarriage. Around 1470 (Waldburg-Wolfegg *Hausbuch,* f° 52b).

pivot on the other. (This appears to be the system depicted on the Carrara seal.) This arrangement permitted only very wide turns, and wagons of this type overturned easily because of their lack of balance. The true artillery carriage appears to have been adopted only in the very first years of the sixteenth century; a remarkable drawing of it by Dürer (1518) is still in existence.

Francesco di Giorgio also drew a kind of "self-propelled wagon." Its steering mechanism included a remarkably presented rack system; the motor consisted of manually operated capstans placed on the upper platform of the carriage, while the wheels acted both as power and as steering devices. Maneuvering such a machine must have been difficult, if not impossible, but the idea is interesting, if only for the perfection of the drawing, which incidentally inspired Leonardo da Vinci.

Practical embodiment of all these mechanisms proved to be quite complicated. We should keep in mind that the four-wheeled carriage in which Henri IV was assassinated did not have a movable forecarriage and that the difficulty of maneuvering it facilitated the task of his assassin.

There was also the problem of suspension not only for passenger-carrying carriages but for all wagons, which were quickly put out of order by jolting over the roads. At first the boxes were suspended with chains or straps, as can be seen in a drawing of 1405; perhaps these are the "swinging wagons" mentioned after 1398. Not until the middle of the sixteenth century do we see further improvements: the chains and straps were then attached, not to the framework of the carriage, but to large springs placed on this frame. A very clear picture of such an arrangement appears in a German manuscript of 1568. Henri IV's carriage had no suspension of any kind. The Italian engineers devoted much attention to these problems; the research of Francesco di Giorgio, Leonardo da Vinci, and Dürer can be studied. Leonardo drew the first (before Cardan) picture of the suspension system that bears Cardan's name.

FIG. 5. Suspended carriage (Singer, *A History of Technology,* II, 549, Fig. 501).

Metal bands expanded by heat and then shrunk onto the wheels also came into use late in the sixteenth century; there is no example prior to the middle of this century. Judging by the engineers' drawings, nailed plates were still the most widely used method of protecting wheels from abrasion.

Progress in wagon transportation was thus slow, although visible. In any case it made possible a considerable increase in the importance of carting services, as can be seen in the pictures of certain painters and engravers of the first half of the sixteenth century.

The ship and its development

The evolution of the ship is equally difficult to trace. Our knowledge is very fragmentary because of the lack of exact sources and detailed pictures; moreover, this evolution itself was undoubtedly somewhat vague: there were no abrupt changes, and the historian who attempted to establish an exact date for the appearance of a given type of ship would probably make grievous errors. The transition from one type to the next was unquestionably imperceptible; there were experiments (not all of them are known to us) with various types of sail before the sails were finally standardized. Nothing would be more incorrect than to draw chronological boundaries and to attempt to divide the history of the ship into a series of steps.

All the elements that were to form the ship of the great discoveries and their exploitation had long been in existence. While the Nordic type of ship was now adopted by the Mediterranean fleets, certain characteristics of the southern navies, particularly as regards the sails, were in turn to exert an influence on future shipbuilding. A fifteenth-century Venetian manuscript depicts cargo vessels that are not very different from the ships of the thirteenth century. Their hulls are very flaring; the castles have become part of the planking; the stern is a heavy, square box whose flooring fits over the large round hull, and which soon increased in height, while the bow is a triangular platform with overhanging prow, commonly called an overhanging forecastle. The sheer is very pronounced; there are no straight walls, and the tops have decreased in size. The strength of this hull posed problems of resistance that were difficult to solve. Shipbuilders gave a hostile reception to compartmentalization and the closer spacing of the timbers. External reinforcements were therefore utilized; they protected the ship's wall against shocks, but also impeded its progress through the water.

At the beginning of the sixteenth century, a certain evolution in the shape of the hull can be seen. Around 1520, several English ships abandoned the overhanging forecastle; the prow was dropped to the level of the bridge, and the forecastle dominating it became square, while the external ribs also tended to disappear. The square stern gradually made its appearance. The sterncastles, which continued to increase in height, also became smaller.

The origin of the caravel is shrouded in obscurity. The term had already been long in existence at the time when Christopher Columbus set out on his first voyage; moreover, it was applied indifferently to all kinds of vessels. Gradually it came to designate a ship with clearly defined characteristics, but it is

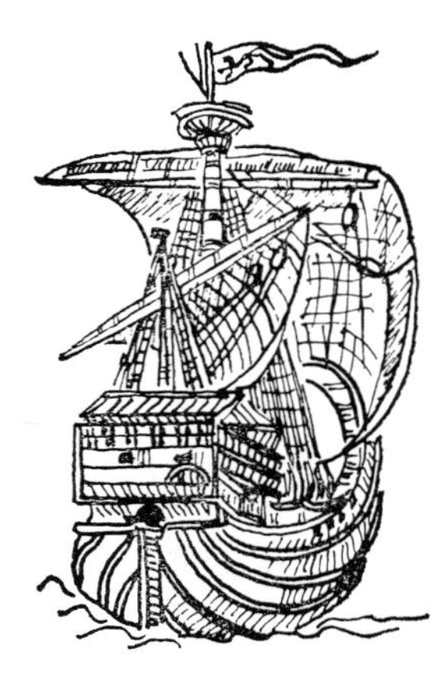

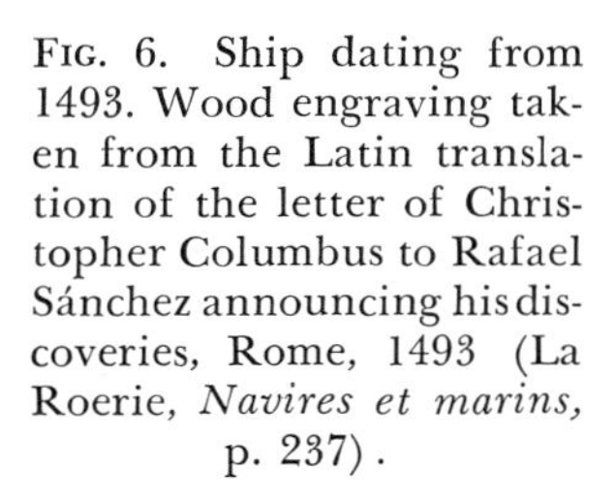

FIG. 6. Ship dating from 1493. Wood engraving taken from the Latin translation of the letter of Christopher Columbus to Rafael Sánchez announcing his discoveries, Rome, 1493 (La Roerie, *Navires et marins,* p. 237).

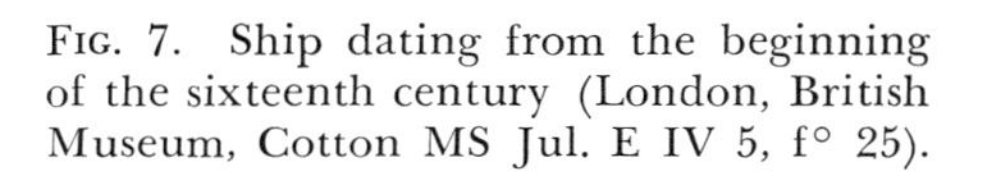

FIG. 7. Ship dating from the beginning of the sixteenth century (London, British Museum, Cotton MS Jul. E IV 5, f° 25).

nevertheless dangerous to list these characteristics, on which no two authorities seem to agree, and the attempted reconstructions are quite hypothetical. Definite pictures of the caravel do not appear until the end of the sixteenth century, by which time this ship was well developed and hybrid forms had already appeared. In any case, it was a small ship, with a capacity of approximately 150 tons. The caravel was easy to maneuver, and thus its shape was particularly successful and sufficiently fine to resist drifting; it was also speedy. The *Santa María* of Columbus probably had an overhanging forecastle.

The transformations relative to rigging and sail appear to have been more decisive, and it is not impossible that these modifications may have brought about a certain development of the hulls. The essential factor is their combination of Nordic and Mediterranean sail types. The ordinary ship — as for example Columbus's caravel — now had three masts, not including the bowsprit (which we shall discuss later). The foremast carried a square sail; the large mainmast, the heir of the primitive single mast, also carried a square sail, while the mizzenmast on the aft castle carried a lateen sail. Only the enormous mainmast was of unquestionable importance; it was very tall, and was constructed of jointed pieces. Some authorities believe that the other masts and sails served as counterweights, acting by lever arm; together they kept the ship on course, and, when acting separately, caused it to alter course.

This new system of sail was to change again between the second half of the fifteenth and the middle of the sixteenth centuries. To begin with, the divided-sail plan came into use; it had been known to the Romans, but was now to be systematized. Above the top appeared a small square sail, the topsail, which gradually increased in size (Christopher Columbus had one, but it was still small). On Anthony Roll's ship, the *Harry-Grâce-à-Dieu* (1509–1547), this sail was larger, but not yet very well developed. The mizzenmast was doubled, and

Plate 3. Oceangoing three-master getting under way. Pen drawing by Holbein, 1532.
Photo Städelschen Kunstinstituts, Frankfurt-am-Main

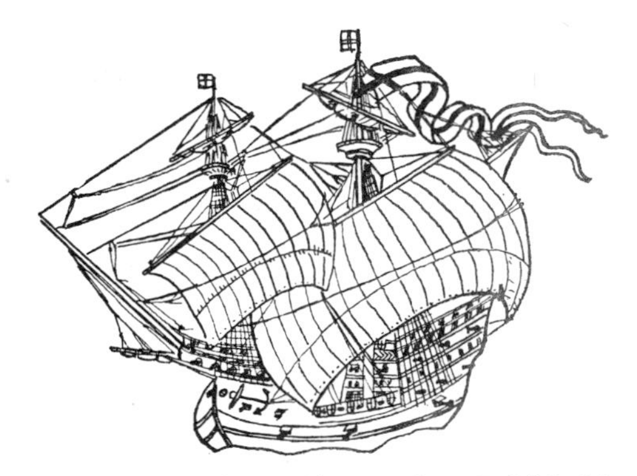

FIG. 8. English ship, 1545 (London, British Museum, map of Calais by Pettyt; La Roerie, op. cit., p. 259).

the after mizzenmast appeared; it can be seen in Pinturicchio's painting depicting the return of Ulysses. The after mizzenmast also carried a lateen sail; a spar hanging from the bowsprit projected over the bow to hold the sheet of the after mizzenmast. Historians are not in agreement on the origin of the spritsail, the small square sail suspended from the bowsprit; in any case, it can be seen in Hans Holbein's pretty drawing of a small ship, done around 1532.

Ships of the same type could each have very different sail equipment. Columbus's caravel had three masts, the last with a lateen sail. The smaller *Niña* had three lateen sails; the *Pinta,* originally rigged with lateen sails, was remodeled during the voyage.

As one sailor states, these ships had a tendency to sail to windward, which is characteristic of ships that drift. Thus they had little need of sail aft. In any case they were not very suitable for tacking, which perhaps led to the use of the after mizzenmast. Only the caravels could be maneuvered in this way; Columbus held course at six "quarters" of the wind (67.5 degrees) and even at five "quarters."

A great number of types of ships were in existence during the fifteenth and sixteenth centuries. While carracks were already in use in the first part of the fourteenth century, their shapes must have gradually changed; the galleon, destined to such a splendid future, appeared in the fifteenth century. However, we are unable to say exactly what were the characteristics of these various ships, and these names were probably applied to greatly dissimilar ships. Nor do we know what exactly is meant by the various local names, which have come down to us without details. Moreover, the transition to finer hulls must have been completely imperceptible.

Problems of navigation

The history of navigation has given rise to intense controversies; incorrect interpretation of texts and arguments over priority have greatly obscured its development. Apparently we must consider the various elements one by one in order to attempt to clarify our present knowledge.

The preceding period had seen the birth of geographical maps. We know that the Majorcan and Catalan schools were beginning to produce veritable masterpieces of mapmaking, which, however, were less useful than has been claimed. Beginning in 1354, however, Pedro IV of Aragon ordered that all his ships carry two charts. The discoveries of 1492 and Magellan's voyage of 1522 were to revolutionize early cartography. The globe of Martin Behaim (1436–1506) is the last step in medieval cartography. On Johannes Schöner's globe (circa 1523) Europe and Africa are correctly represented, but Asia and America are drawn as one continent; mapmakers did not begin to show them as separate areas until 1592.

While the voyages of discovery made it possible to clarify the general form of the continents, and while descriptions of the coasts proliferated, thus becoming of greater assistance to navigation, the major part of the problem remained insoluble, since it was impossible to mark a route on a chart. Not until further progress was made in scientific cartography could the use of these documents become practicable. To Gerhardus Mercator (1512–1594) we owe the first achievements in the transformation of flat maps into maps that detail the earth's contours; he published the first of these "reduced" maps in 1569.

The application of rudimentary trigonometric tables to navigation dates from the thirteenth century. As soon as this aspect of mathematics began to be further developed in the universities, its use spread. The so-called Marteloio or Martelogio Tables appeared at the beginning of this period; examples can be found in the atlas of Andrea Bianco (1436) and in a manuscript of 1444. This table, according to Guy Beaujouan, made it possible to calculate the distance in which a sailing ship driven off course by the wind will regain course by following a given rhumb; it also made it possible to calculate the difference in latitude between two ports.

There remains the delicate problem of sighting instruments. The use of a quadrant by the navigator Diego Gomes during his second trip to Guinea (1463) appears to refer to a sighting made on land; notwithstanding unfounded assertions, navigation by the stars does not appear to have existed before 1480. The nautical astrolable makes its first attested appearance on a chart dated 1529 and preserved in the Vatican. While we may with good reason place its invention slightly earlier, we cannot, however, trace it beyond the beginning of the century. The use of the quadrant at sea was difficult because of the lack of stability of the plumb line. Pedro Nuñez, describing (in 1533) the instruments for finding height, had already replaced the line by a small rule, undoubtedly to the detriment of its precision. The Jacob's staff or *balestilha* (the cross-staff) does not appear to have been used before the beginning of the sixteenth century. Consequently navigators had few sighting instruments, and greater precision would in any event have been purely academic, since there was no way of noting latitude on the chart. Land sightings, especially on the coast of Africa, appear to have increased, and apparently points of orientation were established that permitted the navigator to determine where he was.

Thus, as far as latitude is concerned, the difficulties were primarily material. The tables of declination of the sun were printed in 1483 in Venice, and reprinted in 1488 and 1492; Francesco Bianchini's tables were published in 1495. It still remained to determine correctly the position of the sun, and errors were both numerous and important. The problem of longitude appeared almost insoluble. Columbus and Vespucci attempted to employ lunar methods and the conjunctions of planets. In 1493, at Haiti, Columbus sought a safe port from which to observe the conjunction of the sun and the moon. In 1520, Magellan's better-educated pilot observed conjunctions. The use of the eclipses and the lunar distances were recommended by Alonzo de Santa Cruz, the cosmographer of Charles V (Charles I of Spain). The ephemerides of Regiomontanus (Johann Müller) were used for the years 1475 to 1506, and the *Calendarium eclipsium* for the years 1483–1530.

The astronomical problem of longitudes, which theoretically had been solved, in actual fact was still unsolved, since its solution required the simultaneous realization of a certain number of material conditions. Thus navigation continued for some time to be a matter of guesswork. It still remained to solve the two questions of direction of course and speed, and then to find a method to utilize this information to determine position, in other words to solve the loxodromic problem. As for speed, almost the only method of determining it was sight approximation; the ship's log is mentioned for the first time only in 1577. Ideas such as that of magnetic declination were still denied by certain navigators and even certain scientists: the *Arte de navigar* of the mathematician Medina, published at Valladolid in 1545, is an example. Consequently, it was not until the first half of the sixteenth century that a solution was about to be found.

CHAPTER 3

THE GROWTH OF MECHANIZATION

THE DEVELOPMENT of the machine and a belief in its almost infinite possibilities—a belief that was to be shared by the engineers discussed earlier—are certainly among the most essential characteristics of the technology of the Renaissance. This evolution was most probably due to a certain number of causes that it is quite difficult to define, the most important of which are undoubtedly the invention of new systems of transmission of movements and the more widespread use of a new material, metal.

It was inevitable that the machine, now constructed with vastly stronger elements and more perfected systems, should make decisive and rapid progress. To these material improvements must be added a taste for mechanization on the part of the most advanced technicians of the period. They utilized machinery both for attacking and defending strongholds, for the preparation of the "games" that were so greatly developed by the fifteenth and sixteenth centuries, and for purely industrial purposes (purposes that, it is true, were quite foreign to the mentality of certain technicians). The growth of mechanization also appears to have been in large measure the result of a speculative type of thinking; Leonardo da Vinci was not the least of its representatives. Before their practical applications were discovered, machines were created for intellectual amusement, for the pleasure of making them, and for the solution of difficult problems. All these engineers were painters or sculptors, and most of them undoubtedly regarded their mechanical research as an aspect of their art. This is evident in certain drawings, where mechanization and anthropomorphism are perfectly allied, and in which we become aware of a mechanization of nature that will continue to increase until the beginning of the seventeenth and even into the eighteenth centuries.

The flywheel and the crank-and-connecting-rod system

The invention of the crank-and-connecting-rod system is perhaps the most important mechanical acquisition of the early fifteenth century. This system, which was unknown in the medieval period, permits the transformation of a continuous circular movement into a straight-line, alternating, back-and-forth movement, or vice versa. Its absence, as we have seen, had greatly limited the development of mechanization.

While the crank had been in general use for centuries, the same was not true of the bitbrace, no picture of which is found in the medieval period. In Meister Franck's "Carrying of the Cross" in the Hamburg Museum, which dates from 1424, we see a bitbrace for drilling holes in wood in a basket carried by one of the soldiers. We find it again the triptych of Le Maître de Mérode (Flémalle), which is slightly later.

The combination of the crank and the brace forms the crank-and-connecting-rod system. The jointing of movable pieces obviously posed problems that were difficult to solve, particularly the problem of wear, at a time when the pieces were still made of wood. The invention of this mechanism is still difficult to date, as are all the inventions of this period. At most we may discover its first appearance in practically utilizable form after long inconographic research that is not yet complete.

In Kyeser's manuscript, at the extreme end of the fourteenth century or in the very early years of the fifteenth, we find a hand mill operated by the crank-and-connecting-rod system. It was by now understood that the device had two dead centers. The flywheel that appears in Kyeser's drawing was created to compensate for this disadvantage, and so the principle of inertia, the theory of which was not to be established for almost a century and a half, had at least been discovered, if not clearly understood. Very similar drawings appear in the so-called "Hussite War" manuscript; they too include flywheels.

While nothing similar appears in the work of Taccola, the engineers of the second half of the fifteenth century seem to have completely adopted the new mechanism, and its use was spreading rapidly. Solutions to the problems of dead center and regularity of movement, especially in the hand mills, were now perfected. The ideal solution can be seen in a drawing in Francesco di Giorgio Martini's treatise on architecture, where the flywheel has a governor with fly balls. Here we have passed the stage of a simple device. This picture reveals a great deal about the mechanical genius of our engineers.

After this, applications of the crank-and-connecting-rod system occur rapidly and are easily to understand; Francesco di Giorgio's treatise presents the most striking examples. First there was the problem of the saw. We have studied Villard de Honnecourt's drawing of the saw (middle of the thirteenth century), with its pole that acted as a spring. Francesco's drawing shows us a modern saw of the type in existence until the end of the nineteenth century; it has a wooden frame sliding between two poles, and is operated by a crank-and-connecting-rod system powered by a hydraulic wheel, while another device moves the piece of wood forward. Another difficulty solved was the problem of pumping water with lift-and-force pumps, an application that was to undergo considerable development in various activities.

A last and equally interesting application of the crank-and-connecting-rod system is depicted in the von Waldburg-Wolfegg *Hausbuch*. Here we find a spinning wheel with flyers inaugurating a series of treadle lathes, which were to develop considerably in the coming centuries.

The application of the crank-and-connecting-rod system was still impeded by difficulties airsing from the materials used and the jointing of the moving parts. The attachment of the connecting rod generally consisted of a simple ring which it was difficult to maintain in position. Between Francesco di Giorgio's water pumps (around 1470) and those that appear in Agricola treatise on metallurgy (1556), no improvements appear to have been made. The en-

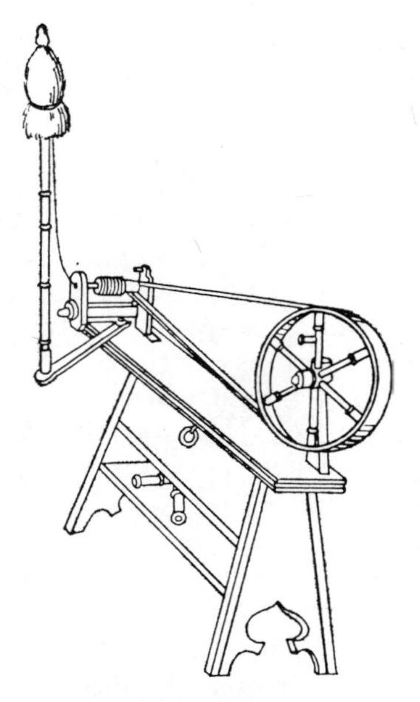

FIG. 9. Spinning wheel with treadle. Around 1470 (Waldburg-Wolfegg *Hausbuch*).

FIG. 10. Lathe. Around 1500. Leonardo da Vinci (Milan, *Codex atlanticus,* f° 381, r⁰b).

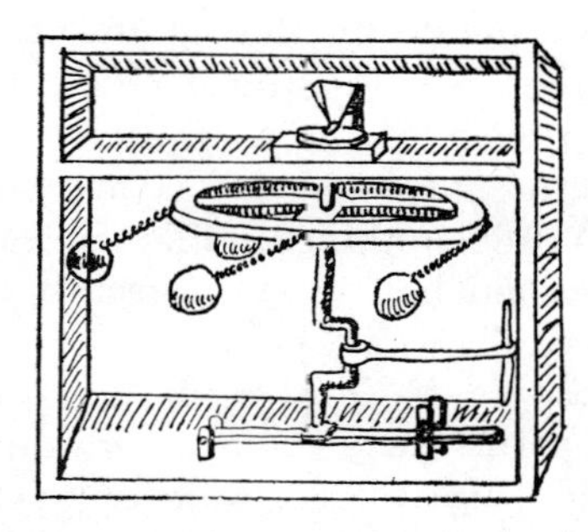

FIG. 11. Governor with fly balls. Around 1475 (Turin, Duke of Genoa 148. MS of Francesco di Giorgio).

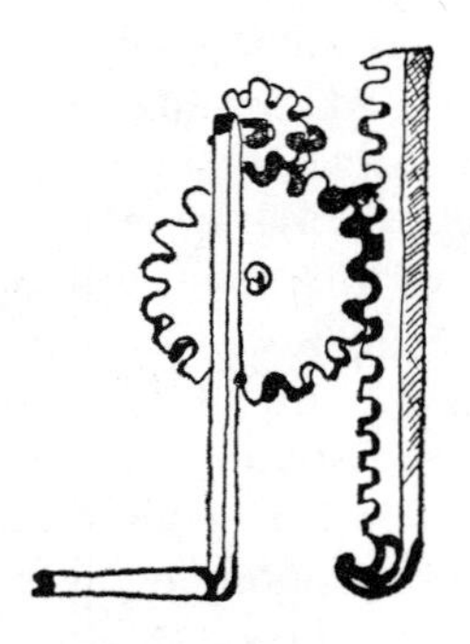

FIG. 12. Lifting jack. Around 1500. Leonardo da Vinci (Milan, *Cod. atl.*).

FIG. 13. Type of link chain. Around 1500. Leonardo da Vinci (Milan, *Cod. atl.*; Uccelli, p. 75, Fig. 237).

gineers themselves were aware of these difficulties, and in Francesco di Giorgio's work pumping systems with bucket chains are as numerous as the use of the improved device. Two treadle lathes, probably dating from the sixteenth century, can be see in the Deutsches Museum in Munich and the *maison des brasseurs* at Antwerp, respectively; both of these machines are still extremely rudimentary. The engineer's drawings are sometimes more advanced than their practical achievements. This aspect of the technology of the Renaissance will be discussed later.

We also find tentative solutions in the notebooks of Leonardo da Vinci. His early solutions to the problem involve a simple, relatively useless device: a cylinder on which is carved a double helicoidal groove to engage a tooth on the shaft of a piston, a weighted beam being utilized in the case of the two pumps. Leonardo later had an idea that was adopted by certain sixteenth-century technicians. A continuous circular movement is first transformed into an alternating circular movement. A wheel with teeth on half of its circumference gears into two lantern gears attached to the end of an axle that transmits the power. The first lantern therefore turns in one direction, the other in the opposite direction. After this it was a simple matter to transform this alternating circular movement into a straight-line alternating movement by means of a third lantern gear and a rack. Ultimately Leonardo achieved a true crank-and-connecting-rod system, but only after experiments that are indicative perhaps not so much of lack of imagination as of the difficulties involved in constructing the device, especially in the case of large projects.

Gear trains

We shall quickly summarize this problem that in theory rather than in fact occasioned fairly detailed studies on the part of the Renaissance engineers, chiefly Leonardo da Vinci. One of the keys to the problem, however, continued to be the use of metal for stronger gears; this would have required metal-cutting tools that did not yet exist. Only in Agricola's work do metal gear trains appear somewhat more frequently. However, for a long time wood continued to be the principal material used for the making of machine parts, as Diderot's *Encyclopédie* proves.

The most frequently encountered gear train is the type that consists of toothed wheels and lantern pinions; it had long been in existence and undoubtedly had been perfected. The principal problems had been solved in antiquity, in particular that of multiplication; for others, no satisfactory solution was then possible. Some of his drawings reveal that Leonardo was interested in the problems of friction and wear of the gears of toothed wheels; he successfully demonstrated how important it was for the teeth of the two pinions to catch more or less against each other to avoid excessive wear. The *Codex atlanticus* also contains drawings of profiles of toothed wheels and the manner of tracing and making them. Leonardo's research went even further than this: in Manuscript "B" we find projects for toothed wheels with trapezoidal cross sections.

The problem of irregular gearing with a regular movement was much more difficult to solve, because it was not known how to calculate the necessary curves. The sketches in Leonardo's notebooks are only very distant foreshadowings of later achievements. Toward the middle of the sixteenth century, Jerome Cardan made the best contribution of this period to the theory of irregular toothed wheels. The first achievements are shown in a work of the French mechanician and mathematician Jacques Besson, in the last quarter

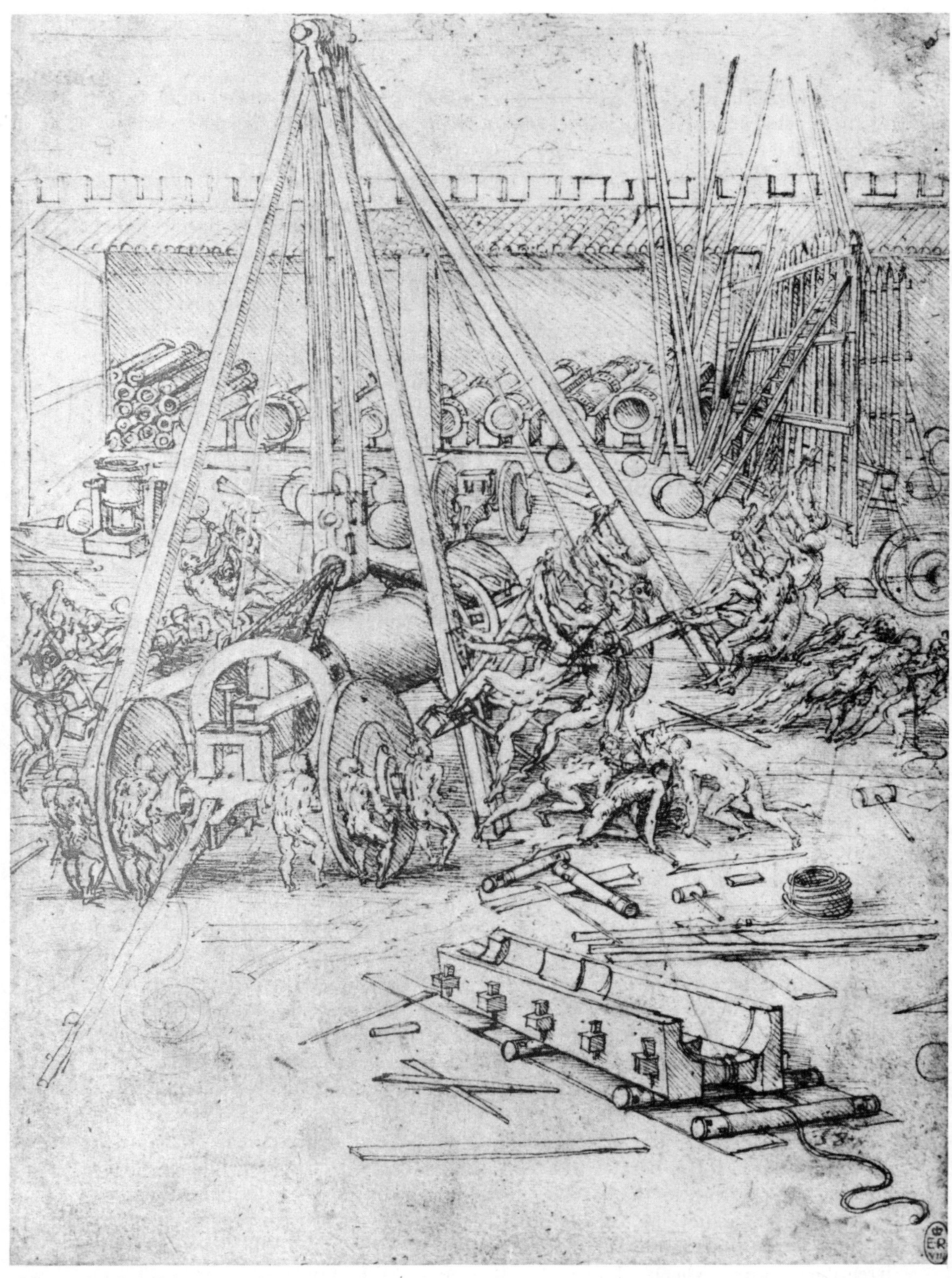

Plate 4. Emplacement of a gun barrel on its carriage. Drawing by Leonardo da Vinci, Windsor Castle. *Photo Royal Library.*

of the sixteenth century. Not until the Danish astronomer Olaus Roemer do we see the use of toothed wheels with epicycloidal profile, independently of the French mathematician Gérard Desargues.

Probably because of the difficulties we have just mentioned, Francesco di Giorgio seems to have favored gear trains composed of endless screws, which were also frequently used by Leonardo da Vinci in his projects for machines. The chief defect was their considerable tendency to wear, at a period when the oiling of turning parts was still very rudimentary. These gears with endless screws are widely used in power devices. In addition to the classic screw jacks, which were already in use in the Middle Ages, we find machines for moving heavy weights, columns, and even pyramids; we also find them combined with jointed parallelepipedic joinings. Some of these drawings were adopted by Leonardo da Vinci (to whom all the credit is given) from Francesco di Giorgi. In addition to raising monolithic columns (a practice brought back into use by the architects of the Renaissance), these machines were used in particular for moving certain structures. Undoubtedly it was the devices of this type that Aristotele Fioravanti straightened the tower of the Podestà at Bologna, and in 1455 moved the tower of the church of the Magione, which weighed 407 tons and was more than 59 feet tall.

Leonardo da Vinci raised another problem, for which he sketched several solutions: the changing of speed by substituting one gear for another. In his screw-cutting machine, in which he could act simultaneously on the speed of rotation of the object being cut and on the speed of the forward movement of the tool-holding device, the movement could be modified only by dismantling the pulling gears and replacing them with others. Later he tried to combine the various speeds that could be utilized into a single mechanism; this resulted in a superposition of toothed disks on a single axle and a lantern gear in the shape of a truncated cone that could by changing position gear into one or the other of the disks. In the cam type of hydraulic machine, the solutions of changing the crown bearing the cams and changing the flow of the stream in order to obtain varying speeds continued to be used for a long time.

Other devices for the transmission of movement were perfected. In lathes that did not yet use the crank-and-connecting-rod system, leather straps rather than ropes that slipped on the wooden axle were used. In Taccola's work (written before the middle of the fifteenth century) we find a drawing of an ordinary

FIG. 14. Gear train with trapezoidal cross section. Around 1488–1489. Leonardo da Vinci. (Paris, Bibliothèque Institut, MS B, f° 73 v°).

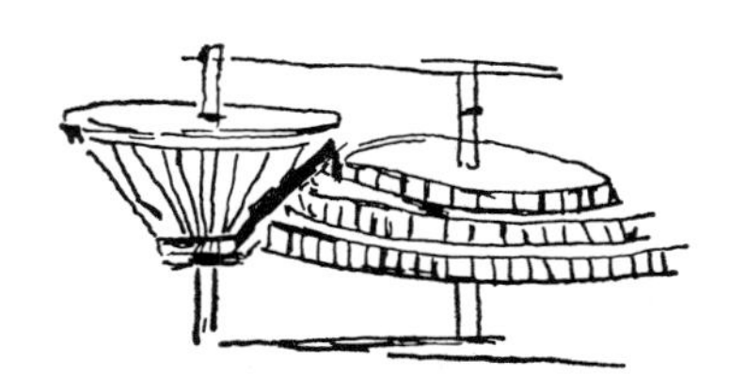

FIG. 15. Cone gearing. Around 1490–1500 (Milan, *Cod. atl.*).

Plate 5. Chain-type lifting machine. Engraving taken from Ramelli, *Le diverse e artificiose machine*. Conservatoire des Arts et Métiers. *Photo by the Conservatoire.*

chain ultilized as the organ of transmission. But this is only an isolated example. The *Codex atlanticus* contains a series of drawings that seem remarkably advanced for that period; they are drawings of jointed chains that are specially conceived to be used for the transmission of movements, and bear a strong resemblance to modern chains. The difficulty of obtaining metal elements for the chains depicted leads us to suppose that they were probably never made, and in fact such chains were not finally constructed until the first half of the nineteenth century.

The utilization of hydraulic power

There were several ways of improving the efficiency of the hydraulic wheels, and all of them were used concurrently. The first concerned the wheels themselves. Here we realize the interest of certain of Leonardo's studies: his spirit of observation and undoubtedly his feeling for experimentation are better revealed in these projects that relate to a precise technological question than in invention. In regard to mill wheels, Leonardo envisioned a series of cases, depending on the flow of the stream, and the height of the fall, in order to determine both the position of the wheel and the shape of the paddles. It is even possible that he constructed models in order to perform the necessary experiments. He quickly decided on the use of inclined paddles (which may, however, have been in use before him), and devised trough paddles to utilize the weight of the water contained in these troughs. Such research can have had little influence, since Leonardo's notes remained buried in his notebooks. They are perhaps a reflection of isolated empirical attempts noted by Leonardo in the course of his travels. While we find no very striking improvements in the wheels depicted in artists' drawings of this period, we do find an increasing use of inclined paddles.

The second problem had to do with the very principle of the wheel. We have already discussed horizontal wheels, for whose existence we have no proof before a fairly recent period. The engineers' notebooks give us not only the first evidences of these horizontal wheels, but even drawings of genuine hydraulic turbines. In Francesco di Giorgio's treatise we see these first attempts at turbines, with supply pipes and suitable nozzle to obtain the greatest possible power. The paddles were also carefully constructed, and their curve perfectly

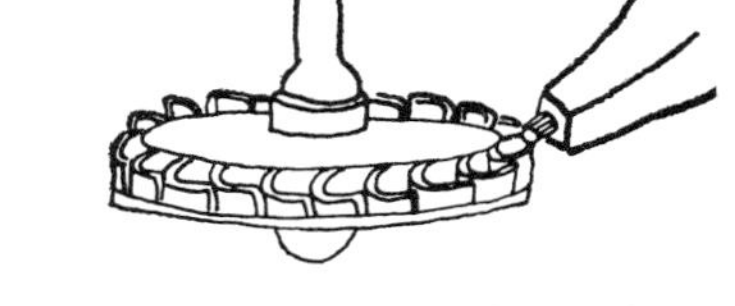

Fig. 17. Project for turbine. Around 1475 (Turin, Duke of Genoa 148. MS of Francesco di Giorgio).

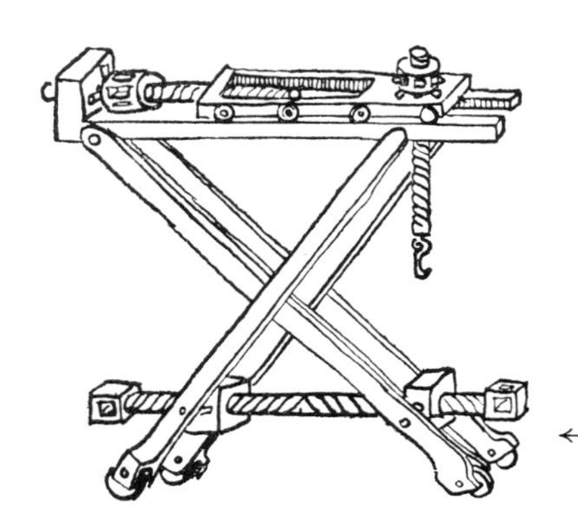

← Fig. 16. Lifting device with various positions. Around 1475 (Turin, Duke of Genoa 148. MS of Francesco di Giorgio).

adapted to the use to be made of them. The idea of the turbines was adopted by Leonardo and all the sixteenth-century engineers. It is nevertheless true that these wheels were seldom used prior to the eighteenth century; the Bazacle wheels, of which we have already spoken, are the best example.

Thus, while traces of undeniable progress can be seen in the engineers' notes, their practical achievements were probably not so rapidly accomplished as is frequently claimed. If instead of examining these engineers' projects we were to study carefully the paintings of this first half of the sixteenth century (some schools of painting are very rich in works that are perfectly useful for the historian of technology), we would find that mill wheels had changed very little since the medieval period. The paintings of Joachim de Patinir (1480–1524) and Henri Met de Bles (1490–1550) show no major changes, and the same can be said for the pictures in Agricola's treatise. However, the latter work does contain a picture of a wheel with reversible movement, intended for transporting ore extracted from a mine to the surface. This machine consists of two wheels back to back, whose paddles are inclined in different directions. A reservoir has two water pipes, one on each wheel. By maneuvering the proper vanes, the winch can be turned in either direction.

The utilization of wind power

Windmills, in contrast, made considerable (and undoubtedly immediately applied) progress. We have already noted that the old windmills were all conceived along the same lines; a drawing in the "Hussite War" manuscript is an excellent representation of such a windmill. The mill is raised on a tripod, and the entire structure turns in order to orient its wings to the wind. Thus the instrument had to be built of wood.

The windmill with turning cap was undoubtedly born in the fifteenth century. The mill that appears in a manuscript of Guillaume de Machaut (end of the fourteenth century) may be the first of this type, but the drawing is not sufficiently clear to be conclusive. Not until the fifteenth century, in the work of Leonardo da Vinci, do we find the next, this time definite, specimen. In the very first years of the sixteenth century this type of windmill, now constructed of masonry, began to spread very rapidly. Photographs of a few extant sixteenth-century specimens appear in handbooks on archaeology.

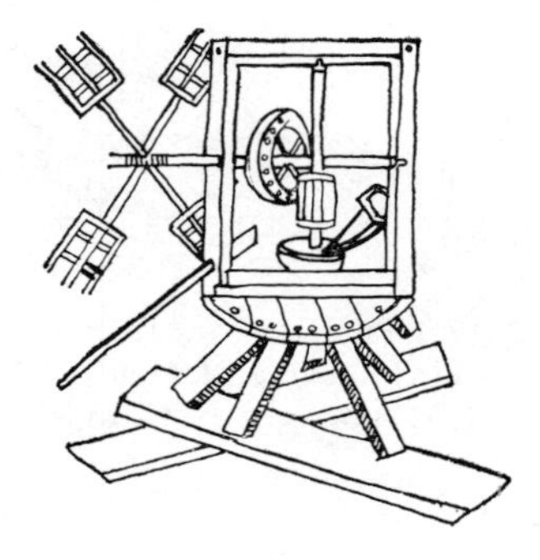

FIG. 18. Windmill. Around 1430 (Munich, SB, lat. 197, "Hussite War" MS).

FIG. 19. Windmill with pivoting cap. Around 1490–1500. Leonardo da Vinci (Milan, *Cod. atl.*).

Plate 6. Windmill with turning cap, operating a hydraulic ball pump. Engraving taken from Ramelli, *Le diverse e artificiose machine.* Conservatoire des Arts et Métiers. *Photo by the Conservatoire.*

Plate 7. Windmill with turning cabin operating a wheat mill. Engraving taken from Ramelli, *Le diverse e artificiose machine*. Conservatoire des Arts et Métiers.
Photo by the Conservatoire.

Birth of automation

The idea of automation certainly was not present in the minds of the Renaissance builders and engineers; and yet we can see that some of their research and their mechanisms ultimately led up to it, thus laying the groundwork for ideas that were to appear in more detail in the works of Descartes and, after him, in specialized literature.

The regulation of movement is certainly an ancient idea. It is present as soon as we see the slightest intervention of a flywheel or substitute for a flywheel; we find it in the first mechanical pendulums, which in general have two systems of flywheels, one continuous and the other alternating, to regulate the movement. We find the same idea with the appearance of the crank-and-connecting-rod system, for which this regulation was indispensable because of the dead center. In the first extant pictures of it (dating from the early years of the fifteenth century) we note the presence of the indispensable flywheel. With Francesco di Giorgio we go a step further, since he preceded James Watt by almost three centuries in imagining a governor with fly balls. Though nothing came of the idea, at least not immediately, its appearance should be noted, as much for the effort of thought behind it as for its application to the problem of the mill.

In the first half of the sixteenth century, we find another application of automatic regulation of movement. The funnel-like device (the hopper) that distributes the grain over the millstone of a mill is joined to a piece that rubs against the square portion of the axle of the millstone; in this way the hopper is shaken and lets the grain fall. Thus it is the turning movement itself that regulates the distribution of the grain, causing the millstone to work more when the energy available is more abundant. The hardness of the grain may also play a role, causing the millstone (and therefore the distribution) to slow down. A simple invention, to be sure, and one whose author is completely unknown to us, but an interesting one nevertheless. It is difficult to date it, for this governor appeared under a multitude of forms, and it is not impossible that it is depicted in the drawing of a mill mechanism in the manuscript of Herrad of Landsberg. In this case, it would be possible to say that the first example of automatic regulation of movement dates from the twelfth century.

Other automatic devices also existed, those used in clockmaking being probably the most highly perfected. Their influence in the construction of machines was certainly very great; the utilization of metal gears and cams led to the development of increasingly complex mechanisms that were certaintly more fruitful than the "toys" of the School of Alexandria. An additional complexity in clocks of this period was the automata that sounded the hours. Our only information about these early automata comes from vague texts, as in the case of the cock of Cluny, (mid-fourteenth century), or through still-extant models that have, however, been much repaired and probably modified in the course of centuries. The oldest examples of the latter are the automata of Orvieto, which date from 1351, and those of Courtrai, which were brought to Dijon in 1381, after the Burgundian victory.

In these devices it was not, strictly speaking, a question of regulation, but rather a mechanical release on a regular movement intended to produce effects at a given moment. An example of its industrial use can be seen in Leonardo da Vinci's spinning machine designed to wind thread evenly on a bobbin. These devices are of interest insofar as in the second half of the sixteenth century they were to produce more precise and more effective results.

Later in this work we shall study the various machines imagined, or actually created, by these engineers. We should again stress the fondness for mechanization that was particularly characteristic of the end of the fifteenth century — a tendency favored by the appearance of these new elements of new mechanisms, and by the more widespread use of metal. The essential fact, however, is not so much the still relative proliferation of machines as this systematic research in certain areas. It was finally realized that the use of machinery constituted a gigantic step forward and that, thanks at times to a somewhat excessive imagination, this progress was now within reach. By the second half of the sixteenth century, the interest taken in machinery had become, with a few exceptions, a game. This was the period of the innumerable "Theatres of Machines," which contributed nothing and merely repeated the old ideas *ad infinitum;* the work of the Italian Ramelli (1531–1600) is an excellent example.

Our very rapid study of hydraulic power reveals the theoretical character of this movement, whose practical realization continued to be hampered by numerous material difficulties. A careful examination of the machines depicted in Agricola's treatise (middle of the sixteenth century) reveals their imperfections; their functioning was probably equally inefficient. Only in very slow stages were really perfected and usable mechanisms obtained.

Undoubtedly the importance of certain research must not be exaggerated. While Leonardo da Vinci brought about great progress in technology — rather, while we can justly claim, based on an examination of his notes, that technology was now on the road to progress — he did not, however, invent everything possible in the domain of mechanics and power. Hydraulic turbines already appear in the notebooks of his predecessors. The information (also known by his predecessors) on the power of expansion of steam that he utilized apparently does not suffice to place him among those who aided in the invention of the steam engine. Certain intuitions — if a number of them are not simply a heritage from earlier generations — do not, strictly speaking, constitute genuine progress. Progress lies rather in the generalization of a few problems (gears, mill wheels), and it was his presentiment of these problems that reveals the superiority of his mind over the minds of the other engineers of this astonishing fifteenth century.

CHAPTER 4

TECHNIQUES OF ACQUISITION

THE RENAISSANCE did not revolutionize the techniques of acquisition to the extent claimed by certain historians. Certainly some of these techniques were greatly perfected, for example mining techniques, the development of which was completely unrelated to the great discoveries. In contrast, only insignificant transformations occurred in agriculture. (It is easy to understand, however, why this extremely dispersed activity, involving a multitude of workers, was confined within a fairly stable traditionalism: technical progress is quicker to reach the sectors of limited production, where the centers of exploitation are concentrated in a small number of hands.)

AGRICULTURE

Agriculture remained one of the essential economic activities that continued to occupy most of the population of western Europe. Technical transformations in this field proved to be almost imperceptible. The epidemics of the fourteenth century, the trouble of the Hundred Years' War, and the political instability did not favor a rapid evolution until the end of the fifteenth century. The discovery of America contributed several novelties, especially in the botanical field, but their consequences were felt only very slowly, in certain cases being fully developed only at the end of the eighteenth century.

Even the science of agronomy does not appear to have produced works of great interest. Until the beginning of the fifteenth century, only the works of already known authors were popularized. Pietro de' Crescenzi's work was printed in every language: between 1471, the date of its first printing at Augsburg, and 1550 there were six Latin, nine Italian (the first in 1478), one German (1518), and seven French (the first in 1533) editions. The Latin writers Marcus Cato, Marcus Varro, Lucius Columella, and Rutilius Palladius were printed in multiple copies, sometimes separately, sometimes together; there were thirty-one editions in the same period, in Latin, French, Italian, and German. The first edition dates from 1472, in Venice, grouping the other texts.

Original works did not appear until the sixteenth century, and most of them were satisfied merely to repeat notions and ideas already widely accepted. Two French sixteenth-century manuals, the *Maison rustique* of Charles Estienne, published in Latin in 1564 (several portions had, however, been published earlier), and the *Théâtre d'agriculture* of Olivier de Serres, printed in 1599, clearly lack originality. M. Grand regards Olivier de Serres's work as an end rather than a beginning, in that he summarizes the body of knowledge handed down through centuries of practical experience. At most we find in his work a timid intuition of the role of artificial meadows. The only genuinely important work was that of Bernard Palissy, *Recepte véritable par laquelle tous les hommes pourront apprendre à multiplier et augmenter leurs thrésors* (True guide by which every man will be able to learn to multiply and increase his treasures), published at La Rochelle in 1563. Palissy attempted to introduce scientific ideas into agricultural techniques. He was not widely read, and remained without any genuine influence; Parain notes that he had a vision of the future whose assurance and perspicacity his contemporaries were undoubtedly incapable of grasping.

The same situation prevailed in the neighboring countries, which produced a handful of modern works whose value is completely relative, although their success is attested by the number of editions in the course of the sixteenth century. Such is the case of *The Boke of Husbandry,* by an English author named John Fitzherbert, which appeared in 1523 and went through eight editions before 1550. Other examples are the *Libro de Agricultura* of the Spaniard Alonso Herrera, which was published for the first time in 1539, and *La Coltivatione* of the Italian Luigi Alamanni, published in Padua in 1546. These are anthologies of mediocre value that offer no evidence of important modifications in medieval agriculture.

It is true that scientific achievements which could be of direct use in this field were as yet few in number. The science of botany was still in embryonic state; the botanists studied Pliny the Elder (in print since 1469) and Pedanius Dioscorides (printed in 1498, with commentaries by Mathiole written in 1548). The German author Otto Brunfels (1489–1534), of Mainz, a doctor at Strasbourg, published the first work on botany in 1530. Other doctors, including Leonhard Fuchs (1501–1566), Rembert Dodoens (1517–1588), and Hieronymus Bock (1489–1554), established the first (though still vague and incomplete) plant descriptions. Of undoubtedly greater importance was the role of the gardens, which merits a serious study. Italy in particular had innumerable botanical gardens, the most famous being those of Ferrara (established in 1528 by Alfonso I d'Este), Pisa (1544), Padua (1546), and Bologna (1548).

The governing authorities encouraged, although without a clear-cut plan, a vaguely perceived technical progress. Charles V of France had a French translation made of the works of Pietro de' Crescenzi and Bartholomaeus Glanville, and ordered the composition, by Jean de Brie, of a treatise on sheep raising. The Sforza library contained several manuscript copies of the Latin authors and Pietro de' Crescenzi. Louis XI encouraged the rehabilitation of French agriculture, which had been devastated by the war, and tried to introduce several improvements.

The arrival of new plants

The principal consequence of the discovery of America could only be the introduction of new plants into Europe. Europe's contribution to America was in this regard undoubtedly more important. However, not enough is yet known about these botanical exchanges that resulted from the great discoveries. The historians are not in agreement on certain individual cases; some believe that a number of plants are of American origin, while others deny this and believe that America was able to furnish Europe with several new varieties that were of greater interest than the European species.

Discussion continues on two plants, with conclusions being difficult because of the lack of precise information. Nevertheless, the most modern historians of agriculture believe that the white bean is not an arrival from America. The case of maize is still doubtful. Those who believe in an American origin cannot say whether this plant was brought by Columbus after his first trip in 1492, or by Cortez in 1529. Others feel that America simply contributed a better variety of maize. In any case its cultivation spread widely through western Europe during the first half of the sixteenth century, where it replaced panic grass, sorghum, and the little remaining millet.

What remains of the American botanical contribution? The potato did not really become known until the second half of the sixteenth century. The tomato may also have come from America, but we are not absolutely certain. Coffee, sugarcane, and certain cereals are European importations into America.

Thus the American contribution may be quite small, and all the more unimportant insofar as other, non-American, crops appeared or developed considerably at the end of the fifteenth and in the course of the sixteenth centuries. Improvements in the plants that furnished a dietary supplement came about for the most part through a slow transformation of the various species, apparently achieved in Italian gardens. Not until the Renaissance, when the gardeners made it less woody, did the carrot become a valued vegetable. The beet may have been developed from Swiss chard improved upon in the Italian gardens. Lettuce was not enjoyed and cultivated in France until the reign of Charles V, at the end of the fourteenth century. The artichoke was the product of a modification in the thistle by the peasants of southern Italy; we find it in Florence in 1466, at Venice in 1493, and after that on the tables of the wealthy all over Europe. The melon also reappeared in Italy in the fifteenth century, and was brought to France by Charles VIII. The cantaloupe was imported from Armenia in the fifteenth century.

Progress in the various slowly transformed species of fruit was also evident. In the second half of the fourteenth century the strawberry, which until then had been picked wild in the woods, appeared in gardens; in 1368 it reached the lands of Charles V, in 1375 those of the Duke of Burgundy. The raspberry and the currant also developed from wild berries into cultivated plants.

At the end of the fifteenth century there also appeared the plants that were to assure the development of artificial meadows. Sainfoin arrived in the south of France at the beginning of the sixteenth century, probably from Italy, where it had been under cultivation for some time. Red clover and alfalfa began to spread in the middle of the sixteenth century; they are noted in southern France around 1550.

Other plants came from the East. There were, for example, cauliflower, known to the Arabs in the twelfth century but not found in Europe until the sixteenth century; cloves and cinnamon, brought back by Vasco da Gama from the Spice Islands in 1498; the white mulberry, introduced from the Levant into Tuscany around 1434, into France around 1495. Buckwheat, undoubtedly cultivated in eastern Europe in the Middle Ages, continued its progress westward: it appeared around 1460 in Normandy, around 1500 in Brittany. Under Henry VIII hops became common in England, causing a very visible retreat of barley.

Thus there was no revolution in the geography of cultivated plants. Undoubtedly more attention should be paid to the important mutations obtained through an extremely detailed market-garden cultivation and in the botanical gardens.

Techniques of cultivation Transformations in the techniques of cultivation and improvements in the medieval methods are even less apparent and more difficult to discern; in this regard a reading of the treatises on agronomy mentioned above is significant.

During the Renaissance no new techniques of soil preparation were added to those already in practice. Drainage and irrigation also continued to be practiced with the same methods. The irrigation of pasturelands appears to have been developed, which perhaps explains to a certain extent the development of livestock, although the proportion of meadows to cultivated lands at the beginning of the sixteenth century was approximately the same as it had been in the ninth century, at least in the Paris area. In the Vivarais in the fifteenth century, the meadows were plowed and cultivated every twenty years, to ensure the renewal of the grass. No new fertilizers were developed, and the existing varieties were still insuffiicient to return vigor to the soil. Tree planting made little progress until the beginning of the nineteenth century, the principal revolution of the eighteenth century being, thanks to the artificially created meadows, the development of cattle breeding. All the regional studies relating to the sixteenth century stress the lack of manuring of cultivated land, a defect that continued to exist despite clauses in the contracts of tenant farmers.

The techniques of cultivation as such required few modifications. At most we find better utilization of certain tools, probably as the result of great specialization of the various types of instruments. For example, the use of a light plow for plowing the vineyards appeared in the Bordelais in the fifteenth century, whereas in the Middle Ages this operation had always been done with the spade.

There were few perceptible changes in agricultural equipment—a stability commended at the end of the sixteenth century by Olivier de Serres ("Do not change your plowshare, for fear of the danger of loss that all change brings with it.") In the face of this excessive traditionalism, Bernard Palissy noted that while the military engineers were making major efforts to improve weaponry, their ingenuity disdained agricultural equipment, which still continued "in its accustomed fashion." Note should perhaps be taken of the improved construction of agricultural tools, a consequence of the progress in metallurgy. But the two quotations just cited clearly demonstrate that despite certain efforts, in most cases the practitioners rejected change on the grounds of their very experience. Both Olivier de Serres and Charles Estienne recommended the plow (*charrue*) as superior to the *aratrum,* without differentiating between the nature of the types of soil — a waste of effort, since the *charrue* was unsuitable for hilly areas, narrow fields, and the light soil of the Mediterranean world.

Had any improvements been made in this *charrue?* The two major types of plow were already in existence: plows with fixed moldboard, and the two-way plow with raised forecarriage. Local specialization was already complete. The only improvement made in this instrument appears to have been the mount, perhaps introduced at the same time as the gun mount. The *Livre des femmes nobles et renommées* given by Jean de la Barre to the Duke of Berry in February of 1404 depicts an admirable specimen (moreover, the miniature is astonishingly detailed). The two-wheeled forecarriage had a mount made of a forked tree branch; the beam rested on this mount, whose exact point was determined by a wooden peg that could be inserted into a series of five holes arranged along the beam. The closer the peg was to the colter, the tighter the harness and the less deeply the plowshare penetrated the earth; this was the system of regulation of cut by gauge. According to Maget, the author of this description, this was one of the most highly perfected types of plows in France prior to the revolution in agricultural technology in the nineteenth century. A miniature in a work by Egidio Romano (early fifteenth century) shows a different system of regulating depth.

Other agricultural activities remained unchanged. Estienne and Olivier de Serres recommended the use of the roller and the harrow with wooden or iron teeth to "break up the weeds" and cover the seeds, but these instruments had long been in existence; southern Europe seldom utilized them before the middle of the nineteenth century.

As for hand tools, few changes could be expected. Spades, formerly made of wood with a metal band, were now made completely of metal, thanks to improvements contributed by the rolling mills. In addition, these metal tools existed in greater numbers and were utilized over a wider area.

A reading of the writings of the agricultural technicians and an analysis of pictures as detailed as those of Brueghel amply demonstrate that the few achievements that have been noted are still very relative, and that the vast majority of the western-European peasantry did not modify their primitive practices in any way.

The forest

The same stability was characteristic of the techniques of exploitation of the forest. The only tree whose exploitation appears to have been placed on a rational basis is the poplar; a stronger and more useful variety may perhaps have been imported from America. In any event, the plantations of poplars in low-lying areas and damp places date from this period.

The great geographical discoveries did, however, have an indirect influence on the techniques of forest exploitation, which until then had admittedly been quite primitive. Larger ships were needed in larger numbers, and the work of naval construction inevitably drew the attention of the ruling powers to the problem of wood. This problem had gradually become very serious in a number of regions that were threatened with a shortage. Certain historians have emphasized the fact that the decline of Spain coincided with the impoverishment of her forests, which were stripped by the demand for wood for naval construction. This forest destruction may in addition have caused climatic changes harmful for agriculture.

These were certainly not the only problems. At the end of the fourteenth century, during the reign of Charles V, the first reforms attempted to establish sectioning in order to combat the rights of usufruct, the greatest scourge of the French forests during the Middle Ages. Only very gradually, however, did French legislation relative to forests become oriented toward a more logical exploitation of forest resources. While a royal ordinance of 1346 forbade any new concession of usufruct in the royal forests, the first really effective measures were not taken until the sixteenth century. The ordinance of 1544 prescribed rational cuttings to replace the random selection of the best trees for felling. An ordinance of 1520 had already intervened in the private forests, but not until the famous ordinance of 1573 were exact techniques prescribed, in a series of measures — staddlings, the sowing of acorns, the closing off of exhausted cuttings, weedings, the sparing of seed trees — which organized both a logical exploitation and especially an effective restocking of the forested areas.

Such measures were useless, however, if excessive depredations were not reduced. The authorities had struggled against rights of usufruct, but there were other, equally important, dangers. The appearance of water-powered saws, thanks to the crank-and-connecting-rod system, contributed to an increased use of large trees. Equally distrusted were iron furnaces, glassworks, tileworks, and carbonization shops. The establishment of these factories and shops was the object of increasingly close control by both the royal authority and the provincial governments. A declaration of 1543 strictly forbade the creation in France of new wood-burning factories. In Franche-Comté the local authorities were equally concerned with limiting the number of forges, essentially in order to ensure a supply of fuel for the cities. The practice of floating logs has often been claimed for the end of the fifteenth century or beginning of the sixteenth; as we learned in the first volume, this was not so, but it is true that with the increase in size of the cities this technique was practiced on a hitherto unknown scale. It is also possible that the impoverishment of the forests in the immediate vicinity of these large centers of consumption made it necessary to search for wood at a greater distance and to utilize the most economical method of transporting it, namely, floating.

ANIMALS

As far as the animal kingdom was concerned, there were no modifications in the ancient techniques that had been laboriously perfected in the course of earlier centuries. The most characteristic event was the introduction of the horse into America, where it had been unknown before the discovery of that continent by the Europeans. Here, again, then, it was Europe that made the greatest contribution to America. Few studies have been made of problems relative to the animal world, and we are therefore limited to presenting the few basic facts that have been pointed out by the historians.

Fishing

Fishing appears to have been less important in this period. The consumption of meat had probably increased, making such dietary supplements less indispensable. Nevertheless, because of religious requirements fishing continued to be a very important element in the economic life of the coastal areas of western Europe.

Little need be said about river fishing, since its techniques had changed very slightly. At most we find an attempt at systematization that appears to be the major characteristic of this period. The surveillance of the royal fisheries was entrusted to the same service that performed this function for the forests. Certain ponds were set aside as special preserves, some for carp, others for barbel, perch, and so on, depending on the nature of the water and the bottom. The first attempts at pisciculture appeared in this period. According to a manuscript of 1420, a Benedictine monk of Burgundy is thought to have invented long boxes with wooden bottoms and sides, the ends being rush grilles. The bottom of the box was covered with a layer of fine sand which contained small cavities filled with previously fertilized eggs. These boxes were then placed in a fairly shallow stream for the eggs to hatch. This method probably did not produce very good results; in any case the monk had no imitators. The experiment is valuable, however, for the motives that inspired it.

Fishing methods and equipment did not change perceptibly. However, an Act of 1430 attests to the use of drugs in certain ponds in Languedoc; for this purpose the milky juice of the Euphorbia plant was used.

Innumerable regulations continued to forbid the catching of certain fish at specified periods of the year (generally in spawning time), and the use of certain devices (in many cases we do not know what kinds of devices exactly are meant). The size of the fish sold in the market was sometimes regulated; an ordinance of 1350 specified the number and size of the trout that could be brought in baskets to the Paris markets.

There were no modifications in the practice of sea fishing, at least insofar as the specific operations of fishing are concerned. There was one important development, however, due essentially to the migration of fish and the shifting of the banks. In the course of the fifteenth century shellfish and certain other fish (for example, cod) gradually disappeared from the western coasts of Europe. Whale hunting became increasingly rare. In contrast, the fishermen attempted to travel greater distances in their search for cod. Unfortunately, the texts are not very explicit, and the historians are not in agreement, on the subject of deep-sea fishing. This type of fishing posed numerous problems: in addition to the question of ships and navigation, there was also the problem of preserving the fish, which had to be brought back to land after long journeys, and which therefore had to be more carefully prepared on the fishing vessels. In the fourteenth and fifteenth centuries a press was invented for pressing salted cod into barrels. There are tales of the presence of Basque fishermen in the waters of Newfoundland before the discovery of Canada. In 1497 John Cabot, sailing in this direction, may have crossed abundant banks of cod; the Basques may have preceded him in 1454 or even as early as 1443. Undoubtedly it will never be possible to determine the truth.

Herring fishing had also retained the ancient techniques, one of which proved perhaps more important in its social consequences than in its actual technical implications. Shortly before 1447, a certain Guillaume Beukels is supposed to have perfected the curing and barreling of herring. This art consisted, not of piling the herring in barrels filled with an ordinary brine, but of placing them in the containers with the scrupulous care and special methods described in detail in the ordinances of Amsterdam. Upon arrival in port, all that remained to be done was to ship the barrels to their distant destinations. The entire job being done at sea, the coastal population thus found itself deprived of part of its activity, and mass migration followed.

Cattle breeding

We find the same almost total stability in cattle breeding. However, two new barnyard animals arrived in western Europe. The guinea fowl was introduced into France by merchants coming from Guinea at the beginning of the sixteenth century. Once again the historians disagree on the introduction of the turkey: some claim it is an import from America, while others claim that it came from the East, perhaps in the fifteenth century, and that its breeding developed rapidly in the sixteenth century.

Aside from these innovations, cattle-breeding techniques changed very little. The treatise by (or attributed to) Jean de Brie, written around 1379, gives us a description of the activities of the shepherds as they had been practiced for centuries and were to continue to be practiced for a long time to come. This treatise is known to us only through an abridged edition dating from the beginning of the sixteenth century. It lists the various attentions to be given to these animals, as well as the manner of caring for their illness, and advice on feeding and pasturing. It is quite improbable, however, that this manual, which was not printed until a century and a half later, and no manuscript copy of which is extant, had great success in the rural areas; moreover, it is merely a repetition of the customary techniques. The same is true of the more numerous treatises on the care of horses.

Cattle breeding appears, however, to have become widespread toward the end of the fifteenth century. More, stronger, and better fed and cared-for animals made possible a greater consumption of meat in western Europe. Certain techniques that had already appeared in preceding periods became more widespread, for example transhumance, which was now widely practiced in the Alps and in the Pyrenees. In the latter area we find a dual form of transhumance – toward the plains and toward the mountains.

The principles of hunting had remained the same. We know at least certain types of hunting thanks to the magnificent miniature hunting scenes by Gaston Phoebus, Count of Foix, who himself possessed fifteen or sixteen hundred dogs from all over Europe. The handbook of Guillaume Tardif was the first treatise on hunting to be printed (it appeared in 1492), but this treatise on falconry was more the work of a teacher of rhetoric. No truly original work on hunting to hounds existed until the publication, in 1561, of *La Vénerie,* by Jacques du Fouilloux. The old works were also reissued: the *Chasse du roi Modus* (early fourteenth century) was published in 1486.

Texts and images prove that hunting techniques had changed very little.

Falconry was tending to disappear, while hunting to hounds was spreading. Crossbows were still preferred to firearms, and a multitude of devices, nets and traps being the most common, were still in use.

Human nourishment It would have been interesting to see whether diet had changed during the period 1450–1550, but we possess only very fragmentary and not very extensive information on this subject.

Undoubtedly there was a slow improvement in diet. This improvement can be summarized as an increase in the consumption of meat, which explains to a certain extent the decreased importance of fish; the expansion of the richer cereal grains, particularly wheat; a greater variety of plants for general consumption. In this way a more balanced, and consequently more beneficial, nourishment was obtained.

THE EXPLOITATION OF UNDERGROUND RESOURCES

The Renaissance made its most decisive progress in the exploitation of mines. This expansion of mining appears to have begun around the middle of the fifteenth century, and to have progressed very rapidly until the middle of the sixteenth century, by which time it had acquired the form it was to retain until the eighteenth century.

The medieval period had produced no treatises on the art of mining. Only the lapidaries supply very vague, limited, and imprecise information. No works on the art of mining appear among the first printed books of the end of the fifteenth century. The first short treatises (*Bergbüchlein*) appear in the first half of the sixteenth century; they are devoted to the discovery of veins and deposits rather than to mining techniques as such. Not until the appearance of the famous work of Agricola, published posthumously in 1556, did we possess a genuine mining and metallurgical encyclopedia. The sometimes expressed belief that Agricola's work is indicative of a considerable progress occurring around the middle of this century is quite inaccurate. This work, which is essentially a descriptive work by a man who is not a technician, in fact reflects the end product of a slow evolution that had begun almost a century earlier. Fortunately, there are artists' works prior to or contemporary with Agricola, which furnish us with a multitude of details on the conditions of exploitation of the mines. Thanks to these sources it is possible to attempt to draw as concrete a picture as possible of mining techniques – something it is impossible to do for the medieval period because of the lack of sources.

The solution to the problem of the search for veins and deposits and the knowledge of their general direction was dependent solely on the advancement of geological knowledge, probably still very rudimentary in this period. The *Bergbüchlein,* which appeared at Augsburg in 1505 and enjoyed sufficient success to warrant six editions before 1540, proves that this technique was still very empirical; there is still a belief in an unquestionable influence of the planets on the formation of metal-bearing and mineral deposits – a centuries-old belief of Eastern origin that was not to disappear until the eighteenth century. The use

FIG. 20. Horse-driven treadmill for raising ore from mines. End of the fifteenth century (Vienna, Bibliothek National, frontispiece of the *Kuttenberger Kanzionale*).

of the divining rod, a double hazel branch, appears to have been extremely widespread. The few extant narratives, most of them German and all of them certainly more or less legendary, show that the discovery of the richest veins was in most cases a matter of chance discovery, luck being the miner's principal aid.

This search was useful especially in the case of metal-bearing deposits. Coal was known only by its outcroppings; as for iron, the miners continued to exploit alluvial bodies of ore in massive formations, which were particularly abundant in certain regions.

The mining industry was still very far from achieving its present-day degree of concentration. In many places there existed only small individual exploitations, the techniques of which had remained unchanged. Each had its shaft, with a small hut surmounting the opening; a simple hoist lowered the miner to his vein and brought up the basket of ore. Shallow levels were dug at the base of the shaft. A perfect example of these individual mines can be seen in the retable of the church of Annaberg in Saxony, executed in 1520–1521 by Hans Hesse; in it we can distinguish all the shaft entrances with their ladders, surrounded with fill, and crowned by windlasses and huts. We also see several airshafts with vanes, and open levels in slopes. A mining charter of 1395, for the region of Allevard in the Dauphiné, shows that these galleries were sometimes joined, and caved in when they were located one below the other.

In these small mines the shaft was circular, like an ordinary shaft. In the larger mines square or rectangular shafts were used, which permitted a more rapid timbering. The two bases were oriented in the direction of the slope of the rock layers. Various extant drawings show that the shafts were generally timbered. In the Sainte-Marie-aux-Mines manuscripts, as in Agricola's work, we note that the timbers were arranged vertically rather than horizontally. The levels were also timbered, when the terrain was not solid and when the water level was near. This timbering was generally done with ash timbers; the Sainte-Marie-aux-Mines manuscript shows the cutting of the props. In circular shafts the timbers were jointed like the staves of a barrel. In certain areas, however, trapezoidal pieces and even masonry were used. When necessary, recesses with

Plate 8. Working the mine. Painting by L. Gassel, sixteenthcentury. Musées royaux d'Art ancien, Brussels.
Copyright A.C.L., Brussels.

FIG. 21. Mine shaft and small loading wagons on rails. End of the fifteenth century (Paris, *École des Beaux-Arts,* MS of the mines of La Croix).

buckets were installed for the collection of the water. The shafts were always dug to a slightly lower level than the last gallery to act as a reservoir.

There were also air shafts and safety shafts, topped with a truncated cone-shaped chimney surmounted by a vane; this arrangement appears in both the Annaberg retable and Agricola's work. At the base of this shaft a fire was installed to activate the circulation of air, a necessary condition for ease in working and, in the coal mines, a preventive measure against the terrible pit gas. For this ventilation canvases waved either by workers, hand-, foot-, or hydraulics-operated bellows, or even windmills were used, as can be seen in Agricola's work. To determine the presence of gas in the coal mines, two miners, their heads covered with canvas bags, explored the unlighted level before work began.

In contrast to the modern practice, the deepest levels were exploited first. Water could then flow into the worked-out areas, and in this way the intermediate levels were drained.

In coal mines fire had to be avoided as much as possible. In rock layers and metal mines, in contrast, the old method of lighting a fire was often used to break up the rock. This was because mining equipment was limited, the princi-

FIG. 22. Working face of a mine. End of the fifteenth century. (Paris, *École des Beaux-Arts,* MS of the mines of La Croix).

pal tools being picks and clubs, together with wedges. Examples can be seen in Agricola's work. In his drawings of Sainte-Marie-aux-Mines, Heinrich Gross shows us the miners collecting tallow for candles, picks, and spare iron tips before their descent into the mines. In the same manuscript we see the miners chopping an enormous block with the help of wedges and clubs. Tradition claims that powder was utilized for the first time at Chemnitz in 1527, but perhaps a later date should be postulated. Once a level was exhausted, the empty areas left between the pillars were filled up with sterile fill.

Small exploitations had only simple windlasses; this arrangement can be seen in most of the shafts pictured in the Annaberg retable and on the borders of the Gradual of Saint-Dié and of the Museum of Siegen, which range in date between 1480 and 1520. This mechanism was ample for individual exploitations. When the shaft was too deep or the production too abundant, a kind of treadmill was used – the *hernaz* of the Liège area, the winding pulley in the French mines. A drum unrolled a chain on which boxes or baskets were hung. Two shafts in the Annaberg retable seem to have treadmills. A very good example is seen in a manuscript in Vienna, where the treadmill, operated by four horses, is clearly apparent. Agricola's treatise depicts a perfected machine that we have already mentioned, in which there are two hydraulic wheels, back to back, whose paddles are inclined in opposite directions. Two conduits with movable vanes carry the water over one wheel or the other; thus the windlass can be turned in either direction. The miners descended to the bottom either by ladders, poles, steps cut in the rock, or by sliding down a leather carpet.

The coal or ore was brought up to the surface, frequently in baskets (in the Vienna manuscript numerous baskets can be seen lying on the ground). In the levels, either wheelbarrows (Annaberg retable, Agricola) or trucks rolling on wooden rails (Sainte-Marie-aux-Mines, the painting by Gassen [1544] preserved in the Brussels Museum) were used to bring the loads to the shafts. This system appears to be of German or Belgian origin. Several authors have even attempted to date this practice from the twelfth century, but without convincing proof. In any event it appears to have been in general use by the time of Agricola, at a period when mine levels had already been considerably developed.

The problems of evacuation of water were much more difficult to solve. In mines situated on slopes, as was the case, for example, in the vicinity of Liège, matters were easier. The water was collected in the levels and ran out of the hill through a channel dug across the seams; it was then used to feed a network of public water distribution. Elsewhere a pumping device had to be installed to bring the water up from the bottom, or at least as far as the beginning of the drainage channel. This was done at first by windlasses and treadmills, with vats. A miner from the Liège country named Antoine Gentil is supposed to have invented (in 1521) valve vats that permitted automatic filling. Later, other methods borrowed from other techniques were attempted. The simplest were the norias, either treadmill norias (most often the case) or sometimes hydraulic norias. These devices generally consisted of a system of endless ropes from which were hung vats or various vessels. A perfect example can be seen in a tapestry of Salins that dates from the beginning of the sixteenth century, and which is preserved in the Louvre museum; here the device is being used to raise salt water. Bucket chains are also depicted in Agricola's treatise, where we also find a system of chains with balls passed through pipes, the whole plunging into the collecting basins at the bottom of the mine. The chain was pulled either by treadmills, squirrel-cage wheels, or hydraulic wheels.

Not until Agricola's treatise do we first see the use of lift-and-force pumps, operated by various sources of power and naturally equipped with a crank-and-connecting-rod system. We have no exact information on the first use of lift-and-force pumps in the draining of mines; an unverifiable tradition gives the date of 1531, in the Liège country. The construction of these pumps appears to have been very difficult. Pictures in Agricola's treatise depict the drilling of wooden pipes. The joints shown are still very primitive.

In any event, thanks to these various improvements the exploitation of mines developed considerably, especially in Germany. The fame of the German miners later spread throughout Europe: their services were everywhere in demand for the development of this industry and the introduction of the new techniques.

CHAPTER 5

THE TRANSFORMATION OF MATTER

THE TECHNIQUES for the working of raw materials were to undergo very profound changes in the two centuries between 1350 and 1550. This evolution was very clear-cut and even rapid, since in actual fact it occurred for the most part between the mid-fifteenth and mid-sixteenth centuries.

These changes took place in every area of activity. While certain of the thermal techniques, which are generally the most stable, experienced radical departures in technique, all the mechanical processes, in contrast, were literally revolutionized by the considerable extension of mechanization to which we have already alluded.

However, it is evident that there were limits to this technical flowering of the Renaissance, about which little is known. A machine depicted in an engineer's drawing, or even in a technical treatise, is perhaps often an idea rather than an actual invention. A highly developed process of mechanization could be achieved only by an "age of iron," since wood does not lend itself well to the construction of complex machines. Whatever the progress of metallurgy, western Europe in the middle of the sixteenth century had not yet reached this "age of iron." And even if these machines had actually been constructed, we would still be justified in inquiring to what extent they were utilized. The historian of the nineteenth century discovers that the practice of the old craft techniques, utilizing very rudimentary tools and procedures, was still widespread in this century. What, then, can have been the situation in the sixteenth century?

THERMAL TECHNIQUES

As regards thermal techniques, the problems of the preceding period, still almost insoluble, persisted into the sixteenth century. The temperatures utilized are unknown, and we have little information on the construction of the furnaces, their shapes and functioning. Fortunately, pictorial representations and a few technical treatises nevertheless permit us to answer several questions.

The same fuel was still being employed, under the same technical conditions; wood was still by far the most widely utilized combustible, whether directly or in the form of charcoal. There are several extant pictures that depict the making of charcoal; the manuscript of Sainte-Marie-aux-Mines contains a perfect depiction of the millstones utilized by the woodmen in accordance with methods which were to remain unchanged for centuries. The expansion of the coal mines proves, however, that coal was the fuel most frequently used. While a date at the beginning of the fourteenth century is mentioned for the utilization of coal in a forge, in the Belgian village of Marchiennes, undoubtedly we must not regard this date as absolute; there were probably many earlier trials of the new material.

Glass

More information on transformations in the glassmaking techniques is supplied by archaeological evidence than by detailed texts. There were various innovations, especially between 1450 and 1550. First there was the more widespread use of soda, which gradually replaced potassium. Soda produced a fusible glass that was easy to handle, even and clear; it thus became possible to make flatter, clearer glass.

A new type of glass appeared in the second half of the fifteenth century that was destined to be greatly developed. This was Venetian glass, or crystal. Here, again, its origins are known only through a tradition that attributes its invention or perfection to one Beroverio, a famous glassmaker of Murano, in 1463. In 1623, however, Norman glassmakers claimed this innovation for their own; their ancestors had apparently been practicing it for 250 years in Languedoc. Venetian glass was a silico-alkaline glass (silicate of potassium and lead), later replaced by a lead glass (potassium silicate and lime), which since the nineteenth century has represented (at least in France) genuine crystal. To obtain this transparent glass the Venetians employed purer products, soda from Egypt and Syria, produced by the combustion of a special type of grass (*kali*) and pebbles from the Ticino.

Fig. 23. Glassmaking. End of the fifteenth century (London, British Museum, Add. MS 24189).

Plate 9. Glassmaker's shop. Engraving taken from Agricola, *De re metallica.* Conservatoire des Arts et Métiers. *Photo by the Conservatoire.*

Few changes occurred in the production of plate glass. As we have noted, clear glass appears to have been less tinted than in earlier centuries. To obtain colors, new types of glass were produced by "casing" (the superimposing of layers of glass). The colored sheath overlaying the clear glass was then worn down by mechanical methods in order to produce certain effects. Some plate glass of this period consisted of several superimposed "casings." Larger pieces of glass (two to two and one-half feet) were also produced; they were cut with diamonds.

In the sixteenth century glass production underwent a tremendous development because of the expanded use of glass. Especially noteworthy was the use of plate glass. Windows, formerly quite rare and generally reserved for special buildings, became far more common. There was also more widespread use of mirrors (lead-coated sheets of glass), many of which appear to have been still cut from balls. The Venetian mirrors, which were beginning to become famous, were made by blowing crystal into a cylinder. This production has been attributed to the del Gallo brothers, who are supposed to have mentioned it in a petition in 1503.

It is evident, then, that there were no major modifications in the technique of glassmaking. The most important event was perhaps the considerable development of the use of windows in private dwellings, a use that perceptibly transformed the way of life.

Metals: Iron

The evolution of metallurgical techniques is undoubtedly one of the basic factors in the history of technology in the Renaissance. Unfortunately, we have little knowledge about its origins and stages of development, and the historians give very differing evaluations of the few pieces of information that can be found in the texts.

Fig. 24. Refining hearth, 1452 (Florence, Biblioteca Nazionale, II.1.140, f° 127, Filarete's treatise).

Fig. 25. Reducing furnace for ore, 1540 (Biringuccio, *De la pirotechnia,* Venice, 1540).

We have mentioned that the blowing of air into the ore-reducing furnaces by means of a hydraulically operated bellows appears to be mentioned in the first half of the fourteenth century, but the first pictures of these bellows do not appear until the first half of the fifteenth century, in the notebook of an Italian engineer. This more powerful blasting technique undoubtedly made it possible to increase the size of the apparatus. Step by step the furnace was thus perfected. German authors have assigned specific names to the various steps in this transformation: first the *Stückofen* (the shaft furnace) and then the *Flussofen,* or classic blast furnace. By increasing the size of the structure, thus permitting the burning of a larger amount of fuel, a more carbonized iron — cast iron — was obtained, which had the considerable advantage of flowing easily. This required furnaces of from twelve to fifteen feet high and the indispensable hydraulically operated bellows.

The question of the place and date of origin of the blast furnace is a highly controversial one, and it is even probable that this problem will never be solved. Some authorities argue for the Netherlands, more precisely the Liège region, around the very end of the fourteenth or beginning of the fifteenth century. Others, more specific, though without good reason, claim that the first blast furnace was constructed at Nassau in 1474. Actually, the blast furnace must have appeared in the second half of the fourteenth century, either in the Liège region or on the banks of the Rhine. As soon as the furnace began to rise aboveground, and a sufficiently powerful bellows existed, no further difficulty stood in the way of the production of cast iron. It must have been accidentally produced in isolated instances.

The geographical development of the blast furnace followed easily explicable routes. A technical discovery, at least in this period, was disseminated almost exclusively by migrations of workers, and it is to the travels of master smiths that we must look for the lines of expansion of the new machine. In the principality of Liège, where it was probably born, the blast furnace became the exclusive instrument of iron production only in the course of the sixteenth century. It slowly reached the flat country of the Hainaut and Namur regions, and by the end of the fifteenth century several French provinces seem to have known of it: Lorraine, Champagne, Nivernais, Normandy. In the middle of the sixteenth century it probably reached Alsace, Franche-Comté, and Brittany. Only later did it begin to travel south. It was a Liégeois, one Louis de Geer, who brought the new technique to Sweden around the middle of the sixteenth century. While it was already being utilized in the Rhine Valley in the middle of the sixteenth century, the blast furnace was not yet known in all areas of Germany. From France it traveled to England with French workers.

Cast iron had one tremendous advantage: it could be poured, and thus could be used for a multitude of new projects, in particular for cast-iron ware for domestic and industrial use. However, it was necessary to obtain iron, and therefore to burn the carbon that was present in an excessive amount in the reduced metal. For this purpose the reduced metal was treated in the metal refinery. By the end of the eighteenth century there existed a certain number of methods for refining this metal. Though these methods are generally designated by geographical names, we can nevertheless affirm the common origin of these various processes, which then became differentiated both by the arrangement of the refinery fires and by local particularities of production. The Walloon method appears to be the oldest.

Oxidizing the cast iron was not an easy operation. The Walloon method consisted of melting the pigs at one end over a fire made with charcoal. The drops, falling into the fire, were oxidized, and the impurities (carbon, silicon, and manganese), upon burning, gave rise to the formation of particles of iron that gathered at the bottom of the crucible, forming a mass of iron called a "bloom." Phosphorous was only partly removed by this method; thus it was necessary to eliminate phosphorous metals and use only white metals. The Walloon method undoubtedly originated as an attempt to utilize this by-product of a poorly functioning furnace, which was called cast iron. To shorten the operation, the pig was first heated in another fire, the chafery.

As for molding, we have few details on this process. Only the cast-iron plates on the chimneys bear dates that would permit us to establish the origin of iron casting; a scant handful from the end of the fifteenth century are still in existence. The first iron casters molded their objects from earth, in a molding box, following the technique practiced by the copper and bronze casters. These molds, however, had to be adapted to higher temperatures. In addition, cast iron could be utilized only for the casting of massive objects that would require no further working. It is probable, therefore, that this technique remained for a long time in an embryonic state.

Our knowledge of the construction of the furnaces is quite limited. For the period prior to the time covered by the treatises of the mid-sixteenth century, we are reduced to several works executed by painters who worked chiefly in the Liège region. In a painting by Patinir (circa 1480–1524) we see a furnace constructed in stone, with a stairway leading up to the mouth; it appears to measure between 16½ and 19½ feet in height, and has a hole in the bottom through which the metal flows out. The furnace is attached to a thatch-roofed building that undoubtedly sheltered the refinery and the forging hammer. The opening of the furnace is obviously a figment of the artist's imagination.

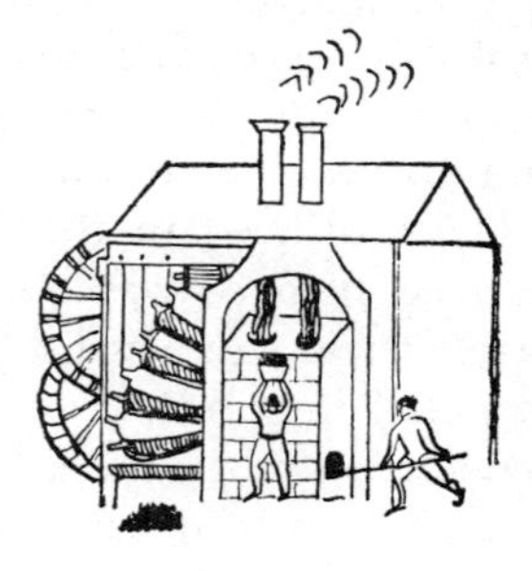

FIG. 26. Blast furnace. Sixteenth century (drawing of Public Record Office).

The works of Henri Bles (circa 1490–circa 1550) appear to depict the same factory. We can distinguish three mine shafts topped with simple hand-operated windlasses; and the buddle, where a worker with an open lamp is shaking the ore in running water. In the background is the blast furnace, whose flame-crowned mouth towers over the roof of the thatch-covered casting shop. The water is carried over the furnace wheel by two wooden gutters; one of them, in the classic model, overhangs the wheel and pours the water over its top, while the other is in three-quarter position, in which the paddles have already completed a portion of their descending course. A similar arrangement is seen again in the furnaces of the Van Walkenborch brothers, shortly after 1550, with the addition of large poles that, acting as a spring, ensure the filling of the bellows.

As for the refining furnaces, we are reduced to mere suppositions. It is,

Plate 10. A blast furnace. Painting by J. Brueghel. Sixteenth century. Galleria Doria Pamphili, Rome. Ph. del Gab. Fot. Naz. Roma, *Photo Alinari.*

however, a known fact that these furnaces were not modified until the second half of the eighteenth century.

Nonferrous metals

While we are relatively well informed on iron metallurgy, we know very little about the other metallurgies; by and large we obtain no information on this subject until the appearance of the treatises of Biringuccio and Agricola. In truth, metals other than iron were relatively few in number. Tin was frequently found allied with copper; its extraction presented few difficulties, given its low fusion point. Gold was in most cases found in a free state. As for silver, except in the case of veins of native silver it was difficult to separate it from the lead in whose ore it was usually found.

Lead and copper ores often existed in the form of sulfur. Before any reduction operation could be performed, it was necessary to calcine them to extract the sulfur and to transform them into oxides, easily reducible in furnaces of very primitive appearance that bore a strong resemblance to the early furnaces utilized for iron. The silver found in lead and copper ores was probably frequently abandoned. According to a persistent legend, in 1451 one Johannsen Funcken perfected the method of separating silver from lead and copper.

The devices utilized for copper apparently underwent important changes between 1450 and 1550. Copper, until then employed in a small number of products, especially in bell metal and copper utensils, came into great demand because of the development of bronze artillery (to be discussed later). New factories, which the Germans called *Saigerhütte,* were created; they were larger and more efficient thanks to the general use of hydraulic power for the bellows of the furnaces and the hammers.

Thanks to the development of these techniques, great fortunes could be made in supplying a metal for which the demand was constantly growing; Jacques Cœur (mid-fifteenth century) and the Fuggers (between the end of the fifteenth and middle of the sixteenth centuries) profited greatly from this demand. However, the instruments of this production are not well known. A picture in the manuscript of Sainte-Marie-aux-Mines clearly depicts a foundry that, according to inscriptions on the drawings, was used for both lead and copper. Only the exterior of the furnace, with its three draw holes and the bellows operated by a hydraulic wheel, can be seen. Next to it is the indispensable refinery: round furnaces with metal calottes that could be lifted off by means of a chain. On the Annaberg retable, completed around 1521, we see the stoking of the masonry shaft furnace, from the base of which the incandescent metal is flowing. To the right, in the background, a workman is standing before a furnace with a soleplate very similar to that of Sainte-Marie-aux-Mines, and is using a hook to pull the gangue of a silver casting that is flowing down the furnace wall. These shaft furnaces and furnaces with soleplates correspond exactly to those described in the works of Biringuccio and Agricola.

Copper was hammered with hydraulic drop hammers similar to those used for iron. It was generally shaped and beaten by hand, using techniques that continued in use in certain craft centers until the end of the nineteenth century. Lead was used for numerous purposes, including the making of pipes, which continued to be made by soldering the lips of sheets of lead rolled around cores. The soldering was done, not with molten lead, as in the Roman period, but with a soldering iron and tin; the lips were first beveled. The jointing of the pipes was done by melting a lead casing, which was then hammered, or by a tin solder casing.

The most commonly used metal, after iron, was bronze, which could be cast easily; the development of artillery, as we have noted, ensured its extensive development. Cast-iron cannons, which appeared around the middle of the fifteenth century, did not enjoy a popularity equal to that of the bronze cannons, which were the customary weapon of land-based armies beginning in the second half of the fifteenth century and especially at the beginning of the sixteenth. Thanks to Biringuccio's writings and to a certain number of other still-unpublished treatises, we possess very detailed information on the casting of bronze cannons from the end of the fifteenth to the middle of the sixteenth centuries.

The bronze used in cannons was of a very different composition from that used in bells. Whereas the latter contained approximately 23 to 26 percent tin, the bronze used in artillery contained only 8 to 12 percent. The most difficult part was not so much the casting itself (which, however, was not completely free of obstacles) as the making of the mold. Sandy soil crumbled into powder, while argillaceous earth shrank in the fire; the two varieties had to be combined in order to arrive at a satisfactory product. First a model of the proposed object was made of wood or terra-cotta, in most cases of wood. It was bound with ropes, and then covered with a preliminary layer of "cement earth" (beaten earth, pulverized brick and tile, and water), then with a second layer of "fire clay" (loam, wadding, and dung). This block was scraped and shaped into an exact model of the object to be cast; the surfaces were smoothed and the raised areas carved. This model was then dried, while a separate mold was made for the breech. The model was coated with oil, and a thick layer of pulverized earth, well beaten and moistened, was applied to it; it was then wrapped in iron bands and hoops. A second layer of earth was applied, and was covered with iron strips placed lengthwise; the work was terminated by a third layer. The mold was fired on the outside, and the model was removed. The mold was then placed upright in order to be fired internally, after which it was placed in vertical position in the pit. Its bore also had to be formed; the core used for this purpose, an

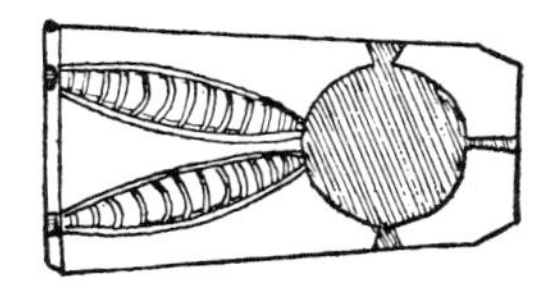

FIG. 27. Reverberatory furnace for casting bronze, 1540 (Biringuccio, op. cit.).

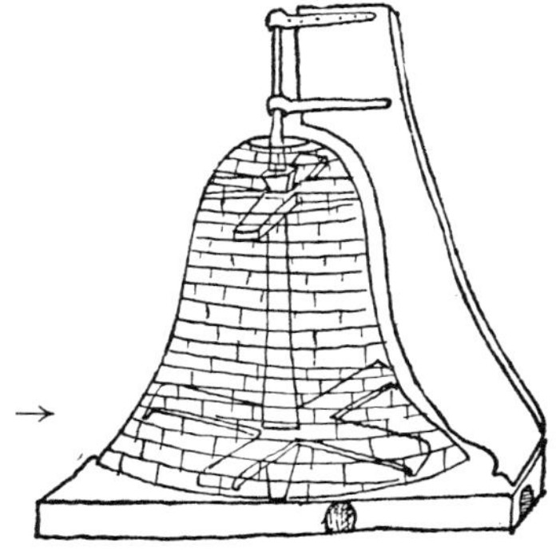

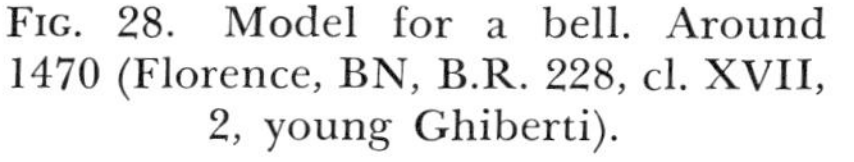

FIG. 28. Model for a bell. Around 1470 (Florence, BN, B.R. 228, cl. XVII, 2, young Ghiberti). →

iron core coated with earth, had to be able to withstand the pouring of the metal, and in addition it had to be held in the center of the mold, a task that created numerous difficulties.

Casting was not done at as high a temperature as in the nineteenth century, but care had to be taken that the metal was completely melted; to determine this, a bar of iron thrust into the bath had to come out clean. At regular intervals samples were taken to ensure that the alloy was successful. Finally, the casting had to be done very slowly. The furnaces utilized for the smelting of metal were reverberatory furnaces constructed of baked or dried bricks or refractory stones, with two or three openings for the escaping flames. These furnaces were not very large, and could hold only a limited quantity of metal. Probably it was impossible to cast more than one large piece at a time. Frequently the metal must have been insufficiently liquid; in such cases the inside of the object was inspected with a candle or a "cat," a clawed instrument that revealed hollow areas. Next came the polishing of the surface and the drilling of the priming hole.

The production of large pieces of sculpture cast in bronze was done with similar methods. Failures, however, must have been quite numerous. In one of his manuscripts Leonardo da Vinci has left notes that, while not as complete as those of Biringuccio, give similar information on bronze casting, both for guns and for large pieces of sculpture. Moreover, Leonardo had begun his career in a caster's workshop.

From these brief remarks we can see that between the middle of the fifteenth and the middle of the sixteenth centuries metalworking made decisive progress. These techniques were preserved almost intact until the great Industrial Revolution of the eighteenth century. This perfecting of metallurgical methods was to have important consequences that were manifested particularly in a much more general use of metal in numerous areas of activity.

Pottery

Improvements in pottery were far less spectacular than the progress of metallurgy. The few innovations occurred less in the firing techniques as such than in the preparation of the material.

G. Fontaine correctly believes that the art of pottery, at least in France, was formed in the thirteenth and fourteenth centuries and that these techniques were handed down by tradition, without striking modification, until the eighteenth century and even until modern times. Progress was most apparent in Italy, where the potters passed from baked, lustered, and varnished pottery to faïence, the fundamental distinction being the nature of the glaze covering the terracotta. The material, which varied considerably, still consisted of a more or less pure clay, sand, and calcareous marl, while the glaze was now a stanniferous enamel. The decoration was placed on the unfired enamel; the metallic oxides used as dyes became part of the enamel during the firing (*grand-feu* decoration). As in the case of glass, the metallic oxides that could sustain high temperatures were very few in number, being reduced to manganese (violet, brown, black), iron (red), copper (green), antimony (yellow), and cobalt (blue). This technique, which was known to the Arabs in the eleventh century, was adopted in Italy in

Plate 11. Roman lime kiln. Painting by Sébastien Bourdon, seventeenth century. Munich Pinakothek. *Photo Bayerischen Staatsgemäldesammlungen.*

the fifteenth, and became so famous in the region of Faenza that the technique took its name from that city. It was not introduced into France until the sixteenth century.

In France, around the middle of the sixteenth century, Bernard Palissy was utilizing similar although slightly different techniques; his procedures, however, were often unorthodox. His enamel was a plumbiferous enamel that produced a yellowish white after firing. Only then did he lay his metallic oxides, with a brush, on the still-damp flux. Not being a sculptor, he modeled his decoration after nature.

We have little information on the making of bricks, tiles, and pottery for ordinary use. The brickmaking process is depicted in a Flemish miniature of the middle of the fifteenth century; we see the earth being mixed with water, the forming of the bricks with a mold, then the piling up of the bricks, probably for drying. Unfortunately, this small picture does not show the kiln. The use of brick, while still infrequent, was becoming more common with the passage of time.

MACHINERY

Here again we find a striking change occurring between the mid-fourteenth and mid-sixteenth centuries, and especially after the second half of the fifteenth century. This technical evolution can be characterized essentially as a period of major development of the machine, which in many cases was to replace equipment that had become particularly outmoded. We have already noted the limits that had to be imposed on this development; it was as yet impossible to hope for an unlimited development of machines that for the most part were still being made of wood, despite the development of metallurgy.

It was not only machine tools (although these were the most numerous) that were to appear in the various crafts; there were also machines for the refining of raw materials, and these were to revolutionize several industries, including some of the most important.

The iron industry The birth of the blast furnace had already revolutionized the conditions of metal production in the course of the fifteenth century, and the appearance of a complete and perfected hydraulic apparatus was to speed the transformation of this industry. It was through this door that industrial capitalism entered the field of iron metallurgy.

Hydraulic hammers had been in existence since the twelfth century. It was probably at the end of the fifteenth century or in the first half of the sixteenth that the lateral and frontal types appeared in the forges; until then only the terminal hammer, the lightest of all models, appears to have been known, and it is the only model that appears in the drawings of Leonardo da Vinci, often with imaginative details it would be difficult to translate into reality. We find these small tilt hammers, with their crudely constructed housing, in Brueghel's painting *The Fire;* they undoubtedly give us a good idea of what the ordinary

Plate 12. Various types of workshops, beginning of the sixteenth century: the smith, the printer, the ceramist (?). Satirical drawing by Dürer. Musée de Bayonne, coll. Bonnat. *Photo Bulloz.*

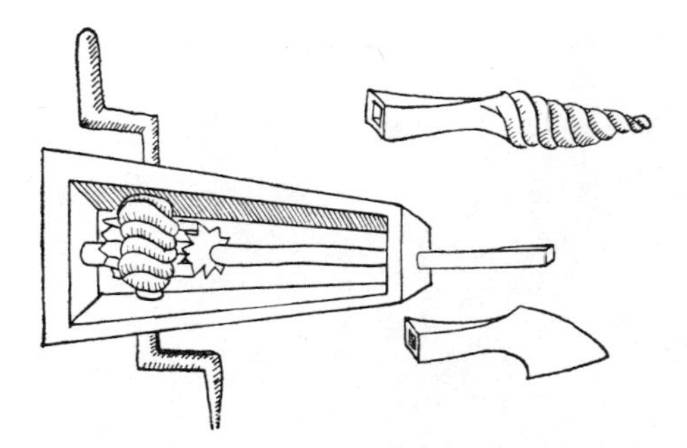

FIG. 29. Trepan with two bits. Around 1475. (Venice, Marciana, ital. 86).

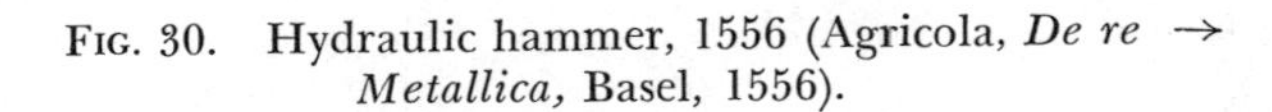

FIG. 30. Hydraulic hammer, 1556 (Agricola, *De re Metallica,* Basel, 1556). →

tools of this period must have been like. Lateral hammers, already heavier, appear for the first time in Agricola's treatise.

There were three other machines that were greatly to facilitate ironworking: the rolling mill, the splitting mill, and the hydraulic wire mill. As can be seen, all these machines utilized hydraulic power, which made it possible to concentrate industry along a river.

In Leonardo da Vinci's works we find the drawings of two machines that appear quite perfected but that undoubtedly were never built. Their purpose was essentially to shape a heated bar by passing it through a hole of the desired size and shape. The bars were pulled by means of a mechanism operated by a hydraulic turbine. To a certain extent these devices were similar to those utilized in hydraulic wiredrawing.

When we pass from Leonardo's drawings to actual machines, we are much poorer in precise information. The building of rolling mills posed difficult problems, if only for the cutting of the cylinders, which at this period could only be made of cast iron. The casting of such masses, their machining to give them the desired size, and their polishing suppose very advanced techniques about which we know absolutely nothing. The same was true of the splitting mills, composed of cylinders with cutting edges that acted on the principle of shears. While few rolling mills appear to have existed until the eighteenth century, splitting mills are often mentioned, and do not begin to disappear until the time of the great expansion of the rolling mills.

Only very brief references to rolling mills and splitting mills can be found. The two instruments are mentioned in the Liège region for the first time in the middle of the sixteenth century. Thus they clearly seem to have been, if not invented, at least perfected there during the first half of the century. Not until 1615 do we find the first picture of a rolling mill, in a work by the famous Salomon de Caus.

As for wiredrawing, more extensive investigations would have to be carried out. To what extent did wiredrawing exist prior to the sixteenth century? We do not know. While mention is frequently made of gold and silver wire, whose production posed no difficulty, given the great malleability of these metals, no reference is made to iron or brass wire. From this we must perhaps deduce that these products came into existence either at the end of the fifteenth century or the beginning of the sixteenth. Apparently it is to Germany that we owe the hydraulic drawing mill. A detailed source proves that this technique was not introduced into France until the very first years of the seventeenth century. A picture by Biringuccio, dating from 1540, gives the first representation of a mechanical drawing mill. The hydraulic wheel ensured the spooling of the wire after its passage through the drawing device.

The difficulty of making tinplate was perhaps due less to a mechanical problem than to the methods of tin-plating iron. Certainly it first had to be learned how to make sheets of metal sufficiently flat so that a layer of tin could be poured over them; in addition, the tin had to be strong. For a long time these techniques were kept secret. Judging by eighteenth-century manuscripts, which describe the making of tinplate at that period, it is probable that completely empirical methods were utilized. The making of tinplate is believed to have been perfected in Nuremberg in the second half of the fifteenth century; its use was still limited to utensils. It was the food-canning industries that gave this product its greatest expansion.

On the basis of a drawing, the invention of the drop hammer has been attributed to Leonardo da Vinci. It is a fairly complicated device in which cams and counterweights act successively. The hammer itself is a simple pounder, which does not appear to be unusually large, and is of a type that existed in much simpler forms until the end of the nineteenth century. The operating of such an instrument would have been much too complicated, and it would have offered no marked advantage over the hydraulic hammers, drawings of which appear on the same page.

Thus it was probably as a result particularly of the discovery of the blast furnace and smelting that metallurgy began to develop. This event inevitably caused the development, undoubtedly in the first half of the sixteenth century, of more or less perfected machinery, which was to persist almost unchanged until the dawn of modern metallurgy and which made possible the production of more directly usable components, thereby contributing to the growth of the use of metal in all areas of activity.

Metalworking

Metalworking continued to be a very difficult art, not so much because of lack of machinery as because of the scarcity of the indispensable machining equipment. For example, practically no metal-turning machinery existed before the very end of the eighteenth century. The only operation relating to these techniques is the boring of bronze cannons. We possess two very sketchy drawings, dating from the middle of the fifteenth century, of devices that were undoubtedly intended less for boring cannons than for giving the bore a more or less round shape. The problem of boring did not begin to exist, however, until the calibers were strictly standardized, and the bore of the cannon had to be of precise dimensions, failing which the ammunition would have been useless. These two fifteenth-century devices worked the cannons in vertical position. Another vertical system appears in Brueghel's painting *The Fire;* here the tool is operated by a horse-driven treadmill. In Biringuccio's work, published in 1540, we see a horizontal system, which certain poorly informed historians have attributed to the seventeenth century; in this case the tool is operated by a squirrel-cage wheel. A drawing of the various tools utilized is still extant. In one case the fraise is made of pieces of steel mounted in wood; in another, steel strips are held in a bronze crown; in the last drawing the entire head of the tool is made of steel. What was only a simple boring operation became a genuine drilling process in the eighteenth century, when the Maritz brothers discovered the method for boring cannons that had been cast solid.

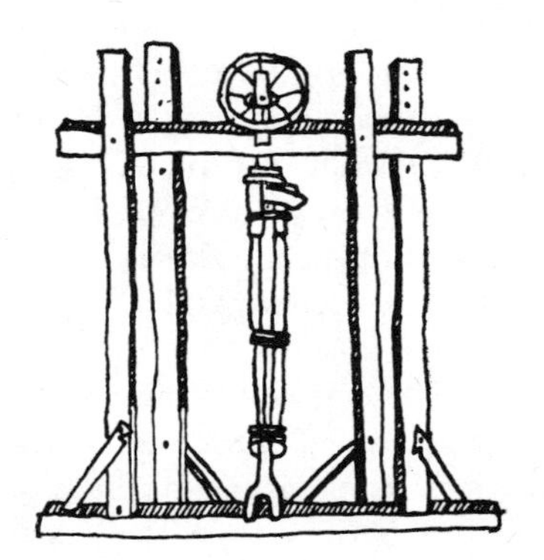

FIG. 31. Machine for boring cannons. Around 1450 (Uccelli, p. 61, Fig. 186).

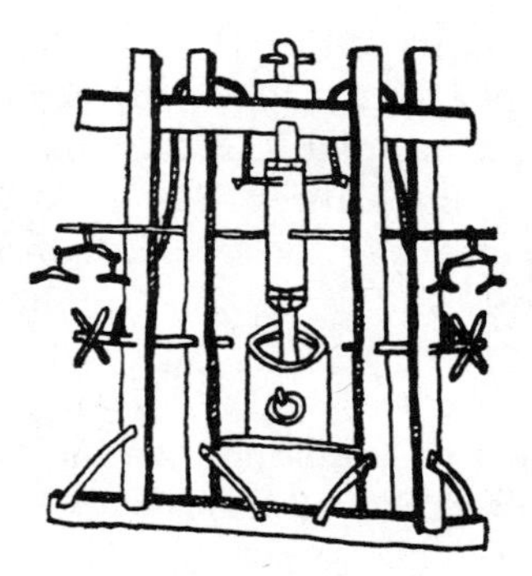

FIG. 32. Machine for boring cannons. Around 1450 (Uccelli, p. 61, Fig. 185).

Strictly speaking, the other metals were worked only by hammering. However, several machines were invented for this work, most of which undoubtedly never progressed beyond the engineers' notebooks. A drawing in a manuscript in Nuremberg, dating from the end of the fourteenth century, shows a workman cutting a file with a chisel and a hammer; he created a checkerboard pattern of grooves in the metal strip. Leonardo da Vinci made this the work of a machine in which the chisel was moved mechanically by means of a camshaft; a gear train ensured the forward movement of the object to be cut. Leonardo's desire to replace the workman by a machine as soon as the movements of execution of the work are sufficiently simplified and regular is again evident. As for screw-cutting machines, so often depicted in Leonardo's drawings, we can apparently be certain that they cut only wooden screws. In the latter machine it is the toolholder that moves, thanks to two endless screws on which it slides. The object to be cut is placed in the center, and turns by means of a crank that also gears into the screws of the toolholder. These gears could be changed to obtain screw threads of varying sizes.

The evolution of minting is more worthy of note. The striking of the coins was ordinarily done with a wedge and a hammer; various pictures dating from the beginning of the sixteenth century give excellent depictions of this operation. In the manuscript of Sainte-Marie-aux-Mines, as in the Annaberg retable, this is the only method shown. Obviously the coins thus obtained were not without defects. When the artist-medallists demanded more precise strikings of their masterpieces, the striking operation had at times to be replaced by molding, a technique that had been used in antiquity. To prevent the coin from moving at the moment when the blow was struck, Leonardo imagined an improvement on the system by inventing a casing that held the two wedges in an exactly parallel axis. Not until the middle of the sixteenth century did Benvenuto Cellini take a decisive step in the minting of coins by inventing the coining press, which was more or less copied in the printing press. Since then the instrument has changed very little in its general conception. Cellini's invention dates from around 1530. Henri II purchased his first coining press from a German in 1550.

Machines for drilling and boring

The drilling and boring operations were undoubtedly the easiest to mechanize, since they utilized essentially circular movements, but their mechanization was only gradually achieved. The so-called "Hussite War" manuscript (circa 1430) shows an early and still very rudimentary machine for boring the wooden pipes that were often used in urban piping systems, a purpose for which this machine was utilized by the inhabitants of Nuremberg. The drawing shows the housing of the machine and the wheel, which was undoubtedly turned manually, although some authorities believe they see in it a hydraulic wheel. Leonardo also drew a machine for drilling beams, in which the beams are held in vertical position. In the "Hussite War" manuscript we see another drilling machine (it may also be a machine for boring cannons); its most striking feature is the very advanced shape of the boring tool, which can be clearly distinguished.

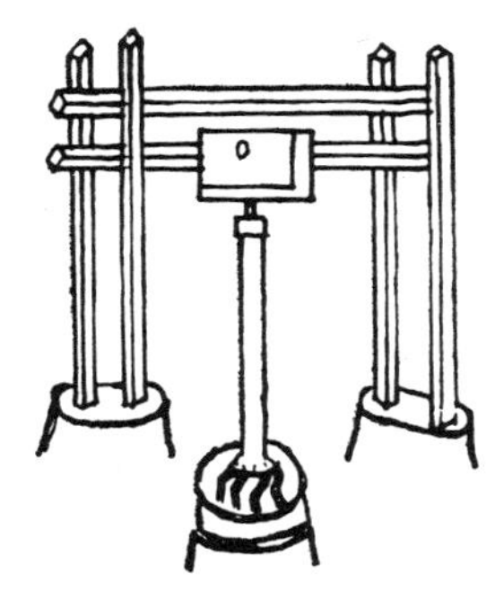

Fig. 33. Boring machine. Around 1430 (Munich, SB, lat. 197, "Hussite War" manuscript).

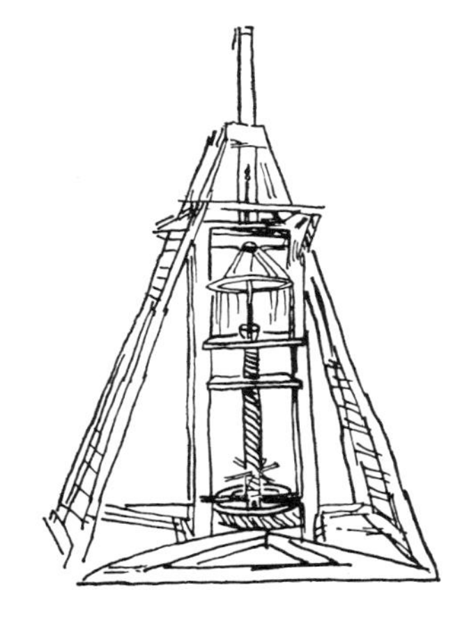

Fig. 35. Machine for boring beams. Around 1488–1489. Leonardo da Vinci (Paris, Bibliothèque Institut).

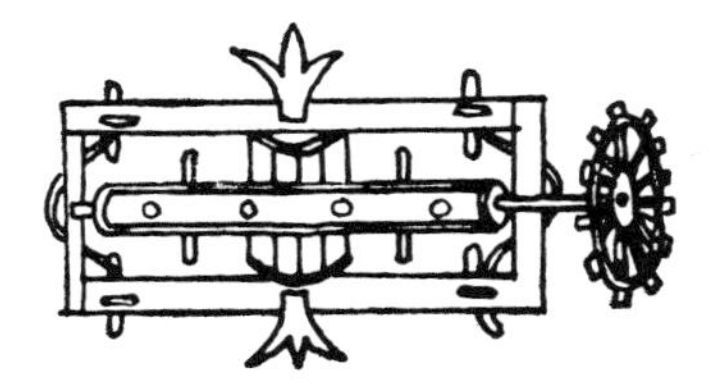

Fig. 34. Machine for boring means. Around 1430 (Munich, SB, lat. 197, "Hussite War" manuscript).

In most cases the traditional instruments, from the auger to the various types of drills, which had been in existence since prehistoric times, continued to be used. One fact should be mentioned, however. We have noted that neither pictures nor texts mention a bracebit, although this instrument may possibly have been known in antiquity. The first specimens appear in the last quarter of the fifteenth century. Francesco di Giorgio shows relatively perfected models with double cranks and movable, replaceable bits.

There appears to have been no major changes in saws. With the appearance of the crank-and-connecting-rod system, the problem of the hydraulic saw was completely solved. Very clear examples can be seen in a drawing by Francesco di Giorgio (later copied by Leonardo da Vinci) of a frame saw that slides within a housing, as is sometimes still seen today. Here, too, a gear pulls the piece being sawed. Another drawing by Leonardo da Vinci shows a manually operated saw with two parallel blades held in a frame.

Polishing machines

We have very little information on the polishing methods utilized in the Middle Ages; we know only that the polishing was done on a tin table and was finished on leather. The invention of the stone-polishing machine was a great achievement. It consisted of a revolving disk turned by a crank and an endless belt. The stone was held in a small vise whose position could be regulated, and was set, as is done today, in a material that it is difficult to identify. This machine is described by Benvenuto Cellini in 1568, and according to most historians described and pictured for the first time in a work of André Félibien (1676). These dates apparently should be revised. The "Hussite War" manuscript depicts a machine that according to the accompanying text is a stone-polishing machine. It is evident that this drawing is unfinished, but we can nevertheless recognize some of the components of the classic machine: the pivot of the disk, and around it the crank-turned belt (the top of the machine is missing, perhaps because of lack of space). The complete machine is depicted in a manuscript in the Bibliothèque Nationale of Paris, exactly as it appears in Félibien's work. Dating this manuscript, which is composed of a number of parts, is difficult; most of the notebooks included in it appear to

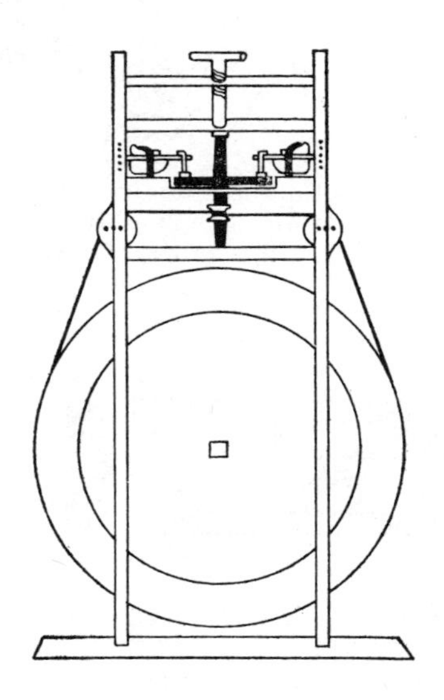

FIG. 36. Machine for polishing precious stones. End of the fifteenth century (Paris, Bibliothèque Nationale, lat. 7295, f° 137).

date from the first half of the fifteenth century. The handwriting in the identifying text must apparently be dated between the end of the fifteenth and the beginning of the sixteenth centuries. The polishing disk, with the endless belt wrapped around its axle, and the supports for the stone (which can be regulated to fit the stone) are clearly distinguishable.

The disk was not the only technical invention. The other essential improvement was the polishing of diamonds with their own dust. Here our chronological information is much more vague. A legend claims that this technique was invented in 1476 by a German named Louis Berken, but no other exact text can be quoted in support of this claim. Several recent historians believe that the modern system of polishing precious stones was perfected in the last quarter of the fifteenth century, in the northern portion of the Netherlands.

Eyeglasses had been invented in, or introduced into, western Europe in the second half of the thirteenth century (around 1285). At first they were made of cut rock crystal; later they were made of glass, when a method of producing a sufficiently transparent glass was discovered. Originally they were produced only for the farsighted. Diverging lenses for the nearsighted do not appear to be earlier than the sixteenth century. Mirrors were made in most cases of polished metal. When glass mirrors were created, an attempt was made to polish curved surfaces that could be used for all kinds of purposes. This paved the way for the appearance of the telescope. In Leonardo's notebooks we find drawings relating to machines of various types for polishing the surfaces of flat or concave mirrors, either with curved tools on which the mirror is turned or with revolving grindstones, or by a combination of both movements. If these mirrors were really attempted, their production must have given rise to almost insurmountable difficulties, despite the reconstructions that have been attempted.

Lathes

For turning objects of wood, bone, or other material that could easily be worked, there existed primitive wheels and lathes equipped with the crank-and-connecting-rod system. This new equipment does not appear, however, to have spread very rapidly. Most of the extant pictures of lathes prior to the sixteenth century depict spring-type lathes (pole lathes or bow lathes), which were in use in the Middle Ages. The only treadle instruments known are the spinning wheel with flyer in the Waldburg-Wolfegg *Hausbuch* (circa 1470), a drawing by Leonardo da Vinci, and the two tools already mentioned. There were also crank-operated lathes operated by an apprentice, who turned a large wheel attached to the lathe by a belt.

The driving mechanism, however, was less important than the tools themselves. We have very little information about the tools that were utilized, but undoubtedly they had changed very little over the centuries; they consisted principally of shears and other cutting tools. The equipment familiar to us from the miniatures, drawings, and engravings that depict, for example, St. Joseph's workshop, is still relatively modest; the Merode tryptych, now in the United States, is a good example. At most we note the presence of the bitbrace, which led to the gradual disappearance of the primitive drills.

Although the second half of the sixteenth century witnessed the development of more advanced equipment, both mechanical and manual, we should not lose sight of the fact that the great majority of workers made use of methods most of which were hardly an improvement over those of their medieval ancestors. Abundant proof is found in the inventories contained in extant notarial documents.

CHEMICAL PROCESSES

While the discovery of certain devices could contribute more or less major innovations in the area of mechanical processes, the same was not true in the field of chemical methods. Not until the scientific discoveries of the eighteenth century was truly decisive progress made in this area.

Dyes

For all practical purposes the textile industry was still using the same dyes it had already used. (Cochineal was a later development.) The phenomenon of a "decline" toward simpler techniques continued in the textile industries, and thus certain strictly prohibited (at least in luxury fabrics) products make an official appearance in the regulations in the second half of the fourteenth century. (This phenomenon will be more fully discussed later.) One change, which was mechanical rather than chemical, should be mentioned: the appearance of the wheel on the dye vats. It appears at the end of the fourteenth century in the record of a community of Nuremberg, and in another work written around 1540.

As for stained-glass windows, it was taste rather than the techniques that evolved. There was a gradual turning away from color, a process that reached a peak in the second half of the fourteenth and the first third of the fifteenth centuries. Bright colors were avoided as much as possible. Multicolored glass was reserved, writes Marcel Aubert, for backgrounds and a few accessories, and its brilliance was lessened by the use of diaper patterns (which appeared at the end of the thirteenth century at Troyes) painted with the brush, in grisaille lines, on a colored background. In the sixteenth century they were in most cases executed with a stencil, in order to simplify the work. A new color, silver stain (glass stained yellow with oxide or chloride of silver), was widely used. Its execution was very simple. On the back of the glass a thick layer of ocher and chloride was laid with a long-haired brush. This was fired, and the ocher, which was only a support, was then removed with a brush. The silver salt impregnated the glass, turning it yellow in the areas on which it had been laid. The first timid appearance of this color occurs in Parisian stained-glass windows of around 1310–1315.

This evolution undoubtedly resulted from the fact that the glassmakers were by the end of the fifteenth or the first quarter of the sixteenth centuries turning out clear glass made with vastly improved materials, instead of the old slightly tinted glass. "Flashed" glass appeared in the second half of the fifteenth century; it was obtained by blowing colored glass, then dipping the parison into the crucible of clear glass and continuing to blow until a cylinder that was colored on the inside and clear on the outside was obtained. This was then worked and stretched. In this way colored glass was encased in a colorless sheath. Red was always "flashed," and frequently other colors as well (the process was rarely used for yellow or green). Several sheets of glass could be superimposed in this way, thus providing a great range of colors. This sheathed glass was then carved with emery, a cutting wheel, or a fraise to remove the clear (or colored) glass in a pattern (end of the fifteenth and first quarter of the sixteenth centuries).

At the beginning of the sixteenth century color returned, with a vastly expanded palette. Light colors were given a more violet tone with the addition of cobalt oxide; flesh tints were obtained with a vitreous enamel pigment and sanguine crushed into a powder (this color appeared around 1490). Fusible and translucent enamels, and vitreous pigments, colored by metallic oxides fired to adhere to the glass, then came into use, permitting the elimination of leading.

Explosives

The production of explosives was still, if not in its infancy (at least in the mid-fourteenth century), at any rate quite crude. The development of artillery was probably due as much to the development of powder as to the evolution of methods of casting. In the "Hussite War" manuscript, composed around 1430, we find an excellent, though brief, description of the method used:

"Powder consists of fine blended sulfur and coal, with niter salt. If you wish to make it more combustible, add *eau-de-vie,* for one pound four ounces, then dry thoroughly in the sun. If you do not have *eau-de-vie,* use vinegar for your mixture, and then dry the powder. Camphor is better than vinegar. If [sulfurated] arsenic is combined with the powder, the stone is hurled farther." Constant improvements in this process were made throughout the fifteenth century. A long series of manuscripts, chiefly German, give numerous descriptions of the methods used, beginning around 1420. The object was to obtain a nitrate of mercury, $Hg(NO_2)_3$, then a fulminate of mercury, Cy_2O_2Hg. Later, the proportions of the mixtures changed. An English historian gives the following compositions:

	Saltpeter	*Sulfur*	*Charcoal*
1380	1	1	1
1410	3	2	2
1480	8	3	3

In France the early powder contained 75 percent saltpeter, the rest of the mixture being divided equally between sulfur and charcoal. This powder burned slowly but did not detonate.

It was a long time before the saltpeter workers were organized; in France their statutes date from 1487. They were assigned both an area to exploit and a quantity to be furnished. In the houses scrapers and sieves were used, elsewhere picks, hammers, shovels, chisels, and mattocks. The product gathered was carefully washed, then refined in caldrons in two or three boilings. The saltpeter remained at the bottom in the form of crystals when the water was evaporated; it was then melted and molded. When the powder was to be made, it was crushed, as were the sulfur and charcoal. The three ingredients were blended in the proportions just mentioned, then moistened, kneaded, strained, and dried. The charcoal was made in most cases of willow and hazel ashes.

The three materials were crushed and pulverized separately. For this purpose water mills or hand-operated pounders were utilized, as is shown in the manuscript copies of Kyeser's work.

Distillation

No decisive progress was made in the art of distillation, but an attempt was made to perfect the devices that had been utilized in the Middle Ages. Leonardo da Vinci distilled various substances, for example azotic acid, alcohol, and perhaps terebenthinate. He was struck by the disadvantages of the devices available, and attempted in particular to improve the chilling of the head of the alembic. The usual method of passing the vapor through a tube chilled in water is found in the thirteenth century in the works of Thaddeus Florentinus. Leonardo tried to keep the entire head of the alembic continuously chilled by a flow of running water. In an early drawing we can sense his desire to avoid mixing the boiling liquid with the product gathered in the receptacle, in order to have a clearer, more brilliant distillate. In other drawings, the heads are soaked in basins with flowing water. The problem was to obtain sufficiently watertight joints. A late drawing shows an alembic whose head has a double compartment in which water circulates; the water probably enters through a hole at the top of the device. This device is almost the equivalent of the Liebig condenser. But could a glassblower have made such an object? Treatises of the period give us little additional information. The distilling of a considerable number of plants was attempted, in most cases for medical purposes; in any case we find the first example of the distillation of beet sugar. Distillation remained for the most part a scientific rather than a technical operation.

We shall not discuss the techniques that utilized the properties of certain bodies. The chemistry of salt and fats had evolved very little, except by a slow, almost imperceptible improvement in procedures. Methods of refining (salt, sugar), about which we know practically nothing, were to make it possible to obtain purer products of better quality. However, our research is still too incomplete to permit us to give very detailed information about these methods, although texts on these subjects are fairly numerous.

While chemical processes were not revolutionized to the extent that some authorities have supposed, it is nonetheless true that a certain amount of progress, not easily discernible, can be seen. One proof of this is the relatively large number of treatises printed between 1500 and 1550. We have mentioned the works of Brunschwygk – the first was published in 1500 – numerous editions of which were printed. There were also two treatises on dyeing, published in Italy. The anonymous *Mariegola dell'arte di tintori* appeared in 1510; the other, by Rosetti, was printed in Venice in 1548 and had several editions. We also find several references to these problems in the famous treatise of Cennino Cennini, *Libro del arte,* which, however, remained in manuscript form until the middle of the nineteenth century. Although they were not disseminated as extensively as the other technical treatises, these works do appear to have been widely known. It must, however, be noted (and Leonardo da Vinci is not an exception to this rule) that the engineers of this period did not devote much attention to this phase of technology: gunpowder was their sole interest.

CHAPTER 6

TECHNIQUES OF ASSEMBLY

THE PROGRESS MADE in the techniques of assembly consisted for the most part of a series of minor, incomplete inventions rather than of radical transformations. Traditions were very strong, especially in the guild industries, and major changes were impossible. But while the continuation of deeply rooted traditional techniques appears to have been one of the basic characteristics of agriculture, we do find a certain change of orientation and a new flexibility in the industrial techniques. The new and tremendously significant technique of printing, whose importance lay more in its intellectual consequences than in its techniques, naturally stands in a class by itself.

TEXTILE TECHNIQUES

A few minor improvements were made in the textile industry. The most interesting transformations were due particularly to the necessity of turning out cheaper products, especially in the second half of the fourteenth century, principally to meet the competition from other textiles, chiefly cotton. On the whole, however, the textile operation retained its medieval appearance. The very powerful economic and social organization of the textile industries certainly contributed to the continued existence of procedures that were to remain in use until the dawn of the great technical revolution in the eighteenth century. This revolution began precisely with the renovation of the textile equipment.

A large part of the products of the textile industry were still of family origin, especially in the rural areas — a confirmation of the stability already noted. This simple fact of the absence of a wide market in itself set limits to the progress we might have expected to discover.

The yarn Wool remained by far the most widely used textile, although cotton was beginning to compete with it. The economic crisis of the second half of the fourteenth century had seriously affected the manufacturing of luxury fabrics. The cloth manufacturers therefore turned to less refined techniques, and the numerous prohibitions contained in the medieval regulations were abandoned, beginning with those concerning the types of wool and their preliminary preparations. Lamb's wool was generally prohibited; it is found only at Valenciennes, beginning in the early fourteenth century, sometimes alone, sometimes as a supplement. In the last quarter of the same century other woolen centers, whose products were even more famous, also began to permit its use. In 1377 Arras introduced a new type of fabric made from lamb's wool; Douai followed suit in 1390. The use of carcass wool was also permitted at Valenciennes; it began to be used in Arras in 1344. The practice of oiling the wool to make it easier to comb was generally frowned upon. A new fabric mentioned in 1343 was made with wool coated with butter. Oiling of the wool appears to have been definitively adopted in 1394 at Arras.

The oldest regulations are completely silent on the subject of carding; the first references to this process are never earlier than the beginning of the fourteenth century. As in the case of the spinning wheel, its use must have long been forbidden or reserved for the coarser grades of wool, mixed fabrics, and wools died with prohibited products. At Valenciennes, which appears to have been a center for the practice of techniques forbidden elsewhere, carding gradually spread during the first half of the fourteenth century, although in 1358 it was forbidden for certain qualities of fabrics. The new textile factory of Arras employed it in 1343, and the regulations of Douai mention it in 1352. At Troyes this technique was forbidden in 1359 and 1361, but protests were so numerous that permission for its use had to be granted in 1377. The argument in favor of these prohibitions was that the action of carding favored the mingling of foreign bodies in the fabric. The primary goal of carding was to untangle the wool, to multiply the strands by breaking them slightly, and to place them lengthwise side by side. But carding also served to combine the different types and colors of wool, and to form a new wool or new color by the precision and closeness of the combination.

The principles involved in spinning had not changed. Most spinsters continued, and were to continue for a long time, to spin with the spindle and distaff. In 1475 the spinsters of Leyden used the spindle for spinning a long, carded type of wool called *étaim,* and not until around 1496 and 1527, respectively, did the use of the spinning wheel become general in The Hague and Leyden. The new instrument made very slow progress; it appeared in Douai for the first time in 1305, at least according to the texts, and the second reference to it dates only from 1362. Beginning in the second half of the fourteenth century, its progress became more rapid and widespread. The closing years of the fifteenth century saw a major improvement: around 1470, in the Waldburg-Wolfegg *Hausbuch,* we find the first picture of a spinning wheel with flyer and automatic winding system. The system of the flyer made it possible to give an extra twist to the yarn. Thus this invention, which was long attributed to Leonardo da Vinci, is in fact considerably earlier, if we consider this first picture as being probably later than the actual appearance of the flyer. Leonardo da Vinci designed a fairly complicated mechanism that gave the spindle a back-and-forth movement, permitting an even winding of the yarn. After the sixteenth century, in any case, we see from numerous paintings that the flyer was now customary.

The manufacture of cloth No major modification occurred in the preliminary operations for weaving before the middle of the sixteenth century. As for weaving itself, we have little knowledge of the development of the loom, which must have been slow. Two-shaft (that is, plain) weaving was certainly still the most frequently practiced type. Three-shaft looms are attested in 1403 at Douai and Valenciennes, and also at Bergues. More complex structures were probably nonexistent.

Fabrics continued to be prepared with the same techniques utilized in earlier centuries. However, there was a return to more primitive techniques, undoubtedly regarded as better in accordance with a habit of thought that has not yet completely disappeared. Thus at Nogent in 1403 and at Chartres in 1444 there was a return to fulling by foot, after mechanical fulling had been in use for some time. Here, again, products that previously had been strictly forbidden were now permitted for the fulling process.

If we wish to characterize this period in a few words, we may say that carding and the adoption of the flyer for the spinning wheel were almost the only innovations. Leonardo da Vinci may have dreamed of applying mechanical principles to various textile operations, but it is unlikely that he succeeded in translating his ideas into reality. Nor were they utilized by succeeding centuries. The mounting of several bobbins on a single driver did not suffice to solve all the problems of mechanical spinning; Leonardo's drawing is not sufficiently detailed to show the details of the mechanism. The same is true for the mechanical weaving loom that appears in the *Codex atlanticus,* a drawing it is difficult to interpret. Obviously the idea of operating a loom mechanically may very well have entered Leonardo's mind; by carefully studying the workers' actions, a habitual activity of his, he imagined such a device. But translating the idea into reality could not be done. The same was true of other, less complicated, operations whose mechanization it was easier to imagine, for example teaseling with the card and shearing. Leonardo has left drawings of the machines he would have invented for these purposes. It would have been difficult actually to construct them, especially in the case of the machine for the preparation of the cloth hose so often worn by the common people.

Silk

Silkworking techniques had developed little during the two preceding centuries. We have already mentioned the appearance of the white mulberry, which gave products of better quality, but aside from this innovation the techniques already known by the middle of the fourteenth century appear to have continued in use. At most we

Fig. 37. Weaving loom, 1387 (Nuremberg, *Hausbuch der Mendelschen Zwölfbrüderstiftung;* Uccelli, p. 129, Fig. 10).

Fig. 38. Silk mill (Singer, II, p. 206, Fig. 171).

find an expansion of machines that until then had been concentrated in Italy, where production secrets were carefully guarded. Hydraulically powered spinning machines may not have crossed the frontiers of this country until some time during the fifteenth century; silk weaving followed in their wake. Not until the last quarter of the fifteenth century do we see numerous attempts being made to install these industries almost everywhere in western Europe. The Bolognese machines and the so-called "Jean le Calabrais" loom then appeared everywhere.

A treatise on silk dating from 1487 includes a certain number of miniatures relative to the manufacture of silk fabrics. We find no innovations in the material and processes it describes; however, it does show a vertical loom.

The other textiles certainly continued in use, especially flax and hemp, whose techniques had undergone few modifications. The great expansion of cotton as a textile for garments should be noted, together with flax, generally reserved for the upper classes.

Leonardo's inventive nature is evident in his ideas for the manufacture of hemp rope — assuming that he did not simply reproduce a drawing or an actual machine. His drawing shows a machine for making rope, which it does by twisting a certain number of threads.

Thus there were no major transformations in textile techniques. However, our knowledge of these industries is still very incomplete. One industry in particular has resisted all attempts at investigation: knitting, which appeared at the beginning of the sixteenth century, in the form in which it is still practiced today. No picture earlier than this period, and no archaeological evidence, exists for this technique, which makes it possible to weave with as it were a single thread, and the statements that have been made on this subject are purely gratuitous. In any case, the theory that techniques used for coats of mail were similar to those used for knitting wool has no basis in fact.

JOINERY

The same stagnation is seen both in the equipment and the methods used in the techniques of joinery; we find at most certain perfections of detail.

Cabinetwork was perhaps the most stable of all techniques; construction had no need of transformation. Furniture itself, however, changed radically. The chest, which since antiquity had undoubtedly been the principal storage place, almost completely disappeared, taking refuge in the rural areas, where it persisted for a long time. The armoire in two sections, which theoretically is made by placing one chest on top of another, made a timid appearance around the middle of the century. The table was also completely transformed; it was attached to its legs, and became monumental in size. The chair, in contrast, became lighter, and finally became an armchair. The large poster bed appeared later, and led to the Renaissance bed with steps at the sides and a chest at its foot.

The same types of joints found in medieval furniture were used in all these articles of furniture, except that beginning in the fifteenth century tongue-and-groove joints were used for joining the boards.

The techniques of roof trusses remained the same, at least in stone buildings where the truss was not visible. In France, most of the churches or large halls were vaulted, which considerably limited the role of the truss. In England the case was completely different, and the practice of open-timber roofing continued to develop until the first half of the sixteenth century. Viollet-le-Duc noted its striking similarities to the techniques of naval construction. The essential problem of the open-timber roof was to eliminate the tie beams without lessening the strength of the structure. In order to prevent accidents, strong, thick, wide beams had to be used. The purlins then acted to strengthen the structure, thereby acquiring an importance they did not have in, for example, the French trusses.

At Malvern Abbey (Worcestershire), whose wooden roof dates from the middle of the fourteenth century, the builders used a kind of square framework made of individual pieces. The purlins are supported by ties or curved struts that support the rafters and prevent the truss from giving way under the horizontal thrust. It was a complex, heavy structure, an expensive price to pay for the elimination of the tie beams – and its span was not very great. Ely Cathedral was roofed at the end of the fourteenth century. The builders used large curved braces cut from a single piece of wood, whose bases were fixed in a strut while the top was jointed in the queen posts. The purlins were fixed between the principal rafter and this brace; the intervening interval was filled with timbers to strengthen the structure.

The masterpiece of English roofing is uncontestably Westminster Abbey, with its span of almost 69 feet. Large wall posts resting on brackets supported the curved braces, which met at the base of the queen post. Here a hammer beam cut in a single piece rested on a post, which rested on the end of the protruding strut. The pressure on this end was extremely great, since both the curved braces and the posts met at this point. Part of this stress was met by large arched timbers that joined the hammer beam, the post, the brace, and the wall post that supported the brace. In this way the strength needed in order to avoid warping was obtained.

It is difficult to find roof trusses of this type outside England: most of the European countries preferred the vault. However, a few examples of roof trusses built on the same principle can be found on the Continent, for example the main hall of the City Hall of Saint-Quentin.

It was perhaps in the construction of the wooden house that the most visible modifications were made. The system of construction utilized in the Middle Ages required long, square planks. The growing scarcity of wood forced the builders to use other methods in the fifteenth and sixteenth centuries, methods based on the dividing of the posts, which thereafter were cut to fit each story. The horizontal pieces could then be placed on top of the vertical pieces. The post, resting on the template, carried the girders that supported the joists and, over all, the sill that separated the two stories on the outside. Thus the girder rested on top of the post instead of fitting into it, as in the earlier system. The fear was that the tenon, the only organ that held the two, would break under the strain, causing the post to collapse. The girder was therefore wedged by extending the side of the (frequently enlarged) post (half-open-mortise joint). The sill is the old brace, inserted between the uprights of the old system. In addition, the two horizontal pieces had to be prevented from sliding on each other. To obtain a better joint, the girder and the sill were dovetailed, and the upper post was jointed to the sill by a tenon. A kind of triangular bracket attached to the post and placed under the girder replaced the brace that had formerly supported the post and the sill; it strengthened the angle in which it was placed.

The half-open-mortise joint, which was frequently used at Rouen, does not appear to have been utilized in Germany or England, but in contrast was used in both western and eastern France. The studwork common in Rouen consisted essentially of studs and a few oblique struts that were always jointed in horizontal pieces. To the west of the Seine we find groups of struts jointed in the sills and posts, a system which gave birth to an oblique studwork. The same concern for ensuring triangulation is found in southern and western Germany, where it is manifested in the practice of supporting the posts with pairs of large braces placed on each side. Curved planks form a network of twisted forms which are often complicated with pieces crossed to form a latticework. In England we find either vertical studs evenly spaced and close together, or intersecting curves. In Sweden the studwork forms panels in which triangulation is ensured by several large braces. The walls are often covered with pieces of wood or tiles to protect the framework from infiltration of rainwater.

Short planks and complex systems of joints thus predominated in construction. By the end of the fifteenth century the art of the carpenter had achieved a perfection that was demonstrated in corbeling. The simplicity of the studwork was derived in particular from the increased use of windows, which is linked to the question of lighting. It is difficult to establish a date for this new system of construction, which, for a long period, coexisted with the old system, sometimes even in the same buildings.

The corbeling of the medieval wooden houses is one of the famous characteristics of this type of construction. Corbeling was easier to achieve with the techniques we have just briefly described than with the construction methods of the earlier system. Several reasons have been given for its origin, one being in-

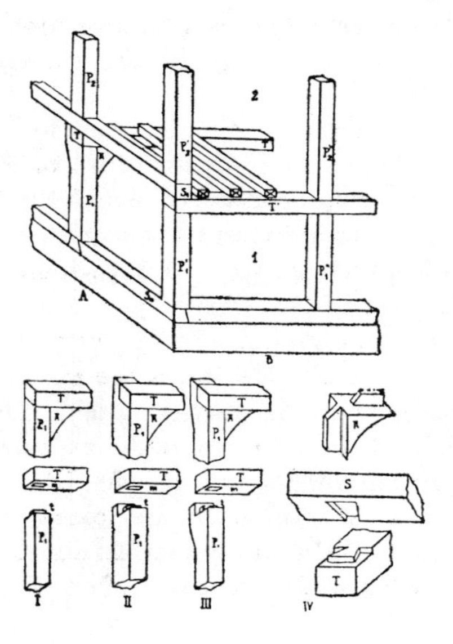

FIG. 39. Joints used in the houses of Rouen. End of the fifteenth century (from Quenedey).

creased space, which despite the opinion of certain authors was not negligible. Another reason was that it protected the lower stories from the rainwater, and served to shelter the passersby. It has been claimed that in Germany corbeling was due to a constructional arrangement, additional stories having been gradually added on that utilized the protruding girders laid to receive the roof truss. This formula is supposed to have been born in northern Germany and to have then spread to Hesse, the Harz region, Westphalia, and the Rhineland. From here it may have entered France, but this is much less certain.

There are numerous methods of constructing corbeling. In the earlier type of house, with its posts rising straight from the ground, corbeling could be created only by bypassing the joists, an arrangement common in Germany, England, and eastern France, or with triangular brackets and false joists, a much less common arrangement. In the later model corbeling on joists was naturally invented. To bear the lateral thrusts on the tenons of the joints, braces were inserted between these vertical pillars. This system made possible the creation of projecting braces called "corbeling braces." In this way extremely strong frameworks were obtained. This system made for a good distribution of the joints, so that the tenons were not all accumulated at the same point. The method appeared at Rouen in the first half of the fifteenth century, and became common in every area where projections resting on joists were to be found.

The art of the framework was thus a manifestation of a certain originality, which showed the extent to which simple craftsmen had reflected on the factors involved in their work.

PRINTING

The dissemination of ideas and knowledge presented serious difficulties as long as manuscript copies were the only material that could be utilized; copy work was long and costly, and by no means completely accurate. By the beginning of the thirteenth century the growth of the universities had brought about the appearance of broader needs that it was difficult to satisfy. The professional copyists organized into numerous workshops, and a system of verified copies that were rented to copyists was perfected. So that these master manuscripts, which were written in a fairly large hand (a certain number of examples are still extant), could be kept circulating as rapidly as possible, they were divided into separate books (the *pecia* system). In this way the texts of the classical authors were disseminated. More than two thousand thirteenth- and fourteenth-century copies of the works of Aristotle are still extant, which, taking into account the number that have disappeared, represents a much larger number of copies actually made. Nevertheless, the difficulties of dissemination were still considerable.

A number of large workshops had been opened that were organized either under the control of the universities or in connection with the major religious institutions, or even for the benefit of important public figures. Assembly-line production made it possible to turn out not only pious works but even nondidactic works written in the vernacular. Two hundred and fifty copies were made of *Le Voyage de Jean de Mandeville* (1356): seventy-three in German and

Flemish, thirty-seven in French, forty in English, and fifty in Latin, to say nothing of a lesser number of manuscript copies in Spanish, Italian, Danish, Czech, and Gaelic. The growing need for books, which was unquestionably one of the causes of the invention of printing, thus found an organization to which the new techniques were to be adapted.

Papermaking in Europe In various chapters of the preceding volume we have seen how papermaking appeared in the Far East. Paper had been introduced into Europe in the twelfth century by Italian merchants, who imported it from the Middle East. The first paper mills in Europe were operating in Spain around the end of the thirteenth century, but the real center for the dissemination of the paper industry throughout Europe in the fourteenth century was Italy. This dissemination was governed by various economic and technical factors.

The raw material consisted of linen and hemp rags, and a sufficiently abundant supply did not become available until the cultivation of these two textile plants was well developed and the use of garments utilizing these fibers became quite common. The location of paper factories was also determined by the problem of water supply. Water was used in large quantities in the making of the pulp, and it had to be free of mineral salts, which might discolor the paper, and of refuse from the urban agglomerations. In addition, water supplied power needed to operate the mill. The latter consisted of a hydraulic wheel whose camshaft raised the mallets; these fell of their own weight into the troughs where the rags were crushed. These mills, which as we have seen had been in use in Europe since the early Middle Ages, were not intended solely for the making of paper pulp; in the early stages wheat mills were transformed to be used in the papermaking industry.

Production procedures remained practically unchanged for almost six centuries. The rags were torn up in the rag cutter, and were then placed in the "rotting room," that is, in a damp place where fermentation began. They were then shredded in troughs containing soapy products, and were beaten with pounders equipped with metal blades. The pulp was then beaten again in skimming troughs, after which it was lifted in brass molds into vats and diluted in warm water. The sheet was formed by the draining of the water when the mold was lifted out of the vat. The sheets, separated by layers of felt, were piled up, and were passed through a press. They were then dried in the open air, after which they were plunged into a vat of sizing made from gelatin and alum.

Between 1475 and 1560, paper factories spread throughout Europe. The paper centers in the Champagne and Vosges regions tripled in importance during this period, while those of Normandy, Brittany, Angoumois, Beaujolais, and Auvergne increased with still greater rapidity. In Germany the first paper mill was constructed in Nuremberg in 1391, followed by those of Kleve (1431), Augsburg (1460), Ulm (1469), and the eastern part of Germany (1480–1496).

Whether through local production or importation, the European countries now had a sufficiently large supply of paper to permit the realization of one of the indispensable conditions for the appearance of printing. The use of parchment remained common, but it could spread no farther. It was expensive, its production was limited, and the distrust with which paper had been greeted and the prohibitions against its use for official and commercial documents (out of fear that it was not durable) had almost completely disappeared.

The only problem concerned the supply of rags, and solving this problem continued to be difficult until the nineteenth century, because of the continually growing demand once printing was invented. Moreover, it was the activity of the printing plants that favored the development of papermaking and paper selling toward the end of the fourteenth century.

The origins of typography

Methods of reproducing a drawing had existed prior to the fourteenth century. The bookbinders utilized sheets of metal with sunk engravings to decorate the leather. For rapid reproduction of large, ornate initial letters, the copyists' workshops used relief stamps made from wood or metal. The method of printing fabrics had already been imported from the East. The first xylographic prints appear to have been prints on paper of woodcuts intended for fabrics, some seventy years before the invention of printing, in the last quarter of the fourteenth century. Wood engravings seem to have originated in the Rhineland and the Burgundian States. Thanks to crude equipment, religious pictures could in this way be disseminated in large numbers, and they enjoyed great success. The first woodcuts were pictures without text; later, short captions appeared. From this essentially religious imagery developed the production of cards, satirical posters, and calendars in which the text played a much more important role. Small xylographic books quickly followed, and persisted even after the appearance of typography.

It now appears certain that the origins of typography are not to be sought in these woodcuts. It was impossible to create groups of wooden characters, which could not have the necessary precision. Moreover, metalworking was completely unknown to these craftsmen. In fact, it was the printing industry that gave printing ink and the press to the wood-engraving industry.

Three inventions were indispensable for the appearance of printing: movable characters, printing ink, and the press. The first two posed technical difficulties of obvious importance. A punch that had to be engraved in relief in a hard metal struck a matrix in a softer metal. The character was cast in this hollow matrix, placed in a mold, with a metal that was fusible at low temperature (tin or lead). The technique of engraving the punch had long been in use by the goldsmiths. By the end of the fourteenth century, the Italian goldsmiths had even brought the techniques of casting medals back into use.

Sources for the dates of the first metal movable characters are rare and difficult to interpret. The first references appear in the lawsuit between the goldsmith Gutenberg, the son of a family of minters, and his moneylenders, in Strasbourg in 1439. Though the terms are rather vague, they seem to refer to an early attempt at printing. This was not the only attempt: another goldsmith, Waldfoghel, a native of Prague, was working along the same lines in Avignon around 1444–1446. This source mentions metal alphabets for writing artificially, but it remains uncertain whether this refers to isolated characters, entire pages, or other methods. The text of the Chronicles of Cologne for 1499 attributes to Gutenberg, at Mainz in 1440, the invention of printing or of a typographic method that may thereafter have been utilized on a wider scale in Holland, where around 1441 a certain Coster probably perfected the art of assembling the movable characters. An examination of certain Dutch incunabula has confirmed that the characters were cast in sand molds, perhaps prepared with wooden punches. Numerous attempts were probably made, and solutions proposed, in the years immediately prior to the middle of the fifteenth century. These experiments were undoubtedly encouraged by the immense success of the woodcuts.

Gutenberg had returned to Mainz before the end of 1448. In October 1457, after many financial difficulties, the first dated work appeared: the *Psalter of Mainz,* printed by the shop of one Schoeffer, and absolutely perfect in execution. Had Gutenberg printed other works before his break with his moneylender? We shall probably never know. The legend of the genius robbed of his invention is a common one. By 1450–1455 several printing shops were already in operation in Mainz, and it is here that the famous forty-two-line Bible, traditionally considered the first printed book, was composed.

The growth of this new technique was rapid and considerable, despite the secrecy that surrounded the first editions. By 1458 the King of France was making inquiries about it. In 1459 a Bible was printed in Strasbourg. The first Italian books date from 1470, by which time the number of printers was already considerable. Henri-Jean Martin estimates that by the end of the fifteenth century, some fifty years after the appearance of typography, at least 35,000 editions, representing at the very least 15 to 20 million copies, had already appeared. Of this number, 77 percent were in Latin, and 45 percent were religious works, and 236 cities had shared in this production.

The making of the characters

The technical difficulties involved in making the characters were numerous. In particular, they involved the discovery of metals that would make sufficiently strong punches, matrices that would not wear out too quickly, and characters that could be thoroughly inked and that would last for a fairly long period of time. The first punches appear to have been made of brass or bronze, and the first matrices or matrix molds were obtained by pouring lead around the punches. Schoeffer is believed to have been the first printer to utilize steel for the punches and copper for the matrices. Lead matrices were still in existence at the beginning of the sixteenth century. Abbreviations and ligatures required a much larger number of characters than is the case today.

The problem of the characters themselves was the most difficult matter. By the fifteenth century very sturdy characters must already have been developed, but even these wore out quickly: in 1570 Aldus Mannucci made a new set for each volume printed. The printers probably utilized a tin alloy without too much lead, which would have damaged the matrices. A small quantity of antimony was also added, as can be seen in characters from Lyon dating from the end of the fifteenth or beginning of the sixteenth century. Determining the correct proportions of the metals used was, however, difficult.

The cutting of the punches quickly became a craft for specialists, engravers or goldsmiths. The standardization of types after the almost universal adoption of Roman type made it possible for a small number of shops to acquire a quasi-monopoly of this production. The standardization of the ensemble of the typographic material was achieved slowly; however, the type height was not standardized until the number of producers had decreased. For a long time the size of the characters also was not standard.

Composition

Composition by hand, which is now tending to disappear, was universal. The characters were arranged in the case, a wooden rack divided into small pigeonholes. The characters selected for use were placed in a wooden composing stick; the completed line was placed in the composing galley, a small tray in which the lines were mounted. The lines were grouped in pages, which were then placed in the form. At first the case was in a very low and slightly inclined position, which made the work painful; not until the second half of the sixteenth century was the arrangement still in use today adopted. There was no easy method for determining the correct position of the characters, which resulted in numerous errors. The arrangement of the letters in the case was not standardized, which was particularly disconcerting for workers going from one shop to another.

The press

Before they had realized the usefulness of, and perfected, the press, the early printers perhaps began with the method of the rubbing that was used by the artisans of woodcut engravings. The fact is that we have very little precise information on the method used by the early printers. Martin considers it quite mysterious from certain points of view, and perhaps even quite different from what might be supposed, particularly as regards the form and assembling of the characters. Why was a hole drilled in the stem of the character? Why was its end cut in the shape of a bevel or a chevron? Some technicians believe that the inverted form was placed on the sheet of paper, in contrast to the position later adopted. The instrument must therefore have been infinitely more simple. These problems will be considered in more detail in the chapter on improvements in printing, in the second part of this volume.

The reconstruction of Gutenberg's first press remains quite hypothetical; it was a simple wooden press very similar to those utilized for wine and oil. The printing was done by page, since this was the size of the form, with resulting shifting. After 1470 the double-printing system appears to have been used, the form being composed of several pages or even an entire sheet. The printer had to be able to move the form rapidly and with precision. The form was now placed on a movable carriage and moved (beginning before the end of the fifteenth century) by means of a system of cranks and pulleys.

The press utilized in Lyon in 1500 by the German printer Husz still consisted of a large wooden screw. Between the vertical posts was a table that held the form; this form had a hinged cover in which the sheet was inserted. It was then lowered, and the bar was pulled over to cause the sheet to adhere to the composition. At the beginning of the sixteenth century three types of presses were in use: French (or Lyonnais), German, and Flemish. The trademark of a printer from Ghent established in Paris shows us a considerably improved

FIG. 40. Printing press, 1507 (*Prima pars operum Baptiste Mantuari,* published by Judocus Badius Ascensius in 1507. Title page).

model that may be that of the middle of the sixteenth century. The platen, placed beneath the screw, is still small. The movable table slides on two rails, and is able to hold a much larger form. The composition is easier to ink, and it is less difficult, between two turns of the bar, to introduce the second part of the form.

Numerous problems would have to be studied in order to exhaust this subject, but they involve indirectly related techniques. An oil-based ink suitable for this work had to be created, but we do not know much about it; a carefully sized paper had to be used. Printing was undoubtedly less important for its internal technological developments than for the influence it exerted on intellectual development.

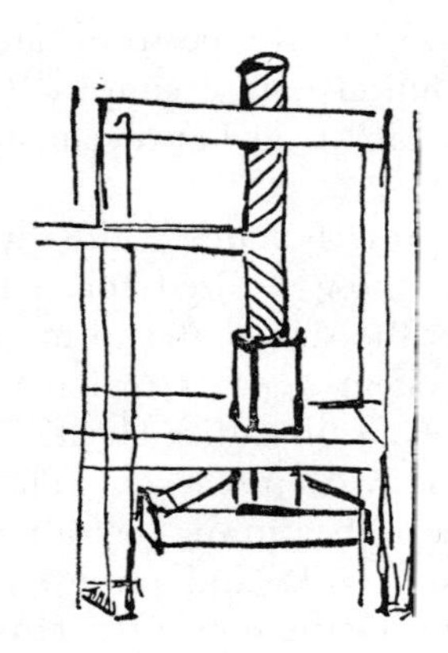

FIG. 41. Project for printing press. Around 1490–1500. Leonardo da Vinci (*Milan, Cod. atl.*).

CHAPTER 7

MILITARY TECHNIQUES

RECENT WORKS have demonstrated that the development of the art of warfare has often been the basis of technical and even scientific progress. It was the profound modifications in military techniques that brought about a sudden awareness of the importance of technical problems. The military techniques of the Middle Ages offered no material for further major development; in contrast, the invention of and development in firearms led to much research in numerous fields.

Firearms dominated military technology, at least after the second half of the fifteenth century. From the middle of the fourteenth century to the middle of the fifteenth, progress was slow, and firearms were used only on a very small scale. After 1450, cannons very slowly became an essential element of the art of warfare, modifying not only combat techniques but also all the systems of fortification. It was in the besieging of strongholds that the cannon proved to be most quickly utilizable. In Normandy in the second half of the fifteenth century, the French were able to reconquer sixty strongholds from the English in a year and six days; in 1515 they captured the castle of Milan thanks both to the artillery of Galiot de Genouillac and to the mines of Pedro Navarro. In 1509, despite his artillery, Maximilian failed to capture Padua. The role of artillery did not dominate ordinary combat until the second quarter of the sixteenth century. At Fornovo, the only real field combat of Charles VIII during his Italian expedition, the French cannons fired at little aside from the enemy batteries; the artillery of both sides combined killed fewer than ten men. The first serious use of artillery occurred at Ravenna in 1512, and artillery played an important role at Melegnano in 1515.

WEAPONRY

At the very end of the fourteenth century the trebuchet had been utilized at the siege of Rennes, and throughout the fifteenth century the handbooks on military technology (or, more precisely, the collection of drawings of war engines) still show the medieval counterpoise artillery; we find it even in the works of the great Francesco di Giorgio Martini in the last quarter of the fifteenth century. But counterpoise artillery was soon to become only a memory of a bygone age; the last reference to it seems to be the use of trebuchets at the siege of Burgos in 1475–1476, in a country where military traditions were slow to disappear.

Transformations in mobile weapons occurred much more slowly. It was quite difficult to perfect portable firearms; for a long time the slowness of their fire and their lack of precision prevented their inclusion in military equipment. This is why, in the first half of the sixteenth century, we find the most modern armies equipped not only with picks (which still had a long life ahead of them), but also with bows and crossbows. These last vestiges of medieval armament did not disappear until the second half of the sixteenth century.

Firearms

In the middle of the fourteenth century the cannon was a very crude instrument, dangerous for its users and of limited military interest. However, we have very little information on this primitive artillery, of which practically nothing remains.

According to certain writers, the first cannons were long and thin, and were made with iron staves or were cast in iron and copper, reinforced at intervals by rings of iron, and carried by mules or on carts. (A later age called them *"veuglaires."*) The tube was open at both ends; a box containing the powder charge and the projectile was fitted to one end. This breech was completely independent of the tube, and was attached to it by means of a movable stirrup. This system permitted a large part of the gases to escape, and thus considerably reduced the power of these weapons. On the other hand, it had the advantage of being easy to transport.

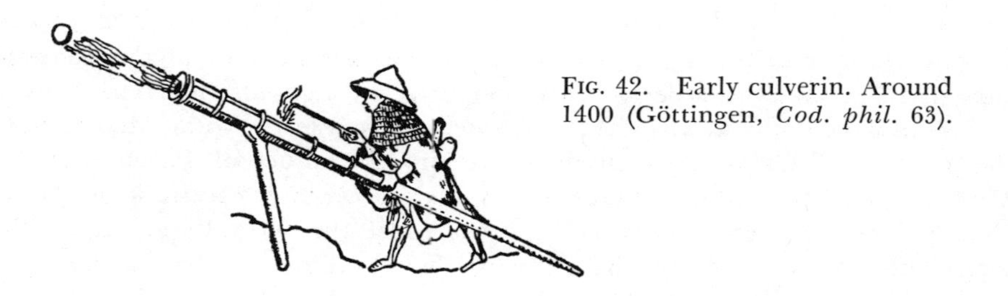

FIG. 42. Early culverin. Around 1400 (Göttingen, *Cod. phil.* 63).

There also existed very short, wide guns, in all calibers, closed at one end. These "bombards" were utilized particularly in siege warfare for hurling very heavy stone balls.

This primitive artillery was not very easy to maneuver, and posed little danger for the adversary. The problem was to transform it into a genuine weapon. We are able to study the transformation of these engines in the numerous military manuscripts of the fifteenth century. But the drawings, except for several very special features, are difficult to interpret precisely. Numerous guns of this period can be found in the museums of western Europe, however, and by combining the two sources we can briefly reconstruct the evolution of these techniques.

Several problems had to be worked on simultaneously in order to achieve a satisfactory result. The first difficulty was the manufacturing of the guns. Throughout the fifteenth century cannons continued to be made of forged iron.

But progress in metallurgy made it possible to obtain guns infinitely superior to the early models. The arsenal of Basel contains a remarkable gun of forged iron that dates from the third quarter of the fifteenth century. The breech is forged in a single piece. The bore is composed of a staving of iron strips 1.2 inches thick by 2.4 inches wide, held in place by thick iron rings. The diameter of the mouth is 13 inches; the priming hole is very narrow. A similar gun of much larger caliber is preserved at Edinburgh Castle. Tradition claims it was forged at Mons in 1486. These were, therefore, despite improvements, guns of the old type. There was one important modification: the chamber, fitted in by means of a groove and attached by a strap that very often broke, had been abandoned. In the first half of the fifteenth century the powder chamber was embedded in a traplike opening in the cannon; it still had the disadvantage of wearing too rapidly at the muzzle, and single-piece guns loading at the mouth were therefore gradually developed. When these guns became very large and consequently very heavy (the Edinburgh gun is a perfect example), transporting them became difficult. The breech was therefore separated from the gun, the two pieces screwing into each other. The places left for the wooden levers that served to raise the gun can still be seen.

A change in technique was in the offing. While some cannons appear to have been made with cast iron after the appearance of this metal, there was some hesitation about using it, perhaps because the techniques of casting were not yet well known. Copper was therefore used for cannons. Specimens dating from the first half of the fifteenth century are still in existence. A better solution, very soon reached, was the bronze cannon, which remained in existence until the following century and which, until the appearance of steel, constituted the customary type of artillery. The arsenal of Basel houses a copper gun six and a half feet long, dated 1444. Considerable progress was made in techniques of casting bronze cannons. Cast-iron cannons, references to which occur in the middle of the fifteenth century, became known in Italy around 1460–1464; it very soon became evident that they were too fragile to be useful. Bronze gave the gun manufacturers a stronger metal that made it possible to make the single-piece guns that had come into existence through the transformation of the breech. By about 1535 England had brass cannons. After 1540 the casting process for cast iron was perfected to the point where it could be used for artillery, which considerably decreased its cost. Ships were now equipped with cast-iron rather than bronze cannons. The dazzling success of the French artillery beginning in the reign of Charles VIII, in contrast to the frequent bursting of the Italian cannons, was due to the quality of the French bronze.

Until the middle of the sixteenth century the priming hole was simply a hole opened in the tube. It had the serious disadvantage of becoming quickly enlarged, to such an extent that it made the gun useless. Between 1550 and 1561, the idea of lining the priming holes with iron or steel tubes inserted in a hole drilled in the cannon became common. This represented considerable progress; previously a cannon could not shoot more than sixty to seventy shots in succession, and even then it had to be cooled. In 1561 a Venetian ambassador speaks of this method as a novelty peculiar to the French artillery. Unfortunately, these vents rusted quickly.

The transportation and mounting of cannons

The new artillery posed a delicate problem that had been unknown to the medieval "master gunners," who had constructed their instruments at the battle site. Artillery had now become permanent, and it had to be transported. The first guns were mounted on gun carriages without wheels; they were embedded in a trough cut in large pieces of wood and held together with bolts, iron clamps, or even ropes. The gunners were so distrustful (with good reason) of their guns that they attempted as much as possible to literally surround the weapon with large pieces of wood, a valuable protection against bursting, which was a frequent occurrence. In order to transport them, the guns had to be hoisted on carriages; numerous examples of the gin, the hoisting device used, can be seen in the military manuscripts. The idea of genuine gun carriages armed with light cannons, which can be seen in the oldest of these manuscripts (German, end of the fourteenth and beginning of the fifteenth centuries), probably originated in this device.

The laying of these guns could be done only with levers and wedges, or by sinking one of the ends. Aiming was only approximate, since in siege warfare the first bombards did not fire against the walls but hurled their balls over the walls into the besieged camp. In the second half of the fifteenth century two wheels were placed on the forward section of this gun carriage. The carriage was then divided into two superimposed pieces, the upper portion being able to turn in a portion of an arc. The cannon was embedded and held in jointed pieces of wood pivoting on a horizontal pin placed under the mouth. The effort required to lift the gun was still considerable, and the lower carriage was therefore equipped with racks — iron arches with holes that held the metal ties supporting the upper carriage (the one supporting the cannon). Examples of these racks can be seen in the treatise of Kyeser, in the early years of the fifteenth century.

Instead of moving the entire upper framework on an axle placed under the muzzle of the cannon, the lower portion of the carriage was made mobile, and the pin, instead of being placed at the head, was placed to the right of the breech. The entire weight of the cannon was thus borne on the axle. The same system of racks was used for aiming. The idea of fitting trunnions to the tube to make it pivot more easily is believed to have been conceived in 1460. These trunnions probably were an addition to the tube, until the period of Charles VIII, when they were cast in one piece with the gun. The gun was now aimed not by raising the carriage but by using first wedges, then a screw, under the breech of the cannon. Around 1530 a third wheel was added to the two already present in the rear carriage, which was then separated into two strong "cheeks." The third wheel, which facilitated the return to battery, was mounted between the cheeks.

The birth of the two-wheeled gun carriage, which could be attached either to a carriage or, later, to a movable forecarriage, solved to a great extent the problem of transporting cannons. The cannons of Charles the Bold, effortlessly seized by the Swiss at Morat in 1476, and preserved at the arsenal of La Neuveville, in this sense are already modern cannons. In 1515, when François I ordered his army to cross the Alps, transporting the artillery gave rise to serious difficulties: the cannons had to be removed from the gun carriages and hoisted over the escarpments by means of windlasses. (Bas-reliefs on the king's tomb depict this delicate maneuver.) Elmwood was the wood used in gun carriages. The idea of planting rows of elms along the highways of France, not so much to shade travelers as to supply the necessary wood for the artillery, has been attributed to Sully. Similar measures were taken in 1553 by Henri II. The mounting of these gun carriages also required a considerable number of pieces of iron of all types, to give them strength. According to authors of the period, a gun carriage with its wheels contained more than one hundred pieces of iron. Also for purposes of strength, the spokes of the wheels were inclined over the hub.

Ammunition and firing The early firearms were used only for hurling the same projectiles that had been used for the counterpoise engines, that is, stone balls. The difficulty was to cut stones sufficiently even to fit into the cannon without leaving too large a space between the ball and the walls of the gun; it was necessary to avoid loss of power by leaving too much play. When hurled against walls, these balls had the additional disadvantage of bursting or breaking wthout producing the desired effect. However, it is almost certain that these stone balls were utilized until late in the fifteenth century. In many cases they could be made on the site, which resulted in important economies in transportation.

In the second half of the fifteenth century, cast-iron and even bronze balls were invented, which, however, were very fragile; perfected casting techniques later produced stronger projectiles. This weakness of the ammunition explains to a certain extent the unsuitability of artillery for the siege of strongholds. It was the use of artillery for the defense of, rather than in offensives against, the strongholds that explains the transformation of the fortifications. We shall see that fortification walls were lowered to provide the artillery with the possibility of firing on the assailant by utilizing the paved platforms that inevitably topped the towers and made overhanging galleries unnecessary.

The military experts very soon thought of making these balls hollow and filling them with powder, which, by scattering the pieces created by the bursting of the ball, made them much more deadly. It is difficult to learn where the bomb was born, but its birthplace appears to have been Italy. It is possible that the first bombs were made of wood banded with iron. Projectiles of this type may even have been utilized with counterpoise artillery. In his treatise published in 1472, Valturio seems to be one of the first writers to give a description of bombs made of molded cast iron. It is possible, however, that Francesco di Giorgio perfected a technique that was still only in its infancy.

Firing case shot was also known. When lead was substituted for cast-iron bullets and balls in small-caliber cannons, it was realized that small pieces in a container scattered and became more deadly. However, lead ammunition very quickly became the appanage of mobile weapons.

The appearance of metal ammunition permitted a perfect correspondence between the ball and the cannon. The idea of the caliber, which made it possible to manufacture reserve stocks of ammunition, was soon conceived. It seems, however, that the ammunition stocks were quite small. In most cases the balls were made in the country, by requisitioning metal found on the spot; church bells were the chief victims of this practice. Another superior feature of the French artillery was its early realization that the number of calibers must not be increased. However, the definitive establishment of the calibers does not appear to have occurred before the sixteenth century. It was probably around 1525 that the castings were reduced in France to six or seven calibers — the double cannon, the serpentine, the large culverin, the bastard, the cannon-moyen, and the falcon. In the same period, Niccolò Tartaglia enumerates twenty-six calibers for the Italian artillery. Under Henri II the double cannon was eliminated, and the famous "six calibers of France" came into existence, the length as well as the diameter of the muzzle being regulated; this reform appears to have been introduced by the ordinance of 1544. The Italians had probably realized the obvious advantages of the reduction in the number of calibers. In the third quarter of the fifteenth century Francesco di Giorgio had listed the sizes of ten carefully defined calibers. His example was not followed, and the variety of Italian calibers seems to be due more to a lack of organization than to a lack of imagination.

As for the firing techniques, we have very little information. Purely ballistic problems are beyond the scope of this study, and belong more to a history of science than a history of technology. Some notion of ballistics did exist, however, in the absence of solutions that would not be found until much later. Tartaglia's writings already herald (due allowance being made) the research of those engineers who, beginning in the second half of the sixteenth century, were to begin to build this science.

At a very early period there existed, at least in the imaginations of the military engineers whose notes are still extant, small combination cannons, called "organs," which could fire simultaneously or be pivoted either with their bases or on themselves. Leonardo da Vinci adopted the idea, which certain authorities claim as the prototype of the machine gun. Such guns can be seen in the manuscripts of Kyeser in the early years of the fifteenth century. These notebooks teem with ideas and notes on the most varied and sometimes most unexpected uses of cannons or, more generally, firearms. By the second half of the fourteenth century, firearms became installed on ships.

As in the case of counterpoise artillery, the users of this artillery probably acquired through gradual experience empirical ideas that made it possible for them to achieve, if not systematically satisfactory results, at least an honorable proportion of successes. In 1495 we find Charles VIII and his soldiers, on the beach of Moncalieri, attempting to fire on canvases stretched between the masts

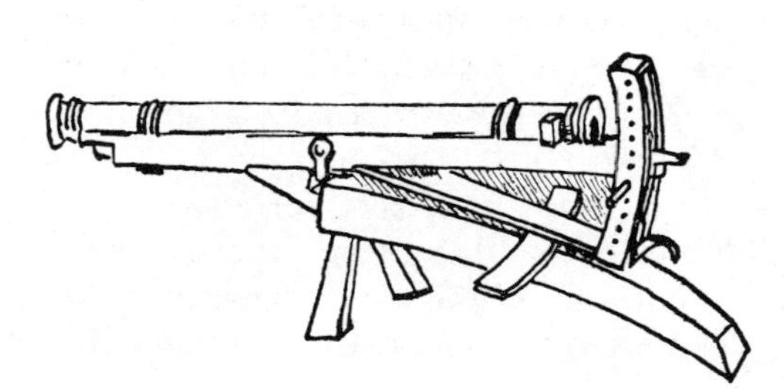

FIG. 43. Cannon with mount. Around 1490–1500. Leonardo da Vinci (*Milan, Cod. atl., f° 9, r° b*).

FIG. 44. Portable weapon, 1475 (*Rudimentum novitiorum,* Leipzig, 1475).

of ships. We can therefore suppose that repeated exercises helped to train the gunners of this period, and also led to increased theoretical knowledge.

Mobile weapons

The history of mobile weapons is much more difficult to trace, because of the lack of information. We have noted the slow elaboration of a system of firearms, paralleling the continued existence of the old weapons.

The crossbow had not remained this long in existence without undergoing some modifications. In order to obtain an increasingly powerful and precise fire, the spring of the bow was constantly being reinforced. At the end of the fourteenth century the windlass crossbow appeared. A small box was attached to the stock, in which there was a small windlass. With the help of the windlass the bowstring could be brought into the notch of the slider. All that was needed was to turn two cranks located at the sides. The stock was approximately 3 feet long, and the span of the bow was only $2\frac{1}{2}$ feet; it was much thicker than earlier models.

The fifteenth century saw the birth of the jack-type crossbow. A small jack fitted to the weapon made it possible to bend the bow. This made the weapon easier to handle than was the case with the windlass system of ropes and pulleys. On the other hand the stock was slightly shorter, ranging from $23\frac{1}{2}$ to $25\frac{1}{3}$ inches. At the strongest part of its center the bow was approximately $1\frac{4}{5}$ inches by $\frac{2}{3}$ inch. The military engineers achieved great power and rapidity of fire, which explains the persistence of this weapon in the face of guns that were for a long time dangerous and difficult to handle. Crossbows persisted for a good part of the sixteenth century, being used as hunting weapons.

The appearance of mobile firearms is still a subject of discussion. The dimensions of the *veuglaires* could certainly be reduced to produce a weapon of very small caliber, and consequently mobile, and this is exactly what was done. The first pictures of portable weapons date from the end of the fourteenth century – Kyeser, around 1390–1400, depicts a very good example. Historians have assigned the dates of 1364 in Perugia, and 1381 at Augsburg, for the first mobile firearms mentioned in texts.

The first weapons of this type known are tubes approximately 2 feet long, in six or eight sections, extended by a stock about $3\frac{1}{4}$ feet long. This stock was placed under the arm, and the gun was lighted with a wick through a priming hole located on the side. With this model as a base, the more easily handled weapons of the end of the fifteenth and beginning of the sixteenth centuries were gradually created. The tubes of these weapons were mounted on wood, but were still held under the arm. Gradually the iron stock was replaced by the wooden butt, which was rather large, for the backfire of these weapons was extremely violent.

There were two types of portable firearms: the harquebus and the *hacquebutt.* One was held in the hand, the other was rested on a stand, which was replaced in 1520 by a bracket; the *hacquebutt* then became the musket. One of the superior features of the army of Charles V (Charles I of Spain) was that it was equipped with these weapons; it was the Spanish harquebusiers who decided the Battle of Pavia, where the supremacy the French Army had held for several decades vanished. In 1523 Monluc succeeded in persuading François I to substitute firearms for crossbows.

FORTIFICATIONS

The development of weaponry unquestionably brought about profound modifications in the systems of fortification. In the beginning, as we have noted, artillery facilitated the defense rather than the besieging of strongholds. The increase in firepower and the appearance of more efficient ammunition nevertheless led to new techniques that were slowly perfected.

It was necessary to envisage the role of the artillery from a twofold point of view: on the one hand the defense of strongholds against the new machines, which at least in the beginning were not very dangerous, and on the other hand the appropriate use of this artillery in defense. It is this latter aspect of the problem that clearly predominated during the last years of the fifteenth century and led up to the essential features of modern fortification. The problem was not only the emplacement of this defensive artillery but also the necessity of storing within reach of the batteries considerable stocks of ammunition that had to be protected (particularly in the case of powder) from unfortunate accidents. All these difficulties and problems made it necessary to develop a new type of military architecture. Although this development had not yet been completed by 1550, much progress had nevertheless been made since the middle of the preceding century.

Attempts at adaptation The first idea that naturally came to mind was to adapt the existing fortresses by making several modifications in their layout. Certain fortresses were also constructed along modified (but not novel) principles. This work of adaptation continued for a long period. We find traces of it as late as the first half of the sixteenth century: François de Mondon added new defenses to the fourteenth-century walls of Vence, and Giuliano Sangallo surrounded the Castel Sant' Angelo with new outworks, a method also employed at Ostia by Baccio Pontelli. However, the attempt to adapt new ideas to the old formulas seems to have been the specialty of the French engineers, while the Italians as a general rule tried to find new solutions.

Between 1450 and 1470, a certain hesitation is evident. Several dates and several types of works permit us to establish very precise dates for the origin of this development. The very feudal-looking castle of Langeais, constructed around 1465, remained unfinished; by this time military experts had realized that it was becoming useless to continue an outmoded work. The castles of Nantes, begun in 1466, and Saint-Malo, which dates from approximately the same time, certainly look like medieval fortresses, but they are as if sunken; the builders tried to place an efficient artillery on top of the towers, and very high walls were not suited to this use. J. Richard has recently discovered an important document: a memorandum from the pen of François de Surrienne ("l'Aragonnais"), the Duke of Burgundy's master gunner and one of the most famous captains of the Hundred Years' War. According to Richard:

"This document — this 'counsel,' as its author calls it — is of additional interest by virtue of its date (June 25, 1461). It was written in the middle of a period of transition, for it was at this period that the impossibility of preserving the medieval system of fortification in its existing form was realized." Surrienne set forth his ideas on the fortifications of Dijon, which dated from the twelfth century and had been almost completely rebuilt after 1358, the only improvements being the construction of *berles,* or ramparts, before the gates. New towers were later built, and in 1461 the Saint-Nicolas tower was begun. Surrienne advised completion of the works begun and the filling in of the second moat, which, although a logical defense against a medieval army, now offered a very safe shelter for the attacker's artillery. This was the first attempt to construct a glacis. This glacis was to be beaten by the traditional plunging fire from the top and by a grazing fire from the foot of the wall, preventing access to the ramparts, which were utilized to beat the glacis parallel to the curtains. The theory of the bastioned front was beginning to be defined. In order to give better protection to the curtain, François de Surrienne suggested low, solidly built, horseshoe-shaped half towers between the towers, with very narrow openings equipped with light pieces of musketry.

The keep of Ham, in Picardy, probably dates from between 1470 and 1475. It was 108¼ feet high, and had the same diameter; the walls, built somewhat in the form of a truncated cone, were 36 feet thick at the base. All the openings and the paved terrace were equipped for the installation of artillery. The overhanging galleries had disappeared.

With the addition of these various components in the second half of the fifteenth century, practically all the ideas that were to govern the building of French fortifications in the first half of the sixteenth century — a defensive system, we repeat, that was an adaptation rather than a revolution — were now complete. The high walls had been completely abandoned, and fortifications were constructed whose general design was not greatly modified but which were lower and thicker. The platforms were now equipped for the installation of artillery, and the interior was divided into vaulted casemates. The walls were extremely thick, often more so than in earlier periods; machicolations were now unnecessary. This is the form of the circular bastion of Langres, built at the very end of the fifteenth century. Completely similar is the great tower of Toulon. Begun in 1514, under the direction of an Italian engineer named Giovanni Antonio Della Porta, it was completed in 1524. The fort is almost 197 feet in diameter; the walls are 16½ feet thick at the base, almost 10 feet thick at the top. It contains 16 vaulted casemates separated by cisterns; the casemates

FIG. 45. Bastion of Langres. Around 1495 (Viollet-le-Duc, *Dictionnaire d'architecture*).

FIG. 46. The Great Tower of Toulon, 1514–1524 (Arch. Génie à Paris, plan of 1707).

have embrasures for grazing batteries. It served as inspiration for Du Chillou in his construction, in 1517, of the great tower of Le Havre, unfortunately now destroyed.

The Spanish fortifications were slower to adapt to the new conditions of siege warfare; for a long time to come the high walls of earlier centuries were all that could be seen in Spain. Here, again, the first transformations were felt much more in the construction itself than in the traces. The new forms appeared at Coca, around 1473, and at La Mota de Medina del Campo (1479–1480). At Coca a very deep, masoned moat is still in existence, but the walls are very low in relation to the level of the surrounding ground, to permit a grazing fire.

The first complete major fortification of this type is the Spanish castle of Salses in French Roussillon, begun in 1497 under the direction of the engineer Ramírez. Gerhard Ritter notes that it is the first large fortification protected for the most part from the direct fire of the opposing batteries by its low position. However, the trace is still very medieval: a rectangle flanked by four round towers. On the east and south fronts, strange works whose flanks are rounded on the outside and become part of a triangular projection rise isolated from the moat, prefiguring half-moons. The paved platforms held artillery that shot through the embrasures of a very thick parapet constructed in a glacis. In addition there was a network of subterranean passages, listening galleries, and casemated galleries. Salses is undoubtedly one of the first great examples of a new fortification system.

In France, however, the Italian engineers employed by François I continued to build structures of the same type we have just mentioned. The castle of If, near Marseille, is an excellent example: a powerful square redoubt flanked by three large cylindrical towers whose platforms are equipped for artillery. The techniques of the great tower of Toulon are repeated here.

Italian innovations

The chronology of the Italian works, which have not been as throughly studied, is more difficult to determine, especially in the latest aspects of military architecture of the period. It seems certain, however, that the definitive transformations, both in the trace and in the structures themselves, occurred in Italy. To increase the flanking surfaces, Camille Enlart tells us, the Italian builders first gave the bastion the form of an elliptical tower attached to the curtain by a narrower portion. Then, since the front of this work was weak, it was given the form of a pointed caponnier; lastly, the bastion was simplified by turning it into a simple pentagon. The end result was Jean Errard of Bar-le-Duc's polygonal system, which was adopted and perfected by Vauban (Sébastien Le Prestre).

It would be interesting to be able to attach names to the principal stages in the elaboration of the modern fortification. Notwithstanding the opinions of certain German authors, Taccola does not appear to have played an important role in the transformation of these techniques. In contrast, we would have to devote more study to the work of such engineers as Lorenzo di Pietro ("il Vecchietta") (1412–1480) who, beginning as a painter, later became a leading military engineer at Sartaneo, Urbetello, and Monte Argentoli. His most famous pupil was Francesco di Giorgio Martini, whose talent has been mentioned several times in these pages. This school also included the Sangallo brothers, Giuliano (1445–1516) and Antonio (1455–1534), the Florentine Baccio Pontelli, a pupil of Francesco di Giorgio, and Michele Sanmichele. The decisive steps seem to have been taken within this narrow circle of military architects.

The tenaille trace undeniably appears in the writings of Francesco di Giorgio. But while he definitely conceived a system of a trace in which the curtains are broken in their center, it is useless to try to see in the salients a new type of fortification: there are only machicolated towers. On the west side of the city of Rhodes, the Saint George gate was protected by a rectangular tower, originally isolated (1421–1431). It was preceded, on the side facing the open country, by a projecting work which is an amplification of the *fausse braie*. In 1496 Pierre d'Aubusson built a powerful rampart around the outworks; it followed their contours, although separated from them by a moat, and its purpose was to put the defensive artillery as far as possible to the front of the curtains. In 1521, on the advice of Basilio della Scuola, of Vicenza, the inner moat of the rampart was filled and connected to the escarpment wall; the rampart was now one with the curtain and overlooked the *chemin de ronde*. Between 1496 and 1521 the Italian fortifications had indeed made decisive progress.

We find the first forms of the caponier bastion in the last years of the fifteenth century (1494–1497), at Civita Castellana, where the Sangallo brothers were working, but they are combined with the traditional formulas. The same is true of the citadel of Ostia. Although it belongs to the same period, its overhanging galleries and the height of its walls give it an older appearance.

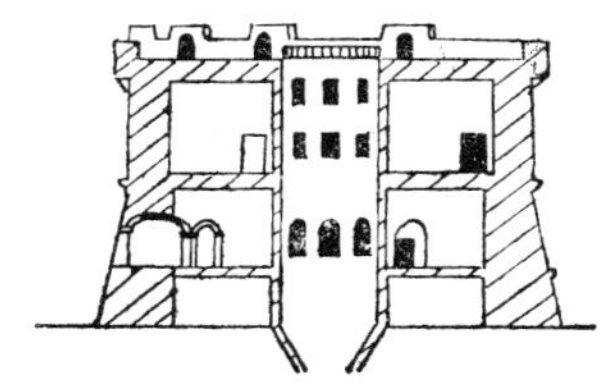

FIG. 47. Cross section of the San Michele Tower at Ostia, by Michelangelo (Uccelli, p. 480, Fig. 11).

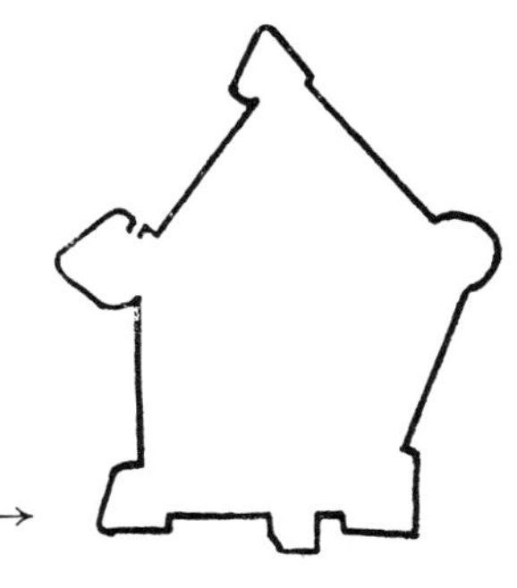

FIG. 48. Plan of the fortifications of Civita Castellana, 1494–1497 (Uccelli, p. 477, Fig. 2). →

The trace of Nettuno (1501–1502) also has caponier bastions flanking a traditional rectangular plan still preferred by the Sangallo brothers. The polygonal plan was finally adopted in the fortification of Civitavecchia, conceived by Antonio da Sangallo in 1515. He repeated the same ideas, better adapted and better conceived, at Ancona in 1527, and in the defensive projects of of Rome in 1537.

The stage was set for the appearance of the first bastions with orillons, naturally in Italy. The Maddalena bastion at Verona was constructed by Michele Leoni, engineer of the Republic of Venice, in 1527. Sanmichele imitated

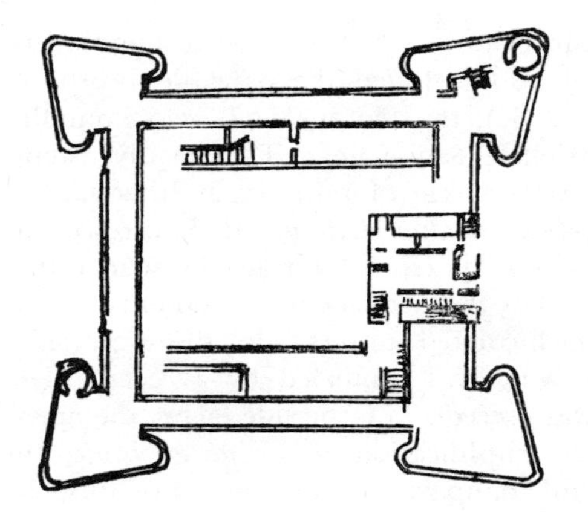

Fig. 49. Plan of the fortress of Nettuno, 1501–1502 (Uccelli, p. 478, Fig. 3).

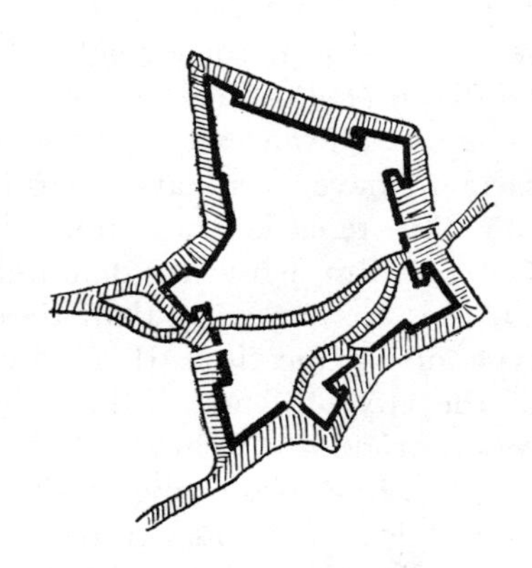

Fig. 50. Plan of the fortifications of La Fère, 1540 (Uccelli, p. 482, Fig. 22).

this example in Venice, Sebenico, and Crete; the same was true of the Verde bastion in Turin (1536). In France the Italian engineers very soon constructed, at Troyes (1524–1529), Saint-Paul-de-Vence, and Navarrenx (1543–1569), works so similar that some authorities have questioned in which country this new formula was perfected, or indeed whether it may have been perfected in both simultaneously.

Siege techniques

Thanks to all these modifications, the art of fortification was certainly in advance of the art of siege warfare. It can safely be said that between the middle of the fifteenth and the middle of the sixteenth centuries no stronghold was captured by means of a general assault or an intensive bombardment. New techniques, therefore, had to be invented, and in order to perfect them the authorities naturally turned to the same men who had just revolutionized the notions of defense. The great military architects were employed as much to take strongholds as to construct and defend them; Francesco di Giorgi is a remarkable example.

But while artillery proved to be of little help in this aspect of warfare, the same was also true for other techniques formerly used against the high walls of the medieval fortresses. A good sapping operation could easily bring down high walls, but this technique was useless now that walls were low and thick. The engineers then thought of combining simple sapping with the technique of setting fire to the supporting timbers, and the effects of gunpowder, for a mining operation. The name of the inventor, and the date and place of the first use of mines, will undoubtedly continue to be topics of discussion. One possibility is Taccola (Jacopo Mariano), who included several very crude pictures of the mines in his treatise completed around 1440. The mine was certainly in existence at this date, there is no certainty that Taccola invented or even utilized it. We have formal proof of its use in 1495 at Castel Nuovo in Naples, and in 1501 at the siege of Cephalonia.

Thus it is almost certain that miners were utilized in the second half of the fifteenth century, in particular by Francesco di Giorgio, and that by the beginning of the sixteenth century certain engineers were already past masters in this technique. François I made use for this purpose of the talents of an engineer who was undoubtedly of Spanish origin: Pedro Navarro (1460–1538). His successes were brilliant. Between 1500 and 1503 his experience in minelaying in Sicily and southern Italy for the Spaniards led him to produce a type of explosive device that remained almost unchanged until the nineteenth century.

These few aspects of military technology that we have just briefly reviewed show that extremely rapid progress was made in less than one century; the engineers abruptly passed from a medieval heritage to the first manifestations of modern techniques. Even the participants in the drama realized the rapidity of the change. In 1559 Cardinal de Granvelle (Antoine Perrenot) replied to an agent of Queen Elizabeth of England: "Your men are courageous and valiant, but what training have they had of late? And the art of warfare is now such that every two years it must be completely relearned." War on the seas was also to be profoundly modified; in the last years of the reign of Henry VIII (1540–1547), a great number of English ships were equipped with cannons of various types. Just as the cannon had revolutionized the art of fortification, it was to have an equally profound influence on the art of naval construction.

Like the great geographical discoveries, however, the development of military technology does not appear to have influenced technological development as a whole. Progress in one area cannot produce a general movement. The new military needs that resulted from the new methods of combat of the sixteenth century, notes John U. Nef, were not a factor of capital importance in the technical achievements that represent an essential part of the first "Industrial Revolution." The growth of metallurgy that in large part conditioned the revolution in military technology was completely independent of any military policy. Only the expansion of certain discoveries was in many cases based on the desire of a state or a prince to be able to supply armies with the best equipment available.

CHAPTER 8

THE ORGANIZATION OF SPACE

IT WAS undoubtedly in the techniques of organization of space that the men of the Renaissance best demonstrated their originality and genius for adaptation. They succeeded in combining a keener sense of real needs and a superior logic, in addition to more powerful methods, with a conscious empiricism that had been one of the most brilliant successes of the Middle Ages. Thinking in terms of an overall picture seems to have imposed itself more extensively, a process in which the formation of centralized monarchies was an invaluable aid.

The advent of more effective governments, supported by a vigorous middle class, and their control over wider geographical areas that had formerly been parceled out into feudal holdings, unquestionably made it possible to derive greater benefit from experiments and to put the best formulas into general use. This period also saw the birth of those great works of public utility that until then had been difficult to accomplish because they surpassed both the geographical framework of the existing political units and the financial means at the latter's disposal.

Urban planning and major public works of all kinds were therefore to claim most of the attention of the Renaissance engineers. To them, and particularly to the Italians, we owe most of the achievements of this period. Their notes and projects constitute abundant sources that can easily be utilized, at least for the very rapid study we are able to attempt here.

THE CITY AND ITS PROBLEMS

The city was both the center of the great intellectual activities of the Renaissance and the principal base of centralized power. Thus it is easy to understand the large urban expansion of this period, and, thanks to this very expansion, the progress made in all the techniques connected with it. The influence of antiquity was not without importance; the gradual rediscovery of ancient Rome encouraged the most brilliant minds to become interested in urban problems.

Birth of theoretical urban planning

Urban planning offered to theoreticians and technicians alike a wonderful field for experiment. The great architects of the Renaissance understood this perfectly, and several of them set forth their ideas on the subject. To be sure, they were still far from deliberate functionalism; a certain systematization, still medieval in spirit, was very slow to disappear. But, as P. Lavedan notes, these men had already left far behind them the medieval period with its diversity, expediency, sense of the relative, and application of a different solution to an individual case. The very notion of urban planning, with all that it involves in the way of general concepts, marks an undeniable progress.

Throughout his *Re Aedificatoria,* composed in 1452, and printed for the first time in 1485, Battista Alberti often returns to the problems of urban planning, though without any particular order. While he was an innovator in architecture, Alberti was not revolutionary: he was still very much a part of the Middle Ages. In Book IV he examines the problem of the site, in order to establish immediately that there are neither general rules nor universal principles but only individual cases. Everything should be determined by the site: "The circumference of a city and the distribution of its parts must be changed depending on the diversity of the sites." That is, it is a question of adaptation. After these judicious remarks, Alberti divides the city systematically into twelve sections, following Plato's system, and attempts to make it conform, like the cosmos of Pythagoras, to the law of numbers. He appears as a partisan of rigid division, which had been one of the principles of the Middle Ages, especially as regards the various trades; there are separate markets for each type of business and residential quarters. As for the layout of the streets, he accepts the twofold principle of straight lines and curves. He warmly defends curved streets, an inexhaustible source of variety if not of military utility, but also gives us for the first time a theory of the straight avenue, with strict alignment of houses all of which are of the same height, terminating in an important building. When he comes to describe the ideal city, Alberti chooses the radial-concentric plan, which produces precisely these long straight lines and circular, concentric streets. According to Lavedan, the appearance of a system of urban esthetics is in fact Alberti's most important contribution.

Filarete, whose treatise is slightly later and remained unpublished despite the large number of manuscript copies, arrived at the same conclusions. His ideal city consists of a sixteen-sided polygon forming a star, with the city gates in the recessed angles. In the center is a large rectangular square, from which radiate sixteen straight avenues; these cross sixteen secondary squares on which are eight churches and eight markets and which are linked by a great circular avenue. In Filarete's work several buildings, such as the theatre and the hospital, occupy eccentric positions.

Francesco di Giorgi Martini is also very close to the classical tradition; his *Prolegomena* mention Aristotle and Vitruvius, as well as Pliny and Palladio. But he is equally close to the spontaneous medieval practices. While he, like his predecessors (to whom he owes much), proposes a very rigid radial-concentric model, he seems much more sensitive to individual cases and to the inevitable irregularities of terrain.

In his work, published in 1527, Albrecht Dürer suggests a square city, divided somewhat in the manner of a Roman camp, and organizes this "chessboard" in an extremely detailed order. This leads him to a very rigid specialization of the various areas, and even to a primitive form of zoning. He places the church at an angle and leaves green belts by eliminating construction from certain sections of his chessboard.

All these treatises had a certain amount of influence. More precisely, we are able to measure the importance of the ideas contained therein, ideas that seem to have been generally adopted by the builders. There is first of all an organic linking of the various parts of the city, and the subordination of the whole to a center, the main square — something unknown or almost unknown in the medieval cities, which had at most a marketplace. It is for this reason that the radial-concentric plan appeared to the Renaissance planners to be the most logical one. A second element of the theory was the need for monumental perspective, which had hitherto been nonexistent. The monument was not conceived solely for itself, like the Gothic cathedral, but also in order to be placed in a certain context; there is often a mutual influence of the monument on the plan or of certain perspectives imposed on the appearance of the monument. Lastly, the urban planners introduced the notion of a program, which left nothing to the arbitrary, individual case. The street is no longer an empty space between houses; it has become the determinant of individual structures. The house must respect the alignment; it must be of a certain height, which height must be in proportion to the width of the avenue or square it faces.

Actual accomplishments A great distance often separates a theory from its applications. Nevertheless, the century between 1450 and 1550 attempted, when means and circumstances permitted, to put into practice the principles so clearly defined by the theoreticians. No systematic redevelopment of existing sites was done; the medieval cities continued to exist until the early years of the nineteenth century. The urban planners could act only in the case of new sections of the older cities, as occurred in northern Italy, or in the building of new cities, particularly frontier towns in northern Europe. We shall limit our considerations to a few examples that are indicative of a new spirit.

There is practically no evidence of any remodeling of old cities. One of the few examples, which is also one of the oldest and the most significant, is that of Pienza, in Italy, the successor to the small city of Corsignano. This city was fortunate enough to be the birthplace, in 1405, of Enea Silvio Piccolomini, the future Pope Pius II, who in 1462 ordered Bernardino Rossellino to transform his humble native town. The architect's principal creation was a public square, the first example of the public squares built during the Renaissance. Around it he placed the cathedral and various palaces, which gave it an appearance never before seen.

Urban planners no longer indulged in individualistic fancies. The new sections built in various cities now corresponded to norms that were sometimes very rigidly established. Such was the case, for example, at Turin, where the checkerboard plan was adopted. In France, although such spectacular cases were nonexistent, we do find strict rules: the new sections of Amiens, begun in 1479, and Chalon-sur-Saône, begun in 1466, generally followed a set of rules. The alignment of buildings was established, and deviations were forbidden and punished. Thus in 1432, encroachment by storefronts was forbidden at Amiens, and in 1474 limits were placed on corbeling.

Very few completely new cities were built in this period; in the few such examples, however, the builders were careful to give their plans the same precision. At Corte Maggiore, constructed between 1470 and 1481, the checkerboard plan was adopted; the same was true at Gattinara, which was enclosed within a rectangular wall. The first great name in French urban planning was Le Havre. Its first builder, Guyon Le Roy, created the first portion of the city on ceded lands; he constructed it without a general plan, producing a remarkable disorder which it was difficult to correct. The king then called upon the Italian Bellarmato, in 1541, who conceived the Saint-François quarter of the city on a preestablished, regular plan, with a plan for the street layout and even an architectural program for the houses. Included in it were direct-to-sewer drainage, a cistern for each block of houses, and systematic street paving. The first regular, fortified frontier city was Vitry-le-François, constructed in 1545 on the plans of the Bolognese Girolamo Marini, in a very strict checkerboard plan. However, the church was placed in an angle of the main square, and the marketplace was slightly off-center. This was the first model of those very regular, fortified cities, somewhat severe in appearance, which proliferated at the end of the sixteenth and during the seventeenth centuries, and with which the name of Vauban is frequently connected.

The activity of the urban planners was exercised almost exclusively in Rome, in a still timid and erratic fashion. Paul II (1464–1471) had begun to open up the extension of the Via Flamina, which, under the name of the Corso, connected up with the Piazza del Popolo at the Piazza Venezia. Sixtus IV (1471–1484) drew up a construction plan, and in 1473 ordered the Ponte Sisto built by Baccio Pontelli. Alexander VI (1492–1503) made improvements on the Borgo, Julius II (1503–1513) opened the Via Julia, and Leo X (1513–1521) the Via Leonina, which linked the Piazza del Popolo to the Piazza Navona. This was the beginning of a series of great avenues that eased the increasingly congested traffic to a remarkable degree. The sack of Rome in 1527 interrupted these projects. In 1540 Michelangelo drew up projects (completed in the second half of the sixteenth century) for the construction of the Capitol and the surrounding area.

There were few other transformations of ancient cities. While Leonardo da Vinci conceived multilevel highways to solve Milan's traffic problem (traffic was a major concern almost everywhere), the project was never realized.

In 1499 the Roman authorities took the most effective measure: they forbade the use of carriages to all but cardinals. At Ferrara the new city was linked to the old city, the castle serving as the meeting place of the two. Here we find the first very large highways: the Corso is 52½ feet wide, 39 feet of which is the actual road. The Via Po was 59 feet wide.

The true originality of the Renaissance appears to have resided in the layout of the public square, around which the city, or at least a neighborhood, sprang up. The width of the square, according to the most widely accepted canons, had to be from three to six times the height of the surrounding buildings. In Florence the squares of Santa Croce and especially the Annunziata were created by Sangallo in 1518. The great statues placed in the centers of these squares were also an innovation of the Renaissance. Donatello's Gattemelata was the first; it was ordered in 1446 and executed in 1453. In 1538 Paul III had an ancient statue of

Marcus Aurelius placed in the Capitoline square.

Problems of municipal administration

In many instances the administration of the Renaissance cities made considerable progress, but this was far from being generally true. It was not easy to correct the practices inherited from earlier centuries, which had left much to be desired in this area. In certain countries we find a series of public works that, although not concerted, nevertheless revealed a mentality that was to continue to develop. The most symptomatic of these was the paving of the streets. Dijon was completely repaved between 1383 and 1386, Rouen in 1406, Albi in 1418, and many other French cities during the fifteenth century. In Paris in 1366, either large blocks of sandstone or small blocks of hard limestone from the deposits at Gentilly were used. At Douai in 1350 and at Amiens in 1406, this paving was sunk in a bed of fine sand. In 1452 the flagging (not paving, which was felt to be unhealthy) of the streets of Rome was begun.

Sewer services spread slowly because of the material and financial difficulties of such projects. In most cases they utilized a stream that crossed the city; the gutters of the streets, which sloped down to the stream, drained into it. In Amiens in 1459 the general collecting stream was dredged. The Suzon of Dijon was straightened between 1359 and 1404, and in some places covered over; in 1450 its bed was leveled by Jehan de Mousterret, master builder to Philippe le Bon. When he revised the plan of Le Havre in 1541, Bellarmato made provision for general direct-to-sewer drainage. City cleaning was also part of the program of urban improvement; Tournai, Valenciennes, Saint-Omer, and Lille organized sweeping services, which were imitated at Dijon in 1443 and Amiens in 1462.

Fire was the most dreaded plague of these cities, which were constructed almost entirely of wood. As can easily be imagined, fire-fighting methods were extremely limited, and the better course was to attempt to eliminate as far as possible the occasion of and opportunities for fires. Thatched roofs were accordingly forbidden, for example at Abbeville in 1466 and in Dijon in 1524. Lighting nevertheless continued to be limited and haphazard in all the European cities.

The water supply was one of those basic problems whose solution was extremely difficult. Despite visible mechanical progress, manifested particularly in the increased use of lift-and-force pumps like those in use or coming into use in the mines, the piping of drinking water into the large urban agglomerations made little progress. It is true that numerous difficulties existed. It was not sufficient to draw the water from a pure stream: it had to be distributed, and the techniques of piping, like those of measurement, which are indispensable for economical use, did not keep pace with the progress made in the other techniques. Naturally the authorities thought first of repairing and improving preexisting or disused facilities. It was not by pure chance that one of the first Latin authors printed before the beginning of the sixteenth century was Sextus Julius Frontinus, the famous author of the book on the aqueducts of Rome. By the middle of the fifteenth century the Aqua Virgo was the only section of the ancient piping system of Rome still functioning in approximately normal condition. It was restored by the great architect Alberti in 1453, at the same time that the famous Trevi Fountain was constructed. Not until the end of the sixteenth century were two new aqueducts constructed, the Aqua Felice and the Aqua Paolo. Once repaired, the Aqua Virgo supplied 63,000 cubic yards of water per day. The situation in Paris was much the same. The Belleville aqueduct was repaired in 1457; a new waterworks was installed, and more manholes were opened to ensure its efficient functioning. Until the beginning of the seventeenth century, the aqueducts of the Pré-Saint-Gervais and Belleville supplied all the needs of the city of Paris. In 1499 the city proper had only twelve fountains; five additional fountains were located outside the city walls. Nineteen seigneurs and religious orders had obtained concessions. In 1481 the Roman aqueduct of Segovia in Spain was repaired, a project that was the work of the monk Juan de Escobeto.

At Liège the water piping, which had impressed foreign visitors as early as the fourteenth century, came from a special source that supplied an abundant, widely distributed quantity of water. This water came from the drainage channels of the mines in the surrounding hills. The Richefontaine channel was harnessed in the second half of the fourteenth century and diverted into storage fountains where the inhabitants could come to draw water. Thirteenth-century Rouen still had only three fountains, until, in the fifteenth century, the Gaalot spring was harnessed to feed three new fountains. At the end of the fifteenth century major projects were begun to supply each section of the city with water — so successfully that by about 1550 there were twenty-five fountains in the city.

Around the end of this period we see the appearance of those great lift-and-force pumps, operated by hydraulic wheels, which in the coming centuries were to constitute the principal method of piping water. The pump of Toledo, which to a great extent replaced the old lifting wheel, was installed in 1526 by an Italian named Della Torre. The Gloucester pump in England was built in 1542. In Germany the first great pumps were those of Augsburg (1548) and Bremen. In the very large cities, however, pumps did not appear until later; the first pump in London was installed in 1582 by a German named Peter Maurice, and the Notre-Dame pump of Paris was installed by the Fleming John

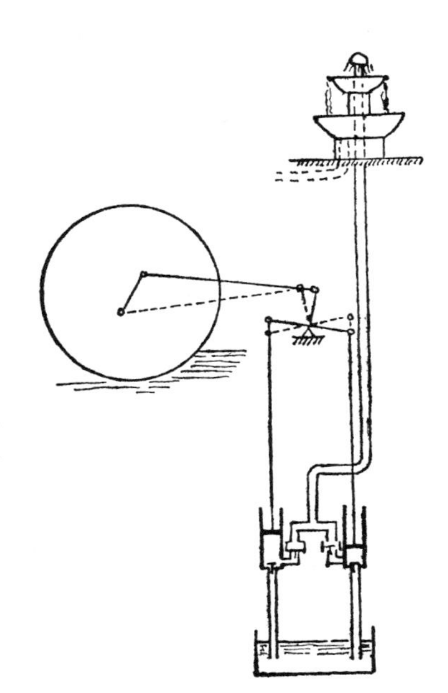

FIG. 51. Diagram of the Nuremberg pump. Mid-fifteenth century (Uccelli, p. 332, Fig. 9).

Lintlaer in 1608. In his commentary of 1548 on the machines of Ctesibios, Rivius speaks of pumps operated by hydraulic wheels as if they were fairly common. All these machines are known to us by rather vague texts, and it would have to be ascertained that they (or at least some of them) were not simple lifting wheels or, as in the case of the English city of Poole, bucket chains.

The technical problems of distribution were equally difficult. The only pipes available were fragile, costly terra-cotta pipes or wooden conduits that rotted quite quickly. The technique of using terra-cotta pipes had not changed since Roman antiquity. Wooden pipes had also been used in antiquity, but the extension of the piping systems caused some change in the techniques, which by the time of the Renaissance had achieved a high degree of perfection. These pipes were generally made by boring out tree trunks or large branches. Pieces of wood still covered with their bark were preferred to shaved pieces, because the bark gave better external protection. Oak, elm, and the alder were the preferred woods. We have already briefly discussed the methods of drilling; as for joining wooden pipes, there were two methods. In the first, one end of the pipe was whittled down, and the opening at the other end was enlarged into a cone; the narrow end of one pipe was then inserted into the flaring mouth of the next. The flared end was sometimes banded with iron to prevent it from bursting when jointed. To ensure watertightness the joints were encased in moss or tow and a cold mastic composed of mutton fat and crushed brick. Another method utilized conical iron rings fitted into notches prepared for this purpose. The pipes were then held at an angle and forced into each other by hammering with a mallet. The dampness of the wood, once the piping was put into use, caused the wood to swell, ensuring a watertight joint. Diversions were equally difficult; forked branches of trees were the preferred method.

Once the blast furnaces were producing a cast iron sufficiently fluid to be cast in relatively thin sheets, the hydraulics engineers undoubtedly thought of cast-iron pipes. Only a few specimens, which, moreover, are poorly dated, are still in existence. Lead pipes were made by soldering with a soldering iron and tin the edges of sheets of lead rolled around cores.

Faucets were practically nonexistent: fountains, both private and public, flowed continuously. Ultimately, however, some kind of rationing had to be effected. In Paris, in 1529, a water concession was granted only on condition that the individual maintain "the said fountain, so that it opens onto the street for the assistance of the inhabitants of the neighborhood." It was therefore necessary to install tanks whose overflow poured into the nearest public fountain. Tanks with valves and a floater system to stop the flow of water were very soon devised. Two bronze faucets used between 1515 and 1539 in a lead conduit in London, and now preserved in the Science Museum of that city, are an exact copy of Roman faucets. The cones were polished with emery, or sometimes even mechanically. Measuring the water was practically impossible. In the few known private concessions of this period, only the width of the faucet or the delivery pipe was regulated, their dimensions being ensured by a copper or silver ring of the diameter permitted. The water pressure at the junction was not taken into account, because it was believed that the volume of water supplied by an opening was in proportion solely to the diameter of this opening.

We have several examples of these water-distribution problems in the cities. At Dijon, in 1445, wooden conduits were installed by a talented carpenter and fountain-maker, P. Belle; they were jointed by means of "iron screws." The waterworks or reservoirs were of masonry. In 1509 the entire piping system had to be renewed and lead pipes installed.

TRANSPORTATION

The circulation of merchandise and people unquestionably increased during this period. On the one hand the return of peace in the middle of the fifteenth century led to an undeniable development of commercial relations, a movement that was also favored by the formation of the great centralized monarchies. In addition, the discovery of the New World brought about the need for hitherto unheard-of travel, on a scale and over distances never before known.

The technical revolution in wagons and ships, whose importance we have already discussed, naturally had an influence on the growth of transportation after the end of the fourteenth century and throughout the fifteenth.

This conjunction of phenomena was to cause a parallel development of roads and methods of communication. To facilitate the expansion of maritime commerce, it was necessary to improve and develop port installations. To handle the increased traffic, and also thanks in part to new political and military conditions, the highway networks had to be considerably developed and improved. For the first time we see the birth, on a much greater scale, of the idea of a more consistent and complete system of canals.

Port installations

Navigation, along with the growth of warfare on the high seas and the increase in tonnage, inevitably caused the transformation of poor installations — or at least their evolution in the direction of progress.

The old problems nonetheless continued to exist, and particularly the difficult problem of silting. Between the mid-fourteenth and mid-fifteenth centuries the port of Bruges gradually died. The inhabitants mobilized every supply of fresh water to flush out the silt, and dredged constantly, but nothing could stop the constant and rapid silting. In 1544 the Hanseatic League dropped Bruges from its list of ports of call. In 1564 the Brugeois, in a last heroic effort, constructed a new canal from Bruges to Damme; in 1572 the Dutch filled it in, thus ensuring the death of the port that in the Middle Ages had been one of the greatest ports of western Europe. At Harfleur silting was equally menacing. Between 1391 and 1398, locks were installed on the Lézarde to create a flushing reservoir. The port of Granville was fortified in 1437; shortly thereafter, Admiral de Coëtivy created a seawall with a tower and a mill at each end, which also formed a flushing reservoir. Port-en-Bessin was fitted out in 1475; the end of the port was ensured of a constant water level by the construction of two wharfs that faced each other and were linked by a bridge that formed two basins. The arches were equipped with grooves for sluice gates. Here, too, the upriver basin formed a flushing reservoir.

Dredging the ports was a much more delicate operation. Everything was tried: cofferdams, "otter boards," flushing systems, locks or "bars," dragging, cleaning out with shovels and picks, and even paving and planking. When necessary, the stream was diverted and the port of Harfleur was cleaned out with shovels and wheelbarrows or hods. The port was nevertheless inoperable in 1388, 1389, 1446, and 1453. It was then necessary to drop anchor in the roads.

The outer harbor of Harfleur began to be fitted out in 1474. The so-called "Spanish" cove had two indentations and two projecting necks of land separated by a marsh called the Hommet. The Lézarde was diverted at its exit from the *clos des galées* and brought to the Spanish cove through a palisaded ditch more than two miles long, dug through the sand and mud. On one side the stream carried away the sand; on the other, the channel communicating with the sea

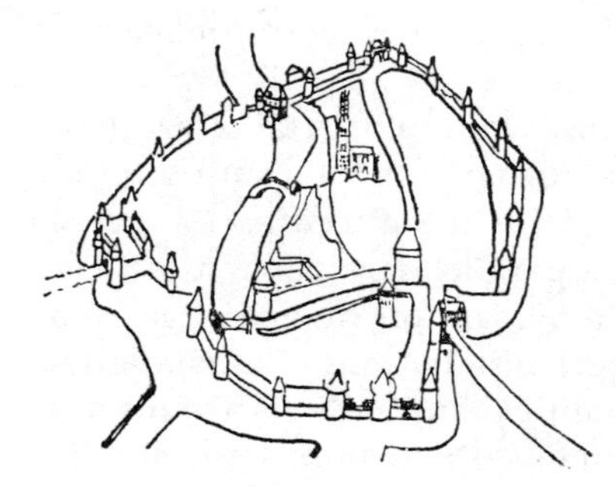

FIG. 52. Bird's-eye view of Harfleur in the fifteenth century (Hérubel, *L'homme et la côte,* p. 115).

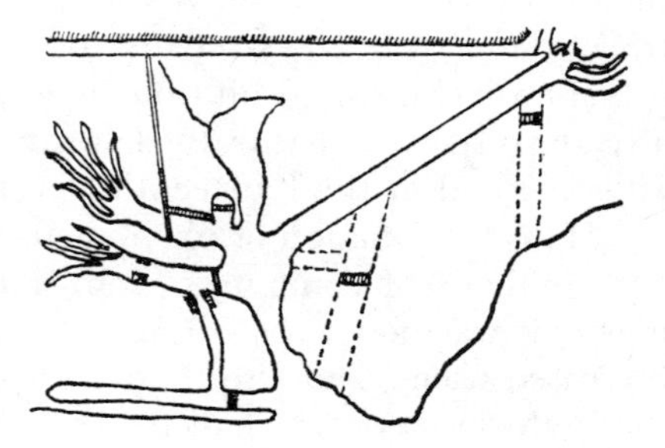

FIG. 53. Plan of the port installations of Harfleur. End of the fifteenth century (Hérubel, *L'homme et la côte,* p. 120).

permitted the entrance and withdrawal of the tide. A wooden bridge linked the banks of the northern neck of land and formed a wharf, and a corduroy road joined up with the Harfleur road via Granville. The southern neck was fitted out in the same fashion; it was linked by a channel with a masonry lock to the moat of Leure, isolated from the sea by a cofferdam. This moat was generously supplied with water by the swamp, and formed an additional harbor. In this way a rational system was gradually developed. However, it was necessary more or less to abandon the earlier installations and create a kind of outer harbor. The knowledge gained from experience made it possible to create ports where none had ever existed. The project of Le Havre de Grâce, located in a cove that had to be completely equipped, was begun in 1517. This cove had a single opening on the south; this was closed off by manual labor, and a new opening was created in the angle formed by the coast. A canal 4¾ miles long was to link the end of the port with Harfleur (it was not completed until the following century, by Vauban). The plan called for two wharfs and two jetties terminated by towers, the whole to be constructed of masonry; the tower of the south jetty was never constructed. A tributary stream had been closed off by a lock, and formed a flushing system. A bridge was then built across the narrow portion of the large stream (1526) and five supplementary flushing tanks were planned. In addition to the jetties and three defensive towers, the *"hable neuf"* included an outer and an inner harbor extended by the "Grande-Barre" cove and two other, smaller coves. Stone jetties were joined to the Great Tower. Wooden wharves bordered the north side of the *hable* as far as the entrance to the large cove. The sixteenth-century *hable* was still far from perfect: it was dry at low tide. In some ways it was (to use M. Hérubel's expression) a *portus* in a basin. Many other ports were still no more than simple beaches, as for example Saint-Malo. This city had been constructed on a cliff rising from a vast beach that was completely covered at high tide, and which was crossed by a small river, the Routhouan. In 1509 the city was linked with the tip of the coast by the seawall of Le Sillon, thus closing off a part of the beach.

The problems of the Mediterranean region were different, since there were no tides and rocky coasts were more numerous. Very often it sufficed to erect a large dike to supplement some natural feature of the coastline. Thus the Mediterranean ports were characterized by large moles whose existence was threatened only by storms. The first port installations of Naples had been destroyed in 1343 by a violent storm. Queen Giovanna II then began the construction of a new basin between the mole and the small port; in 1447 Alfonso I of Aragon completely rebuilt the large mole, which in 1495 was extended and equipped with a lighthouse. At Palermo in 1541, at Livorno at the same period, at Catania in 1438, similar projects were begun to create mooring basins intended to accommodate larger merchant fleets. We have little knowledge of how the installations of the Spanish ports were developed after the discovery of America. Some of them had been so well endowed by nature that practically no additional work was needed to prepare them for considerable ship traffic; Cadiz is the best example.

However, several Mediterranean ports did have a problem of cleaning; it was one of the principal preoccupations of Marseille. The municipal authorities gladly welcomed all the suggestions that were made by inventors of various nationalities; each suggestion was tried, and a decision then made. In 1413 the syndics experimented with a new device suggested by a Genoese; the device proved unsatisfactory, and the contract was canceled at the end of one month. We have little information about the devices actually utilized; they were undoubtedly large vertical bucket wheels placed on a pontoon, which poured the debris brought up from the bottom into special boats that dumped it at a distance. The results were neither speedy nor perfect, which explains the constant appeal for inventors.

Roads and bridges

No major modifications in the techniques of road construction were possible. However, the formation of centralized governments did help to facilitate development of the highway network. Our information is very fragmentary, and to date we still lack badly needed studies of the medieval and Renaissance roads. From the little information available, we can deduce the royal concern for this problem, at least in France. In Beauce, in 1475, wide stone roads were constructed between Chartres and Blois. Louis XI took an interest in the maintenance of "wide roads paved with stone, very beautiful and remarkable" between Orléans and Sercotte, and saw to it that the inhabitants of the riverbanks did not encroach on the public highway. From these two examples we note two facts that had also existed in the preceding period: the defense of the public highways against the encroachments of the river dwellers, and the assurance that, thanks to paving, a suitable road for wagons was always at hand. In the same spirit, Emperor Charles IV recommended that bushes be kept at a certain distance from the highway, for vegetation was sometimes as harmful as human beings.

However, we do know of several examples of new techniques that continued to develop. A German Dominican of Ulm describes several highways in a work in which he relates his pilgrimage to the Holy Land between 1480 and 1484. In this work we see for the first time the use of mines to break up rock for the construction of a road in the Tyrol, near Bolzano, under the supervision of Sigismond le Riche. At the express desire of Louis XI, the construction of the Mount Viso tunnel, built to link the Dauphiné and the marquessate of Saluces, was made possible. The king wrote to the Parliament of Grenoble, which was charged with attending to the matter, that he desired that the work be done, that it would be "of great benefit to the country and a great honor to all of you that in your day such a great deed should be done." The people of that time therefore realized to a certain extent the novelty and daring of this project. The tunnel was drilled under the pass, more than 6,560 feet above sea level. It is 6½ feet wide and 8¼ feet high, with fairly regular walls; because of an error in the plan, it curves slightly. The tunnel tends to slope toward Italy. The work was done between 1478 and 1480, under the direction of Martino d'Albano and Baldassare d'Alpiasco, and was a great event at the time.

However, the importance of the highways and the traffic they carried must not be exaggerated. The merchandise they transported was still relatively low in volume, and the development of maritime navigation had, moreover, helped to limit this traffic considerably, the ships having absorbed a major part of the growth of trading.

The same comments hold true for bridges, whose technique had little need of major transformations. In the middle of the fourteenth century the bridges began to lose the appearance of the very sharply curved Gothic bridges that were so characteristic of the preceding period. The Middle Ages had constructed few large stone bridges; most of those that are still extant cross small streams. Wide spans were generally bridged by wooden bridges, to which the texts make numerous references. The foundation techniques of piling by now seem to have progressed sufficiently to permit the construction of bridges with several arches over wide rivers. The Ponte Vecchio of Florence, whose arches are completely modern in appearance, was constructed in 1345; the Pont Notre-Dame in Paris was constructed in 1413 and rebuilt in 1499. These two bridges definitely represent very clear progress in construction. Another perfect example is the fortified bridge of Verona, a masterpiece of brick construction. Its arches range in span from 78¾ feet to 159¾ feet; it was begun in 1354, under the direction of Giovanni da Ferrara and Giacomo da Gozo. The now destroyed bridge of Trezzo, constructed between 1370 and 1377, was still very Gothic in appearance because of the wide span (239½ feet), which made it famous. The bridge of Prague, constructed in 1357, also has a very classical appearance.

Information about the methods of construction of these bridges is very fragmentary. It is evident that as the span increased, the builders were obliged to build the piling in the water. It was therefore necessary to construct coffers with piles dovetailing into each other, as can be seen in engineers' drawings. (These techniques were also used for certain major hydraulics works.) To bury the piles, pile drivers moved by hand-operated windlasses (Francesco di Giorgio has left us several drawings of this operation) were used. In addition, the water in these coffers had to be pumped out. In Amiens at the end of the fifteenth century, devices with chains and buckets were utilized for this purpose.

The conception of the light temporary bridges utilized by armies in the field was completely different. The engineers' notebooks are filled with drawings of these bridges, which were usually constructed of wood and for which the most ingenious systems were invented. Leonardo da Vinci's wooden bridges, many of which are simple rafts on supports, are of no major interest. There were, however, several turning bridges, which became useful when canals proliferated. In his drawings of fortresses, Francesco di Giorgio depicts not drawbridges but sliding bridges, traces of which were found by Viollet-le-Duc. We find, however, that the most interesting examples are the early works of this type, those of the very end of the fourteenth century: folding bridges for crossing moats, prefabricated bridges sketched by Guido da Vigevano around 1330, and floating bridges. This military technique for crossing rivers and moats already appears highly perfected.

It seems, therefore, that by the first half of the sixteenth century perfect mastery of bridge construction had been achieved. Leonardo da Vinci's designs for the coffering of the vaults of bridges, the foundation techniques of piling, and all the equipment we find in other engineers' notebooks, were utilized in succeeding centuries without notable modifications.

Navigation canals

We have already had occasion to note the valuable and important contribution made by the waterways of northern Europe to commercial relations during the Middle Ages. It was very soon realized that this communication network could be corrected and even supplemented by artificial waterways. However, the creation of these canals was still hindered by numerous difficulties. Some of these were due to features of the terrain, and required, on the one hand good leveling, which was still difficult to accomplish, and on the other hand a system of locks that long remained a major disadvantage because of the lack of proper methods. In addition, it was absolutely necessary to feed water into these artificial waterways. The techniques for creating a canal and ensuring its watertightness constituted a difficult problem. The Renaissance appears to have made a fundamental contribution in all these areas to the progress of technology.

We have little exact information on the material available; however, the engineers' notebooks, especially those of Leonardo da Vinci, do furnish us with several interesting elements. Leonardo appears to have been greatly interested in all these techniques; under the direction of his supervisors he worked on all the construction sites begun in the plain of the Po to obtain both navigable waterways and irrigation channels for the Milan region. The first problem was that of excavating a channel sufficiently wide and deep. However, it was not so much the techniques of terracing themselves that attracted the attention of the Tuscan engineer (it seems that only the spade and wheelbarrow were used) as those of the removal of the fill. Relatively numerous drawings depict devices, generally of the squirrel-cage type, for removing the excavated material from several construction sites installed at the various points of attack. We have much less information on the techniques of the slopes and of the watertightness of the beds thus created; we discover, however, that Leonardo planned to plant trees on these slopes in order to hold them in place, a technique that had long been in use, especially on the levees of drainage dikes. Some of the slopes may have been retained with pile planks driven in with the help of the pile drivers mentioned above.

The locks still pose difficult problems of interpretation for the historians. As we have already noted, locks with movable gates were already in existence at the end of the twelfth century, on the canal that linked Bruges with the sea, but it is almost impossible to say how these locks functioned. We still do not know whether they had movable gates or merely large sluice gates. It seems, however, that in 1394 John, Duke of Berry, installed such locks on the canal that runs from Niort to the ocean. Others were probably constructed on the Juine, in the county of Étampes, at the same period. The system of locks was probably perfected in the first half of the fifteenth century. Shutter-type stop gates now came into general use; Filippo Visconti made use of them in 1440, and the Venetians utilized them on the Piovego in 1481. Alberti describes this system in his treatise on architecture, composed in 1452. A document of August 3, 1447, shows that

FIG. 54. Lock gates with vertical control (Singer, II, p. 663, Fig. 626).

similar gates had already been utilized in 1435. But it was not so much the gates themselves (whose construction created no major difficulties) as the methods of opening and closing them that were perfected in the first half of the fifteenth century. In the engineers' notebooks we still find frequent examples of the ramps on which the ships were hoisted, with the help of windlasses, to pass them through the various levels.

It was these improvements that made it possible to dream of the creation of more complete and consistent networks of internal navigation. The first very large projects in this field were begun within one century after this: the Briare Canal during the reign of Henri IV.

In this period, however, we begin to see major works that go beyond the elementary level of improvement of natural navigable waterways — not, however, that this improvement was abandoned. (Certain achievements in France even lead us to believe that it occupied a leading place in official plans.) In particular, we find considerable activity around the Loire and its small tributaries. In 1468 Guy Farmeau and Guillaume Baudet were charged with continuing the work, begun a long time before, of regulating the flow of the Loire, in order to eliminate silting and dangerous passages, and also to remedy the disadvantages and disasters of the excessively frequent floods. In 1480 these projects were continued in the valley of Cisse, where they can still be studied, under the direction of Hardouin de Maille, Antoine de Galles, and Jacotin le Mercier. Another project was to make the small tributaries of the Loire, which acted as access routes into the back country, navigable: the Auron led to Bourges; the Brenne made it possible to transport wood to Tours, the favorite sojourn of Louis XI. The Brenne was made navigable from Château-Renault to Tours by Nicolas d'Aubigny. The Clain was to link Poitiers with the Loire network: after the mud and weeds had been cleaned out, locks were installed. In 1460 Louis XI brought visitors to view the Loire, the Maine, and the Sarthe, all of which had recently been made navigable. On the Clain, when the Nouâtre lock broke, one Antoine Martin (who was probably an Italian) was hired to fix it.

In 1455 Charles VII had ordered the deepening of the Eure, between Nogent and Chartres: lock gates were installed and a towpath created. Streams were diverted around Étampes in order to channel the stream, which, by way of the Juine and the Essonne, permitted the passage of ships. The Sèvre was improved between Niort and Marans. In 1462 the canal from Luçon to the sea was deepened for the same purposes of navigability. It is in this spirit that the works entrusted to Leonardo da Vinci, to create a canal in Berry around the Romorantin, must be envisaged.

France was not the only nation to undertake these major works of practical utility, whose development unquestionably made possible the progress of technology. In the plain of the Po, projects had been under way for centuries, both to improve navigation and to facilitate irrigation. We have seen that in the fourteenth century the diversion of the Ticino had been extended as far as Milan, with the Pavia Canal, and that work had been done to control the water level of the Po from Ponte Alberto as far as the junction of the Lambro. The next problem concerned the linking of the marble quarries of Verbano with the construction site of the cathedral of Milan, for which Verbano was supplying the stone. This problem was solved in 1395 with locks that made it possible to bridge a five-fathom difference in level. The diversion of the Martesana was begun in 1457; it was originally intended to ensure a supply of water for Milan, and it drew its water from the Adda. This project was the work of Bertola da Navata, Francesco Sforza's engineer. The next project was to make this canal navigable, in order to spare boats the long twistings and turnings of the Adda. Leonardo da Vinci undoubtedly participated in this project, under the supervision of Bartolomeo della Valle.

At the beginning of the fourteenth century the Stecknitz was made navigable from Lake Mölin to Lübeck, a distance of about eighteen miles, with a drop of forty feet. A canal then made possible the linking of the Elbe at Lauenburg, forming the first bridge over the watershed between the basins of the Baltic and the North Sea. This work was completed in 1391–1398.

The techniques employed in all these projects, except for a few minor details, appear to have now been sufficiently advanced to permit large-scale projects; although the sixteenth century provides no examples of this type of undertaking, succeeding centuries accomplished such projects.

THE GREAT HYDRAULICS WORKS

In the field of major hydraulics works, the projects inherited from the Middle Ages were already numerous, and the contribution of the Renaissance was to be no less considerable. The sources furnished by Leonardo da Vinci in connection with the draining of certain Italian swamps are well known, but we are less familiar with similar efforts begun or completed almost everywhere in Europe in matters of both drainage and irrigation. Here, again, political centralization and the fondness of the period for large-scale works probably provided strong support for the development of these techniques.

Drainage projects

There were no major modifications in the technique of drainage, and especially coastal drainage. If a natural flow of water could be ensured, the difficulties were not insurmountable; the establishment of a checkerboard network as closely spaced as possible was more than sufficient. When work was resumed on the swamps of the Sèvre around Niort, after the depredations of the Hundred Years' War that in many places had caused the abandonment of the conquests of the preceding centuries, the engineers were satisfied to repair the holes in the *bots* (filled-in lands) and to clean out the drainage canals. The Le Roi Canal, completed in 1283, had been the last major work of the medieval period. More than a century later it was necessary to repair the disasters caused by the war. In 1409 the royal government became concerned, and between 1438 and 1443 partial projects were carried out to put the drainage network back into use. Not until the first half of the sixteenth century (in 1526, to be exact) do we find the authorities thinking in terms of an overall plan for the swamps. It was then discovered that the sinking of the dried alluvions would entail a considerable raising of the *bots,* if their protection was to be continued. The efforts made by the Italian cities in the same period were equally important. Here, too, a long tradition existed, and certain abbeys already had major projects to their credit. It was a question not of reclaiming lands from the sea, but of draining and protecting the low valleys periodically invaded by overflowing rivers. At the beginning of the sixteenth century, Venice appointed officers to supervise the canals and swamps; Verona established a school for the Adige River, and in 1549 Florence also appointed officers to superintend rivers, bridges, and moats. At Venice it was a question of improving the entire swampy region that extended between the Brenta and the Piave. The water level at the mouths of all these rivers, which caused catastrophic floods upriver by becoming choked with mud, had to be regularized. At the same time the navigability of all the streams that linked the port of Venice with the very rich mainland was improved. This was a long-term

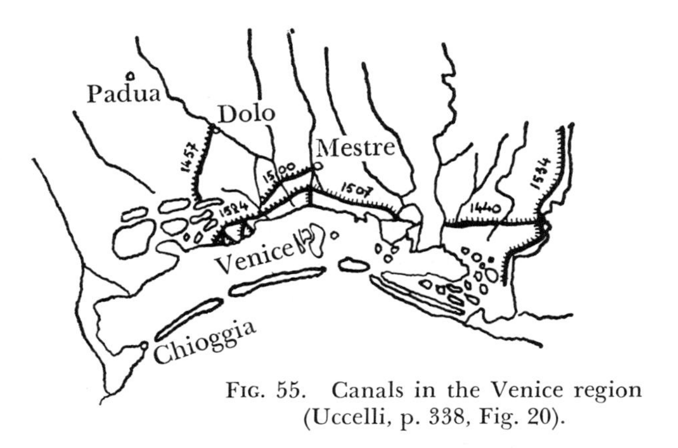

Fig. 55. Canals in the Venice region (Uccelli, p. 338, Fig. 20).

project, as is indicated in Figure 55. The first part of the work was executed between 1440 and 1460, doubling the outlets of the two main rivers. Then, between 1500 and 1530 a network at right angles to the natural slope was established. Around 1400 the passage of Fusina was constructed; here ships passed from one level to the next by means of a ramp and an enormous windlass operated by a horse-driven treadmill. Two centuries later this mechanism was replaced by the lock of Moranzano.

In 1514 Pope Leo X ceded to his brother, Giuliano de'Medici, the drainage projects of the Pontine marshes to the south of Rome. Leonardo da Vinci's project (now preserved in the Windsor collections) dates approximately from this period: he had in mind a great transversal canal, to be called the Fiume Giuliano, parallel to the Via Appia, opening onto the sea at Terracina, and a canal at right angles opening into the Gulf of Asturo. This was in large part a repetition of the Roman achievements. It was actually executed only later, by Dutch engineers.

During the two centuries here under discussion, the Dutch had also been continuing work on projects that were already ancient. Several errors were made, and a major catastrophe demonstrated that work must not be done without an overall plan. Just as the Zuider Zee had been formed around 1300 by a sinking of the alluvions that had been dried either naturally or by human labor, in 1421 it was the turn of the Dordrecht region. On the night of November 18–19, 1421, 65 villages were submerged, swallowing up some 10,000 people. The entire region between Bergen op Zoom and Moerdijk, that is, the lands lying between the mouths of the Schelde and the Meuse, had been protected in the fourteenth century by a dike, but the neighboring region of Dordrecht had been neglected. It is possible that the installation of locks on the Meuse River was one of the causes of the disaster of 1421.

While some protection against invasions by the sea and the overflowing of the rivers had been achieved since the twelfth century, and although a portion of the rich alluvious behind the dikes had been drained, more powerful methods of drainage were needed if the work was to be continued. The idea of utilizing windmills to operate what undoubtedly began as simple Archimedean screws apparently appeared at the beginning of the fifteenth century (some historians advance the date of 1408). The proliferation of drainage points would permit a more rapid and more complete reclamation than had been attempted up to now.

Work on the Biesboch, the region inundated by the catastrophe of 1421, could therefore be resumed quite rapidly. Before 1430 and 1460 the region was surrounded by dikes; beginning in 1435 polders were gradually installed. The method consisted of surrounding the region to be drained by high dikes, on top of which the drainage canals were dug, and of installing a checkerboard drainage system in this area; windmills were used to raise the water from the interior into the surrounding canals, which were above the level of the sea or river. Ultimately this became the method systematically employed.

It was necessary, however, to continue the construction of a protective system of powerful dikes, the sole guarantee of the drainage work being done inside, in the lowlands. Work was therefore continued without interruption until the end of the fourteenth century, and throughout the fifteenth. In the region of Alkmaar an attempt had been made to construct dikes; until 1388 it met with small success. At the end of the fifteenth century the dikes of the island of Walcheren, about 2½ miles long and 328 feet wide, were constructed; they dominated the high tide by 15¾ feet and the low tide by more than 27 feet. This did not prevent the dike from being broken in 1530 by a powerful storm; it was immediately repaired. The dikes of Frisia are later than the first half of the sixteenth century; they were the work of a Spaniard, Casper de Robles, around 1570.

The use of watermills made it possible to consider draining the internal lakes that still covered a large surface of the country. This project was to be the work of a pleiade of great engineers who were being trained at this period and who were to endow these projects with universal fame. Andrew Vierlingh, who in 1578 wrote the first treatise devoted to these techniques (it remained unpublished until 1920), had already begun his work before 1550, for in 1552 he was named master of dikes at Steenbergen. Between 1542 and 1548 a first group of small internal lakes were drained and put into cultivation: Dergmeer, Kerkmeer, Kromwater, Weidgreb, Rietgreb, and still others in the northern region of the country. The reputation of the Dutch engineers began to be valued everywhere, and many European countries called upon them to create similar projects. These early Dutch projects included the draining of the mouths of the Vistula, between Elbing and Danzig, at Tiegenhof, in the period 1528–1562.

Irrigation

Irrigation techniques had remained unchanged for several centuries; no attempt was made to utilize the more powerful methods now available with windmills and lift-and-force pumps. There was, however, a slow improvement in certain techniques, in particular for the tapping of mountain streams, a project that was always difficult of accomplishment. Thus, to take only one example, in the Pyrenean canals these diversions had to be constantly replaced, since they were ruined by the erratic courses of the torrents. Such was the case with the Ille Canal, created at the end of the thirteenth century, which was fed by the Têt river. Not until 1550 was the project linking the canal with the stream begun; it was done by means of a subterranean gallery, in a place where the presence of rock created the feeling that the bed of the torrent would not change. But a number of other small canals continued to be supplied directly. At the end of the fifteenth century it was clearly specified that the dams established for these diversions must not be constructed of masonry, to ensure that they would not constitute dangerous obstacles in case of flood or endanger the interests of the riverbank inhabitants downstream in periods of drought. The dams, made of piles and stone, must in effect have permitted a large quantity of water to filter through.

Techniques employed by the Arabs in earlier centuries, for example dams with collecting reservoirs, were to develop considerably. Leonardo da Vinci had already outlined a project for a collecting basin above Florence that would distribute water from the Arno River in periods of drought. It was in the southern part of Spain, however, at the very end of this period, that the basic projects were constructed. By 1586 the Almansa reservoir had already been in existence for an unknown period of time; it may have been constructed around the middle of the century. The dam, which is built completely of masonry, is almost 68 feet high and 292 feet long, and forms a square reservoir about 4,920 feet long. This reservoir, which is now badly silted up, was 262½ feet deep. The Tibi reservoir, also of masonry, may have been constructed in the third quarter of the sixteenth century; some authorities attribute it to Juan de Herrera, the architect of the Escurial.

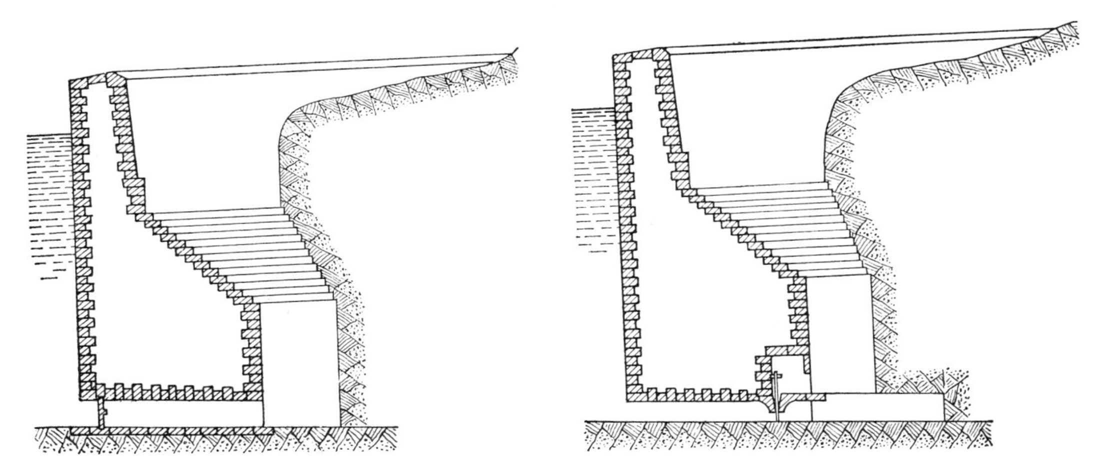

FIG. 56. Cross sections of the Almansa dam. Around 1550–1560 (Aymard, *Les irrigations du midi de l'Espagne,* Paris, atlas, pl. V).

The great Renaissance contribution to the techniques of organization is thus a very visible one: the very spirit of this period was inclined toward those great works of public utility that formed part of a new idea of the State, an idea whose continuing development began in this period. The birth of the idea of great overall projects, and a certain spirit of systematization, furnished technology with an infinitely wider field of application that was consequently very favorable to innovation. The vigorous middle classes, who were very conscious of material problems — we have seen that urban planning had made great progress in the cities of northern Europe, where the system of municipal magistrates was well organized — and the appearance on the shores of the Mediterranean of dynasties of enlightened princes, some of whom sprang from several generations of statesmen and who were seeking a certain monumental order that perhaps had its roots in a rediscovery of antiquity, favored this movement to a great degree. The onset of economic policies and more extensive military concerns all helped to orient the action of the public powers toward the construction of a better-organized world.

None of these phenomena, which were outside the realm of technology, could have borne fruit if the political powers had not found those remarkable men, the "engineers of the Renaissance," to serve their purpose. It was to answer to this new desire (or these new desires) that these men became military engineers in addition to being talented architects or experts in hydraulics. They were not satisfied to draw up projects and execute fortifications, great buildings, dams, or canals: since they were given the opportunity, they attempted to raise problems in general terms, an example of which is Leonardo da Vinci's very important notes on the erosion of riverbeds. This research probably led him to construct experimental models with wooden planks and sand in an attempt to envision every possibility, just as modern hydraulics experts do. Thanks to this work and research, technology was extended beyond individual cases. Going beyond the particular instance, and sometimes even its own objectives, Renaissance technology was to envision more formal problems and to assist a science still seeking its working methods.

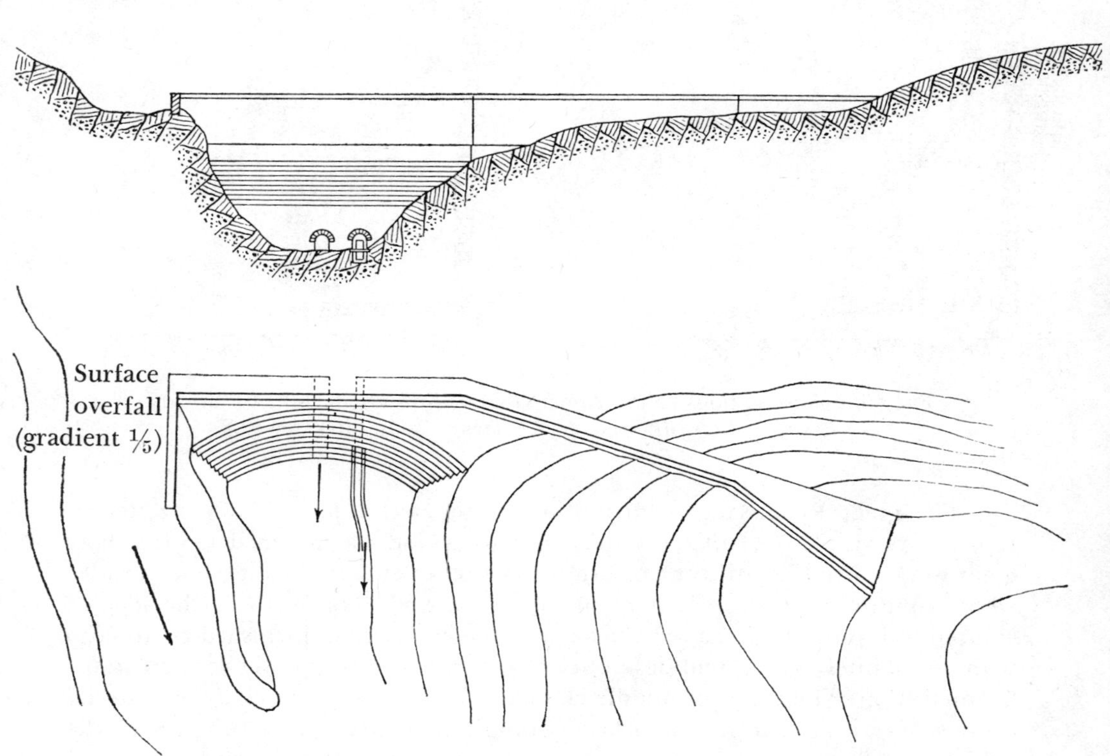

FIG. 57. Elevations of the Almansa dam. Around 1550–1560 (Aymard, loc. cit.).

CONCLUSION

THE EVOLUTION OF THE TECHNICAL CIVILIZATION

WITH THE CLOSING WORDS of the last chapter, we have now returned to the problems we outlined at the beginning of the history of the technology of the Renaissance. This technology is perfectly in accordance with a change in orientation of the entire civilization, a change that was the concrete manifestation of Renaissance originality. Between the end of the fourteenth and the middle of the sixteenth centuries, technology succeeded in capturing a place it had never before occupied. This importance is revealed by a series of facts that perhaps have not been sufficiently studied from this point of view. The first, and one of the most symptomatic, is the inclusion of technical knowledge in the body of knowledge, which resulted in an undeniable interrelationship between technology and science — an extremely important phenomenon whose origins date from the time of Archimedes. On the other hand economic development, like military requirements, helped to place technology among the great political problems. Thus there was as it were a second entrance of technology into the life of nations, at the highest level. This importance of the technical evolution was not without influence on social problems, although this was only the beginning of an influence that was to become increasingly great in succeeding centuries.

TECHNOLOGY AND SCIENTIFIC THINKING

The interrelationships between technology and scientific thinking are still obscure. The problem is made particularly difficult by the fact that, as we have already noted, the scientist and the technician were often one and the same man. While it appears that technical progress was not yet able to benefit from the scientific development, it is difficult to determine to what extent scientific thinking was able to receive sustenance of any kind from technical problems.

The "mathematization" of technology

There is no doubt whatsoever that mathematics was regarded as the indispensable instrument of the "mechanical arts" and that this was considered its true and only use. This had been the position of the medieval thinkers, and was universally held until the time of Descartes. "Mechanics is the paradise of the mathematical sciences, because it is in mechanics that the latter find their realization." This sentence of Leonardo da Vinci could have been written in the thirteenth century and also at the beginning of the seventeenth. What was new was the growing importance of mathematics. "When none of the mathematical sciences, nor those that are based on mathematics, can be applied, then there is no certitude." This sentence, again by Leonardo, attests to a new type of thinking. It was, moreover, in the millieus of the technicians that Leonardo da Vinci was able to build up his mathematical knowledge: in Andrea del Verrochio's studio, and on the construction site of Milan Cathedral, where his geometrical knowledge was perfected. He studied the problems of perspective, as defined by the architect and engineer Alberti benefiting especially from his friendship with the monk Luca Pacioli (1445–1514). Pacioli was a disciple of Piero della Francesca who was closely linked with the technical milieus of Urbino, which taught the calculation of irrational numbers. Leonardo's mathematical procedures are interesting in the sense that they are a perfect revelation of the technician's desire to find in mathematics a working tool that would permit them to perfect their methods. The painters had cherished the same hope with regard to perspective, that technical intrusion into the domain of art. This entire period, notes P. Francastel, considered the destiny of art to be, not realism and the pure simple recording of sensations, but the depiction of a new sensory experience illuminated by a new scientific knowledge. The technicians attempted to utilize mathematics along the same lines, not as a simple measurement of the real, but as an aid to a logical construction of the universe.

There are symptomatic examples of struggles between partisans of the old techniques and those who were recommending a more systematic recourse to mathematics. Proof of this is seen in a construction diagram from the pen of one Stornacolo, one of the architects of the Duomo of Milan, around the end of the fourteenth century. The construction of this cathedral gave rise to a debate between old master builders and modern engineers over the question of interpreting and utilizing geometry for the construction of buildings. The medieval builders had generally utilized flat projections of geometrical figures in order to calculate relationships and to explain how one could go from the plane to the elevation by evaluating surface relationships between the geometrical figures. Beginning very early in the fifteenth century, gometrical figures were utilized in relation to numbers, to establish a link between abstract mathematical speculation and the drawings, without limiting oneself to empirical relationships. Thus, concludes Francastel, a transition was made from a fragmentary, empirical, concrete rationalism to a universal speculation. In a sense, the builder was no longer a stonecutter utilizing flat models to draw volumes from them: he had become an engineer.

Little more sustained information can be found, it must be admitted, in the still extant engineers' notebooks of the first seventy-five years of the fifteenth century. Mathematical ideas are extremely rare in the treatise and hasty notes of Francesco di Giorgio, and not until Leonardo da Vinci do we find a more modern position. His algorithmic calculations, notes P. Sergescus, are usually correct, and the numerous pages devoted to calculations of bastions, piping systems, and fortifications of all types testify to the correct use of modern rules. It is very evident that Leonardo was not the only engineer of his time to practice mathematics in this way. We are able to ascertain that his immediate predecessors, and particularly Francesco di Giorgio, did the same, although their works are no longer in existence.

It is obvious that this intrusion of mathematics into technology did not occur at the same time in every phase of this technology. There were some areas in which "mathematization" was practically impossible, and others where it did not become possible until much later. In naval construction, for example, we know that the first mathematical elements did not appear until the work of the Englishman Matthew Baker, in the second half of the sixteenth century, and that not until the end of the eighteenth century did a correct mathematical theory of the ship appear. In contrast, the construction and techniques of hydraulics systems undoubtedly called for an increasingly greater use of mathematics at a much earlier period. The trace of a fortification, a canal, or a dam obviously could not be done without mathematical equipment, which continued to develop. In some cases complaints were even made about the technicians' lack of mathematical knowledge. In 1534 the Senate of Venice deplored the small number of ships' captains who took the trouble "to learn to put the compass on the chart in order to know how to estimate the distance covered and measure the time when they are drifting and to recognize the shores, important things a sailor should know." This reproach could certainly be made of numerous technicians, but it should be noted that the last quarter of the fifteenth and the beginning of the sixteenth centuries marked a decisive turning point in this regard.

Technicians and physicists

While some of the technicians' observations could be of help to the men interested in physics, so that they were able to pose problems in a completely new manner, it is obvious that there is not necessarily a connection between the two, and it would undoubtedly be an error and an exaggeration to say that the first faltering steps of modern physics began from technology. There is no doubt that the scientific curiosity of architects, hydraulics experts, and gunners led the way and that they were among the first people to become aware of the new problems; we find, however, that the same problems were very soon taken up by laboratory workers who were never technicians and who carried science far beyond the point it had reached around the end of the fifteenth century. Scientific construction uncontestably belonged to a system of logic different from that of the pure technician.

Theoretical knowledge, despite its importance (at least for some techniques), was still largely a matter of oral and empirical tradition. This is perhaps the reason why we find among Leonardo's predecessors none of the notes that a quarter-century later were to make the great Florentine thinker famous, at a time when he himself had become more of a laboratory worker than an engineer. We should remember, however, that Taccola was called the "Archimedes of Siena" and that his reputation as a scientist and investigator was at least as great as his knowledge of fortifications. Francesco di Giorgio was certainly a very well-educated man, and his trip to Urbino, where he knew both Leonardo da Vinci and Luca Pacioli, just as surely turned his mind to theoretical problems. While we probably do not possess all the engineers' notebooks of the middle or third quarters of the fifteenth century, it is nonetheless true that Leonardo da Vinci unquestionably represents a step forward in this regard over his immediate predecessors. The very contradictions of his thinking, Maurice Daumas has remarked, for example on the notion of energy and weight and his early ideas on kinetic energy, reveal the attitude of a constantly alert mind that was no longer satisfied with the teachings of the medieval masters. Leonardo studied the phenomena of shock and percussion, and those of statics (center of gravity, heavy bodies turning on an axle and moving on an inclined plane, composition of forces); ballistics interested him, and especially problems of hydraulics. Not all this, however, was completely new, and he was certainly following in the footsteps of his predecessors. The gap lies perhaps more in the blanks in our information than in the facts. While technology often progresses by leaps, no activity is more continuous than research.

An accurate history of ballistics would illustrate in undoubtedly quite striking fashion the appearance of certain ideas of classical physics between the middle of the fourteenth and the middle of the sixteenth centuries. Not, however, that this science was exclusively the work of technicians. At both the beginning and the end of this history we find pure theoreticians, like Jean Buridan and Niccolò Tartaglia. It would be appropriate, however, to study (if this is possible in the present state of our information) the contribution of the technicians as well, and paricularly Francesco di Giorgio, whom Leonardo Olschki regarded as the inventor of experimental ballistics. There were certainly fruitful confrontations between the technicians, who were beginning to realize the value of their repeated experiments, and the mathematicians who were trying to translate them into a more universal language. "If Benedetti, by far the most remarkable of Galileo's predecessors, sometimes surpasses the level of the Parisian dynamics, it was thanks to his study of Archimedes rather than to his work as an engineer and gunner," notes Alexander Koyré. But was the scientist alone involved? The gunner and the engineer posed exact problems to the scientist (it is very difficult to dissociate the two aspects of the man), and in their hands lay the possibilities of experimentation.

At the beginning of the sixteenth century, however, the scientists had not yet passed the stage of the definition of concepts. Leonardo was still hesitating between Buridan and Aristotelian dynamics, even if his understanding of the subject was better than has been claimed for his predecessors. A scholastic argument, to be sure, but one that was to acquire a remarkable scope when men like Francesco di Giorgio and later Benedetti made available to scientists a great number of experimental observations that undoubtedly were already quite well "classified." The time was now ripe for Tartaglia, who was to be the true originator of the ballistics theory. In 1537 he published his *Nova Scientia,* and in 1546 a second book, entitled *Quesiti et inventioni diversi,* a portion of which was devoted to the various aspects of artillery: combustion of the powder, aiming, casting. In 1531, while Tartaglia was living in Verona, a highly experienced gunner from the fortress asked him in what position he should lay his gun so that it would have the greatest possible range. Tartaglia states that he based his conclusions "on the nature of objects and on geometry." It seems plain, therefore, that he did not take into consideration experience that undoubtedly had led the simple gunner to the same result. We shall not even briefly analyze Tartaglia's theories, which are certainly indicative of the state of knowledge of that period; Pierre Duhem did not have a very high opinion of them, but since then they have been rehabilitated by other historians. Tartaglia's division of the trajectory into three sections does not correspond to the actual facts. It is nonetheless true that he succeeded in bringing about the recognition that the trajectory of a cannon ball is a continuous curve and that to him we owe two interesting theorems of minimum speed and angle of maximum range. It is to his second book, infinitely more practical than the first, that we must look for the origin, if not the theories, of internal ballistics. Tartaglia also considered both the length and diameter of the guns, the weight of the charges needed for maximum range, the causes of bursting, the various proportions of powder, and so on. Here we have the beginnings of a treatise on artillery, whose genesis is contained in the preface of Tartaglia's first work. He had thought about it, but had retreated before the horror of the effects of artillery, finding it detestable to write a work on the perfection of such a dreadful sinew of war. Taraglia's works were studied and plagiarized for almost a century. The work on geometrical artillery by Rivius (published in Nuremberg in 1547), whose title was very promising, was actually only a translation of Tartaglia's writings. The Spanish and English works of the second half of the sixteenth century were also directly inspired by the Italian writer.

We have very little information about other problems of physics raised by the evolution of technology. It seems evident that Leonardo's studies on hydrodynamics were begun by his predecessors, who accomplished a number of projects that demonstrate their theoretical knowledge. Leonardo's merit is perhaps that he carried out certain research in a systematic manner, for example studies on whirlpools and the erosion and formation of riverbeds. Everything concerning the calculation of water flow had certainly been known before his time. As for the theory of pumps, it unquestionably originated with the School of Alexandria, and work on it definitely continued during the first half of the fifteenth century.

The dynamics of Leonardo and Benedetti is, in its main outlines, that of the Parisian Nominalists, closer to experience than Galileo's system. This phenomenon is perhaps similar to the one found in the evolution of technology in ancient Greece, the formation of an excessively pure science. The development of an experimental, and therefore technical, science was also hampered by difficulties in experimentation, which requires a perfected equipment singularly lacking in this period.

The problem of measurement and observation

We have already noted how diversified the system of the ancient measures had become in the Middle Ages. Basically this was due to the fact that despite a common origin, for political, economic, and technical reasons it had been impossible to maintain a single system. Beginning in the second half of the fourteenth century, this unification of the systems of measurement was to be one of the concerns of the civil authorities, for technical as well as political purposes. In the middle of the fourteenth century an archbishop of Lyon named Guillaume de Thurey decided "that in future only two-balance weights would be used, that the weight of Lyon would be a quintal of 102 pounds, a pound being 14 ounces; that all the merchants would regulate their weights on this one, and that the ancient weights would be broken and abolished." In France it was especially the measures of Paris that, because of the growth of the royal power, tended to become the rule throughout the kingdom. In 1501 the city of Amiens ordered the goldsmiths of the city to create a standard equivalent to those of the Chambre des Monnaies of Paris. In 1540 an ordinance imposed the exclusive use of the Parisian ell. In England an attempt had been made long before to establish uniform measures, since the Great Charter of 1197 recommended it, but the same confusions had been introduced, for the same reasons. In 1340, 1430, and 1495, attempts were again made to unify the systems of measures.

Measuring instruments had been little improved. The same balances were still being used, and not until the appearance of the Frenchman Gilles Personne de Roberval, inventor of the Roberval scale in the seventeenth century, was decisive progress made. However, gradual and barely perceptible improvements in weighing made it easier to verify measures, which explains why the ordinance of March 1540 was able to ordain standardization and punching by the minters for the manufacturers of weights. Henri II continued this process of unification, and in 1557 reduced the weights and measures to a single system for the city of Paris, in 1558 for the entire kingdom. It was only a *de jure* unification, to be sure, but it is nevertheless indicative of a clear-cut desire for an indispensable precision.

Until the sixteenth century almost the only scientific instruments in existence were instruments for observing the stars or making topographic surveys. Leonardo's pictures of certain instruments — hydrostatic balances, pedometers, hygrometers, anemometers — are very fragmentary, and for the most part are not original ideas; moreover, these instruments, about which many people had been thinking, were not to be effectively created until later centuries. Instruments for observing the stars had been widely known for centuries. Special attempts were made to simplify them and make them more exact. The Davis quadrant dates from the end of the fifteenth century. In most cases it was constructed of wood, and the scales were engraved on encrusted ebony or ivory plaques. One of the most simplified expressions of the astrolabe was the return to the brass circle, retaining only two orthogonal diameters, one of which represented the horizon. An alidade with sights pivoted around the center. The quarter circle and graduated sectors did not become the customary instruments of the astronomers until the end of the sixteenth century. The square or geometrical quadrant, a solid copper quadrant with a graduation in 90 degrees on its sector, is described at the end of the fifteenth century. Around the middle of the sixteenth century a horizontal circle and a vertical semicircle were combined into an instrument that Leonard Digges calls a theodolite. Rules, squares, and compass supplemented these instruments. We find them utilized in large numbers to measure the heights of towers, trees, and mountains, in the treatises of Alberti and Francesco di Giorgio.

We may thus say that at the end of the first half of the sixteenth century methods for measuring and instruments of observation had made only negligible progress. Major innovations in these techniques were not made until the second half of the sixteenth century.

The influence of antiquity

In spite of a new scientific spirit, the extent to which the reigning spirit of humanism caused the weight of classical antiquity, and especially of Roman antiquity, to weigh upon even the technicians is very evident. The influence, from the architectural point of view, of the Italian cities of antiquity need not be demonstrated. This influence was exerted not only in the domain of esthetics but even in the methods of construction. The reverberations of the discovery at Saint-Gall of a manuscript copy of Vitruvius in 1414, when the text of this author was already well known, indicates the hope that the Renaissance placed in the teaching of the ancients. The first editions of Vitruvius, after 1480, were merely reprints of the Latin text. The first commentaries, written by Fra Giocondo, who belonged to Bramante's circle, and Cesare Cesariano, a pupil of Bramante, appeared in 1511 and 1521 respectively. Not only do the treatises written in the second half of the fifteenth century show the influence of Vitruvius; they also show that a number of well-known architects, like Francesco di Giorgio and the Sangallo brothers (to mention only the most important ones), have left abundant drawings of ancient monuments, accompanied by important technical notations. The influence of other technicians of antiquity — Pliny and the agronomists, Vegetius, Frontinus — has been noted elsewhere in this book.

There is another aspect of antiquity that was investigated with equal ardor. A Leonardo da Vinci who eagerly searched the manuscript copies of Archimedes has recently been discovered. Archimedes' works were undoubtedly investigated both for their scientific value and for their technical interest, the memory of his military inventions being still vivid. Archimedes is the writer most frequently quoted (sometimes even incorrectly) by Leonardo. However, he had been well known long before Leonardo's time, through translations. Taccola himself, although "illiterate" like Leonardo, probably had indirect but genuine knowledge of him, and Benedetti was later to discover the same sources.

This technical aspect of the influence of antiquity at the time of the Renaissance has been little studied, but it is evident that here again technology is part of an entire intellectual context. We even possess rather strange proofs of this fact, discovered by Lucien Febvre. In 1525 and 1526, when the Senate of Venice was deliberating on a type of ship capable of destroying the pirates, Matteo Bressano, an old master craftsman with long experience, presented a project for a round ship. But Vittorio Faustus, a public reader of Greek rhetoric and a humanist trained in Greek mathematics and Aristotelian mechanics, ventured into the practical realm and submitted plans for a quinquireme. The marvel was that in the contest the memory of ancient Greece won the victory over modern tried and tested techniques.

To be sure, all this is only a straw in the wind: the contacts between book learning and practical knowledge were undoubtedly .very weak, although they definitely existed. Perhaps the Renaissance men were in search not so much of instruction as of justification.

BIRTH OF TECHNICAL POLICIES

A discussion of the origin of official policies on technology would entail research that has not yet been begun. We must therefore be satisfied with a few fragmentary remarks that in one way or another will indicate the outlines of the problem.

As early as the Middle Ages certain political powers had attempted to attract scientists, technicians, and poets; we have mentioned several cases, ranging from Alfonso the Wise to the Hohenstaufen. Frequently, however, this was done for intellectual rather than utilitarian purposes, and similar attempts of several small fifteenth-century Italian courts like Urbino and Rimini, whose military preoccupations were, however, much more definite, must still be explained in this spirit. The formation of the great modern monarchies implied a twofold military and economic effort in which technical progress suddenly acquired a preponderant role. Powerful and well-equipped armies, and a system of frontier defenses both terrestrial and maritime, and as inviolable as possible, became necessary. After the disorders of the Hundred Years' War, it was equally indispensable, in order to maintain an active policy, to carry out an economic reorganization and to eliminate the outflow of wealth, in particular the wealth lost through costly imports. The second half of the fifteenth century thus witnessed the birth of the first industrial mercantilist systems, which required large numbers of foreign technicians. This economic policy was to be one of the most powerful factors in the expansion of technical progress, much greater than the diffusion of books or of treatises, which were usually vague.

The military effort The military effort is undoubtedly the most thoroughly studied aspect of this movement, but there are nevertheless large blank areas in our knowledge that merit more thorough research. The problem has two aspects: armaments and fortification. In addition, the universal knowledge of certain technicians, who were nevertheless hired for a certain specific job, constituted a source of enrichment that went beyond the purely military stage: specialists in fortifications, called to France, worked as urban planners, like the architect Ridolfo Fioravanti who worked in Russia as a caster of cannons.

The weapons manufacturers, armorers, and casters seem to have traveled very little. These activities, whose progress was slow and almost imperceptible, could very easily be adapted to the traditions of a local labor force that was everywhere in abundant supply. However, we find Louis XI hiring Spanish crossbowmen and inventors of devices for digging moats and trenches. The evolution of artillery also appears to have occurred independently in every country. Few Italian casters were hired for service in France or Spain: the French casters seem to have been clearly superior to those of the rest of Europe.

Such was not the case (especially in the first half of the sixteenth century) for fortifications, in which the Italians were past masters. A large number of Italian engineers went abroad to teach their art to foreigners. The great engineers of François I had such names as Marini, Bellarmato, and Castriotto. All or almost all of the chiefs of the engineers' corps until the end of the sixteenth century were Italians. Not until Jean Errard of Bar-le-Duc do we see the formation of the school that was to give birth to Vauban. Girolamo Marini gradually built up the fortifications along the frontier of Champagne: La Fère, Laon, Soissons, Épernay, Vitry-le-François, Château-Thierry, Troyes, and Joinville. Bellarmato worked at Langres, Dijon, Besançon, Nuits, Vesoul, and especially Le Havre. Another Italian, Donato Buono dei Pellizuoli, constructed, at the command of Charles V (Charles I of Spain), the remarkable fortress of Antwerp, modeled after the fortifications of Verona. Shortly after 1550, Italians also brought the new fortification techniques to Spandau and Düsseldorf, before the growth of the German school (which had already begun with Dürer), which was more rapid than any other school, thanks to Daniel Spekle (1538–1589).

Technology and economy All the European countries sought to reach the level of the most advanced countries, especially northern Italy and, to a certain extent, Germany. Foreign experts were therefore called in to advise not only on industrial secrets but also on artistic programs and sometimes even an administrative organization completely different from the one that had formerly been in existence.

Russia is the best possible example of this policy. In 1472 Ivan III married Princess Sophia (Zoë) Palaeologus, a niece of the last emperor of Byzantium, who had been brought up in Italy. An Italian minter, Gian Battista della Volpe, who had played an important role in this marriage, was already working in Moscow. This union led Ivan III to bring numerous technicians from Italy to modernize his country. One of the first was Ridolfo Fioravanti of Bologna, a man with an encyclopedic brain who was simultaneously an architect, engineer, caster, medallist, and an expert in hydraulics, fortifications, and pyrotechnics. In 1488 another Russian mission was sent to hire architects, goldsmiths, casters, armorers, and another great architect who also worked on the Kremlin – Pietro Antonio Solario of Milan. Other and similar missions were sent to Italy in 1493, 1499, and 1527. After this period, however, the flow of Italians, who were suspect to the Russians for reasons of orthodoxy, ceased. This "invasion" explains to a large extent the resemblance of certain buildings in Moscow to the architecture of northern Italy.

Russia also called on experts from other countries. In 1484, the czar asked the King of Hungary to send him miners; the same request was addressed to the emperor in 1488. In 1550, printers were requested from Denmark. Around the same time a group of Englishmen brought with them artisans and miners. In 1554 a German named Hans Sitte, from Goslar, arrived in Russia with 123 German masters. All this activity is indicative of the continuity and importance of the movement.

We can reverse the problem and seek in the relocations of workers the explanation of the expansion of certain techniques. Such was the case of mining, a technique in which the Germans appear to have held first place. We have just seen that twice within a fairly short period Ivan III called in Hungarian and German miners (the former were probably also of German origin). In 1450 Charles VII made use of a German, Klaus Smerment, to exploit the confiscated mines of Jacques Cœur. In 1452 Henry VI of England delivered safe-conduct passes to Bohemian and Hungarian miners who were coming to settle in England. Louis XI also called upon German printers and glassmakers. It was only natural that it was the Germans who disseminated the art of printing, which they had discovered, throughout Europe.

Other industries experienced the same phenomena. There was, for example, the case of the silk techniques, which for the most part had been perfected by the Italians. For purely mercantile purposes, Louis XI attempted to introduce this industry (especially milling) into France; he sent for Italian workers, whom he installed at Lyon and Tours. This policy became exceptionally widespread. In 1478 we find the Estates of Brittany subsidizing the City of Vitré for the installation of a mill to be operated by Italians; another attempt was made in 1483. Cologne and Marseille followed this example in 1470 and 1474 respectively. Other instances could certainly be discovered.

Simultaneous with these movements organized for a clearly defined purpose, migrations of workmen, which we mentioned in connection with the Middle Ages, naturally continued. It would be very interesting to study them more attentively than has yet been done. We find them in certain industries, such as the metalworking industry; at the beginning of the fifteenth century numerous workers from Germany and Lorraine were working in the forges of Berry. Piedmontese and Styrian workers had been working in Dauphiné since the thirteenth century. It was probably workers from Liège who helped to disseminate the blast furnace and certain methods of refining cast iron throughout western Europe. In England the technical revolution in iron metallurgy appears to have been due to French workers around the beginning of the sixteenth century; a large number of the terms employed in this craft are of French origin. The considerable number of French workers employed in this industry is attested by the letters of naturalization demanded of them in 1544, when Henry VIII was preparing for war against France.

It was these migrations of workers, whether spontaneous or organized, that made the greatest contribution to the expansion of technical progress. This situation persisted until the end of the eighteenth century, and even into the first half of the nineteenth; only very recently have oral traditions ceased to be the customary vehicles of technical progress.

Technical progress and the social milieu

Here again the gaps in our information are considerable, and it is hardly possible to treat even superficially of a subject whose scope can easily be comprehended. Thus we must inevitably limit ourselves to several very general considerations and to the few examples discovered by the historians.

Technical progress is revealed by an increase in investment: it is necessary to renew the material equipment, whose acquisition becomes increasingly burdensome as this progress asserts itself more vigorously. This evolution is in turn revealed by changes in social relationships within the industrial classes. If there was no revolution in the social aspect of agriculture during the two centuries under discussion, it is because there were no major modifications in agricultural techniques. In industry the case was different. These social phenomena occurred, however, on completely different planes. In some industries in which production was of necessity concentrated while a wide distribution was needed, certain forms of large-scale capitalism were born. Such was the case, for example, in mining. The heavy demand for several metals, particularly copper, created by the development of bronze artillery and linked with the technical development of methods of exploitation (drainage, the drilling of deeper shafts, timbering, transportation), resulted in the accumulation of large fortunes whose best representatives are the Fuggers in Germany and the Hochstetters in England. The formation of these large operations in turn permitted a more rapid development of certain techniques utilized in these activities. In more localized industries, other types of organizations were formed to improve already existing techniques or to transform the material used in production. We know, for example, of numerous associations for the construction or exploitation of metal-working factories formed between the beginning of the fifteenth and the middle of the sixteenth centuries. Few of the Dauphiné nobility were sufficiently rich or sufficiently daring to invest single-handedly the enormous sums involved in the construction of drop hammers and complementary instruments; they formed associations to which they admitted rich members of the middle class. Here again, then, technical improvements brought in their wake a new organization of capital and therefore new social formulas.

But this influence of technical progress was not felt only in the domain of capital and a certain middle-class, industrial society: a reaction was also felt in the social hierarchy. The growing complexity of technology gradually led to distinctions between workers, and a new social stratification was henceforth to correspond to a hierarchy of knowledge. This phenomena had certainly been long in existence, but it was now to become increasingly pronounced. A more clear-cut position was now occupied, at the top of the scale, by the great technicians, whose role until then had been ill-defined and in particular poorly integrated into a society that had little place for the technical occupations. The ephemeral age, in the last quarter of the fifteenth century, of the omniscient technicians and the great engineers courted by the various political powers was succeeded by a series of specialists, military engineers, organizers of cities and communicatons, who were gradually integrated into a centralized administration whose technique had been considerably modified. Great "officer" technicians appeared, great master gunners, officers of bridge and road, water and forestry services, naval officials, and many others, who attest to the scope of the technical problems. However, their training was undoubtedly still very empirical, and only haphazard use was made of their abilities. Great schools were hesitantly developed (rather than created), heralding the higher technical instruction whose organization was to require more than two centuries longer; by 1506 Venice had a school of artillery, an example that was soon imitated by Burgos.

No study has been made of the working class in the sixteenth century. Such a study would undoubtedly reveal a hierarchization that can be glimpsed in several professions: the history of the saltworks of Franche-Comté is revealing in this regard. At the beginning of the fifteenth century we find a salt expert visiting the saltworks of a large number of western European countries to perfect his knowledge. Technical progress unquestionably brought about the growth of an unskilled labor force, directed by technicians who were now better educated than their predecessors. This growth of the industrial labor force was in part caused by a demographic expansion evident in the first half of the sixteenth century. Again, then, technical progress and demographic progress appear to go hand in hand.

Social reactions to progress or to certain achievements of technology are not known to us; we find at most a few references whose veracity it is often impossible to verify. In 1529 a certain Anton Muller saw in Danzig a machine that could weave four or five fabrics at a time. The invention of this machine is supposed to have caused such disturbance among the working class that the city magistrates did away with the machine and drowned the inventor. The great printers' revolts in France and in the middle of the sixteenth century were undoubtedly caused by modifications of the printing press that brought about a reduction in the number of workers. Isolated and inconsequential events? Or the first manifestations of an opposition that would be more definitely apparent and wider in scope at the end of the sixteenth century and the beginning of the seventeenth? It is difficult to say with the scanty testimony that has been preserved.

In any case it would be unthinkable that the major technical development occurring during this period should have had no repercussions on the social structures. The evolution of technology must have been manifested, on the social level, by geographical distribution of the population, social hierarchization, and phenomena of social psychology.

Between the middle of the fourteenth and the middle of the sixteenth centuries, technology had acquired an extremely important place, and was tending to become one of the dominant factors of human existence. The philosophical and scientific current, and the social evolution that placed classes of society whose importance was especially economic in nature in the forefront of activity, in a sense caused the birth of the Renaissance. This early Renaissance reveals, in all domains, the importance that was now attached to the problem of technology. Everything about the Renaissance was technical, including its art (geometrical perspective) and its political conceptions, which reveal a logical, practical organization unknown until then. Evidence of this fact ranges from the engineers' notebooks to the activity of a triumphant middle class and the national policies of various countries of western Europe.

Western man was thus beginning to become conscious of technical progress and of its possibilities, which some regarded as infinite. This progress inspired invention and the renewal of both material and spiritual equipment. Invention even seemed to be a new idea, an idea so important that a system of law was to be built around it: the institution of the patent, in still-primitive form, appeared for the first time in Italy, in the second half of the fifteenth century, and rapidly spread to all the countries of western Europe in the first half of the sixteenth century. It is surprising to learn that more patents were granted in Germany in the sixteenth century than in the eighteenth.

It is obvious that not all the projects envisioned in this immense movement, which developed especially in the century between 1450 and 1550, were translated into reality. Much remained in the state of virtuality: Leonardo da Vinci's projects, both adopted and original, are visible proof of this. Not everything could be accomplished, and, in the fact of a progress that was conceived and defined at least in its major outlines, we can see that real achievement continued to be slow until the eighteenth century, as is evidenced by the *Encyclopédie*. We have already noted the extent to which the use of wood as a basic material had retarded the development of mechanization.

We almost have the impression (equally difficult to verify) that on the whole it was the material impossibility of achieving everything that was promised and of maintaining continuous growth, and undoubtedly certain fears of various kinds, which halted this second period of expansion and technical progress around the middle of the century. Economic decadence set in during the second half of the sixteenth century, together with a deterioration in the political and religious situations. There was another demographic decline, and once again Europe endured the desolation of national and civil wars. The rise in prices, following upon the influx of precious metals from America, the effects of which were generally felt throughout Europe after 1550, was an additional cause of the halt in technical progress.

The Middle Ages witnessed a continual progress between the middle of the twelfth and the last third of the thirteenth centuries, coinciding with the demographic growth. The same phenomena occurred between 1450 and 1550, with equally precise points of rise and fall and with the same combination of material progress and increase in population.

BIBLIOGRAPHY

See also the bibliography for medieval technology (Volume I, Part V).

GENERAL WORKS

GILLE, BERTRAND, *Engineers of the Renaissance* (Cambridge, Mass., 1966).

NEF, JOHN U., *The Conquest of the Material World* (Chicago, 1964).

———, *Cultural Foundations of Industrial Civilization* (Cambridge, 1958).

PARSONS, WILLIAM BARCLAY, *Engineers and Engineering in the Renaissance* (Baltimore, 1939; reprinted, Cambridge, Mass., 1968).

ROSSI, PAOLO, *Philosophy, Technology, and the Arts* (New York, 1969).

WOLF, A., *A History of Science, Technology, and Philosophy in the 16th and 17th Centuries* (New York, 1950, paperbound ed., 2 vols., N.Y., 1959).

SOME SPECIALIZED WORKS

DRAKE, STILLMAN, and DRABKIN, I. E. (eds.), *Mechanics in Sixteenth-Century Italy* (Madison, Wis., 1968).

GOODMAN, W. L. *The History of Woodworking Tools* (London, 1964).

HALL, A. R., "The Military Inventions of Guido da Vigevano," *Actes du VIIIe Congrès International d'Histoire des Sciences* (1957), pp. 966–969.

HART, IVOR, *The World of Leonardo da Vinci* (New York, 1962).

HEYDENREICH, LUDWIG H., *Leonardo da Vinci* (2 vols., New York, 1955).

KYESER AUS EICHSTATT, CONRAD, *Bellifortis* [1405], 2 vols., trans. into German and edited by Georg Quarg (Düsseldorf, 1967).

MACCURDY, EDWARD, *The Notesbooks of Leonardo da Vinci* (2 vols., 1938).

RETI, LADISLAO, "The Codex of Juliano Turriano (1500–1585)," *Technology and Culture,* 8 (1967), 53–66.

———, "The Leonardo da Vinci Codices in the Biblioteca Nacional of Madrid," *Technology and Culture,* 8 (1967), 437–445.

———, "Francesco di Giorgio Martini's Treatise on Engineering and Its Plagiarists," *Technology and Culture,* 4 (1963), 287–298.

SPENCER, JOHN R., "Filarete's Description of a Fifteenth Century Italian Iron Smelter at Ferrière," *Technology and Culture,* 4 (1963), 201–206.

THORNDIKE, LYNN, "Marianus Jacobus Taccola," *Archives Internationales d'Histoire des Sciences,* n.s., 24 (1955), 7–26.

PART TWO

THE MAJOR STAGES OF TRANSITION

Section One

The Exploitation of Raw Materials

CHAPTER 9

AGRICULTURAL TECHNIQUES

At the close of the medieval period, during which only minor progress had been made in agricultural techniques, it was to be hoped that the sixteenth century (especially its last quarter) would be a period of renewal for agriculture. The international wars had ended; the civil wars were drawing to a close. The monarchies were being organized, and their first concern was to bind up the wounds suffered by the rural areas. They undoubtedly had other tasks as well; but is not the feeding of people who are returning to a state of peace the best means of reestablishing the political and social order? In France, King Henri IV and his ministers were proclaiming that "rural renewal" was their most urgent task: "plowing and pasturing . . ." in the famous words of Sully. In England, where the disasters of the Wars of the Roses were equally extensive, the desire of the Tudors to restore agricultural prosperity to the country matched the hopes of the French monarchy. Spain, in the midst of rebuilding after the Reconquest and the reign of Charles I, had not yet been dazzled by the gold and silver from America. In Italy, while the cities still presented the spectacle of a power that had long been in the forefront of activities, it was in the active silence of the countryside that the reservoir of a large population was being reestablished. Even in the Balkan Peninsula, the Turkish power established itself by organizing the rural areas. Everywhere the life of the fields was regaining its strength. The moment of liberation from the weight of tradition was perhaps at hand. Advisers for this purpose were available to the farmers and those who could guide them.

Writings on agronomy

The activity of the ruling powers was in fact based on a new series of writings on agronomy. This was a direct consequence of the Renaissance, which reexamined a large number of traditional ideas, attempted to break with the past, and sought out innovations with joyful enthusiasm.

The "return to antiquity" brought about the discovery, or rediscovery, of the authors who had assembled the works regarded by the Greek and Latin authors as the expression of agricultural wisdom. Cato, Varro, Columella, Palladius, and even Theophrastus, as well as Hesiod and Virgil, found enthusiastic readers. Columella was translated into Italian by Pietro Lauro in 1554, into French by Charles Cotereau in 1555; Palladius had already made a verse translation in 1420 (which, it is true, remained in manuscript form). Thanks to this zeal for the past, Pietro de' Crescenzi's *Opus Ruralium Commodorum sive de Agricultura,* with its description of Bolognese agriculture at its best, was discovered.

Still more significant, and richer in possible influences, was the appearance of original writings in Italy, France, and England. The catalog of books on agriculture printed between 1471 and 1840 (published in 1927 by S. Aslin under the auspices of the Rothamsted Experimental Station) gives a list of works for the sixteenth and seventeenth centuries that would be impressive but for the fact that many of them were simply compilations. Some of them nevertheless enjoyed significant success. The success of John Fitzherbert's work (if it is not from the pen of his brother Anthony) was so great that Lord Ernle (Rowland E. Prothero), the historian of English agriculture, was able to say that the *Boke of Husbandry* (1523) "became and remained for more than fifty years a classic work on English agriculture." Hardly less famous was Thomas Tusser's *Hundreth Good Pointes of Husbandrie,* which was published in 1557; the book went through several editions, and as late as the eighteenth century no writer concerned with agriculture failed to quote from it.

A similar movement led to the birth of the Renaissance in Italy and France. After Michelangelo Tonaglio's *De Agricultura* (1490), the ideas contained in Camillo Tarello's *Ricordo de Agricultura* were (1567) discussed throughout Italy and in foreign lands, as well as by the Venetians, for whom he had written it. Nor were the French idle: in his *Recepte véritable par laquelle tous les hommes pourront accroître leurs trésors* (The true guide whereby every man can increase his treasures) (1553), Bernard Palissy demonstrates a completely novel knowledge of the types of soil, which the art of pottery had taught him to observe. The *Praedium Rusticum,* the poorly arranged compilation of Charles Estienne and Liébault, was a more direct result of the return to antiquity; although it was translated into French in 1564, it never had the power of penetration that would have enabled it to inspire new agricultural practices.

There was Olivier de Serres, a Huguenot "gentleman," courageous fighter, and skilled diplomat, who upon his return from the war settled down at his farm at Le Pradel, in the Vivarais region of Switzerland. Here he created an oasis under the black basalt peaks of the Coiron, at the foot of the wild garrigues of Les Gras, through his hard work, application, and love of the land. He was not only a good technician, however; he was also extremely observant, an experimenter who was simultaneously prudent and daring. Being a learned man, he had read the Latin agronomists, knew Tarello, had traveled in France, Italy, and Switzerland. His book, the *Theâtre d'Agriculture ou le Mesnage des Champs* (1600), is a summary of the best methods of cultivation as conceived by the sixteenth century. De Serres tells us: "There are people who ridicule all the books on agriculture and send us to the unlettered peasants, who, they claim, are the only competent judges in this matter, as based on experience, the sole and safe rule for cultivating the fields." De Serres, in contrast, wants the *"ménager"* (husbandman) to keep in mind that working the land is a science that is "more useful than difficult, provided it is understood in its principles, applied with Reason, guided by Experience, and practiced with Diligence."

Did this agronomical zeal have the effects expected by its proponents? The talent of an Olivier de Serres undoubtedly sustained the royal power's activity in favor of agriculture. His magnificent work went through eight editions during his lifetime. It is said that Henri IV had several pages read to him each day; he undoubtedly took great pleasure in this evocation of the life of the fields that is simultaneously direct and poetic, in the manner of the master of Le Pradel, and enjoyed its prose, which by its firmness and suppleness is a forerunner of the beautiful language of the classical period. But those familiar with the *Theâtre d'Agriculture,* like those who were reading the Italian and English agronomists, were not genuine farmers; these continued to be "unlettered peasants." The nobility had just laid down its arms; it took refuge at court, or, if it did return to its castles, led a modest, sometimes even poverty-stricken life. The middle class, which had for more than a century been in the habit of investing money in real estate, was not anxious to court disaster by experimenting with agricultural innovations; it was generally satisfied to have "property in the sun," receive profits from it, and organize its tenant-farming and métayage (sharecropping) systems while hoping to find in its landholdings and the venality of public office an opportunity to be ennobled. Neither the nobility nor the middle class had any ambition to be the yeast in the amorphous dough of the peasant world.

The routine of the farmers

As soon as peace returned, the peasant class turned all its energies to putting the fields back into production and to attending to its most pressing need, which was to feed itself and to produce enough, if possible, to sell the remainder as surplus. In both the rural districts and the cities, years of suffering had accustomed the people to demanding no more than their daily bread. What they wanted, then, was the traditional, tried and true crops, and especially the bread grains. The only renewal possible at the end of the sixteenth century — and for a long time to come — was a renewal of the past.

This becomes very apparent when we note the advice given to the farmers by the most progressive of the writers on agriculture. As late as 1727, Richard Bradley, a professor of botany at Cambridge, declares in his *Complete Body of Husbandry* that the farmers whom he was able to advise concerning improvements in crop cultivation first asked him whether he knew how to drive a plow, "because they believe that the entire secret of a plot of ground resides in this skill." Even Olivier de Serres, who was so attentive to plowing and so concerned with good plowing tools, observes that "those who have invented new plowshares are admired rather than imitated, so revered is the ancient manner of working the land." And he repeats after Cato, "Do not change your plowshare."

The *aratrum* and the *charrue,* each in its own domain, continued to open the ground in the old ways, in some places by flat plowing, elsewhere by plowing so that the soil of a furrow was pushed onto an unplowed adjacent strip, as the agronomist from the Vivarais observed in the Beauce, where the farmers attempted by means of these curved ridges to "drain the rainwater from the sides and the low areas." Plowing tools undoubtedly acquired forms adapted to the local conditions of work for which they were intended, but the method of work was fixed by habit, and it was only the skill of the plowman that could distinguish one field from another field in the same locality.

An attempt was thus made to work the land with the traditional methods established by experience. Olivier de Serres describes what this entailed in France, and what he says is undoubtedly valid for a large part of Europe. Work continued to be adapted to the agricultural systems, and the latter to the climatic characteristics and the qualities of the land. The best lands, remarks de Serres, can support a continuous grain cultivation, using a rotation system of wheat, rye, or mixed wheat and rye, followed by barley, oats, or "other spring grains." This was undoubtedly an extreme case found only in good alluvial lands, where triennial rotation was habitually practiced. Lands of average fertility were left to rest every other year; this was the typical biennial rotation. The poorest lands had a fallowing period of two or three years. Everywhere the stubble was burned, and uncultivated land was plowed as much as was necessary to ensure that "the weeds growing there do not draw from it its substance." Every ten or twelve years the "good husbandmen" turn over the land with "wooden shovels tipped with iron" (the *besse* or *beche, luchet* or *lichet* in Provence and Languedoc); the mattock in its various forms was utilized in stony ground. The spade was widely used; it made possible a "mixture of old and new soils," and the field "was cleaned to perfection." If the farmer was careful to divide the plot of land into ten or twelve equal portions, each being turned over in turn, "the entire domain remains at the highest degree of bounty."

Plowing was first and foremost, and the old, dying plowman advises his readers to "Plow, burrow, dig; leave no spot unturned by human hand." This was the solemnizing of the marriage of man and the earth, regarded as indissoluble, completely similar, in its perenniality, to seemingly eternal nature herself.

MEDITERRANEAN AGRICULTURE

Climatic deficiencies and their consequences

No area of the Western Hemisphere seems to have been more constrained to fidelity to its traditional empiricism that the Mediterranean world. Here agriculture had attained such a high degree of perfection that it had to a certain extent lost its power of renewal. The circumstances of its history, including the Arabic conquests, barbarian pillaging, and even wars were not the only, and undoubtedly not even the most important, causes of its economic decadence. The Mediterranean regions bore within themselves the seeds of their backwardness from the moment that they failed to endow their agricultural techniques with a power that would have made them capable of nourishing as many human beings as the northern regions. Their geographical characteristics destined them to brilliant, but stunted, production.

Their climate had given them fine grains that supplied white flour, the grape for their beverages, and the olive for their oil, lighter and more golden than anything produced by the northern countries — a prestigious trio that had served as a base for the civilizations of antiquity. To this the farmers of the south had gradually added other oil-yielding and sweet fruits — the almond and the carob bean, the apricot and the peach, and the fig, whose variety of appearance and taste had been praised by Olivier de Serres. Fruit, eaten fresh or dried in the sun or on screens, delayed the inevitable exhaustion of the grain harvest, and formed part of the food supply during the period just before the arrival of the new harvest. (In eighteenth-century France, doctors found that the rural inhabitants sometimes ate too much partially ripened fruit.)

The relative scarcity of the food grains — even when supplemented by fava beans, chick-peas, and vetches — was nevertheless a chronic problem. The summer drought limited the varieties of grain almost exclusively to wheat and barley. These were low-yield varieties of wheat (often less than four or five to one), and their cultivation continued to be restricted to biennial rotation because of the lack of a spring grain; this was the "two-part cycle" (Marc Bloch) described by Virgil and the Latin agronomists, evidence of which already appears in the works of Homer and Hesiod. Half of the grain-bearing lands thus lay fallow each year, since wheat could not follow wheat without suffering from foot rot and other diseases. The ancient leases strictly forbade the planting of the same grain two years in succession on a piece of land; it could at most be supplemented by a few heads of grain or vegetables in a corner of a fallow plot.

Immutable practices were associated with this form of cultivation, whose methods dated almost from prehistoric times: plowing with the *aratrum*, harvesting with the sickle, and threshing with the sledge (*tribulum, plaustellum*), the feet of animals, the flail, and sometimes by beating the stalks on a stone. These techniques, which were suited to the low yields, are still found in backward regions of the Near East, North Africa, and even Spain.

It is not surprising, then, that shortages were endemic among the poor and that bad harvests were accompanied by terrible famines. An increase in population, for one reason or another, sufficed to permit terrible ravages by these famines, for example at Naples in the second half of the sixteenth century.

It was perhaps this habitual inadequacy that gave rise to a busy trade between the more productive and the less productive regions; Venice and Genoa, in particular, derived much profit from this activity. From the Danubian and Rumanian plains, the provinces of the Levant, the Greek and Bulgarian basins, Thrace, Egypt, and sometimes even Spain, the grain purchased by small dealers traveled to the ports, where negotiation and speculation then took control of it. This trade gave rise to the practice of laborious sorting (whose techniques have been described by Flückiger) to separate good quality wheat for sale (*grani forti,* in Sicily) from the more mediocre varieties that were either consumed locally or sold at a lower price.

This grain was preserved, following the ancient tradition, in chests, jars, or on the floor of ventilated granaries. The merchants in the ports stored it in genuine warehouses or sometimes turned part of the stock (probably that of poorer quality) into flat cakes. This business, however, was uncertain, and was unable to survive the arrival in the Mediterranean of grain-laden ships from northern Europe, the appearance of which caused, or coincided with, the decline of the grain trade in the Mediterranean area.

The grape, however, was able to continue its advance. It was the crop of the poor soils, of slopes which, though rocky, had good exposure, and which centuries of labor had broken up into terraces behind which stretched narrow fields (*"faïsses* or *"échamps,"* as they were called in the Cévennes and Vivarais regions). An attempt was made to improve the quality of the wines, and the search for good plants had long been a concern of the vine-growing peasants. Complaints were later made about the anarchy of species that sometimes resulted from this patient introduction of new varieties and that the entire Mediterranean basin presented to the curiosity of the growers. It is striking, in any case, that in certain regions the plants that produced abundant but mediocre wine (*gamay, melon,* and so on) were forbidden by regulations as strict as those that protected the local sale of the products.

In addition to the ordinary wines consumed locally, the Mediterranean regions had a quasi-monopoly of the liqueurs. From Greece to Italy, Spain, and Portugal, by way of Provence and Languedoc, each country was proud of its own products and tried to surpass its competitors. The choice of plants and the attention given to wine making were the basis of these products, whose fame resounded throughout the Western world: wines from Cyprus, Marsala, Málaga, malmsey, muscatel, claret, grenache, and so on. Their painstaking preparation remained very localized; it was sometimes combined with special production techniques for which wine or eau-de-vie was the base, as in the case, for example, of those liqueurs whose production, according to Boulainvillers (*État de la France,* 1737), was a secret of Montpellier.

Other wines, which graced the tables of the rich, came from regions of the French Midi that, while they had not yet completely discovered their grape-growing vocation, were nevertheless proud of the reputation of some of their products, as for example the wines of the Rhône region, today known by such characteristic names as Côte-Rôtie, Hermitage, Côtes du Rhône. They were obtained from small plots of land, frequently built up with difficulty on steep slopes, behind dry stone walls. The vine growers remained faithful to venerable varieties of plants and rules for wine making, with each vine grower proclaiming that he applied the best of these rules.

In the vineyards of Bordeaux, the Garonne, and the valleys of the Lot and the Dordogne, the fruit ripened under a more temperate climate. Harvesttime was later, and in some cases the grapes were permitted to overripen in order to obtain naturally sparkling wines (as at Gaillac on the slopes of the Tarn) or sweet wines (at Monbazillac in the Bergeracois). The rights granted to Bordeaux during the period of the English domination hindered the expansion of the vineyards, and not until their abolition in 1776 were the Graves and Médoc wines able to develop.

The Dutch trade, however, which was interested particularly in the white wines, favored the products of the Charente, the Adour and Bigorre, and Armagnac. Even when their alcohol content was low, these wines found knowledgeable buyers who sweetened them or muted them with alcohol. Their producers were led to utilize heavy producing plants like the *picpoul* and the *folle blanche.* Planting was frequently done *en masse,* without the use of vine props. Part of the produce went to the alembic, which especially after the beginning of the seventeenth century was installed in the peasants' wine-making plants.

Fruit trees — pear, peach, fig, and especially plum — were combined with the vines, as well as mulberry trees when sericulture began to progress, and olive trees where the climate was favorable. This trend toward mixed cultivation was fairly general, although in many regions the vines and the olive trees were planted in separate sections. Their combination with the grains created special landscapes in Provence (*"oulières"*), Aquitaine (*"cances"*), and Italy (where they were known by the expressive name of *"coltura promiscua"*). This accumulation of crops on the same fields appears to have been definitely encouraged by the increasing spread of the métayage system, in which the tenant tilled the land in return for a share in the harvest; the tenant's poverty encouraged him to obtain the largest number of harvests possible on the narrow plots on which he earned his bread by the sweat of his brow. It was also an indication of certain technical retardations. For example, in damp areas the grapevine was attacked by diseases for which no cure was known. By planting the vines in widely separated rows the damage was lessened; supporting them on trees (the elm, the maple [*Acer campestre*]) protected them, as the ancient farmers who had practiced cultivation on vine props had already known. The vine shoots hung in festoons from one tree to the next, as is still frequently seen in the plain of the Po and in central Italy; they also clung to the branches of the trees, as in the Portuguese Minho and in corners of the Pyrenean valleys. Another advantage was that the lack of meadows and artificially created pasturelands was compensated for by the gathering of leaves on the supporting trees — genuine "aerial pastures." As has been demonstrated in the case of Italy, in the sixteenth century, with the increase in population and the reclamation of lowlands, *coltura promiscua* began to make great progress. It is an expression, in its own way, of that "narrowness" of the plowlands of which Plato had already complained and which the Italian historian Giuseppe Prezzolini unjustly ascribes to niggardly nature.

How could these disinherited, repressed peasants have dreamed of new techniques? Fidelity to the customs of the past at least gave them the quasi-certainty of being able to exist. In addition, everyone was able to satisfy his few additional needs from his piece of land. The livestock, which was limited to a few sheep, a few goats if there was no cow, and an ass or a mule, sufficed for the needs of the family, the pulling of the *aratrum,* and the limited transportation required (when this was not done by human labor). The animals fed off the scanty woodlands, bare hills, the garrigues, areas of slightly better growth in the calcareous lands, the scrub in the siliceous soils, in addition to the common grazing lands. The old *saltus* (uncultivated area of heath and woodland) was still more extensive than the *ager* (public domain). It was also the domain of the bees, and was sometimes visited by wood gatherers who came to strip the cork oaks and collect fallen branches, oak bark for the tannery, sumac (the Provençal "red"), kermes, wild vegetables and berries, and chestnuts, in the acid soils of the mountainous areas (Corsica, the Cévennes, northwestern Portugal and Spain) in which grain did not grow well. Thus the various forms of food gathering that had been practiced by the hungry peoples of prehistory continued to exist.

When necessary the never distant mountains provided summer pastures for the transhumance of the large collective flocks. In Spain the Mesta (association of owners of flocks) had profited by the devastations of the Reconquest to organize and increase its hold on vast regions covered by the seasonal migrations of its thousands and tens of thousands of merinos, the chief suppliers of wool. By the fourteenth century the association had come under royal protection, and in 1561 its rights became a matter of state law. It possessed pastoral routes, the *cañades reales,* along which agriculture of any kind was practically impossible. This was an extreme case, but even elsewhere, in France, Italy, the Balkan Peninsula, Asian Turkey, and North Africa there were routes for transhumance (called *drailles* in Provence and Languedoc), as well as rules and rights. The struggle between the shepherds and the farmers continued for centuries.

However, the sixteenth century seems to have marked a turning point, if not in the raising of sheep for wool (which now reached its peak), then at least in the breeding of draft animals. The breeding of the horse, an animal for warfare, was everywhere threatened, in Italy as in Spain, in Cyprus as in North Africa: the mule replacing it, and this animal's "victory" over the horse was undeniable, although regarded (in the time of Charles I of Spain) "as a frightful calamity" (Alonso de Herrera, quoted by Fernand Braudel). Plowing, therefore, was long restricted to the scraping action of the *aratrum,* while the stronger horse was conquering the heavy soil of northern Europe.

Peasant individualism and the indifference of the large landowners

This mediocrity of technical equipment was well suited to peasant individualism, which is so common in the southern regions that there is a tendency to consider it as a "racial" trait. The tenants, who were almost everywhere freedmen, regarded themselves as hereditary landowners. The exploitation of the grain-growing areas did not impose upon them — or no longer imposed, with a few exceptions — those traditional collective rules, replete with servitude, that it did in vast areas of northern Europe. Each man cultivated in his own manner a narrow holding composed of bits and pieces of land. Provided that he had a few crops of wheat on the best lands, a corner of vinyard with good exposure, and a few oil-yielding and fruit trees, the southern peasant was content. The division of the land reflected these modest ambitions, which it would have been difficult to increase. The largest farmers possessed holdings whose surface area seems to have been not much larger than could be worked by the immediate family. The *"mas"* (farms) of Provence and Languedoc rarely consisted of more than ten or fifteen hectares, and could support only a "patriarchal" life of the type evoked by Frédéric Mistral in the mid-nineteenth century.

The landscape and the dwelling usually bore witness to this poverty. Almost everywhere the peasants had gathered in squat villages that clung to the hillsides and to isolated hillocks. Fear of pillaging warriors and bands of vagabonds that the armies had turned loose upon the defenseless countryside, as the sea and the rivers leave their foam after their moments of fury, had driven the people into these strongholds. Fernand Braudel has drawn an impressive picture of this banditry, a sequel to the wars, and Lucien Romier has shown the extent and danger of vagrancy in France under Catherine de' Medici. Insofar as it was possible, the people avoided the highways, permitting them to cross the country in hostile indifference.

This form of community imposed by the Middle Ages, the rebirth or maintenance of which had been caused by the wars of the sixteenth century, did not favor the evolution of agricultural techniques. From this point of view, as from many others, the drawing together of the populace was more favorable to the preservation of the old customs than to individual initiative. While it permitted the formation of close bonds, it forced the farmers into the routinism of tradition. This factor alone would have smothered those impulses toward change that differentiate and contrast.

Progressive individuals could have existed among the large landowners; they alone had the material means to liberate agriculture from the rut in which the weight of the past constrained it. There is no evidence that they actually thought of doing so. The best of them were satisfied to manage their lands efficiently, following the customary rules. Good King René, in his castle at Gardane (end of the fifteenth century), demonstrates his attachment to order and economy more often than an inclination toward innovation; his administration remained patriarchal, but his efforts are far from the creation of a "model farm," as is claimed by his historian, Abbot M. Chaillou. Lucien Romier is correct in affirming for France, and particularly for the Midi, that "the laborer made no complaint about his condition because no one dared even to think about means of improving it."

Throughout southern Europe the nobility survived the wars usually impoverished and frequently uprooted, but sometimes, on the contrary, with much land from which it sought only to profit. This was the case in the Turkish Empire of the Balkans, where the seigneurial regimes collapsed through the conquest of the invaders and the pressure of peasant revolts. The Christian *latifundia* (huge rural landholdings) were replaced by Moslem *timars* (military fiefs) which the new seigneurs had much difficulty in defending against the encroachment of the state. Agricultural progress, nonexistent in these areas, perhaps existed in the *tchifliks,* which seem to have been a kind of colonization system, sometimes on still-uncleared lands. Braudel relates this transformation to the development of wheat cultivation, and believes that its equivalent could be found as far as the Danubian countries and even in Poland. The Polish historian S. Malowist has shown that all the Baltic countries participated in this grain expansion, which supplied their foreign trade. But this agricultural progress was based on the labor and sometimes even the servitude of the workers, who were so numerous and so repressed that it was unnecessary to search for new technical methods.

We find the same situation in reunified Spain. Their Catholic Majesties and later Philip II undoubtedly tried to curb the nobility, but did not go so far as to prevent them from acquiring vast estates. The laws of Toro (1505) had sanctioned the institution of the *majorats,* and the great aristocratic patrimonies found their place between the immense holdings of the church and the king; none of them concerned themselves with agricultural progress. Instead of leading to useful investments, the growth of revenues, in correlation with the constant rise in prices throughout the sixteenth century, favored absentee landlordism and a life of idleness in the palaces that dotted the countryside abandoned to its misery, and in the rich houses of the nobility in the cities. The cultivation of wheat, essential on the meseta, and the silk production in Andalusia fed profitable trading operations. Both activities retained their ancient techniques. Sheep raising and the rights of the Mesta supported the maintenance of extensive cultivated areas in which the fallow lands (*barbechos*) occupied as great an area as the lands under cultivation (or an even greater area).

Italy experienced the same triumphs and deficiencies. The nobility proliferated, the *latifundia* became more extensive, but little or nothing was done to make genuine progress in agriculture possible. The population was numerous enough to permit manual cultivation of the land, and this was sufficient.

As for the middle class, in the Mediterranean world it was absorbed in its urban functions of industry and commerce, although it was everywhere acquiring land. Its capital and income did not go to "irrigate" agricultural development, but were used to enlarge holdings and to buy jobs and sometimes titles. This is what Braudel calls the "betrayal" by the middle class. He is speaking, it is true, of social change, which in the sixteenth century was occurring at an "abnormal rate of speed." But the landowning middle class was an even greater traitor to the rural areas over which it daily drew its net more tightly and where it was soon to flaunt its upstart titles of nobility.

Water: miracle and menace

The only way to increase agricultural production in the Mediterranean world would undoubtedly have been, not to extend the centuries-old work of land clearing — all the good land and many less-favored areas had already been conquered — but to harness streams and improve the irrigation systems. They were very rarely neglected wherever peasant labor made it possible to harness some spring or to lead a trickle of water toward a small vegetable plot or piece of meadowland. When we read, in the old land registers, of the extraordinary division of the irrigated lands, we realize the value placed by the cultivators of dry lands on water, the bearer of life. The Arabs were past masters in this art. In isolated places in Asia Minor, large norias created luxuriant areas of green along the Orontes and the Barada. Ingenious *fogarahs* sought out the water hidden deep in the sands and dry alluvions, to cause the flowering of the oases on the edges of the Sahara Desert. Gushing water made the gardens of Grenada and Seville enchanting, and constituted the luxury of the Spanish patios in the houses of the wealthy. In the *vegas* (open plains) of the Iberian Peninsula the conquerors had lengthened the piping systems, and had at least maintained and reinforced (if not created) the systems used in harnessing the water. This activity had had its rules, jurisdiction, and even its tribunals (for example, that of *las aguas* in the *vega* of Valencia) since the early Middle Ages. Thanks to the water distributed by these methods, it was possible to develop cotton, rice, and citrus fruit crops in the coastal provinces of the peninsula.

After the Reconquest, circumstances had either favored or destroyed the extension of irrigated crops in various regions. One Portuguese example was the case of the low valleys of the Sabor and Villarica, tributaries of the Douro, where during floods the dammed-up water deposited fine alluvial soil of exceptional fertility, on which the peasants grew abundant crops. Profiting by the impetus given the ropemaking industry by the maritime enterprises of the fifteenth and sixteenth centuries, they planted hemp crops on these lands, which became and remained famous. Other examples could be mentioned, but on the whole achievements of this type were minor and sporadic. The triumph of the *secano* (dry and unirrigated, though arable, land) over the *vega* and the *huerta* seemed to be the general rule in Spain. We find, at best, the construction at the beginning of the modern period of more water mills, and perhaps fewer windmills, for the crushing of olives and oil-bearing seeds and the grinding of grain into flour. At the same time fulling mills proliferated wherever the cloth industry prospered, as for example in Languedoc.

The damage done by the violence of unharnessed water and the stagnation of slow-moving water was in proportion to the high value of the piped and distributed water. The peasants fled the banks of coleric rivers, but were even more fearful of the pestilence-laden swamps. Low plains that could have become fertile regions remained abandoned. The plain of Macedonia, a portion of the Venetian territory, the Pontine Marshes, the banks of the *étangs* (ponds) of Languedoc and Roussillon, a few *albuferas* (lagoons) in the Spanish Levant, scattered inland bodies of water buried in the hills, remained in a swampy state or returned to that condition as soon as the population ceased to attend to the drainage of the water. Many workers and authoritarian and continuous supervision were needed to cultivate such land. As Braudel remarks, this work could be effectively carried out only if "the rich and the powerful were involved in it." The rich sometimes ruined themselves through this activity, as for example one Claude de Montconis who between 1603 and 1611 drained the Pujaut *étang* in Languedoc, after the failure of Hugues Pelletier de Salon's attempt in 1586; the Carthusians of Villeneuve were ultimately its principal beneficiaries. Elsewhere the large landowners were satisfied to establish large farms for extensive stockbreeding, as did the rich Roman landowners in the Pontine Marshes.

The poor formed labor associations to compensate for their lack of wealth. The inhabitants of Ledenon in the Gard region were the first in Languedoc to carry out the draining of a marsh (between 1592 and 1597). In the Bas-Rhône area, Adam de Craponne, who had already drained the alluvial plains of Fréjus, and was later (1577) to construct the canal across the Crau which bears his name, did not succeed in bringing the communities of Tarascon and Arles to agree to the drainage of their swamps. The major works undertaken were those the Dutchman Van Ems successfully completed in the Arles region, but after his death in 1651 the installations were not maintained. The failure of the projects carried out under the leadership of Humphrey Bradley (1630) to put the Marseillette *étang*, near the valleys of the Aude and the Argendouble rivers, into cultivation suffered the same fate, and the projects had to be begun anew in the eighteenth century. Almost everywhere similar attempts resulted in the same quasi-failures.

The only major works during this period were the Italian projects. At the end of the sixteenth century Venice completed a tremendous effort to increase and improve its agricultural territory. The plain of the Po, whose improvement had begun long before the modern period with the work of the Benedictines and Cistercians in the twelfth century, was finally exploited in the fifteenth and sixteenth century, when the Lombard canals were being completed. Much of this area was given over to rice, perhaps an import from Spain prior to the end of the fifteenth century. This was thanks in large part to the capitalist landlords and to the seasonal contributions of labor supplied by upper Lombardy and the Apennine regions. With its meadows, flax, and rice, the plain became one of the richest areas of Mediterranean Europe — at the price, it must be admitted, of the labor of numerous workers rather than of a search for technical progress. In any event, how would technical progress have been greeted? When in 1566 Tarello suggested a new type of fallowing to the Venetians, he merely justified the system in use in the Brescia area, where flax, millet, wheat, and clover alternated on the irrigated and plowed meadowlands. The combination of cereal crops and forage crops suggested by Tarello contained the secret of the most fertile of agrarian revolutions, but it did not result in a more abundant harvest of grains, a fact that negated his teaching.

Thus, while we are unable to trace a complete picture of the agricultural techniques of the Mediterranean world in use during the sixteenth and seventeenth centuries, our information points to a deficiency in its spirit of innovation. Wherever irrigation was impossible, stability in agricultural practices was the rule. The only recourse open to the peasants of the East, North Africa, and southern Europe was fidelity to the old ways. They continued to drive their *aratrum* over the fallow lands, as Varro and ibn-el-Awan advised them to do. They knew the effects of this "dry farming" long before the existence of the term, even if they did not know its determining reasons.

To shatter the prestige of this tradition, only one resource was open to the Mediterranean countries; an attempt at new, rich specializations. But the time was not yet ripe. The beginning of the modern age permitted them to achieve only a timid simulacrum of this specialization, which was condemned to disappear within a short time.

The minor but brilliant successes of the South

This seemingly rigid system of agriculture did in fact make several efforts to free itself from its traditional techniques. The Mediterranean world, which lay in the path of certain intercontinental trade routes, maintained relations with the countries of northern and eastern Europe, and remained open to influences from the outside world. Urban growth and the expansion of industry (especially the textile industry) gave birth to new needs, some of which could be met only by the products of the sunny south, for example certain dyestuffs — saffron, madder, woad.

The cultivation of woad made progress, particularly in the area of southern France that is closely related to the Mediterranean regions. The initiative appears to have come from the merchants of Toulouse, who were already doing business with Italy and northwestern Spain, as well as with London and Antwerp. These merchants, who were also landowners, may have been responsible for the attempts at woad cultivation on the small métayage farms in the Terrefort Lauragais area. By the fifteenth century this crop was established in the damp areas of valleys, and on the cleared areas of meadows plowed with the spade. The harvest, gathered leaf by leaf, was begun in June and continued until autumn. The preparation of the woad was adapted from the countries that had cultivated it earlier, and bore a sufficient resemblance to the preparation of indigo to justify a belief that it derives from the latter. The leaves, which were crushed in a mill, supplied a pulp that was formed into balls (called *coques or coquaqnes* in the Toulouse area). These *coques* were latter crushed to supply a powder that fermented when mixed with water. The result was a kind of black mastic, designated in the merchants' account books by the name *pastellum agranatum.*

Competing with woad was indigo, which also gave a beautiful blue dye. The Romans had obtained their supply from India. In the fourteenth century the Dutch began to import it in small quantities, and when its use became common (in the eighteenth century) the cultivation of woad disappeared.

Saffron gave a yellow dye, and was utilized for pharmaceutical purposes. Cultivated in ancient times in Asia Minor, and as far as Persia and even Kashmir, it spread to Spain, Italy, France, and even England (Essex and Cambridge counties, before the last Stuarts). The bulbs were planted in ground opened with the spade, and the plant was carefully hoed. The flower was gathered by hand; the stigma were then removed and dried, either in the sun or over hot braziers, as in the Gâtinais region, long famous for the high quality of its saffron.

Madder, which grew wild in Asia Minor and southeastern Asia, was cultivated far beyond its native regions. It had made a place for itself in North Africa and Spain; it was found in Alsace, West Germany, and even in England. Dutch commerce had gained a monopoly of the supply of the beautiful red dye obtained with madder root, and lost it only when the plains of the Vaucluse, which took to cultivating madder in the eighteenth century, became its principal producers.

Silk production was more Mediterranean. Originating in China, it had invaded much of southeastern Asia, then Asia Minor and the Greek regions. The Arabs practiced the breeding of silkworms in Africa, whence it reached Spain and Sicily, thanks to their efforts. From here it became established in Italy, from which country France learned the secret. The first French attempts at sericulture remained "obscure" until the reign of Louis XI, but Charles VIII, François I, Henri II, and Henri IV (especially the latter) all attempted to wrest the silk monopoly from Spain and Italy.

The Italians had not adopted the breeding of silkworms without a certain amount of resistance, but by the end of the Middle Ages they possessed extensive mulberry-tree plantations. These trees were very often planted among the rows of grape plants, between which lay the narrow fields of the *coltura promiscua.* Success had crowned their efforts by the end of the sixteenth century, to the point that in 1602 Jean-Baptiste Le Tellier was able to write that "the Italians have devoted themselves so completely to this work that they are like sponges filled with gold and silver. . . ."

France required a longer period for the establishment of the mulberry. The southern countryside, with its olive, walnut, almond, and fruit trees was already overburdened with trees, and to add mulberry trees might endanger the production of the grains. Royal encouragement, and the writings of Barthélemy Laffemas, Olivier de Serres, Le Tellier, and Nicolas Chevalier were powerless to persuade the rural people to refrain from "resisting such a great good" (Isaac de Laffemas). However, by the end of the sixteenth century Languedoc, after Provence and the Comtat (the modern Vaucluse), was participating in the production of cocoons.

The raising of silkworms, although simple, was in need of improvement. Olivier de Serres criticizes the methods commonly employed to cause the worms to break out, the "seed" being placed "under the armpits or between the breasts of women." But he himself did not know how the cocoonery should be managed in order to be properly heated and ventilated. Not until the eighteenth century was a method discovered to measure the temperature most favorable for the hatching of the worms (86 degrees Fahrenheit).

Once the cocoons had been gathered, the silk had to be unwound. It was the spinning process ("warm water and a woman," in the words of Eugène Rouher) that brought in money. For a long time the unwinding was done in the houses where the silkworms were raised. Not until the nineteenth century (at least in France) did the spinning process become organized in small factories; this was true in almost every country. Travelers writing descriptions of the Asian countries between the sixteenth and eighteenth centuries found only the most primitive methods, including those which made it possible to obtain "with neither water nor fire" the Shantung filoselle spun by *Antheroea Pernyi,* a silkworm raised in the open. The same was probably true among those pioneers of Virginia who planted "an innumerable quantity of mulberry-trees" at the beginning of the seventeenth century (1619), which their neighbors in Georgia and South Carolina sought to imitate: by the end of the century neither breeding places nor spinning devices remained in any of these areas.

Once the silk thread (raw silk) had been obtained, the industry still had only a raw material. In order to weave it, the irregularities had to be removed and several threads had to be twisted together in order to increase their strength: this was the milling operation, which was done with the help of mills operated with a rapid rotating movement. The Chinese had long known of this technique, although some of their fabrics were woven with raw silk, and their methods were used throughout the silk-producing area of Asia in small rural workshops very close to the silk-breeding farms and spinning shops. In Europe, Italy seems to have been the first country to perfect milling. The first silk-twisting mills appear to have been constructed in Bologna in 1272 by a man from Lucca. From here they reached Modena, then Avignon (fifteenth century). Under King Ferdinand, Naples built up a silk industry, but by the time of Charles V (Charles I of Spain), only the spinning operation remained. Around the same time, Lyon and Paris already had silk-spinning factories. Milling next appeared, together with weaving, at Saint-Chamond, then at Lyon, thanks to Italians from Cremona and Florence who arrived in 1539 and 1542, according to the silk historian Natalis Rondot. In the sixteenth century Saint-Chamond became the center of this industry, having progressed beyond the artisan stage. Switzerland was also developing milling factories, as a result of the immigration to Zurich, in 1555, of Protestants from Locarno. The impetus had been supplied. Under Colbert, a spinning mill and a milling factory was established near Condrieu, south of Vienna, and at Aubenas in the Vivarais. Sericulture, now supported by industry, made progress wherever it was possible. The rural districts of southern Europe now found in silkworm breeding, spinning, and milling the supplementary resources they so badly needed.

This search for additional activities and techniques was only a veneer on a background of ancient traditions. Despite its vines, fruit, and silk, the agricultural Mediterranean world remained a world of poverty. From the fellahin of the East and Africa to the *cafoni* of southern Italy and the peasants of Castille and Provence, all these persistent workers remained bent over their hoe or *aratrum*. Gaston Roupnel, although somewhat harsh on the people of southern France, nevertheless admits that their backwardness and routinism were due not so much "to a rudimentary agriculture as to a precarious soil."

NORTHERN AND ATLANTIC EUROPE

Factors in its progress

While Mediterranean Europe was achieving minor progress in the sixteenth and seventeenth centuries, and that in line with its oldest traditions, northern Europe was paving the way for a far-reaching revolution.

The first factor in its agrarian transformation was its abundant population. The regions of the north, which were routes of passage and places of exchange, were populated with such intensity that they far surpassed the regions of the south. Their cities grew through trade and industry, and this urban growth, a picture of which has been drawn for the Middle Ages by Henri Pirenne, was in itself a unique stimulant of agricultural activity. These cities, whose population depended on the rural areas for the satisfaction of their vital needs and their industry of raw materials, were markets that from day to day and year to year required increasingly abundant supplies.

The peasant populace itself was sufficiently dense to meet these needs. What François de Belleforest said of the flat country of the Île-de-France ("abundant in peasants and large towns and villages, a land as pleasant, fertile, and copious as could be desired . . .") could be applied to large areas of northern Europe, at least to its dry plains. When necessary this peasant class enlarged the area under cultivation; it reclaimed new lands from the sea, the mud flats of the rivers, the swamps and peat bogs, dry sands, and forbidding heaths. It made the most unyielding soil fertile, and transformed barren lands into gardens, when they could not be made into fields. This class had long been accustomed to struggling, to constant and painstaking work, to ingenious search for methods most likely to draw from the earth what it was capable of giving — and sometimes even that which it seemed incapable of giving.

Northern Europe had other advantages over the south. Her vast plains were sometimes covered with an alluvial mud whose natural composition was that of grain-growing soil par excellence. Her misty, often cloudy skies, by slowing down the vegetative cycle of the grain, permitted the use of coarser, stronger varieties of higher yield than was possible in the south. In addition to the winter grains, the summer rains permitted the growing of barley, oats, and rye that were sowed in the spring, and buckwheat, planted later in the year. Northern Europe was thus able to have a rhythm of cultivation that was less brilliant but also less dependent on chance than that of the south. Over large areas it practiced a grain rotation system that permitted two harvests per year, with one-third of the land remaining fallow. Thus each summer southern Europe was one harvest behind northern Europe, a deficiency that was overcome only during the nineteenth century.

The northern climate had still other advantages. The frequent and prolonged rainfalls, the heavy dews, and the abundant snowfalls were favorable to grass and woodlands. Everywhere pasturing was easy, and livestock numerous: sheep heavy with wool, pigs that were easily fattened, cows and horses that were large and vigorous. This advantage became more evident when means of better nourishment for this livestock were sought and when artificial pasturelands were created for this purpose.

In the meantime, thanks to the oxen and horses there was no lack of power for pulling heavy loads. While the Mediterranean lands with their mules, asses, and small oxen remained faithful to the *aratrum,* in the north plowing was done with the *charrue,* the only instrument capable of turning over the heavy earth, aerating it, and renewing its nourishing power. The *aratrum* was very soon abandoned, and plowing acquired a new dimension. There was also the spade (the Flemish *louchet*), which was a tool better suited to gardening than to farming, but to which the Flemish farmers nevertheless gave an honored place.

The Flemish example

In the sixteenth century northern Europe began to benefit from the example set by Flemish agriculture, which by then had achieved first rank in Europe.

The transformation of the methods of cultivation in Flanders was achieved "without fanfare, in the midst of general barbarism," in the words of the Count Adrien de Gasparin. It was achieved without preconceived plan or doctrine, and solely through the labor of the inhabitants. Flanders did not even have the advantage of a naturally rich soil. On the contrary, the Flemish soil, which was either sand or clay, lacked the rich alluvial mud of the plains of the Paris basin and a large part of the Hainaut and Brabant regions. Sand covered at least half of Flanders – more than half in the eastern portion. It produced a light soil that was hurt by frost but that warmed quickly; it was generally easy to cultivate, except in places where it had caked near the surface into a kind of tufa (*rokke, rotse*) which the plow could not break. In its natural state this sand, like that of the dunes, is poor in lime and phosphoric acid. It supported woods and moors, and its original sterility is still visible in the savage landscapes that partly cover the territory between Bruges, Ecloo, Ruysselede, and Thourout. Elsewhere, in the maritime plain, a compact clay had been deposited during the last of the sea invasions. Plowing this clay, which stuck to the feet and tools, required powerful teams; it is said that two men could sometimes be seen leaning on the handle of the plow, while a third poured water on the land the plowshare could not penetrate. At least it was of astonishing fertility. All that was required for its exploitation was to dig ditches and canals to lead the stagnant water to the locks, where it flowed into the sea.

The Flemish farmers were rebuffed neither by the lack of fertility of their sandy soil nor by the tenacity of their clay. Even in the Middle Ages they had a reputation for courage in the face of hard work. In areas that it was difficult to plow, the soil was turned over with the spade, like that of a garden; it was said that the spade was the peasant's "gold mine."

It was necessary to add fertilizer to this prepared, mixed, and aerated soil. The supply of manure was insufficient, despite the abundance of livestock thanks to the meadows, which were green all year round. It was supplemented with mud from ditches and canals, debris from industries and private houses, residue from oil presses, and night soil ("Flemish fertilizer") collected even from the cities — techniques related to those of Chinese agriculture, whose application was possible only by the use of hand labor. The astonishing transformation of the Flemish land was accomplished by a veritable debauchery of work.

By the end of the sixteenth century we find fields replacing woods, moors, and swamps almost everywhere, and the land being conquered, worked, and populated, presenting the spectacle of that intensive cultivation which was to astonish travelers and inspire imitators even in modern times.

Next to winter and spring grains — wheat, rye, barley, oats — Flanders utilized buckwheat, and produced more grain for human consumption than any other region of Europe. To these crops it added oil-yielding plants — colza, camelina, poppyseed. It had fields of coarse and fine flax that fed its industries. It practiced large-scale cultivation of fava beans and field beans, vetches and lentils, peas and beans. It bred heavy horses for farm work, and both heavy and small livestock either by pasturing or in the stable. To feed these animals it very soon adopted clover and turnips. This harmonious combination of cattle breeding and farming, and the wide range of cultivated plants, permitted skillful rotation in which the earth was never fallow. With the abundance of its farms and its eternally green fields Flanders offered the appearance of a garden — and in many ways was a garden.

It is not astonishing that the English learned from Flanders the art of cultivating vegetables (with the exception at least of those that could be cultivated in the open fields). It has been claimed that England received onions and cabbages from Flanders at the beginning of the fifteenth century. During the reign of Henry VIII, Queen Catherine ordered varieties of lettuce purchased from the Flemish gardeners, and it was through emulation of the Dutch that the farmers of Suffolk began to cultivate vegetables on a large scale. In 1651 Robert Child declared that in Surrey, where market-garden vegetables prospered, "old men claimed to have known the first gardeners who came . . . to plant cabbages and cauliflowers and to sow rapes, carrots and parsnips, and also early peas." He adds that in certain areas of northern and western England the names for gardening and the hoe, the customary tool of the gardeners, were hardly known. Perhaps we should see in this backwardness a result of the British diet, which was rich in meat, fish, eggs, and dairy products; the same was undoubtedly true of northern and northeastern Europe.

It was from Flanders, too, that the hop came to England, around the end of the fifteenth century. By 1552 its importance was sufficient to warrant legislation by Edward VI on the subject. In 1574 Reginald Scott, in his *Perfite Platforme of a Hoppen Garden,* gave precise instructions on the cultivation, drying, and bundling of hops; no new features were added until around the end of the seventeenth century. Kent was then the English center of hop production, whence it spread to Suffolk, Surrey, and several other counties.

The cultivation of hops was connected with the use of beer, which gradually replaced barley beer in popular consumption. The addition of hops to beer was probably an arrival from eastern Europe; the beverage known as *piana braga* owed its special flavor to hops. In the seventeenth century (between 1656 and 1696) the breweries of Strasbourg began to utilize hops as a general practice, and hop farms appeared in Alsace (until then they are mentioned only in the Boulonnais, Champagne, and Artois regions).

However, the agricultural acquisitions from Flanders were of small import by comparison with the suggestions offered by the Flemish example. Sometimes the Flemings themselves carried the power of their innovations and the courage of their arms to the outside world. Protestants persecuted by the Duke of Alba introduced the clover into the Palatinate, whence it spread to Bavaria and Alsace a century later. It was, however, the ensemble of their agricultural techniques that was to be held up as a model.

The obstacles of the agrarian structure and adherence to routine

Based on what he had seen in the Netherlands, Barnaby Googe insisted in his *Foure Bookes of Husbandry* (1577) on the importance of fertilizer and the value of marling, liming, and the use of wood ash. He recommends the cultivation of the rape, as is done in the Principality of Kleve, not only for its oil or for the feeding of sheep, but also as crude fertilizer. He places a high value on the cultivation of the clover (which he calls "Burgundian Grass"), which was already widespread from the Baltic to the Rhineland, and recommends the cultivation of rapes in the fields. He thus established a formula for a revolutionary agriculture that would have placed the crops planted in rows and the artificially created meadows next to the cereals. Sir Richard Weston also turned to Flanders. After his exile in 1642, he visited the Netherlands and studied the methods of agriculture in use there, especially the use of flax, clover, and rape. On his estates he introduced a new type of crop rotation based on the cultivation of root plants and clover. Weston was no better imitated than Googe was heeded.

Opposition came not only from routinism and the type of national pride that reproached Googe for having written his book "in Germany," but was rooted in the agrarian regime itself. Tenants gained nothing by perfecting their system of cultivation if they lived under the threat of sudden or carefully calculated eviction. In his *Discours of Husbandrie Used in Brabant and Flandres* (1645), Richard Weston recommended that leases be extended for a period up to twenty-one years, as in Flanders, and asked that the landlord be required to pay the tenant the value of the improvements made by the tenant to the farm. Samuel Hartlib also praised the Flemish lease system, and Walter Blith adopted the idea in order that (he said) the proverb current in the County of Berck — that "He who ruins must remain, he who improves must flee" would no longer hold true. However, this was not the principal obstacle to the adoption of new techniques. More important was the system that confined the greatest number of tenants and obliged every one of them to cultivate in regulated sections. Everyone had to sow the same grain in the same year, and put the same section of his lands in fallow at the same time. These communal obligations, to which was added the existence of commons, prevented the enclosure of land. Thus the fields were all open, in what the English call the "open field" system.

This synchronization of sowing and harvesting united the peasants to the point that "the laziness of one was able to paralyze the activity of twenty." The system definitely hindered all individual initiative, which was already difficult to accomplish on holdings that consisted of sometimes numerous but always widely dispersed plots. Lucien Romier's words about "the peasant who is master of a holding, retaining initiative and independence in cultivation," is almost never completely valid for open-field countries. The method for ensuring progress in agricultural techniques would have been, on the contrary, to break this kind of peasant "monolithism" and make every farmer a genuinely free agent.

In the fifteenth and sixteenth centuries England made a by no means negligible attempt in this direction when the authorities tried to develop the enclosure system. Enclosed individual holdings could permit each farmer to benefit from all the fruits of his work and experiment, and each holding thus separated from the open field could be adapted to the capacities of the soil, the climate, and the demands of the market. But achieving this revolution in the structure of the rural districts and their techniques would have necessitated the consent of the majority of the landlords and tenants, and for this the time was not yet ripe.

At the beginning of the Tudor period certain enclosures included portions of the commons reserved for pasturing. Thus livestock was partly or completely excluded from the fields reserved for grain. On other occasions this method was used to close off newly cleared lands reclaimed from the forest, the moor, or the swamp. Finally, lands used for the growing of grain were separated from the open fields and pastures were established on them with a view to raising sheep for wool. These areas for breeding livestock generally belonged to those merchant adventurers who invested capital in and speculated on the land, and expected an income from it, which in a sense made it part of the capitalist system.

These attempts aroused vigorous opposition. The poor lost heavily — sometimes their lands, and frequently their possibilities of employment, since cattle breeding employed fewer people than did agriculture. Such measures caused misery, an increase in vagrancy and begging, and an alarming exodus from the rural areas, especially in the grain-growing districts of the Midlands. The enclosure movement, which was halted in the seventeenth century by the civil war, was not resumed until the eighteenth century, at the time when agricultural improvement became general.

The conquest of new lands

Meanwhile, lacking the power to break the existing agricultural structure, England as well as a large portion of France and northern Europe made an attempt to improve its content. Reclamation of new lands and improvement of the lowlands were the only means of coping with the increased demand in areas of increasing population.

Lands in need of drainage were not lacking along the Atlantic coast, and even along the North Sea. In general they corresponded to gulfs whose submersion dated from the last ("Flandrian") invasion of the seas. These swampy stretches formed pestilence-laden zones in places where it had not been possible to establish crops or at least pasturelands, or even salt marshes. Attempts were made to push back the boundaries of these areas. Along the edges of such swamps, gardens were sometimes squeezed between canals "of slow-moving clear water," as Henry of Navarre wrote upon his arrival one evening at Marans on the Sèvre swamp, during the period when he was waging war in Aunis and Poitou. The remaining territory, which was sometimes wet, sometimes dry, supported a few trees, some coarse grass, and bodies of stagnant water. The work of drainage required was thus considerable and difficult, but it was worth the trouble: the reclaimed land was excellent. In Volume XII of his *Cours d'agriculture* (1788) Abbot Rozier declared, in the article "Swamps," that the "vegetal earth accumulated in the swamps by the perpetual and continually renewed decomposition of animals, plants, insects, and so on," gave a soil that "deserved preference over all others." Between the Loire and the Gironde, the packed mud areas formed a claylike soil (*terre de bri*) whose qualities naturally varied from one swamp to another but which was often of astonishing fertility.

The easiest type of reclamation was reclamation from the sea. Generally it was sufficient to protect the coastal mud judged "ripe" by means of dikes. This construction of dikes continued over several centuries around the edges of the Aiguillon cove, near the island of Noirmoutier, around the bay of Bourgneuf. The victory over the sea was sometimes a precarious one that had to be protected against the sea and also against the water that filtered through the dikes; on the Bouin dikes, Dutch-type windmills were continually pumping out the excess water.

The "wet swamps" were conquered in the same fashion; for a long period the labor that was to put them into cultivation was done especially by monks, with the assistance of teams of peasants. In the drained land, behind powerful dikes, there was a network of canals and improved natural streams. The water collected was poured into the sea by locks.

During the reign of Henri IV, swamp drainage in France increased in scope; it was patterned after the work done by the Dutch in the conquest and improvement of their dikes. In April 1599 Henri IV entrusted Humphrey Bradley, an engineer from Brabant, with the position of "master of dikes and canals." In 1607 Bradley founded an association for the draining of swamps and lakes in France. In accordance with his plans or those of his associates and successors — men with names like Coymans, Van Uffe, de la Planche, Van Ems, Hoeufft, Strada, Coorte — the projects were extended from the swamps of Poitou, Saintonge, and Aunis to every region that had swamplands to be conquered. The seventeenth and eighteenth centuries accomplished tremendous work on the area around the Sèvre, the swamps of Charente, the marshes of the Gironde near Bordeaux, lower Languedoc, and Provence. In the interior of the country, the zeal was equally ardent: the lake of Sarliėves near Clermont-Ferrand in Auvergne, the Bourgoin swamps in Dauphiné, and still others were completely or partially drained; the turn of the marshes of the Comtat Venaissin came later, when the Persian Althen introduced madder to the Vaucluse plains.

Needless to say, these projects were not all equally active, and they frequently had to be abandoned and resumed. Even in modern times their use is continued or extended only through attentive maintenance and the use of large-scale methods. But wherever new lands were opened up for cultivation, colonizers, large landowners, and humble workers established homes.

The techniques of cultivation were not discovered immediately, and varied greatly depending on the nature of the soil, the degree of perfection of the drainage projects, and the management of the property, as has been demonstrated by M. L. Papy for the case of the swamps of the Charente and lower Poitou. On the drained lands of the Sèvre and the Lay, divided into vast *"cabanes"* of 150 hectares and more, large numbers of cows and horses benefited from the eternally green meadows. Wheat, fava beans, and flax occupied plots of land near the villages, and were worked by a large number of tenant farmers and agricultural laborers. In the Poitevin swamp two-thirds of the land was cultivated, the remainder being left for pasture. On richer lands that had been covered by fresh water for a long period, the peasants used spades to combine the topsoil with the land dug up from a depth of three or four feet. In this way they cultivated wheat, flax, and hemp, using ditches to maintain the moist condition of the ground. In the Breton swamp, where the *"métairies"* covered, on the average, only about thirty hectares, cultivation was intensive, and never included fallowing. The workers very often speculated on the natural fertility of the new lands won by drainage. Frequently the soil was quickly exhausted because of the lack of proper care. Sometimes the workers did not even take the trouble to manure; droppings of the heavy livestock, mixed with straw and then molded into flat cakes (*bouzats*) and dried in the sun were used as fuel. Not until the nineteenth century were the swamps, at least in certain areas, put to their most rational agricultural usage: as pastureland for livestock breeding, which had by then become intensive.

England had the same problems. Walter Blith (1649), in the words of Lord Ernle, gives the impression that "the situation of the swamps had become a question of national importance." The Wapping Marsh had already been drained in 1544 by a Dutchman, Cornelius Vanderdelf, and smaller swamps had been brought under cultivation during the reigns of Henry VIII and Elizabeth. However, there remained the Fens, a plain of 700,000 acres covered with water and overhung with a "heavy atmosphere, full of mist and rotting odors." The eight streams that drained into it did not always succeed in carrying their water to the sea, and after rain or at high tide the entire plain became a shallow watery area. Only the islands were inhabited; the monks had transformed them into beautiful cultivated areas with vines, apple trees and other trees.

It was the Flemish example that inspired the conquest of the immense bay known as the Great Level. In 1626 Cornelius Vermuyden began to work on the project with varying degrees of success, although he had the support of the local nobility (which beginning in 1630 formed the Commissioners of Sewers) and the Count of Bedford. In 1652 Vermuyden was able to declare the work finished. This did not prevent the water from occasionally retaking possession of its former domain, either because the dikes broke or because the windmills did not succeed in coping with the floods. Not until the nineteenth century were the swamps completely dried and put into cultivation.

This latter part of the task was not the easiest. The inhabitants preferred "catching pike and plucking geese" to pasturing cows or sheep, and soldiers sometimes had to be sent to oversee them. Basically the resistance of the poor people resulted from their poverty and ignorance, and was related to the resistance encountered by those who wished to enclose lands or put the commons under cultivation. Here, again, this was work for rich men, and it was the rich who seem to have derived the greatest benefit from it.

This attempt to win new lands was, however, of profound significance. It proved that the desire to increase the area under cultivation existed everywhere. Having failed effectively to transform the methods of agriculture, the necessity of gaining more ground was felt. While the great landowners – the nobles, the ecclesiastical authorities, the middle class – accomplished or sought to accomplish the draining of the swamps, the small holders nibbled at the commons, grubbed in corners and forests, and cleared small areas of the moors. Soon these expedients were no longer sufficient. The eighteenth century, enlightened by the science of agronomy, and driven by the necessity of feeding an increasing population, undertook a kind of general revision of agricultural techniques, and paved the way for the most astonishing revolution ever seen in agriculture.

CHAPTER 10

THE EXTRACTION OF CHEMICAL PRODUCTS

THE GENERAL DEVELOPMENT OF INDUSTRIAL CHEMISTRY

THE PERIOD between the end of the sixteenth century and the middle of the eighteenth century was a very important one for the future of industrial chemistry. The methods for the extraction and transformation of the raw materials extracted from the ground or from living beings underwent no major changes, but the traditional practices were perfected, the purpose of these improvements being to create new techniques in order to achieve the most economical possible treatment of larger quantities of materials. Thus the application of chemical techniques gradually lost its "craft" character, and in the course of these two centuries a genuine chemical industry was born.

Factors in this evolution The preparation of coloring matter and dyestuffs, metals, glass, and certain foodstuffs had long served as a basis for the observation of the properties and chemical reactions of elements. In this way the science of chemistry was born. It is noteworthy that until around the middle of the seventeenth century this science contributed nothing to the progress of corresponding industrial techniques. The numerous treatises on chemistry published during this century, for the most part by French and German authors, merely describe the practices then in use by the manufacturers; it is doubtful that their contents could have influenced the techniques of preparation. The authors of the treatises bring these operations down to the level of the laboratory; they suggest no new industrial methods, and at most their writings ensured a certain dissemination of these operations. This is the role played by the writings of "Basil Valentine" (Johann Thölde) (1604), Libavius (Andreas Libau) (1604), Johann Glauber (1658), Tachenius (1668), and Nicolas Lémery (1675), which were to be read and commented upon for more than one hundred years (the sixteenth-century authors, particularly Biringuccio and Agricola, are still frequently quoted).

The principles of the operations of the chemical industry remained almost unchanged. They consisted of calcination either in the open or in a closed area in order to obtain the decomposition of solids (generally oxides, sulfurs, nitrates, and sulfates), dissolutions followed by double decompositions for the preparation of certain salts, decantations, filtrations, and crystallizations. Since the number of known mineral and organic compounds increased only in very small proportions during this period, the products of the chemical industry remained unchanged. However, it must not be thought that the industry itself did not evolve. While the technological language had remained approximately the same for two centuries, the characteristics of this industry had changed completely. The original causes of this evolution may have been economic and political. Until the sixteenth century, all the products (with the exception of the metallurgical products) of what we would today call the mineral chemistry industry were introduced into Europe by the Venetians. The importance of Venetian trade had favored the creation in Italy of an industry for the transformation of the raw materials imported from Asia Minor, which were generally dyestuffs; the Italians refined these products before placing them on the market. This activity had led the Italian manufacturers to undertake the production of certain products such as nitric acid, sal ammoniac (ammonium chloride), borax, spirits of salt (hydrochloric acid), and even alum, either from imported materials or from those available locally.

Naturally, this type of production had gradually spread through all the European countries, and was widely developed, particularly in Germany. Until the end of the seventeenth century the production of alkaline products from a combination of soda and potassium was done chiefly in Spain and Germany, but Italy continued to retain her lead over the other European countries in this field in the sixteenth century. (The situation apparently changed during the second half of the seventeenth century.)

The rise of the maritime powers of England and Holland resulted in the transference of the center of chemical production to these countries, at the period when that industry was declining in Italy. This geographical shift had a salutary effect on the perfecting of production techniques. The chemical industry, which participated in the expansion of industrial production and commercial diffusion, benefited by a new outlook that was tending toward expansion beyond the ages-old routines. During the mid-seventeenth century an attempt was made to modify the old methods with a view to increased production, and to achieve local production of products that had formerly been imported.

This type of industry apparently began to develop in France at the beginning of, or in the first quarter of, the eighteenth century. The production of acids and mineral products spread first into Flanders and Picardy. Factories were opened in Paris and then in several other cities, for example Rouen and Grenoble. The French factory owners adopted the Dutch and English methods, and even became innovators in the preparation of nitric acid.

The improvement of methods

Several scientific ideas clarified by the German authors of this century — for example, the more precise definition of acids, which assigned them a special place outside of the major category of the "salts," and the differentiation between soda, potassium, and ammonia — contributed to progress in this industry. In this way the manufacturers acquired a better knowledge of the products they were preparing and selling, and learned to adapt their techniques in order to obtain better-defined categories of products. The industry was still incapable of producing pure or relatively pure products, as this term is understood in the modern period. However, it was able to supply materials whose chemical and physical properties became relatively standardized and filled the requirements of their various users.

The preparation of pure compounds, for example for laboratories and pharmacies, remained the monopoly of very small enterprises generally created and managed by chemists who left their mark on the history of their science. These men were doctors or pharmacists (chemistry was taught almost exclusively in these faculties) who prepared on demand for their colleagues (who often constituted a large clientele) mineral and organic compounds utilized in the preparation of medicine or in demonstration and research experiments. The discovery of new salts, such as double tartrate of sodium and potassium (by Seignette) and sodium sulfate (by Johann Glauber), were immediately exploited by their discoverers, who jealously concealed the secret of their composition and preparation. Moreover, until the end of the eighteenth century the entire chemical industry was shrouded in secrecy, and this is why it is quite difficult to determine how its transformation was accomplished during the period under consideration.

Other products were less subject to the influence of international trade, for various reasons. The preparation of gunpowder, for example, had become a matter of national necessity during the fourteenth and fifteenth centuries; while saltpeter had to be imported from India, every ruler nevertheless sought to establish factories for its production in his own country. In the sixteenth century the collection and purification of saltpeter began to be practiced in systematic fashion in every country; during the same century the technique of preparing the mixtures was standardized and industrialized. These practices, which were dominated by empirical rules established in rigid fashion for reasons of security, remained almost unchanged during the following centuries.

The preparation of paper pulp, the extraction of oils, waxes, dyestuffs of vegetable origin, and several other products, and the production of glass were carried on in widely dispersed establishments. Some of these which exploited local raw material were established in places favorable to their development by virtue either of the presence of streams to drive the pounders, the proximity of forests supplying fuel for furnaces, regional crops, or climatic conditions. Commercial reasons were also decisive, even if this entailed overcoming some technical difficulties. Despite their dispersion, these establishments were sufficiently numerous to permit the procedures to be transmitted from place to place and standardized during a period of several centuries. They continued to be exploited until the appearance of synthesizing industries and mechanical equipment.

Improvements in equipment

During the eighteenth century, however, the general structure of the factories began to evolve; it was learned how to construct stronger and larger furnaces, and to increase the capacity of the distilling devices, decantation vats, and evaporation surfaces. A rudimentary mechanical device was utilized to effect the flowing and decanting of liquids (formerly done for the most part with large spoons) and the brewing and stirring of baths and mixtures. Crushing mills became more powerful; they were now capable of efficiently operating several batteries of pounders, thanks to cam shafts moved by means of reversing lantern gears. (In the sixteenth century these pounders were still being lifted manually.) Lifting and handling devices were beginning to lessen the labor of the workers in the mirror factories and to make it possible to obtain larger surfaces of better material. Boiling kettles were perfected, and in certain industries drying rooms with constantly regulated temperature were becoming available.

This constant progress in the methods utilized in the chemical factories, which probably began in the second half of the seventeenth century and has continued without interruption into modern times, was the outstanding characteristic of the evolution of chemical techniques until the beginning of the nineteenth century. It appears to have been constantly misunderstood by the historians, who in most cases have concentrated on describing the principle of the operations rather than their execution. For example, a kind of revolution in the chemical industry appears to have occurred somewhat artificially, following the work of Antoine de Lavoisier and the chemists of his generation. In reality, the chemical industry was not influenced by the new theories until fifty years later. The manufacturers, who for centuries had found no inspiration in the various theories propounded, were not to feel the need for recourse to the new principles until another generation had been trained in this school of thought — in fact, until chemical technology had become a subject of instruction and made its effects felt, which did not occur until around 1840–1850. This does not mean that prior thereto industry made no use of the discoveries of chemical science. Thanks to the progress achieved in the preceding century in industrial machinery and equipment, the industry was able to begin commercial production of the elements recently discovered in an extremely short period of time. Such was the case, for example, with chlorine.

MANUFACTURING TECHNIQUES

The Mineral Products Industry

It is customary to regard the utilization of the Leblanc method for the production of soda as the beginning of the large-scale chemical industry. "Beginning in the first quarter of the nineteenth century," writes Paul Baud, one of the first historians of the chemical industry, "the principal soda-works or factories for the production of 'Leblanc soda' tended to present that economic phenomenon of concentration which is today defined as 'vertical concentration or integration' and which involves, under a single management and in workshops located on the production sites of a raw material, in most cases extracted from the earth, a series of mutually derivative productions." There is no doubt that the appearance of the Leblanc method, along with several other events, transformed the nature of the processes of, and gave new impetus to, the chemical industry, but it seems clear that Paul Baud's words could be applied, given the same conditions, to the first quarter of the eighteenth century, substituting establishments to distill acid for the Leblanc soda factories. Speaking of "chemical France around 1806," the same author writes that "the large-scale chemical industry was born with Holker's sulphuric-acid works and Leblanc's soda-works. . . ." Holker, who introduced lead-lined rooms into France, established his sulfuric-acid works in Rouen around

1766. But the production of and uses for this acid were still too limited to permit the new method to bring about changes in the nature of the chemical industry.

In reality it was the preparation of nitric acid in various concentrations that constituted the most important branch of the chemical industry in this period — a situation that was to continue for fifty years.

Nitric acid In the sixteenth century nitric acid was prepared to meet the needs of the assayers, who used it for the separation of gold and silver. It was extracted from saltpeter that had been heat-treated with alum or ferrous sulfate. The treatises of this period indicate that the saltpeter had to be refined. To a mixture of the two elements were added sand, lime, or bits of pottery, and the result was placed in glass phials with heads. These phials were placed in a masonry furnace that was capable of holding two rows of four or six phials; the phials were buried up to their necks in earth and cinders, which distributed the heat and prevented cracking. Heads projected from tubes that led to an equal number of condensing phials arranged on a bench outside the furnace (Figure 1). All joints were carefully sealed. Heating was at first moderate, in order to dry the material; then the temperature was increased every six hours. The nitrous vapors were carried off by the water of crystallization of the salts. When the color of the distilled products indicated that the decomposition was finished, the heat was gradually diminished.

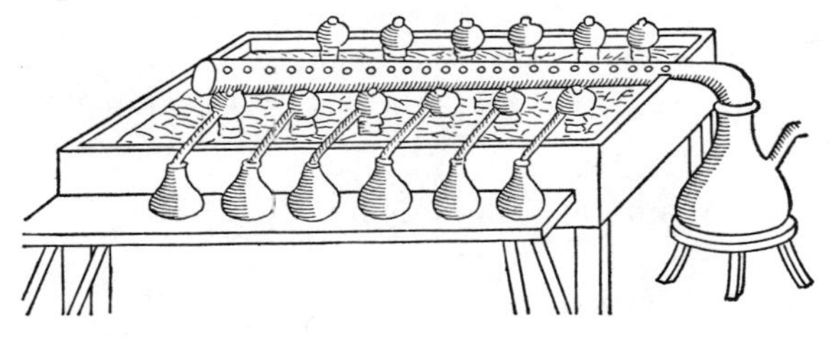

FIG. 1. Furnace for the distillation of nitric acid. Sixteenth century (from Biringuccio).

The ferrous sulfate method again described by Glauber was undoubtedly the only one utilized until around the beginning of the eighteenth century. The English and the Dutch appear to have continued to use it for a long period, but in France the ferrous sulfate was replaced by alumina. One century earlier the manufacturers had felt that the yield from alumina was lower than that obtained with the use of vitriol; slightly later the opposite opinion prevailed, but we are unable to determine what observations led to this change of opinion, except that it definitely did not come about through laboratory experimentation. The treatises on chemistry consisted of formulas collected by the authors from "practitioners" (when the authors were not themselves practitioners). The modification of the traditional method was probably due to the fact that attempts were gradually made to produce a less costly nitric acid in larger quantities.

For centuries nitric acid had been utilized only by the assayers, and for a few industrial purposes. The description of its production is a supplement to metallurgical procedures for the treatment of precious metals. Gradually other industries began to require larger quantities of this substance. In addition to its use by refiners and goldsmiths, it was needed by those who worked certain metals (for example, the casters); tinmen and bookbinders used it for cleaning objects. The dyers used it for dissolving tin shavings to brighten the colors made with cochineal, the furriers, skinners, and hatmakers for removing oil from and polishing the skins and preparing dyes, the engravers as a mordant. It is difficult to say whether the use of nitric acid increased because it was available at a lower price and in more standardized grades, or whether production increased in proportion to the demand. A reciprocal action between supply and demand was probably established during these two centuries, which caused a gradual evolution in the preparation technique and, ultimately, the beginning of the vertical concentration that heralded and paved the way for the structure of the great chemical industry of the nineteenth century.

The use of alumina as an economical raw material was certainly the result of a combination of traditional experience and a great number of experiments. It was known that not all clays were suitable for this purpose, and particularly those that contained pyrites, blue or gray clays from which the potters made portable furnaces. In France clay from Gentilly was used, even by distant provincial establishments, for the preparation of nitric acid.

Refined saltpeter was thus abandoned for large-scale productions. Red saltpeter from the first boiling and mother liquors from the first crystallization were adopted; along with a large proportion of sea salt, they contained nitrates of alkaline earth and magnesium. The mixture of clay and red saltpeter was moistened with this mother liquor; the waters from the distillation of the nitric acid were also reused for this purpose.

The most important changes concerned the equipment of the distilleries; it was these changes that began to give these processes the character of a major chemical industry.

Although the principle of distillation itself was not modified, larger furnaces were constructed; in France they were called crucible furnaces, and their characteristics were clearly defined. The mixture to be distilled was placed in earthenware vessels (*cuines*) equipped with a waste pipe that fitted onto the earthen vats where the nitric acid collected (Figure 2). This material was produced according to fixed standards. Around the middle of the eighteenth century a pottery factory in Savigny near Beauvais had almost a monopoly of this production.

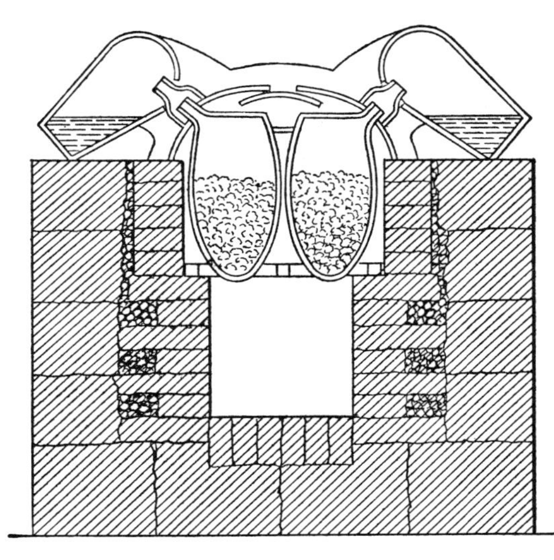

FIG. 2. Cross section of a crucible furnace, *cuines,* and pots for the distillation of nitric acid. Eighteenth century (De Machy, *Art du distillateur d'eaux-fortes*).

By varying the method of performing the distillation and then rectifying the product, three types of nitric acid were prepared, which were differentiated by their concentration and impurities, particularly their hydrochloric acid content.

De Machy, the author (for the *Académie des Sciences*) of *L'art du distillateur d'eaux-fortes* (The Art of the Distiller of Acids), published in 1773, wrote, "The method of distilling nitric acids by means of clay is very modern, and may be a French invention."

The French manufacturers did not completely abandon the vitriol method. Distillation was done on cast-iron furnaces in iron cucurbits surmounted by an earthen head (Figure 3). The nitric acid (which in this case was sold under the name of spirits of niter) was collected in matrasses (glass vessels used in distilling). This was the Dutch method, which was also widely used in northern France.

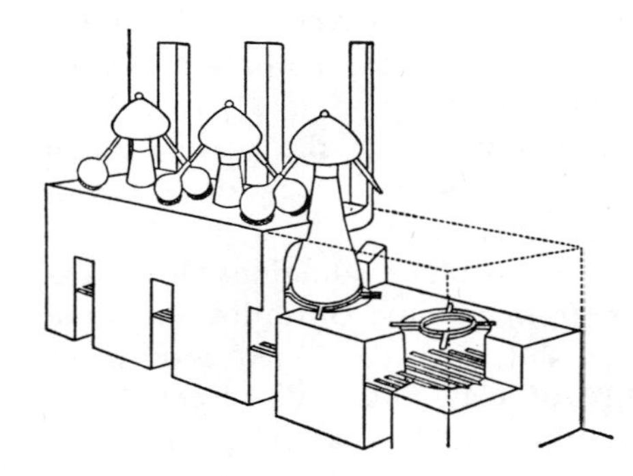

FIG. 3. Furnace and cucurbits for the distillation of nitric acid by the Dutch method (de Machy).

Fuming spirit of niter was prepared in small quantities for limited use, by the action of oil of vitriol (sulfuric acid) on purified saltpeter.

As for the production of nitric acids, around the end of the seventeenth century the manufacturers began gradually to organize a group of complementary preparations that permitted them to utilize the same raw materials or by-products.

Hydrochloric acid

Among the products produced on a large scale was hydrochloric acid. It was obtained by a method similar to the process that produced nitric acids, namely, by treating the mother liquors of the residues from the distillation of nitric acids with clay. This residue was itself sea salt, and its mother liquor was simply a concentrated solution of sea salt and earth chlorides. The manufacturers also purchased from the arsenals the mother liquor of the residues from the purification of saltpeter. Thus the same supplier furnished the raw material for both this product and nitric acid.

The operations were performed in equipment identical to that used for nitric acid, in earthen containers of the same shape, and on crucible furnaces. The various elements of the distillation, collected separately, produced varieties of hydrochloric acid of diminishing concentration. As in the case of nitric acid, hydrochloric acid was also obtained by treating sea salt with either ferrous sulfate or oil of vitriol.

By-products

A number of by-products were treated in the factories, both for sale and use in other preparations.

The calcined clay removal from the *cuines* supplied an excellent material for the making of special cements, obtained by mixing the crude product, crushed, with fresh lime. The pavers utilized this cement for paving courtyards, reservoirs, and all surfaces that had to be watertight. When the colored compound had been washed out of the cement, it was used for the walks in ornamental gardens.

A polishing earth was extracted from the residue of the action of ferrous sulfate on saltpeter. When dried and calcined, this residue, which was rich in ferrous salts, furnished a red substance from which all the soluble salts were removed by washing, so that only the colcothar, naturally mixed with large proportions of foreign bodies, remained. Colcothar was also prepared directly by the calcination of ferrous sulfate.

From the decantation water of this polishing earth, the iron remaining in solution was precipitated by an alkaline solution of potassium, to produce potassium sulfate; Glauber salts (sodium sulfate) were prepared by performing the same operation with a soda solution. As a complement to this series of products, the manufacturers prepared double tartrate of sodium and potassium by treating cream of tartar (potassium tartrate), produced chiefly in Montpellier, or imported from Germany, by a soda solution. All these salts were utilized solely for pharmaceutical purposes, and although they were consumed in fairly large quantities they did not represent large-scale operations for the factories; in addition, other enterprises could easily prepare them cheaply. Thus the potassium sulfate produced in Germany directly from potassium and green vitriol (ferrous sulfate) was sold at such a low price that the distillers of nitric acids abandoned its production. The other alkaline salts and magnesium oxide were also prepared locally by the saltworks of Lorraine, England, and the French maritime regions.

The mother liquor of niter (that is, of the purification of saltpeter) from the first or second boiling, which as we have seen constituted one of the raw materials utilized by the distillers for nitric acid or hydrochloric acid, also furnished magnesium oxide. This compound was extracted from mother liquor diluted with water by precipitation with the help of an alkaline potassium solution.

Saltworks

There is no need to describe the methods of exploitation of the salt marshes, which underwent no special development during the period under discusssion.

Rock salt was extracted from deposits in the Franche-Comté, Poland, and Germany, using the ancient method of preparation of the brine brought up in buckets or by an overshot wheel. In the eighteenth century the traditional raising devices were replaced by the chain pump. Purification was done by means of several successive crystallizations. The methods of keeping the fire going during boiling, regulating the temperature for the crystallization, and decanting the mother liquors grew out of traditional procedures whose perfect execution determined the quality of the products obtained. The salt was left to drain on a ramp, and dried in cakes over a low fire. After the sixteenth century graduated salt pans were also used, and remained in use until the nineteenth century. These were pyramids formed by hurdles made of straw or branches, over which the concentrated solution flowed; in the open air the solution crystallized on this base.

All these treatments reveal a thorough knowledge of salt solutions and an application of the then-unknown laws that govern physical properties. It is remarkable that long before their formulation by the physicists and chemists these laws were well known to the practitioners. Without the experience accumulated by the latter over the centuries, the laboratory workers undoubtedly would never have suspected that they could be developed. The preparation of saltpeter, like that of the various salts mentioned above, required still more complex knowledge, since it involved the phenomenon of, among other things, double decomposition, which was used to excellent effect long before its theory was known.

Saltpeter

The operations for purifying saltpeter were based on the fact that the solubility of potassium nitrate increases with a change in temperature, while that of sodium chloride remains constant; in addition, it utilizes the variation in solubility of a salt in a salt solution, in particular the solubility of sodium chloride in a solution saturated with potassium nitrate. In the sixteenth century purification was done in two successive boilings (dissolutions followed by evaporations and crystallizations). Then the saltpeter from the second boiling was melted in a covered iron or copper receptacle; on top of it was thrown a small quantity of sulfur, which burned with the impurities that came to the surface. It was then left to solidify. The methods of purification varied, but all of them followed the same principles; which became increasingly standardized in the course of the seventh century.

The long wars during the reign of Louis XIV caused a considerable increase in the production of saltpeter in Europe. In France the consumption of saltpeter tripled between 1690 and 1700. This increase in production accelerated the perfecting of its techniques.

For a long time the saltpeter used for the making of gunpowder was imported from India. The niter collected above ground was called saltpeter earth. Because of its price, made still higher by the cost of transporting it, and especially with a view to compensating for the uncertainty of maritime communications in time of war, the local sources of saltpeter were exploited with increasing frequency. In the European countries it was collected by washing all the walls where dampness and the nature of the materials favored the development of nitrifying bacteria (sheepfolds, cellars, damp rooms and caves, old broken bricks and plaster, and so on) with a solution made with cinders of vegetable matter. Not until the eighteenth century, and especially the last third of that century, were artificial niter beds systematically built and maintained.

In France, an ordinance of May 11, 1658, established a statute for the saltpeter workers, which decreed not only general rules for collecting, quality control, and delivery of the products manufactured, but also all the details of production, including even the size of the vats and the characteristics of the various grades of saltpeter, which had to be delivered to the arsenals where the dyers, distillers of nitric acids, glassmakers, and metal casters came to obtain their supplies. All these branches of the chemical industry were thus supplied with raw materials of standard quality, which helped to standardize the methods of production and the equipment of the factories.

The solution containing the saltpeter was given three successive boilings, in the course of which the impurities were skimmed off and the sea salt was eliminated by fractional crystallization. When a certain degree of concentration had been achieved by boiling, the sea salt precipitated, forming what the saltpeter workers called the "grain." This was scraped from the bottom of the boiling kettle and placed in baskets to drain. The solution was further concentrated until the saltpeter, when chilled, was able to crystallize alone. The sea salt that had not precipitated remained in the liquid, since its solubility, unlike that of niter, did not diminish with the change in temperature. The niter crystals were then plunged into boiling water for a second and then a third boiling. At the end of the third the liquid was clarified to eliminate the last remaining impurities before the crystallization.

The saltpeters and the mother liquors from each boiling were used for various purposes. In the making of gunpowder, the saltpeter from the third boiling was further treated by fusion and solidification.

The production of gunpowder consisted of mixing three elements, saltpeter, sulfur, and charcoal, in a series of operations controlled so as to ensure maximum safety during the pounding. The composition of the mixture varied according to the grade of powder produced — cannon, hunting, fireworks, mines. As for powder for military use, there were several sizes, depending on whether it was to be used for cannons, muskets, rifles, or blunderbusses.

The traditional method consisted of first pounding together the sulfur and the saltpeter; charcoal moistened with water was then added. The operation was done by batteries of cylindrical mortars with semicircular bottoms. Above each was a pounder whose head was also semicircular (Figure 4). During the pounding the contents of one mortar were emptied into the next; the mixture thus moved from one end to the other of the battery, depending on its degree of desiccation.

When removed from the mortars, the powder was taken to the granulator, where it was left to dry in *mayes,* or flat containers. It was next passed through a series of sieves in order to separate the various sizes of grains, then through a sifter to eliminate the dust. Polishing was the final step.

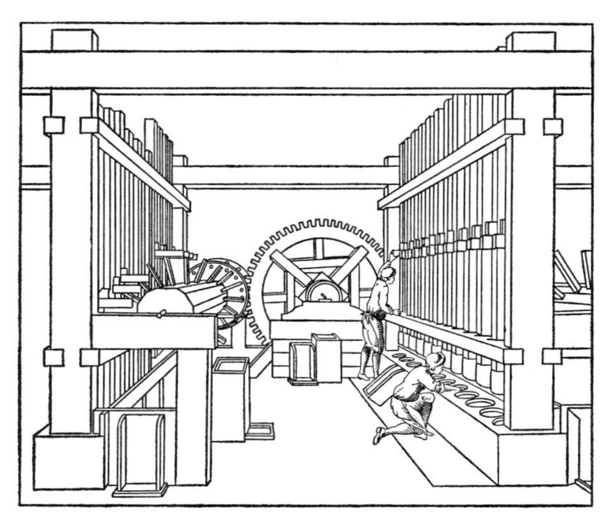

Fig. 4. Pounding mill for the making of gunpowder (illustration from the *Encyclopédie,* Vol. VI).

During the seventeenth century the material and the equipment were improved as production increased. In particular, the batteries of pounders became larger and stronger; moreover, they were all operated by a camshaft driven by a hydraulic wheel. In the sixteenth century most if not all pounders had been raised manually. The progress made in this industry was due essentially to this transformation of the workshops and their more methodical organization. The strict work regulations established may even have had an influence on the organization of the nitric acid distilleries and certain other workshops.

The quality of the powder produced was naturally improved, and various other methods of preparing powders for specific uses were perfected. The mixture for hunting powder was prepared, not with the pounder, but in vats with triturating balls (tin or copper balls) arranged horizontally and turning around their axle. This is perhaps the first example of the rotating cylinders later put to such extensive use by the chemical industry, for example around 1860–1880 for the making of artificial soda. The still-damp powder was then run twice through a mill that contained two vertical millstones operated by a horse-driven treadmill. The finishing process consisted of several granulations, polishing, and drying.

For use in mines, a powder of round pellets was prepared. The powdery material was placed in horizontal cylinders in which a certain humidity was maintained (Figure 5). The powder, crushed by the rotation, collected around the drops of water in perfectly round pellets. Looking at the drawings that depict the details of the production of the material used, we realize that the inventers and constructors of this material possessed a very thorough knowledge of their subject, and an art of solving problems that is much closer to nineteenth-century techniques than to the empirical experiments of their seventeenth-century predecessors.

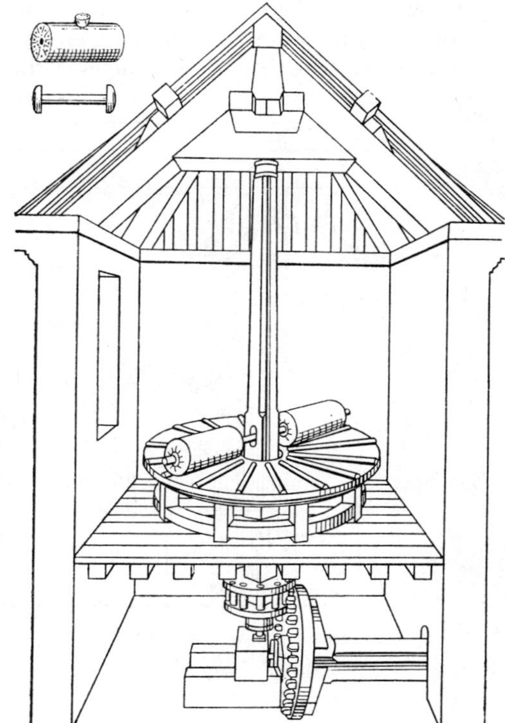

FIG. 5. Machine for making powder into round pellets. Top, left: detail of the mounting of the cylinders. (Illustration from the *Encyclopédie,* Vol. VI.)

Sulfur and sulfuric acid Although it had been in existence since the fourteenth or fifteenth century, sulfuric acid (oil of vitriol, spirit or acid of vitriol) was not produced industrially before the second half of the eighteenth century; for centuries its use was confined to the laboratory or the pharmacy. It was very expensive, and for all practical purposes it was not until bleaching products became known after the discovery of chlorine by Carl W. Scheele and the work of Louis Berthollet that it was needed on a large scale. Its production was then systematically perfected and developed. Previously, sulfuric acid had been obtained by two methods: the combustion of sulfur and the calcination of sulfates.

Sulfur existed over a wide area in natural state, and for a long time no attempt was made to extract it from its compounds. During the seventeenth century, however, an attempt was begun in certain countries located far from the sources of natural sulfur, and first of all in Saxony and Bohemia, to extract this element from pyrites. It was not done in France until the beginning of the nineteenth century, when Dartigue invented a satisfactory furnace.

Until then two quite rudimentary methods, which gave very low yields, were used. The raw material was abundant, however, and contained high proportions of sulfur. The first method consisted of collecting the sulfur by distillation. The pyrites (ferrous sulfide) were heaped on a bed of faggots and logs that was set on fire. The shed under which this operation was done had an opening that deflected the steam into a container filled with water, in which the sulfur condensed. The second method consisted of collecting the liquid sulfur. The pyrites were arranged in a compact heap on top of the fuel, and the melted sulfur accumulated in holes left in the pile of fuel. A worker collected this sulfur from the holes with the help of a long-handled spoon.

It is difficult to determine at what moment sulfuric acid became known, and also the method first used in its preparation. It seems certain that at the end of the sixteenth century both methods (combustion of the sulfur in a damp atmosphere, and pyrogenation of natural sulfides) were utilized but that the chemists did not know that oil or spirit of sulfur and oil or spirit of vitriol were one and the same element. It is true that the complexity of the elements obtained in the making of sulfuric acid is such that the lack of precise ideas on the part of the technicians of the sixteenth and seventeenth centuries is easily understandable.

The combustion of the sulfur was done under large glass bells. Naturally, a large proportion of sulfurous gas was produced that escaped into the air, as well as a very small quantity of sulfur trioxide that combined with the water from the damp air to produce sulfuric acid. The liquid dripped along the edges of the bell, and was collected in a flat basin underneath; the outside of the glass bell was covered with a layer of clay, and was suspended over (without touching) the basin, whose diameter was slightly larger. The sulfur to be burned was placed in a cupel in the center of the basin. Combustion was continuous thanks to a steady addition of sulfur. This method, in its rudimentary form, is the one that led to the modern production techniques of sulfuric acid (Figure 6). Not until 1740–1750 did an English manufacturer, Joshua Ward, introduce the first improvement.

The pyrogenation of vitriols was a longer and more expensive process. Green

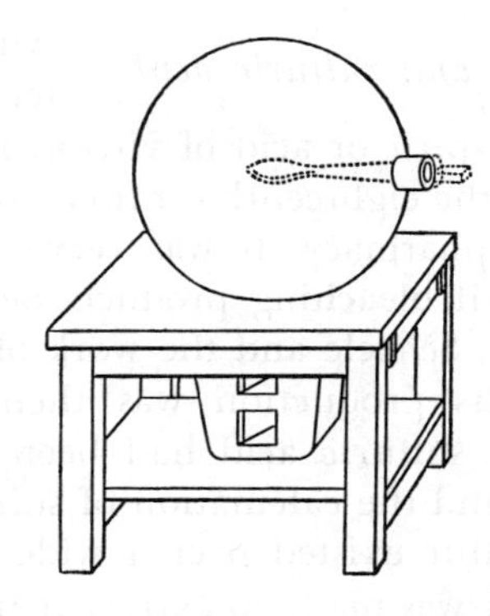

Fig. 6. Combustion of sulfur in a glass balloon in preparing sulfuric acid. This represents a transitional stage, during the second half of the eighteenth century, between the bell method and the use of lead chambers (De Machy, *Art du distillateur d'eaux-fortes*).

or blue vitriol (that is, ferrous or copper sulfate) was calcined after desiccation in a glass retort placed in an oven. The neck projected horizontally from the furnace, and was luted onto the neck of a large receptacle containing water. The fire was maintained for two or three days. In this operation only a small amount of sulfur trioxide was formed. The product dissolved in the water was distilled to obtain a more concentrated acid. Sherwood Taylor has estimated that a week's work was required to obtain two pounds of sulfuric acid. But this method made it possible to prepare concentrated acid and even oleum, that is, acid retaining dissolved anhydride. During the eighteenth century a factory in Nordhausen in the Harz Mountains specialized in this production, and until modern times the name "Nordhausen acid" designated a special type of sulfuric acid.

Alum

Alum (double sulfate of aluminum and potassium) had been used in dyeing since ancient times. Until the end of the Middle Ages it was the object of a very important trade between the countries of the Levant and Europe. When the Turks conquered the lands in which it was found in natural state, alum became scarce in Europe. Exploitation of the deposits of alunite discovered on volcanic lands in Italy was therefore begun. The alunite contained an excess of hydrated alumina which careful calcination transformed into anhydrous insoluble alumina without decomposing the double sulfate. Water was added to the residues to transform them into a paste that was then transformed into a solution, and decanted; the decantation was concentrated and left to crystallize (Figure 7). This industry spread first to Holland and then to Flanders; during the seventeenth century it developed much as the nitric acid industry had done; that is, the equipment was perfected and conceived in a more rational manner.

During the same period in Holland, and later in England, a method for treating schist in order to transform it into alum was discovered. The schist was calcined on a roaring wood fire; the ferrous pyrites in the schist were oxydized into ferrous sulfate. The fire was built still higher so that the decomposition of this element brought about the conversion of the alumina silicate into sulfate. The sulfate of alumina was then extracted from the mass by leaching, and this solution was treated with seaweed ash or the ashes of the wood that had been used for the calcination. Fermented urine, at that time the principal source of ammonia salts, was also used. The product obtained was either a potassium or an ammonium alum, or a mixture of both.

Fig. 7. Extraction of alum. Top: calcination of the alum, and the decantation basins. On the left are the buildings where the concentration and crystallization are performed. Bottom: interior view of one of the buildings shown above (illustration from the *Encyclopédie,* Vol. VIII).

Needless to say, the theory of these transformations was completely unknown; only the alkaline character common to the potassium and ammonia solutions had guided the choice of the technicians. The yield from these methods was less than 1 percent. However, the raw materials were cheap and in abundant supply.

The mineral salts

Similar methods were used for preparing or purifying various mineral salts utilized chiefly in dyeing. We shall discuss only green and blue copperas (that is, ferrous and copper sulfate). We have already noted the use of ferrous sulfate in various operations that have already been described. For example, the grilling of pyrites for the preparation of sulfur left a portion of the oxidized ore in the state of a sulfate. The pyrites oxidized naturally in the open air in the deposits. Leaching, concentration, and crystallization were the classical operations of the industrial chemistry of this period. As in all the other, similar industries, during the eighteenth century the equipment acquired a new character that permitted increased production at less cost of more satisfactory products (Figure 8).

Copper sulfate was purified in the same manner. In the seventeenth century a certain number of mineral compounds useful in dyeing began to be produced, at first in Holland. One example is borax, which until then had been supplied almost exclusively by Italy, and tin chloride, obtained by dissolving the metal in acid, after Cornelius Drebbel had discovered the use of this compound as a mordant for the brilliant scarlet dye of cochineal. Other products were the mercury compounds, vermilion and corrosive sublimate, and lead compounds, min-

FIG. 8. Leaching of pyrites and crystallization of ferrous sulfate (illustration from the *Encyclopédie,* Vol. VI).

ium, massicot (yellow lead), white lead, litharge, and lead acetate (sugar of lead), used by dyers, boilermakers, goldsmiths, and numerous other trades. The method of producing white lead by attacking lead with vapors from the fermentation of fresh manure (called the "Dutch method" and utilized until the beginning of the twentieth century) dates from this period (Figure 9).

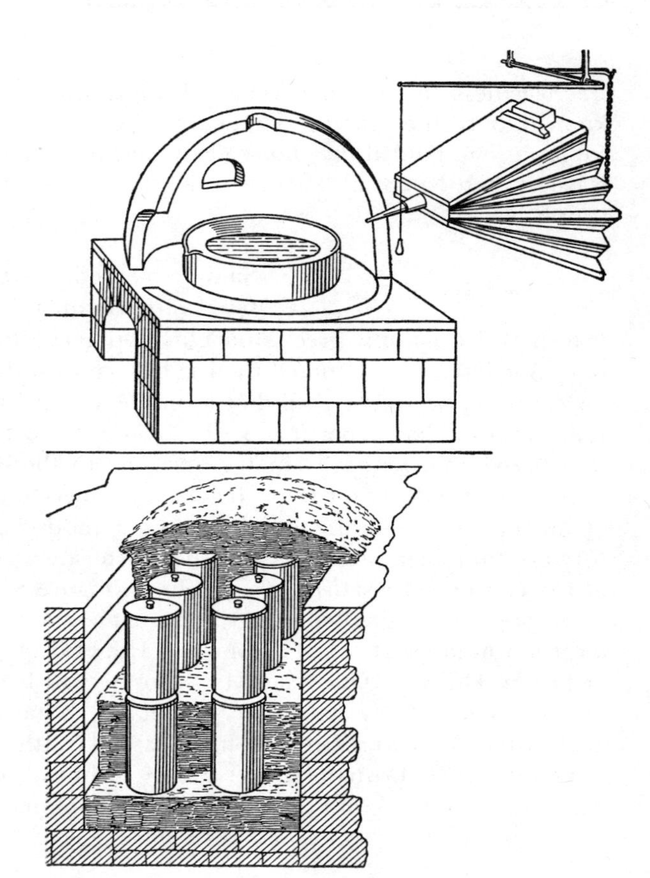

FIG. 9. Production of litharge (top) and white lead (bottom). The methods remained practically unchanged until the beginning of the twentieth century (De Machy, *Art du distillateur d'eaux-fortes*).

Regarding the copper salts, verdigris was produced in the Montpellier area by attacking copper with vapors from fermenting grapeskins. The Dutch in turn found a method of making it directly from acetic acid distilled from vinegar; crystallization was done on small rods split crosswise.

Soda ash and potassium The alkaline products were sold in the form of carbonates, generally obtained through open-air combustion of plants known to be rich in organic salts of these bases. What was actually obtained were mixtures of carbonates and various other salts, such as sulfates and chlorides, from the two alkaline bases, mixtures in which the soda or potassium salts predominated, depending on the origin of the raw materials. The composition of these products remained almost unknown until around the end of the seventeenth century, when the chemists established a distinction between soda and potassium; long before this, however, manufacturers had recognized the differences in properties between the sodas obtained from various products, and had utilized these sodas for specific purposes. The exact nature of soda and potassium in raw state became known in the early years of the nineteenth century, thanks to Sir Humphry Davy's electrochemical work, but the manufacturers did not benefit from this discovery at the time.

The extraction of vegetable soda had probably been in existence since the Gallo-Roman period or the beginning of the Middle Ages, and became increasingly important during the fifteenth and sixteenth centuries when the making of glass, soap, and dyes became more extensive. Pliny's famous natron was a sesquicarbonate that had perhaps sufficed in ancient times for the production of the first vitreous materials; now, however, there was a need for sources closer to hand. In the sixteenth century Biringuccio tells us that the glassmaking industry utilized cinders "made from the grass called Chali," which in France grew along the Rhone and in the diocese of Maguelonne. Various varieties of the salsola soda shrub were cultivated for this purpose in the Mediterranean regions.

In Spain the cultivation of barilla became very important during the seventeenth century. Until the nineteenth century the soda extracted from its ashes was famous under the name of "Alicante soda."

Seaweed ash was also an abundant source of soda; this plant continued to be burned until the beginning of the twentieth century, but by the middle of the nineteenth century its ash was no longer an industrial product.

It was very quickly learned that calcination of the sediment that forms in fermented liquid furnished an alkaline product of excellent quality that was much in demand, especially by dyers and glassmakers. The potassium tartrate decomposed during calcination to produce a carbonate mixed with caustic potassium. The properties of this potassium carbonate were similar to those of the "sodas" imported from western Europe, Germany, Poland (especially "Savary des Bruslons" soda from Danzig), and even Russia.

Mother liquors from the salt marshes or from the treatment of rock salt were also a very abundant source of alkaline salts.

Volatile alkali was identified during the seventeenth century, as was volatile salt (ammonium chloride), but did not become an object of industrial production until the eighteenth century. Until then, volatile salt was imported from Egypt.

The belief was that it was extracted from the soot left by the calcination of animal remains; the leached soot had to be treated with sea salt. After evaporation the ammonium chloride (volatile salt) was separated and purified by sublimation. The first attempts to produce this product were made in England; it was introduced into France undoubtedly around 1760. The process consisted of distilling woolen rags or animal substances in metal retorts of a special shape (Figure 10). This distillation produced a syrupy oil and watery ammonia solution. This mixture was poured into a saline mother liquor that had first been concentrated in the presence of lime in order to fix the hydrochloric acid. The volatile salt was then extracted by crystallization and sublimation.

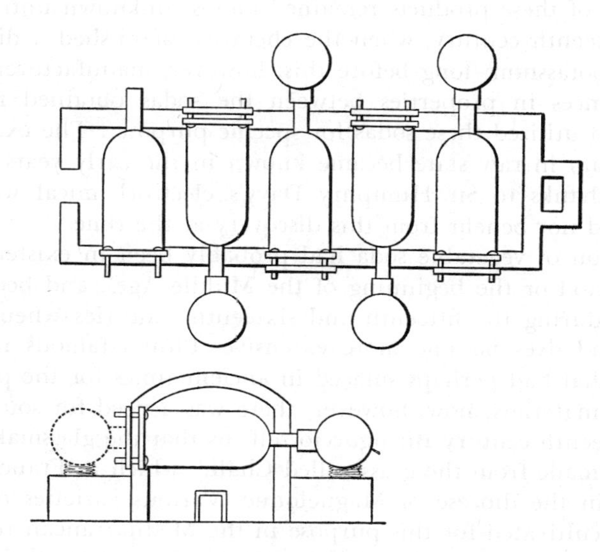

FIG. 10. Furnace and battery of metal retorts for the preparation of ammonium chloride. Top: diagram. Bottom: cross section of the equipment. At the sides can be seen the spherical balloons in which the ammonia was collected (De Machy).

Minor metallurgy As in earlier centuries, lead, tin, mercury, and the precious metals were the object of an industry whose techniques had evolved very little. From Agricola to Diderot, technological innovations in this domain were very few in number. For example, the reproduction of an illustration in the *De re metallica* sufficed for a description in the *Encyclopédie* of the volatilization of mercury from cinnabar. The basic achievements accomplished prior to the sixteenth century, and described in the technological treatises of that period, remained in use at least until the beginning of the nineteenth century.

The only major innovation that can be mentioned is the introduction into Europe of metallic zinc, imported from China during the seventeenth century. European preparation of this metal did not begin until the eighteenth century, and did not become important until a century later.

Beginning in the seventeenth century, tin, which had been known since ancient times, was utilized in Germany to make tinplate, and this practice quickly spread to other countries. The sheet metal was prepared locally with pig iron in bars, which were first refined. The iron was flattened by hammering manually or with a drop hammer. (Iron rolling did not come into practice until the very first years of the eighteenth century.) The sheets of iron were then dipped into vats containing a hot decoction of fermented rye. The plating was done by soaking the sheets in a bath of molten tin to which a small quantity of red copper had been added. At the last minute a small quantity of tallow was added, probably to reduce the tin oxide formed by the heat on the surface of the bath and the sheets, and to make the plating more homogeneous (Figure 11).

FIG. 11. Shop for the production of tinplate. Right: cleaning and scraping (in the vats) of the sheets of metal. Left: dipping the sheets into the tin (illustration from the *Encyclopédie,* Vol. VI).

Heavy metallurgy

Heavy metallurgy included the preparation of copper and bronze on the one hand, and that of cast iron, steel, and iron on the other. The methods employed in both cases were based on traditional knowledge that progressed very slowly until around the end of the seventeenth century. At this time, however, a certain evolution began, marked at first by attempts to substitute coal for charcoal, and the appearance of blast furnaces and industrial methods for making steel. This evolution continued steadily, and its rhythm became much more rapid beginning around the middle of the nineteenth century.

Glass

The production of glass and mirrors was one of the most important "heavy" industries of this period. Production methods had remained practically unchanged for centuries; the composition of the material was determined by empirical rules governed by the quality of the glass prepared and the purpose for which it was intended. The techniques were also regulated by a long tradition. The ovens, equipment, phases in the production of the objects, and the various workers were all designated by a very specialized professional vocabulary perhaps equaled only by the vocabulary of, for example, the printing trade.

The raw materials consisted of silicates, alkali compounds, chiefly sodium and potassium carbonates, and metal oxides, especially lead oxide for the making of crystal. The soda and silicate were first melted together to form the frit, which was then reheated; the proper metallic oxides required were added later. The various phases of the heating, patterned after the methods of the Venetian glassmakers, were popularized at the beginning of the seventeenth century. During this century England and Germany both made great efforts to develop their glass industries. This industry consumed enormous quantities of wood, and it was during this period that coal began to be used to heat the ovens; until the end of the eighteenth century wood nevertheless remained the most frequently used fuel. In France the term "wood glass" was used to designate the objects produced in factories that utilized this fuel. The glassworks that utilized coal, during the second half of the eighteenth century, turned out the same articles. In this period crystal in particular was prepared in coal-heated furnaces. In addition, the quality of the crystal was dependent on the choice of materials and especially on the preliminary purification of the sand used; potassium replaced soda; lead oxide replaced manganese oxide.

During this period, in glassmaking as in the other industries already discussed, we find that progress in production techniques particularly involved the equipment of the factories, and especially the size and arrangement of the furnaces. A comparison of Figure 12 of this chapter with Figure 23 in Chapter Five, Part One, of this volume (which depicts a fifteenth-century glassmaker's oven)

FIG. 12. Interior of a hall for the manufacturing of small glass objects. Wood was the fuel used. Middle of the eighteenth century (illustration from the *Encyclopédie,* Vol. X).

shows the progress accomplished in this area in less than three hundred years. The method of varying the composition of the glass had undoubtedly been known for a century or two, as well as the results that could be obtained from each mixture. But the quality of the articles made was improved only by greater control over working practices; in this way the factories of England and Germany were able to improve the production of crystal during the seventeenth century and successfully compete with the Venetian products. To a certain extent it was this same strictness of organization that made possible the establishment of the great French factories during the reign of Louis XIV.

These enterprises were supported by letters of patent that protected them from domestic and foreign competition. But it seems that in addition the French manufacturers are to be credited with the introduction of the technique of plate glass into the industry.

Until the closing years of the seventeenth century, glass objects, crystal, and mirrors were made only by blowing. This technique, which has so often been described and which undoubtedly had changed little between the Renaissance period and the appearance of mechanization in this industry, is well known. Flat surfaces could be obtained in two ways. The material collected with the blowing iron was blown and rolled in a special manner to form a long cylinder, the bottom of which was detached; the cylinder was split lengthwise and opened out on a flat table. The other method consisted in shaping a circular "plate" at the end of the blowing iron; the diameter of this "plate" was increased by rapid rotation (Figure 13). All these operations were performed with extremely skillful handling on the part of the master-worker (called the *gentilhomme*), interrupted by frequent, adroit reheating in front of special ovens. (For further details, see the chapters entitled "Construction Techniques" and "Techniques of the Decorative Arts."

Mirrors

The production of plate glass is linked with the founding of the famous Saint-Gobain factory. In 1665 Colbert, who wished to create a mirror industry in France, granted a license to Nicolas du Noyer, who established a factory at Tourlaville near Cherbourg. One of his partners, a man named Poquelin, did a wholesale business in Venetian mirrors. He hired several Venetian workers, who brought with them to France the blowing techniques they had learned by tradition. Around 1688 Abraham Thevart in turn obtained a license to make plate glass, "which until then was unheard of in Europe. These mirrors must have been poured in the manner of lead which the lead-workers reduced to sheets, and this new invention made it possible to make mirrors double the size and volume of those blown in the Venetian manner" (Savary des Bruslons).

The new factory was established at first in Paris, in the area called "La Grenouillère," at some distance from the Palais Mazarin (on the site of the present Orsay station). But since the cost of development was too high in the capital, the enterprise was moved to the former Saint-Gobain castle, near Laon, which had the advantage of being close to the La Fère and Oise forest. In contrast to what had been planned, however, the two establishments competed with each other. In 1695 Louis XIV decreed their fusion into a single "Compagnie

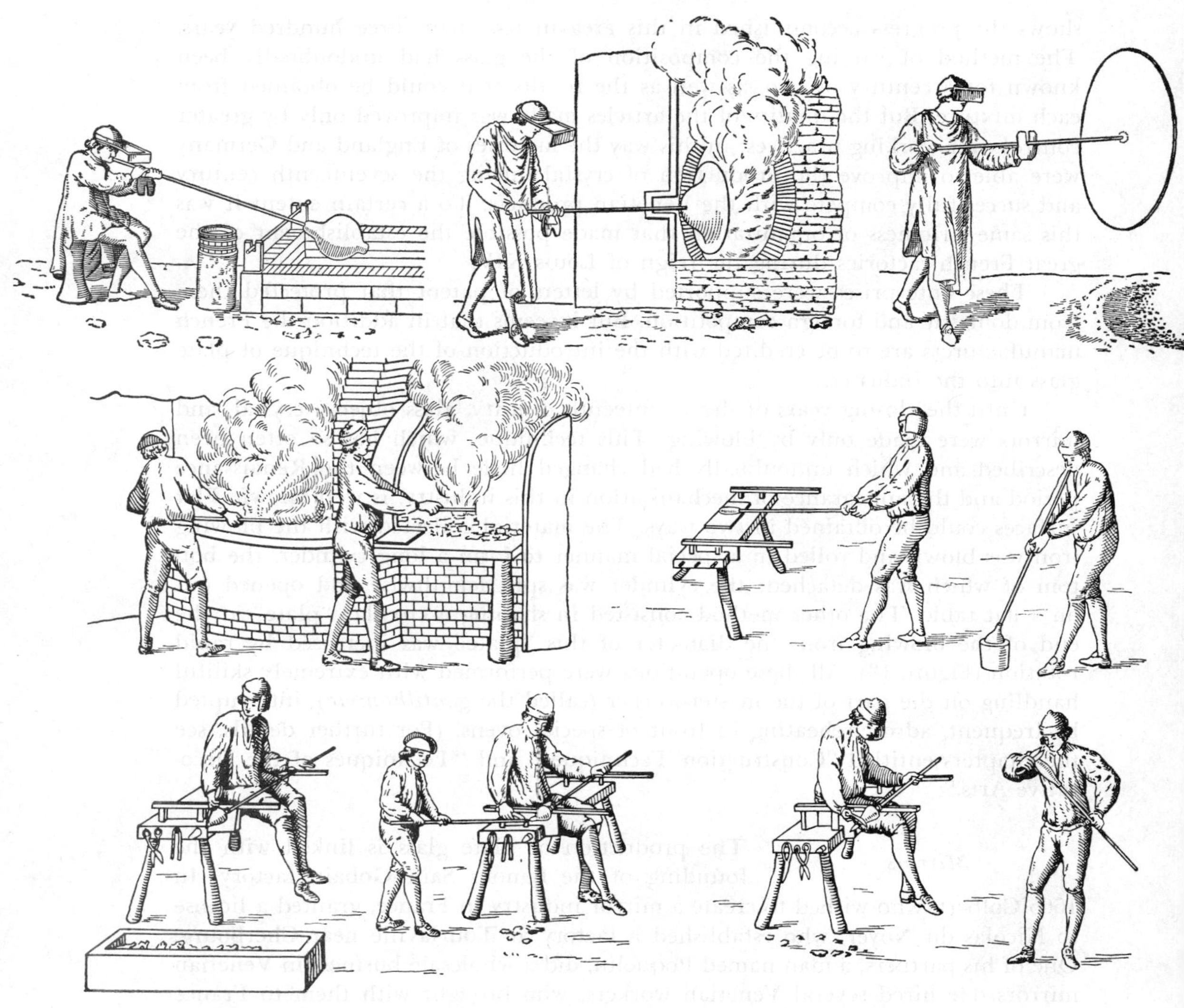

FIG. 13. Various phases of making a glass plate and a drinking glass (illustrations from the *Encyclopédie,* Vols. XV and X respectively).

des Glaces" formed by François Plastrier. The new company had numerous financial difficulties at first, but it is still thriving today. To a great extent it was the activity of this company that brought about the improvement of working methods and equipment in France, through which mirror making gradually acquired the character of a heavy industry.

Blown mirrors continued to be made at Roulaville, but two halls for this method were installed at Saint-Gobain. Each hall included ovens for the preparation of frit, the melting of the glass, and the heating of pieces of broken glass that were added to the frit before the melting, and lastly the ovens for reheating the mirrors. The handling of the melting pots was done manually with the help of special hooks; these pots were about three feet wide and 36 inches high (Figure 14).

The production of plate glass made use of the same raw materials, but in larger quantities. The sand utilized was specially extracted from a pit near Creil. Sand and soda were both carefully purified and transformed into a frit. The oven (*arche*) in which the melting was done had to be surrounded with twenty-four reheating furnaces (called *carcaises*). The ovens for the preparation of frits and the calcination of the pieces of broken glass were also larger. The *arche* contained a melting pot that was capable of holding approximately 260

FIG. 14. Workers removing a pot of fused glass from the *arche* (illustration from the *Encyclopédie,* Vol. X).

quarts, and small basins into which the molten material was placed for refining and in which the melting was repeated three times. The small basins were removed from the *arche* by a team of workers. They were placed on a wagon, and rolled over to a cast-iron (later bronze) table. The material was skimmed; then the basin was tipped so that the material poured out onto the table. A roller regulated to give the desired thickness of glass (Figure 15) was passed over the material. The still-soft glass was pushed into the *carcaise* for a last heating, followed by a slow cooling.

Some of the workers who performed these operations specialized in specific tasks. Because of the weight of the materials, which were handled while red-hot, this work was extremely difficult. Easier methods for most operations were soon sought by equipping the halls with devices for lifting and carrying, which facilitated removal of the basins from the ovens and the pouring of the material onto the table. The table itself was equipped with wheels on rails so that it could be easily moved in front of the battery of *carcaises*. Here we see one of the first (if not the first) examples of heavy-industry equipment for assembly-line firing and removal of components. In this spirit the operations performed in the

FIG. 15. Shop for mirror casting. The workers tip the small basin and pass the roller over the glass (illustration from the *Encyclopédie,* Vol. IV).

glass- and crystal-making enterprises were perfected in the course of the eighteenth century.

EXTRACTION OF VEGETABLE AND ANIMAL PRODUCTS

It was perhaps through the utilization of certain products extracted from vegetable and animal matter that the chemical industry had made its first progress in the early days of human civilization. Plants supplied coloring matter and detergents, fermented beverages, and oil; animals furnished fats, skins, and furs that had to be treated. In the case of the textile fibers, until the advent of the twentieth century the chemical industry was involved only in the dyeing and preparation of the yarns or fabrics.

On the whole, the origin of the processes by means of which these products were extracted is completely unknown. Production began to increase during the Middle Ages, under curcumstances that are unknown to us. The first printed treatises of the Renaissance make no reference to these subjects. Not until around the end of the sixteenth century do relevant technical texts begin to appear. We then note the existence of a genuine industry whose methods have been acquired by experience and perfected in proportion to their transmission, orally and by example, through the ages. This industry was widely dispersed, and was to remain so at least until the end of the eighteenth century. Thus it is difficult to discover exact traces of earlier achievements. Even the equipment and manual techniques appear to have remained fairly stable, in contrast to the situation prevailing in the treatment of materials extracted from the earth.

Empirical knowledge was extensive, however, and it was from this source that the chemists of the seventeenth and eighteenth centuries gained their scientific knowledge (the identification of certain organic acids and the first esters, and the study of fermentations, to mention only a few examples). This empirical knowledge was also utilized by the chemists of the nineteenth century to begin the great adventure of organic chemistry.

However, progress was probably made in the course of the seventeenth century, and here as in other fields it was undoubtedly made in equipment and installations. The same factors mentioned above (and chiefly the increase in production) probably were the cause of this evolution. Thus, by the first half of the eighteenth century we find establishments capable of handling large quantities of material under the best possible conditions.

Dyestuffs

The importance of the production and sale of dyestuffs from the very beginning of civilization is well known. Many of these products were mineral in nature; in particular, the products utilized for the removal of oil from wool, for preparation, and for mordanting formed the principal basis of this industry. Most of them, however, were extracted from vegetable matter. Until the end of the Middle Ages, and even until the beginning of the sixteenth century, there was no way of obtaining brightly colored cloth; the only products available produced subdued colors like crimson, dark red, rose, indigo, and of course black, which for a long time continued to be the color most commonly used for fabrics destined to be made into garments. Ultramarine blue and vermilion were unknown. Sherwood Taylor notes that it is difficult to form a precise idea of the colors then obtained. Depictions of costumes in paintings and miniatures do not reproduce the same colors, for artists utilized mineral pigments that had long been in existence but could not be used in dyeing.

A genuine scarlet was not obtained until the use of tin chloride as a mordant for cochineal was popularized at the beginning of the eighteenth century by Cornelius Drebbel. From Holland the use of tin chloride spread, first to England, then to the other European countries. The development of the dye industry, or rather its transformation into a genuine industry in the modern sense of that word, is linked with the spread of the silk industry and, later (beginning around the middle of the eighteenth century), with the increasing use of cotton cloth.

Vegetable dye products and animal products such as cochineal, which were utilized in the period under discussion, had all been known to earlier centuries. They were very numerous, and sometimes products of varying origin and nature were designated by the same name.

The most widely utilized products included woad, extracted from the *Isatis tinctoria L.*, for blue. Kermes was formed by an insect of the Coccidae family gathered from the kermes oak; much discussion has been devoted to the question of whether it was a worm (whence the name "vermilion," also applied to mineral pigments) or a seed; the word "grain" was also frequently used for this color, while the name "kermes" was of Eastern origin. The dye extracted from the madder root contained alizarin, whose synthesis in the nineteenth century was to bring about the decline of this crop. Archil (*oricello*) and yellow-weed (weld), which replaced dyer's greenweed, gave a yellow dye; spurge flax and a matter extracted from the bark or root of the walnut tree supplied a fawn color. Galena and nutgall gave the dark brown color that was still so common. Use of the sunflower became more extensive during this period.

All these products were extracted from plants that had been cultivated in the European countries at least since the Middle Ages. Their area of cultivation developed according to the historical periods and the fluctuations of the dye (that is, the cloth-manufacturing) industry. Thus the extraction of dyes was a very active industry in the seventeenth century in Holland, Flanders, Artois, and Picardy. At the end of the seventeenth century and until the beginning of the nineteenth, the cultivation of woad, sunflowers, and madder, among others, was developed in the south of France. A great number of other products were imported into Europe from the Mediterranean countries, the Levant, and India, including indigo (whose consumption increased in Europe after the second half of the seventeenth century), brazilwood, saffron from the Indies, the tumeric from the Antilles, cochineal from the Spanish West Indies (often confused with the kermes of the Arabic countries and sumac from Portugal), and others.

The Gobelin brothers The dye industry, which was widely dispersed, was naturally linked with the textile industry. Its establishment depended on numerous general circumstances, most of which, however, were local. In Paris, dyers settling in the thirteenth century in the Faubourg Saint-Marcel (or Saint-Marceau), along the Bièvre stream, which was then outside the city walls, were able to evade certain guild rules. In 1450 the brothers Jehan and Gilles Gobelin founded an establishment in this suburb that was to become famous. The brothers imported the formula of a famous scarlet dye that was prepared with kermes or cochineal, woad, and a mushroom called agaric; this mixture was boiled with alum and tin salts. The Gobelins constructed buildings (known to the seventeenth century under the name "Folie des Gobelins") which in 1667 became the Hôtel royal des Gobelins, while the Bièvre itself acquired the title of "the Gobelins' stream." As early as the beginning of the seventeenth century, Henri IV lodged Flemish tapestry makers at the Gobelins' establishment; by an edict of 1667 Colbert restricted the Hôtel des Gobelins to workshops for the painters, tapestry makers, sculptors, and cabinetmakers whom he brought to France.

Colbert took a great interest in the reorganization and control of the textile industry, and naturally also established rules regulating the dyers. In 1669 a statute was established for master dyers in permanent and temporary dyes. At the same time Colbert published instructions on the manner of dyeing woolen fabrics, and hanks of silk thread, wool, and yarn. These instructions are extremely detailed technological compendiums. Exact rules must be followed in dyeing each grade of fabric or yarn, depending on the use for which it is intended. It would be impossible to attempt to analyze them here; we shall merely mention one example taken from the instructions for dyeing expensive fabrics black. Savary tells us that the best, darkest woad must be used:

"The good quality of this woad results from the fact that it is made with only six pounds of fully prepared indigo for each ball of woad, when the vat is low, that is, when the woad is beginning to give off its blue flower. . . . Next it must be boiled with alum, tartar of bitartrate of potassium, and after that maddered with ordinary madder . . . and finally completed in black with nutgall . . . , cuperas and sumac toned down, passing it over the yellowweed to give it the perfection of black. So that the black will be completely fixed . . . the fabrics must be well scoured in the fulling mill before being placed in the woad. And after being put through the woad, they must be fulled by foot in water, then put through the madder; after they have turned black, they must be thoroughly washed until no color comes from them."

In these instructions we find the essence of the dyers' methods in the second half of the seventeenth century, as they were to remain for more than a century and a half. In particular, we note that the terminology appears vague to the reader who was not a member of this profession in this period — whence the considerable difficulty experienced when we try to determine exactly the nature of the products utilized by the dyers.

Dyeing methods

Several examples will permit us to give an idea of the manner in which these operations were performed. The so-called "Gobelins dyeing" consisted of dyeing fabrics piece by piece. (The name originated from the place where it was done, namely, in the workshops located on the banks of "the Gobelins' stream.") These shops were equipped with large boiling vats heated by furnaces installed on the floor beneath. Using long poles, the workers pushed the pieces of fabric, which unrolled behind them, into the vats. The fabric was first passed through a cleaning bath. Then it was spread over a kind of large crank-operated reel that was placed over the dye vat and permitted the fabric to be easily dipped and removed (Figure 16). After dyeing, the fabric was washed in the stream, whose banks were lined with masonry platforms, and which flowed under the building. The surface of the fabric then had to be brushed smooth with the help of a special brush (Figure 17), and put to dry on a frame so that it would remain taut.

FIG. 16. A Gobelins dyeing shop. Dipping the fabric in the dye (illustration from the *Encyclopédie,* Vol. X).

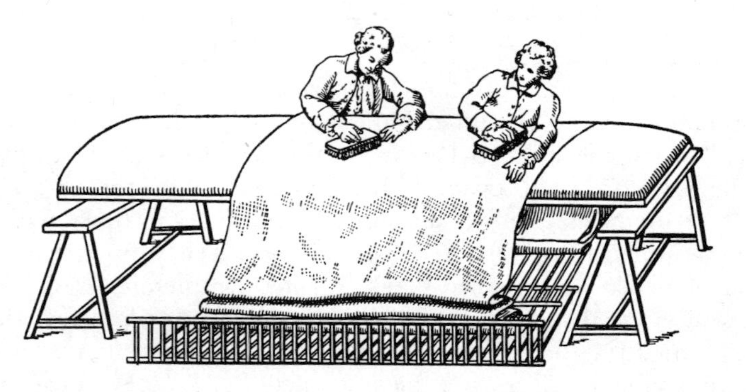

Fig. 17. Brushing the fabric after dyeing (illustration from the *Encyclopédie,* Vol. X).

Fig. 18. Shop for dyeing hanks of yarn (illustration from the *Encyclopédie,* Vol. X).

The dyeing of hanks of silk or wool yarn was called "stream dyeing." The hanks were hung on rods in the cauldrons, and were soaked in the dye baths by being turned on the rods (Figure 18). For certain operations the hanks were enclosed in cloth bags that were plunged into the bath. The hanks were washed in the stream, on pontoons; then they were hung on racks in the drying room, where a sheet-metal furnace maintained a constant temperature.

The extraction of colors With the development of the dyers' art the industry of the extraction of dyestuffs underwent an evolution similar to the one we have already mentioned in other contexts. In this particular case the local resources were increased by the increasing cultivation of the plants involved. The workshops of this industry were naturally located on the site of cultivation. Its basic procedure consisted of crushing the parts of the plant that contained the dye — the leaves, for woad and sumac; the flower, for the sunflower; the roots, for madder; as well as the bark of the walnut tree, and lastly nutgall for galena.

The leaves were dried under a shed, which protected them from the rain and sun. They were then crushed and, if necessary, the oily residue was removed; the crushed material was made into a paste. This operation was performed in a mill that generally consisted of two vertical millstones turned by a horse (Figure 19). Woad was also made into a paste, and kneaded with the feet and hands until perfectly homogeneous. The paste, made into bars, was left to dry for two weeks. The crust formed on the bars was broken, then carefully blended with the mass inside, the result being a homogeneous material of properly dry consistency.

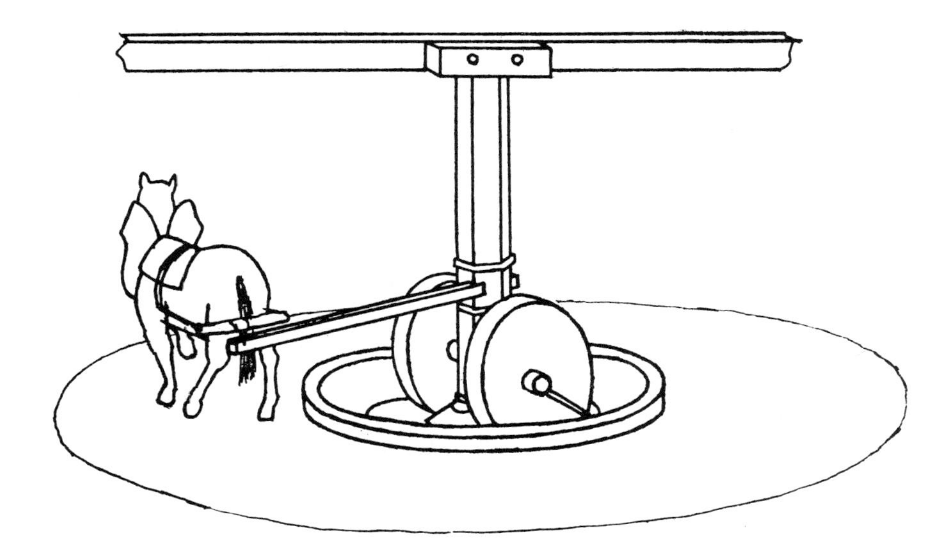

FIG. 19. Mill for crushing the dried leaves of plants that furnish dye (from Beauvais-Raseau, *L'art de l'indigotier*).

Sunflowers were crushed in oil mills and placed in bags so that their juice could be pressed out by a press. The extract was exposed to the sun for about an hour; then rags were soaked in it until they began to be impregnated. The rags were dried in the open air, and exposed to the ammonia vapors of urine to which quicklime had been added. This gave it color. The operation was repeated once. A large portion of this product was purchased by the Dutch manufacturers, who knew how to transform it into paste and bars ready for sale.

The indigo factories

The most characteristic factor in the dye industry at this period was the considerable increase in the production and use of indigo. This product had been in use for centuries, and was imported from the East, probably from India. The product was also known under the name "blue inde," which, however, does not appear to have designated exactly the same variety. As late as the beginning of the seventeenth century, inde was undoubtedly extracted only from the leaves of the anil (indigo plant), while indigo was extracted from both the stems and the leaves. A decree of Henri IV forbade the use of inde and indigo by the dyers, under pain of death. The establishment of the French in Santo Domingo resulted in an increase in anil cultivation and the extraction of indigo, which at the beginning of the eighteenth century acquired an excellent reputation in the European countries. But until 1730 the use of indigo as a dye in its own right was forbidden in France; it was permitted only as a supplement to woad. Its use became still more general after the Englishman Barth discovered (in 1744) the method of heating indigo with Nordhausen fuming sulfuric acid to produce indigo sulfonic acid. This was used as a wool dye, under the name "Saxe blue."

By the second half of the eighteenth century the equipment of the indigo factories of Santo Domingo had achieved a degree of perfection characteristic of the spirit of organization that during this period conquered every branch of the chemical industry (Figure 20). Basically the factory consisted of three masonry or wooden vats arranged in such a way that each vat poured into the next. The first was the "soaking vat," in which the plant was placed to macerate and ferment. The quality of the water was carefully controlled; only stream water or water from certain wells, clean and clear, could be used. The material was transferred into the second vat, called the "beater" because it was topped with mechanical beaters (*buquets*) operated by a horse-driven mill or a hydraulic wheel. Beating was done with a small quantity of fish oil to cause the foam to fall. The liquid changed from green to dark blue; it was decanted, by means of conduits with large pipes, into the third ("resting") vat. The bottom of this vat had a special shape so that the starchy grains fell into a small side basin (*bassinot, diablotin*), from which they were removed and placed in bags hung on the wall of the "resting vat" in order to drain.

When reduced to a paste, this was dried in small boxes placed on benches under a shed. The last operation consisted of letting the starch ferment in a large barrel for three weeks, and then drying it completely in the open air.

What is particularly noteworthy in these production techniques is the arrangement of the equipment, which permitted the workers to perform their work with the greatest economy of motion and handling. Even before the end of the eighteenth century the factory owners had already perfected the methods of exploitation thanks to which their nineteenth-century successors were to build large enterprises.

Oils and soaps

We find the same characteristics in the other branches of the industry of raw materials, particularly the branch that dealt with oils and fats and their by-products. If we can rely on the treatises on chemistry and descriptions of methods, these techniques appear to have remained unchanged from the Middle Ages until Eugène Chevreul, French chemist, did research on animal fats and soaps during his long career (1786–1889). Actually, a slow but certain evolution took place that gave rise to equipment better adapted to the handling of the raw materials and to the development of products of superior quality.

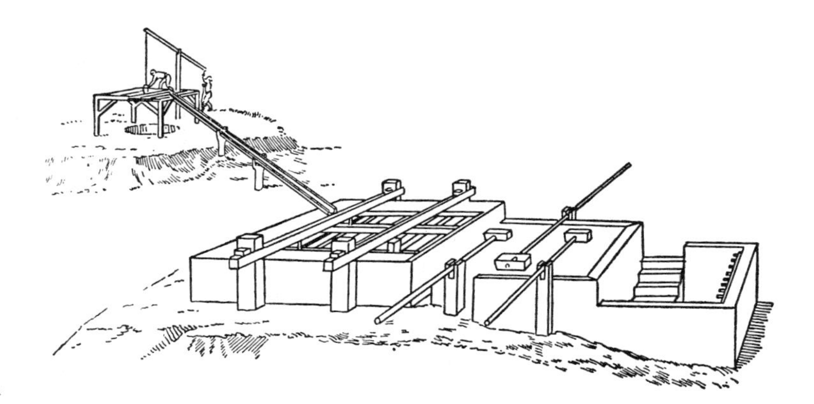

FIG. 20. Indigo factory on Santo Domingo. Below: Plan and elevation of the three vats—the "soaking" vat, the "beater," the "resting vat." The *bassinot* appears at the bottom of the third vat (from Beauvais-Raseau, *L'art de l'indigotier*).

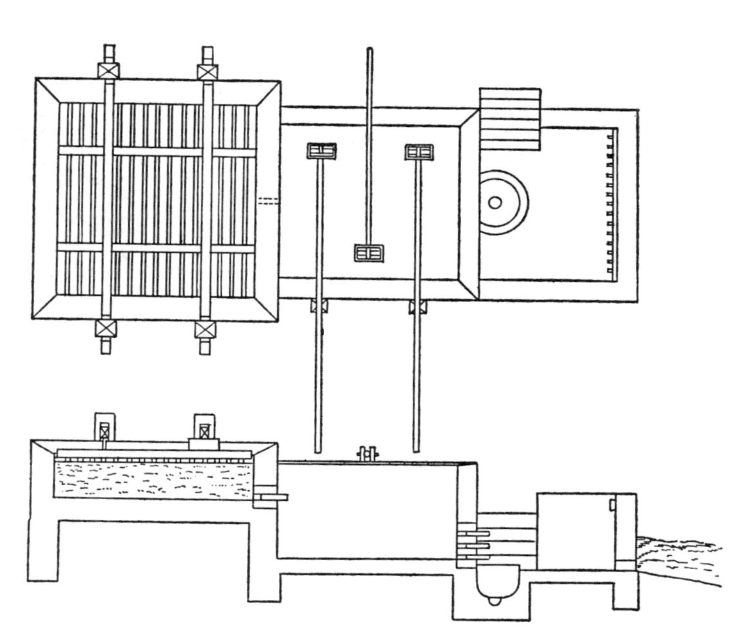

As regards the extraction of oils and fats, no particularly striking characteristics draw our attention. The extraction of olive oil was done following the traditional methods. The olives were crushed by a millstone rolling in a circular trough, in order to remove the pits. The pulp, collected in baskets or mats, was put through a classic screw press, and sprinkled with boiling water. The separation of the water and the oil was done in stone decantation vats, and the oil was stored in underground reservoirs called "hells" (*enfers*). Another separation removed the poor-quality oil used only for lighting. Around the end of the eighteenth century it became the practice to perform a second pressing of the pulp from the first pressing.

We shall not list all the plants whose fruit or seeds supplied oil for food, lighting, and especially soapmaking. Those most widely used for industrial purposes included hempseed, colza, rape, flax, and, beginning early in the eighteenth century, poppy. Tallow was another raw material used for the making of candles and soap products.

Hard soaps were generally manufactured in the Mediterranean regions, at Toulon, Marseille, Cartagena, and Alicante, where the alkali compounds made from soda were utilized. England, Holland, Germany, and northern France produced liquid soap from potassium. Soapmaking was done with quicklime that decomposed the alkaline carbonate (Figure 21). In addition, the manufacturers added various ingredients, such as copperas, starch, and pigments or dye products like cinnabar, red ocher, and indigo, depending on what qualities of soap they wished to obtain. The dyeing industry employed only white or marbled soap. Liquid soaps were used in the fulling and felting operations.

FIG. 21. Operations in preparing soap. Left to right: a worker pours a bucket of lye into the vat; two workers empty the vat; a worker pours oil into a vat (illustration from the *Encyclopédie,* Vol. IX).

Tapers and candles

Tallow candles must have already been in production by the thirteenth or fourteenth century; the production of wax candles and tapers undoubtedly began to become common in the sixteenth century. Domestic habits changed, and street lighting developed; thus the use of candles began to increase rapidly and steadily in the seventeenth century.

The tallow candle was at first prepared by dipping; that is, the wicks were dipped several times into liquid tallow until they reached the desired thickness. Here again we find that the manufacturers succeeded in establishing production equipment perfectly adapted to its goal.

The tallow utilized was a mixture of mutton fat and ox fat, cut into pieces (to avoid excessive heating, which would have adulterated it) and carefully melted in a copper boiling kettle. The melted fat was then skimmed, and the remaining impurities were precipitated by a small quantity of water. The fat was kept hot in a wooden tub, from which it was removed through a copper spigot.

By the seventeenth century the wicks were being made of cotton yarn. The hanks were reeled on a small wheel. Several threads were combined, depending on the thickness of the candles to be made. They were then cut to the desired length with a wick knife, which had a fixed blade and a frame whose distance from the blade depended on the size of the candles. In this way a frame with the number of wicks needed was obtained almost automatically. The yarn for each wick was hung on the frame, and the two strands were twisted together with the fingers.

Dipping was done in a vessel with a special shape (called the "abyss"), which held the melted tallow. All the wicks on a frame were dipped, removed, and drained; the tallow was left to harden while the worker worked with other frames. He thus repeated a certain number of dippings — the *"plingure,"* the *"retournure,"* and lastly the *"remise."* In this way the diameter of the candle increased to a predetermined size. Then came several finishing operations.

A candlemaker named Brez, installed in the Faubourg Saint-Antoine under a "Royal factory" license granted by Colbert, perfected (if he did not invent) the method of making molded candles. The mold was a tin tube closed at it lower end by a movable collar, at the top by a small soldered funnel topped with a hook. The collar had a hole in it through which the wick was threaded by means of a long metal needle; the wick came out through the funnel, and was attached to the hook. The molds were arranged in a row on a special table with two shelves, the first of which had holes the size of the molds, while the second had a trough to catch the overflow. After the candles had cooled, the collar was removed. The candle was now ready for use; there was no need of further finishing.

Tapers and candles were made only with wax, which was a relatively expensive product but which gave much more satisfactory light than the tallow candle. Later, thanks to the work of Chevreul, the stearin candle combined the advantages of the wax candle with the low price of the tallow candle, which disappeared quickly with the improvement of oil lamps beginning at the end of the eighteenth century.

The wax collected had to be purified and bleached. This was done in a series of operations repeated three times, which consisted of melting the wax and letting it drop into cold water, or spread out in thin strips. These strips or drops of wax were put through a large copper sieve to separate the water, then left for about ten days on canvases exposed to the sun. In the last operation the wax was made into bars by means of cylindrical molds. Considerable improvement was made in the equipment probably beginning in the first half of the eighteenth century. Since the melting vat was in a raised position, the wax flowed into tubs topped with a small crank-operated mill, on which the wax was automatically made into strips (Figure 22).

Candles were made in two ways, either by spinning or by throwing. The spinning method may have been invented around the middle of the seventeenth century by a Parisian waxmaker named Pierre Blesimare. It consisted of working the wick back and forth in the melted wax. The wicks, which were made from flax yarn, were stretched between two large rollers turned with the help of cranks; they became impregnated with wax, and were passed through reels to give the candles the desired diameter. A length of 400 or 500 ells was spun in this way. This was then cut and sold in round or flattened rolls similar to modern wax tapers (Figure 23).

Slightly later, probably at the beginning of the eighteenth century, it was learned how to make thicker candles by throwing. The wicks were suspended from a circular crossbeam, and the worker threw over them melted wax, dipped up with a copper spoon; this operation was repeated several times. During this process the candles were passed under a roller to make them perfectly cylindrical. The

FIG. 22. Shop for the purification and bleaching of wax (illustration from the *Encyclopédie,* Vol. III).

FIG. 23. The making of candles and tapers. Right: "spinning" candles. Left: the "throwing" method (illustration from the *Encyclopédie,* Vol. III).

same operations were repeated for the finishing. Finally the clipped and dried candles were bleached in the bleaching plant.

Whale oil and sperm oil To give some idea of the variety of the materials whose exploitation became important during the seventeenth century, and which it is impossible to discuss completely, we shall mention only the material extracted from whales and sperm whales; whale oil and sperm oil. Whale hunting was being practiced around the middle of the seventeenth century by both the Basques and the Dutch, but the latter ultimately acquired supremacy in this activity. A fisherman of Ciboure, François Soupite, discovered a method for melting whale oil on the open sea (until then it had been done on land). The melting furnace on the ship was fueled with the scraps and debris from earlier melting operations. The whale oil was used for various purposes: refining sulfur, treating hides, preparing fabrics and caulking products for ships, and the manufacture of soap. As for the spermaceti, which was extracted from the head of the sperm whale, it was purified by several heatings and washings, and was utilized in medicine and the perfume industry for the preparation of unguents and cosmetics.

Leatherworking

The preparation and working of leather required the use of large quantites of very ordinary chemical products. Like many other industries, the tanning activity was widely dispersed; the factories had to be built on the edge of streams (in the case of Paris, on the Bièvre, in the lower portion of the Faubourg Saint-Marceau) because the hides had to undergo long soaking during the various operations. Throughout the sixteenth and seventeenth centuries the products of the French tanning industry held the first rank in Europe. Tanning included three basic stages for the purpose of cleaning the hides, removing the hairs, and tanning them.

Scouring was done by the stream. The hides were submerged for several long periods in the water, and were worked with a round knife on a wooden trestle. The second operation was liming, during which the hides were dipped in large vats filled with lime. They remained for several days in each vat, going from the vats of slaked lime to those that held quicklime. Acid baths prepared with the fermented liquid of barley flour were also used, particularly in England in the middle of the seventeenth century. After this long treatment the hides were worked to remove the hair. Depending on the type of hide, liming might consist of exposing the hides to the heat of a fire and to smoke; the hides were then soaked in a liquid made from bark.

The tan was made from the bark of a young oak, ground into coarse powder in a special mill. "English-style" tan included genista. The actual tanning was done in ditches. The hides were covered with a mixture of hot water and tan, and soaked and stirred for a long period, after which they were rinsed in cold water, then returned to the ditch. This operation was repeated numerous times. The length of the treatment and the various techniques depended on the nature of the hides worked and the qualities desired.

Leatherworking was still practiced by various separate trades: the curriers, the "Hungarian" tanners, the "Morocco" tanners, the chamois-leather dressers, the tawers (Figure 24). Each was responsible for special operations, either for refining the hides or for preparing certain qualities of leather. The curriers refined ox, cow, calf, and sheep hides for harnesses and saddles by fulling them by foot and with the anvil on a hurdle after dampening, scraping off the bits of flesh remaining after the tanning, coating them with tallow, and working them with special tools. "Hungarian" dressing replaced tanning when a strong oxhide for saddles, harnesses, and carriages was needed; the tan was replaced by alum and sea salt. The last operation consisted of flaming the tallow-coated leather over

Fig. 24. A tawer's shop. Washing, scrapping, fulling (illustration from the *Encyclopédie,* Vol. III).

burning coals, the flesh side being turned toward the fire. The Morocco tanners, the chamois-leather dressers, and the tawers prepared the hides used in glove-making and fancy leather goods. Morocco leather was made of goatskin tanned with sumac wood. The hides were made into a bag by placing their outer sides together and sewing along the edges; this bag was filled with sumac and submerged in a sumac bath. The chamois skins or lambskins (from which imitation chamois was also made) were treated with lime and fulled; after the washing and scraping operations, they were treated with oil.

The food industries

The food industries comprised only the preparation of fermented beverages from grains, and the distillation of wines and alcohols (to which the distillation of vinegar for the preparation of acetic acid can be compared), as well as the roasting of coffee and cocoa beans and the production of chocolate, and the extraction of cane sugar.

As for the fermented grain beverages, their production is very ancient in origin, and probably dates from the Gallo-Roman period. In the course of the eighteenth century, industrial breweries were established in those European countries that are still famous for this type of production. Barley was fermented on the paved floor of large storerooms, where it was sprinkled and turned regularly. It was then heated in a vat on an inclined grill. The brew was prepared by heating a mixture of this malt and another grain (wheat, rye, or millet) in a vat filled with water. During this operation the starch was hydrolized and made soluble. Once drawn off, it was cooked with hops; then the liquid stuck to the gelatin was left to ferment.

The techniques of distillation of wines and alcohols changed little, except that the alembics were better conceived and were built in larger sizes (Figure 25). As for the other foodstuffs, they had just been introduced into Europe at the beginning of the period under discussion in this chapter. Their production was still on the craft level, although the methods were perfectly known and consumption

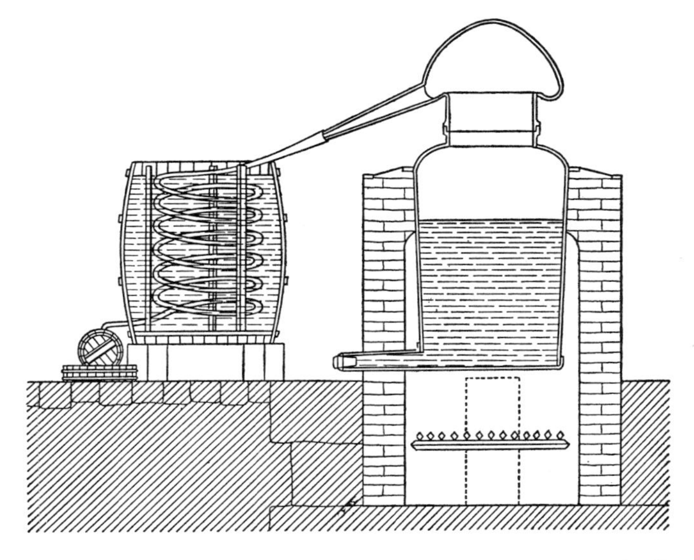

FIG. 25. Burner for spirits. Around 1775 (De Machy, *Art du distillateur-liquoriste*).

continued to increase rapidly. Between the sixteenth and the end of the eighteenth centuries this minor industry perfected its relatively simple techniques and its production equipment. These remained unchanged until the end of the nineteenth century, with the exception of the appearance of new foodstuffs, for example beet sugar, whose extraction did not acquire the appearance of an industry until the middle of the nineteenth century.

Papermaking

Only the activity of papermaking remains to be discussed. The long history of paper has been traced in several chapters of this volume and Volume I. Until the end of the eighteenth century the evolution of production techniques did not involve the principles of the methods themselves; here, again, progress meant simply progress in the organization and equipment of the factories, and in the use of power to operate the pounding mills, which, like the vats, presses, and other pieces of equipment, were constructed with better materials and in more rational shapes.

The papermaking industry, like a great many others, was the subject of controls by Colbert in 1671. Actually, these controls merely codified the traditional operations that had undoubtedly been brought to their current degree of perfection in the first half of the seventeenth century, in keeping with the technical level of the period. The raw material utilized for the preparation of paper pulp consisted solely of hemp and flax rags. These rags were washed, then shredded on cutting blades attached to tables. They were next placed in vats filled with water, where they began to decompose. Next they went through the rag-cutting machine, a workshop where the strips of rag were torn by the workers on vertical blades. After this they were put through the mallet mills. In the first of these mills the bottoms of the mortars and the heads of the mallets were covered with sheets of iron (Figure 26). The resulting pulp next went through batteries of mallets that did not have this iron frame. Next the pulp was put into boxes, and dried. When ready for use it was put through another mallet mill.

Plate 13. Paper mill. Engraving taken from Zonca, *Novo teatro di machine et edificii,* Padua, 1656. Conservatoire des Arts et Métiers. *Photo by the Conservatoire.*

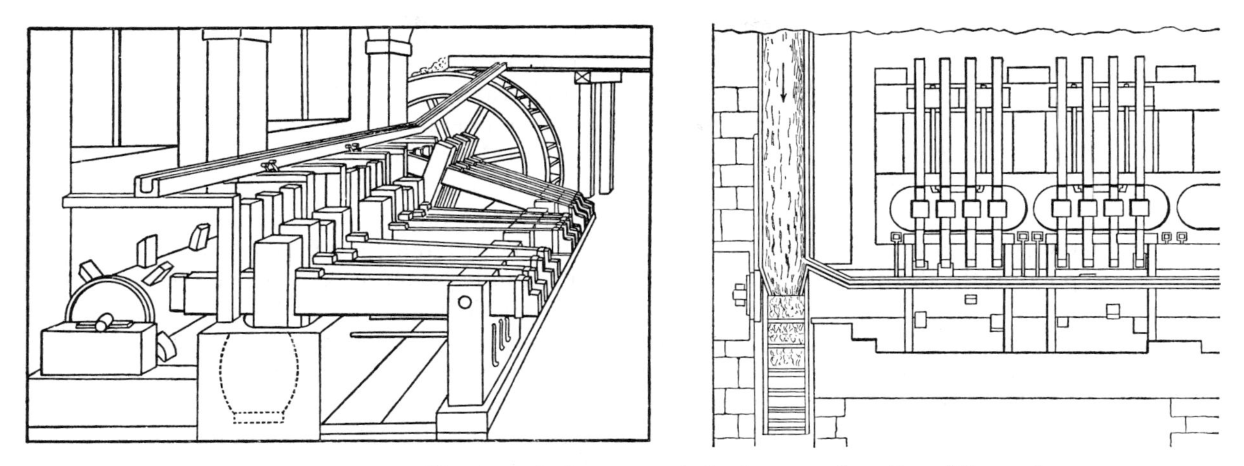

FIG. 26. View of a paper mill; plan of the vats and the battery of mallets (illustration from the *Encyclopédie,* Vol. V).

The paper itself was made in forms. The pulp was moistened in vats of warm water into which the molds, or forms, were dipped. These forms were wooden frames whose bottoms were made of very fine brass screening; the whole was stretched by larger brass wires. This screen also bore the distinctive marks of the manufacturer and the size of the sheet of paper; the wires and marks appeared as a filigree in the body of the sheet. After the form had been lifted out, various operations were performed; the drained sheet was turned over onto a piece of felt and covered with another piece of felt so that the next sheet dipped up by the "plunger" could be placed on top of it. The completed pile of sheets was placed in the press. Then the sheets were separated from the pieces of felt and pressed a second time (Figure 27). The paper was dried on ropes, then dipped into

FIG 27. Papermaking shop. Left: worker lifting a sheet in the form. Right: passing piles of paper and felt under the press (illustration from the *Encyclopédie,* Vol. V).

a clear warm sizing prepared with scrapings of leather or parchment to which was added alum or white copperas (aluminum sulfate). Once again the paper was pressed, dried, and then packed in quires and reams.

The decree of 1671 that controlled all these operations also fixed the traditional numbers of sheets at twenty-five sheets to the quire and twenty quires to the ream. These standards, together with the various sizes of paper, are still in existence. Although the names of the sizes have gradually fallen into disuse in the past few decades, the sizes themselves have been retained. The names correspond to the distinctive signs originally required to be watermarked on the sheets; the "raisin," "pot," "large eagle," "sun" were marked with those designs, while the "jesus" showed the name of Jesus, the "crown" the arms of the General Comptroller of Finance, the "tellière" paper the arms of Chancellor Tellier. Certain grades of special papers still retain their names and characteristics. The "serpent" was an extremely fine paper marked with a serpent. Joseph paper, blue paper for wrapping sugar, gray paper for patterns, and so on, were probably being produced by the beginning of the seventeenth century, if not already during the sixteenth, in exactly the same form they retained even after the major transformations in this industry at the end of the eighteenth century. The persistence of these standards is not a chance occurrence, or even the result of a sentimental respect for ancient traditions. It proves that the techniques of this and many other industries had achieved a sufficiently high level during the seventeenth century to create conditions of production that continued to be satisfactory for several centuries.

A period of transition The chemical industry presented in this chapter developed during a period that has too long been regarded as unimportant. The fame of the work of Lavoisier and the chemists of his generation led to the belief that until then chemistry did not exist. The historians of chemistry have now recognized their error. But they have continued to believe that the major chemical industry was born spontaneously at the beginning of the nineteenth century, and they make little distinction between the production methods of chemical products in the time of Agricola, for example, and in the period of the *Encyclopédie.* The fact is that this period of transition was extremely important. While this transitional era did not witness the appearance of many original methods, it did develop the foundations for a modern large-scale production technology that was to be utilized by nineteenth-century chemistry to create new industries.

BIBLIOGRAPHY

SOURCE MATERIALS

AGRICOLA, GEORGIUS, *De re metallica* (1556), transl. by Herbert C. and Lou Henry Hoover (London, 1912; reprinted, New York, 1950).

Bergwerk-und Probierbuchlein (1524), transl. and ed. by Cyril Stanley Smith and A. G. Sisco (Chicago, 1949).

BIRINGUCCIO, VANNOCIO, *De la Pirotechnia* (1540), transl. and ed. by Cyril Stanley Smith and Martha T. Gnudi (Cambridge, Mass., 1942; paperbound 1959).

EDELSTEIN, SIDNEY M., "The Allerly Matkel (1532): Facsimile Text, Translation, and Critical Study of the Earliest Printed Book on Spot Removing and Dyeing," *Technology and Culture,* 5 (1964), 297–321.

ERCKER, LAZARUS, *Treatise on Ores and Assaying* (1580), trans. and ed. by Cyril Stanley Smith and A. G. Sisco (Chicago, 1951).

SMITH, CYRIL STANLEY (ed.), *Sources for the History of the Science of Steel,* 1532–1786 (Cambridge, Mass., 1968).

SECONDARY WORKS

EDELSTEIN, SIDNEY M., "Dyeing Fabrics in Sixteenth-Century Venice," *Technology and Culture,* 7 (1966), 395–397.

GUTTMAN, OSCAR, *The Manufacture of Explosives* (2 vols., London, 1895).

MULTHAUF, ROBERT P., *The Origins of Chemistry* (London, 1966).

SINGER, CHARLES, *The Earliest Chemical Industry* (London, 1948).

TAYLOR, F. SHERWOOD, *A History of Industrial Chemistry* (New York, 1957).

WILKINSON, NORMAN B., "Making Powder, by Jean Appier Hanzelet," *Technology and Culture,* 6 (1965), 633–635.

Plate 14. Spinning, weaving, and finishing. "Clothmakers' Window," Semur-en-Auxois (Côte-d'Or), fourteenth century. *Photo Archives photographiques d'Art et d'Histoire.*

CHAPTER 11

THE TEXTILE INDUSTRY

SPINNING, WEAVING, AND FINISHING

DURING THE seventeenth century several original inventions and a number of improvements in the various phases of fabric production came to herald the technical evolution that, beginning in the middle of the eighteenth century, was to increase in speed and finally lead to the Industrial Revolution. The seventeenth century was not a period of major changes. For the time being, the only changes were to be fragmentary, isolated innovations without a unified economic stimulus; in a word, they were palliative measures intended to ameliorate individual technical disadvantages that had become too acute.

The principal innovations of this period were made in areas other than the nature and use of the fibers involved. Only a single innovation occurred in the types of fibers available: to wit, a certain renaissance of cotton. This fiber had been well known in the Middle Ages (it is attested in Ulm, Venice, Greece, and in the islands of the eastern Mediterranean). Around the fifteenth and during the sixteenth centuries, the West continued to import it from the Levant, but for secondary purposes such as candle and lamp wicks, and as stuffing material. In the seventeenth century its use for more serious purposes began to become common again. The first definite reference to cotton during this period occurs in 1641, in connection with a factory in England.

As for the treatment of fibers, there were no innovations in the handling of either flax or wool, except that in the latter case the preparation of slivers of combed wool before spinning continued to be important (Figure 28). This tech-

FIG. 28. Combing wool (*Encyclopédie méthodique*).

nique dates from antiquity. It is mentioned by Homer in the *Odyssey,* and the medieval methods are depicted in a miniature of Boccaccio's series of *Noble Women* and in a sculpture of the royal portal of Chartres Cathedral. Combed wool yarn was utilized in the manufacturing of light woolen fabrics and in knitting stockings *(bas d'estame);* it thus spread with the use of the knitting machine.

The principal innovation in this period in the treatment of fibers concerned silk thread, and related to the mechanical twisting of the thread from the cocoon, which in the fifteenth century began to be done in Italy with the throwing mill. (This important machine will be discussed at the end of this chapter.)

When we consider technology in its "craft" stage, we find that certain very minor and incessantly repeated operations are as it were beneath human level. For this reason the idea of performing them by mechanically operated devices was soon conceived and put into operation — for example, the twisting of silk (fifteenth century, Italy), the weaving of ribbons (Holland, beginning in the seventeenth century), and the knitting of stockings (France, same period).

In silk twisting, 100 to 200 million twists are required for the preparation of two pounds of thrown silk; the knitting of a man's sock requires the execution of 150,000 to 175,000 stitches. As for weaving very narrow fabrics, the value of silk ribbons made the high price of the labor required bearable, but this was not the case for the utilitarian ribbons made of flax and later of cotton.

Other innovations were adopted, not for quantitative reasons (increase in productivity) but rather for qualitative reasons: customers wanted homogeneity of surface and unity of treatment in fabrics. These innovations included the machine for friezing cloth, the calender for polishing light Flanders serge, and the napping machine. In machine work the area covered by the functional parts far surpassed the possibilities of the human hand; in addition, the use of the machine made the operation continuous, which made it possible to obtain the desired uniformity and regularity.

While the shortage of labor may to a certain extent have stimulated the activity of the innovators, their interest was not the result of a feeling of pity for the labor of the women and child workers, but rather the pressure of labor shortages, which followed an annual seasonal rhythm. Each year, at the approach of the great fairs, there was a disparity between the productivity of the labor supply and the volume of the demand. Manual labor was becoming incapable of satisfying the needs of entrepreneurs.

The search for qualitative results antedated the search for quantitative effect. Except in the production of stockings and ribbons, these innovations did not yet bring about a transition to an industrial stage of production, especially since the other phases of fabric manufacturing continued to be performed manually.

The bar loom

The bar loom was a permanent loom that made it possible to weave a certain number of ribbons simultaneously, on a single loom operated by a single worker. It was a forerunner of mechanical weaving, which it preceded by exactly two centuries.

In weaving a ribbon, which is simply an extremely narrow piece of fabric, the width of the warp is less than the length of the shuttle. Thus, the shuttle does not need to be thrown in order to cross the warp; a slight impulse is sufficient to transfer it from one edge of this miniature fabric to the other. This explains why the mechanical weaving of ribbons developed long before that of fabrics of normal width.

Thus it was the narrowness of the back-and-forth movement of the shuttles that distinguished the bar loom from the mechanical loom for fabrics. On the latter, and on the hand looms equipped with John Kay's fly shuttle, the shuttle actually "flies" from one edge to the other across a fairly long space varying from 4 to $6\frac{1}{4}$ feet.

The bar loom had, and has retained, approximately the same width as a loom for weaving wide fabrics. But it wove twenty-four or more ribbons, side by side, by the following method:

Along the entire length of the beater was a slat with metal fingers. A cam pushed this slat before each stroke of the beater, alternating between left-to-right and right-to-left. The extent of this push corresponded exactly to the trajectory of the shuttles; it was the metal fingers that transmitted the impulse to each shuttle (Figure 29).

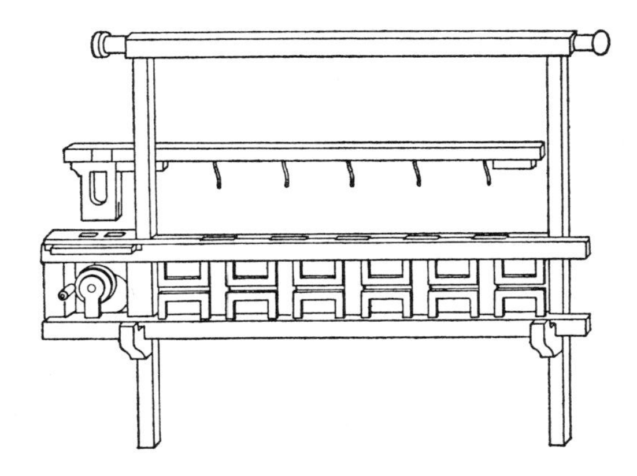

FIG. 29. Principal parts of the bar loom. Below: the beater, with slots for the shuttles. Above and behind: bar with iron fingers that move the shuttles.

The four characteristic functions of mechanical weaving are the raising and lowering of the heddles, throwing of the shuttle, beating by the reed of the weft left by the shuttle, and the unrolling of the warp and the rolling up of the woven fabric. On the bar loom these functions were already synchronized, and derived from the movement of a rotating main shaft, heralding the achievements to be made in fabric manufacturing after 1796.

We know that the inventor of the bar loom was named Willem Dierickzoon van Sonnevelt, an inhabitant of Hondschoote in French Flanders (now located in the Nord Department), but we know nothing of his life. The first reference to his invention dates from 1604.

Sonnevet's loom was preceded by a small multiple treadle loom called the Danzig loom, which seems to have been introduced at Danzig around 1586 and which was capable of weaving from four to six ribbons simultaneously. Its use is supposed to have been immediately forbidden by the Municipal Council, which according to legend went so far as to drown the inventor. This loom was also prohibited at the end of the sixteenth century in Bruges and Antwerp. As it could be useful in small-scale production, it continued to exist to a limited extent even after the spread of the bar loom.

As we have already mentioned, the oldest extant reference to Sonnevelt's loom dates from 1604. On July 10 of that year the inventor granted licenses for the use of his large "ribbon mill" to several Walloon ribbon makers at Leyden. The following year Sonnevelt received from the Estates General an inventor's patent valid for ten years.

Until 1623 the documents speak only of Leyden. By 1638 the loom had reached London; around 1664 it is found in Germany and Switzerland. In France the administrator of the Poitou Barentin region transmitted to Colbert in 1666 a request for a license made by the ribbon makers of Poitiers, concerning a bar loom that wove ten ribbons simultaneously and was capable of weaving twenty. In 1706 François Prevel introduced bar looms from the Spanish Netherlands into Lille. His factory was wrecked during the hostilities of the next few years. In 1716 he presented a petition, transmitted to the Académie des Sciences, which drew the attention of the company to the bar loom. René-Antoine Réaumur and the Abbot Bignon, who were at this time preparing the *Description des Arts et Métiers,* had drawings made of the bar loom. Still unpublished they are preserved in the archives of the Institut.

The city of Basel became a very important center of ribbon-making. The promoter was J.-H. Hummel, who in 1730 established a hydraulic installation. By 1754 there was 1,225 bar looms functioning in Basel; in 1786 there were 2,246.

Naturally, however, the development of the use of the loom was subject to prohibitions and workers' revolts, the enumeration of which would be too lengthy. In 1791–1792 the Parisian workers, in their protest to the National Assembly against the use of the bar loom, make use for the first time of the term "mechanical loom."

The invention of de Gennes

Though a project for a mechanical loom published by de Gennes in 1678 in the *Journal des Savants* has often been presented as an actual achievement, this is not the case, although de Gennes, a naval officer and governor of the island of Saint Kitts, was a distinguished mechanician. In addition to a cylinder that climbed a ramp, he had, apparently long before Jacques de Vaucanson, constructed an automaton in the shape of a duck that could walk, eat, and digest.

De Gennes's loom project is indisputably the forerunner, by perhaps fifty years of the mechanical loom. The inventor may have been inspired by the Dutch "ribbon mills," which were already known by the middle of the seventeenth century. Unfortunately, his suggested mechanism, which was a purely intellectual conception, was completely impractical. To give a brief idea of this device, the loom had a shaft bent in the shape of a double bitbrace. Ropes led from its angles to the treadles, which were similar to those of an ordinary loom, ensuring the opening of the warp. This shaft had four cams. Two of them, in the form of a quarter circle, caused the action of the reed. Two others in turn moved two sliding arms that held the shuttle; the latter was grasped in alternating rhythm by pincers at the end of each arm, the opening of the pincers being caused by the advance of the arm. De Gennes thought he could accomplish three of the four operations essential to mechanical weaving by means of a rotating main shaft.

De Gennes ends his description with these words: "The author also has a very simple method for making the fabric fold of itself as it is produced." This device, however, is not described.

The inventor was inspired principally by the cloth loom, in which the operator throws the shuttle from one hand to the other; it was his intention to imitate these gestures. Vaucanson later adapted this limited conception of the mechanization of weaving. It was the mechanical throwing of the shuttle (John Kay, 1733) that opened the path for further developments. Without this "fly shuttle," as Kay's device later came to be called, mechanization was impossible.

Preparation and Finishing of Fabrics

Teaseling machines (gig mills)

Teaseling is the operation that consists of passing a tool equipped with thistles over the surface of a fabric in order to raise the nap. The first reference to teasling is found in the writings of Pliny the Elder.

The oldest drawing of a teaseling machine is found in the manuscripts of Leonardo da Vinci; it is completely impractical. Vittorio Zonca's drawing (1607) of a teaseling machine is primitive, to be sure, but it is definitely an actual machine. However, the various movements are not yet synchronized; the coordination between the forward movement of the fabric and the speed of the two cylinders equipped with thistles require the attention of the worker and his assistant (Figure 30).

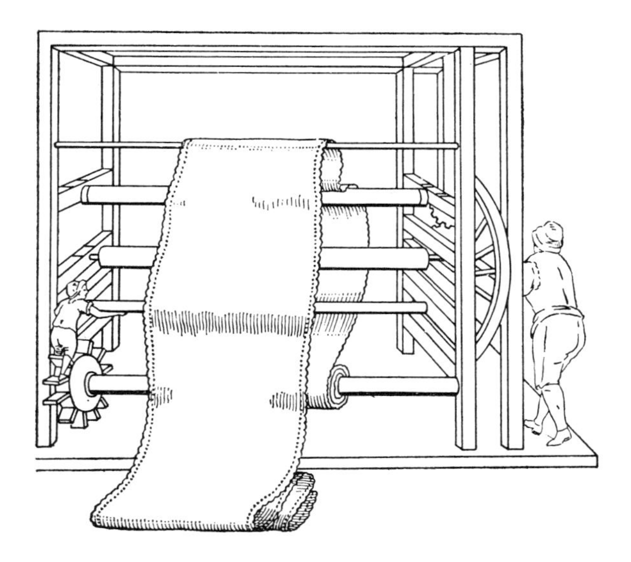

FIG. 30. A gig mill (Zonca).

This machine must have made possible more uniform results and a productivity higher than that of handwork. However, the still incompletely perfected coordination of the movement precluded the use of this device for teaseling high-quality cloth, even in factories where the machine was utilized despite the rules.

The functioning of this teaseling machine of 1606 is very simple. The movement of a wheel turned by a worker turns two cylinders, equipped with teasles, at a high speed. The fabric moves forward slowly as a result of the action of a pulling cylinder, operated by an assistant; after the operation the fabric winds onto this cylinder. The two movements are completely independent. Teaseling machines, or "gig mills," the name by which they are designated in works on economic history, for example those of Karl Marx and Paul Mantoux — were forbidden in England during the reign of Edward VI (statutes of 1551), along with the use of metal cards, and this prohibition was renewed by his successors. Metal cards and gig mills were nevertheless utilized before 1640 in Gloucestershire for coarse fabrics. In 1758 the first completely mechanized gig mill, installed at Heytesbury, Wiltshire, caused a famous riot. In France a regulation of Sedan, dated 1666, vaguely alludes to these machines.

The cloth press

The pressing of fabrics is depicted in one of the frescoes of the "Fuller's House" in Pompeii. All the cloth centers practiced this technique. The pieces of fabric were piled, accordion-pleated, under a press; a flat board (later a piece of cardboard) was inserted between each fold.

Heat-pressing considerably increased the sheen of the fabric. It was known in Western Europe by the fifteenth century, but was prohibited by all the codes. In one such regulation we read that "this manner of pressing fabrics hides their irregularities and defects, a situation which could give workers and manufacturers an opportunity to overlook and facilitate business frauds." Although this assessment seems to be correct, by the seventeenth century heat-pressing had become an essential phase of the finishing of high-quality fabrics, and has remained so until modern times.

The heating of a pile of several pieces of fabric placed under the press was done by means of the intercalation of sheets of cast iron first heated in an oven to the proper temperature. These cast-iron sheets, which were perfectly flat and approximately 4.5 millimeters thick, were not placed directly in contact with the fabric, but were separated from it by planks of alder wood (which does not react to heat) or cardboard (Figure 31).

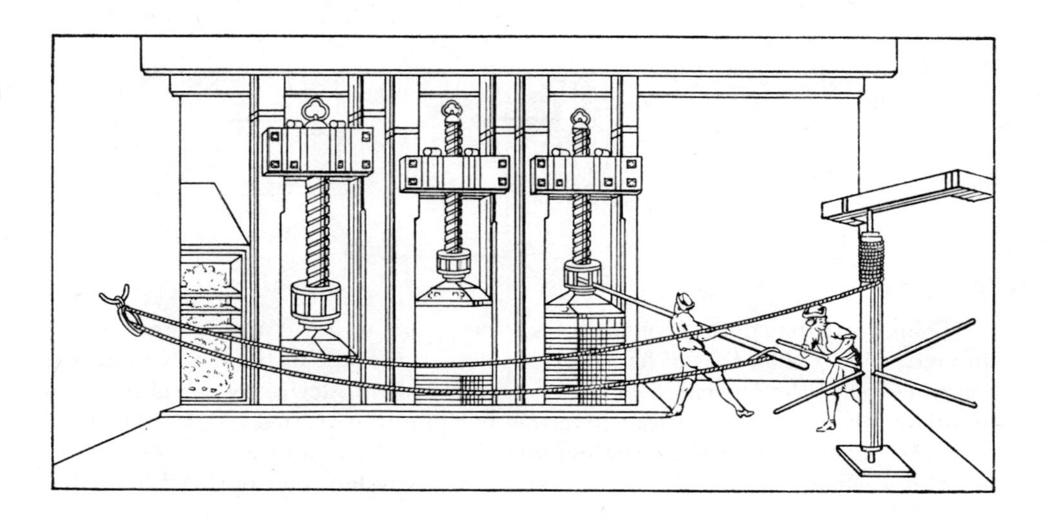

By about the first quarter of the eighteenth century, heat-pressing had become an authorized preparatory operation in France and Germany; undoubtedly it had long since become legal in England and Holland.

Fulling of fabrics

Fulling is the basic operation of fabric finishing. It is done in a warm damp place, using specific ingredients. Under these conditions the threads of the fabric are "soldered" to each other, owing to the fact that a portion of the wool fibers becomes separated from the thread. The fabric thus acquires consistency and strength, but decreases in size.

Until the eleventh century fulling was done only with the feet, in vats. Then the first fulling mills, which are also the first example of the use of natural power in the textile industry, appeared near Grenoble. They spread rapidly across Europe; examples can be seen in the Technical Museum of Göteborg and the Open-Air Museum of Stockholm. In this primitive device, small wooden beams that slide vertically are raised by cams. When they fall they knead the piece of fabric in a vat whose bottom is lined with round pebbles or kernels to soften the shocks. The action of these mills was very harsh; they could be used only on rough homespun for monastic usage and similar coarse fabrics whose fulling was very difficult. In the seventeenth century fine fabrics were still being fulled with the feet.

In the sixteenth century, however, hammer-type fulling mills appeared almost everywhere in Europe. Zonca's work contains an engraving of them. These mills were less harsh, and their hammers had carefully constructed profiles, thanks to which the fabric in the vat was not only kneaded but also gradually rolled and stirred, which helped to distribute the fulling action more evenly (Figure 32).

Around 1850 these hammer mills were replaced in woolen-goods factories everywhere by fulling cylinders, but they are still utilized for fulling articles made of small pieces, for example knitwear, berets, and certain types of felt used for technical purposes.

← Fig. 31. Cloth press, with furnace for heating the sheets of iron. From an engraving of about 1720, showing the operations in the making of fabrics at the factory of Horni-Litvinov (Bohemia). With permission of the Narodni Tecnické Museum, Prague.

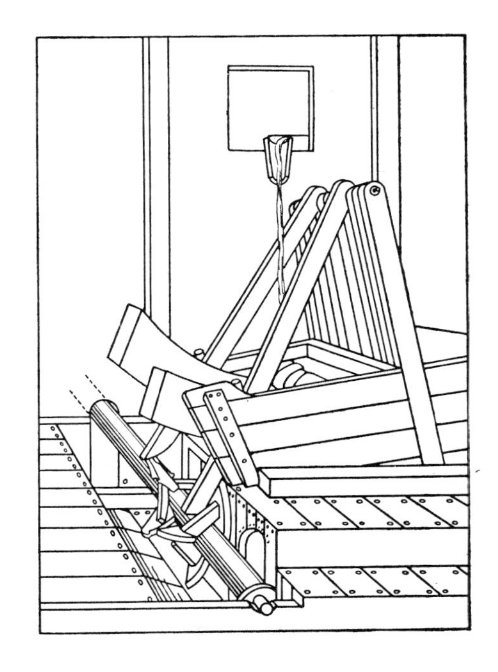

Fig. 32. Fulling mill (Zonca). →

Cropping of fabrics Since the days of antiquity, fabrics had been cropped with manually operated shears. The first improvement in the shearing operation was a device of English origin that resembled an inverted mallet. It had a groove at the top to receive the back of the lower blade of the shears, and exercised traction on the upper blade by means of a cord. This facilitated the closing of the shears, which opened spontaneously as soon as the traction ceased, the shears acting as a spring (Figures 33 and 34).

The appearance of this device caused sometimes violent resistance on the part of the shearing guilds in France and Holland. In several centers (Leyden, Sedan) this resistance continued for almost a century. The first echoes of the trouble caused by this innovation dated from 1680, at Leyden, and continued throughout the first half of the eighteenth century. Not until around 1760 did the guilds resign themselves to its use in most of the fabric centers. By making the maneuvering of the shears easier, the new device facilitated the recruiting of apprentices. Without it, the young boy standing between two journeymen and obliged to grasp the shears directly collapsed after a few hours or days of work.

We possess a detailed medical report by a doctor who witnessed cropping with the simple shears, analyzing the effect of this work on the muscles and tendons of the hand and wrist. Another witness reports that the hands of the shearers were bleeding by the end of the week, but they persisted, although they knew that elsewhere the shears were handled with the help of the more modern device.

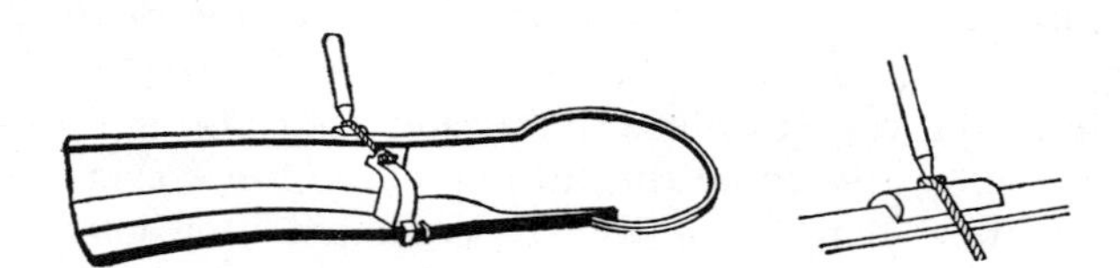

FIG. 33. Shears for cropping fabrics, showing device for aiding in their opening and closing. The detail shows the attachment of the strap (*Encyclopédie méthodique* and Tolson Memorial Museum, Huddersfield).

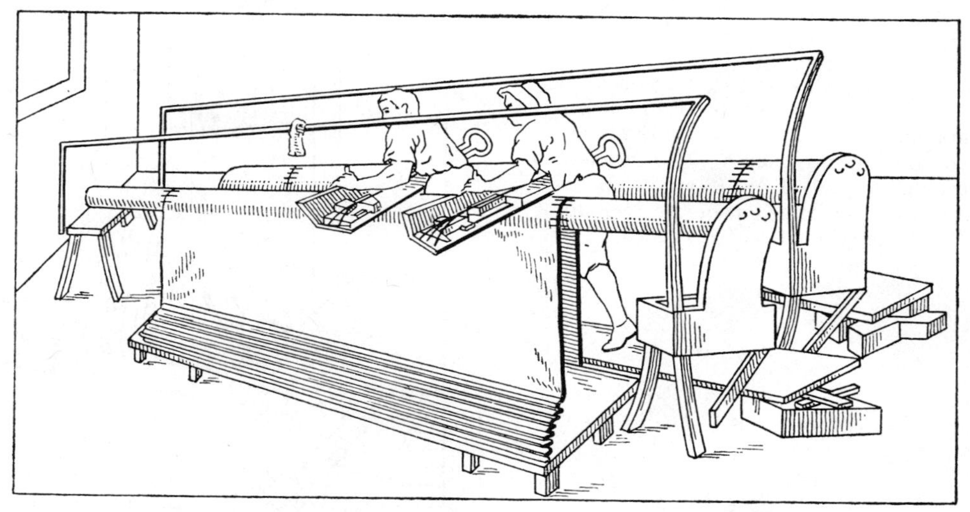

FIG. 34. Croppers working on a fabric (factory at Horni-Litvinov).

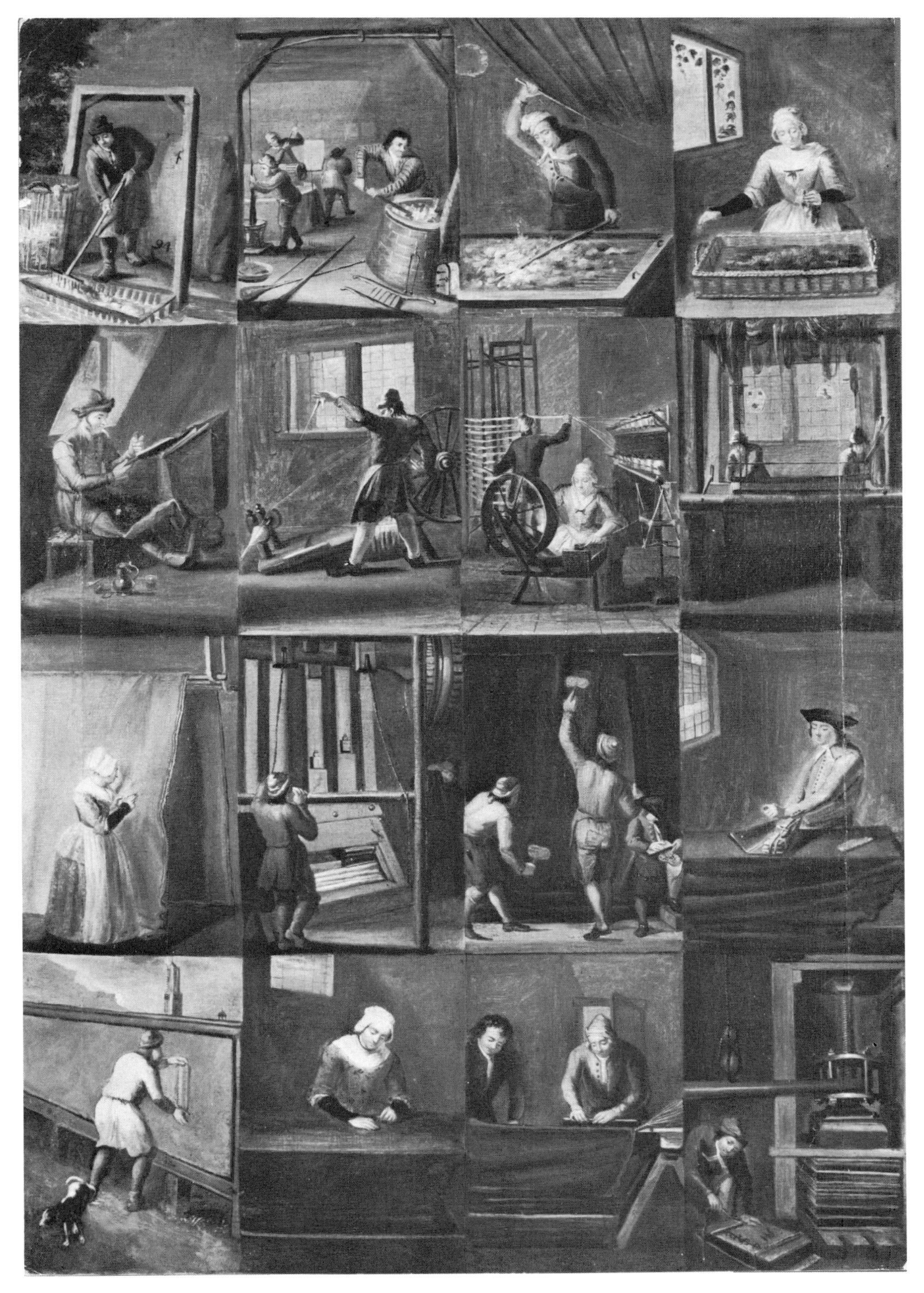

Plate 15. Various scenes depicting the working of wool and the finishing of fabrics. Anonymous painting, sixteenth-seventeenth centuries. *Photo Centraal Museum, Utrecht.*

This stubborn resistance is explained by the fact that the ease of recruiting apprentices for work with the new type had harmful repercussions on the slender earnings and the very relative advantages of the journeymen shearers. The device also made it possible to increase the rhythm of the shearing. Without this accessory, cropping with shears could not have continued as it did into the middle of the nineteenth century, that is, fifty years after the appearance of the first shearing machines.

Friezing The vogue for ratines and friezed fabrics continued practically without interruption throughout the *Ancien Régime*. In addition, the wrong side of ordinary black fabrics was also friezed.

Friezing was done on the pieces of fabric after the nap had been raised by teaseling. The prepared fabric was stretched on a table. To agglutinate the tufts of hair, the fabric was rubbed, using a single motion over a surface as wide as was possible with manual labor.

The tool employed was called the "tile"; it was a slab of sandstone or a plank whose bottom face was coated with coarse sand. The tile was covered with a sticky substance: egg white, honey, or a gelatin solution. The rubbing was done by moving the tile gradually from one edge of the fabric to the other with a small gyrating movement. This monotonous work required skill and, especially, close attention constantly maintained for the entire interminable workday customary during that period.

In order to eliminate irregularities, the manufacturers very soon thought of mechanizing the friezing. This was a daring idea, for as yet there existed no machine having the slightest resemblance to what was desired. To obtain the effect of the friezing, the friezing mill had to have two synchronized movements: first the gradual advance of the tautly stretched fabric under the vibrating plank, then the slightly gyratory vibration of this plank in order to reproduce the rubbing exercised by the tile in the hand operation. No automatic device for spreading the agglutinating substance was conceived (Figure 35).

FIG. 35. Machine for friezing fabrics (from Duhamel du Monceau, *Description des Arts et Métiers* of the Académie des Sciences).

It is at Troyes, an important fabric center during this period, that we find (in 1678) the first reference to friezing mills. They had been brought to Troyes from Toulouse by F.-J. Puget, a member of an old Toulouse family of clothmakers and fullers. These mills rapidly spread to all the clothmaking cities of France. As in the case of the gig mill, the friezing mill acted energetically on a larger surface, friezing the nap with a single movement over the entire width of the fabric. In addition, the synchronization of the forward movement of the fabric with the vibration of the plank assured the uniformity of the friezing along the entire piece of the fabric.

Two workers were needed to do the friezing, and the machine's essentially qualitative effect was thus very apparent. This is why the friezing mill, unlike other innovations in fabric finishing, met with no opposition anywhere from workers or people opposed to innovations.

The first friezing mills that appeared in Troyes were manually operated, but soon a horse-driven treadmill or hydraulic wheel came into use. Since the archives of the mostly highly developed fabric centers of England and Holland mention only manual tile friezing during this period, it can be accepted that this machine did in fact originate in Toulouse.

The friezing mill was a remarkable machine tool for that time, by virtue of the synchronization of its two movements and by the fact that the horizontal gyratory movement made within a very short radius by its principal working organ was regulated by an eccentric vertical axle.

Currying (fulling by boiling)

Étamine (worsted warp and weft or wool/silk mixture) and say (cheap, thin cloth resembling serge) were light fabrics made from combed wool yarn, in plain weave. The word *étamine* comes from the Latin *stamen,* meaning "warp thread" (in contrast to "weft thread"). The manufacturing of this fabric had been in existence since the Middle Ages at Leyden, Brussels, Hondschoote, Lille, and Le Mans. These two fabrics were utilized for women's clothing and ecclesiastical garments, and were exported to all corners of the globe — the Near East, Italy, Spain, America.

The essential phases of say making depended on the dyers and curriers. Currying was one of the most delicate methods of finishing light fabrics of combed wool yarn in the sixteenth and seventeenth centuries. This operation, which was both physical and chemical in nature, remains practically unchanged in modern factories, where it is known as "fulling by boiling." Its purpose is to improve the appearance of the woolen goods and to ensure the permanence of their dimensions.

To avoid the formation of wrinkles, the unbleached pieces of fabric are rolled tightly on rods, which are stood in a boiling kettle and boiled for at least an hour. They are then drained and cooled, after which the operation is repeated by rolling the fabric in the opposite direction on other rods.

A drawing by Otto van Veen (Leyden, 1602) that depicts a currying shop is the oldest extant evidence for this operation. Slightly later, an innovation made it possible to effect an economy in labor. To ensure the tautness of the fabric while it was being rolled, a worker applied pressure to it. At the beginning of the

Fig. 36. Currying shop (from a drawing by Van Veen, 1602; Archives of the city of Leyden).

seventeenth century the worker was replaced by rope brakes weighted with stones (Figure 36).

Not until after the Second World War, thanks to research done on the structure of wool fiber, did a scientific explanation of the effects of currying become possible. Under the combined effect of dampness and high temperatures, the internal tensions in the fiber are released. Within the keratin molecule, the polypeptide chains are detached and then reunited, but this time under the pressure of the rolling. Thus the fabric is fixed in size and flatness, every wrinkle having disappeared. As long as it does not undergo another, similar treatment, it is protected from breaks and wrinkles.

The flatness obtained by means of currying made it possible to polish fabrics on a mill for polishing serge. This machine, the first reference to which dates from 1631 (Brussels and Hondschoote), may have been an ordinary but more carefully constructed calender with alternating movement, or perhaps a polishing machine with a rotating cylinder.

Véron's shop

The combination of the series of inventions just described ensured the prosperity, at the end of the seventeenth and beginning of the eighteenth centuries, of the Le Mans factory of Guillaume Véron. This shop acquired a certain notoriety thanks to the fame of the lawsuits Véron had to sustain against the various guilds that, being jealous of his success, energetically pursued him.

This establishment included a scouring mill similar to a fulling mill, and a mill for removing grease on the same principle but with pounders terminating in beaters, whose action imitated that of laundry beaters. Muslins were also polished in this shop, probably with the help of the mill for polising serge already mentioned.

The silk-throwing mill

The cocoon of the silkworm supplied a thread that was continuous but much too fine to be used singly. A certain number of these threads were therefore combined as soon as the cocoon was unrolled (the reeling process), and the result was a thread called "raw silk." The next step required was milling, or the twisting of the raw silk. Executing this operation with a spindle rolled under the palm of the hand, or even with the spinning wheel, was beneath the level of human activity, and in the fourteenth century a mechanical solution was therefore devised that utilized a large number of spindles simultaneously; a number of these could be attended by a single worker.

The housing of the throwing mill formed a cylindrical cage approximately 16½ feet in diameter and of the same height, the axis of which was occupied by a vertical shaft. "The spindles, approximately 336 in number, are arranged in a series of circular rows (*vargues*), and are turned by a very primitive mechanism. Inside the mill, opposite each *vargue,* felted semicircles (*strasins*) are attached to the central shaft that gives them a rotating movement; rubbing against the spindles, they cause them to turn with a rotating movement that may reach 600 to 800 revolutions per minute" (Ballot). The extended vertical shafts moved a series of reels in a second row.

Montaigne saw these circular mills in operation in Florence in June 1581, and reported that one worker attended to 500 spindles — a figure that seems exaggerated. In 1607 Zonca made a fairly crude engraving of such a mill. A drawing from the same period, originating in Leyden, gives a much more exact idea of the equipment. Given the provenance of this drawing, it is probable that these machines were functioning in the Netherlands long before their introduction into England in 1717 (Figures 37 and 38).

By 1717 the throwing mills were in operation at the factory in Derby founded by the Lombe brothers. This was the first factory to operate in England, and its activity later covered many operations is addition to silkworking. The factory had several floors, and power was transmitted by a vertical shaft (which at this period was the best method of distributing the power from a hydraulic wheel) to a great number of spindles. Each drum had three rows of 96 bundles, an arrangement that later inspired all the eighteenth-century inventors of spinning machines. A drum of a throwing mill from Lombardy is preserved in the Technological Museum at Winterthur.

With this machine, we are still far from a genuine mechanization of spinning, which was not to occur until it was learned how to regulate and orientate short fibers, draw out the casing thus obtained to decrease its diameter and increase its length, and then fix this refined bit of material by twisting it.

KNITTING

Hand Knitting

Its origins

The origins of knitting are extremely obscure. It is impossible to determine what period and what civilization witnessed the birth of the worker who first had the idea, genial in its simplicity, of making a fabric, not by crisscrossing a warp and a weft, but by using a single thread to form row after row of loops continuously picked up to create new loops until the net had reached the desired length.

The weaving loom had certainly been used in ancient Egypt, and was far

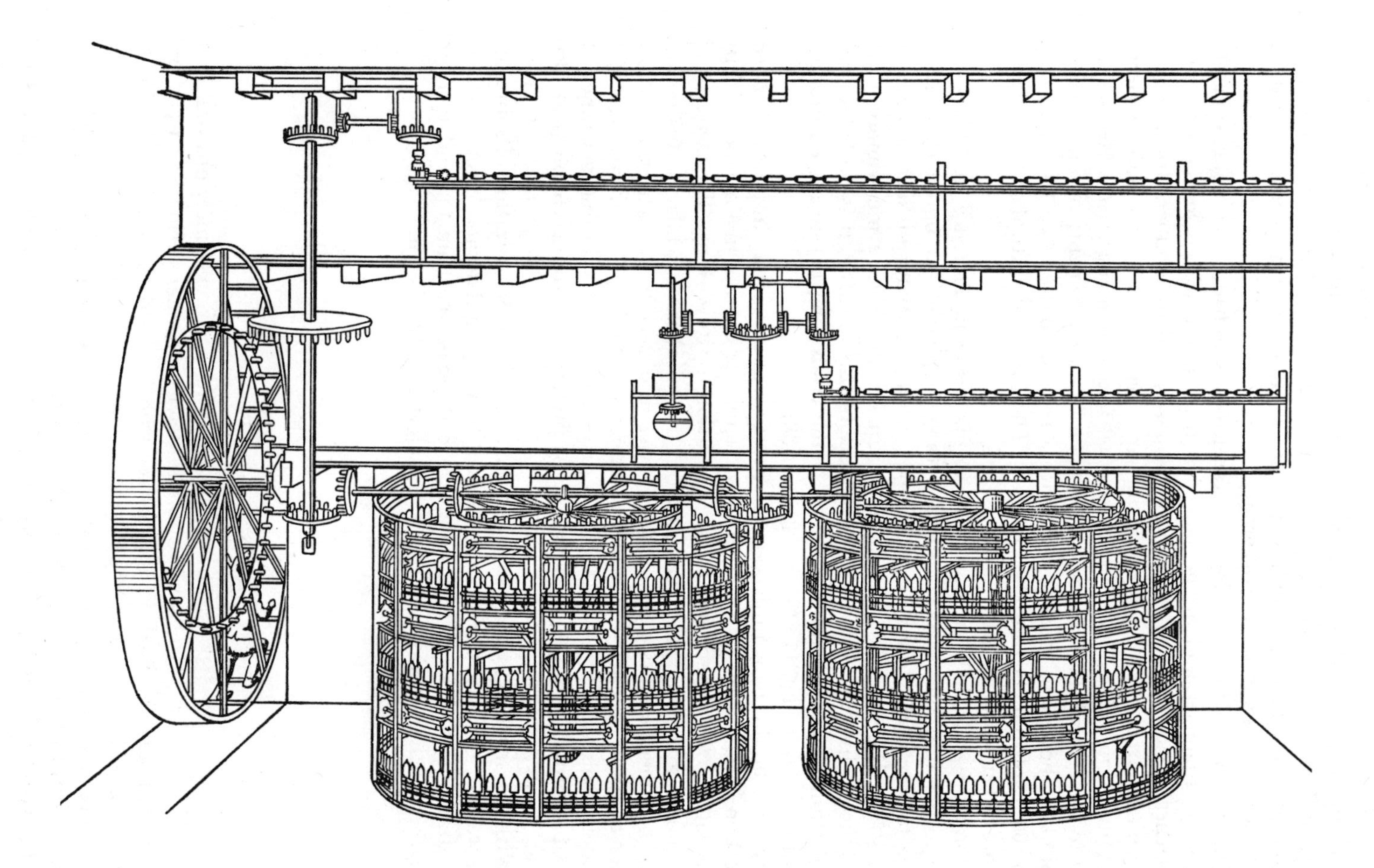

Fig. 37. General layout of a silk-throwing mill. (From a drawing by I. V. Verven, beginning of the sixteenth century. Archives of the city of Leyden.)

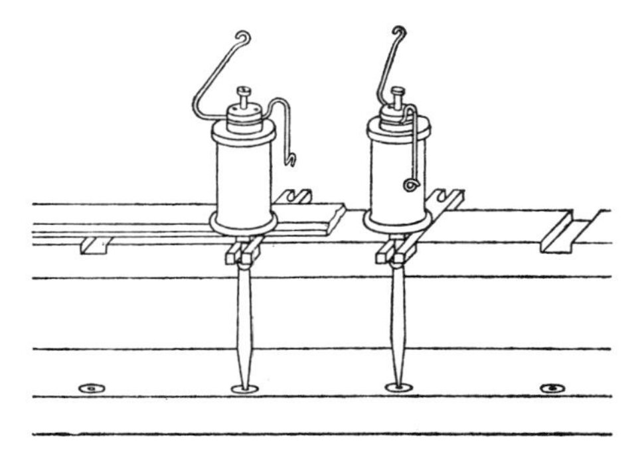

FIG. 38. Detail of the spindles of the silk-throwing mill (*Encyclopédie*).

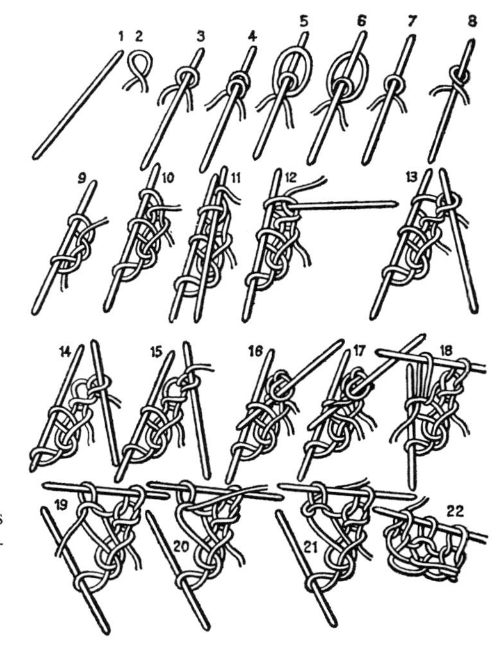

FIG. 39. Steps in hand knitting. From Hindret's notebook (Paris, Bibliothèque Nationale, Cabinet des Estampes).

earlier than the invention of knitting. It has been thought that knitting might have originated in a simplified and ingenious use of the wooden frame that formed the essential part of the weaver's equipment. A reduced model of this frame may have been equipped with pegs around which the yarn was wound to form stitches. Once the first row had been obtained in this manner, each stitch was picked up with a hook to form the next row, and so on. In this way a tubular net was formed, similar to what was done, beginning in the nineteenth century, on circular looms. This technique appears to have been practiced as early as the first century of the Christian era by a Christian sect, which is supposed to have learned it from the nomads of the Egyptian desert. Woolen bonnets from North Africa, believed to have been made by framework knitting, are still extant.

James Norbury, to whom we owe this attractive and ingenious hypothesis, has attempted to reconstruct the transition from the frame to needles. He believes that someone had the idea of replacing the frame with pegs by a rod around which the yarn was wound to form the first row; each stitch was then picked up with a hook. This method was still being used at the beginning of this century by many knitters in the rural regions of central Europe. Undoubtedly the primitive rod quickly became a genuine needle, fixed in a kind of wooden handle often attractively decorated. When this handle disappeared, two needles (three or four when a seamless fabric was desired) were used, as is still done in modern times.

This explanation (which, however, is given by very few authors) takes into consideration all the stages in the development of the technique of hand knitting, beginning with the frame (square, rectangular, circular) inspired by the frame of the weaving loom. The handles of the primitive needles are sometimes seen in museums, which furnishes a considerable support for Norbury's interpretation.

The most fantastic hypotheses have been proposed with regard to the great antiquity of knitting. According to some of these hypotheses, it may have been known in the time of Homer, and Penelope's famous cloth may have been a knitted fabric, which would have made it easy to unravel the work quickly and begin it all over again. It has also been suggested that the sheath garments that clothe Isis and Hathor in the Egyptian bas-reliefs seem too narrow to permit these goddesses to walk — unless the garments were made of a very elastic fabric. The next step is the idea of a dress knitted in ribbing, permitting the slightest movement to stretch the fabric. This step, however, is reached a little too quickly: there is nothing in the numerous pictorial remains of antiquity to indicate that knitting was known in Greece or Egypt. The Egyptians of the pharaonic period, who were so fond of depicting all kinds of work, show not a single person knitting, either with frame or needles. The small slipper found by Champollion and given by him to the Louvre Museum probably does not antedate the Coptic period; the same is true of the one found in a Christian tomb of the third century, and preserved in the Victoria and Albert Museum of London.

The oldest known knitted objects

The latter object is a very important piece of evidence. Done in knit-one-purl-one ribbing (that is, by knitting alternately one stitch in front and one stitch in back), it proves that hand knitting existed by the early centuries of the Christian era; we are unable, however, to determine how widespread this type of work was. Is the rarity of extant specimens due solely to the fragility of textiles?

It is believed that the art of knitting had never been lost in the Near East and that it spread to the West in the wake of the Crusades. The Metropolitan Museum of New York and the Detroit Institute of Arts possess specimens of two-tone stockings or slippers with high uppers done in stockinette stitch. According to the information supplied by the curators of these museums, they may date from the eleventh or twelfth century, and are of Eastern origin.

The Treasure of the Saint-Sernin Basilica in Toulouse houses a very beautiful pair of bishop's gloves knitted of linen thread in stockinette stitch. Although they are known as "a pair of gloves belonging to Saint Remy," these gloves do not appear to be earlier than the thirteenth century.

A major piece of evidence reveals that knitting was being practiced in Germany by the beginning of the fifteenth century; the Buxtehude Retable (Kunsthalle, Hamburg) depicts the Blessed Virgin knitting, on four needles, a garment for the Infant Jesus.

Knitted articles from the fifteenth century appear in slightly larger numbers in our museums; they would probably be even more numerous but for the fragility of the materials utilized. The Cluny Museum and the Musée des Arts Décoratifs in Paris possess very beautiful purple wool gloves knitted in stockinette stitch, like those of Toulouse, but decorated with motifs knitted with a gold thread, which forms a kind of encrustation in the fabric. Still more luxurious is an Italian glove that can be seen in the Musées Royaux d'Art et d'Histoire in Brussels; it is of red silk with gold, silver, and green silk designs, and dates from

Plate 16. The Virgin knitting. Painting by Buxtehude, fifteenth century, Kunsthalle, Hamburg. *Photo Kleinhempel.*

the end of the fifteenth century. All these gloves belonged to church prelates. Since the ritual required that liturgical gloves be seamless, they could only be knitted.

First evidence of knitting in England

We have little information on knitting in France in the Middle Ages, but the situation is much better in the case of England, which seems to have had very powerful and well-organized knitting guilds and corporations in the fifteenth and sixteenth centuries. Everything had been done to ensure perfect organization of the knitters. The apprenticeship lasted for three years, after which the newly formed journeyman had to go abroad for an equal number of years to learn the knitting techniques practiced in other countries. After six years of experience the journeyman could petition for a license as a master craftsman. For this he had to complete four specimen projects: a "rug measuring 8′ x 12′, a woolen shirt, a woolen cap, and a pair of woolen hose." The "rug" was actually a kind of tapestry that had to depict a very complex design composed in accordance with a conventional stylization of foliage, birds, and flowers, and executed with numerous shining colors (the use of twenty or thirty colors was not rare). These specimens had to be completed in thirteen weeks.

In Tudor days knitting was given a great impetus by the fashion for caps. This headgear, whose manufacture was similar to that of the Basque beret, was worn by all classes of society. Some of them were highly complicated, being rolled, draped, and even pleated. The more simple versions formed the traditional headgear of the apprentices, for whom they were a kind of insignia.

We shall later see how the development of knitting during the reign of Queen Elizabeth I quite naturally led to the invention of the first knitting machine. For the moment, let us merely point out that this invention did not immediately deal a deathblow to handwork, which had by then reached a rare degree of perfection, even in the very difficult art of knitting stockings, provided that they were of wool or coarse yarn. Mention should be made of the very strange pair of horseman's stockings preserved in the Victoria and Albert Museum in London.

In the valleys of Yorkshire hand knitting became an extremely prosperous craft in the seventeenth and the first part of the eighteenth centuries. The finished articles were collected by merchants, who sold them at the Kendal market. But hand knitting soon bowed before the machine competition, and from then on only small local groups who were satisfied with small markets continued to exist in England.

Here, again, we owe this information on knitting in England to James Norbury.

Appearance and development of knitting in France

As for France, not until 1505 do we discover statutes for brotherhoods of knitters or, more precisely, "hosiers," in the strict sense of that word. They are found at Troyes, which with the advent of the "machine age" was to become one of the international capitals of the knitting industry.

At the beginning of the sixteenth century the principal activity of these early "hosiers" of Troyes was undoubtedly the manufacturing of hose, almost always of wool (J. Ricommard), but there was no question of stockings, which now as in the Middle Ages were cut from cloth, and sewn, following a very ancient production technique. The Musée Bargoin at Clermont-Ferrand possesses such stockings (or rather "hose") found in a Merovingian tomb. In the sixteenth century in France, as in England, the most important people wore hose that were cut, frequently from extremely rich fabrics, and sewn. As we have seen, however, as early as the twelfth century (and even earlier) the Arabs were already knitting stockings; in this activity the East was more advanced than Western Europe.

In any event the hosiers' guilds seem to have been extremely prosperous by the sixteenth century, for in 1554 the statutes were revised and supplemented. At this time mention is made of the manufacturing of stockings, apparently by hand. By 1698 the number and size of the workshops justified the renewing of the earlier statutes and the addition of five new articles.

The regulations of 1505 were primarily religious in nature; those of 1554 and 1698, in contrast, are indicative of technical preoccupations along with an evident desire to protect the guild. Thus, Article I outlines the requirement of the submission of a specimen ("masterpiece") "for anyone who wishes to be a master hosier in this city of Troyes and its suburbs."

Article 5 stipulates that "all persons other than the masters of the said craft are forbidden to work for anyone and to order hose, stockings, slippers, woolen wristlet gloves, caps, brodequins, and other articles, whether of wool, yarn, cotton or worsted, on large or small needles, and even with the loom under pain of three pounds fine." Worsted, which is often mentioned in seventeenth-century texts relating to hosiery, is a combed wool.

Mention of the loom in Article 5, at a period when mechanical production of hosiery was not yet in existence in Troyes, is evidence of a desire to leave the door open to a possible future introduction of the knitting industry (which as we shall see was already prospering in other centers) into the capital of Champagne.

Ordinary and luxury knitwear

It is unfortunate that no knitted articles of the sixteenth century have survived. They were certainly cheap, and intended for a clientele of modest folk who got rid of slippers and hose when they were worn out. Knitted stockings were worn by the rich only when they were of silk. According to a tradition (unverified as to origin) related by all the authors, the Spanish became internationally famous for their hand knitting of silk stockings, a feat accomplished thanks to the use of very fine metal needles. In France, as in England and Italy, wooden or bone needles were used, which produced a rather crude article. The price of these hand-knitted silk stockings was so high that only kings and the most powerful nobles were able to own them; King Henry VIII of England had only one or two pairs in his wardrobe. On this subject the chronicler John Stow tells us: "He wore only slippers from cloth or satin 45 inches wide. . . . Unless by great good luck there came from Spain a pair of Spanish silk stockings." (*The Annals or General Chronicles of England,* by John Stow, London, 1615.)

Speaking of Edward VI, Henry's successor, the same author notes that "he had a pair of long Spanish silk stockings, sent as a sumptuous gift." And in the margin he notes that "Sir Thomas Gresham gave them to him."

Such a gift must therefore have been an important and valuable present, if a chronicler saw fit to mention them about sixty years later in *The Annals.* Stow also tells that in the second year of the reign of Queen Elizabeth, her first lady-in-waiting, Mrs. Montague, presented Her Majesty with a pair of knitted black silk stockings. "They had been knitted by the donor herself, and the Queen was so enchanted by them that she no longer desired to wear any others." It seems that from that moment on, the Queen's ladies were ordered to knit her silk stockings for her.

No author gives us any information on the degree of strength of these delicate masterpieces, one of which (or at least a similar specimen) is still preserved at Hatfield House.

At this point the question of productivity must have become extremely important. At the beginning of the seventeenth century, at the very moment when, as we shall later see, the inventor of the loom was beginning to build his first machines, efforts to solve the problem do not appear to have made great progress, since King James I, who apparently wished at the moment of his accession to

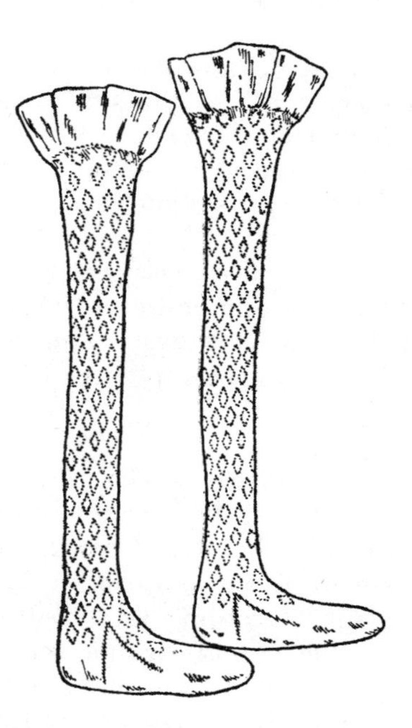

FIG. 40. Hand-knit stockings of Queen Elizabeth I of England (Hatfield House, Hertfordshire).

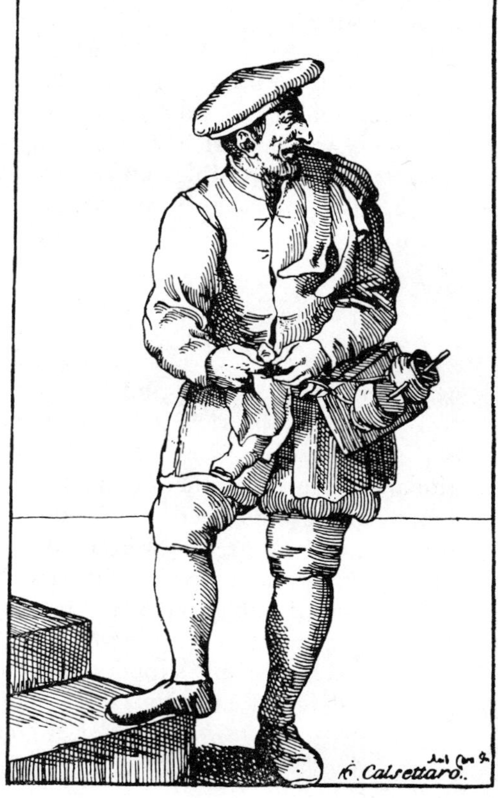

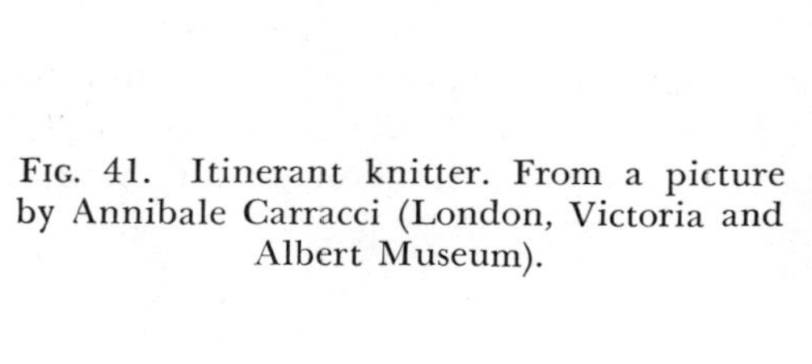

FIG. 41. Itinerant knitter. From a picture by Annibale Carracci (London, Victoria and Albert Museum).

the English throne to make a good appearance before the foreign ambassadors who had come to congratulate him, had to borrow a pair of silk stockings.

Around the same period, this time on the Continent, we find a reference to the price such articles could command. In 1614 the Register of the Councils of the City of Avignon notes an expenditure of 14 crowns authorized in order to offer a pair of Milan silk stockings "to the secretary of the Lord High Constable." These were surely hand-knitted stockings, since, as we shall later see, not until fifty years later did the stocking machine begin to spread through continental Europe. By now, however, this gift already appears to have lost its former character as a very rare item reserved only for royalty. It is impossible to say whether the labor (and the raw material) were less expensive in Milan than in Spain.

Knitting appears to have been practiced by the twelfth century in certain regions of Italy. Later, at the end of the sixteenth century and the beginning of the seventeenth, the Italian markets had itinerant knitters. Proof exists in a strange engraving that reproduces a detail from a painting by Annibale Carracci: A man dressed in a craftsman's or peasant's costume, in a standing position, is knitting a stocking with needles (Figure 41). This stocking must have been of two colors, for the man is using two balls of wool attached to a small board that he has slung by a strap over his shoulder (Victoria and Albert Museum, London, Department of Engravings, E. 2669–1928). The itinerant knitters (who were also salesmen) undoubtedly made ordinary articles, but in certain regions of Italy (for example in Venice, a rich city where the most costly garments always found purchasers) the manufacturers produced admirable tunics of multicolored silk, embellished with gold and silver threads, which date from the seventeenth and eighteenth centuries (Victoria and Albert Museum, London).

In the middle of the eighteenth century there still existed guilds of hand knitters who had preserved the tradition of a long, conscientious, rigorous period of training terminated by tests similar to those which two or three hundred years earlier had crowned the years of apprenticeship and journeymanship. One proof is furnished by the strange "rug" bearing the date of 1740, executed for the purpose of obtaining the diploma of master knitter of the Colmar Guild (Musée d'Unterlinden, Colmar).

After the invention of the hosiery loom, and when its exploitation had reached the industrial stage, hand knitting continued for more than a century, but only in connection with two types of articles: on the one hand, luxury garments and beautiful tapestries; on the other, ordinary, moderately priced woolen hose and worsted stockings. Even after the establishment of the mechanical knitting industry in France, Colbert continued to encourage the establishment and maintenance of hand-knitting shops, in order not to deprive the poor of this breadwinning activity. They had to be satisfied with low wages, however, and also work at dizzying speed. According to authors of that period, the record is supposed to have been eighty stitches per minute.

Not until around the end of the eighteenth century did hand knitting disappear completely, or at least cease to be a craft activity and become limited to a domestic task.

Invention and development of the stocking loom

Two important factors promoted the invention of the stocking loom: first, the necessity for rapid production because of the continually growing demand, and, second, the fashion for silk stockings, which it was difficult and time-consuming to knit by hand.

The same question asked earlier in connection with hand knitting now arises in connection with this second stage in the development of knitting. But here, despite the fact that the origins of the invention of the stocking loom, absolutely prodigious for its time, are still to a certain extent enveloped in mystery, we are now on more solid ground. Despite the somewhat legendary nature of the circumstances that aroused the spark of genius in the inventor, it is now definitely known that the inventor of the machine was the English clergyman William Lee, a graduate of Cambridge University, where he received a Master of Arts degree around 1582–1583. The procedure of his work can apparently be reconstructed, with a good chance of probability, as follows:

We have seen that in the sixteenth century silk stockings were a luxury article extremely rare on the market. The knitting of stockings was being practiced in this period (it was particularly widespread in England), but these were coarse woolen stockings knitted in one day by thousands of hands. "Bonnets" and garments that could be knitted quickly could permit a worker to earn his living provided he worked rapidly. Such was not the case with stockings, which took much longer to knit properly and could not be sold at a high price, since the rich felt they were unworthy to be worn with the splendid and sumptuous court costumes about which the sixteenth century was so enthusiastic.

Under these conditions it was natural that an inventive spirit should try to solve both phases of this problem by creating a machine capable of doing the work faster and at the same time producing a finer article. In the loom a series of steel fingers by moving simultaneously formed an entire row of stitches with a single action, thus permitting much faster execution.

The circumstances of the invention

Certain authors, more inventive than well informed, have surrounded the somewhat mysterious personality of William Lee with a charming legend. Lee was the Vicar of Calverton in his native Nottinghamshire, though we do not know whether he was born at Calverton or in Woodborough; his birth is registered in neither place. No registers were kept prior to 1568.

Being engaged to a poor girl who earned a meager living from her knitting needles, Lee wished to free her from this painstaking work and give her the possibility of earning greater profits. According to another version of the story, he is supposed to have married while still a student at Cambridge, in violation of the rules, which led to his expulsion. The young couple, reduced to quasi-misery and with an infant to support, owed their subsistence solely to Mrs. Lee's courage and nimble fingers. It is difficult to choose between these two versions, which are probably equally imaginary, but it seems certain that Lee conceived of his invention by watching hand knitting.

We do not know exactly what type of education corresponded to the title

"Master of Arts," but it probably included a certain amount of technical drawing, without which its holder could not have embodied his idea in an actual machine, or, more precisely, ordered the construction under his direction of a prototype by a locksmith or blacksmith, with the assistance of a cabinetmaker for the construction of the wooden housing.

Milton N. Grass, the modern historian who has made the most complete existing study of the English origins of the hosiery loom, gives a detailed history of the tribulations of the unlucky inventor, who, it is believed, could not succeed in obtaining from Queen Elizabeth the license indispensable to anyone who wished to start a factory. The reason given by the queen was the coarseness of the sample Lee made in her presence. It was undoubtedly made of thick wool and could not be compared to the delicate masterpieces made by Mrs. Montague and her imitators. But if the inventor should succeed in making silk stockings on his machine, the question could then be reconsidered. Apparently Lee was not discouraged, and went back to work.

The authors of that time give us scanty, but very important, information. The first of these references is found in a work entitled *A Transcript of his Majesties Letters Patents, granted unto Simon Sturtevant, for the said Metallic Business, for one and thirty years,* and printed in London, by "George Eld, on May 22, 1612." Its author was not satisfied to reproduce patents, but also described various mechanisms, mentioning their origin. With regard to William Lee's invention, he says that ". . . in hose and stockings, there is a progression of three [steps]. Cloth or kersey, stockings with needles, thirdly and lastly, in knit stockings with loome, which is a late invention of one Maister Lee."

In his *Annals,* published in 1615, John Stow devotes several lines to the history of the invention:

"In the year 1599 the art of knitting or weaving silk stockings, vests, and various other articles with steel machines was invented and perfected by William Lee, Master of Arts, of Saint John's College, Cambridge.

"He then went to France, where he obtained patent letters from the King and taught this secret.

"His companions were welcomed to several foreign countries, such as Spain, Venice, Ireland, and still others, where they taught the secret of their art, and in this way it spread through the world."

Here we have two definite statements on which to base our discussion, since they are contemporary evidence. However, several of Stow's statements have been contested, beginning with the date, which all the modern historians place in 1589, not 1599. We are thus forced to admit that the first date must be based on an oral tradition, and in our opinion the unanimity of the testimony gives the date of 1589 a certain credence.

However, both dates can undoubtedly be retained, for William Lee could not have constructed a perfect loom on the first attempt. His first results were undoubtedly too crude, and perhaps too loosely knitted, and he may have worked for long years to perfect its mechanism by giving it thinner needles and a finer gauge.

As for Lee's emigration to France, the only early references to this subject are those of Stow and all the authors who repeated the story after him, often

Plate 17. Workshop of an English hosiery maker. Engraving taken from *The Universal Magazine*, VII, 1750. *Photo Conservatoire des Arts et Métiers.*

embroidering on and embellishing the theme. According to this author, Lee is supposed to have enjoyed great success in France and other countries. In contrast, all the other historians tell the story of his disappointments. He went to Rouen with his brother James and six workers, bringing with him an equal number of machines. Shortly he was a fellow victim of the assassination of Henri IV in 1610, before he had had the time to install his factory and profit by the patent letters the king, who perhaps was interested in him because of his Protestant religion, is said to have granted him.

We are uncertain whether he died in 1612 or in 1620, after a life of misery in France where the knitting machine seems to have been regarded, as in England, as a fantasy and a utopian idea. We have not been able to find in the Archives of Rouen the slightest trace of William Lee's passage. Nor do we have any confirmation of James Lee's return to England. It has, however, been claimed that he brought the machines back to England, sold them to various London merchants, and returned, without the hoped-for success, to his native Nottinghamshire. Here, thanks to the collaboration of one of his former friends, a man named Aston, he again revised the system and, having brought it to the level of industrial productivity, finally succeeded in founding the first factory for mechanically knitted silk stockings.

In any event, one fact is certain: in 1657 a group of framework knitters requested from Cromwell a charter similar to those that then governed the manual professions. This charter was granted to the "Masters, directors, assistants, and Company of framework knitters," and the group was later known under the name "Worshipful Framework Knitters."

These regulations, which were reinforced by a new charter after the death of Charles II, were aimed at preventing abuses and frauds and also (given that the machine was of purely English origin) of forbidding the export of the invention. Thus they were attempting to maintain both the privileges of the masters and a genuine state monopoly that was very profitable to the English crown.

It has too often been repeated that the death penalty was incurred by anyone who revealed the secret of the hosiery loom to a foreign power. According to the texts, the punishment actually appears to have been no more drastic than a very high fine.

The success of the machine was tremendous, and factories proliferated throughout England. By 1669 England possessed a total of 660 looms, 400 of which were in London; three-fifths of them were silk looms (W. Felkin, in *An Account of the Machine-Wrought Hosiery Trade* [London, 1845]). By 1782 the country had 20,000 looms. By now, however, England's monopoly was no more than a memory, for despite a veritable "hunt" for smugglers of the machines, the secret soon spread over the Continent.

First appearance of the stocking machine in France

In 1656 Louis XIV decided to establish a factory for "the production of stockings, canions, dressing jackets, and other silk objects . . . both to give employment to several workers who would find in this factory an honorable means of supporting their families and to prevent the flow of various sums of *deniers* into the foreign countries where these objects are manufactured. Which production was very successfully begun through the efforts of Sir Jean Hindret, whom we established in our Château de Madrid, and who was so successful in this enterprise that, its progress having become considerable, in order to increase it we decided to establish a company which has since then made every possible effort to perfect it."

Thus in 1656 the stocking loom, which approximately fifty years earlier had made a furtive entry into France followed by a rapid retreat, returned, this time in triumph, with the blessing of the royal protection. Jean Hindret, who at Colbert's instigation brought back the secret from England, was given the title of "the first instructor in France of silk stockings made on the machine." In January 1656 the king granted him an exclusive license (confirmed by the Parlement on May 13, 1659) for all France.

Apparently overwhelmed by the extent of the task, Hindret soon had to request the assistance of partners, and by 1666 the factory of the Château de Madrid at Neuilly was being administered by a genuine company that included, in addition to Hindret, Lords Babbert, de Rotrou, Vanganguelt, and de Ny.

The factory seems to have begun as a kind of school; it was first necessary to show the students how to construct the looms, then how to use them. Despite a stunningly successful (in the opinion of contemporaries) beginning, its early years were difficult, as is attested by the patent letters granted by Louis XIV in 1672, establishing the factory as a master school and community. The first two hundred masters who graduated received from the king the sum of two hundred livres to assist them in paying for the loom, which was to be supplied by the factory.

The hosiers of Avignon At the same period, however, "English-style" stocking looms appeared in the French Midi, introduced by English workers who settled in Avignon. These workers were Catholics who had left their country for this reason, and perhaps also because they hoped to find better living conditions in the Comtat. Their hopes apparently were not disappointed, and in 1662, while Hindret was struggling with difficulties, we find them installed as master craftsmen, training apprentices, building, selling, and renting machines, signing contracts, buying houses, and engaging in similar activities. At the beginning of their activity the price of a machine was approximately five hundred livres, but this price had dropped to three hundred livres by the end of the century. The metal parts – needles, sinkers, presser bars, and so on – were made by locksmiths and even clockmakers, the wooden housing by carpenters or cartwrights.

At first only silk stockings were made; soon, however, wool, cotton, and filoselle (a kind of low-grade silk) were also being used. Production was not limited to stockings; it included all the articles mentioned above in connection with Hindret's factory. By 1666 the contracts mention "gloves, chemisettes, and hose."

The Archives of the Vaucluse have preserved the names of these manufacturers of stockings and other articles knitted by machine: Chapman, Brent, Tillee, Hill (called "Montagne"), Ware, Poulton, and others, who had come to France with their families, and established themselves there. Some of them brought the knitting industry to southern cities like Toulouse, where, in 1688, William Ware settled as a "merchant of stockings for men and women."

The success resulting from the establishment of English-style mechanical knitting caused similar attempts to be made throughout the entire region, a factor that was to bring about rapid progress.

In 1667 one Louis Boucherat, an important person at least by his rank and fortune, obtained through patent letters established at The Hague the authorization to "order the construction in this city of Orange of all the structures, English-style looms, and other items necessary for the establishment of the manufacturing of stockings, slippers, pants, ribbons, camisoles, petticoat breeches, canions, and other similar objects of silk, thread, and wool, with exclusive license for the Principality, under pain of three thousand livres' fine and confiscation of looms and merchandise for fifteen years."

Numerous contracts of apprenticeship, journeymanship, renting of services, and so on, attest to the activity of Boucherat's factory, but all these items are signed by his representative at Orange, and he does not appear to have ever come to inspect his factory. Although he is sometimes called a "silk-stocking worker," he does not appear to have been a technician, but a rich bourgeois with a sense for "a good business" – for the success of the English at Avignon had taught him that the time was ripe for such a venture.

Even at the time when the English were settling at Avignon, a few machines were installed at Oppède. Soon there were several in Toulouse, introduced by William Ware; Montauban was next. This aroused continual protest on the part of those who had every intention of preserving their monopoly over the hosiery trade.

Beginnings of the prosperity of Nîmes

At Nîmes, which was to become an important center, the first machine seems to have been introduced by Cuvellier, who had been trained by Hindret at the Château de Madrid. According to J. Boissonnade its introduction was due to Louis Félix, who had the metal parts made by the locksmith Timothée Pastre in 1680.

A memorandum of 1764 mentions the year 1640 for the establishment of the knitting industry at Nîmes. In the absence of other testimony, however, it is impossible to be certain of this date.

The success of the first stocking-manufacturing shops at Nîmes was astounding. This industry was making progress almost everywhere, but the principal centers were very soon established at Lyon (the Fournier factory), Caen, Granville, Saint-Lô, Valognes, Coutances, Louviers, and Bayeux. The Château de Madrid factory, after a very promising beginning, did not develop, and the competition created, despite Hindret's license, by the factories just mentioned did not help matters. Moreover, Hindret's monopoly was more theoretical than actual, as is proved by the foundation of the Fournier factory at Lyon, which in 1667 received from the councilors an extremely strict set of rules whose highly detailed text shows that this was a genuine factory (Bibliothèque Calvet, Avignon, fds. Chobaut, No. 00349).

After Hindret's death in 1697, the centers of the knitting industry multiplied to such an extent that the truth had to be recognized: the monopoly was no longer justified, and the demand was such that the proliferation of factories was not dangerous. In 1700 a royal decree authorized the "Master manufacturers of stockings" to exercise their craft in the cities of Paris, Dourdan, Rouen, Caen, Nantes, Oléron, Aix-en-Provence, Nîmes, Toulouse, Uzès, Romans, Lyon, Metz, Bourges, Poitiers, Amiens, Orléans, and Reims (Arch. Nat., F. 12, 1396). Most of these cities had not waited for the royal authorization.

One year later other cities, such as Toulon, Marseille, and Montpellier, as well as smaller places, also entered the competition. The Midi continued to prosper, as is proved by the statistics. For example, in 1706 Nîmes had 870 machines; in 1711, 1,100; in 1743, 3,200. By 1782 the region had 9,000 machines.

Troyes

It is noteworthy that the decree of 1700 does not mention Troyes, which later became the great international center of the hosiery industry. The city had long had an important trade guild of hand knitters, but it did not become interested in mechanical knitting until the middle of the eighteenth century.

Until then the weaving industry had been very prosperous throughout the region. By 1745, however, its markets were becoming scarce. Knowing of the prosperity brought to many other cities by the production of stockings, Mayor Jean Berthelier conceived the idea of establishing this sister (or at least cousin) industry of the work to which the workers were already accustomed. Encouraged by a first successful attempt at Arcis-sur-Aube and the environs, he ordered several machines to be tried out at the Hôpital de la Trinité, an orphanage for poor children. In this way a source of cheap labor was available, and at the same time these unfortunate children could be taught a trade.

The Musée de la Bonneterie in Troyes is the fortunate possessor of one of the first stocking-knitting machines in the city. It is a very beautiful and very unusual object, whose delicate decoration proves that it must have been constructed by a craftsman accustomed to making beautiful furniture. However, we may ask whether this machine, and particularly its metal parts, was built at Troyes. Was it perhaps ordered from a center like Nîmes, or from abroad, for example from Weimar? Was a plan obtained, and if so where? These questions have not yet been solved.

From Arcis, and later from Troyes, the machine knitting of hosiery spread rapidly through the rural area, providing the peasants with additional sources of income that they were to retain for almost two hundred years and that were not to disappear until very recently in the face of the enormous development of the monumental factories which now supply the livelihood of a large part of the population of Troyes.

In the eighteenth century, however, southern Champagne was simply one hosiery-manufacturing region among many in France, and its prosperity could not be compared with that of the cities of Languedoc. Only at the beginning of the nineteenth century, thanks to the genius and perserverance of its loom builders, did Troyes assume one of the leading roles in the international knitting industry.

Technical Development and Productivity

Operation of the old hosiery loom

The ordinary stocking loom consists of a series of hooked needles mounted parallel to each other in a single horizontal line and spaced in accordance with the closeness of the mesh to be obtained. This mounting is called the "needle bar" or "needle plate."

The extended hook of the needle does the work of the loom. The hook is sufficiently flexible to permit its outer portion (the spring beard) to be closed by pressing it against the stem of the needle. At the point of contact is a groove that holds the beard of the needle so that it cannot catch the loop of thread sliding toward the bottom of the needle. The beards of the needles are closed by lowering a horizontal bar called the "presser bar."

Between the needles are vertical blades called "sinkers." Each sinker has a right-angled hook, and forms a convex curve beneath this hook; they are capable of moving both vertically and horizontally in a straight line.

When a piece of work has been begun, it is suspended from the needles by a series of stitches (Figure 42, Diagram 1). The knitting yarn is laid loosely on the horizontal row of the needles. A descending movement of the sinkers causes their

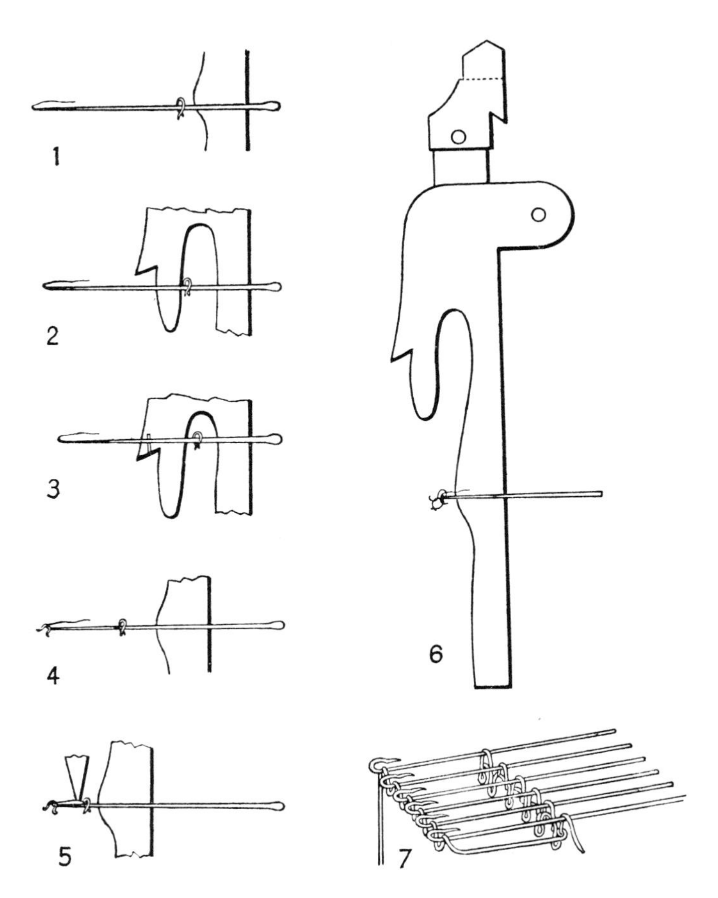

FIG. 42. Diagram of the operation of the ordinary knitting loom.

hooks to press on the thread and force it into the empty spaces between the needles. The straight thread is thus changed into a kind of zigzag line (diagrams 2 and 3). The sinkers then push the thread toward the front end of the needles, and the thread slides under the curved portion (Diagram 4). The presser bar is lowered, and the yarn is locked in the beards of the needles. The sinkers are raised, thus freeing the row of knitting begun; the lower portion of the sinkers pushes it toward the end of the needles until it comes out above the closed beards. The presser bar is lifted when desired, that is, when the stitches reach the bottom of the beards (Diagrams 5 and 6). Obviously the row of stitches first formed passes under the beards, while the row being made is still engaged. This is what forms the stitch (Diagram 7).

After the execution of these various movements necessary for the formation of a row of stitches, the functional parts need only be made to resume their original positions.

Naturally, in order to begin a piece of knitting it is not necessary to start it with hand needles; a first row of stitches need only be cast on by forming a series of closed loops of yarn around the needles of the machines.

Questions about Lee's mechanical loom

Many documents that would permit us to determine exactly what William Lee's loom was like are missing. The sketch of his instrument illustrating the charter granted by Cromwell in 1657 is extremely rudimentary, and cannot be used to aid us in the description of the device as it came from the hands of its creator. Moreover, this sketch is known to us only through an engraving dating from the beginning of the nineteenth century that reproduces an older document, now lost, and there is nothing to permit us to affirm the precision of the copy.

The Technical Museum at Falun in Sweden possesses a knitting machine that, according to a tradition communicated by the director, is supposed to correspond exactly to Lee's model. However, this machine was not introduced into Sweden until 1723. It had been constructed in England and was smuggled in, together with two others of the same type, by one Jonas Alströmer, who thereby succeeded in introducing the stocking loom into Sweden.

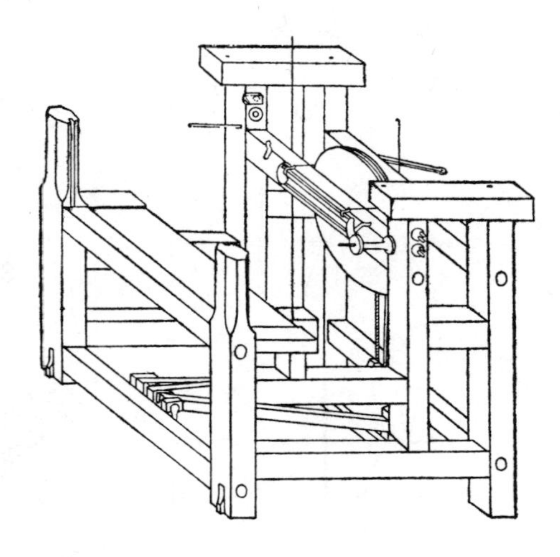

FIG. 43. Housing of the early knitting loom (Hindret's notebook and the *Encyclopédie*).

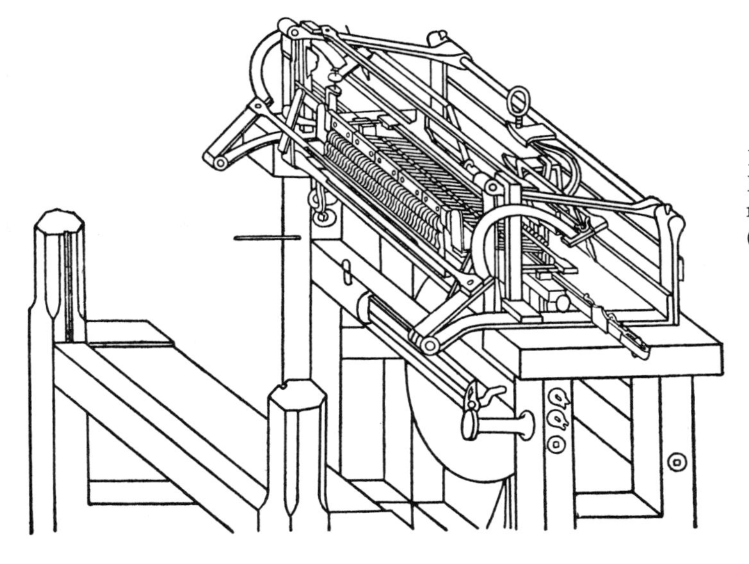

FIG. 44. Early knitting loom. The mechanism is mounted on the housing (Hindret's notebook and the *Encyclopédie*).

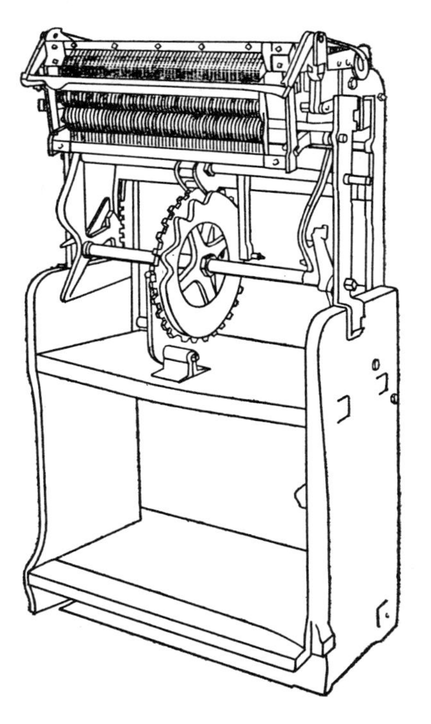

FIG. 45. Knitting loom constructed in Sweden by Polhem, around 1730 (Bergslagets Museum, Falun, Sweden).

Hindret's loom

This loom is completely classical in type, and in essence is no different from the loom drawn by Jean Hindret himself for the construction of the prototype at the Château de Madrid (Cabinet des Éstampes, Paris, LH, 32, 1571). This series of exceptionally interesting drawings is to the best of our present knowledge unquestionably the oldest extant document concerning the construction of a hosiery-knitting device. It was operated by treadles, and the movement was transmitted to the working parts by complicated systems of ropes. Its working parts — that is, the needles, levers, and sinkers — are arranged in the same manner in both Lee's loom at Falun and Hindret's drawing, and we find the same arrangements and characteristics pictured in the plates of the *Encyclopédie* of Diderot and d'Alembert ("Métier à faire le bas" and "Faiseur de bas au métier").

The authors of the *Encyclopédie* were confused by the extreme complexity of the machine. They write: "It came from the hands of its inventor almost in the state in which we see it; and as this circumstance must add greatly to our admiration, I preferred the loom as it was before to the loom as it now exists, observing only to indicate their minor differences as they arise" *(Encyclopédie,* Vol. II, p. 98, Col. 1).

Fig. 46. Spinning wheel; from Hindret's notebook (Paris, Bibliothèque Nationale, Cabinet des Éstampes).

Thus the machine as it is drawn and described in the *Encyclopédie* is that of the beginning of the eighteenth century — Hindret's machine, with the exception of a few details. In this way we can retrace our steps back to the inventor, and are able to form a fairly precise idea of what the model was like after its revision by Aston. It would be foolish to believe that the original device was a simple machine that became more complex as it was improved. The system could not be simple: it had to be capable of making a stitch.

The text divides into five operations the movements required for the complete formation of the stitch around spring-beard needles. The inventor's first idea was certainly to imitate the phalanxes of the fingers with the help of the play of the levers and sinkers, the needles being placed between the latter as if they were placed between the fingers. This mechanism must have been the embryo of the system. Once it had been found and assembled, together with the transmission of the movement to the treadles, the device was able to knit. Later improvements and perfections had only two purposes: to make the fabric finer by bringing the needles closer together and by using thinner sinkers, and to increase productivity.

Gauge and productivity The gauge, a unit of measure still in use, was undoubtedly the principal concern of the first constructors of looms. It corresponds to the number of needles contained in a given space (the English inch). The fineness of the work was in direct proportion to the number of needles in this space.

An eighteenth-century technician tells us that "for silk stockings, looms 24, 25, 28, 30, and 40 are the most favored looms, the first being used for working striped, chine, spun silk, and brocaded stockings. The 28 and 30 are for the solid-color black, white, and other colors of stockings. The 22 looms are used to make woolen, cotton, filoselle, and floss silk stockings.

"There are not many gauge-40 mills; few workers perform this work, which is harmful to the eyesight, and few consumers would pay the price for it" (*Mémoires sur la bonneterie du Département de Nîmes,* 1782, Arch. Nat., F. 12, 1398).

It may be asked what was the output of the looms of the seventeenth and eighteenth centuries. We would be singularly mistaken if we imagined that at the end of the day dozens of pairs of stockings were piled up beside each loom. In 1667, at the Fournier factory in Lyon, a good journeyman was expected to produce three pairs of silk stockings per week, working twelve to thirteen hours a day. These were straight stockings. If the owner called for stockings shaped at the corners and along the leg, the number was reduced to two. Taking into account the fact that this Fournier was particularly demanding and that his employees were subjected to a genuine "forced labor" regimen, we are forced to conclude that this was the maximum output.

One century later the output had increased slightly; at Nîmes a good worker made four pairs of silk stockings and six pairs of filoselle stockings per week. Productivity was higher at Marseille: in 1779, forty looms made sixty pairs a day, and in the same period ten looms made fifteen pairs of cotton stockings.

Thus there was visible progress, perhaps due to better construction of the mechanism and also undoubtedly to the increased skill and dexterity of the labor force. It must be understood that while the knitting was done mechanically, the worker was nevertheless obliged to work the needles by hand in order to form the leg and foot, and these operations were long and tedious.

The contribution of the nineteenth-century mechanicians was to be their conception of a device that made it possible to multiply on a single loom the sets of needles and sinkers (or "heads"), thus permitting the knitting of several stockings simultaneously.

BIBLIOGRAPHY

Encyclopédie, ou dictionnaire raisonné des sciences, des arts et des métiers (17 vols. text, 11 vols. plates, Paris 1751–72, plus supplements). This is the famous *Encyclopédie* of Denis Diderot and Jean d'Alembert. A selective condensation, including reproductions of some 485 of the original 3,100 plates, has been published under the editorship of Charles C. Gillispie, *A Diderot Pictorial Encyclopedia of Trades and Industry* (2 vols., New York, 1959).

FREUDENBERGER, HERMAN, *Waldstein Woolen Mill* (Boston, 1963). Description of an eighteenth-century textile complex in Bohemia.

GRASS, MILTON and ANNA, *Stockings for a Queen: The Life of the Rev. William Lee, the Elizabethan Inventor* (London, 1967).

HEATON, H., *The Yorkshire Woolen and Worsted Industry from the Earliest Times up to the Industrial Revolution* (Oxford, 1920).

HOWITT, F. O., *Bibliography of the Technical Literature on Silk* (London, 1946).

LAWRIE, L. G., *A Bibliography of Dyeing and Textile Printing Comprising a List of Books from the Sixteenth Century to the Present Time* (1946), (London, 1949).

LIPSON, E., *The History of the Woolen and Worsted Industries* (3rd ed., London, 1950).

USHER, ABBOTT PAYSON, *A History of Mechanical Inventions* (rev. ed., Cambridge, Mass., 1954; paperbound ed., 1959).

Section Two

The Mechanical Arts

CHAPTER 12

INDUSTRIAL MECHANIZATION

THE CONCEPT of the Industrial Revolution, which has been so carefully fostered and utilized by historians, is usually accompanied by another, vague and often unformulated, idea: the notion of a technical revolution. If it does continue to have some significance in the modern world, the Industrial Revolution of the end of the eighteenth and early nineteenth centuries can find it only in its economic and social aspects. In the history of the actual technological achievements, however, it did not distinguish an age as exceptional as has for so long been claimed. The technological history of the two centuries that now separate us from the period regarded as the beginning of this phenomenon, as well as of the history of the two centuries that preceded these same fateful decades, will assist us in understanding this fact.

The myth of the technical revolution

For a long time the myth of the technical revolution concealed from historians and their readers the fact that this progress had been a continuous process. It seemed to them that industrial mechanization had been born spontaneously of that age, considered as crucial, which allegedly corresponded to the Industrial Revolution. The fact is that mechanization had by then attained a very advanced state of development, which was to lay the groundwork for the age of the machine tool.

Not until the first quarter of the nineteenth century was the machine tool actually made available for use by industrial technology. The first attempts to adapt the lathes and drilling and boring machines then available for industrial use took place during the second half of the eighteenth century. The fact is that several of these machines had already been set up to manufacture large objects and to obtain results that could not possibly be achieved with the hand tool. This implies that other mechanical methods were also available to the metalworking industry for the handling and movement of these large objects.

In industries of this type, progress in mechanical adaptations was in fact achieved in steady, rhythmical fashion. An innovation, to say nothing of an invention, is of interest only if it can take its place within an apparatus that has already attained a sufficiently high level of development to accept it. For example, the perfecting of a precision borer or a slide lathe was of genuine interest to factory owners only if the time gained on one operation formerly done by much slower methods was not wasted through inability to accelerate the complementary operations. Examples of the application of this law of progress can be seen constantly around us, and it does not apply only to our own age; it has governed progress at least since the beginning of modern times, that is, since the end of the Middle Ages.

The first stages of industrial mechanization

The metallurgical industry was not the only one concerned with the perfecting of mechanical methods. Until the eighteenth century wood remained the most widely used material, even for the construction of machines. Machines for the production of power were constructed essentially of wood, especially until the advent of the steam engine in its early forms, and it was perhaps for the cutting and working of this material that the first major mechanical techniques were utilized. The working of other materials, such as stone, marble, and then metals, in turn benefited by the adaptation of these methods. Thus the human race acquired a long experience with these problems, and at last realized the necessity of solving them in ways other than by haphazard empirical experimentation.

Most of the chapters in these volumes demonstrate the increasing preeminence of mechanization in human labor. In this way we learn how industrial mechanization appeared in a form appropriate to it by virtue of the methods available to its creators in each of these periods. During the fifteenth and sixteenth centuries, particularly in connection with water mills, the first models of power hammers, grindstones, drilling machines, and mechanical saws evolved from the primitive forms given them by the Middle Ages into stronger and more effective structures. Once this age of development was over, the evolution continued very steadily and with more assurance during the next two centuries. These centuries have been unduly neglected in this regard, and it is precisely for the purpose of drawing attention to their importance that we wish to combine in this chapter a certain number of facts that are dispersed through the preceding and following chapters.

HEAVY INDUSTRY

It is possible to apply the term "heavy industry," even in its modern acceptance, to the centuries of transition that separated the age of the Renaissance from the end of the eighteenth century. By their combination of various methods of production, the workshops that produced cannons, ships' anchors, statues, large bronze objects, and brass and lead sheets constituted a "heavy industry" before its actual invention, by comparison with the other contemporary industrial fields —glassworking, ceramics, industrial chemistry, food and textile chemistry, and those industries that utilized metal products (sand-molders, blacksmiths, locksmiths, cutlers, and manufacturers of cutting tools, clockmakers and watchmakers, and others). The illustrations in the numerous technical treatises of the sixteenth, seventeenth, and eighteenth centuries demonstrate (to a much greater degree than the reading of economic statistics on the tonnage of raw materials and combustible substances used and the quantity of manufactured products) the existence of this industry and the constantly improving technological foundation that supported its activity.

Evolution of artistic representation

It is undoubtedly impossible to trace the stages by which it was perfected during this period. The texts tell us little, and the art of technological description was rudimentary in the sixteenth and seventeenth centuries; not even by the eighteenth century had it achieved the clarity and precision of style that would permit the reader to follow, step by step and without ambiguity, the development of the various operations. The illustrations speak more clearly to our imagination, but here, again, in order to interpret them we must take into account the skill of the artists of each period. The illustrations of sixteenth-century works that have maintained their reputation down to our own times — those of Agricola, Biringuccio, and Ramelli, for example — perhaps do as great a disservice to their authors as the literary style. The naïveté and primitive technique of their draftsmanship certainly lagged behind the mining, kiln firing, and weapons techniques they were intended to describe. (The awkwardness of the thirteenth-century Villard de Honnecourt, who was, however, an architect, is a case in point.) The illustrators of the technical works of the sixteenth and early seventeenth centuries were obviously not great masters of perspective. The teaching of technical draftsmanship was not begun until the middle of the eighteenth century. It is undoubtedly for this reason that Leonardo da Vinci's thousands of sketches seemed all the more astonishing when they were revealed to our contemporaries. Between Jacques Besson (end of the sixteenth century) and the end of the eighteenth century (including Salomon de Caus and the Fathers Chérubin and Plumier, and even Bélidor) the draftsman's art improved together with the machines and tools it depicted. The eighteenth-century artists employed by the *académiciens* to illustrate the *Description des Arts et Métiers* published by their Order, or by Diderot for the illustrations of the *Encyclopédie,* possess perfect mastery of their craft. Possibly the distinction between a sixteenth-century and an eighteenth-century workshop, as regards machinery and mechanical equipment, was not as great as might be supposed from their artistic representations.

Even after taking this uncertainty into account, however, it is undeniable that a considerable amount of progress was achieved during these two centuries.

Iron working and the drop hammer

This is clearly illustrated by the example of the methods of working and shaping (one could almost say manufacturing) large metal objects. In metallurgy, ships' anchors together with cannons were unquestionably the largest objects being manufactured at the beginning of the eighteenth century. Their production was described in detail by Réaumur (in *L'art de forger les ancres . . .*), and the authors of the *Encyclopédie* used the drawings commissioned by him as their model.

Fig. 47. Shop for forging anchors. Forging of the stock (illustration from the *Encyclopédie,* Vol. IV).

The principal piece of equipment in the large forge was certainly the drop hammer powered by a hydraulic wheel, which had been known for perhaps eight or ten centuries (Figure 47). By now, however, it was no longer a primitive instrument; the accumulated experience of many generations had developed it into a sturdy machine that functioned dependably. Little by little the structure of the horizontal shaft, its axle and the pivots on which it rests, and that of the lugs acting as cams had been as rationally perfected as the form of the anvil and the hammerhead falling heavily on the object to be forged. The various types of drop hammers had been so perfectly adapted to the uses for which they were designed that they were to remain unchanged until the second half of the nineteenth century, and even, in the case of some models, almost to our own day. When the steam engine was substituted for the hydraulic wheel, they continued to be used for a long time after the appearance of the first forging presses.

The layout of the shop, providing for the worker to permit him to accomplish his heavy work with the greatest economy of time and expense, as depicted in eighteenth-century illustrations, is in striking contrast to the disorder and turbulent activity that seems to have characterized all the workshops depicted in, for example, the *De re metallica.*

The rolling mill

The same arrangement is found in all the other metallurgical shops in which heavy materials, large sheets of brass, table castings, and sheets of lead were worked or handled. Beginning in the last years of the seventeenth century, mills for lead rolling had come into use in England, and this industry continued with uninterrupted success; it was introduced into France a half century later (around 1730). Despite the resistance of the guilds, which violently opposed all new developments of this sort during that period, several enterprises sprang up in succession, first in Paris and then in several other cities.

Plate 18. Smith's drop hammer and bellows operated by undershot hydraulic wheel. Böckler, *Theatrum machinarum*. Nuremberg, 1662. *Photo Conservatoire des Arts et Métiers.*

FIG. 48. Lead-rolling shop. The furnace and the casting table (illustration from the *Encyclopédie,* Vol. VIII).

The lead-rolling factory established at Romilly near Rouen (probably modeled on the factory that had been in operation in the Faubourg-Saint-Antoine since 1730) is an example of the equipment found in contemporary workshops. We are struck by the soundness of judgment displayed in the placing of the casting table in relation to the furnace containing the molten lead, so that the channel could be filled with the molten metal (Figure 48). It was no longer necessary to utilize a natural slope in the ground, as has been the case two centuries earlier. Moreover, the shops were now installed in covered, well-enclosed halls. Two horizontal lever rams permitted the tilting of the channel at the end of the table. Each of the chains supporting it was attached to the edge of a half disk of wood; the channel fell into place by its own weight, the weight of the lever arms being distributed on each side of the suspension point. The sheets of metal were removed from the table with the help of a crane equipped with a winch and a hook held by a movable pulley. Since this operation could not be performed by hydraulic power or by a horse-driven treadmill (the only source of power then available), every mechanical resource was utilized to lessen the expenditure of muscular energy. Thus a single worker handled the heavy lead sheets to be rolled.

The same crane, pivoting on its vertical axis, was used to place the sheet of lead on the rolling table. The latter, with its two rollers slightly inclined toward the center, already presented all the characteristics of a modern rolling mill (Figure 49). The mill housing had the best improvements that could possibly be achieved with the mechanics of the time. The bottom roller was drawn by a hydraulic wheel or a treadmill, the movement of which was transmitted by a wooden lantern gear; there was a shift device for disengaging and reversing. The distance between the rollers could be regulated to the height required for each passage of the sheet being rolled. This was done automatically by a simple turn of a crank, which was transmitted by a gear train to the supports of the axles of the vertical roller.

Until the end of the eighteenth century, lead was the only metal rolled industrially. Zinc was also rolled, but its consumption became important only with

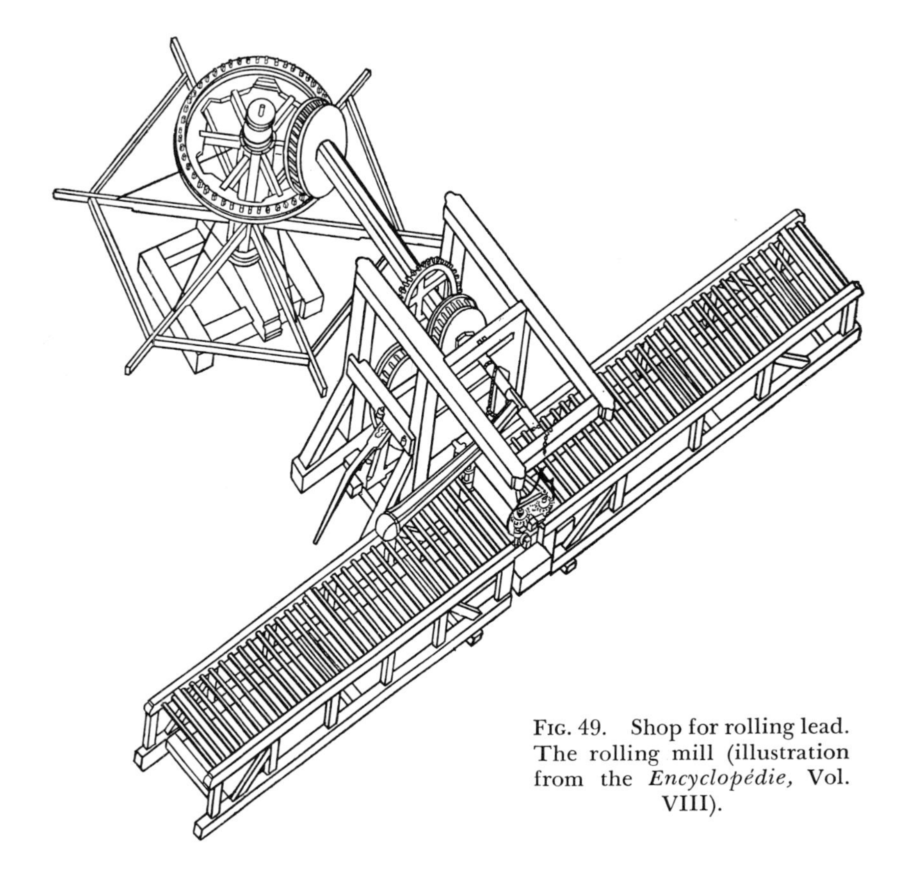

FIG. 49. Shop for rolling lead. The rolling mill (illustration from the *Encyclopédie,* Vol. VIII).

the manufacture of tinplate. Copper and brass were rolled, by hammering or with small hand rollers, by mechanicians and instrument manufacturers, precious metals by the goldsmiths. Mints were perhaps the only establishments that utilized larger rolling mills, sometimes operated by horse-driven treadmills, to roll blanks. Copper rolling was practiced on a large scale only after 1770 or thereabouts. A large establishment for the rolling of copper was created in 1782 at Romilly. Toward the middle of the eighteenth century iron rolling was begun in England and Germany for the production of sheet metal; the method began to be utilized by several plants for the requirements of tinplate production. The largest factory in France was established in 1777, near Saint-Omer at Blendecques; copper rolling was also begun here, five years later and with greater success. The technique of rolling was already highly developed. While the French factories met with various failures until the end of the eighteenth century (whereas their English and German predecessors and competitors in this field were prospering or at least successfully continuing their activity in satisfactory fashion), the reasons for this lack of success were in no way connected with deficiencies in the technical methods being used. The French establishments began to flourish after the wars of the empire.

Mechanical saws

The sawing and drilling operations were mechanized very early. Volume I contains a sketch by Villard de Honnecourt of a thirteenth-century mechanical saw; its blade, raised by a flexible pole, was lowered by the lugs attached to the horizontal shaft of a hydraulic wheel. The system was perfected and apparently more widely employed beginning in the sixteenth century. The mechanics of these devices remained unchanged until the nineteenth century: a framework holding several parallel blades was propelled in an alternating, vertical movement, while between each stroke a ratchet wheel pushed a rack pulling the carriage to which was attached the tree trunk to be cut into planks.

This method was applied to harder materials, for example to stone and marble. In Bélidor's *Architecture hydraulique,* beginning with its first edition in 1737, we find a picture of a double saw for cutting marble, which was by no means a novelty at that time (Figure 50). The movement was supplied by two cranks that were turned by hand — for these machines had to be installed right at the quarry, for reasons that are readily understandable.

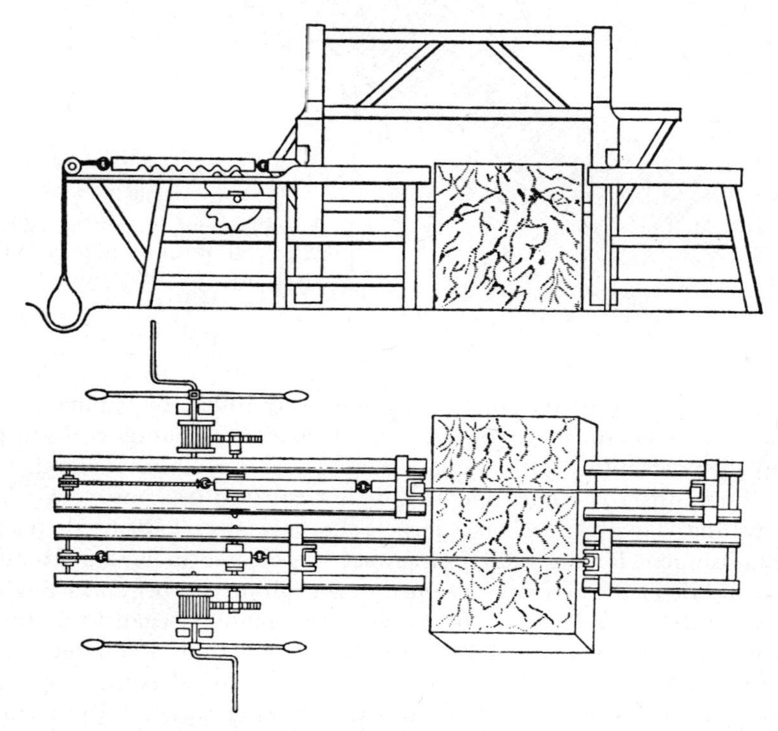

FIG. 50. Mechanical sawing of marble (Bélidor, *Architecture hydraulique,* Vol. I, Pl. IV). The two blades function simultaneously in opposite directions. Each saw is pulled in the direction of the teeth by a weight, and brought back by a rack and a partially toothed wheel.

Boring of tubes

Woodworking also provided workers and engineers with the opportunity to experiment with mechanical methods of boring tubes. Wooden conduits drilled in tree trunks had been in use perhaps since antiquity, particularly in the mines and for hydraulic installations. Figures 33, 34, and 35 of Part One of this volume show several examples of the method used for boring beams. It seems that mechanical drilling was seldom employed at the beginning of the sixteenth century. Agricola shows us a workman

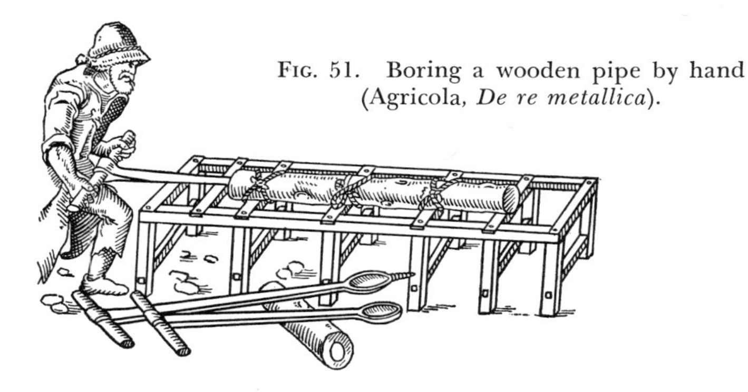

Fig. 51. Boring a wooden pipe by hand (Agricola, *De re metallica*).

hollowing out a pump barrel in a tree trunk with the help of a hand drill (Figure 51). Only fifty years later, however, Beringuccio shows us this drill already being driven by a hydraulic wheel. Lastly, Bélidor describes a complete water mill for drilling wooden pipes (Figures 52). The drawing is interesting for more than one reason; the hydraulic wheel utilizes a waterfall, and is an ancestor of our modern Pelton turbines. Despite the mediocre perspective, it is easy to perceive the use of lantern gears for reversing. The pressure against the drill of the object to be drilled is ensured by a wheel whose ratchet operates at each turn of the toothed wheel, to which it is connected by a stationary spindle.

Cast iron and lead pipes were molded – drawing of lead pipes did not come into use until the end of the eighteenth century. For about two centuries the metallurgical industry used drilling only for the production of cannons. For several centuries, from the birth of firearms until the end of the sixteenth century, cannons were cast in a mold buried in earth and supplied with a central core for the bore (see page 474). Printed treatises of every age explain in detail the construction first of the forms, then the molds.

In Biringuccio's *Pyrotechnia* (1540) we find one of the first (if not the first) descriptions of a machine for boring cannons cast in one piece (Figure 53).

The gun was placed on a wooden platform that formed a carriage and that could, with the help of a winch, be shifted longitudinally on a heavy base, also made of wood. Its height was adjusted so that its axis was in line with that of the drilling tool. This consisted of a forked drill fastened to the end of a long

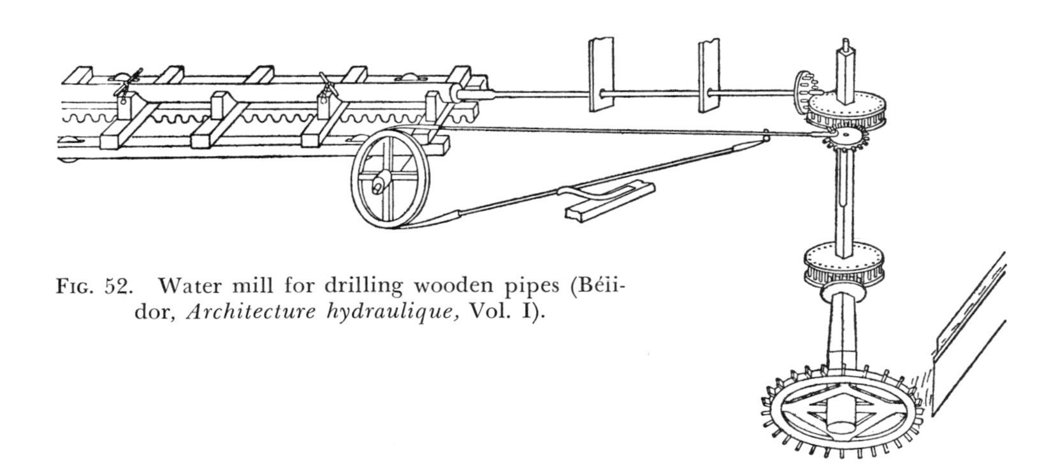

Fig. 52. Water mill for drilling wooden pipes (Béiidor, *Architecture hydraulique,* Vol. I).

horizontal bar, which was turned either by a manually operated device or by a squirrel cage, a water mill, or a horse-driven treadmill. The forward movement was supplied by the winch; the drill performed the preliminary work, after which it was replaced by one or several borers for the finishing work.

In 1713 the Swiss Jean Maritz began to utilize a vertical boring machine (reproduced in the *Encyclopédie*) that was undoubtedly not of his own invention. The pressure of the drilling tool on the metal was obtained by the weight of the cannon, balanced by counterweights to avoid breaking the tool or the bar supporting it (Figure 54).

This method was utilized for a long period of time, for Gabriel Jars, writing about his travels between 1757 and 1769 for the purpose of visiting the major European metallurgical installations, describes methods of drilling and boring quite similar to the foregoing. He writes:

"At the iron forges of Moss, the cannons are cast in molds in which a core has been placed, so that there remain only a few millimeters to be drilled. The machine used for this operation is the one that was first known. It operates perpendicularly; that is, by means of levers and iron chains, which hold the cannon in this direction; the latter drills itself by its own weight, by pressing against the drill, which is propelled by a pulley wheel and a lantern. These in turn receive their propulsion from a large 'waterwheel.' "

It is clear that the work thus performed, whether on this machine or on the preceding one, could not be very precise, for there was nothing to prevent the drilling tool from deviating.

However, around 1730 Jean Maritz invented a horizontal drilling machine in which the object to be drilled was propelled before the stationary tool by a

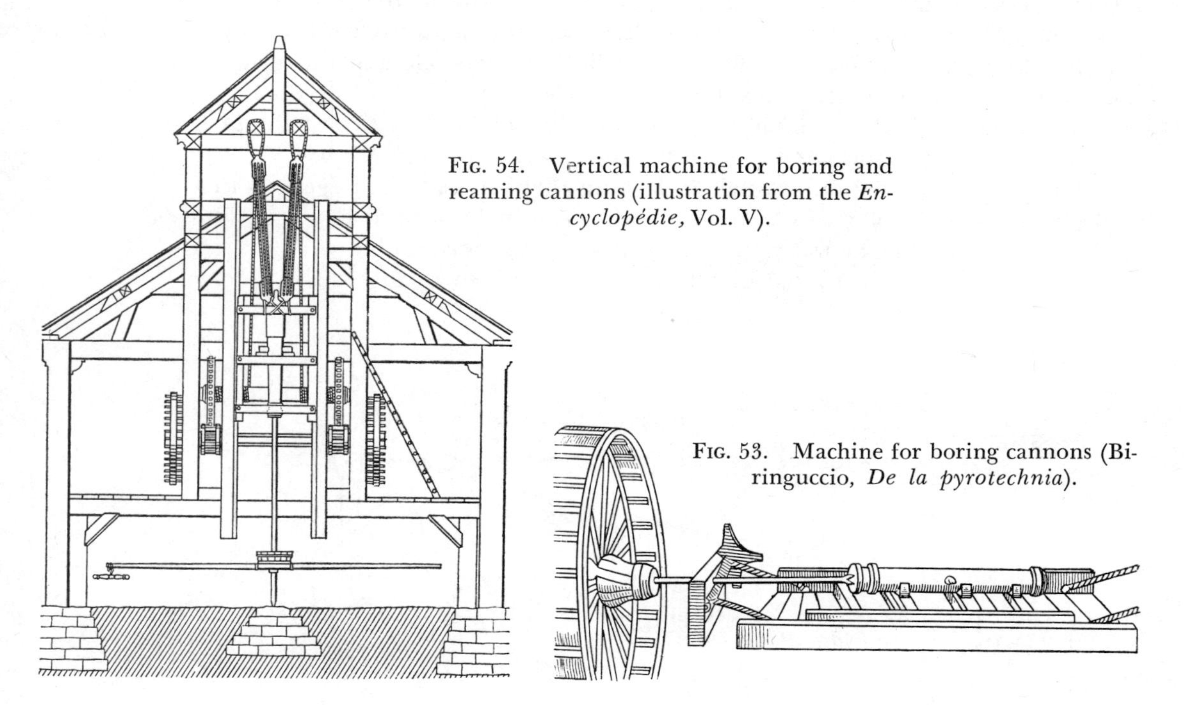

Fig. 54. Vertical machine for boring and reaming cannons (illustration from the *Encyclopédie,* Vol. V).

Fig. 53. Machine for boring cannons (Biringuccio, *De la pyrotechnia*).

rotating movement. In this way a deviation of the tool was much less important than on the other machines. This method, which at first was not much used, ultimately became universally accepted. Maritz and his son were placed in almost complete charge of artillery manufacturing plants in France.

John Wilkinson's machine, which was later modified for the boring of steam-engine cylinders with a precision ensuring a minimum of tightness, was also used at first for the boring of cannons. This was a horizontal machine whose conception and functioning, as explained in the patent of January 27, 1774, were based on completely different principles.

Similarly, the necessity of melting down old, worn-out, or defective cannons led to the invention of a simple machine that cut a cannon into three or four pieces in a single day, depending on its diameter. This process has been described as follows:

"This machine consists of a small, toothed wheel, about a foot in diameter, made of wrought iron but equipped with steel teeth. It is firmly attached to a long, thick iron bar which is fixed on one side, and fitted into the hub of an axle on the other. On the opposite side, and in the same direction, is another machine which supports the cannon on the same line. A single workman performs the entire operation.

"Having placed the part of the cannon he wishes to divide on the toothed wheel, he sets the large waterwheel in motion and, as it turns, he gradually lowers the cannon, until it is perfectly cut, and continues in this way, moving the gun forward on the toothed wheel."

Musket barrels were formed by hot-rolling iron strips drawn by hammering. To give these strips proper thickness, they were milled in the same way that a piece of wood is planed on a planing machine. A machine built for this type of work was described in 1716 by Villons.

These inventions belong to the period in which mechanicians gradually broke away from the empirical methods of their predecessors and, using the methods inherited from the latter, created the modern machine tool.

MECHANICAL EQUIPMENT FOR LIGHT INDUSTRY

Using more modest equipment, numerous industries increased their methods of production by gradual mechanization.

Splitting mills and wire mills

The splitting mills and wire mills utilized rolling techniques at an early period, and at least beginning in the first years of the eighteenth century. In the splitting mill, iron rods were reduced to thin strips, which were then split by being passed under rollers with suitably profiled cutting edges (Figure 55). The work was done by a series of pairs of rollers mounted next to each other on the same shafts. Power was supplied by a horse-driven treadmill or a hydraulic wheel. The distance between the axles of the rollers was uniform for each pair, but diminished from one pair of rollers to the next. The iron rods, heated red-hot in the furnace, were passed between each pair of rollers in succession; the last pair had a cylinder with a cutting edge. The thin rods cut by the chopper

FIG. 55. Splitting shop (illustration from the *Encyclopédie,* Vol. V).

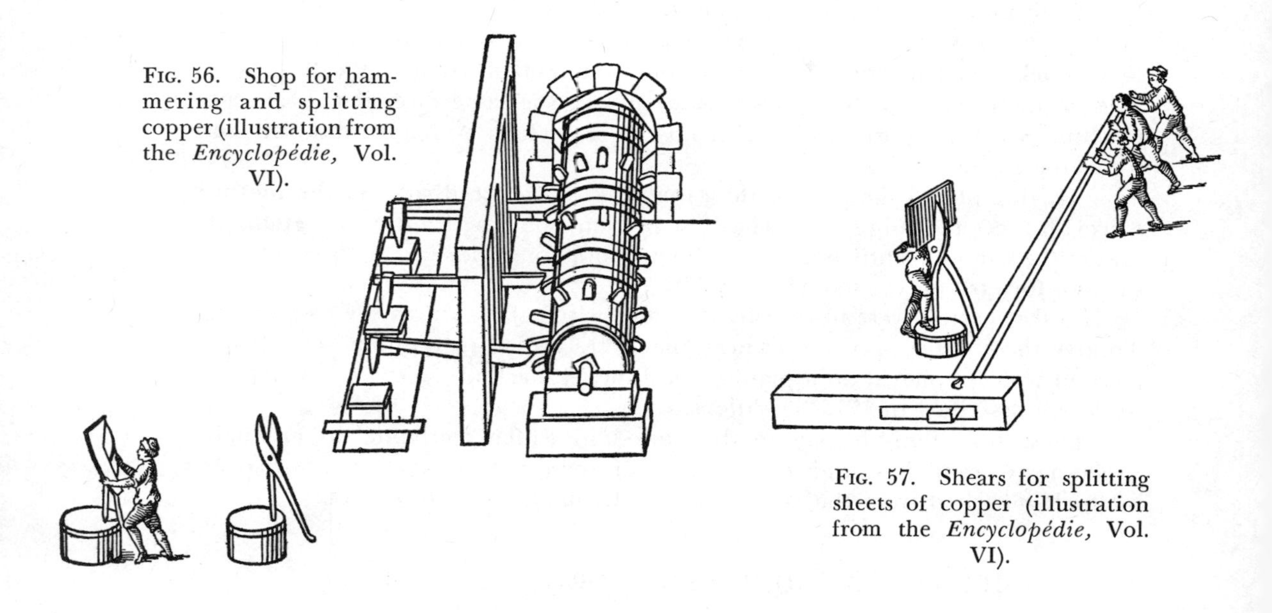

FIG. 56. Shop for hammering and splitting copper (illustration from the *Encyclopédie,* Vol. VI).

FIG. 57. Shears for splitting sheets of copper (illustration from the *Encyclopédie,* Vol. VI).

were gathered together with a tongs and a hook and made into a bundle.

Copper wireworking utilized a different, more traditional type of equipment. Until the end of the eighteenth century, as we have just seen, thin strips of copper or brass were obtained only by hammering. Large drop hammers were substituted only for small ones with a single hammer, and in the sixteenth century even the latter were undoubtedly employed in conjunction with hand hammering. At the beginning of the eighteenth century a battery of two drop hammers could drive six hammers working simultaneously. Comparing those in Figure 56 with the type used in an anchor-working forge (Figure 47), for example, we note that their shapes are different, being adapted to the operations to be performed. The bronze plates were cut into strips with an enormous shears fixed vertically on a block (see Figure 56). Such shears were to be found in all the shops in which zinc and sheet iron had to be split. These shears were operated by means of a lever moved by several men (Figure 57).

The actual wiredrawing was done by passing the thin, heat-softened rods through a set of dies. The wire, seized with a tongs as soon as the end projected, which had been forced through the hole, was drawn out by the worker as he moved away from the workbench.

Although mechanization economized to only a small degree on human muscle power in these operations, the improvements achieved in the construction of the traditional machines were not negligible. The structure of the treadmills, hydraulic wheels, batteries of pounders and drop hammers, and in general the construction of the "mills" that formed the nerve center of almost all the factories in the eighteenth as in preceding centuries, suggest the distance that had been traveled. These improvements, which were made possible by a better choice of materials and increasing experience on the part of the manufacturers, were necessitated by the growth of production and the necessity of increasing efficiency. In numerous shops devoted to the production of small articles, the hand tool was gradually mechanized. It is not possible to trace the stages in these mechanical adaptations; while we find several examples of them in sixteenth- and seventeenth-century treatises, their method of propulsion by machine does not seem very convincing. However, this must have been gradually perfected, and the results must have become sufficiently satisfactory to justify the spread of their use.

Grinding shops

We can understand by what methods this early mechanization of equipment was gradually achieved when we watch the cutlers and mirror polishers at their work. These are branches of production in which every increase in efficiency was easily turned to account.

In the cutler's workshop the grindstones were now worked neither by hand nor by treadle (Figure 58). A hydraulic wheel furnished the power which by means of a shaft with pulleys turned the shaft to which the grindstones were attached. The combination of various diameters made it possible to obtain through a series of pulleys the necessary speed of rotation of the various grindstones. Some of these grindstones and pulleys were interchangeable, in order to increase the range of possible combinations.

FIG. 58. Sharpening machines (illustration from the *Encyclopédie*, Vol. VI).

Polishing and polishing machines

Mirror polishing did not require a great deal of energy, but it had to be exerted steadily and for a long period of time. The pressure that had to be placed against the wheels and rollers was obtained from stretched bows (Figure 59), as in the case of the woodworking lathes that will be discussed later. The upper end of the bow (which was placed in vertical position) was hinged to the ceiling, the other end to the polishing roller; the length of the bow was calculated so that it remained taut, and thus exercised a strong pressure on the polisher's tool. The operator had only to move the tool steadily over the face of the mirror attached to the workbench. With the growing prosperity of the glass factories, such devices made possible the proliferation of mirror-working shops.

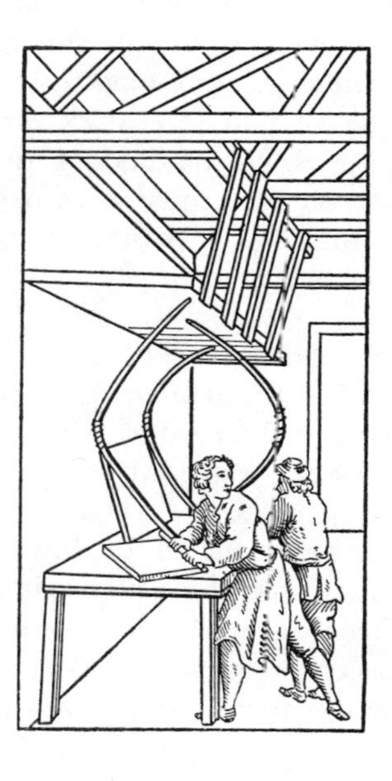

FIG. 59. Mirror-polishing shop (illustration from the *Encyclopédie,* Vol. VI).

The conception and creation of more complex machines, which automatically performed the polishing of mirrors with large surfaces, soon followed. In Part Two, Section One, Chapter 10, we saw that in the course of the eighteenth century these factories acquired equipment comparable to that of the metallurgical workshops. The polishing shops attached to these factories must have benefited from the same technical progress which utilized the mechanical resources of the age with great ingenuity. The large polishing machine at Saint-Ildefonse is a singular example of this (Figure 60).

Punches and balance presses

The minting of coinage was the basis for the creation of several machines that worked metal by shaping. Until the middle of the sixteenth century coins were made by hand hammering. In 1550 Henry II purchased from Marx Schwab, a goldsmith at Augsburg, the three machines indispensable for their mechanical production: the roller for flattening the metal strips after their preliminary working on the anvil, the cutting press (or punch), and the balance press for striking the coins.

Drawings for a cutting press and a roller appear in the manuscripts of Leonardo da Vinci, and even before Schwab, at the beginning of the sixteenth century, Benvenuto Cellini was striking medals by means of a balance press in which the diameter of the screw was approximately 60 millimeters, as he states in his *Trattato dell' oreficeria*.

In order to realize the cost of such work, it is interesting to note that Gabriel Jars, while visiting the Kremnica Mint in Hungary in 1758, had noticed that the balance press required two teams of eight men working alternately for fifteen minutes in order to mint eleven two-florin coins per minute.

Light machinery Equipment of the same type made it possible to decrease the time required for the production of knickknacks and hardware. What has been said above demonstrates the methods that made it possible to mechanize the work of rolling and stamping thin objects, and the operations of drilling, stamping, and punching. It is not correct to speak of machine tools in connection with this light machinery that at the beginning of the eighteenth century ensured the fortune and reputation of the factories of Birmingham. However, mechanization was applied with great ingenuity to the worker's labor, that is, to making the performance of each opera-

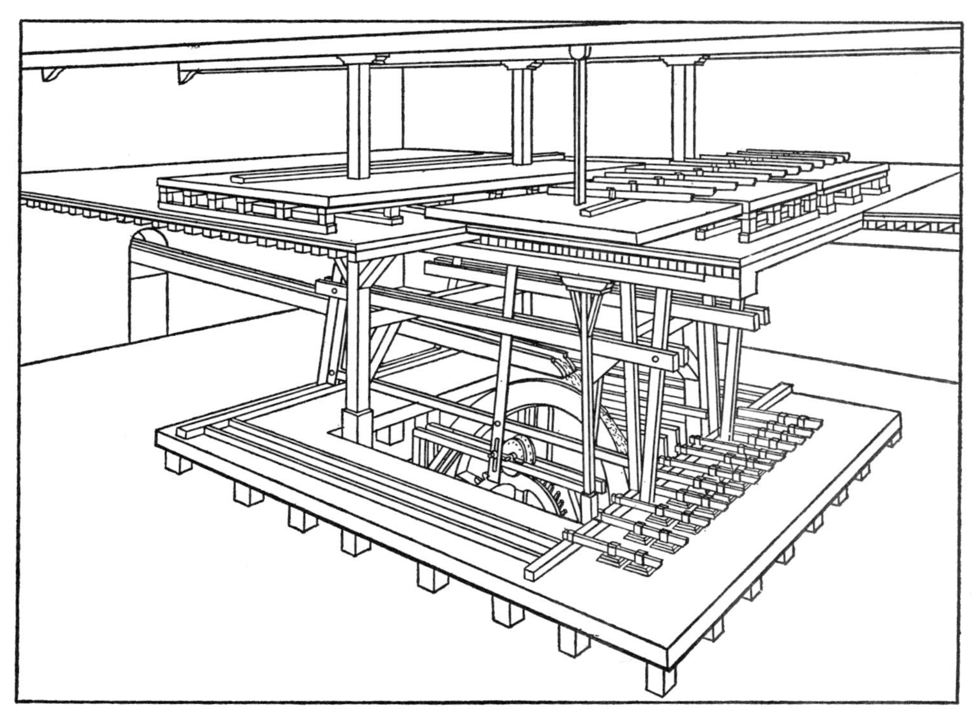

FIG. 60. Large mirror-polishing machine used at the Saint-Ildefonse factory (illustration from the *Encyclopédie,* Vol. VI). This machine is composed of two levels that hold the mirrors to be polished in a horizontal position. The polishers, which are plainly visible on the lower level, are attached to a carriage moved back and forth by a system of levers regulated by the crank of a lantern pinion. The entire machine is operated by a paddle wheel.

tion shorter and less fatiguing. A rigorous division of labor made it possible to obtain the highest possible production from this equipment. The production of small steel chains and buttons in such a shop around the middle of the eighteenth century is a characteristic example of early mass production. Jars writes:

"All the articles of hardware are made in approximately the same manner, using the same machines. For steel chains, for example, each component part is stamped. However, since it would be very costly if the dies used for this purpose were engraved, as in the case of coins, the following procedure is used:

"A single matrix is engraved for each new design, in high rather than sunk relief. The matrices, being made with great precision and well hardened, serve to form all the new dies with which the stamping is done. For this purpose the design of the matrix is imprinted on a red-hot piece of steel. Then it is hardened in a solution, and this procedure is repeated until the dies break or are worn out. The first matrix is used repeatedly for this purpose.

"The steel used for the chains is flattened between two rollers, depending on the thickness desired. Then the design is stamped on it. After this it is given to another workman, who cuts around the design with heavy punches specially designed for this purpose.

"Thus the spaces left in the design are cut out with punches, each following its form. The entire outer portion of each piece is then filed to remove the burrs, after which all the small objects are hardened in a solution. Now all that is needed is to polish them.

"The first polishing given to them is that of the brushing machine, with oil and emery, for the rounded sides. To make this work faster and easier to handle, children thread all the small objects of the same size and shape on coarse wire, large objects being strung on two pieces of wire; they are fastened by bringing together the two pieces of wire. In this state they are given to girls or women, who hold the objects against the brush until they are quite polished. As for the flat surfaces, they are polished on horizontal grindstones by children, using emery.

"These grindstones are operated hydraulically. There are six or eight of them in the same mill, positioned in a circle, with a small boy at each grindstone. To make it easy to hold these objects, pieces of wood are hollowed out in conformity with the designs of the objects; the object to be polished is set into the wood, and since each person always polishes the same design, the work goes very quickly. Damascened buttons are polished on the same grindstones and in the same manner. Then the objects are taken into another shop for the final polishing of the flat surfaces. For this purpose there is a piece of wood, eight to ten inches long and seven or eight inches wide, covered with a kind of tar. A child holds this tarred wood before the fire, places the steel objects on, and embeds them in the tar, and then places them as close together as possible. After which women place a little loam on them, and rub with both hands simultaneously, and for as long as they judge necessary for the objects to be perfectly polished."

THE DEVELOPMENT OF PRECISION INSTRUMENTS

The creation of a genuine precision mechanics followed a rather strange and devious route. Its origin can be fixed in the second half of the seventeenth century, and the efforts pursued, at first unconsciously, for two centuries were the outgrowth of two types of interests whose goals did not begin to become unified until the second half of the nineteenth century.

A great variety of consequences resulted from the invention of compound optical instruments — the telescope, the microscope, and slightly later the reflecting telescope. We have seen in the *Histoire générale des sciences* how scientific knowledge benefited by both the development of theoretical optics and the generalization of experimental methods. At the same time, optical workers were utilizing the mechanical methods then in existence; in particular, they adapted the carpenters' lathes for their own purposes. Following their lead, the builders of instruments for astronomy and surveying developed their industry and created increasingly precise instruments intended not only for scientists but also for navigators and topographers. These workers adopted and perfected certain machines utilized by the clockmakers. In this way the machine for dividing and cutting teeth in gears came into general use.

These activities were directed solely toward creating the methods indispensable for the construction of new and hitherto unknown objects. However, the engineers who were concerned with hydraulic installations and the construction of water mills or windmills were anxious to discover a method for increasing the productivity of the machines they designed, particularly by improving the organs of transmission. This brought up the problem of the profile of gear teeth. No practical solutions to this problem, which was theoretically solved at the end of the seventeenth century, were actually found until a century and a half later, when it became possible to create the first modern machine tools. These machines, originally conceived for the working of large objects, gradually began to benefit from the contributions of light precision mechanics. In their early stage they satisfied the needs of the construction of steam engines and the great metallurgical industries deriving from the expansion of this construction. Later they became the indispensable auxiliaries of increasingly precise machining, when certain types of manufacturing (for example, the internal combustion engine and rapid-fire weapons) in turn began to develop.

In order to respect the chronological limits of this volume, only the first stage in this development will be discussed here.

The evolution of the primitive lathe

It is not possible to say with certainty that the lathe is the oldest known machine tool. From its origin, undoubtedly in earliest antiquity, until the appearance of the machines utilized by the clockmakers to cut fusees for watches and pendulums, it was simply a moving support for the object being worked; the tool was always held in the worker's hand.

Probably until the middle of the medieval period the lathe consisted of a simple pivot rotated in an alternating movement by a small bow or rope operated by the worker; the pivot was the object itself. In this form the lathe is attested by iconographic documents in almost every civilization of the first millennium in the East and the Mediterranean basin.

Around the thirteenth century the lathe appeared in the form it was to retain until the nineteenth century for the working of wood, ivory, and, more rarely, soft metals like tin. It has a horizontal bench on vertical feet, whose housing holds the two lathe centers between which the object to be worked is held (Figure 61). The object is in horizontal position, at a height that permits the worker to work standing (his predecessors had to work in a stooping position). A flexible rod attached to the ceiling of the workshop at one end, and having at its other end a short hemp or gut rope which makes a half hitch around the object being worked and joins a treadle near the floor, permits the worker to move the object while leaving his hands free to hold the tool; with a simple movement of his foot he can rotate the object. It is an alternating rotation, however, which means that the tool can attack the object only at every other turn. In addition, large objects cannot be turned.

FIG. 61. Lathe with continuous rotation (*Encyclopédie méthodique,* Vol. IV). There is one improvement over the early lathes of this type in the sixteenth century: the double wheel at the side makes it possible to utilize two different turning speeds. Before the window in the rear can be seen the classical pole lathe.

The sixteenth century witnessed the appearance of devices that permitted a continuous rotation of larger pieces, and possibly also the turning of metal objects. One of these devices consisted of a large crank wheel at some distance from and at right angles to the lathe; an endless rope passed through the groove of the wheel and a groove in a pulley attached to the axle of the lathe (Figure 61). Here, again, however, the operator had to be aided by an assistant who turned the crank. This device was suitable for large turners' shops, which had to produce quickly and in quantity, and also for carpenters, wheelwrights, and pewtermakers. It remained in use until the beginning of the nineteenth century.

The other device was a simple modification of the treadle lathe, in which the flexible pole was replaced by a heavy crank-turned bronze wheel attached to the left side of the housing (Figure 62). The inertia of this flywheel replaced the flexibility of the pole in pulling up the treadle. The rotating movement, however, was continuous. This type of lathe was suitable for the amateur and for the single artisan producing delicate objects, for example, the maker of chessboards and similar fine articles.

Around the middle of the sixteenth century the art of the turner became a "high-class" occupation that enjoyed tremendous popularity among the wealthy of all the European countries, from Scandinavia to Italy, during the seventeenth and eighteenth centuries.

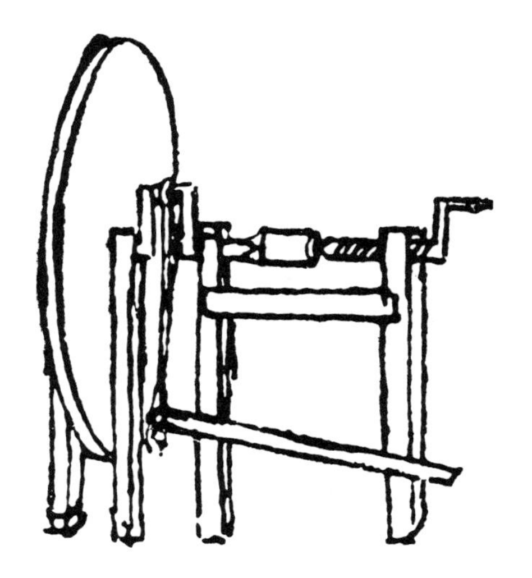

Fig. 62. Lathe with treadle and inertial flywheel (from Leonardo da Vinci, *Codex Atlanticus*).

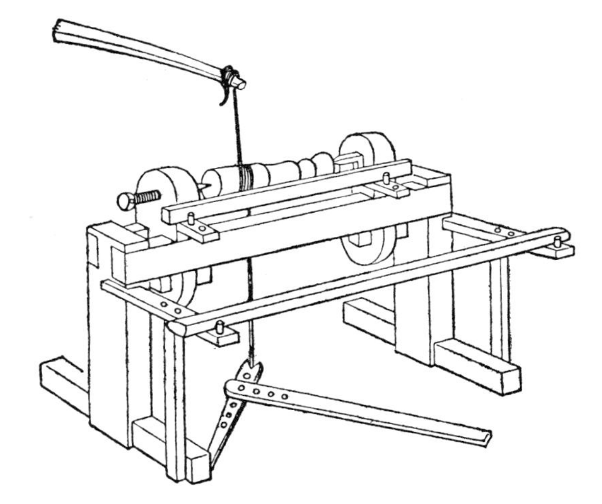

Fig. 63. Wood lathe with support for the tool (from Usher, *A History of Mechanical Inventions,* p. 362).

The lathe with treadle and flywheel represents an important acquisition for future precision mechanics. As we shall see in a moment, it was this lathe that made it possible to begin working metal, particularly steel. It also made possible the development of the lathe with fixed tool, which was independent of the worker and could be guided and moved forward as desired.

Threading of screws on the classical lathes

The type of lathe in which the tool had to be held by the worker continued to exist for decades. Not until around the end of the fifteenth or middle of the sixteenth century did a supporting bar appear; it was placed horizontally on the bench of the lathe, so that the operator could support the tool on it (Figure 63). Ultimately the tool was fixed to the lathe, probably for the purpose of cutting screw threads, and thus became the first attempt at a genuine machine tool. The first efforts in this direction appear to date from the sixteenth century. All that was involved, naturally, was the cutting of threads in wooden shafts, using lead screws. In Leonardo da Vinci's notebooks we find several drawings of similar machines (Figure 64), which suggest that threads of various pitches could be cut with the same lead screws; similar drawings are found in Jacques Besson's books (Figure 65). The first mandrels and fixed screw-cutting dies appeared on lathes with continuous rotation. Sliding headstock dies can be distinguished on the lathes described by Besson in his *Teatrum Machinarum* of 1578.

The professional lathes of the seventeenth century did not yet permit the working of iron. In his famous work of 1693 entitled *L'art de tourner ou de faire en perfection toutes sortes d'ouvrages au tour* (The art of turning, or the perfect construction of all types of objects with the lathe), published at Lyon in 1701, Father Plumier writes that he carefully sought out "the workers who knew how to turn and quickly cut iron," and adds: "But in all my searches in the

Fig. 64. Screw-cutting machine (manuscript B of Leonardo da Vinci). The two lateral screws lead the tool that cuts the center screw. The drawing suggests that several combinations of toothed wheels can be utilized.

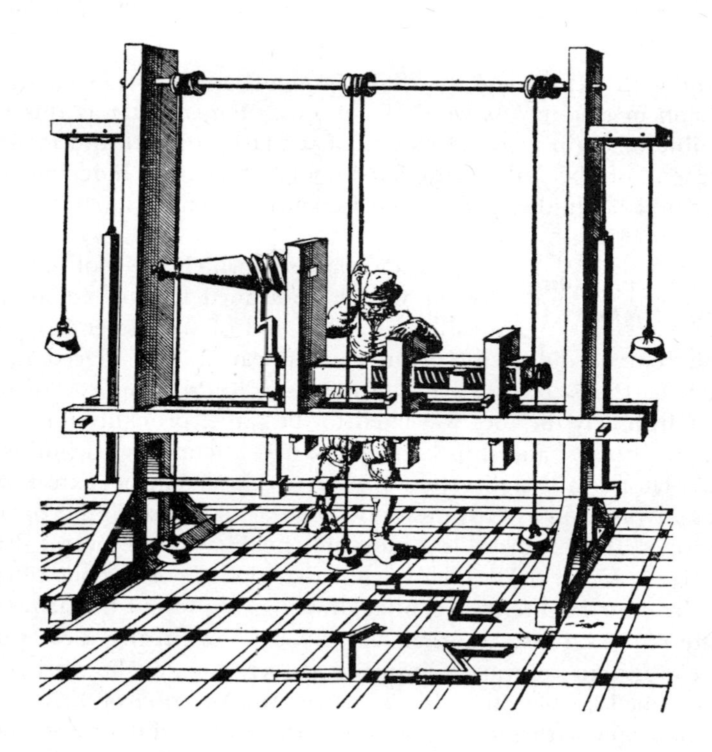

Fig. 65. Mounted lathe for cutting threads (from Jacques Besson, *Théâtre des instruments mathématiques*). The lead screw is placed in the center of the lathe; it pulls the vertical tool that cuts the object to the right of the worker. Note that all the movements are assured by weights. The weights on the ends ensure the pressure of the tool by means of the lower crosspiece.

course of my voyages, I met only two workers capable of satisfying my requirements: one in Rome, a German called *Il signor* Guillelmo, employed at the mint, and the other in Paris, called *sieur* Pierre Taillemars, a mathematician, whose name was well suited to his virtue and skill, since without hooks nor wheel, but simply with his foot and the pole, a mortise chisel, and a lathe with two centers, or with a mandrel, he cuts iron and steel in as large a cut and as quickly as the *sieur* Maubois, the famous turner of the king in the Louvre, cuts ebony and ivory" (Figure 66).

It seems, however, that ironworking on the lathe began with the cutting of helicoidal threads. At the end of the seventeenth century Father Plumier described the method of threading screws, which had undoubtedly been in existence for some decades. For this purpose the turners used a fairly small lathe with a headstock and a tailstock as well as a toolholder fixed to the bench by means of a bolt and a wing screw. An iron rod was placed on two screw-cutting dies to turn freely, its end being first machined to the desired diameter. A screw stock and dies could be used to cut the thread, but this method was not good, for in most cases it involved a decentering of the axis. Not until a century later was the method of making screws with the screw stock perfected, as can be seen in Bergeron's *Manuel du tourneur,* the first edition of which appeared in 1792. In Plumier's time the most skilled turners utilized another method. They soldered to the opposite end of the bar a previously threaded screw tap held within the pincers of a wooden screw-cutting die (Figure 67). The screw tap traced its thread in the die and acted as a lead screw. The spindle to be threaded was first worked with a graver or a sharp-edged chisel; the threading was then finished with the file, or, better yet, with a comb with three teeth, the spacing of the teeth corresponding to the thread to be cut.

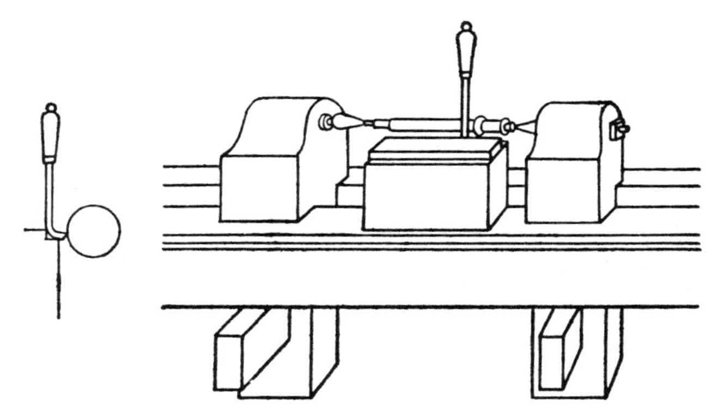

FIG. 66. Lathe for working iron. Seventeenth century (Father Plumier, *L'art de tourner*).

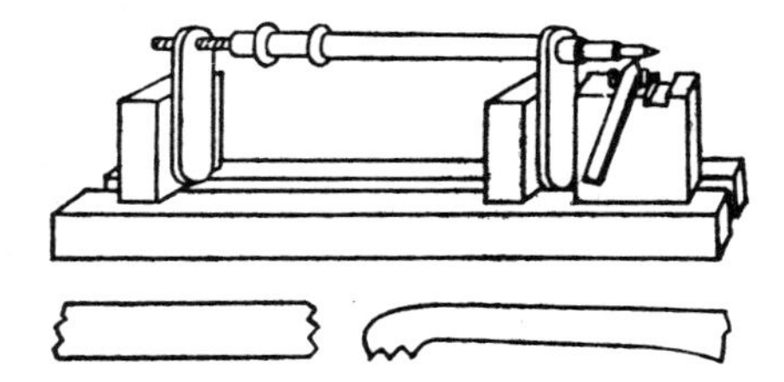

FIG. 67. Small lathe for threading iron screws (Father Plumier, *L'art de tourner*).

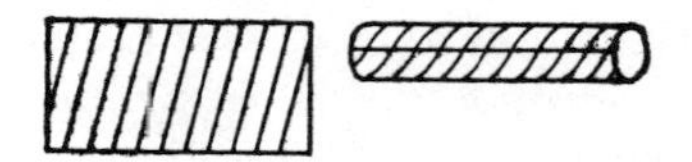

FIG. 68. Method for showing thread to be cut on lead screws.

The method of making lead screws remained unchanged for decades. Oblique lines were drawn on a rectangular sheet of paper; their spacing and angle of inclination corresponded to the thread to be traced. The paper was glued on the spindle to be threaded (Figure 68). The threads were traced, following this pattern, with a sharp file; the cutting was done first with a triangular file, and then with a metal comb.

Plumier's work also contains descriptions of a keyed headstock which made it possible to thread with lead screws of varying pitches. This device, which was intended only for lathes for amateurs, does not appear to have been utilized by the professionals before the middle of the eighteenth century. The keyed headstock was placed on the lathe, enclosing a portion of the mandrel, on which a series of threads of various pitches had already been cut (Figure 69); each key corresponded to one of these threads. When one of the keys was raised, the mandrel was guided along that particular pitch.

Opticians' lathes Corrective glasses had been in use since the twelfth century, yet nothing is known about the origins of the working of optical glass. The astronomical telescope, which had been more or less explicitly mentioned by various authors — Roger Bacon, Leonard Digges, Jerome Frascator, Giambattista della Porta — appeared almost simultaneously in Holland and Italy. Beginning in 1608, the first instruments were constructed by a few skilled men — glassmakers, mirror makers, mechanicians — and in most

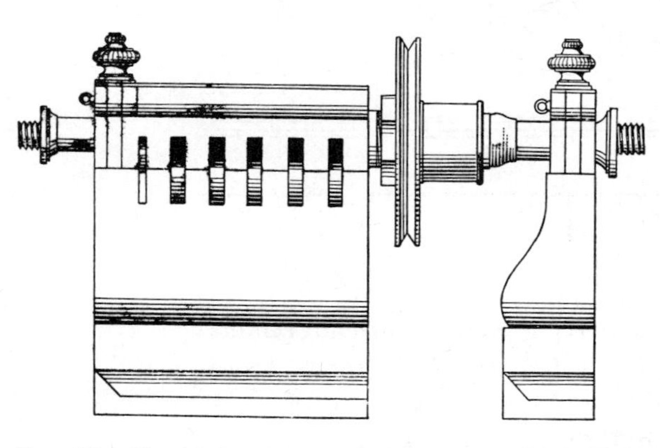

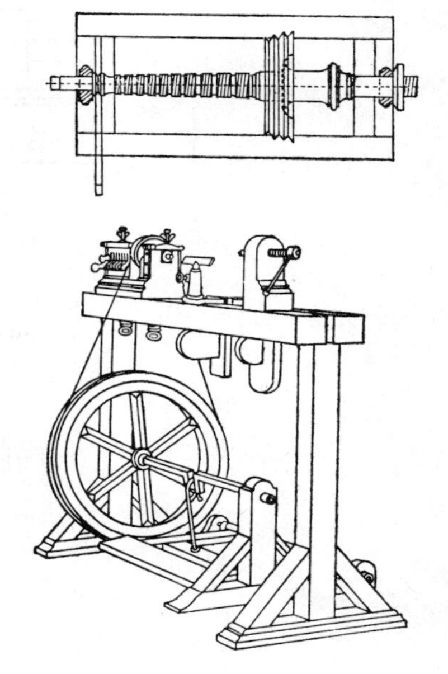

FIG. 69. Keyed headstock for threading on a lathe. Diagram of a mandrel with threads of various pitches, and its mounting on the lathe (from Plumier, *L'art de tourner,* and Holtzapffel, *Turning and Mechanical Manipulation,* 1847).

cases by the scientists themselves as a side activity. Several years later (between 1612 and 1618) the first compound microscopes appeared. By 1625 there existed workshops specializing in the manufacturing of these new instruments; the oldest known is that of a Parisian optician named Chorez. Slightly later, two Italian opticians, Eustachio Divini and Giuseppe Campani, acquired an international reputation.

Campani may have invented the first machine for cutting spherical glasses. It was a small lathe operated by an adjacent wheel and equipped with two fixed dies installed on a sturdy housing (Figure 70). The glass to be cut was attached to the nose of the mandrel — one of the first, and probably the first, instances of a face lathe. The tool was attached to the end of a very long horizontal pole pivoting around a pin through its other end. While an assistant turned the crank, the optician moved the tool over the surface of the glass, which itself was moving with a rapid rotating movement.

In his *La Dioptrique oculaire* (published in 1671), Father Chérubin described various lathes he had adapted for glassworking. The one shown in Figure 71 is quite interesting from various points of view. It is operated by a small bow suspended from the ceiling; the shaft of the crank holds a series of pulleys, thanks to which the speed of rotation of the mandrel can be varied. The operator stood to the left of the lathe, which he operated by means of a stirrup hanging from the end of a strap wound on a pulley. In front of him was a toolholder that slid on a prismatic bench. In addition, the traversing of the tool could be regulated by a tightening screw.

The results obtained from opticians' lathes of this period were apparently not completely satisfactory. The best glasses were worked by hand, with the glassworker utilizing a cutting wheel of suitable shape. Several of the devices on Chérubin's lathe, however, are very far in advance of their time. They are found again in the eighteenth century on the guilloching lathe utilized almost exclusively by amateurs.

Lathes for ornamental work and guilloching

The vogue enjoyed by the amateur's lathe in the seventeenth and eighteenth centuries is attested by the important work published by Father Plumier in 1701, *L'art de tourner ou de faire en perfection toutes sortes d'ouvrages au tour*. Father Plumier writes that "it is an established fact that in present-day Europe this art is the most serious occupation of people of intelligence and merit, and, between amusement and reasonable pleasures, the one most highly regarded by those who seek in some honest exercise the means of avoiding those faults caused by excessive idleness. The variety and surprising delicacy of objects of wood, ivory, gold, silver, iron, and copper, and many other strange materials, made on the lathe by so many industrious people skilled in this art, in France as in Italy, England, and Germany, reveals that in all these countries the exercise of the lathe is so highly esteemed that there are very few intelligent individuals who do not attempt to excel in this art and who do not work through emulation to produce on it something splendid which will reveal its merit." The lathes used by amateurs were perfected especially at the beginning of the eighteenth century.

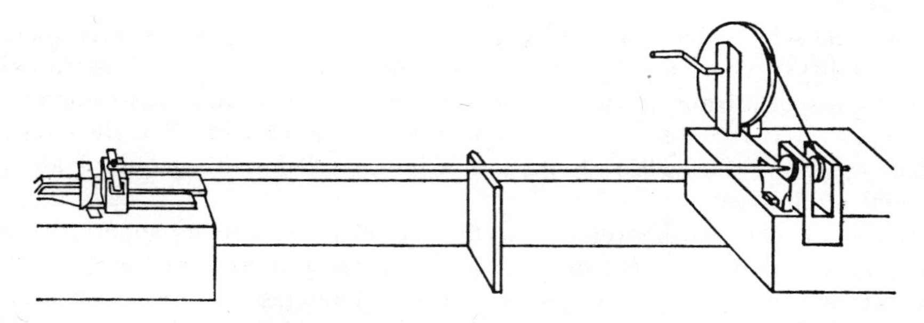

FIG. 70. Campani's lathe for working glass. Around 1664 (from Fougeroux de Bondaroy, *Mémoires de l'Académie des sciences,* 1764).

FIG. 71. Chérubin's lathe for cutting spherical glasses (Father Chérubin, *La Dioptrique oculaire*).

On August 30, 1719, Philippe de La Hire told the Académie royale des Sciences that "we first turned in the round, then in an oval, and it is only later that we found a way of turning ornamental figures." Slightly later, on July 8, 1733, La Condamine wrote that "in the past century, and especially in the present period, the lathe has been brought to a high degree of perfection."

The most famous amateurs' lathes are those of the Emperor Maximilian (circa 1510), which can be viewed at Innsbruck; their principal components are finely sculptured. Then there is the small German guilloching lathe, a genuine goldsmith's masterpiece of around 1750, in the Science Museum in London. The Musée du Conservatoire national des Arts et Métiers in Paris houses the portrait lathe with portraits of Nartov given by Peter the Great to d'Ons-en-Bray after his trip to France in 1717, and the guilloching lathe constructed for Louis XVI, which bears the inscription, "Made by I. T. Mercklein, Saxon, mechanician of the Furniture-Repository of the Crown in Paris, in 1780."

While the artisans' lathes were generally simple lathes used only for turning in the round, the amateurs' lathes were figure lathes or combination lathes which included the portrait and guilloching lathes just mentioned.

The guilloching lathe (named after the turner Guillot who is supposed to have invented it) "permitted the tracing of intertwining curves, with symmetrical volutes, suitable for ornamentation, as for example the shallow engraving of cases for watches and clockmakers' and goldsmiths' work." This was done by means of disks with contours cut in the desired design (called rosettes). It differed from the ordinary lathe in that "the center of the circle described at each moment by each point of the surface on which you are working is no longer a fixed point, for it has a slight oscillating movement that creates curves. These curves differ from a circle in that the oscillations are more frequent for a single rotation and are wider in relation to their distance from the center." In 1730 La Condamine constructed a small device that made it easy to trace the rosette corresponding to a given curve. This object seems to have been an intermediary between the cam and the reproducing device.

Another very beautiful guilloching lathe, which also belonged to King Louis XVI, is preserved at Saint-Germain-en-Laye, near Paris, in the pavilion where Louis XIV was born. It bears the name of its builder: Wolff (porte Saint-Martin). The chasing was done by Gouthière, one of the best creators of the Louis XVI style. The machine consists of a massive mahogany bench on which the various components are mounted. The regulating device is a gilded iron column that supports the small crank-operated flywheel and the pulley, of mahogany and gilt bronze, which moves the headstock by means of a strap. As in all the guilloching lathes, the headstock consists of a frame on which the spindle rests; the latter can be shifted longitudinally as well as transversally. It turns in two bearings, and on its "nose" is mounted the mandrel that holds the object. On the spindle, between the two bearings, are the series of rosettes whose profile corresponds to the turns to be given to the object in order to obtain the desired designs. The toolholding carriage faces the mandrel; it is of gilt bronze, chased and inset with the royal coat of arms.

A small, simple, unornamented guilloching lathe, preserved in London and built around 1740, is probably of French origin; it belonged to the second Count of Macclesfield, a famous astronomer and president of the Royal Society from 1751 to 1764. On the whole it is similar to the machine we have just described, but it does, however, present two interesting details. On the one hand it can reproduce medals, when the object to be reproduced is mounted on the rear end of the spindle; it can also thread by reproduction, since one of the sections

Plate 19. Guilloching lathe constructed by Mercklein in 1780 for Louis XVI. Musée des Arts et Métiers. *Photo by the Conservatoire.*

of the spindle has three short, separate threaded surfaces that can be reproduced on the object.

Lathes for portraits and for "reducing" medals were generally intended for the production of medals from a considerably larger model. A palpating device went over the relief of the model, and a tool reproduced it on a smaller scale on a metal disk.

Similar work could be performed on other machines as well, using a small milling cutter.

Several reproducing lathes can be seen in the Musée du Conservatoire national des Arts et Métiers in Paris, particularly Mercklein's lathe for reducing medals (1767), Wohlgemuth's lathe for engraving and reducing medals and cameos (1820), and Maire's machine for reducing medals with a milling cutter.

In the *Encyclopédie* we find a picture of a simple lathe utilized in this period (1770) by the goldsmiths to outline dishes or plates in precious metal by reproducing, by reduction, the contours of a model. The same source gives a detailed drawing of the two-directional carriage whose upper plate, which slides freely in its grooves, follows the details of the model by means of a roller pressed by a spring.

The toolholder

It was the improvements made on the toolholder that rendered perhaps the greatest service to the progress of mechanics toward precision instruments. The toolholder depicted in Figure 72 can shift longitudinally on a prismatic bench thanks to a groove cut in its plate. The angle of attack of the tool can easily be modified, the pivot being locked and unlocked by a locknut. The lateral movement of the tool is obtained by a threaded spindle regulated by a crank. By 1760 the latter device, which was already in use on clockmakers' lathes, had become one of the characteristic principles of the first industrial machines constructed by Vaucanson: his drill and his lathe with a sliding toolholder (to be described in the next volume).

The toolholder of the lathe made by Mercklein for Louis XVI is even more remarkable. The tool is rotated by means of a threaded spindle that gears onto the threaded groove of a circular plate. Graduations on a sliding gauge permit the tool to be operated with the greatest precision. In this particular case the

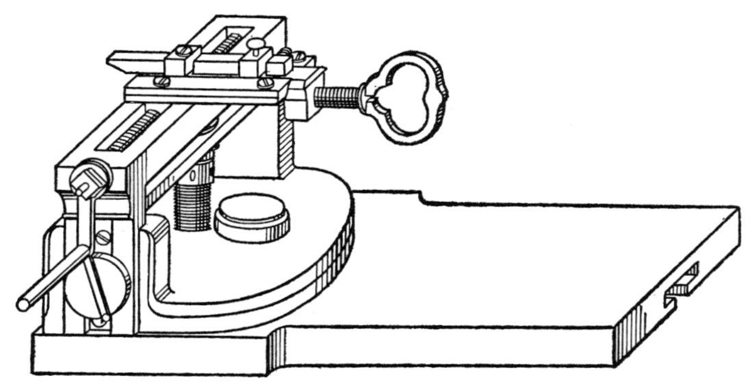

FIG. 72. Pivoting toolholder of a guilloching lathe (illustration from the *Encyclopédie,* Vol. X).

desire for perfection appears to have been only an unnecessary luxury, despite the remarkable execution of these devices. We discover them again, this time with greater justification, on the machines created by the builders of scientific instruments of the same period for the needs of their workshops. These great masters of the eighteenth century borrowed the principles of their machinery from the clockmakers' equipment.

The clockmakers' lathes For five centuries the clockmakers were to be the builders of the most precise mechanisms then known. They very soon invented tools that made it possible to perform with confidence the most delicate operations of their work: machines for dividing, graduating wheels and lantern gears, cutting spindles, calibrating gears, and so on. They began to take an interest in the properties of the brass and steel they used, and invented and perfected machines for making even their simplest tools.

The results were not always satisfactory; such was the case, for example, with the machines for cutting the files that were made with chisel and hammer. Jars writes that "several attempts were made to construct machines or mills for cutting files, but without success." This means that the machines made by du Verger in 1699, P. Fardoil in 1725, Thiout in 1741, Chopitel in 1750, and Durand in 1762 did not produce good results. This very failure, however, reveals a spirit of investigation that was still quite rare among ordinary mechanicians.

By the end of the sixteenth century the clockmakers had conceived a division plate for dividing toothed wheels, a device that eliminated the necessity of using the compass for each individual case. This plate had on its upper face a series of concentric circles, each divided according to one of the divisions usual in clockmaking. Each division was marked by a point incised to a sufficient depth to stop the point of a cutting needle. The wheel to be divided was attached to a shaft that passed through the center of the disk. With the help of an alidade, the divisions of the plate corresponding to the number of teeth divided were marked out on the circumference of the wheel, which was held stationary in each position by the needle. For a long time the clockmakers also utilized the division plate for cutting the teeth of lantern gears on a small lathe (Figure 73).

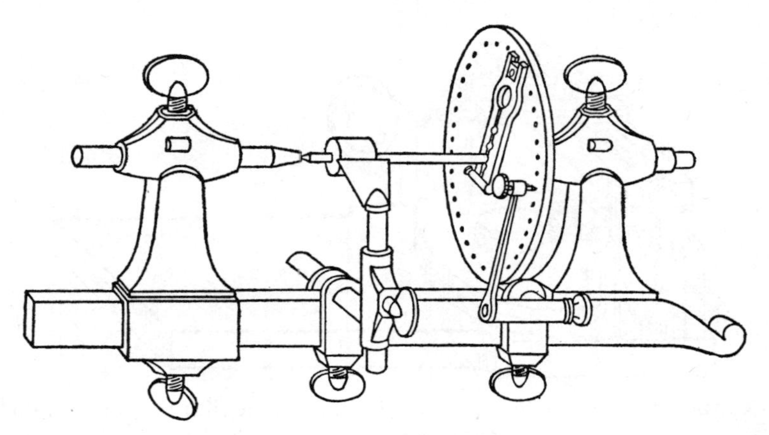

Fig. 73. Division plate mounted on a clockmaker's lathe (from F. Berthoud, *Essai sur l'horlogerie,* Vol. I). The plate has only a single divided circle.

Plate 20. *a*. Machine with division plate built by Hulot to cut the wheels and pinions used in clockmaking. Around 1760. Conservatoire des Arts et Métiers. *Photo Chevojon.*
b. Lelièvre machine for cutting fusees. Constructed by Hulot around 1760. Conservatoire des Arts et Métiers. *Photo Chuzeville-P.U.F.*

Around the end of the seventeenth century, the division plate was equipped, undoubtedly by Robert Hooke, with a small milling lathe, and was thus transformed into a machine for dividing wheels and lantern gears. A half century later, Ferdinand Berthoud invented a machine that automatically rectified the teeth cut with the milling cutter. Even before the advent of the builders of amateurs' lathes, Fardoil seems to have been one of the first (if not the first) to utilize an endless screw to ensure the movement of a toolholder carriage. This device is seen on a small machine, dated 1715, for cutting fusees for a watch or clock. At this period, however, it was practically impossible to trace a perfectly regular helicoidal thread, especially on threaded shafts of such small size. Although the device also appears in Thiout's *Traité d'horlogerie* written in 1741, it does not appear to have been utilized by the clockmakers; they preferred systems that had either a jointed frame whose shape could be changed, or a triangular, inclined guide to ensure the movement of the toolholder along the fusee to be cut, which was placed between the centers of a small crank-operated lathe (Figure 74).

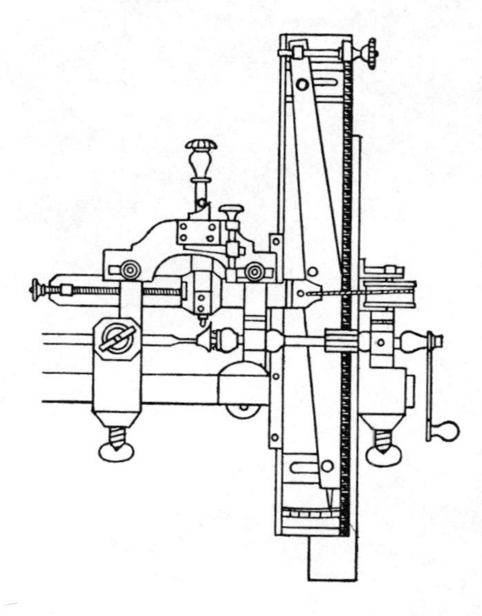

FIG. 74. Machine for cutting fusees (from F. Berthoud, *Essai sur l'horlogerie,* Vol. I). The fusee is placed between the centers of a small crank-operated lathe. The shifting of the rack causes a balancing movement of the triangular piece and consequently the shifting of the toolholder toward the left. The descent of the graver (applied manually), which rests on the upper handle, is regulated by the profile which can be distinguished below this handle. The drum with chain pulls the toolholder back toward the right at the end of the operation.

Ramsden's inventions

The second half of the eighteenth century saw the appearance, first in England and then in France, of extremely skilled mechanicians who were able to adapt all the then known techniques in order to create the first genuine precision machines. The unquestionable leader of this generation was Jesse Ramsden, the great English builder of scientific instruments, and the first man to solve the problem of the mechanical division of limbs and divided circles. Various attempts had been made before him, particularly by Henry Hadley in England around 1739, and the Duke of Chaulnes in France in 1765, to substitute mechanical division for the geometrical methods which at that time were the only ones in use.

Ramsden first attempted to create a machine for dividing circles. As it had to be operated by a gear with a tangent wheel, he was obliged first to construct a machine for cutting helicoidal screws, and another for threading the groove of a circular plate. The former consisted of a lead screw leading a plate by means of a threaded collar; the tool of the plate attacked the screw to be cut, which was positioned parallel to the lead screw. This wheel was turned by a crank whose rotating movement was transmitted to the two objects by a series of gears that could be changed, so that it was possible to cut threads of different sizes with the same lead screw (Figure 75). (Leonardo da Vinci had had the same idea two and a half centuries earlier, but it is doubtful that at that period the means for making such a machine existed.) Ramsden used his first machine to cut the tangent screw destined to pull the division plate and an identical screw whose thread had a cutting edge. This latter screw was intended to cut the thread in the groove of the division plate by pressure.

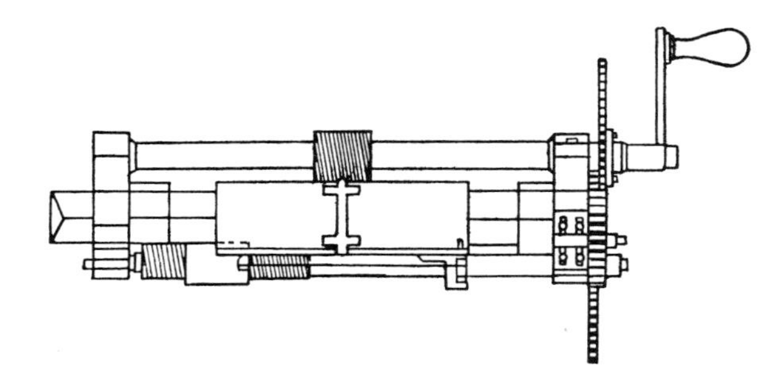

FIG. 75. Ramsden's threading lathe. (*Description of a dividing engine.*) The object to be threaded can be seen in the upper portion. The tool that attacks it has a diamond tip which can cut tempered steel. This tool is held by a carriage with a collar, pulled by the lead screw in the lower portion of the drawing.

The method used by Ramsden to construct the plate reveals the exceptional qualities of a mind that knew how to analyze all the difficulties of such an undertaking. Briefly, his method was to eliminate the absolute quality of the irregularity of the threading by distributing these irregularities as evenly as possible over the entire circumference of the plate. As for the threading of the tangent screw (that is, the lead screw used to cut it), the perfect regularity of the thread was the basis of the precision of his machine. He probably achieved this regularity by trial and error and repeated corrections. This work was possible for a man as highly skilled as Ramsden, with satisfactory grades of steel available to him. In addition, the work was facilitated by the fact that this threaded shaft worked only on a small part of its length, both to cut the thread of the circular plate and then, on the completed machine, to pull this plate. After fourteen years of effort, in 1773 Ramsden succeeded in constructing his first machine for the mechanical division of graduated circles. In a few years this invention transformed the construction of instruments for astronomy and surveying. It marks the beginning of modern small precision instruments.

Shortly after this, Ramsden constructed a machine for dividing straight lines, utilizing an endless screw to pull the platform holding the rule to be divided under the scriber. The making of this screw was a more delicate operation than constructing the machine for dividing limbs because the screw worked along its entire length. For this purpose Ramsden conceived a threading machine that also consisted of a circular plate pulled by a tangent screw. The rotation of the plate regulated the movement of the cutting tool by means of a central pulley (Figure 76). Thus the regularity of the various threads of the tangent screw was distributed over the entire length of the bar to be threaded.

In France two exceptional mechanicians, J.-L. Lenoir and Nicolas Fortin, created dividing machines shortly after Ramsden. We do not know by what method they succeeded in obtaining bars threaded with a continuous thread. Success in this work depended on a personal skill that could not yet be replaced by the available mechanical methods.

The problem of gearing For all these machines created around 1780, the profile of the gear teeth was not yet a major problem. In general the toothed wheel was very rarely utilized on the various types of lathes then known. When it intervened in the operation of certain tools, in the clockmakers' machines, and in the clockworks themselves (watches, pendulums, and so on), it was operated only with slow movements. In addition, in the clock mechanisms the gears always turned in the same direction, and the teeth exercised only slight pressure on each other. Under these conditions it was quite easy for the technicians to discover satisfactory solutions, in empirical fashion, without having to solve the problem of the geometrical shape of the teeth.

This was not true of large mechanisms, for example those involved in the construction of horse-operated treadmills, hydraulic installations, and windmills. A certain number of these devices are depicted in various chapters of this volume. It will be noticed that gears were constructed of wood, following models whose origin dated at least from the beginning of the Middle Ages. Figure 77 recalls the form of the toothed and pegged wheels that did not completely disappear from use until the middle of the nineteenth century, as well as the effects they could produce.

But the engineers who were seeking to increase the productivity of the power machines they constructed or maintained were beginning to become interested in the question of the best shape for the teeth of the gear wheels. A drawing by Leonardo da Vinci depicts wheels whose teeth have a modern profile, but there is no note to tell us whether the shape was seen or simply imagined by the artist.

Early geometrical research Drawings of conical gears in the works of Jacques Besson and Ramelli attest that at least for certain purposes an attempt was made, toward the end of the sixteenth century, to find a substitute for the classic lantern gear. The seventeenth-century mathematicians were to pave the way for geometrical research. Around 1650 or 1660 Gérard Desargues, the creator of geometrical perspective, was undoubtedly the first to

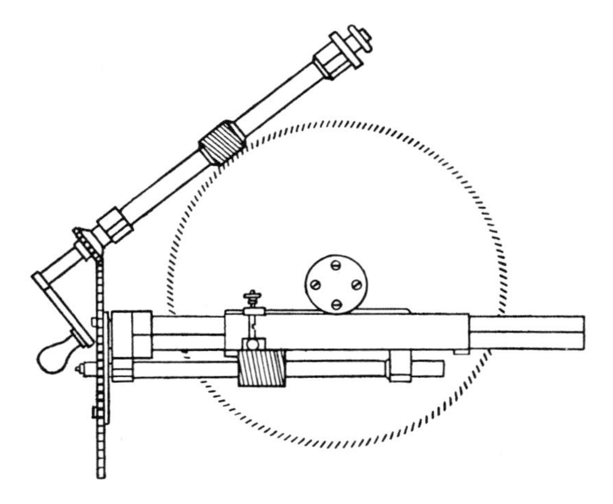

Fig. 76. Ramsden's threading lathe pulled by tangent screw. (*Description of a dividing engine.*) The lead screw pulls the plate with threaded groove, above which the toolholding carriage, operated by a central pulley, moves. The object to be threated is positioned parallel to the carriage.

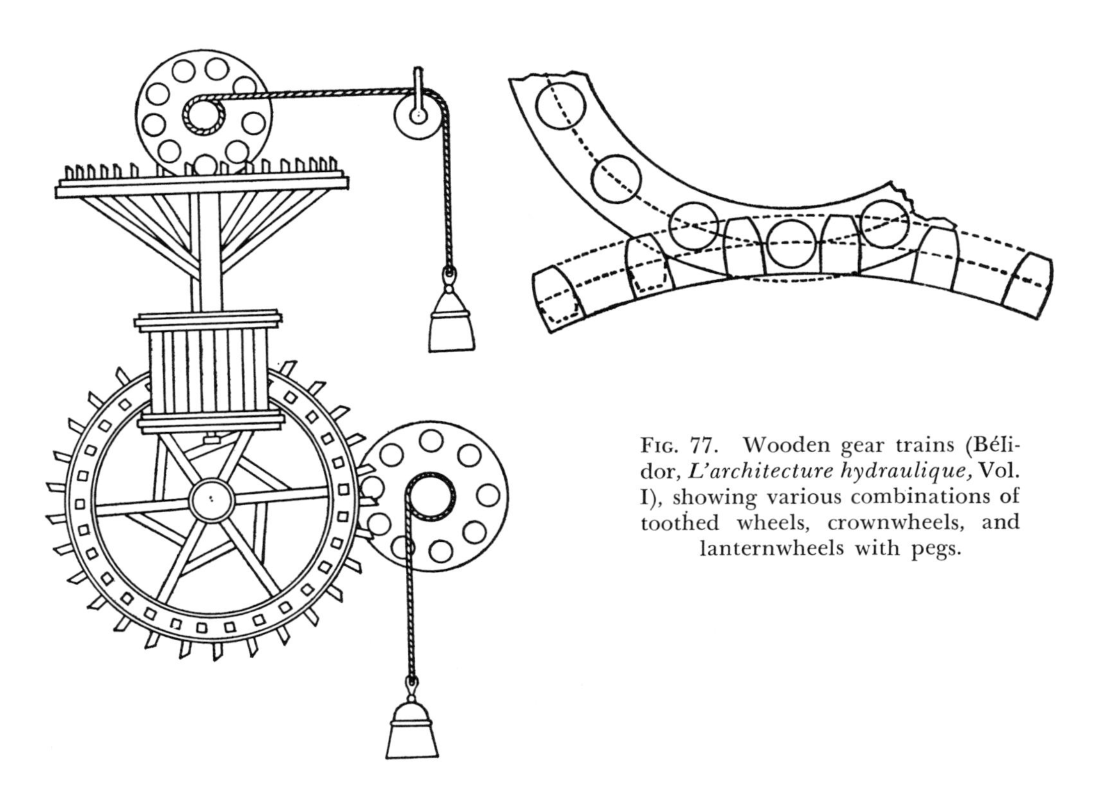

Fig. 77. Wooden gear trains (Bélidor, *L'architecture hydraulique,* Vol. I), showing various combinations of toothed wheels, crownwheels, and lanternwheels with pegs.

give an epicycloidal profile to the teeth of gear wheels. We do not know whether this accomplishment was preceded by geometrical research on the part of Desargues, but the choice of a cycloidal profile was certainly not happy; curves of this type were then interesting the most renowned scientists of every country, and Desargues's achievement was not an isolated event. The astronomer Roemer may also have created such gear trains in 1674; and in 1694 Philippe de La Hire published a *Traité des épicycloïdes et leur usage en mécanique,* followed in 1695 by a *Traité de mécanique* in which the utilization of cycloidal curves for tracing the profile of the teeth is studied in detail.

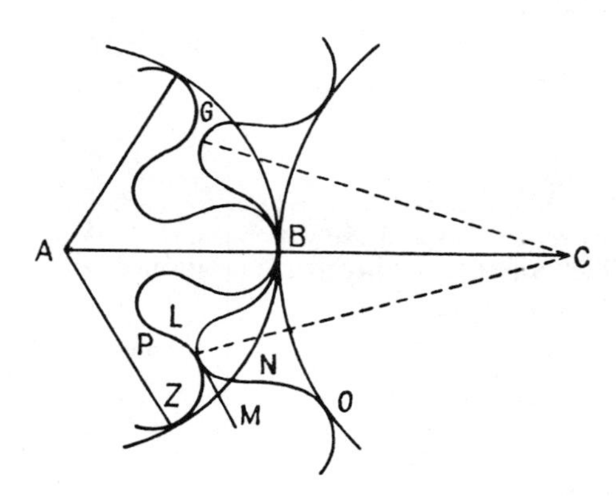

FIG. 78. Extension and arrangement of the teeth of wheels gearing into each other. When tooth ONL ceases working on tooth PZ, tooth BG must begin to work on the corresponding tooth of wheel A (De la Hire, *Traité des épicycloïdes.*)

It was La Hire's intention to determine the shape for the surfaces of the teeth best suited to ensure uniform pressure and movement. He realized that in order to avoid rubbing, the teeth in contact should roll upon each other, and he demonstrated that if one of the teeth has the profile of an epicycloidal external portion defined by a given generating circle, the profile of the tooth that corresponds to it should be that of an internal epicyloid defined by the same generating circle. His discussion introduces for the first time the idea that the pressures should be continuous so that the movement will be uniform. After reviewing the various curves that can be used, he concludes that the involute represents the most satisfactory external cycloid. La Hire's name is connected with a gear train formed of a toothed wheel gearing with a circular rack lined with teeth.

Forty years later, in 1733, the geometer Camus presented to the Académie des Sciences a memorandum, "Sur la figure des dents des roues et des ailes de pignons," which was included in 1766 in his *Cours de mathématiques.* Camus's work is more practical in nature than that of La Hire, which he used as his base. Studying the gears in existence at that time, he investigated the shape of the teeth required for the various combinations. He envisaged only the epicycloidal shape, and neglected to discuss the involutes. But he shows that the best effect is obtained when the tooth of the driving wheel engages on the driven wheel when its own surface has passed the line that joins the centers of the two wheels

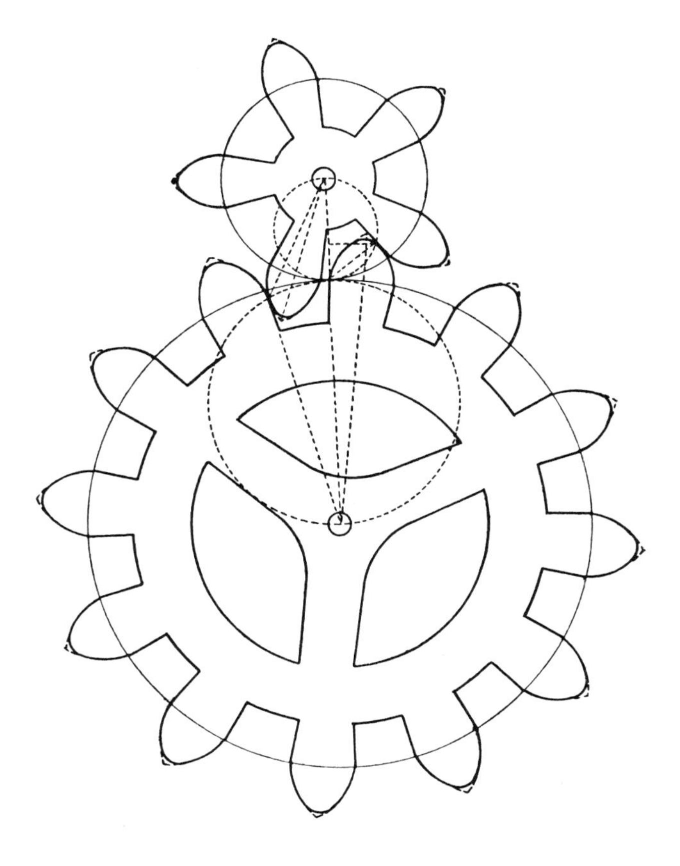

FIG. 79. Establishment of the shape of the teeth of a pinion and a wheel in relation to the respective number of their teeth and the distance of the centers (Camus, *Cours de mathématiques,* Vol. II).

(Figure 79). In addition to this important result, Camus also established the first data on such diversified problems as the minimum number of teeth on a wheel, the best profile for the end of the wheels, and so on.

Between 1754 and 1765, several theoretical works were devoted by Leonhard Euler to the problems of gear teeth. In particular, Euler gave a mathematical demonstration of the fact that teeth with involute or epicyloidal profiles eliminate rubbing, and he determined that the angle of pressure should be 30 degrees.

All the theoretical stages had been achieved, but they remained unknown to those who would have been able to make use of them; the contents of the mathematicians' papers were not accessible to the technicians. Not until treatises on applied mechanics, written for a wider public than the relatively limited circle of academicians, made these results available to the technicians were the latter able to benefit by the mathematicians' ideas. This did not occur until the first quarter of the nineteenth century. In addition, it must be admitted that La Hire, Camus, and Euler were ahead of the needs of their time. Fifty years after Euler the traditional methods were still sufficient for the construction of the machines in use. The scarcity of steel forestalled any idea that it could be used for extensive production of large machines. Gradually, however, the evolution of mechanics (which has been summarized in this chapter), the iron industry, and the steam engine (to be discussed in the third volume) paved the way for the period when all the technical conditions combined to permit the genuine birth of the machine tool.

BIBLIOGRAPHY

SOURCE MATERIALS

AGRICOLA, GEORGIUS, *De re metallica* (1556), transl. by Herbert C. and Lou Henry Hoover (London, 1912; reprinted New York, 1950).

BÉLIDOR, BERNARD, *Architecture hydraulique* (1737–1739).

BESSON, JACQUES, *Théâtre des instruments mathématiques et mechaniques* (1578).

BIRINGUCCIO, VANNOCIO, *De la Pirotechnia* (1540), transl. and ed. by Cyril Stanley Smith and Martha T. Gnudi (Cambridge, Mass., 1942; paperbound, 1959).

CAMUS, C., *Cours de mathématiques* (Paris, 1766).

CHÉRUBIN, P., *La Dioptrique oculaire* (Paris, 1671).

A Diderot Pictorial Encyclopedia of Trades and Industry, ed. by Charles C. Gillispie (2 vols., New York, 1959).

LA HIRE, PHILIPPE DE, *Traité des épicycloïdes* (Paris, 1694).

——— *Traité de mécanique* (Paris, 1695).

PLUMIER, P., *L'art de tourner ou de faire en perfection toutes sortes d'ouvrages au tour* (Paris, 1701).

SECONDARY WORKS

ABELL, G. S., with LEGGATT, JOHN, and OGDEN, W. G., *A Bibliography of the Art of Turning and Lathe Machine Tool History* (London, New York, 1956).

BATTISON, EDWIN A., "Stone-Cutting and Polishing Lathe, by Jacques Besson," *Technology and Culture,* 7 (1966), 202–205.

DAUMAS, MAURICE, *Les instruments scientifiques aux XVIIe et XVIIIe siècles* (Paris, 1953).

JOHNSON, WILLIAM A. (transl.), *Christopher Polhem: The Father of Swedish Technology* (Hartford, Conn., 1963).

KELLER, A. G., *A Theatre of Machines* (New York, 1965).

——— "Mechanical Linkages," *The Chartered Mechanical Engineer* (July 1967), 322–327.

ROLT, L. T. C., *A Short History of Machine Tools* (Cambridge, Mass., 1965).

USHER, ABBOTT PAYSON, *A History of Mechanical Inventions* (rev. ed., Cambridge, Mass., 1954).

WOODBURY, ROBERT S., *History of the Gear-Cutting Machine* (Cambridge, Mass., 1958).

———, *History of the Grinding Machine* (Cambridge, Mass., 1959).

———, *History of the Lathe to 1850* (Cleveland, 1961).

CHAPTER 13

THE BUILDING OF CLOCKS

THE QUESTION OF ORIGINS

WHATEVER MAY BE the importance accorded to its known mechanical antecedents, the appearance of the crown-wheel escapement in the second half of the fourteenth century was a decisive event that marked the beginning of mechanical clockmaking.

This invention, whose origins are unknown despite the considerable amount of historical research done on the question, influenced the construction and proliferation first of large clocks on public buildings, then table models, and finally watches. There has been much discussion of the circumstances that ensured such rapid success for this new industry. In the entire history of technology prior to the seventeenth century, there is apparently no other example of an invention as clearly localized chronologically that in the space of some fifty years was as widely adopted.

Its appearance can correctly be situated in western Europe, perhaps in southern Germany or in England, and the date that can reasonably be assigned to it is around the first third of the fourteenth century. The new invention was quickly transmitted to France and northern Italy — if it was not actually born in this area. We are not absolutely certain, however, whether it was the fruit of an individual effort or a product of the rapid evolution of an already existing device applied to the governing of a clock mechanism. No trace remains of the primitive and poorly developed models that probably preceded and led up to the verge escapement with pallets and crown wheel; the first definitely known specimen is the clock constructed by Giovanni de' Dondi around 1350, of which we have an authentic description (see Figure 84). Still more remarkable is the fact that this type of mechanical construction was reproduced for centuries in a considerable number of models without undergoing major changes; the verge and crown-wheel escapements of the seventeenth and eighteenth centuries are identical in structure if not in workmanship to those of the fourteenth and fifteenth centuries.

The historical factors An invention is able to fulfill its destiny to the fullest extent only when it encounters circumstances that make its exploitation possible — a fact that is true in all phases of the evolution of technology. Historians are still uncertain, however, about the favorable factors that ensured the success of mechanical clocks. The use of public clocks around the end of the Middle Ages does not appear to have filled a specific need; the hypothesis that they originated in the monasteries in answer to the requirements of the schedule of the monastic life has been abandoned, for the succession of the Offices was regulated by the length of the day and the night, and did not require the measurement of regular intervals of time. Moreover, the rhythm of daily life and work outside the monasteries was not yet linked with an exact reckoning of hours. Perhaps the development of urban life was beginning to cause the appearance of certain needs which the first public clocks filled. It is more likely, however, that these clocks, which not only told time by means of chimes and the appearance of automata but also supplied information relating to astronomy and the calendar, were regarded as curiosities and symbols of affluence of the communities that had undertaken to build them. Intercity rivalry for prestige would explain the rapid proliferation of clocks, which at that time were of no practical use. It is possible that at a later date the functioning of the public clocks gradually caused the rhythm of urban life to become adapted to the regular division of the day into hours. It must be admitted, however, that these hypotheses do not satisfactorily explain this phenomenon.

The birth of the escapement

The same uncertainty obscures the origin of the crown-wheel escapement and the manner of its invention. We are almost certain that the weight-driven clock was in existence or at least in the experimental stage by around the end of the thirteenth century. The use of a heavy weight suspended from a cord wrapped around a horizontal drum, for example to operate certain lifting devices, probably dates from the Alexandrian period if not earlier. Without discussing all the written and iconographical evidence that has been indexed for the period under consideration, we need only note the reference to the weight-driven clock in the *Libros del Saber de Astronomia,* which dates from 1276–1277. This work, which includes the famous *Alphonsine Tables* prepared in 1272 at the command of Alfonso X of Castille, contains a drawing of a clock whose mechanism is operated by the falling of a weight; this movement is regulated, not by an escapement, but by the flowing of mercury contained in a annular drum turning around a horizontal axle (Figure 80).

Mr. H. Alan Lloyd correctly observes that if the mechanical escapement had been known in the time of Alfonso the Wise, it would have been described in the *Libros del Saber de Astronomia.* He further deduces from the fact that this escapement appears on Giovanni de' Dondi's astronomical clock, whose description dates from 1364, that this mechanism must have already been known for some decades. It is a fact that Dondi does not claim the invention as his own nor does he refer to it as an exceptional novelty. It can thus be assumed, with Lloyd, that the invention of the crown-wheel escapement took place approximately between 1280 and 1330.

Hypothetical predecessors

Several predecessors have naturally been suggested in an attempt to discover the line of ancestry that led to the invention. Some have claimed to see in a drawing of Villard de Honnecourt, architect of the second half of the thirteenth century (manuscript

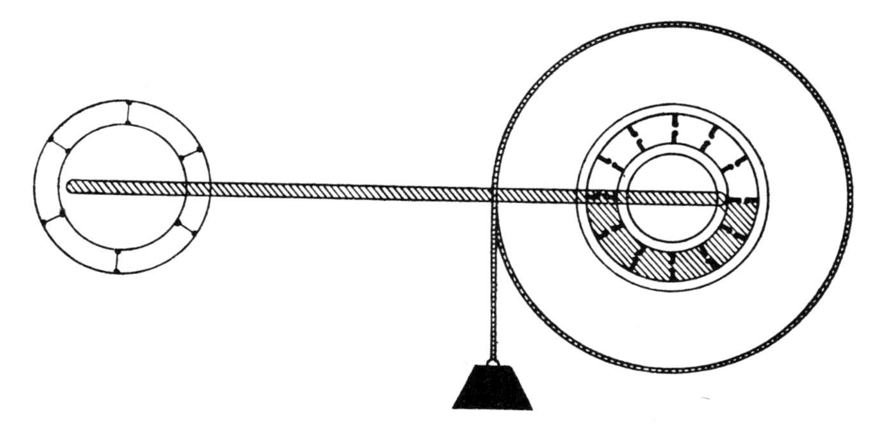

Fig. 80. Clock with weights (Alfonso the Wise). On the right are the driving weight and the regulator drum; the drum contains mercury, which flows from one compartment to the next during the rotation. The horizontal shaft turns a pinion with six teeth (diagram on the left).

of approximately 1260), the presence of an escapement in the mechanism that caused the statue of an angel pointing to the sun to pivot automatically. Actually, this device was simply a flywheel whose wheel was also that of a winch that raised the driving weight (Figure 81). It is difficult to understand how this drawing could have been incorrectly interpreted; Lassus, the first editor of the manuscript, did not make this mistake. Villard de Honnecourt's "escapement" has been mentioned by a series of historians, who perhaps never took the trouble to look at the manuscript or one of its reproductions. This idea must definitely be eliminated from the history of clockmaking.

The Sinologist Joseph Needham and Messrs. Wang Ling and Derek J. Price have made studies of a Chinese astronomical clock described by Su Sung, a high-level official of the empire, in 1090. (The general structure of this clock is pictured in the first volume of this work.) Attention has been drawn by the authors to the governing device, in which they claim to see an antecedent of the mechanical escapement that is supposed to have preceded the appearance of this device in Western Europe by three centuries. It consists of a bucket wheel that advances intermittently; the weight of a bucket filled with water to a certain depth by a constant flow of water releases a locking device. This enables the wheel to turn at the one angle that permits the next bucket to take the place of the first (Figure 82).

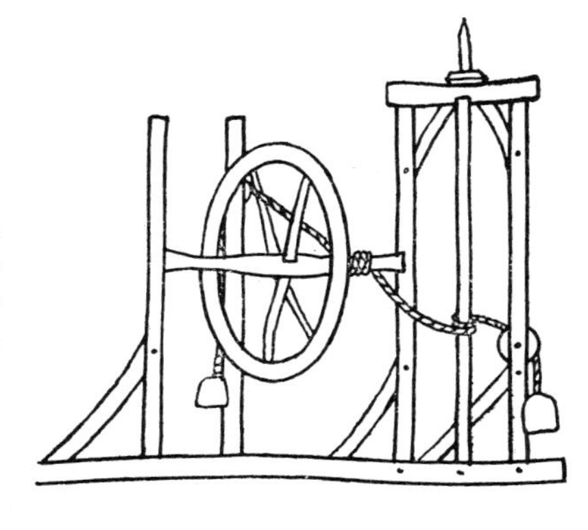

Fig. 81. Drawing by Villard de Honnecourt: countergearing from the winch to a vertical axle. Some writers have incorrectly viewed it as an early escapement.

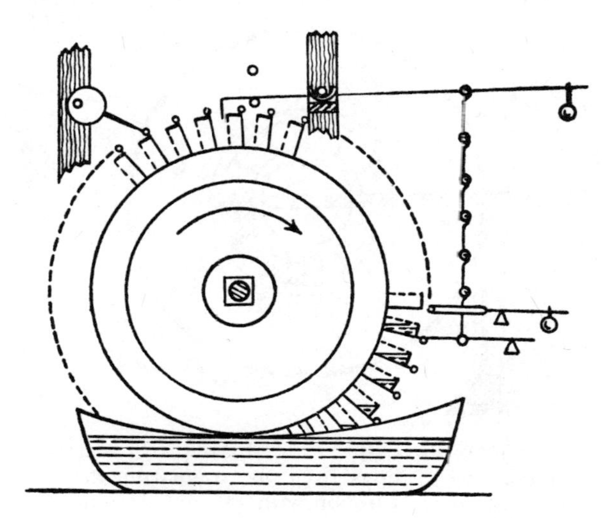

FIG. 82. Principle of Su Sang's clock (from J. Needham, W. Ling, D. J. Price), showing the bucket wheel and the stopping and starting device.

This hydraulic device already contains an early model of the logical continuation of the functions of an escapement: the establishing of a given interval of time by filling a fixed volume at constant speed, and the automatic resumption of movement as soon as the interval chosen has passed.

Dr. H. von Bertele, from whose work we have taken this analysis, has made a thorough study of the "rolling-ball clocks," whose principle presents a striking analogy with that of the Chinese clepsydra. The extant rolling-ball clocks were constructed as curiosities in the seventeenth century, but it can be considered probable that they were a resurrection of an old idea. When an attempt was made to shorten the "chronometrical unit," the rolling ball proved to be too slow, but the continuation of the striking mechanism (which was certainly already known) with the system of release and arrest, combined with a "mover," was able to furnish the escapement in its typical form.

It is nevertheless true that this hydraulic device of Su Sang's clock had no connection with the simple and effective structural principles that characterize the foliot and crown-wheel escapement and that formed the basis for the entire development of mechanical clockmaking after the beginning of the fourteenth century.

Whether it is a question of a purely fortuitous convergence or of an influence (by way of one or several "missing links"), there is nothing to justify an underestimation of the originality of the escapement invention, which was a totally new step in technical progress. It created a decisive break in the continuity of the traditional endeavors, and truly opened the modern period of progress in clockmaking.

The maturity of the technical environment

It would be reasonable to conclude from the foregoing that mechanical clockmaking flowered rapidly at the beginning of the fourteenth century in a technical milieu that was ready to receive it. The art of the mechanician was by then sufficiently developed to utilize the new invention, as is evidenced by a certain number of objects containing toothed wheels and sometimes quite complex gear trains.

We learn from many early documents that the Chinese were probably past masters in this art by the tenth or eleventh century. Another tradition, which may also date from the eleventh century, is attested by the Arabic authorities on the astrolabe, but there is nothing to confirm that knowledge of this technique was transmitted from the Far East to the Middle East. It appears more certain that the West benefited directly, and in a relatively short period of time, from the experience of the Moslem scientists and mechanicians.

In any event, several extant descriptions and instruments show examples of movable calendars for which the problems of the division, cutting, and arrangement of toothed wheels had already been satisfactorily solved by the end of the thirteenth century. But none of these devices constitute, strictly speaking, a clock mechanism, for they are operated manually; the problem of the use of a driving force and its regulation had not been solved in these specific cases.

The first genuine clockworks did not actually appear until the half century between 1280 and 1330, that is, at the time of the invention of the mechanical escapement.

FIRST STAGES IN MECHANICAL CLOCKMAKING

The well-known evolution of mechanical clockmaking extends over a period of some five centuries, from the end of the fifteenth century down to modern times. It is a magnificent development from every point of view, in the continually ascending curve of its continuity, the harmonious relationship of the instruments and techniques, and in the beauty of the inventions that were milestones along its path. The men who worked in this field ranged from craftsmen, so skilled and completely trained that they deserve the name of artist (still applied to fine clockmakers), to universal geniuses. Mechanical clockmaking flowered into the powerful contemporary industry that has produced great wealth while remaining, on a human scale, an industry in which the knowledge and inventiveness of the technician and the skill and conscientiousness of the craftsman triumph on the same plane. The beautiful products of the clockmaking technique, both past (sumptuous pendulums and complex astronomical globes) and present (chronographs of multiple functions, miniscule jeweled watches) have always excited, and continue to excite, admiration and the desire for possession. The men capable of making and repairing these marvels have always enjoyed great prestige. In earlier times they acquired the questionable glory of the magician, today the consideration given to the specialist, the supreme representative of the technological age.

Fourteenth to sixteenth centuries

Mechanical clocks appeared "suddenly" at the end of the Middle Ages in the various distinct yet so "undivided" countries of western Europe. We find evidences of it as early as the second half of the fourteenth century in Italy and France, in the fifteenth century in these countries and in Great Britain and Germany. Of the objects still extant, however, very few antedate 1500 (to use a round figure).

As explained earlier, this "sudden" appearance is certainly an illusion. The early specimens have disappeared, victims notably of the wars that have constantly ravaged Europe. Written evidence is rare, undoubtedly in part as a result of the disdain of the educated people for the "mechanical arts," which in their eyes retained something of the servile character attributed to them by the ancients. It is nevertheless possible to make fairly precise conjectures on the stage reached by the technique shortly before the advent of the oldest clocks still extant, keeping in mind that a possible discovery of a hitherto unknown document would, like certain fossils, temporarily reopen the question.

The development

With the opening of the sixteenth century, when material evidence becomes abundant, we are surprised by the number and perfection of the achievements of, and the wealth of methods and ideas available to, the clockmaking industry. This industry then proceeded to develop at a very erratic pace, sometimes slowly perfecting one detail after another, sometimes with all the characteristics of a revolution, especially after a major invention. In the forefront of the major inventions stand the innovations of Christiaan Huygens: the pendulum (1656–1657) and the spiral balance spring (1675). These two inventions clearly divided the history of clockmaking and the history of chronometry into two periods; the first period was the phase of formation, of trial and error, while the second witnessed the development of the pendulum system and its posterity, which covered two and a half centuries. Finally, modern physics and electronics came to revolutionize clockmaking as they have revolutionized all the sciences. This is why this study will terminate with the inventions of Huygens and the closing years of the seventeenth century, a limit that is dictated, so to speak, by History.

Geographically speaking, for a long time clockmaking remained confined to a limited area of western Europe. At the beginning of its history it was the courts — the courts of England, France, and the imperial court — that appear to have been the principal points of crystallization. We use the word "appear" because the artists, who were great travelers and were often bound by fairly flexible connections to a great noble and followed him quite freely in his movements, exercised their activity in every city of any importance, in many cases training pupils as well. Thus the "original" center comprised England, Flanders, central and southern Germany, Bohemia, France, and northern and central Italy. (The clockmakers of Blois and Augsburg were particularly renowned.) Later a kind of coalescence of the production centers occurred, with London, Paris, and Geneva occupying a predominant place.

The clockmakers' profession

From its origins down to the end of the eighteenth century, the construction of clocks developed within the framework of the guild system. There was a strict hierarchy — apprentice, journeyman, master — and strict control of the members, which successfully ensured the maintenance of quality, but constantly acted as a brake on quantitative increase. The length of the apprenticeship varied from five to eight years, depending on the guild mastership; and accession to master status was dependent on the making of the "masterpiece" (specimen), which almost everywhere was an alarm clock. Among the most famous guilds

Plate 21. Mechanism of an iron clock, end of the sixteenth century. Musée des Arts et Métiers.
Photo Chuzeville-P.U.F.

are those of Paris (first statutes 1544), Blois (1597), Geneva (1601), and the Clockmakers Company of London (whose records are unbroken from 1632 to modern times). While the condition of the "masterpiece" required of the would-be master the complete execution of the clock, from the drafting of the "caliber" (general arrangement of the movement) down to the chasing of the box, in actual practice a well-developed division of work was the general rule. Workshops that were often family affairs, in which each person worked on one piece of equipment, furnished the "rough" movements to the master, who "reworked" them, had the case made, and signed it, usually both on the movement and on the dial face.

Importance of the clockmaking techniques

What place should be assigned to clockmaking in the general picture of technology? It contributed to the general progress of science, and particularly astronomy, and the formation of metrology; it played a role in the discovery of the world by navigation, and led to the formation of the clockmaking industry, which satisfies an essential need of modern life. All these contributions are important, to be sure, and would justify the interest in this subject. Above and beyond this, however, the importance of the development of clockmaking in the general picture of technical civilization went far beyond its direct consequences. Clockmaking, the inexhaustible breeding ground of inventors, inquisitive minds, and skilled artists, was a kind of experimental laboratory, a collection of models in which most of the kinematic components that constituted machinery, in the widest sense of the term, were developed and perfected. Moreover, it was in the hands of the clockmakers that automation was born, first in the elementary form of striking mechanisms of clocks, then as automata with more complex cycles, and finally "control" in the modern sense of the term. Through the men, ideas, and models it furnished, clockmaking played a catalyzing role in the genesis of the world of machinery, a role whose influence was completely unrelated to its own importance.

CONSTITUENT ELEMENTS OF MECHANICAL CLOCKWORKS

The driving force

Almost everyone has at one time or another taken apart an old watch or "repaired" an alarm clock; the basic component parts, which are always the same, are therefore familiar to most people. The driving force is derived from a spring that occupies approximately half of the volume of the spring box in which it is enclosed. This spring is "wound" by a series of revolutions, the number of which varies between five and six and twelve and fifteen, and as it unwinds it transmits the "driving couple" to the works. The driving couple is not very constant; it is very powerful when the spring is tight, but gradually weakens as it unwinds, and on the whole retains half of its initial value at the end of the useful part of the "running down." The weight found in old clocks, which is still used on occasions, is less convenient and vastly more cumbersome than the spring, but the constancy of the driving power it supplies has never been equaled by any device (Figure 83).

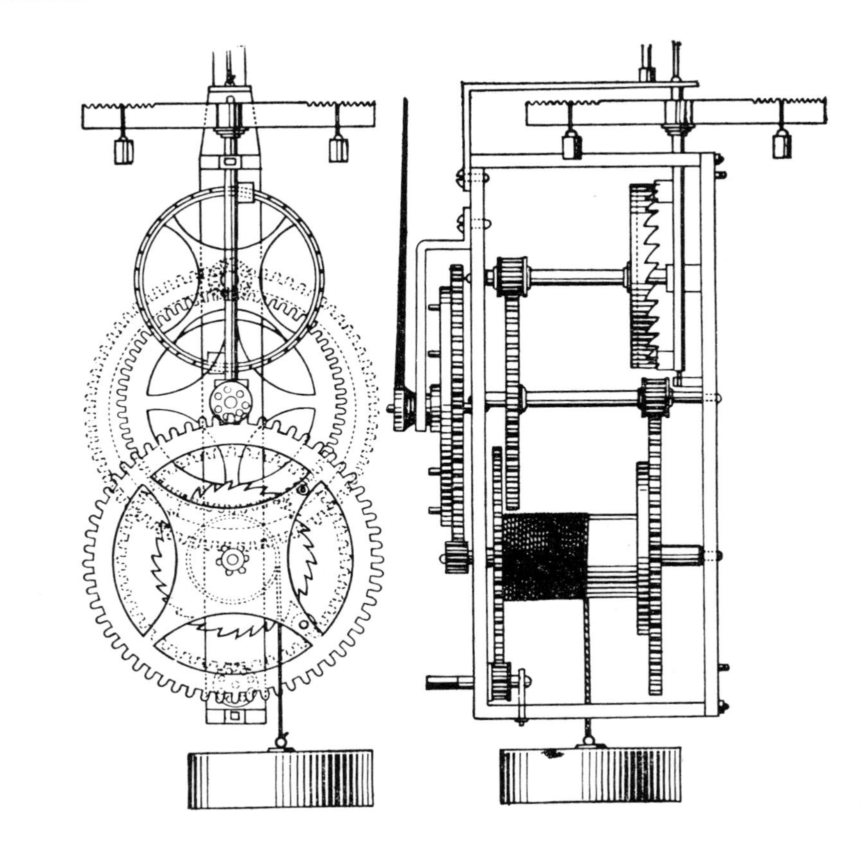

FIG. 83. General organization of an early weight-driven clock (reconstruction by F. Berthoud). Top: The regulator, below the escapement. Bottom: The winch and the driving weight. Center: the works.

The works

The works consist of a series of "movers," each of which has a toothed wheel and a gear. They transmit the movement to the hands, transforming the five or six turns made by the spring box in the course of the unwinding into some three thousand revolutions for the second hand (approximately forty-eight hours of functioning). The inquisitive mind will not fail to note that from the point of view of energy the clock is a completely "unproductive machine"; all the energy stored in the spring is completely dissipated in the frictions, minimal but never nonexistent, of the works, and it performs no useful work. However, it is not work that is required of the clock, but rather "information." The constantly pursued improvement of the works and springs, which means that a spring of a given volume makes the clock run for a longer period of time, produces no increase of "external" work, but does give more "information."

In order for each revolution of the "last mover" to be performed in a clearly determined time (one minute, if this mover is the minute wheel), a governor is needed. This organ always has an alternating movement, because this type of movement is the only one so far known that is connected with a period of oscillation relatively independent of the amplitude of the oscillation (a property known as "isochronism"). We are now using (and undoubtedly will continue to use) the governors invented in the seventeenth century, within a few years of each other, by Huygens: on the one hand the pendulum, whose "restoring couple" toward its position of equilibrium is gravity, and on the other hand the circular balance wheel, whose "restoring couple" is the spiral spring.

The movement governor The works, pushed by the driving spring, tend to adopt a continuous rotation; the movement of the governor is essentially alternating. A special mechanism called the "escapement" resolves this apparent contradiction by giving the works an intermittent movement consisting of brief periods of movement separated by periods of rest. The length of each period is fixed by the governor, which in particular is responsible for the moments when the works "escape." Another function of the escapement is to remove a small amount of the driving force at each oscillation. This ensures the maintenance of the alternating movement of the governor, which if left to itself would be condemned by the friction to slow down and stop. The amplitude assumed by the governor depends on the equilibrium between the driving power and the dissipating powers. When the driving power diminishes with a steady dissipation, the amplitude diminishes. The same is true when the frictions increase, the driving power remaining invariable. If the isochronism were invariable, these changes of amplitude would have no influence on the period of oscillation, but this is merely an intellectual concept. In actual fact there is, and with our governors there could be, no perfect isochronism (a theorem demonstrated in 1933 by J. Haag). This is why the improvements made on the balance wheel and spiral spring system since its creation have consisted basically of a closer approach to this inaccessible ideal. Naturally, a closer approximation than that achieved previously is much more difficult, and involves new phenomena or factors that were hitherto unsuspected.

Additional functions Functions other than the simple indication of the time were then added to the basic component parts of a mechanical clock, whether a monumental structure or a tiny jeweled watch. The most common were striking mechanisms and alarms. Today the striking mechanism is generally simple. Such was not the case in earlier times, when the "great striking" (repetition of the hour just passed after each quarter hour, either for a twenty-four-hour period or only during the night) and "striking by request" (repetition on demand and as often as desired of the hour and the quarter hour, or even the minute, just passed) were highly regarded.

Next came the calendars, some indicating simply the date, others the day of the week and the month, astronomical clocks (age and phase of the moon, and even highly complex planetary systems), and, most modern of all, chronographs (sweep-second hands that can be started and stopped when desired, whether simple mechanisms or the more complex devices in which one hand

Plate 22. Table clock with astronomical information. Constructed by Jost Burgi, end of the sixteenth century. Staatlich Kunstsammlungen, Kassel. *Photo by the museum.*

can be stopped to permit a reading to be taken, and then made to catch up with the other), and self-winding watches.

Some of these complexities have become so firmly rooted in our daily lives that we scarcely notice them; others have become more refined; many have fallen into disuse. All bear witness to the inventive and never-satisfied mind of the creative clockmaker, as is evident from the antiquity of their appearance. It is indeed paradoxical to note that among the oldest known specimens of clocks and watches are some that exhibit the strangest astronomical complexities.

ORIGINAL ELEMENTS OF THE MECHANICAL CLOCK

The component parts of the clock can already be recognized in the earliest extant models. This remark is intended not to minimize the importance of the development and progress achieved, but to illustrate the genius of the early inventors.

The first clocks

These inventors are almost totally unknown to us, as is true of so many other areas of technology; they undoubtedly range over a great number of generations, beginning in the distant past. It is probably impossible to date the mechanical elements, such as levers, detents, locks, click-and-ratchet works, and toothed wheels; they were certainly known to the Greek and Alexandrian engineers, who had received them in part from Egypt or the East. The transmission of the Alexandrian tradition to the Western world, either directly or through the intermediary of the Arabs (who were great constructors of water clocks and astrolabes), and the question of possible exchange of information between the Arabs and China are, as we have already noted, unsolved problems. To limit our discussion to the definitely established data: several fourteenth-century clocks are known, notably the "Gros Horloge" of Rouen (1379), a clock in Salisbury Cathedral (1386), and one in Wells Cathedral (1392). Unfortunately, it is difficult to say what parts of these objects actually date from their original construction: apparently little more than the forged iron frame of the device. The works have certainly been rebuilt more than once; these clocks are now equipped with pendulums that were added in the seventeenth century. We possess a very detailed description, written in 1364, of the astronomical clock built at Padua by Giovanni de' Dondi (here, exceptionally, is one precursor whose name is known); several manuscripts illustrated with numerous drawings have been preserved, and an exhaustive study of them was made in 1955 by H. A. Lloyd. The clock itself seems to have been carried off to Spain, and to have been destroyed during the Napoleonic wars. It had extraordinarily complicated works that reproduced the apparent movement of the moon, the sun, and the planets. The text and drawings of this manuscript tell us little about the hourly movement itself, as if this part of the mechanism was already quite ordinary and well known in Italy at this period (Figure 84). The best extant image of a fourteenth-century clock may be considered to be an object from Dover Castle (preserved in the Science Museum in London); it too seems to have been partially rebuilt, but with greater fidelity and without any attempt at improvement.

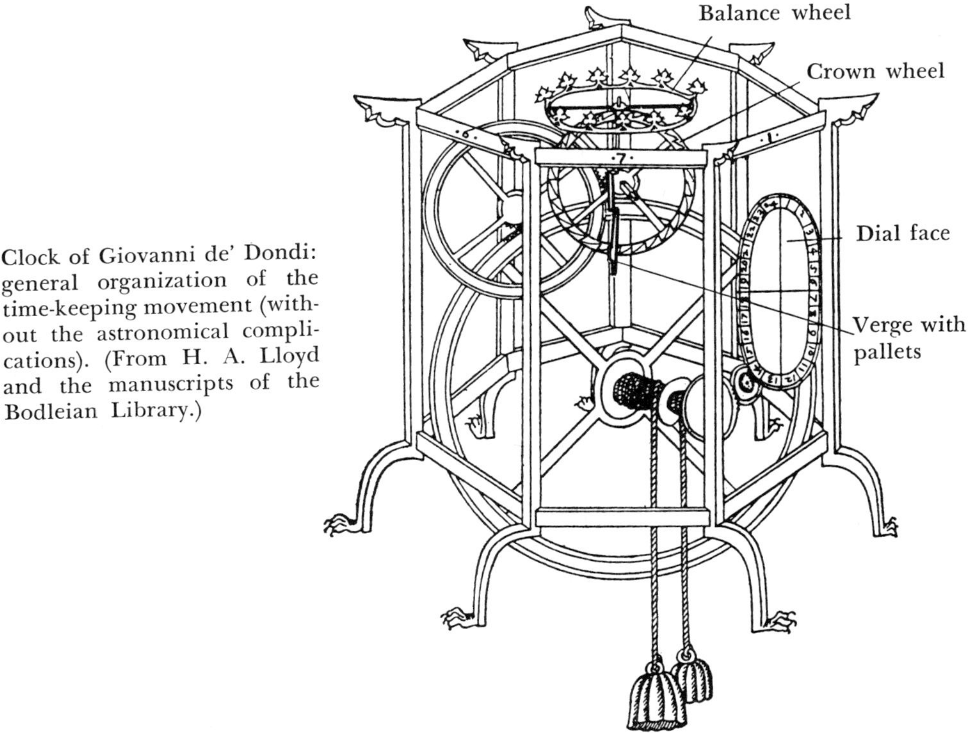

Clock of Giovanni de' Dondi: general organization of the time-keeping movement (without the astronomical complications). (From H. A. Lloyd and the manuscripts of the Bodleian Library.)

The crown-wheel escapement

These clocks have a driving weight whose cord winds around a winch and pulls the works, which in turn operate a single hand on a twelve- or twenty-four-hour dial. The organ corresponding to the governor is scarcely deserving of this name, and only remotely fills this function; it resembles a kind of scale pointer pivoting around a vertical axle. The foliot is thrown alternately to right and left by a peculiarly shaped toothed wheel, the crown wheel, acting on pallets on the axle or "verge." This device has no period of oscillation and therefore no isochronism; the heavier the weight, the more rapid the swing, and the foliot serves more to slow the fall of the weight than as a governor. As it appears, the invention of this mechanism must be regarded as a stroke of genius, and it marks the true birth of mechanical clockmaking (Figure 85).

The second invention, whose date, place, and inventor are all unknown, is that of the mainspring, which made possible the transition from the fixed clock to the portable clock, and from the portable clock to the individual watch, in a series of steps it is impossible to pinpoint. Another major invention was the installation of the spring inside a spring box, which protected it and to a certain extent made its unwinding action more regular.

The appearance of the first table clocks may be assigned to France during the reign of Louis XI, the first watches to approximately 1500. Where were the latter constructed? Perhaps in the Loire Valley at the Valois court, perhaps in

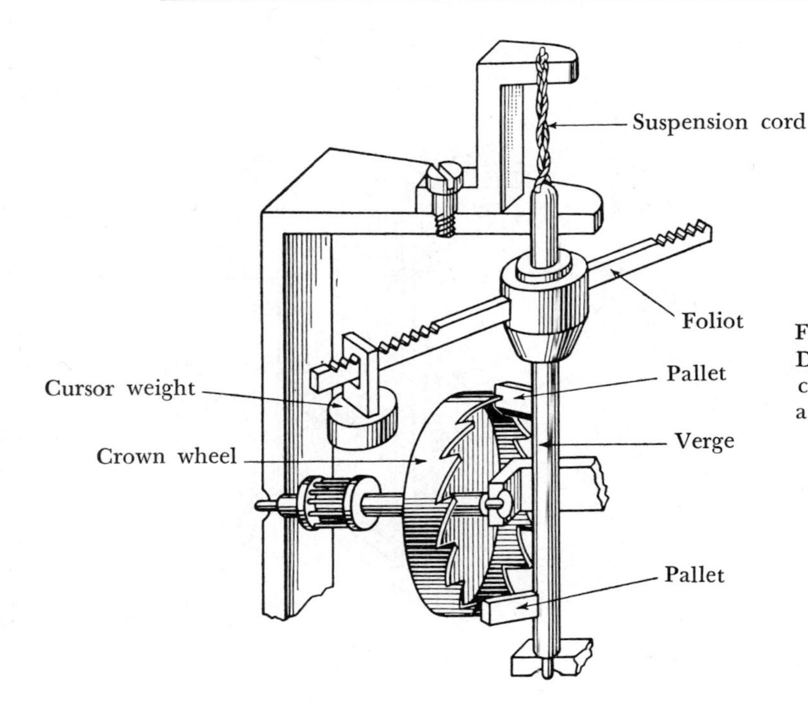

FIG. 85
Detail of the foliot and the crown-wheel escapement (from a drawing by Mr. R. Chaléat).

southern Germany, at Augsburg or Nuremberg, perhaps elsewhere; the answer will probably never be definitely known. It is more likely that the transition from small clocks to very small clocks, and thence to the watch, occurred independently and almost simultaneously in several clockmaking centers, where competition among numerous artists was the rule.

Stackfreed and fusee

The absence of isochronism was a serious disadvantage in a public clock, despite the constancy of its weight and stability of conditions of functioning. The disadvantage was even more serious in a portable clock. As soon as the portable model had been invented, a remedy for this fundamental defect had to be found at any price. An early device furnished a very imperfect solution for the problem by equipping the mainspring with a stop-piece acting on an eccentric shoe, so that the friction was at its maximum on the tightly wound spring, and diminished in proportion to the decrease in driving force. This device is known by the name "stackfreed"; despite the English name, it is found only on sixteenth-century German watches and clocks built shortly after 1600 (Figure 86). The second solution, on the contrary, was theoretically perfect; it made it possible to transmit to the works a constant "couple" during the unwinding, performing this transmission by the intermediary of a pulley with variable radius known as a fusee (Figure 87). When the spring is completely wound, it acts by means of a small cord (later replaced by a small chain) on the portion of the fusee that has the shortest radius; as the couple diminishes, the fusee is attacked at a decreasing point of radius. If the profile of the fusee is well adapted to the spring, the compensation can be perfect. This anonymous (unless it should be credited, based on certain

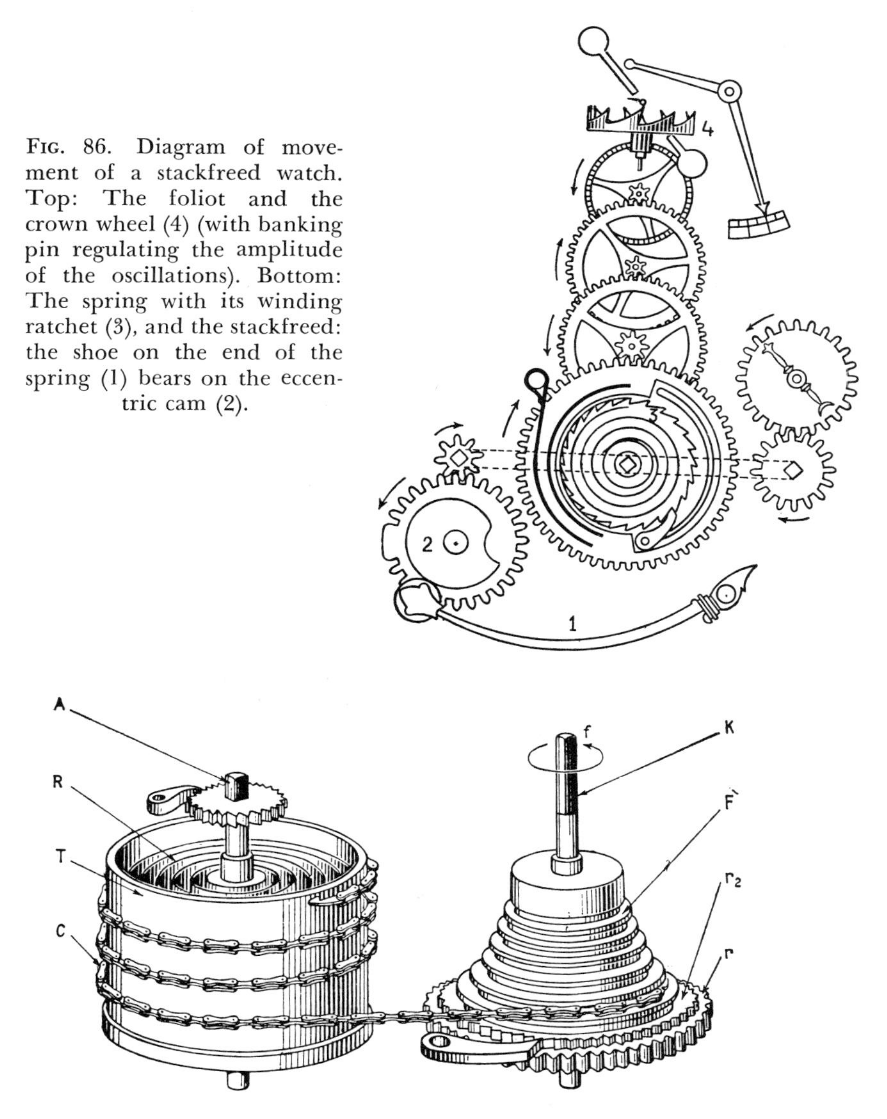

FIG. 86. Diagram of movement of a stackfreed watch. Top: The foliot and the crown wheel (4) (with banking pin regulating the amplitude of the oscillations). Bottom: The spring with its winding ratchet (3), and the stackfreed: the shoe on the end of the spring (1) bears on the eccentric cam (2).

FIG. 87. A modern fusee (drawing by Mr. R. Chaléat). Winding is done by "K," in the direction of the arrow "f." The chain winds around the fusee, "F," and the spring "R" becomes taut. During the unwinding, the chain winds around the drum, "T," pulling the fusee. "A" is used only to regulate the tension of the chain.

drawings, to Leonardo da Vinci [Figure 88]), invention was truly a stroke of genius; it was born between 1500 and 1550, and its use very quickly spread to watches in general, and remained in use until around 1850. Modern isochronism has made it unnecessary (and undoubtedly very expensive), and caused it to disappear from watches, but it is still used in marine chronometers. A splendid example of an invention that has remained unsurpassed for four centuries!

Plate 23. Table clock with astronomical information and engraved crystal cylinder. Early seventeenth century. Musée des Arts et Métiers. *Photo Chuzeville-P.U.F.*

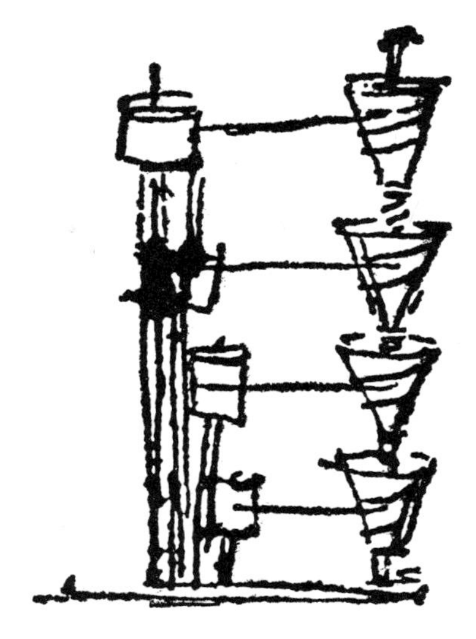

FIG. 88. Drawing by Leonardo da Vinci, probably representing a fusee mechanism. However, it is difficult to understand why there are several fusees.

Constancy of parts

A perhaps more impressive example is that of the striking mechanism. Its origin is unknown, and it is still being constructed in the same manner in which it is found in the oldest clocks, with no change other than in the layout and details of its parts. In our opinion this mechanism should be regarded with respect, not because of its antiquity, but because it is the first example of a programmed device performing a series of movements such that each one of them is regulated by the preceding movement and governs the following movement; sequential automation is the development of this idea.

The invention of the alarm mechanism is also lost in the night of time; alarms and striking mechanisms are found on the oldest extant watches.

Other complex devices were also developed very early; phases of the moon and various dates of the month are found, in the most varied combinations, by the end of the sixteenth century.

On the whole the mechanism remained faithful to this arrangement until the appearance of Huygens. However, interesting details were added; they reveal the ever-alert spirit of research of the master clockmakers, and make it possible to date watches and clocks. Without going into details, mention should at least be made of the devices for regulating the watch by the amplitude of the foliot action and the mechanisms of tension and regulation of the fusee chain.

Production and decoration

The study of the watches of the sixteenth century is attractive (and infinitely so) essentially from the viewpoint of decorative art. They are truly jewels, objects of luxury and beauty of unimaginable richness and variety. Few watches made of precious materials are extant, but more modest models in copper, silver, and rock crystal offer an extreme diversity of form and decoration. The latter is very luxuriant; the smallest parts of the mechanism are decorated, and the single hand is often a tiny work of art.

A great number of signatures of master clockmakers after the middle of the sixteenth century are known to us. Some of them, it is true, are nothing more than monograms whose signification we have been unable to discover. We are nevertheless quite well acquainted with the lives and work of these artists. In general, the method of constructing watches was as follows: the movements were "roughed out" and delivered "in blank" by clockmakers, among whom there was undoubtedly a certain amount of division of work. These blanks were finished, decorated, and gilded by a master who signed them and under whose direction the case was constructed. Anthologies of decorative motifs were published for the use of the engravers. In sum, a watch represented approximately a month's work by one person, but quite a large number of different hands contributed to it.

How precise were these watches? An embarrassing question! The watch might vary between a half hour and an hour from day to day, the instrument of reference being the sundial. This should not cause us to smile, but rather to appreciate the difficulty of the problem involved.

THE MAJOR INVENTIONS

For the clockmaking craft, the seventeenth century was the period of the great inventions that brought about its transition from the empirical to the scientific level, and placed chronometry on a solid base. It is not without astonishment that we find the two most important inventions coming, with an interval of several years, from the hand of the same man: Christiaan Huygens (1629–1695), mathematician and astronomer but by no means a clockmaker. These two inventions are actually only one, for together they constitute the governor, which has a period of oscillation and consequently is at least approximately isochronous. On the practical level, the invention of the pendulum revolutionized clockmaking; on the theoretical level, it was an important step toward the formation of rational mechanics. It must not be forgotten that this invention long preceded the discovery of its general laws and the invention of differential equations, which today seem so natural and so indispensable to the solution of every problem of dynamics.

The pendulum

The discovery of the isochronism of the oscillations of the pendulum is due to Galileo (1564–1642), who believed that this isochronism was invariable. In order to measure short intervals of time, Galileo made a direct count of the oscillations of a free-swinging pendulum. At the end of his life he appears to have imagined a maintaining and recording device that transformed the pendulum into a clock; after his death his son Vincent undertook the construction of this clock, but did not bring it to a successful conclusion. Only sketches of this attempt remain; a few modern reconstructions have been attempted (see L. Defossez, *Les savants du XVIIe siècle et la mesure du temps,* chapters IV and VI (Lausanne, 1946); see also Figure 89.

All pendulum clocks derive from the prototype constructed by Salomon Coster, clockmaker at The Hague, from the instructions of Huygens, in 1657. This clock, which is preserved in the Museum of the History of Science at Leyden, is described in the treatise *Horologium* printed at The Hague in 1658. It

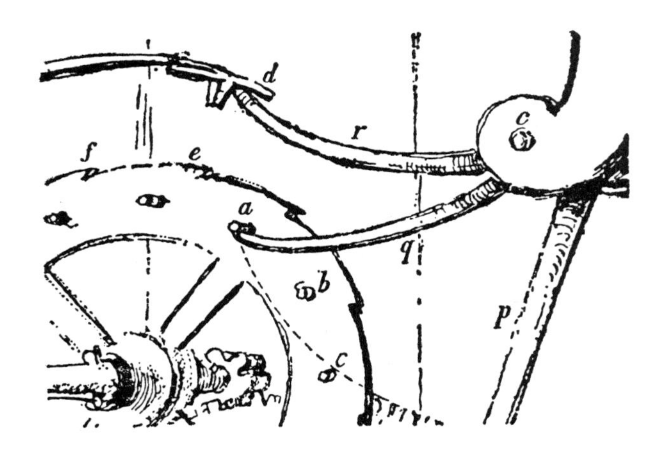

FIG. 89. Reconstruction of Galileo's escapement (from L. Defossez, op. cit.). A fork (r q) forms part of the pendulum (p). When the pendulum approaches the driving wheel, branch "r" raises the locking device (a); the wheel turns, and peg "a" pushes the pendulum by pushing away branch "q."

is noteworthy that the addition of a pendulum to an already existing clock was a quite simple task, the mechanism being retained without any major change, including the crown-wheel escapement. The only modification was that the verge was now horizontal and instead of a foliot had a fork that embraced the pendulum rod. Almost all the existing clocks were remodeled to include the new device, whose immediate effect was to lessen their variation by six or eight to one. This explains the extraordinary rarity of foliot clocks still in their original condition.

The isochronism of the pendulum is only approximate; this means that the period of oscillation is not completely independent of the amplitude of the oscillations, but experiences a slight increase in proportion to its square when the amplitude increases. Huygens was justly proud of having demonstrated that the isochronism becomes invariable when the pendulum mass describes a cycloid, which occurs when the suspension rod of the pendulum, instead of having merely a fixed point, is obliged to remain within the vicinity of this fixed point on guide wires in the form of cycloidal arcs (Figure 90). This is generally the system he used in his clocks, and its theory constitutes the central chapter of the famous treatise *Horologium oscillatorum,* published in Paris in 1673. This improvement, which Huygens considered essential, was to be revealed as being of no practical importance, and the "cycloidal" pendulum was abandoned in favor of pendulums of very small amplitude (two to five degrees) which the clockmakers attempted, in addition, to maintain constant, insofar as this was possible. It is very instructive, for the comprehension of method in the applied sciences, to reflect on these two apparently contradictory procedures when confronted by the property of approximate isochronism. The mathematician seeks to eliminate the defect completely, but the actual construction of the device corresponds only imperfectly to its geometrical definition, and there are other disturbances. The clockmaker reconciles himself to the defect, and tries to minimize it by permitting only small variations of an already small amplitude. In this case experience has shown that the clockmaker's solution rather than the "mathematical" solution was the correct one. With the spiral spring and its progress based on theory, we see an evolution in the opposite direction. A problem be-

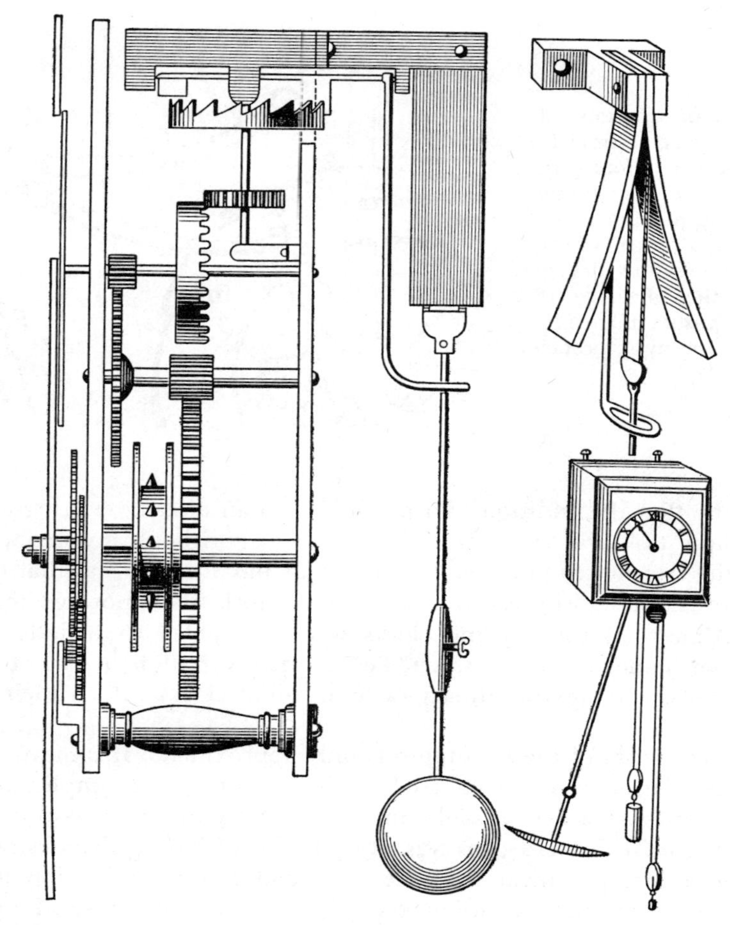

FIG. 90. Pendulum clock of type invented by Huygens. This is a simplified drawing adapted from an illustration in the treatise *Horologium oscillatorum,* showing the suspension of the pendulum by a double rod guided by cycloidal cheeks.

longing within the sphere of the engineer must be considered as successfully solved when the two approaches are able to be given concrete form, and compared.

The balance spring

Could the operation successfully performed on clocks be used for watches? Apparently Huygens immediately asked himself this question, and answered it brilliantly with the invention of the balance spring, which gave the annular foliot of the early watches the "restoring couple" that gravity gives to the pendulum, and transformed it into a governing balance wheel with its own period of oscillation. Here again the isochronism is only approximate, but it is less easy to see, and Huygens was at first able to believe it was invariable. In any event the invention has the same characteristics as that of the pendulum: a major idea with a minimum of actual material work. This is why Huygens was able personally to superintend the construction of the first spring-operated watch by Isaac Thuret (died 1700), renowned master clockmaker, on January 22, 1675 (Figure 91). The immediate result was that the precision of the watch thus transformed was increased by at least five.

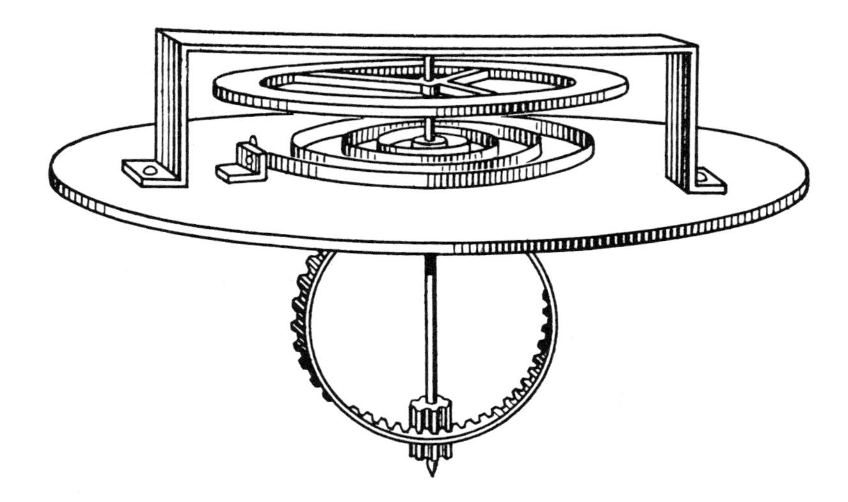

FIG. 91. Balance wheel with spiral spring (from the original drawing by Huygens).

The problem of utilizing the pendulum for regulating clocks had been next on the agenda since the work of Galileo. Huygens was undoubtedly directed onto the path of his research by earlier attempts, but he was the first to find a solution that could be applied immediately and without modification. In similar fashion, the use of a spring to regulate small clocks and watches had led to several attempts before that of Huygens, and he was undoubtedly aware of them. The most vigorous claims of priority were made in France by Abbot Jean de Hautefeuille, in England by the physician Robert Hooke (1635–1703). Both had utilized the principle of the oscillations of a spring, but this was not the spiral spring of Huygens. In addition, his solution of attaching one of the ends of the spring to the balance wheel and the other to a fixed point proved to be, like that of the pendulum, the only practical solution, and it is not prejudicial to anyone's memory to note that all functioning spiral-spring watches derive solely from Huygens's prototype. Thuret himself attempted to present the invention as his own. He later retracted this claim, but Huygens was obliged to realize the usefulness of the patent of "exclusivity" that he had obtained from the king, since control was impossible.

Forgotten precursors

The brilliant success of the inventions of Huygens cannot be overestimated, and there is no doubt that they opened the precision era. However, it would be incorrect to believe that nothing of value had been done before him toward this goal and that the great astronomers of the end of the sixteenth century remained idle in face of the inadequacy of their clocks. Today at least we are aware of a major effort made in this period to create the astronomical timekeeper, an effort due to the Swiss master Justus Burgi (1552–1632), who worked chiefly for Emperor Rudolph II. He envisaged a governor with two flexible foliots operating in opposite directions, which certainly represented a considerable improvement over the contemporary system, and permitted an attempt, with some chance of success, at a minute clock. However, it must be stressed that this device, which was complicated and very difficult to make, was a technological perfection, and not a new principle with almost unlimited promise, as was Huygens's spiral. Burgi's governor was abandoned after his death, and was completely forgotten until its recent discovery by Dr. H. von Bertele while studying and dismantling the clocks of this artist that are preserved in various museums. How many inventions, perhaps of this class, perished without any trace is obviously unknown.

The progress of chronometry

The years of 1657 for clocks and 1675 for watches are therefore genuine milestones in the history of these instruments. But what happened after this? We may claim, with apparently equally good reason, that progress both ceased and did not cease. It ceased, because a movement of a winding watch of 1850 bears an astonishing resemblance to a movement of 1680, except that it is much less carefully constructed and all esthetic character has disappeared from it. Similarly, a "recording" pendulum movement is practically identical to a movement built around 1700. But at the same time progress did not end, because invention succeeded invention without interruption. Only after long delay did they filter down into production as a whole, but they did pave the way for future developments.

It seems fitting to close this chapter by presenting a technological invention of the closing years of the seventeenth century that gave to the watch its familiar, universal face, namely, the indication of the hours and minutes by means of two concentric hands that revolve around the dial face in twelve hours and one hour, respectively — truly a magnificent invention in its elegance and simplicity. We have seen that the watches constructed prior to the time of Huygens had a single hand that made one revolution in twelve hours (rarely in twenty-four); it would have been pointless to expect the indication of the minutes on these watches, which exhausted their possibilities of regulation by marking time within a half hour of precision. The situation charged overnight with the adoption of the spiral spring, and the indication of the minutes thereby became a necessity.

The "solution" was not found immediately, and numerous trial faces were built, each represented by rare specimens in museums. Finally, around 1690, Daniel Quare (1649–1724), a famous clockmaker of London, created "our" dial face, which was not modified until much later by the addition of the sweep-second hand, when technology had made indication of the seconds worthwhile; this was the conquest of modern clockmaking.

BIBLIOGRAPHY

BAILLIE, GRANVILLE HICKS, *Clocks and Watches, an Historical Bibliography* (London, 1951).

BASSERMAN-JORDAN, ERNEST VON, *The Book of Old Clocks and Watches,* revised by Hans von Bertele (4th ed., New York, 1964).

BEDINI, SILVIO A., "The Compartmented Cylindrical Clepsydra," *Technology and Culture,* 3 (1962), 115–141.

———, and MADDISON, FRANCIS R. *Mechanical Universe: The Astrarium of Giovanni de' Dondi. Transactions of the American Philosophical Society,* n.s., Vol. 56, part 5 (1966). Story of a fourteenth-century geared planetarium.

CHAPIUS, ALFRED, and DROZ, EDMOND, *Automata* (New York, 1958).

GORDON, G. F. C., *Clockmaking Past and Present* (rev. ed., London, 1949).

LEE, RONALD A., *The Knibb Family-Clockmakers* (Liverpool, 1960).

LLOYD, H. ALAN, *Some Outstanding Clocks over Seven Hundred Years, 1250–1950* (London, 1958).

MILHAM, W. I., *Time and Time-Keepers* (New York, 1941).

NEEDHAM, JOSEPH; LING, WANG; and PRICE, DEREK J., *Heavenly Clockwork* (Cambridge, 1960).

PRICE, DEREK J. DE S., "On the Origin of Clockwork, Perpetual Motion Devices and the Compass," *U.S. National Museum Bulletin* 218 (Washington, D.C., 1959), 81–112.

———, and BEDINI, SILVIO A., "Automata in History," *Technology and Culture,* 5 (1964), 9–42.

ROBERTSON, J. D., *The Evolution of Clockwork* (New York, 1931).

WARD, F. A. B., *Collections Illustrating Time Measurement.* Part I, *Historical Review* (4th ed., London, 1958).

CHAPTER 14

TECHNIQUES OF MEASUREMENT

ASPECTS OF EARLY METROLOGY

METROLOGY, "the science of measurement," whether commercial or purely scientific, has its own characteristic techniques, which become increasingly delicate and refined in proportion to the metrological qualities required. These qualities include increasing the precision of the results obtained; techniques of preserving and reproducing prototypes to meet the needs of science, industry, and commerce; and operational techniques, which may devolve upon an operator or (in modern times) a machine that verifies its own performance in accordance with standards that are either determined in advance or are transmitted to it (for example, self-correcting machines and ballistic and interplanetary missiles).

This brief description of certain fundamental characteristics of metrology by no means exhausts the list of its applications and methods, which includes the majority of human activities.

The general principles on which the structure of modern and medieval metrology are based are not strikingly different from those applied by the peoples of antiquity; thus the definition of the unit of measure and its representation by an object (the "standard," a stone or metal rule; a container for measurements of capacity; weights for measurements of volume) are what we might call "metrological facts" that go back some fifty centuries, as is indicated by written documents and archaeological discoveries.

Similarly, today as in antiquity and during the Middle Ages the use of measures in business, their conformity to legal standards, and the struggle against fraud have always been the object of technical and juridical regulations whose application has devolved upon specialists: *agoranomoi* in ancient Greece, the master craftsmen of certain guilds until the end of the eighteenth century (apothecaries, scale makers, oil makers, weight makers, and so on), and, in modern times, officials of government bureaus of metrology.

From the creation of the Sumero-Babylonian (third millennium B.C.), Egyptian, Greek, and Roman systems of measurement, down to the institution of the decimal metric system (1790–1799) first on a national, then on an international scale (1889), we can, without risking serious errors, say that in its general structure the physiognomy of metrology has undergone no major changes. In contrast, the framework of this discipline has continued, and continues, to develop in conjunction with the expansion of science and technology that began around the seventeenth century.

While the edict of François I (1540) was aimed at achieving uniformity of measures in the kingdom on the basis of the units of measurement in use at Paris, the success of this operation left no lasting traces. Similarly, in 1557 Henri II followed in the footsteps of François I, without obtaining outstanding results.

Multitude and variety of measures

How could it have been otherwise, given the multitude of metrological systems — communal, seigniorial, provincial — born of the feudal regime and by then traditional, a situation that existed everywhere in the West? From these metrological "varieties" were derived the most diversified expressions for designating measurements, in which a word very often designated entirely different values from one commune to the next.

In addition (and this is true of every age), in evaluating a piece of land the peasant took into account its productivity, geographical location, the effort expended to make it fruitful, and so on. Given equal effort (or, in the modern phrase, given an equal expenditure of energy), the extent of the lands cultivated and their production can be very different. This explains why the terms designating their area and the corresponding values reflect the labor of the worker and the productivity of the soil.

Thus, in Lorraine, the *"hommée"* was the area of land dug each day by one man (approximately 2 *ares,* equal to 8 rods). A *charrée* (approximately 15 *ares*) was equivalent to the area of land from which one wagonload of hay could be harvested. A *fourée* was the area of a field that yielded a *"fourral"* of wheat, or 20 to 26 liters (1/2 to 3/4 bushel).

Thus it is not surprising that the terminology of agrarian measures is frequently identical with that of the measurements of capacity (*boisselée, séterée, quarterée,* and so on). In contrast, the surveyor-geometers evaluated surfaces in terms of basic units *(toise, canne),* their subdivisions (*pan, pied)* or their multiples (*perche, dextre, corde,* and so on). The results were expressed in terms of *perche, canne, dextre, corde, pied,* and square *pan,* that is, in conventional values — not very suggestive for the cultivator — which varied in relation to the unit of local measure.

In another activity, namely, the measurement of fabrics, the size of the ell was far from consistent. In the thirteenth and fourteenth centuries the ell used at the fairs of Champagne and Burgundy (called the "Provins ell") was equal to .86 meter while the Parisian ell measured 118.84 centimeters in the sixteenth century, in accordance with the standard ell dated 1554 (formerly preserved in the Bureau des Marchands grossiers et merciers de Paris, rue Quincampoix, now exhibit No. 3226 in the Musée du Conservatoire des Arts et Métiers). It is probable, however, that the fourteenth-century Parisian ell was approximately the same length as the sixteenth-century ell. The equivalence given by Balducci Pegolotti in his famous *Pratica della Mercatora,* which dates from around 1340, notes that 100 Parisian ells are equivalent to 146 Provins ells, which makes the Parisian ell 120.5 centimeters, or within 1 percent of the standard ell of 1554.

This terminology always creates a certain amount of uncertainty with regard to the value of the measures. For example, in all the local systems of measurement the terms "pound" and "ounce," which designate measures of volume, were

Plate 24. The *perche* (perch) of Frankfurt-am-Main. From the 1633 edition of Jacob Köbel's treatise on surveying. *Photo Bardet-P.U.F.*

commonly employed. In fifteenth-century Paris the pound was 16 ounces, or 489.5 grams; we also find a 16-ounce pound in the southern regions, but there it weighed between 408 and 413 grams, with numerous exceptions. In the north, the pound was equivalent to 428 grams at Saint-Omer, 434 grams at Lille, and so on.

Stability of the measures However, we must emphasize the permanence of the values of the measures, which while not absolute was frequent — a striking fact, in keeping with the requirements of international commerce, the guild spirit of the *Ancien Régime,* and the commercial traditions and customs that under these conditions could only with difficulty have been reconciled with variations in units. In the thirteenth century, for example, the weight standard in minting, the Parisian *poids de marc,* weighed 244.75 grams; this remained practically unchanged until 1800. The *marc* of Cologne, which was 15/16 of the Parisian *marc,* or 229.5 grams, was identical with the *marcs* of the Tower of London, La Rochelle, Portugal, and other countries; the *marc* of Venice was similar to those of Marseille, Montpellier, Avignon, Nîmes, Genoa, and so on. The *canne* of Montpellier — approximately 1.98 meters — retained the same value for centuries. Many examples of this type concerning measures of size or capacity could easily be cited.

For all practical purposes it can be considered that variations in units, when they did not result from the revision of standards accidentally destroyed or worn out by long use, generally appear to have been political or economic in origin.

Another aspect of early metrology claims our attention because it concerns operational techniques. We are referring to those procedures of measurement more specifically concerned with measures of capacity for dry materials. These measures were filled, depending on the nature of the products being measured, to "level," "brimming," "full," "heaping," and so on. With the help of the skill and dexterity of the person doing the measuring, the actual quantity delivered could be different from the quantity due the purchaser — generally to the latter's disadvantage. For this reason the central government, followed by various city governments, decided around the end of the seventeenth century to standardize measuring procedures, so that the measuring implements were filled "level." Even after this standardization, however, numerous exceptions to the rule could be found.

In short, we can say that until the eighteenth century, with the possible exception of major commercial centers such as Lyon, Dijon, Paris, and other cities, metrology completely lacked a national character and a unifying theory. Its structure was based on traditional methods, adapted to local customs. Identical terms masked very diverse units of value. Measuring procedures did not always offer sufficient guarantees, and sometimes had very mediocre metrological qualities.

In addition, the idea of "precision" was quite vague, at least on the commercial level, with the exception, however (and that within certain limits), of precious materials — coins, rare stones and metals — which were very carefully weighed. Moreover, it was in this category of measurement — that of volumes — that the best results were obtained, as will later be seen.

Lastly — and this is of particular interest for the history of metrology —in numerous cases the stability of a fairly large number of units of measure from the fourteenth to the eighteenth centuries has been discovered through experiment. Historically this helps to establish equivalents between the old measures and the metric units. In addition, practical metrology (that is, the various local systems and others) was much more the object of juridical than of specifically technical concerns.

The solution of these technical problems is not always easy. For this reason our analysis will develop this direction, without neglecting the other aspects and applications of metrology. Their renewal (or, more precisely, transformation) was the work of the scientists of the seventeenth and eighteenth centuries.

MEASUREMENTS OF SIZE

The unit of measure of a physical entity

The statement that all measurement of physical entities is a comparison is a commonplace. But comparison requires a previously defined element of reference — a fact that had been realized in the Middle Ages, and even long before, by the "metrologists" of those distant ages. The Moslems defined the "finger's breadth" as follows: it had to be equivalent to six grains of barley pressed front to back, and each grain had to be equivalent in breadth to six hairs from the tail of a mule! In 1324 the King of England decided that one inch was to be defined as three grains of wheat laid side by side.

As for measuring volumes, the Babylonian shekel of 8.4 grams was composed of 180 grains. The term "grain" is found in most modern systems of measurement, since the Anglo-Saxons still use the troy grain, which weighs .0648 gram. In France the grain weighed .053 gram; *the poids de marc* pound (that is, the Parisian pound, which was 489.5 grams) was composed of 9,216 grains, or 384 *deniers,* or 128 *gros,* or 16 ounces, or 2 *marcs.*

Once defined, the unit of measure must be represented by a material object, the prototype or standard. Replicas for public use are established by comparison with this original.

As regards linear measures, which play a preponderant role in metrology, we possess the standard for the Parisian ell just mentioned; it is dated 1554, and is preserved in the Conservatoire des Arts et Métiers. Its length represents the unit of measure allocated for the measurement of fabrics as it was defined by François I, that is, 3 feet 7 inches 8 lignes,* or 118.84 centimeters. This standard was an iron ruler with two square projections at each end. The ell to be compared was to be inserted between these two projections as exactly as the quality of its production permitted.

We do not know how precise this standard was, but we do know that the astronomer Abbot Picard compared it with the standard *toise* (fathom, composed of 6 feet in the seventeenth century), and found that it measured 3 feet 7 inches $10\frac{4}{5}$ lignes; La Hire's academicians estimated it at 3 feet 7 inches $\frac{1}{2}$ ligne, and in 1736 Dufay found 3 feet 7 inches $10\frac{5}{6}$ lignes. It was checked again in 1745 by two commissioners of the Académie des Sciences, Hellot and Camus,

* One ligne = 0.0888 inches (2.2558 millimeters).

who found 3 feet 7 inches 10⅚ lignes. With the exception of La Hire's erroneous measurement, the results of the other comparisons agree very closely (it is usually accepted as 3 feet 7 inches 10⅚ lignes). The values thus obtained surprised the scientists. They felt they could deduce from this experiment that the standard did not correspond to its definition (3 feet 7 inches 8 lignes) and that it was longer by approximately 2.6 lignes, that is, 6 millimeters.

The standard fathom of Le Châtelet

A change had occurred in 1667–1668, however, as the result of the creation of a new "standard fathom" that was shorter by approximately 5 lignes (11 millimeters) than the old model (which was still visible in 1747 on a pillar of the Châtelet building in Paris, "completely warped by the defect of the pillar, which had buckled").

The new standard for the fathom – composed of 6 royal feet (the standard definition) – measured 1.949 meters; it was described by various writers, notably by La Condamine and Paucton. Fixed in 1668 in the wall at the foot of the stairway of the Grand Châtelet in Paris, it consisted of an iron bar ending in two projections. It was crudely constructed, according to Paucton; its angles were round, the inside faces of the two projections intended to hold the fathom had never been polished or squared off and made parallel; moreover, it was exposed to the harsh effects of inclement weather and human destructiveness (people did not hesitate to hammer the fathom in order to straighten it, as happened, for example, in 1758).

The fathom of La Condamine (called the "Academician" or "Peruvian" fathom)

In 1735 La Condamine (1701–1774), an able and farsighted man who together with other scientists had been entrusted with the task of measuring the meridian arc at the equator, ordered the construction, before the expedition's departure, of two copies, as exact as possible, of the Châtelet standard. One of them became the "Peruvian" or "Academician" fathom. It was made by Langlois, the royal engineer, in Paris, under the supervision of the astronomer Godin (1704–1760).

This standard, which was recognized as the national prototype of linear measures in 1766, consisted of a flat rule of forged, polished iron approximately 7.5 millimeters thick and 40 millimeters wide. Each end was notched for half of its depth, so that two projections of 13 millimeters were left on the other half; the notched portion formed a right angle with the projections, and the distance between the angles represented the length of the fathom.

In the prolongation of the angles a line is drawn at each end; it terminates in a sunken point approximately .4 millimeter in diameter. The distance between the axes of these two points is also a standard of the fathom, originally considered as being equal to the other standard. Thus the standard fathom can be determined by reference either to the two points, the two ends, or the length of the rule itself. But its legal length is that of the ends at a temperature of 13 degrees Réaumur (about 61 degrees Fahrenheit).

In order to form an opinion on the precision of this standard, an examination of it was made around the end of the nineteenth century at the Bureau International des Poids et Mesures at Sèvres (Seine). It was found that the fathom measured in relation to the ends at 16.25°C. (13° R.; 61°F.) is equal to 1.949090 meters, while the fathom according to the points, at the same temperature, is equal to 1.949001 meters. Thus these values agree within a variation of less than .09 millimeter. This instrument served as the element of comparison at the time of the creation of the decimal metric system.

La Condamine's fathom was a milestone in the technique of making a standard, not only qualitatively — the graduations of three in three inches are clearly engraved, as is the subdivision of the inch into twelve parts — but metrologically as well. Less than one century after the construction of the first instruments for measuring temperature this fathom was tested at a given temperature, which reveals a certain desire to obtain the maximum precision compatible with the technical methods available at that period.

The conclusions to be drawn from this rapid study are that:

1. In 1554 the standard Parisian ell measured 3 feet 7 inches 8 lignes of the original fathom or of the old Parisian foot (which is the same thing);
2. Expressed in terms of the new Châtelet fathom of 1668, which measured only 1.949 meters instead of 1.96 meters, the standard ell was equivalent to 3 feet 7 inches 10 5/6 lignes of this new fathom;
3. The old Parisian foot (called the "masons' foot"), preserved at the "Escritoire near Saint-Jacques de la Boucherie," was longer than the new one; it measured 32.66 centimeters, while the new one was established at 32.48 centimeters.

Consequently, there would be errors in the evaluation of measurements connected with linear measures in use in Paris if the modification in 1668 of the length of the fathom were not taken into consideration. However, the subdivision of the fathom into 6 feet, the foot into 12 inches, and the inch into 12 lignes has remained classic. The ligne was equivalent to approximately 2.27 millimeters prior to 1668, and to approximately 2.25 millimeters after 1668.

The precision of the measurements

To evaluate fractional parts of a line, scientists and manufacturers of instruments for observation and measurement attempted to discover technical methods capable of solving these problems satisfactorily.

The fields of astronomy, navigation, and topography merit a rapid discussion from this point of view, since the observers were chiefly concerned with obtaining angular measurements as exact as was possible with the delicacy of the division of the graduated limbs, quarter circles, and sextants.

Around 1500 the precision of measurements of angles was hardly within one degree or half a degree on small instruments, 15 minutes or at most 10 minutes on large-scale instruments. An attempt was therefore begun to divide the degree with the help of transversal lines drawn between two concentric graduations, from "*n*" degree on the one hand to "*n* + 1" degree on the other; each transversal was divided into six equal parts. The graduation was read at the point of intersection of the chamfered edge of the alidade and the transversal. It follows that the equal subdivisions of the transversals do not correspond to those of the degree, a fact whose importance was only relative, for the error of reading was less than the error of sighting.

An example of this type of graduation is found on the quarter circle utilized by the astronomer Tyco Brahe (1546–1601); its diameter was 2.50 meters. However, since the degree was divided into "sixths" and the transversals were subdivided by nine equidistant points, the precision of measurement might be less than one minute.

Nuñez

In 1542 a Portuguese mathematician named Pedro Nuñez (1492–1577) suggested describing 46 concentric arcs from the center of the quadrant; the largest was to be subdivided into 90 equal parts, the next one into 89, and so on down to the forty-sixth, which would be divided into 45 parts. The alidade, when oriented toward a star, would according to Nuñez, pass very close to one of the divisions of one of the sectors. If "p" is the number of this division and "q" the number of divisions of the sector that contains them, the measurement of the angle will be equal to $p \times 90$. By means of a table this measurement could be determined within a third! This method, which implied on the one hand considerable technical work in order to build the instrument, on the other hand the use of a table, was of no practical use.

The proportional divider (Pierre Petit)

Methods of this type, for example that of Clavius (Christopher Klau) (1537–1612) made no notable contribution to the solution of the problem. But it was possible to divide the circle with the proportional divider, described notably by Galileo in 1606 in *Le operazioni del compasso geometrico et militare* (The operations of the geometrical and military compass) (Padua, 1606). Jacques Alleaume (1562–1627) was also interested in the division of the circle, as Willebrord Snellius points out in his *Eratosthenes batavus.* Finally, one Pierre Petit of Montluçon (1598–1677), having obtained a license from the king in 1625, published *L'usage ou le moyen de pratiquer par une règle toutes les opérations du compas de proportion* (The use or method of performing all the operations of the proportional divider by means of a rule). Page 18 of the second part of the work, entitled *Construction de la Regle et du compas de proportion* (Construction of the Rule and the Proportional Divider) shows the division of the rule into 100 equal parts.

The transversals (Mersenne)

But the inch can also be divided into 12 parts—hence into 12 lignes—and the ligne into 12 parts. This manner of dividing a ligne into "as many parts as desired," says Mersenne in his *Vérité des sciences,* is very subtle, for it "cannot be done with the point of the divider no matter how small it is." The procedure was simple: it consisted of obtaining the twelfth part of the ligne, or approximately .2 millimeter, by the system of transversals. Figure 92 shows its application to the division of the ligne into tenths; at the right, an inch is divided into 12 parts (that is, into lignes), and oblique lines join one of the divisions of the lower edge to the next division of the upper edge. These transversals cross lines

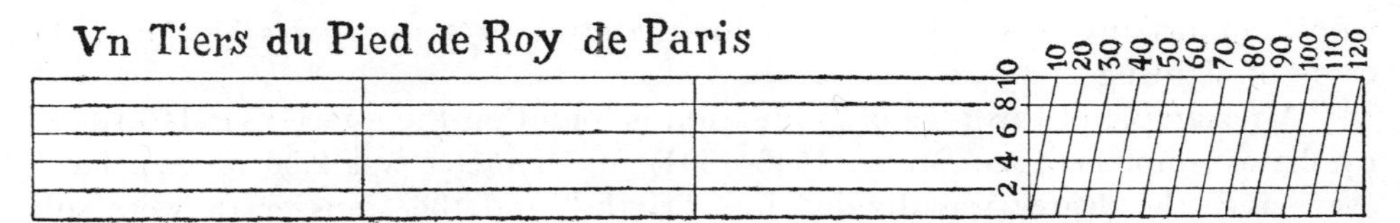

Fig. 92 Four-inch rule divided into tenths of a ligne (from R.P. Mersenne, *Correspondance*).

parallel to the edges of the rule that divide the width of the rule into five equal parts. The points of intersection of the oblique and parallel lines are 2/10 ligne apart, counting along the edge of the rule.

As for compasses and their various technical aspects — plain-hinge compass, beam compass, dividers, and so on — like those drawn by Leonardo da Vinci in his various manuscripts, the reader should consult the extensive and detailed writings by Maurice Daumas on the early French manufacturers of scientific instruments.

Pierre Vernier, inventor of the "Vernier"

In 1631 Pierre Vernier (1580–1637), engineer, native of Ornans in the Franche-Comté, and very interested in cartography, contributed to the latter technique the procedure that he discovered and that bears his name.

His treatise of 122 pages, published in Brussels under the title *La construction, l'usage et les propriétés du quadran nouveau de mathématique* . . . (The construction, use, and properties of the new mathematical quadrant . . .) must have become known fairly quickly, at least to certain specialists, since Jean-Baptiste Morin (1583–1656), astronomer and mathematician, alludes to the vernier in one of his writings (1640). Speaking of the device that a craftsman named Ferrier had constructed for him for his astronomical observations, Morin notes "that it had been very carefully constructed, but that nevertheless the procedure without transversals discovered in 1631 by Pierre Vernier was better."

But the use of the vernier did not appear in cartography until much later, that is, in the first third of the eighteenth century at the earliest. Construction of sighting instruments had made sufficient progress to permit measures to achieve a precision that made their utilization practicable. At this time the vernier was adapted to the first octants of the English optician John Hadley, which were later replaced by marine sextants. In 1699 Newton invented the mechanism of this latter instrument; it was based on the use of two mirrors, one of which could be turned, and which reflected the sighting onto the second, fixed, mirror. The angle of the two directions, which was double that of the mirrors, was read on a divided circle. Latitudes were thus obtained within one minute of an angle, which was a brilliant result (Figure 93).

Precision and extension of the field of observation

Theoretically the vernier was capable of infinite precision; actually, its precision was limited by the thickness of the lines of the graduation. These lines created errors in the reading, and cast a certain amount of doubt on the results of the measurement.

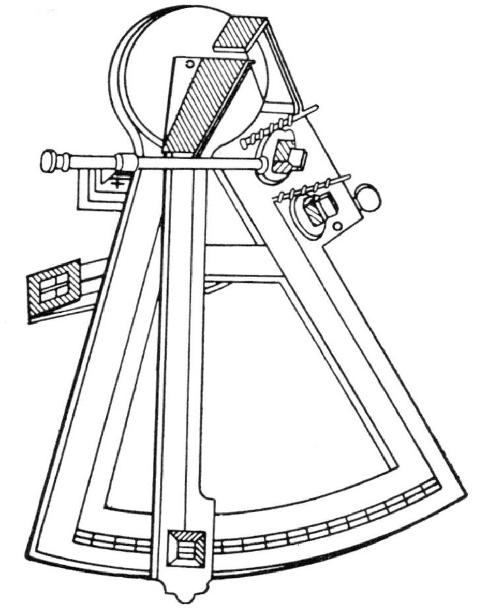

FIG. 93 Hadley's octant. A reflecting sighting instrument, 1731 (from F. Marguet, *Histoire générale de la navigation du XV° au XX° siècle* [Paris] 1931).

This example shows that an increase in precision could be obtained with certitude only if technical procedures adapted to the experiment, as well as to the nature of the physical entities to be measured, were the object of careful development. The scale of observation needed to be increased and expanded. This experimental factor was valid for most areas of metrology. The solution found for the problems it raised came about notably through the development of optics in the seventeenth and eighteenth centuries.

By contributing to the technical improvement of the instruments of measurement and observation – size and purity of the glass; sighting tubes, microscopes, telescopes, and so on – optical procedures enlarged the field of observation in the macroscopic and microscopic worlds. Given the new perspectives opened to the scientists, it was indispensable to invent and create instruments that would offer the possibility of measuring what was seen by the eye, with a minimum of errors.

Auzout and his micrometer

The principle of the micrometer with movable wire – that is, the screw combined with a reticular line – is associated with the name of the French scientist and mathematician Adrien Auzout (1622–1691). It lies at the origin of operational methods and instruments of measurement that made it possible greatly to decrease these errors of measurement. In 1666 Auzout gave a detailed description of a screw-type micrometer combined with movable and fixed wires. The amplitude of the movement of the movable wire in relation to the fixed wires was a function of the pitch of the screw. If a longitudinal movement of one inch corresponds to 20 screw threads, and if the dial is divided for example into 400 equal parts, it is immediately obvious that each division is equivalent to a movement of 2.8 μ * (Figure 94).

* A μ (micron) is the one-millionth part of a meter.

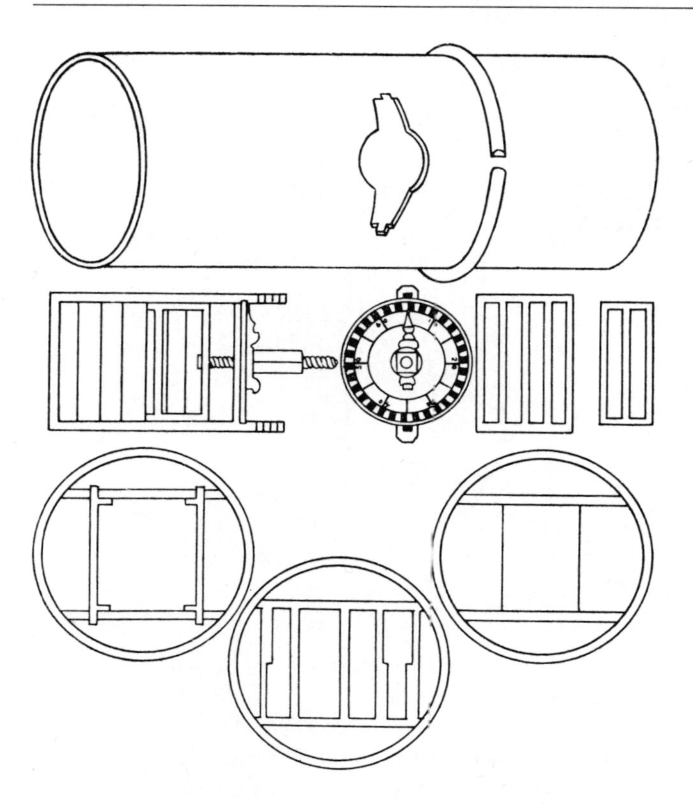

FIG. 94 Diagram of Auzout's micrometer (from *Mémoires de l'Académie des Sciences,* Vol. VII).

When combined with the eyepiece or the microscope, the technique of the micrometer contributed an effective solution to the problems mentioned above, that is, the extension of the field of observation and, correlatively, a method of measurement whose precision was much finer than before.

In Auzout's article "Micromètre" in the *Encyclopédie,* the inch is divided into 2,400 parts. There was also the "skillful theory of Dr. Jurin (James), English doctor and mathematician (1684–1750), which permitted him by the combined use of the microscope and the micrometer to measure the diameter of a globule of human blood, which he fixes at 1/1940 of an inch." Although this result was not completely exact, the approximation obtained was already a very respectable one, for the value given by Jurin corresponds to a radius of approximately 5 μ, while the average value of this radius is on the order of 3 μ.

It is quite surprising to note that, as Albert Pérard has emphasized, the micrometrical microscope, which could have rendered valuable service to metrology, particularly as regards linear measurements, does not appear to have been used before the last quarter of the nineteenth century, at least in this field.

Calibers

Another aspect of dimensional measurements should be stressed: the "calibers" used in industry and notably in the munitions industries. Although no very precise elements of estimation are available, around the end of the fifteenth century the Arsenal of Venice, which employed 16,000 technicians and workers, could in an emergency turn out a galley in one day, by assembling prefabricated parts. Standardization of measurements was indispensable for such assembly-line manufacturing of calibrated objects.

Closer to us in time, in the seventeenth century the production and control of weapons was closely linked to the use of standardized calibers: an approximate standardization, to be sure, but a very fundamental one, for the diameter of cannons had to be matched to the diameter and volume of the cannon balls. In order to check on the precision of the caliber, a caliper was used; this instrument made it possible to check the diameters of the gun and the cannon ball simultaneously. Graduations on one of the legs corresponded to the calibers of the guns and balls, and the inside of the leg had notches that corresponded to the sections of the calibers. The outside caliper was used to calibrate the gun, the inside caliper for calibrating balls.

A comparison of the caliber tables published in 1707 with those included in Nicolas Bion's *Traité de la construction et des principaux usages des instruments de mathématiques* (Treatise on the construction and principal uses of mathematical instruments) reveals only slight differences. We find that the smallest divisions of the calibers expressed in "inches," "lignes," and fractions of lignes seldom descend below 1/8 ligne (in round figures, .3 millimeter.) In the seventeenth century, as we have already seen, it was believed possible to evaluate 1/12 ligne (.2 millimeter) by the use of transversals. Those who had confidence in this approximation were indeed optimistic!

It appears evident that from the second half of the sixteenth century to approximately the first quarter of the eighteenth century, improvement of the precision of the results of measurement was slow; it required much effort on the part of observers in permanent contact with experimentation as well as of the instrument makers. Though these sometimes obscure craftsmen worked with rudimentary materials, their genius for small-scale projects and fondness for careful workmanship permitted them successfully to construct models that even today arouse our amazement.

From the sixteenth century down to the beginning of the eighteenth, precision in the measurement of angles improved from a variation of one degree to a variation of one minute (Hadley's sextant), and even to a half minute. During approximately the same period, the precision of linear measurements improved (under optimum conditions) from a quarter of a ligne (.5 millimeter) to 1/25 ligne (.1 millimeter) with the "Peruvian" fathom of La Condamine's expedition.

At least two inventions, then, with almost infinite consequences, made important contributions to the methods of measurement: on the one hand the "vernier," which made a timid appearance in France, while the realistic English welcomed it and installed it on their marine instruments; on the other hand Auzout's micrometer, which was destined to play a leading role in science and technology.

MEASURES OF CAPACITY

Dry Measures

It is very difficult to form an opinion on the metrological qualities of measures of capacity, whether the hallmarked standards or their replicas intended to meet the needs of business. Their use was determined by rules that defy all idea of measurement.

Traditional practices

To fill a measure "full," "heaping," "brimming," and so on, in accordance with practices left to the good judgment of the measurers, is to rob such operations of all metrological significance. A single example will suffice to illustrate this statement; it is taken from a report of February 15, 1458, on the hallmarking of measures. Copper standards were preserved at the Hôtel de Ville in Paris. In order to regulate the wooden measures by these standards, the official measures, using for this purpose dry grains of rye, had to have "two large, flat wooden basins with holes in the sides, and into which the said grains of rye should be thrown, both into the standards and into the wooden measures; and the grain should be thrown into the standards and also into the measures by two said Measurers, turning their hands before them. As for the standards, the said Measurers should throw as high as they can stretch their arms; and as for the wooden measures, they should throw as high as and near the edge of the wooden measure."

In order to put an end the Measurers' frauds and the repeated complaints of the merchants, the government decided, in a series of regulations established between 1669 and 1671, to cast new standards of the *minot, boisseau, quart, demi-quart, litron,* and *demi-litron,* and to make their content equal to that of the "heaping" measure of tradition. Thenceforth the wooden measures were to be filled with the aid of the hopper, which was already in use for measuring salt, and all measures were to be filled to "level." The sizes of the wooden measures used for retail sale of grain, flour, and so on, had to be identical with the standard vessels.

It should be noted that these vessels did not differ greatly in structure from Roman and modern measures. The wheat *minot* had to measure 11 inches 9 lignes (approximately 3.8 centimeters) high and have a diameter of 1 foot 2 inches 8 lignes (approximately 39.7 centimeters). Also taken into account was the question of what should be put into the measure "before it was priced, to replace the space which the iron support, its pointer, the plate which supports it, and the four brackets which hold the bottom, may occupy."

Beginning with the *minot* (39.27 l.) as a standard of definition (standard in 1800: 39.12 l.), and descending to the *boisseau* (⅓ of a *minot*), the *demi-boisseau, quart, demi-quart, litron, demi-litron,* quarter-*litron* and eighth-*litron,* when we compare them with the values of definition we find relative variations in certain measures of between —. 6 percent and +1. 7 percent. The fact is that it was not always easy for the metal casters charged with the task of making the measures to overcome the difficulties involved in hallmarking, at least the prototypes.

The existence of hallmarked standards is extremely precious, for after they have been gauged and measured they sometimes permit us not only to compare the results obtained with the tables of correspondence established at the time of the institution of the metric system but in addition to verify the stability of certain units. Here is an example of such a chain.

The standard of the *minot* of Châteaudun is dated November 3, 1723; it is 37 centimeters in diameter and 23 centimeters high, and its capacity is 24.75 l. The tables give 24.12 l. One *muid* plus one *minot* of Châteaudun were equal to nine Chartres *setiers.* Taking the Châteaudun *minot* as 24.1 l. (the value given in the tables), we find for the Chartres *setier* a value of 131.3 l., for its *minot* 32.8 l. The standard of Chartres *minot,* which is dated 1283 and is preserved in the municipal museum, is equal to 31.7 l. — a very good correspondence, for the variation, which is infinitesimal given the time span of approximately five hundred years, results not only from wear but from errors of hallmarking that could have been committed earlier.

Stone measures

Measures for dry materials had not always been made of metal; the standards had formerly been cut from blocks of stone, which practice continued to exist down to the definitive establishment of the decimal metric system (1840). Their origin dates from the distant past, for the Greeks and the Romans also utilized very similar measures (the *secoma* of Delos, the *measa* of Pompei, Auch, and so on); the technique of making these measures was known in antiquity, and continued to exist in the West. It consisted of hollowing out small basins in a block of stone, which was often decorated; the stone could be mounted on a frame by means of two pins, and in this way the basin, once filled, could be quickly emptied of its contents.

In some markets merchants from outlying regions sometimes utilized measures whose values differed from those in use on the market; whence the need to establish series of measures that were easy to use and were suited to the requirements of the business. Thus we find blocks consisting of two, three, four, and sometimes more different measures of capacity.

These measures could be permanently fixed in place, for example by being attached to a bridge, as at Billom (Puy-de-Dôme). In such cases they had an opening in their lower portion that permitted the grain to flow into the measure; the hole could be stoppered with a plug. Their capacity was generally large, comparable to the hectoliter, half hectoliter, and so on.

By 1187 Paris had a "standard stone mina," shaped like a miter and kept in the chapel of Saint-Leufroy; it was still there at the time of the chapel's demolition in 1684. The city of Albi has a *"rue à la Piala"* whose route was laid out parallel to the public measures in the market. In the thirteenth century the stone measures were kept in the *place de la cathédrale,* in a spot called the *"Pile."* This type of measure was used throughout the kingdom.

Measures of Capacity for Liquids

Most of the standards for measuring liquids were made of metal: brass or bronze. To meet the requirements of business, the containers were made of sandstone, wood, tin, glass, and so on. In 1635 the city of Lyon made an agreement with the Lyonnais metal caster Boutavau for the production of ". . . six *matricules* [a term equivalent to standard, matrix] of cast iron, namely: two jars, two quarter casks, two half-quarter casks." In 1588–1589, at Cambrai, Michiel Lesecq ". . . made a new casting equal to the municipal jar."

Oil was often sold by weight and measure; the latter corresponded to a given weight. Marked bottles appeared around the end of the seventeenth and the beginning of the eighteenth centuries; witness the large collections of carefully made bottles of this type now housed in the Musée Paul-Dupuy. Prismatic in shape, similar to carafes, these vessels are banded at the neck by a metal band several millimeters wide, held in place by a collar; the band indicates the level of liquid which corresponds to the legal content of the bottle.

Documents and specimens reveal the use of these vessels at Saint-Omer, Boulogne-sur-Mer, Montauban, Cahors, Uzès, Bourg, and other places. In 1292, to mention only one example, such bottles were in general use among the innkeepers of Montpellier.

Measurement of Vats

This rapid summary of the early techniques of measuring capacity must include a few words about the vats, hogsheads, casks, and so on, used for transporting wine, whose actual content was determined by the gauge. If all these "vessels" had been standardized on the basis of certain defined units, such as the Parisian foot, it would have been easy to determine their content and employ a single gauge that could be used everywhere. Such was not the case, however, for the dimensions of the hogsheads were variable. But the tax levies had been determined by ordinances established in June 1680, which fixed the tax in terms of the exact content of the vessels.

To eliminate the disputes that arose between the tax collectors and taxpayers because of the disparity of the vessels and the difficulties encountered with gauges, a scale was published that made it possible to establish the content of vessels once several preliminary measurements had been made. The method used (which was approved by the Académie des Sciences) consisted of comparing the measure with a cylinder whose diameter was equal to half of the diameters of the ends and the diameter of the bulging portion, and whose length was equal to that of the vessel between the ends. To facilitate the work of the tax collectors, the official scale published with the king's approval made it possible, once these two results were known, to immediately obtain "within one pint" the content of the "vessel" in Parisian *muids* (1 *muid* equaled 36 Parisian *setiers;* 1 *setier* equaled 8 Parisian pints). This pint was equivalent to approximately .93 l., a value that appears to have remained unchanged since the fourteenth century. The edict of October 1557 and the patent letters of 1705 fixed the Parisian *muid* at 288 pints (approximately 268.1.) and 37.5 *setiers* (approximately 280 l.). The edict fixed the cask at 1.5 *muid.*

Numerous studies have been devoted to the measuring of vats, for example those of Clarmorgan (1653), Naulot (1668), Pézenas (1742 and 1749), Bion (1752), and so on.

MEASUREMENT OF VOLUME

Scales

Equal-armed scale; the Roman scale

The scale with index pointer and two equal arms has the oldest and most durable claims to nobility, since it has been in existence since antiquity; the so-called "Roman" scale (the *statera*) did not appear until around the second or

Plate 25. Portable armoire containing the royal standards of oil measures of the City of Paris, 1741. *Photo Chuzeville-P.U.F.*

third century B.C. Both types of weighing instruments are still in use.

The equal-armed scale is inseparable from its indispensable complements, the weights. The "Roman" scale, in contrast, needs no weights, since the equilibrium of a burden is obtained by moving a slider (formerly called the *aequipondium*) along the graduated arm of the instrument (Figure 95).

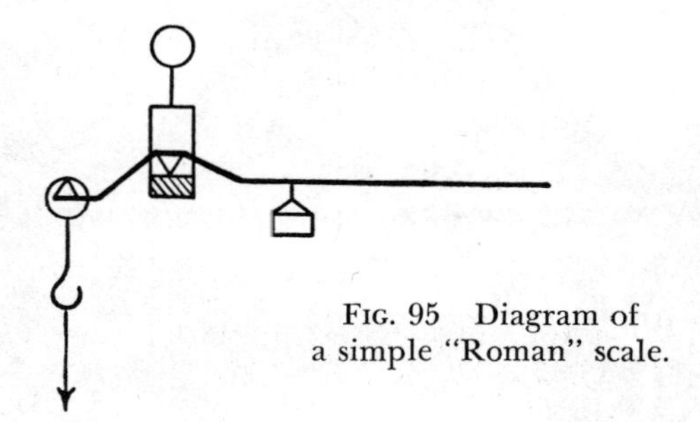

FIG. 95 Diagram of a simple "Roman" scale.

From the technical point of view, the operation of these two instruments is based on the principle of the lever. The mechanical conditions of their equilibrium, which had long since been learned by experience, had been explained as early as the twelfth century by Jordan de Nemore. Even before then, however, the Arabs had amassed a certain amount of practical knowledge in this field, since the patriarch Elyas of Nisibin (died 1049 A.D.) was able in his dissertations on scales to draw attention to the fact that an equal-armed scale with a long, light pointer and a prismatic knife-edge whose edge rests on the fulcrum, is more sensitive than the scale that has a short, heavy index pointer with cylindrical central axle of oscillation.

Techniques of constructing these scales; their qualities

In 1557 King Henri II recommended that the builders of weighing instruments (the scale-makers' guild) make only index scales with a square "nail," that is, with a prismatic center knife-edge. (The term "nail," which designated the instrument's axle of oscillation, was still being used in the seventeenth century by such scientists as Roberval, De La Hire, and others.) At the same time he ordered the destruction or modification of scales that did not meet this requirement. This was a wise decision, since the axle was almost always cylindrical, and sometimes consisted of a simple wooden peg!

The metrological study of several scales of this period (specimens of which are extremely rare) and instruments of the seventeenth and early eighteenth centuries sheds light on the knowledge acquired by the scale makers, undoubtedly in empirical fashion, and on the fact that at least in certain workshops (Paris, Lyon, Rouen, and others, as well as in the Netherlands and Germany) these craftsmen had more or less successfully resolved the problems raised by the construction and perfecting of weighing instruments.

They knew that the principal condition for obtaining correct measurement was the equal length of the arms of the scale; otherwise, they said that the index had "gauge." Formerly, in order to correct this anomaly the balance makers acted on the knife-edge or hammered the pointer in order to diminish the length of the excessively long lever arm. They were aware that the edges of the three knives should be straight and parallel and that the center of gravity of the instrument should be on the vertical line of the center of oscillation, and very slightly below the pointer. There were various methods for suspending the pans from the arms.

Characteristics of the "Roman" scale

The metrological qualities of the "Roman" scale (known to the Arabs as the *quarastun*) by no means achieved as great a degree of perfection as the equal-armed scale. But the possibility of weighing without any need for weights, simply by sliding a runner on the graduated arm of the "Roman" scale, explains the continued existence of this instrument.

In addition, the "Roman" scale could be two-sided ("light" and "heavy"); the instrument need only be turned over in order to change sides. This is why the knife-edge that received the action of the object to be weighed has the form of a lozenge. For example, objects up to one hundred pounds could be weighed on the "light" side; by turning the instrument over, objects of between one hundred and three hundred pounds could be weighed on the "heavy" side (Figure 96).

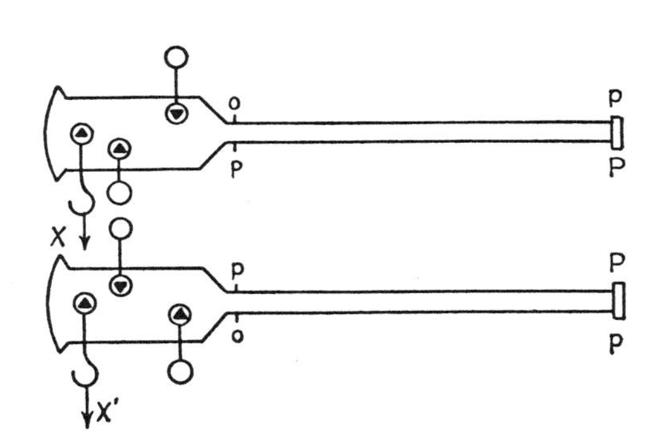

FIG. 96 Diagram of a two-sided "Roman" scale. Top: "light" side; bottom: "heavy" side.

It is difficult to perfect this instrument even with modern methods, and was undoubtedly much more difficult in earlier centuries, taking into account the various mathematical and technical factors.

The graduations on the arm of the "Roman" scale are equidistant, as can easily be demonstrated. Once the unit (for example, the pound) has been established, the marks corresponding to once, twice, or "n" times this unit must be laid out. For this purpose the early scale makers utilized "saber combs," that is, toothed gravers made of fine steel known as "saber steel," and used to engrave these marks on the red-hot iron arm.

It follows that this practice could not match the precision of the dividing engines, which for all practical purposes did not come into use until the eighteenth century (Ramsden's engine). In addition, the inequality of these graduations made the results of the measuring operation and the mechanical conditions peculiar to this instrument (exactitude, precision, and so on) uncertain. It is not surprising, then, to see one Antoine Prévost, a Parisian scale maker, informing the king's councilor (La Reynie) on April 10, 1673, that at Versailles he "seized a Roman scale indicating 25 pounds to the public when actually the weight is only 15"!

Fraud was easy with this instrument. In his study of scales in the *De Subtilitate* (1556), Jerome Cardan notes in connection with the *"traîneau"* and the *"livre"* (the first word designates a "Roman" scale, the second an equal-armed scale) that they are no better than the Roman *traîneaux*.

Given these conditions, it is understandable that that "Roman" scale was disparaged on numerous occasions. It was undoubtedly utilized chiefly in the southern regions, but was particularly widespread in the eighteenth century. It was frequently made of hardwood, and the engraved markings were replaced by brass nails whose heads were leveled flush. Only the suspension axles and the slider, a kind of metal ball hung from the arm by a hook, were made of metal.

These objects, which were for the most part made by turners, were prohibited by the Administrator of Lyon around the second half of the eighteenth century. Three centuries earlier, the Bishop of Rodez had declared that the "Roman" scale was "an instrument of fraud."

Range and precision

The range of the weighing instruments was then as now adapted to their intended use. The weighing of small quantities was done with equal-armed scales, the length of whose index pointer varied between approximately 4½ and 7¾ inches, sometimes more. These scales, called *trebuchets* (assay balances), were used by the minters, changers, alchemists, assayers of metals, and so on.

In laboratories it was sometimes necessary to use instruments of various capacities, in addition to these small assay balances, in order to obtain acceptable results. Long experience had revealed that these comparisons — we could consider them as precision weighings, except for their size — could be effectively carried out only by isolating the weighing instruments and the objects to be weighed from the surroundings.

As early as 1343 an ordinance of Philippe de Valois, relative to assay scales used for minting, prescribed that when weighing "Assays" the "General Assayer . . . should be in a place where there is neither wind nor cold, and should be careful that his breath does not weigh on the scale."

The famous mineralogist Agricola (Georg Bauer, 1494–1555), undoubtedly with a similar thought in mind, used three assay balances, small, medium and large, placed under a glass cage. In order to prevent the knife-edge from becoming prematurely worn, a mechanical system made it possible to raise the index when the scales were not in use. A similar method is used today, but in combination with other modern technical refinements (vacuum, temperature, pressure, hygrometrical degree, and so on) which were practically unknown or could not be used in the sixteenth century (Figure 97).

The Cour des Monnaies, for example, had a series of balances in ascending order of capacity, ranging from the half ounce (approximately 15 grams) to 16 *marcs* (approximately 86 pounds). The fact that any of these instruments were used proves that the early scale makers had realized, through contact with physi-

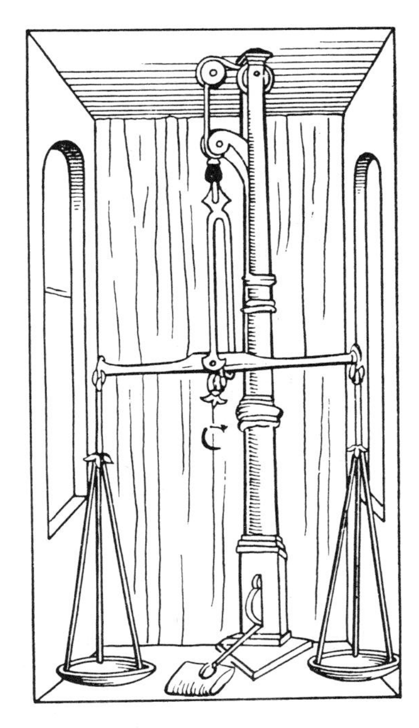

FIG. 97 Agricola's "fine" scale (around 1530).

cal reality, that the smaller the weights to be compared, the more difficult it was to achieve precision. This observation, which is well known to all metrologists, is generally valid whatever the nature of the physical entity being measured.

In the sixteenth and seventeenth centuries the technical level of scale construction, which was already respectable, made it possible to achieve a precision in the neighborhood of 1/10,000, under optimum conditions and when the materials being weighed were around one and two pounds. These figures were determined thanks to the existence of *poids de marc* weights (to be discussed shortly), which are genuine prototypes or copies of prototypes.

For weighing heavy objects, two types of scales were used, notably in arsenals, foundries, wheat mills, and ports. With very large instruments whose arms could measure more than two *toises* (fathoms — approximately six feet in modern terms), it was possible to weigh objects up to 12,000 pounds. The museum of Carpentras houses an eighteenth-century "Roman" scale that was converted to metric units when the new system of measurement was applied; its arm measures no less than 81 inches.

Roberval's scale

Whatever their range, the weighing instruments formerly in use were devices with hanging pans suspended by chains or small ropes to the ends of the arm (or to the head of the "Roman" scale).

In 1669 the mathematician Gilles Personne de Roberval (1602–1675) submitted to the Académie the prototype of the counter type of scale. This innovation had the advantage of stabilizing the pans in a horizontal position, and consequently of facilitating weighing (Figure 98).

While the Academicians were surprised by what they called "Roberval's

Plate 26. Weighing cannons. From the collection of illustrations in the *Encyclopédie*. Musée des Arts et Métiers. *Photo by the Conservatoire.*

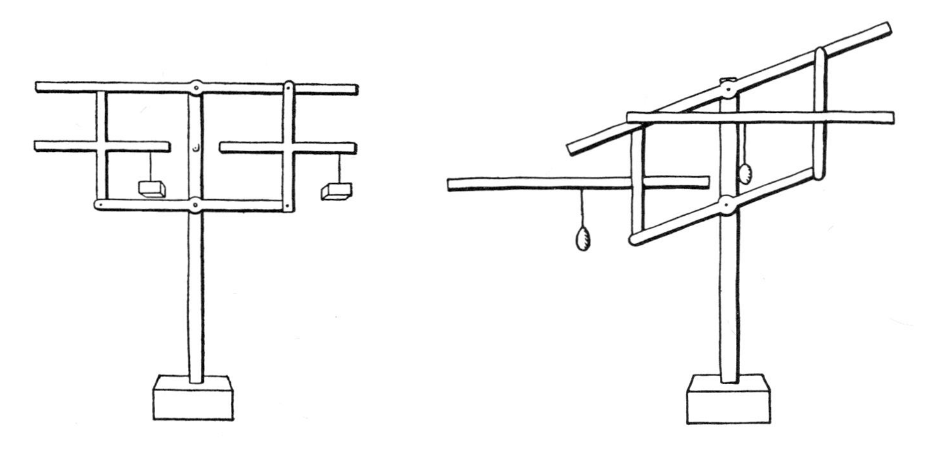

FIG. 98 Diagram of Roberval's scale (from *Mémoires de l'Académie des Sciences,* Vol. X). Left: jointed parallelogram. Right: demonstration of the fact that the pans remain horizontal in any position.

static enigma," the scientist's invention nevertheless remained buried in the Académie's portfolios for more than 150 years before being developed and disseminated by the nineteenth-century French manufacturers (who were, however, preceded by their English counterparts). A reproduction of a scale constructed on Roberval's principle, however, can be seen in Jacob Leupold's *Theatri statici universalis* (Leipzig, 1726).

Automation (Leonardo da Vinci)

In order to utilize a balance of either type, an operator was needed; he placed the objects to be weighed on one of the pans, and balanced it with weights on the other pan, or with the slider in the case of the "Roman" scale. This method is still in use, but it is gradually entering the realm of history, as a result of simplifications due to the generalization of automation in scale making.

The first instrument for automatic weighing — a weighing machine — was invented in 1876 by the French engineer Dujour, who in all probability was ignorant of Leonardo da Vinci's discoveries. The famous engineer-artist left plans — we might almost say blueprints — which reveal beyond any possibility of doubt that the Florentine had conceived the mechanical principles of the operation of measuring instruments that could be read directly. Using these plans, Hugo Rimediotti, verifier of weights and measures at Milan, was able to construct the two types of automatic scales conceived by Leonardo.

Demultiplication of power (Borrel, 1554)

Another invention also helped to modernize methods of scale making and increase its scope of application. This was the principle of the demultiplication of power by series of paired levers, which occupies a prominent place in a treatise on geometry by the French mathematician Borrel (1492?–1572). This principle was "in power" by 1554, the year in which Borrel published his treatise. However, the lack of communication between scientists and craftsmen long constituted an obstacle to any development, a situation exacerbated by immutable craft traditions and the guild statutes, which did not favor the spirit of initiative (Figure 99).

This is why the first weighing bridge with compound levers was born in Great Britain, where it was constructed by an English carpenter, John Wyatt of Weedford (1700–1766), at Birmingham around 1741 (Figure 100).

In 1803, more than fifty years after the birth of Wyatt's weighing bridge, the first bridge with levers placed under the platform was constructed by a man named Merlin, a mechanician at Strasbourg.

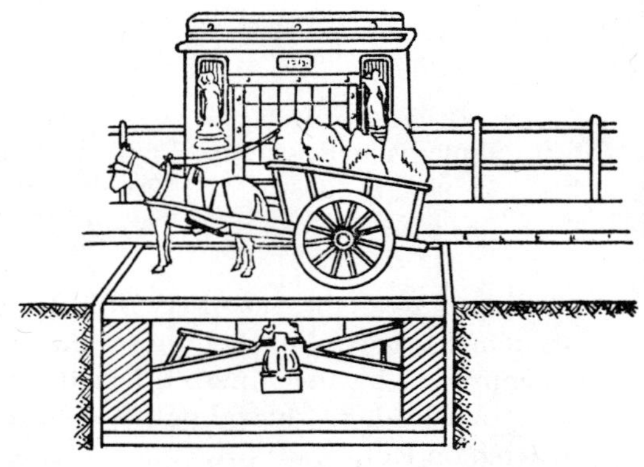

FIG. 100 John Wyatt's weighing bridge, 1741.

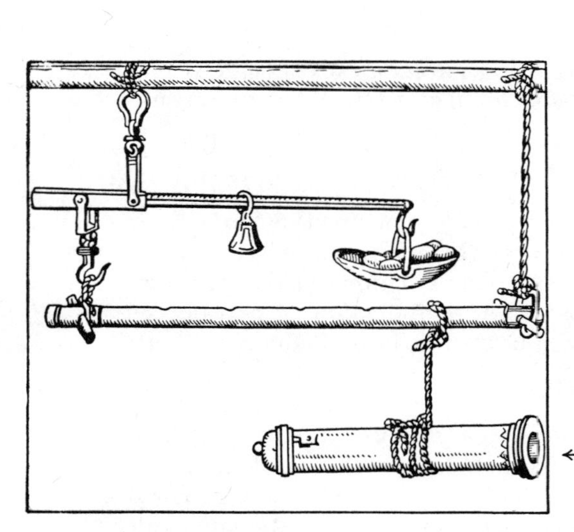

← FIG. 99 Demonstration of the demultiplication of power, by J. Borrel (called "Buteon"). Middle of the sixteenth century.

WEIGHTS

Weights are the indispensable adjuncts of the equal-armed scale. For a long time no state-imposed systematic standardization of weights existed, which explains the variety of shapes given to the weights by the scale makers and metal casters.

The shape varied depending on the caliber of the weight and sometimes also the place where it was to be used. However, the following forms generally predominated: the conical section, the pyramid, cylinder, prism, bell (or similar shape), segment of a circle, disk. Most of the old weights are made of bronze, brass, cast iron, lead, and stone, and can be divided into two types, the marked "coin" type and the unmarked type.

The "coin" type The marked weights are made of bronze, in the shape of a flat disk, and in an unusually elegant style. This model was used basically in Languedoc and the neighboring provinces. It was issued at the latest by 1239, by the municipality of Toulouse, and spread rapidly throughout all the regions of the Midi: in less than a century some thirty cities had adopted this model.

To the best of our present knowledge, the oldest French weights are of the "coin" type; they are dated, their origin is known, and a great number of specimens are still extant. They vary between four pounds and one ounce and its subdivisions; they are between 110 millimeters and 10 millimeters in diameter, and their average thickness ranges between 20 millimeters and 4 millimeters (Figure 101). On both faces of the disk appear the canting arms of the cities and seigniories, surrounded with legends that tell the place and date of issuance as well as the weight of the object *(livra, meia livra, cartaro, meia cartaro, onsa, meia onsa* and so on). The division is obviously binary.

FIG. 101 Coin-type weights of Toulouse. Left: *demi-livre* (190 grams) of Gaillac, 1291. Right: four-*livres* (1,568 grams) weight of Toulouse, 1239 (from specimens in the Musée Paul-Dupuy, Toulouse).

The unmarked type

The term "unmarked" cannot be taken in its literal meaning. At least conventionally, it is to be understood as designating objects marked with a few brief pieces of information: marks and dates of verification, notches or engraved numbers indicating its caliber in pounds, the canting arms of the issuing city. Most of the weights of this type can be grouped under this conventional heading (Figure 102).

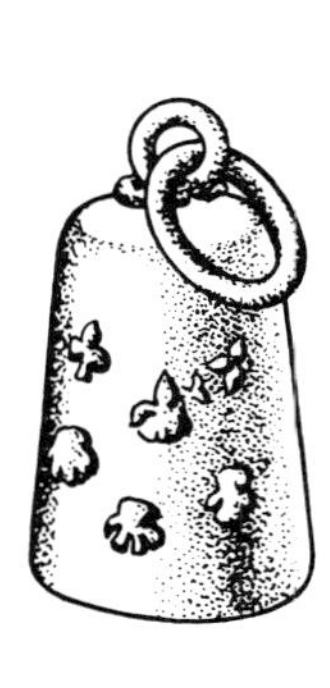

FIG. 102 "Unmarked" weights. Left: Béziers matrix, 1687. Center: bell-shaped weight from Rodez, 1776. Bottom: sixteenth-century weight of Aurillac (from specimens in the Musée Paul-Dupuy, Toulouse).

The flat weight with geometrical forms (square, hexagonal, octagonal), which appears to have been in circulation in southern France around the end of the sixteenth century, ultimately competed with and supplanted (in the eighteenth century) in many cities of southern and southeastern Languedoc the marked coin-type weights that until then had been in use. These weights bore only the escutcheon of the arms of the city, without legends. On the smooth reverse side there sometimes appeared incised letters and marks of verification. The caliber of these objects (as well as their size) is practically identical with that of the coin-type weights.

Various shapes

Other specimens belong to somewhat special categories: voluminous weights shaped like bells or cylinders, such as the standards of Lille, Rouen, Rodez. Sometimes the date of issuance is incised on the side of the object (or on the top, in the case of a round object), along with the indication of the caliber, which ranges between five and one-hundred pounds (quintal). Most of these massive objects have a handle of some sort that is part of the weight, or a ring attached to a holder embedded in the object.

Business people made use of cast-iron weights shaped like truncated pyramids, which were made by the foundries and weighed as much as one hundred pounds. They had a ring firmly held in place by pins that pierced the inner cavity of the weight, where they were embedded in lead. The indication of the caliber, and sometimes coats of arms, were frequently molded with the objects.

Stone weights have been utilized since man has been weighing and measuring; they have had a long career, from the Sumerians down to the creation of the metric system — witness the extant specimens in France and other countries (Belgium, Holland, et cetera). The assistant to the weigher of Poids-le-Roy was still being called "pebble-lifter" in 1434 (the nickname disappeared in 1452).

Small, oddly shaped weights — bottles, amphoras, portions of spheres — also existed. The bottle-shaped and spherical objects sometimes represent medicinal units, which differed from the other, already highly complex units. Amphora-shaped weights are brass standards of the city of Montpellier, which date from around the seventeenth century.

The poids de marc "stack"

The *"marc"* is a unit of measure of volume, originally used in minting, whose circulation is attested by the eleventh century but whose origin does not yet appear definitely determined. In any case, we find *marcs* of various values that had a special vogue, for example those of Troyes or Paris that weighed approximately 244.75 grams — a value that did not change appreciably until the end of the eighteenth century. The *marcs* of Cologne, La Rochelle, London, and Portugal weighed approximately 229.5 grams, those of Venice, Florence, Genoa, and other cities weighed approximately 238 grams.

We do not know whether these *poids de marc* had the same form in the eleventh and twelfth centuries that they had at least by the fourteenth century. At this period they were called "stacks," which name was justified by their shape, since they were composed of hollow cones that fitted into each other, the top one being box-shaped and equipped with a cover, a handle, and a catch, and often artistically decorated.

The *poids de marc* and its subdivisions are described in a fourteenth-century regulation of the city of Albi (Tarn), which sets the rates for hallmarking of the "large" and "small" components of the object. The painting by Peter Christus (1449) showing a moneychanger with his scale and his stack is perhaps one of the oldest depictions of the *poids de marc* stack.

In France and most of the Western states the scale makers imported crude stacks made by the metal casters of Nuremberg, and then hallmarked them in terms of the legal units. For practical purposes these weights were adjusted to the weight of the original *marc* of France deposited at the Cour des Monnaies in Paris; they were used as standards, preserved in the provincial *Juridictions des Monnaies,* by scale makers, goldsmiths, apothecaries, grocers, and moneychangers. They could not be put into circulation by the scalemakers until after they had been verified and stamped at the cour or the Juridiction des Monnaies of the district in which the scale maker's shop was located. Thus these stacks usually bore the marks of the German metal caster, the scale maker, and the official tester.

The uniformity of style and composition of these objects, whose volume ranged from a half-*marc* to sometimes 150 or more *marcs,* justifies their special classification. The stack had at least one predecessor, for archaeology has brought to light several Gallo-Roman objects consisting precisely of nested weights, whose principle is no different from that of the *poids de marc.*

The nesting system was also applied to the standards of measurement of capacity in Portugal around the fifteenth century. The same was done in France (Dijon, sixteenth century) for oil measures (1529).

The Making of Weights

Coin-type weights

The coin-type weights were cast in a mold, as is evidenced by the regulations of the city of Albi, according to which the "matrix weights" (standards) and the "bronze molds with iron pincers" were given each year to "sworn testers." For example, the weights used at Albi in 1673 bear the initials "P.M." for the tester Pierre Mabille. At Rodez, in 1527, the use of molds is attested by the verifier Gransanhe who received them, then in 1722 and 1776 by the metal casters Dubois and Lacombe.

Unmarked weights

The making of unmarked weights falls within the province of metal casting. In 1596 Jehan Gros, a metal caster of Arles, was charged with casting weights for the city of Nîmes on a model supplied by the city consuls; similar information is available regarding the cities of Lyon, Dijon, and Saint-Omer. The museum of Lille possesses a steel mold for casting lead weights.

It is probable that by eliminating the inscriptions that made the coin-type weight an "official" object, the unmarked weight of the same caliber, although it still bore the city's coat of arms, could be made faster and more easily. But it is not impossible that this new formula resulted from the transfer to the Midi of technical procedures in use particularly in the north and in Flanders. Well-preserved specimens of weights, such as those of Ardres, Aire-sur-la-Lys, Saint-Omer, and Malines, have very much the same appearance as the new weights used in Languedoc.

The technique of hallmarking weights varied, depending on whether it was being applied to solid or hollow weights. In the first case, the scale maker could only gradually file down the lower face of the object until it reached its correct weight. In the second case, this adjustment was done by pouring a certain amount of lead into the lower cavity of the cast-iron weights.

To the unmarked weights can be related small objects that represent small subdivisions such as the grain. These weights were strips of metal much the same as those now in use; they weighed a decigram, centigram, or milligram.

The pound poids de marc

The *marc* of 244.75 grams, which as we have already mentioned was originally intended for monetary purposes, became as it were bivalent, in the sense that the pound *poids de marc,* composed of 16 ounces in the fifteenth century (15 prior thereto), and utilized in commerce and industry, was derived solely from the Parisian *marc.*

The prototype of the *marc,* and consequently of the measures of volume, was formerly represented by the stack — incorrectly called "Charlemagne's stack," at least to the best of our present knowledge — preserved in the Cour des Monnaies in Paris, and whose manufacture does not antedate the last third of the fifteenth century (according to Mr. Biancard). This stack, composed of 13 cones forming a total of 30 *marcs,* and a top weighing 20 *marcs,* was thus a stack of 50 *marcs* (12,237 grams). When it became the national standard for weights in 1766 (along with La Condamine's fathom), it served at the time of the creation of the metric system as an element of comparison for the establishment of the new unit, the kilogram. Then, its role completed, the stack of the Monnaie was deposited in the Musée du Conservatoire des Arts et Métiers in 1848 (No. 3261), and become part of history.

Numerically, the pound *poids de marc* was subdivided as follows: one pound = two *marcs* = four *quarterons* = 16 ounces = 128 *gros* or *drachmas* = 384 *deniers* or *scrupules* = 9,216 grains.

The unit of volume for precious materials (stones and metals) was the carat. The origin of this term is the seed of the carob tree — *keration* in Greek, *qirat* in Arabic, *carato* in Italian, carat in French. The metric carat was standardized internationally at .2 grams in 1907.

Inconsistency of units, permanence of fundamental notions

We have now outlined certain aspects of metrology current until around the first third of the eighteenth century. It is to be noted that the diversity of the units, their terminology, and the procedures of measurement, which were poorly defined and sometimes devoid of all metrological sense, did not systematically exclude the permanence of certain fundamental notions.

Plate 27. "Charlemagne stack" weight—the standard for the *poids de marc*. End of the fifteenth century. Musée des Arts et Métiers. *Photo Chuzeville-P.U.F.*

The Sumero-Babylonian standard of dimensional measure, which takes us back forty centuries, was justly regarded as one of the key pieces indispensable to a rational system of measurement — whence the necessity of preserving the prototype. In French villages, how many public buildings had fathoms, feet, and ells fixed to their walls, and sometimes even gauges engraved in the stone to represent the standard, obligatory sizes of daggers (for example at Rouffach in the Upper Rhine area). Even if these standards were not properly kept, everyone understood that it was not possible to do away with them without falling into chaos and arbitrary rule.

In most cities, towns, and villages, the possessors of measuring instruments were subject to controls that history shows us were enforced regularly in some places, more exceptionally in others. But it was always to the standards that parties at litigation referred in case of dispute over the quantities bought or delivered.

However, sometimes the definition of the units of measure was far from precise; they even overlapped, since, as we have already noted, two units of the same size existed side by side — the ell and the fathom! Moreover, the variations in measure were sometimes so great from place to place or over long distances that this startling diversity did not facilitate relations, and in contrast favored fraud.

It is not astonishing, then, that the Estates General received repeated complaints from users, who were anxious to see order brought to this mosaic of units born of the feudal regime. This tremendous task was one of the problems that preoccupied most of the rulers, from Philippe le Bel to Louis XVI. Although the successive governments sought remedies for this evil, which some thought could not be cured, their praiseworthy attempts were almost all destined to failure.

SEVENTEENTH-CENTURY SCIENTISTS AND METROLOGY

In the seventeenth century experimental science encouraged a current of new ideas that effectively contributed at the end of the eighteenth century to the general unification of measures.

Basic characteristics of a system of measurement

However, if it was to bear fruit this unification had to be part of a consistent system of measurement such that the multiples and subdivisions of the units would be in conformity with a fixed scale of numeration.

In ancient Mesopotamia, the Sumero-Babylonian system of measurement was based on a single basic unit, the "cubit," a unit of linear measure. From this unit were derived the units of surface, volume, and mass measures. The unit of mass, for example, was equal to the weight of water occupying a volume equivalent to 1/240 of the cube of the cubit. The system of numeration, then sexagesimal, was perfectly adapted to that of the measures.

In France and elsewhere, particularly in the seventeenth century, the measuring scale was highly erratic. The *toise* (fathom) consisted of 6 feet (sometimes more), the foot of 12 inches, and the inch of 10 or 12 lignes. This was the duodecimal method of numeration. In contrast, the scale of weights was binary: the pound, half-pound, quarter-pound, and eighth-pound; the ounce (1/16 of a pound), half-ounce, quarter-ounce, and so on, down to 1/32 of an ounce. In the case of the ell, the subdivisions were inconsistent: it was divided into 1/2, 1/4, 1/8, 1/16, 1/32, and also into 1/3, 1/6, and 1/12. As for measures of capacity, the *muid* consisted of 32 *setiers,* the *setier* of eight pints, and so on.

Simon Stevin: measures subject to decimal division

The French scientists, who were working in close relation with their foreign colleagues, deplored this anarchy common to most of the Western states. In 1585 one of these scientists, the mathematician Simon Stevin of Bruges (1548–1620), recommended the use of the decimal numeration in his work entitled *L'arithmétique de Simon Stevin de Bruges.* This work included a treatise to which he gave the name *Disme,* and in which he proposed to unify measures and money and make them subject to decimal numeration.

This numeration was already being applied to measures, especially in the provinces of Lorraine and the Barrois, which were not attached to the French crown until 1766. The Lorraine fathom (called the *verge*) measured 2.859 millimeters and was divided into 10 feet (.2859 millimeters); the foot was divided into 10 inches (.0286 millimeters), and the inch into 10 lignes (.00286 millimeters). The Musée Lorrain in Nancy houses series of old measuring standards, represented by octagonal metal bars. The old Parisian foot is reproduced as it was before the creation of the new Châtelet fathom in 1667–1668 — .3265 millimeters. The same gauge shows the Lorraine foot of 284.5 millimeters, divided into ten inches, the inch into ten lignes.

Along the same lines, the Rhineland rod, approximately 3.766 millimeters (12 feet) long, used by the Dutch scientist Snellius (1591–1626) when he measured the meridian arc linking Alkmaar with Bergen op Zoom, was divided into 100 equal parts. Many scientists used the decimal numeration, which possesses the advantage of simplicity, and facilitates calculations.

This rapid and summary analysis of the basic structures of a rational system of measurement makes the enormous gap that separated the numerous local pseudo-metrologies from a genuine, scientifically constructed metrology immediately apparent.

A universal, natural unit of measure

The standardization of measures had already preoccupied certain scientists of the seventeenth century whose attention was attracted by the observations of Galileo (1564–1642) on the pendulum and the isochronism of its oscillations. This phenomenon had inspired them with the idea of choosing the length of the pendulum that beats the second as a "natural and universal" unit of measure. From the Flemish engineer Isaac Beeckman (1588–1637), who proposed this choice in 1631, to the Abbot Picard (1620–1682), creator of the Observatory, and including Father Mersenne (1588–1648), Huygens (1629–1695), and the Abbot Mouton (1618–1694), Lyonese mathematician and astronomer, the conceptions of these pioneers were oriented toward this solution.

They all deplored the diversity of the "Western" units, which made comparisons very difficult, and devised methods of dealing with these difficulties; for example, they reproduced the length or a subdivision of a given unit in the form of a printed line. Although these reproductions were not precise, notably because of the contraction of the paper during drying, given the lack of genuine standards, the lines constituted a useful method of comparing measurements.

In certain cases scientists or travelers undertook to bring back copies of local standards in use in foreign states. Pierre Petit even thought of making cubical vases one foot long (local feet). By weighing the water contained in this vessel, the mass of the cubic foot of water would be known, and could easily be expressed in terms of Parisian, Roman, or other units.

It must be admitted that these procedures were quite complicated and not very precise and that the recourse to a unit of linear measure derived from the seconds pendulum was attractive. It was necessary to ascertain that the amplitude of the oscillations did not vary from place to place; it was also wise to elaborate a system of units of rational measurement in perfect agreement with the method of numeration.

Abbot Mouton's system of measurement (1670)

Abbot Gabriel Mouton was the scientist-craftsman of this painstaking construction. The permanent priest of the Lyonese church of Saint-Jean outlined his plan in 1670 under the title *Novae mensurarum geometricarum idea et novus methodus eas et quascumque alias mensuras communicandi, etc.,* printed at the end of his *Observationes diametrorum solis et lunae apparentium.*

The structure of the system of measurement proposed by Mouton was based on the unit of linear measure which he related to the size of the earth. He borrowed the definition of this unit from the thousandth part of the angular minute, that is, from the thousandth part of the nautical mile of 1,852 millimeters that is 1.852 millimeters. This unit, represented by a standard, is the *virga* (the term which designated the pendulum rod); its value, although less than that of the Parisian fathom (1.949 millimeters) was close to it.

In addition, Abbot Mouton subjected the multiples and submultiples of the unit to decimal division. The scope of this idea was considerable, for the introduction of decimal numeration into the metric system played a most important role that effectively contributed to its expansion throughout the world.

Although the terminology – the nomenclature – of the measures envisaged by Mouton was not adopted in the metric system, it is not inappropriate to give a few examples, which reveal a certain hesitation between a new terminology and a nomenclature composed of traditional terms. (The founders of the decimal metric system were in the same position.) It is noteworthy that the nomenclature given in the table below is drawn exclusively from Latin:

Centesima						
10	*Decima*					
100	10	*Virgula*				
1,000	100	10	*Virga*			
10,000	1,000	100	10	*Decuria*		
100,000	10,000	1,000	100	10	*Centuria*	
1,000,000	100,000	10,000	1,000	100	10	*Milliare*

The terms known were, for example, the *stadium, funiculus, virga, virgula, digitus, granum, pontum.*

Another special feature of Mouton's system attracts our attention: the entity that defined the unit (the nautical mile) was independent of the one that was to verify it, that is, the pendulum. For this reason the scientist performed experiments on a pendulum 1.852 millimeters long (the *virga*) in order to determine the number of oscillations in a half hour. Theoretically, this number should be around 1,252.26. Mouton achieved a remarkable result, since the average of his observations varied between 1,251.8 and 1,252.1 oscillations per half hour.

For the *virgula,* the tenth part of the *virga,* the theoretical value was 3,960 oscillations in a half hour; Mouton found an average of 3,959.2 oscillations. Whether by *virga* or *virgula,* it would always be possible to verify the stability of the standard representing the unit of linear measure, or even to reconstruct it if it disappeared by constructing a pendulum beating 1,252 vibrations in a half hour. Once the linear unit had been defined, the units of measurement of surfaces, capacity, and mass were easily derived from it. This modest scientist, unsung and worthy of his place near the academicians who created the metric decimal system on the foundations he laid, deserves to be honored.

Between 1670, the date of the publication of his system of measurement that was almost an integral predecessor of the metric system, and 1694, the date of his death, the ideas, as well as the experimental result, seemed to reconcile with difficulty the desire to discover a unit of measure of cosmogonic origin and the physical possibility of its existence.

Father Mersenne had begun to study the pendulum long before Picard became interested in it. In his *Harmonie universelle,* published in 1636, he fixed the length of the seconds pendulum at 3½ feet (1,137 millimeters instead of 994 millimeters). However, in his *Harmonicorum Libri,* he describes his pendulum, declares that it has a small lead ball ⅔ ligne (1.5 millimeters) in diameter and that its length is equal to 986.5 millimeters. In one of his works, the *Methodus vitandorum errarum omnium, qui in arte medica comitti possunt* (Venice, 1603), Santorio (Sanctorius) (1561–1636) claimed to have invented a *Pulsilogium* that he later described in his *De instrumentis medicis.* The same doctor specialized for a great part of his life in studying his own variations in weight depending on the absorption of food, transpiration, and so on. For this he utilized a large "Roman"-type scale holding a kind of cage in which he sat (Figure 103).

In 1618 the scientist and physicist Beeckman envisaged the use of the pendulum for observations. Like Santorio, he was seeking the cause of isochronism, and in 1631 suggested the use of the length of the cord pendulum, which makes one oscillation per second, as an international unit.

Thus Abbot Picard was quite out of date when in 1669 he determined that the length of the pendulum beating the second at the Observatory was 36 inches 8½ lignes (.9935 millimeter), a conclusion he reached through comparison with two large pendulum clocks (balance wheel invented by Huygens in 1657).

In 1671 the astronomer proposed the name "Astronomical Radius" for this pendulum; its double would be the "universal fathom," the third the "universal foot." The universal rod would be equal to four "radii" and the "universal mile" would contain 1,000 rods.

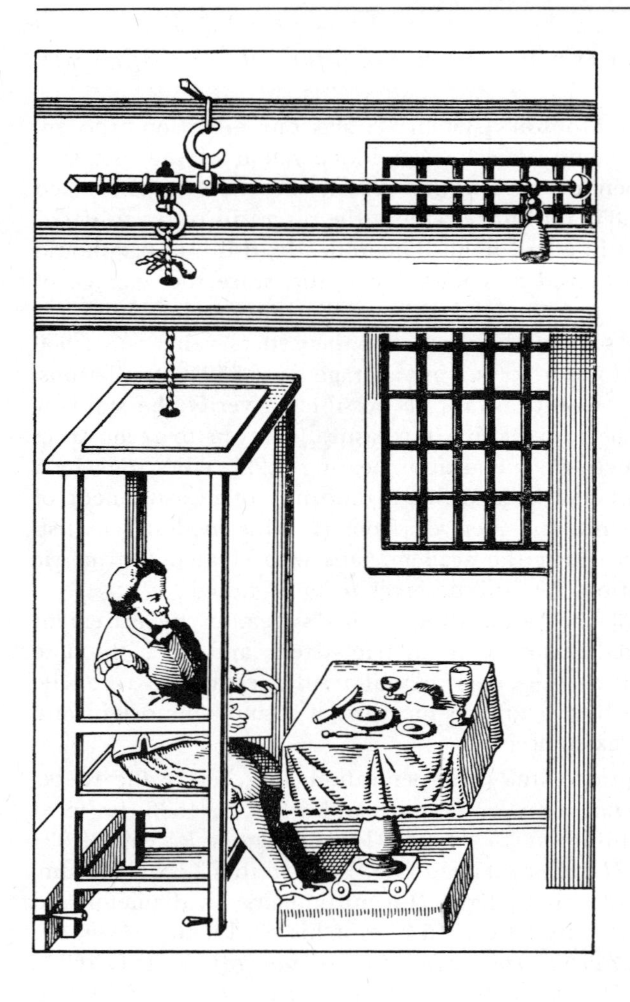

FIG. 103 Sanctorius' use of a "Roman" scale for medical purposes (from the *Commentaria . . .*, 1625).

However, the scientist did not hesitate to express his reservations about the possible variations of the pendulum. (We may ask whether Abbot Mouton did not also have his doubts.) He believed that if experience confirmed them, the idea of a universal unit derived from the pendulum would have to be given up, which would not prevent the existence of a "perpetual and invariable" measure in every place.

Variations of the pendulum (Richer)

Two years later, doubt was raised on this question, which was preoccupying the men of science, by the academician Jean Richer (1630–1695), who made a scientific mission to Cayenne in 1672, chiefly for the purpose of carrying out various astronomical observations. Richer notes that his seconds pendulum, hall-marked in Paris by comparing it with the pendulum clocks of the observatory, lost 2 minutes 28 seconds per day in Cayenne. In order to bring it back to its original period of oscillation, the rod had to be shorter by 1.25 lignes (approximately 2.8 millimeters) at Cayenne than in Paris.

This experimental observation annihilated all hope of discovering a universal unit in nature. It was a long time before Picard would admit the existence of these local variations of the pendulum. But while the search for a unit appeared compromised, these experiments with the pendulum contributed elements of information that were of great interest for the geophysicists.

Geodetic measures

Moreover, scientists and governments were becoming increasingly interested in the shape and consequently the measurement of the earth, in topographical factors, and in the creation of geographical charts indispensable to government and navigation.

France, which was not a leader in this activity, did do considerable work beginning in the seventeenth century, and succeeded in acquiring a leading role in the field of geodetics. One of the first French achievements was that of Jean Fernel (1497–1558), who in 1525 measured the meridian arc linking Paris with Amiens. To attain his goal, Fernel counted the number of revolutions made by one of the wheels of a carriage traveling over the route. At each revolution, a device caused a small bell inside the carriage to tinkle. Taking into account the distance covered with each revolution he concluded that the arc degree in question was equal to 56,746 fathoms 4 feet (Parisian measures). At this period the fathom measured approximately 1.96 millimeters, and thus the meridian degree was equivalent to 111.168 kilometers.

To be sure, Fernel's procedure seems to us archaic, and renders the results of his measurements somewhat uncertain. But at that time there existed no method scientifically applicable to geodetic measurements.

Snellius and the method of triangulations

The credit for creating a method that offered great possibilities for measuring a meridian arc belongs to a Dutch mathematician and professor at Leyden, Willebrord Snellius (1591–1626). His method is based on the close collaboration of the disciplines of geometry and trignometry, an original idea that in 1615 permitted him to set up a series of triangles along the meridian arc between Alkmaar, near Leyden, and Bergen op Zoom, starting from a base of known length. Thanks to this method it was possible to calculate the length of the arc in question, as follows:

Given, on a meridian arc, points *AZ* between which the distance in a straight line is to be determined, and points *m, b, c,* and so on, along this straight line. To the right and left of this line *AZ,* we select points *B, C, D,* which are raised, so that from any one of them it is possible to see the surrounding points. It now becomes possible to measure the angles of the visual triangles, *ABC, BCD,* thus formed. This group of triangles forms what is called a triangulation, or geodetic network (Figure 104).

Suppose that we measure:

1. The angles of triangles *ABC, BCD,* and so on, or simply two angles of each triangle;
2. The angle that any one of the sides — for example, *AB* — forms with *AZ;*
3. The length of any side of one of these triangles. The side thus measured is is called a base. Let *BC* be the base measured; it is then possible to calculate

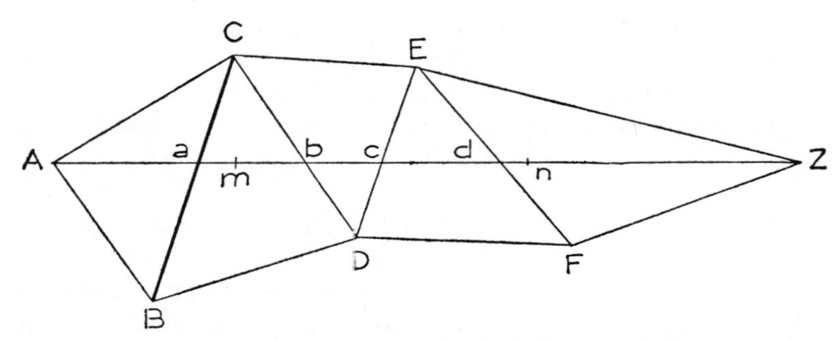

FIG. 104 Diagram of the principle of geodetic measurement by triangulation.

one by one all the sides of the triangles *ABC, BCD, CDE,* and so on. Then, in *ABa,* in which we know *AB, ABa, BAa,* we calculate *Aa, Ba,* and *AaB.* Continuing this procedure, we find $Aa + ab + bc$ . . . ; thus $Aa + ab + bc + cd + dz = AZ$.

It should be noted, however, that the point of departure of the chain is the base *BC.* For this reason the measuring of this base should be done with the greatest precision, to avoid successive repercussions of an initial error that would risk compromising the entire operation. Verification can and should be done by reconciling one or several secondary bases with the chain of triangles.

In addition, various azimuths must be measured. (The azimuth is the angle formed by any side of a triangle with the meridian line *AZ'.*) The usual practice is to measure several azimuths in order to verify the operation.

Lastly, the latitudes at the ends of the meridian arc must be measured, and the difference determined. By comparing the difference with the length of the arc measured, the length of a degree of the meridian in question can easily be calculated.

To summarize: basically a geodetic measurement includes four operations: measurement of one or several bases, construction of a chain of triangles from one base, measurement of the latitudes at the end of the arc to be measured, and determination of the aximuths.

Snellius had deduced from his triangulation, which was composed of 33 triangles, that the degree of the arc measured was equal to 28,500 Rhinelander rods. This rod, which Snellius divided into 100 parts, consisted of 12 Rhinelander feet, and measured 3.766 millimeters. Expressed in sixteenth-century Parisian fathoms (approximately 1.96 millimeters), these 28,500 rods were equal to 54,760.7 fathoms; expressed in Parisian fathoms after the reform of 1668, they equaled 55,026 fathoms (the fathom measuring 1.949 millimeters). It was concluded from the work of Snellius that the arc degree was 55,021 Parisian fathoms, an evaluation that is thus posterior to the reform of the fathom in 1667–1668.

Jean Picard's geodetical measurement

Around 1670 the astronomer and abbot Jean Picard undertook to apply the method of Snellius to a new measurement of the arc between Malvoisine in the Gâtinais (to the south of Paris) and Sourdon, near Amiens in Picardy.

After Auzout's creation in 1666 of the micrometer with movable wire, Picard adapted a reticular eyepiece to his quarter circle, whose radius was 38 inches. The limb, he said, is divided into minutes by the transversals, and it is possible to distinguish "the quarter-minute with a magnifying glass." In addition, he notes that thanks to his divided instrument the measuring of the angles was done with such "correctness" that "on the circumference of the horizon taken in five or six angles, only about one minute more or less than the necessary was found, and that very often it came within five seconds of the correct figure." This was perhaps an excessively optimistic evaluation, but this in no way detracts from the value and intellectual honesty of the scientist.

Picard measured two bases. The first extended from the mill of Villejuif to the pavilion of Juvisy, and measured 5,663 fathoms (11,042 millimeters); the second base, established near Sourdon, measured 3,902 fathoms (7,609 millimeters) and was to serve for the verification of the precision of the calculations. A chain of 13 triangles was formed. The distance included between the ends of the meridian arc proved to be 78,850 fathoms. The difference between the latitudes being 1° 22′ 5″, the arc degree was equivalent to 57,060 fathoms (111.267 millimeters).

Around 1683 the continuation of Picard's arc, on the one hand as far as Dunkerque, on the other to Collioure near Perpignan, was entrusted in particular to the academicians Jean-Dominique Cassini (1625–1712) and Philippe de La Hire (1640–1718). This project was not completed until around 1718. The "mean" value of the degree of the meridian arc was concluded to be 57,061 fathoms.

Thanks to these and later measurements, the problem raised by the "form of the earth," on which subject the scientists had long been divided, was solved.

From geodetic measurements to the creation of scientific metrology

Geodetic measurements are actually no more than practical applications of dimensional metrology (angles and distances). The French scientists excelled in these projects, which brought about an increased "technification" in the construction of instruments of measurement and observation.

The conception of a universal unit was not relegated to the attic of antiquities, however. In his work *De la grandeur et de la figure de la terre* (On the size and shape of the earth) (1718), J. Cassini partly adopted Mouton's project. He suggested a base unit equal to the 1/6000th part of the minute of the degree, that is, a geometrical foot of .3086 millimeter, instead of the fathom of 1.852 millimeters envisaged by Mouton. Ultimately, however, it was La Condamine who reintroduced the idea of the universal unit, whose progress was to continue until the creation of the decimal metric system in 1790.

This rapid sketch of certain aspects of metrology from the sixteenth century to the beginning of the eighteenth century shows the static nature of this science and the techniques it comprised. Methods of production, which were bound by centuries-old traditions, remained the same. The link between scientists and constructors of instruments seems very weak, if indeed it ever existed; and their separation was at the origin of this stagnation. As Maurice Daumas has shown, not until the eighteenth century do we witness the gradual disappearance of the tight compartmentalization created by social and ancestral prejudices between the French scientists and technicians.

It was undoubtedly due to the creation of experimental science by the seventeenth-century "pioneers," and the absolute necessity of technical accomplishments for its further development, that the builders of scientific instruments began to develop more freely.

Persistence of the notion of unity of universal measurement

A scientific metrology was thus created by the scientists who were conscious of its role in their research and of the need to increase the precision of the results of measurement. Then, too, they had to define new units of measurement hitherto unknown: measurement of temperature, pressure, hygrometrical degree, and so on. However, in order to express these measurements they could do no more than borrow the inconsistent language of current metrology, with all its attendant difficulties of exchange of information, on both the national and international levels.

Under these conditions it is easy to explain why the investigators of the seventeenth and eighteenth centuries, thanks to their discoveries, clung to the possibility of finding in nature a universal unit of measure and of creating a rational system of measurement, equally useful to scientists and technicians, business and factory owners — and, in fact, to people in general.

These broad, revolutionary views gradually gained ground in successive, converging approaches. But not until unification of measurement was achieved was metrology able to become the effective basic instrument of the scientific and technical evolution.

BIBLIOGRAPHY

BERRYMAN, A. E., *Historical Metrology* (London, 1953).

BLANCARD, L., *La pile de Charlemagne* (Paris, 1887).

BOOKER, P. J., *A History of Engineering Drawing* (London, 1963).

DAUMAS, MAURICE, *Les Instruments scientifiques aux XVIIe et XVIIIe siècles* (Paris, 1953).

A Diderot Pictorial Encyclopedia of Trades and Industry, ed. by Charles C. Gillispie (2 vols., New York, 1959).

GUNTHER, R. T., *Early Science in Oxford,* vols. I and II (Oxford, 1923).

KIELY, EDMUND R., *Surveying Instruments* (New York, 1947).

KISCH, BRUNO, *Scales and Weights: A Historical Outline* (New Haven, 1965).

MACHABEY, A., *Memoire sur l'histoire de la balance et de la balancerie* (Paris, 1949).

MADDISON, FRANCIS, "Early Astronomical and Mathematical Instruments: A Brief Survey of Sources and Modern Studies," *History of Science,* 2 (1963), 17–50.

NICHOLSON, EDWARD, *Men and Measures: A History of Weights* (London, 1912).

RICHESON, A. W., *English Land Measuring to 1800: Instruments and Practices* (Cambridge, Mass., 1966).

Section Three

Land and Water Transportation

CHAPTER 15

LAND TRANSPORTATION

CARRIAGES AND HARNESS

THE SIXTEENTH century was a period of important progress in the development of carriages.

Wheels and axle In the fifteenth century, crude wooden circles (see discussion of cannons, Section Five, Chapter 19) or wheels consisting of two sets of wooden crossbars crossed at a 90-degree angle (Figure 105) were still in use (these types can still be seen in certain rural districts). But spoked wheels came into universal use in the sixteenth century; improvements were made in their construction, and they were decorated with turned baluster spokes.

In antiquity the wheels had been attached to the axle, which turned with them; the axle had been supported on pieces of wood fixed under the chariot to form platforms. In the sixteenth century this system, which has been retained in modern times in railroad rolling stock, began to disappear from carriages, and wheels turning freely on a fixed axle bound to the frame of the carriage came into use.

The functioning of the wheels was improved by "dishing" them, the adoption of which practice dates from Galiot (end of the fifteenth century), the great artillery master (see Section Five, Chapter 19). It consists of inclining the wheels toward the axle of their rotation (Figure 106a). When the spokes are perpendicular to this axle, a shock or strain (represented by *R*) on the felly causes a sideways thrust (*f*) that is dangerous for the spokes (Figure 106a). The inclination of the spokes by dishing gives better resistance to shocks and sideways thrusts; in addition, tightening of the felly is facilitated by the elasticity of the oblique strokes. When the axle of the wheel (*a*) is inclined (Figure 106b), the thrust from the road is exercised directly on the spokes and the hub, without a sideways thrust that would be dangerous for the spokes.

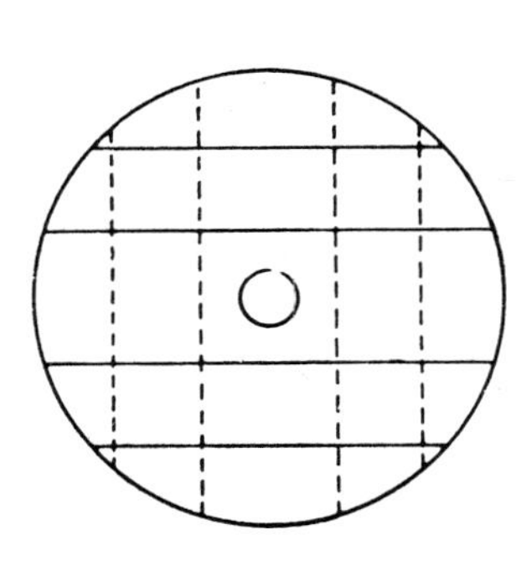

FIG. 105 Solid wheel formed of two sets of crossbars. Fifteenth century.

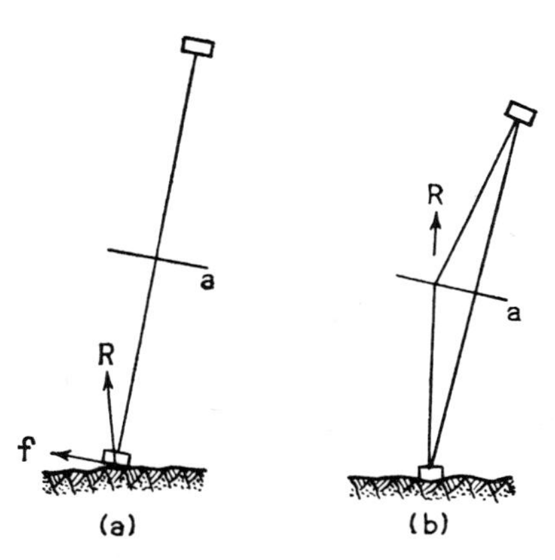

FIG. 106 Dishing (inclination of the wheels on the axle of rotation).

The inclination of the axle was of course more easily done with a metal axle. These improvements were introduced gradually; the number of spokes varied from six to fourteen, depending on the diameter of the wheels and the size of the vehicles. The fellies of the wheels were made of pieces assembled by beveling or scarf joints; they were bound to each other and to the spokes by metal braces, and the rim was banded with iron. All these pieces of metal were held by nails or screws.

Not until much later, in the nineteenth century, was the heat banding of a single piece of metal adopted; the wooden rim was tightened by the cooling and contraction of the metal band.

The box and the suspension

By the Middle Ages the primitive wagon had been replaced by the "traveling carriage," whose box was "suspended" in order to avoid jolts. (In 1610, however, the "coach" in which King Henri IV was murdered was still placed directly on its four wheels.) Not until the fifteenth century do references to these stagecoaches appear. One of the oldest coaches may be that of Isabelle of Bavaria, in which she is supposed to have entered Paris in 1433; Frederick III entered Frankfort by the same method in 1474. During the reign of François I there were only three such coaches, and their use was reserved for ladies; a seigneur of Laval may have been the first man to make use of one, and then only because his obesity prevented him from riding a horse.

When carriages first came into use, royal permission was necessary in order to own a carriage, but rivalry made every member of the court anxious to have its own. The number of carriages in Paris increased to such an extent that Charles IX, Henri IV, Louis XIII, and Louis XIV issued ordinances limiting their number and sumptuousness — a useless step, however, for in a famous satire Boileau describes the Parisian traffic jam:

> "Twenty carriages soon arriving in line
> In no time at all are followed by more than a thousand."

In 1598 François de Bassompierre (1579–1646) introduced the windowed car-

Plate 28. *a.* Sixteenth-century carriage known as the "swaying" carriage.
b. King Henri IV's coach.
c. Carriage from the time of Louis XIII.
d. Carriage from the time of Louis XIV.
Musée national de la Voiture. Palais de Compiègne. *Photo Editions d'Art A.P.*

riage, which was already known in Italy, to the court, but even in the time of Louis XIII the carriage was still closed only with leather curtains. In the course of the seventeenth century carriages closed with windows became prodigiously luxurious, rivaling each other in decoration, sculpture, painting, and gilding. As for the suspension of the box, it was slightly improved by the use of leather straps; at the same period light carriages known as *"sédioles"* utilized the elasticity of the shafts in order to improve the suspension. A Dutchman named Boonen is believed to have invented steel springs, which gave a much better, genuinely elastic suspension.

Public carriages

The public carriage for hire appeared at the beginning of the seventeenth century in the form of the sedan chair (1617); then carriages and coaches for hire were put into service in Paris by a company whose office was in the Hôtel Saint-Fiacre — whence the name given to such carriages. Next came *chaises roulantes* (also called *brouettes*), sedan chairs mounted on wheels and drawn by a single horse. All these vehicles were licensed. In 1660 the only city transportation available to the wealthy was the private carriage; other people had to use fiacres or sedan chairs. Some people preferred the horse.

In 1661 Pascal and a few friends (including the Duke of Roannez) organized a company for the purpose of creating a public transport system, "public carriages after the fashion of the coaches in the country" but with fixed routes and a schedule. The Council gave a favorable response (January 19, 1662), but the Parlement accepted the request only on condition that "soldiers, pages, lackeys and other servants, and laborers, will not be able to enter the said carriages."

The licensees ordered the construction of carriages suitable for this service, and tried them out on February 26, 1662 (letter of the Marquis de Trenan, one of the licensees, to Arnauld de Pomponne, one of the stockholders). Each carriage, which seated eight, was marked with the city's coat of arms and information regarding its route. The box was suspended by thick leather straps (*"soupentes"*) from four pieces of wood (*"moutons"*) projecting from the axles (Figure 107). The trial runs, carried on for two days in succession with rented horses, were satisfactory: four routes in the morning from six to eleven, four in the afternoon, beginning at two and ending at six, with the same pair of horses, which traveled very slowly.

There were five routes. Five days after the inauguration of the first route (March 1662) it was found that seven carriages per route were insufficient, since they could accommodate only half of the would-be passengers. The details of this first carriage service — various incidents, protests by the lackeys and laborers who were the victims of discrimination, the difficulty of paying the 5-sous fare (later raised to six) in small change, and so on — are of no interest for the history of technology. In 1679, seventeen years after its creation, the service was still in existence.

These carriages appear to have later reverted to the status of carriages for hire, in Paris and the rural areas, since the license granted to the Duke of Roannez and his friends was repurchased in 1681 by the beneficiaries of the hackney-carriage license.

Harness

In an earlier chapter we described the modifications in the horse harness which permitted better utilization of the animal's tractive power. The power that could be supplied by a horse with a collar around its neck was approximately the same as that furnished by a donkey:

14 kilograms at a speed of .80 meters/second = 11.2 kilograms-meters/second

The shoulder collar, in contrast, made possible:

45 kg. at a speed of .90 m/s = 40.5 kgm/s

Thus the modern system of harness made it possible to obtain a power multiplied by 3.6.

Major transportation in the rural districts and on large construction sites, however, was done especially with oxen, who could put forth an effort of 60 kg. at .6 m/s, or 36 kgm/s, that is, 9/10 of the power supplied by a horse wearing a shoulder collar. Even taking into consideration the increase in power made possible by the shoulder collar, during this period the horse continued to be used only for light wagons and, later, carriages, while the ox continued to be the customary means of heavy transportation. The horse was, however, also used as a pack animal, which practice was, as we shall see, an important factor in the use of the roads, until the time when regular transportation services by horse-drawn carriage could be organized.

The system of the shoulder collar was utilized, but for harnessing animals to carriages and coaches the principal method was the breast strap (Figure 108). In addition to the headstall required for guiding the animal, the harness included a leather strap passing around the belly of the animal and forming a breast strap on the chest, on which the power was exerted, and breeching on the rump, which made it possible to hold back the weight when descending. This harness was adapted both for the single horse harnessed in the shafts and for two animals harnessed on each side of the shaft.

Carriage teams consisted of two horses harnessed to the shaft, both in the

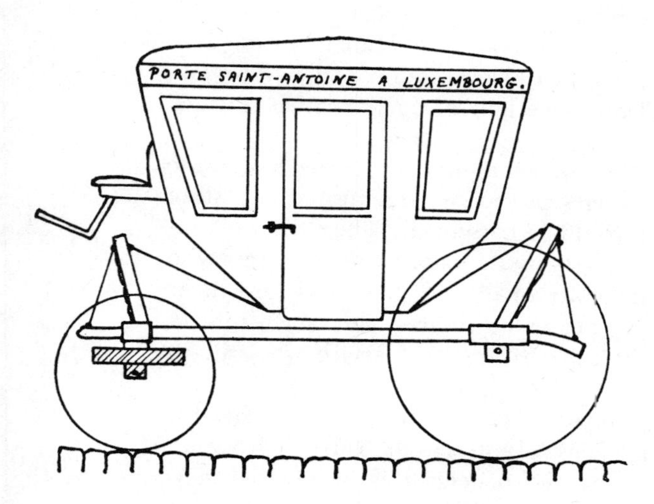

FIG. 107 The five-sous carriage.

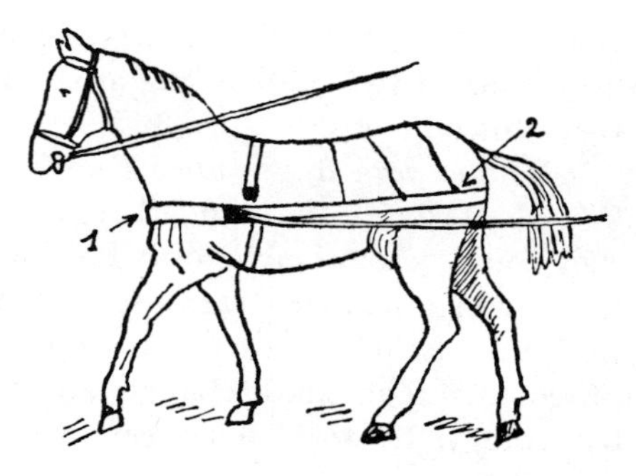

FIG. 108 Harness with breast strap (1) and breeching (2).

city and for the five-sous carriages. For longer journeys in which trotting was utilized in order to save time, the effort demanded of the horses was less (25 kg. at 2 m/s); a pair of horses was harnessed in front, and they combined their efforts on the traces linked with those of the horses in the rear. The driver, seated in front of the carriage box, guided the horses by means of the reins he held in his hand.

For heavy carts the collar was used for the teams, and additional horses — as many as eight, either in pairs or in single file — were used. When they were harnessed in single file, the last horse was in the shafts of the carriage. But in this case it was out of the question to use reins, which would have been impossibly long: the driver walked along beside the team.

THE HIGHWAYS

As Rome had extended her domination, she had created highways that enabled her to communicate with the various parts of her empire.

The Roman roads

The oldest of these highways was the Via Appia, which had already been in existence for 312 years when Christ was born; it linked Rome with Capua. In the eighteenth century it was the object of the admiration of Charles de Brosses, one of the collaborators on the *Encyclopédie*. The Via Aurelia (242 B.C.) linked Rome with Genoa; the Via Flaminia followed the coast of the Adriatic to Rimini. By the time of the Punic Wars, seven main highways radiated from Rome toward the south, east, northwest, and northeast.

In the course of their conquests the Romans developed their highways as methods of penetration; 312 main roads linked Italy with Aquitania, the Narbonnais, Provence, Germany, Spain, Epirus, and Macedonia, constituting a network of 46,200 miles.

From Lyon, the capital of the Lugdunese tribe, a road network of 2,800 miles radiated out to every region of Gaul. In some sections the old roads used by the Gauls must have been used in the establishment of the Roman roads; elsewhere, in the form of tracks, they must have served for internal traffic, which had long been in existence.

The ancient "tin route" up the Seine and the Yonne must have been linked with the Saône by a road that made possible, by means of portage between the rivers, the linking of the English Channel with the Mediterranean. Another route permitted portage between the upper Saône and the Moselle. At that time the journey from the estuary of the Seine to Marseille required thirty days.

The Romans constructed these roads with a great deal of care, strength, and even sumptuousness; the care given to their construction depended on the importance of the highway and the nature of the subsoil. A. de Caumont tells us that their construction (Figure 109) included the *statumen* (the base, formed of flat stones, sometimes cemented together) topped with the *ruderatio* (a layer of crushed stones); next came the *nucleus,* a kind of concrete made of sand, lime, and pieces of broken tile; and lastly the surface layer, the *summa crusta,* formed either of stones with a layer of sand or plaster, debris, or sometimes of paving

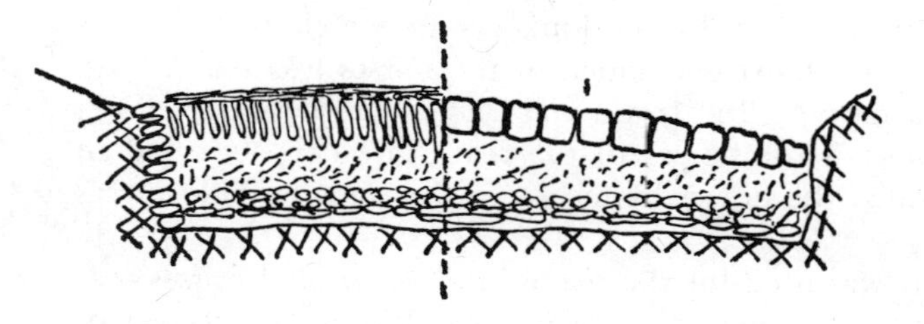

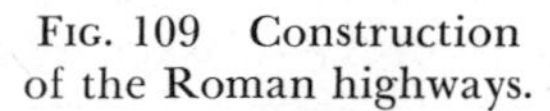

FIG. 109 Construction of the Roman highways.

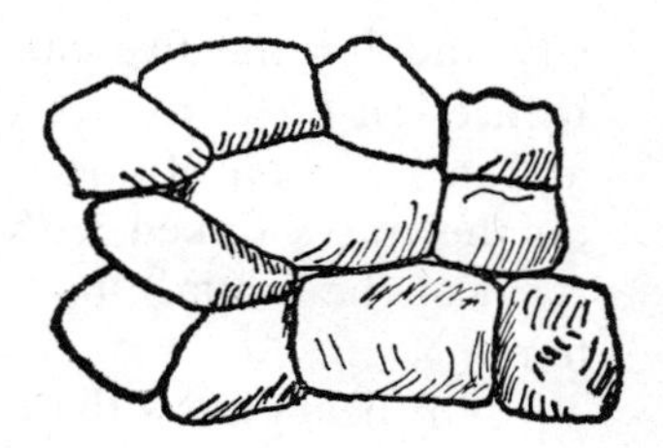

FIG. 110 Paving stones for road.

stones; the latter might consist of well-fitted, small squares or polygonal stones in *opus incertum* (Figure 110). Paving was limited to areas of major traffic (major roads and highways like the Via Appia).

Bergier, who carried out archaeological excavations in the environs of Rheims, and in particular on a road into the swamps of the Vesle, lists the thicknesses of the layers he found:

1. A *statumen* of flat cemented stones (ten inches) set in limestone mortar cobbles or slabs (one inch);
2. A *ruderatio* of round cemented stones (eight inches);
3. A *nucleus* of puddled sand and cement (one foot);
4. A top layer *(summa crusta)* (six inches).

The total thickness of the road was three feet. At Mouzon, six miles from Rheims, the technique was similar, and produced a road three and a half feet thick.

Their disappearance The strength of the Roman roads permitted them to last for centuries; despite invasions and continual unrest, they survived until the time of Charlemagne, and permitted transportation by heavy wagons, as well as the faster journeys of light two-wheeled, one-horse carriages.

Eginhard's story (*Life of Charlemagne*) of the journeys of the Merovingian kings in their heavy, ox-driven wagons, which implied that as late as the eighth century the Roman roads were capable of carrying traffic, has often been repeated. But wear and tear ultimately conquered the Roman work. Between the ninth and twelfth centuries the use of wagons disappeared – according to the chroniclers, because of the bad condition of the roads.

Transportation and travel were then carried on only by horseback and muleback; teams of oxen continued to be used for the heavy transportation needed in the rural areas and for carrying stones to the construction sites, since they alone were capable of the tractive power needed on bad roads.

Mail and communications A network of roads for communication was also needed for the mail service. First the royal mail, later that of the monks, cities, and universities, began by employing foot messengers, but the increasing development of needs led to the creation of a special group, the *chevaucheurs* (riders).

In 1350 the court and had thirteen foot messengers, in 1380, only eight, and by 1421 only two. In 1239 there were twelve *chevaucheurs,* in 1383 at least fifty, not counting the *chevaucheurs* of the queen and the king's sons. In addition, heralds, *"gentilshommes,"* and secretaries had special missions; the great seigneurs also had their mounted messengers.

Louis XI is credited with the creation of a regular mail service, on June 19, 1464. This date has been contested, but in any event the postal service is definitely attested in 1487. This organization included a system of relays — a return to the well-conceived Roman organization.

By the Middle Ages the roads were therefore being utilized for communication, but after the death of Charlemagne the management of the highways fell into the hands of the seigneurs, who found them a particular source of profits by way of tolls levied for passage through their domains.

It was the great religious orders, and particularly Cluny, which undertook the responsibility of maintaining the roads. The pilgrimages contributed to the organization of a network of roads oriented toward the most famous places — Saint Jacques de Compostela (Figure 111), Saint Martin of Tours, and Jerusalem.

In the thirteenth century the royal power began to become interested in highways. Louis ordered the classification of the public highways, and forced the seigneurs to maintain the roads leading to the Holy Land.

The Hundred Years' War interrupted the work of maintaining the highways. When at the end of the fifteenth century Louis XI organized the mail system by creating relays, nine major highways were covered by the king's messengers: Paris to Irun via Blois, Bordeaux, and Bayonne; Blois to Nantes via Tours and

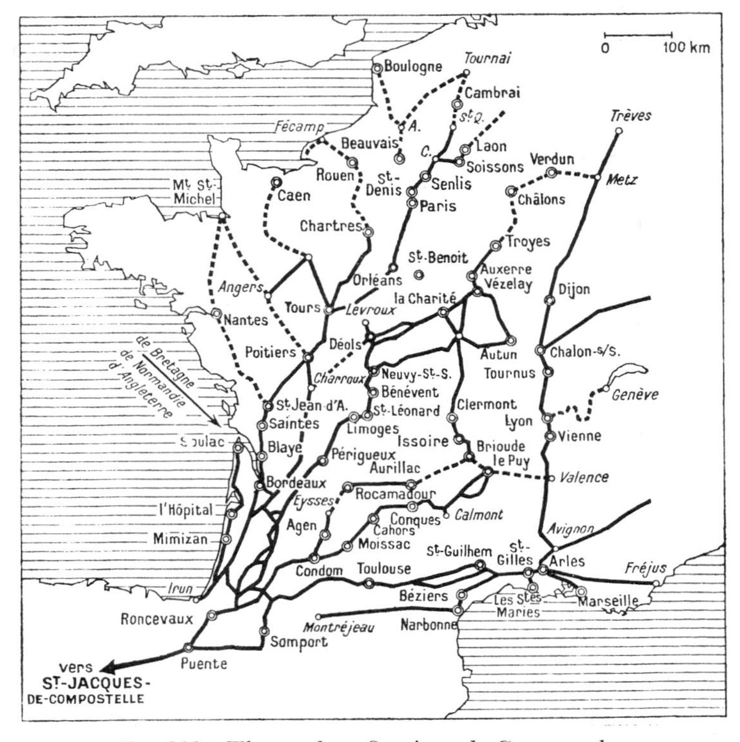

FIG. 111 The roads to Santiago de Compostela.

Angers; Paris to Calais; Paris to Péronne (later to Brussels); Paris to Metz via Château-Thierry, Châlons-sur-Marne, and Toul; Paris to Lyon via the Bourbonnais; Lyon to Saluces; Lyon to Marseille via the Rhône Valley; and Bagnols to Toulouse via Nîmes, Narbonne, and Carcassonne. Maintenance was provided for by tolls at the entrances to cities, bridges, and ferries.

Improvement of the highways

On October 15, 1508, Louis XII confirmed these provisions, which were complemented by edicts and ordinances of François I (1536), Henri II (1552), and Henri III (1579). Until the end of the sixteenth century, however, the French highways remained simple tracks with neither foundations nor strong surfacing.

At this point there appeared carriages with mobile axles and springs, and coaches, the use of the latter being imported from Italy. These developments brought four-wheeled wagons to the roads. Heavy but mobile cannons required stronger and wider roadways and bridges. Roads were no longer used exclusively by pedestrians, horsemen, and a few rustic wagons; they became an important means of communication, competing with the waterways.

In order to get the roads into a condition in which they would be able to satisfy these needs, Henri IV created the post of "Grand Voyer de France," and appointed Sully as its first officeholder. Sully's edict of January 13, 1507, set forth his powers and those of his subordinates, and provided for annual programs of roadwork. However, the extensiveness of the work of repair and construction of the French highways did not permit rapid completion; the reigns of Henri IV and Louis XIII did not suffice for the implementation of the program.

Colbert, who realized the advantages of communication for commerce and relations between Paris, the provinces, and foreign countries, had to reform old practices and attitudes. He supplemented the provincial administrators with commissioners of bridges and roads (1669). He ordered a general investigation, and drew up a very ambitious highway program for the Paris region, the eastern frontier network, and the Cévennes Mountains, the latter for political and military purposes (this region was the last stronghold of Protestant resistance). In the Paris region, communications between Versailles, Marly, and Saint-Germain required new, paved roads on which carriages and military convoys could easily travel. By the time of Colbert's death in 1683, the program was practically completed.

However, while the paved highways still in existence are well laid out and durable, the technique used for ordinary roads was still very mediocre. Paving was not done methodically, and many roads had only a sanded roadway. The picture drawn by La Fontaine in his fable "Le coche et la mouche" is typical:

> "Dans un chemin montant, sablonneux, malaisé,
> Et de tous les côtés au soleil exposé,
> Six forts chevaux tiroient un coche.
> Femmes, enfants, vieillards, tout était descendu,
> L'attelage suait, soufflait, était rendu."

(On rising, sandy, difficult road/ Exposed on all sides to the sun,/ Six sturdy horses were pulling a coach./ Women, children, and old men had all descended;/ The team was sweating, panting, and exhausted.)

This vivid picture can be translated into statistics, as follows:

With speeds of between 2½ and 9 miles per hour (trotting), the carriage traveling over a good pavement required a tractive power of only 44 pounds per ton of weight; on a good road surface of crushed stone, 55 to 66 pounds, on a mediocre road of this type 88 to 110 pounds; on sand or earth, 176 to 220 pounds (figures taken from Clément Colson).

Thus, if a carriage going at a trot to Saint-Germain, on good paving, needed only four horses (total weight two tons), six horses were required on sand or ordinary ground, at a walking pace, and even then it was very difficult, especially if a slope increased the tractive effort. In the case of a road of claylike soil soaked into a mud, matters were still more difficult; mired carriages were numerous, and could be got out only by the use of additional horses.

We shall see that in the eighteenth century only an improvement in highway construction techniques could transform the rapidity and safety of communication by horse-drawn vehicles.

BRIDGES

The technique of bridge construction evolved little between the Middle Ages and the end of the seventeenth century, and we may safely say that it was simply a continuation of the Roman techniques. The Romans had employed the round arch almost exclusively, and it continued to be used frequently in later ages.

Heritage of the Romans

The spans of the Roman arches are not very large, whether we consider the bridge of Alcántara (Toledo), restored without major changes by the Arabs and again in 1258; or the most beautiful of the ancient bridges, the Ponte Sant Angelo (the Roman Pons Aelius), constructed in 136 to provide access to the mausoleum of Hadrian (known to the Romans as P. Aelius Adrianus, whence the bridge's Roman name), and finally the famous Pont du Gard (Figure 112).

When the two banks to be linked were too far apart, the round arch necessitated a ridged roadbed (for example, the Arta bridge at Epirus). This ancient arch still has the original, quite crude springing stones; more evenly cut springing stones, probably dating from the Middle Ages, remain from its later reconstructions. But the necessity of solid foundations brought about the construction of an arch with a wide opening, whence the ridged arrangement of the road it supports.

When they wished (as in the case of the Ponte Sant Angelo) to maintain the access route at a rather low, almost horizontal level, the Romans did not abandon the round arch, but were obliged to reduce its opening and construct several arches, that is, to lay intermediate piers, despite the difficulties imposed by the very irregular flow of the river (117 cubic yards at low water, 5,232 cubic yards at high water). At this period no builder dared to construct a depressed arch, which would have been beyond the technique in use at that time.

In the Middle Ages, and until the seventeenth century, when circumstances permitted, the bridgebuilders continued to be guided by the ancient customs, and the round arch was retained. The bridge of Céret (1321–1336) has an opening of 144 feet; that of Vieille Brioude (1340), 177 feet; Nyons (1351–1407), 161 feet. The same technique was retained in the seventeenth century; the bridge of Claix over the Drac (1606–1611), with an opening of 152 feet, made a ridged roadbed a necessity.

All these structures have an arch of varying thickness, which therefore appears to have been neither calculated nor determined by uniform rules. The arch of Céret has two separate rings; one is 3 feet thick from one abutment to the other, while the upper ring is 1⅔ feet at the haunches, and decreases to 1 foot at the keystone. The thickness of the ring thus constructed varies from a total of 4¾ feet at the haunches to 4¼ feet at the keystone.

The construction of a single-span arch required a temporary wooden centering, about which we have no information. The arrangement of the bridge of Tournon (fourteenth century) reveals the construction procedures used (Figure 113); the arch ring is single at the haunches, but double along a certain portion, starting from the keystone.

From this we can deduce that the individual voussoirs of the springings of the arch were put in position, then locked in place by a masonry filling up to a certain height. At the starting point of the double ring, the upper ring, which gave greater rigidity to the arch, was laid on the centering. The arch thus became thicker, and its span less, reaching a total thickness of 5½ feet. It must have been noticed that warping or difficulties of discentering occurred in single rings constructed on centerings in a single arch. Moreover, the compression of the rings under the weight of the filling brought about distortions and a drop in the keystone, estimated by Séjourné at almost 17 inches for the bridge of Céret.

However, by the end of the Middle Ages depressed arches had been adopted, in order to ensure a better passage of the water, and also in order to eliminate

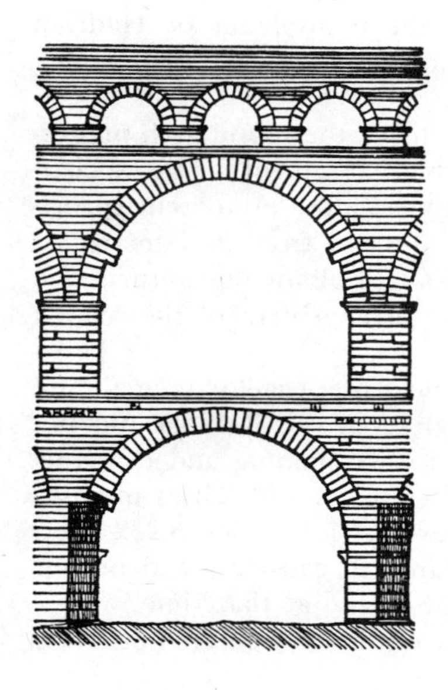

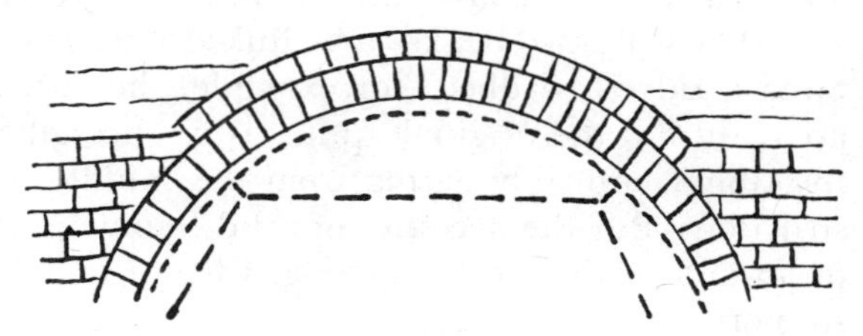

FIG. 113 Single-span bridge at Tournon. Fourteenth century.

← FIG. 112 Arch of the Pont du Gard.

the ridged roadbed. Typical examples are the picturesque Ponte Vecchio at Florence, constructed in 1345 by Neri di Fioravante, with overhanging arches to support shops, and the Ponte Scaligero in Verona, constructed in 1354 to give access to the fortress (1355).

Until the nineteenth century the bridge of Trezzo over the Adda was the largest masonry bridge in existence; like the bridge of Verona, it was constructed to provide access to a fortress. A springing of its arch is still in existence, and permits us to reconstruct the bridge. Built between 1370 and 1377, and destroyed in 1416, it was a circular arch with a span of 236 feet; it was almost 69 feet high at the top of the arch (Figure 114). These factors correspond to an angle of 120 degrees, the center of the arch being 69 feet below the opening, the radius being 137¾ feet. The springing of the arch, which is still in place, indicates a vault 6¾ feet thick, covered with flat cut stones that cannot be regarded as springing stones of a ring; their role might have been decorative, but at the same time they could have distributed the strains of the filling of the tympanum onto the springing stones of the haunches (Figure 115).

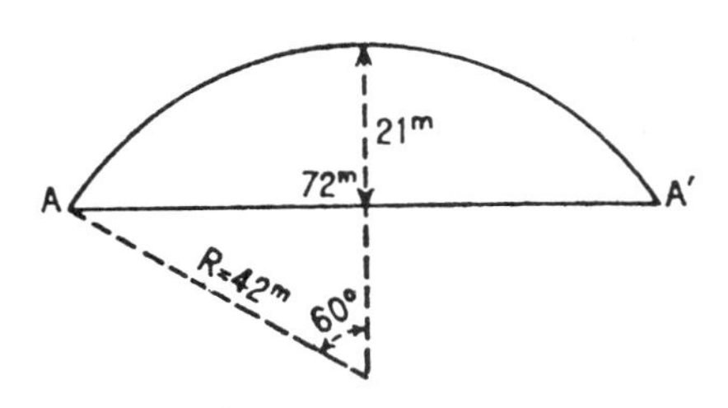

Fig. 114 Arch of the bridge of Trezzo. End of the fourteenth century.

Fig. 115 Trezzo, springing of the arch.

Bridges with several arches

The crossing of a large valley, in which the main riverbed was filled with alluvions, required a bridge of great length. Examples could be seen on the Loire, the Seine, and even in cities (for example Rome [the Tiber at the Aelius bridge], Paris, and some of the cities in the Loire Valley). A single-span arch was not possible, and so the builders were led to construct bridges with multiple arches; in most cases the latter replaced wooden bridges.

When there were several arches, it was possible to lower the level of the road, but pilings had to be built, and consequently their foundations had to be established on alluvions. The difficulty of constructing a solid foundation resulted from the inadequacy of the methods then in existence for pumping out water and working under dry conditions. There were several ways of pumping out water, but their effectiveness was limited, since the Archimedean screw permitted pumping only at shallow depths, with very low-powered "motors" (horse- or squirrel-cage-operated treadmills worked by one or two men).

The piers of the Aelius bridge have large footings, which indicates that the Roman builders intended to distribute the pressures over a large foundation area and to make this area as secure as possible.

An exploratory excavation made under a pier of the old Pont d'Orléans (Figure 116) shows how the foundation could be laid under these conditions. On the sandy bottom was laid a mass of unmasoned small stones, to a level of four feet below low-water level, which appears to have been the maximum possible with the pumping methods utilized. On this base was established the foundation of small masoned stones within a wall of pile planks that rose to a level of 2⅓ feet below low-water level. The actual piling was established on this base.

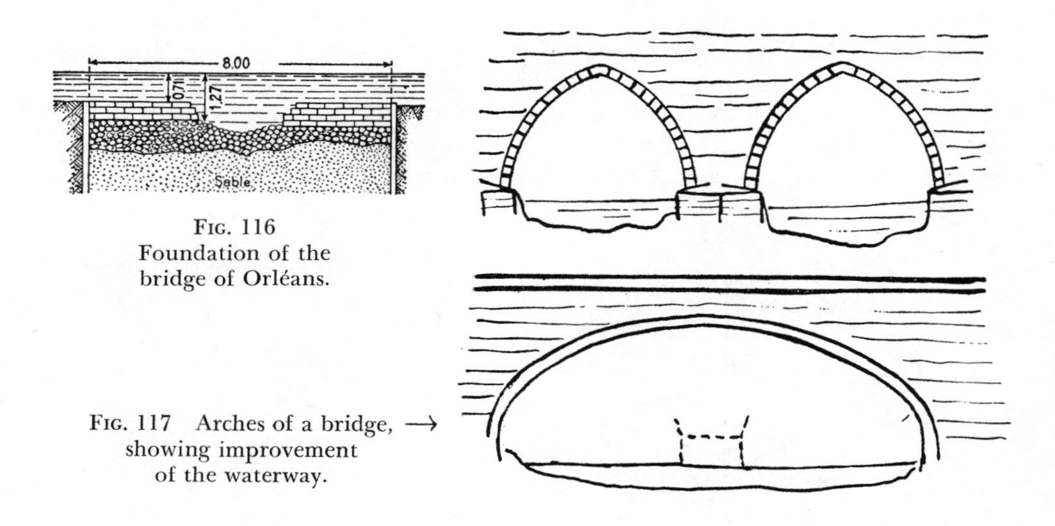

Fig. 116
Foundation of the
bridge of Orléans.

Fig. 117 Arches of a bridge, ⟶
showing improvement
of the waterway.

It is these foundations, which often dated from the eleventh century (Tours, 1037), that supported the piers of the Loire bridges for centuries, generally until the eighteenth century. The shape of the arches is practically the same almost everywhere (Figure 117); they are ogival, which has the effect of reducing the horizontal thrust on the foundation (see Section Six, Chapter 22, "Construction Techniques"). In addition, the piers are very thick, and can support the thrust of the vault alone, without the help of the contiguous arches.

This system of thick piers had the double advantage of permitting the arches to be constructed one by one, without any concern about finding a support for the outside of each arch; in addition, in case one arch broke, the neighboring arches, which held up by themselves, remained unbroken. This is what happened when floods or an accumulation of ice against the piers carried away one or several arches, or when one or two arches had to be broken for the defense of a passage (as at the siege of Orléans in 1429).

However, the thick piers considerably reduced the waterway available in case of flooding. Examples are the old bridge of Nantes, the thickness of whose piers was equal to the openings between the arches. Some of the piers of the Beaugency bridge were 26¼ feet thick, while the arches were 29½ to 36 feet in span; the overall relationship of solid to void was .75. In times of flood such an obstacle created a difference in level between upstream and downstream of two feet, causing stresses on the piers that succeeded in carrying them away. Ice breakups in 1434 and 1435 carried away several arches of the Pont des Tournelles at Orléans.

As bridges were replaced, the waterway was increased by replacing two arches with a single arch (Figure 117) of larger span. Maintaining the road at its previous level then required the adoption of the depressed-arch profile, and we see the appearance of elliptical or almost elliptical vaults, formed of arches of different radii, increasing from the springing up to the keystone (arches of 10 to 16½ feet).

The history of the bridges of London furnishes much the same information. Wooden bridges made it possible to cross the Thames near the site of the present London Bridge, perhaps as early as Roman times but certainly by the period of the Saxons and Normans. These bridges were frequently destroyed, either by floods or fires, until 1175, the date of Henry II's command to construct a stone bridge. This stone bridge was the work of one Peter, chaplain of St. Mary Colechurch, at Cheapside. A chapel dedicated to St. Thomas of Canterbury was built on the bridge, which was lined by a row of houses on each side, forming a street. A drawbridge took the place of an arch, permitting the passage of boats. The bridge retained its primitive appearance until 1761. In 1921, an arch of this bridge was discovered in the course of some work being done, and was transported to Wembley Park.

The builders

All the knowledge of the activity of the eleventh- and twelfth-century builders that has been retained is the name of the "Brothers of the Bridge," architects and builders of such projects as the Pont Saint-Esprit and the Pont d'Avignon; however, we have no precise details on their affiliations or methods. We can only suppose that by analogy with the projects executed by the religious orders they could have been connected with the latter. Later centuries do not seem to have continued the activity of such brotherhoods or construction services connected with the religious orders.

Later, names of individual builders began to appear, particularly in Italy, where remarkable projects of the fourteenth century (Verona, Trezzo) reveal the eminent predecessors of the fifteenth and sixteenth centuries. In contrast, France appears to have had no well-known bridgebuilders. Many wooden bridges had to be replaced, for example, the Pont Notre-Dame at Paris. The bridge of the "planks of Mibray" was rebuilt and named "Pont Notre-Dame" under Charles VI; the latter, accompanied by his son the Duke of Guyenne, the Duke of Berry, the Duke of Burgundy, and the Duke of La Tremoille, came to strike the symbolic blow on the first pile of the new structure. This bridge, which is supposed to have been partially destroyed by a fire, was carried away by a flood, together with its overhanging houses, in 1499.

A decision was then made to send for "master masonworkers" from Orléans, Tours, Amboise, Lyon, Amiens, Nantes, and other "cities where it is known that there are the best workers in masonry, and similarly in bridge-building"; it seems that builders were even brought from Blois and Auvergne. But they do not appear to have succeeded in establishing "the shape and form of the said bridge" being planned. Louis XII then sent to Italy for Fra Giocondo, whose fame in Italy must have been known to him.

Fra Giocondo (1435–1520), whom one of his contemporaries (the humanist Julio Cesar Scaliger) called "a profound mathematician, a wise physician, a prince of architects, a unique model of sanctity and of every type of erudition," had reconstructed the old Roman bridge of Verona with round arches; its type could easily be adapted to the construction of the Pont Notre-Dame. He seems to have constructed it carefully, starting soon after 1500 and continuing until 1513. He laid solid foundations on pilings, which may have been an innovation, and which continued to support the foundations of the bridge reconstructed in 1853. When the bridge was again reconstructed in the twentieth century, the old foundations finally disappeared.

In the sixteenth century we discover a number of Italian architect-engineers known by works that are still in existence. The very beautiful Trinità bridge at Florence, with its three elliptical (and therefore depressed) vaults, was constructed between 1558 and 1570 by Bartolomeo Ammannati (1511–1592), who also left numerous statues to posterity.

Mention should also be made of a bridge that is as interesting as it is famous: the Rialto bridge in Venice. It was built to replace an old, low wooden bridge, one section of which had a drawbridge to permit the passage of boats. The reconstruction project was the object of a competition open to the most famous artists of Italy. In 1513 Fra Giocondo had suggested a project that was not accepted; in 1587, twenty-four projects were examined, including one by Palladio and one that may have been suggested by Michelangelo. The conditions imposed included the possibility of permitting the passage of an armed galley under a solid stone bridge.

Palladio suggested a bridge of the "Roman" type, with three round vaults, which would have necessitated placing two piers in the canal. The project built (1588–1592) was that of Antonio da Ponte, with a single circular arch slightly less depressed than the bridge of Verona; the span was 88 feet, the rise of the arch 24½ feet. The bridge was set on several thousand piles; it is exceptionally wide (72½ feet), and is a genuine street lined with shops in arcades that give the bridge its characteristic, elegant appearance (Figure 118).

The Italian wars had brought the French technicians into contact with the works of the Italian engineers; many of the latter, as we have seen, came to work in France, and the French architects did not fail to to participate in the projects achieved by the Italians.

The Androuet du Cerceau family

In this connection, one of the first members of the famous Androuet du Cerceau family may have worked on the Château de Gaillon with Fra Giocondo in 1505, at the time when the latter was directing the work on the Pont Notre-Dame.

The family lineage of the du Cerceau family is particularly interesting, for it shows how from father to son to grandson, including the nephews, an entire family devoted itself successfully to architectural projects.

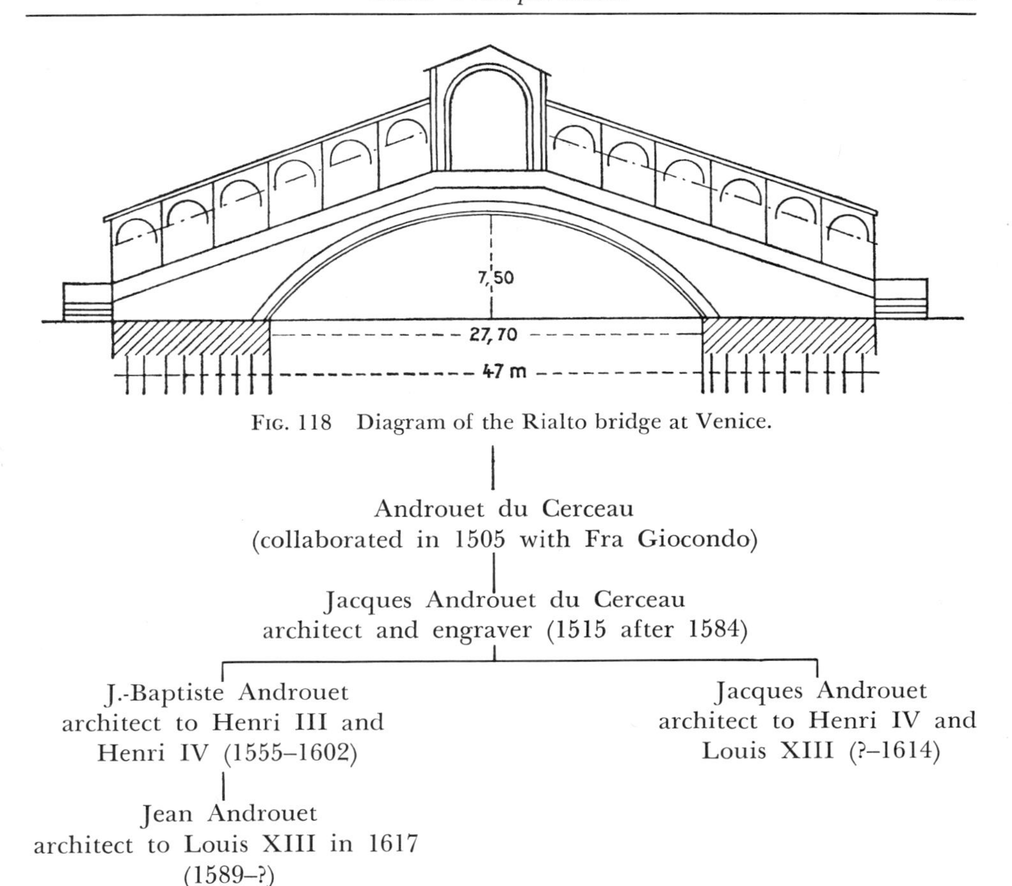

FIG. 118 Diagram of the Rialto bridge at Venice.

Androuet du Cerceau
(collaborated in 1505 with Fra Giocondo)

Jacques Androuet du Cerceau
architect and engraver (1515 after 1584)

J.-Baptiste Androuet
architect to Henri III and
Henri IV (1555–1602)

Jean Androuet
architect to Louis XIII in 1617
(1589–?)

Jacques Androuet
architect to Henri IV and
Louis XIII (?–1614)

The Pont-Neuf in Paris was constructed by J.-Baptiste Androuet du Cerceau on the downstream spur of the Île de la Cité, formed by the union of several small islands. Henri III came out in a boat to place the first stone on a pier raised to water level. The construction of the piers was pushed quite rapidly, to the point of the springing of the vaults, but the serious disturbances of the religious wars caused the abandonment of the spans, and not until the reign of Henri IV was the bridge completed. The king passed under the still-unfinished bridge on June 20, 1603; it was completed in 1604 by Guillaume Marchand, municipal architect, to whom we owe the beautiful cornice that crowns the bridge.

The Pont-Neuf was an innovation in Parisian bridges: for the first time there were no houses on a bridge, and the pedestrians had a view of the Seine. Another novelty was the sidewalks, still nonexistent in the streets of Paris.

Jean, the son of Baptiste Androuet, was in turn charged with the task of constructing the Pont au Change (1639–1642) with Denis Laud and Mathurin du Ry.

The other bridges in seventeenth-century Paris were medieval structures, such as the stone Petit-Pont constructed by Maurice de Sully in 1185 to replace the old wooden bridge. This bridge was destroyed at least ten times by floods or fires. Until 1718 it was lined with houses.

The wooden Pont Marie was reconstructed several times; a portion of the stone model, constructed between 1618 and 1635 fell in 1658. This bridge was the continuation of the Pont de la Tournelle, which continued to be built of wood from 1369 until 1656, when it was reconstructed in stone.

The Pont Saint-Louis, which opened the Île Saint-Louis to traffic, was then called the Pont Rouge, being constructed of wood and painted red (1627). Slightly downstream, in 1685 the Pont Royal replaced the wooden Pont Barbier (also called the Pont Rouge) (1632), which was burned in 1654, then carried away by floods in 1654 and 1684; the stone bridge was constructed between 1685 and 1689 by Gabriel and a Dominican named Roman, using the plans of Mansart.

The seventeenth century thus made a contribution to and a major effort in bridge construction, while numerous bridges of the twelfth and thirteenth centuries (Pontoise, Vernon, Mantes, Pont-de-l'Arche, Orléans, Tours, and others) and even wooden bridges continued to exist. The last of these – the Pont de Grenelle – did not disappear until 1875.

The sixteenth and seventeenth centuries witnessed the establishment of the depressed arch to lower the level of and straighten the ridge of the roadbeds, while at the same time increasing the waterway. But it was not until the eighteenth century that the technique of calculating the vaults and a more scientific layout of the masonry works were introduced.

BIBLIOGRAPHY

Boyer, Marjorie N., "Rebuilding the Bridge at Albi, 1408–1410," *Technology and Culture,* 7 (1966), 24–37.

Gregory, J. W., *The Story of the Road, from the Beginning down to the Present Day* (2nd ed., London, 1938).

Home, Gordon, *Old London Bridge* (London, 1931).

(Rose, Albert C.,) American Association of State Highway Officials, *Public Roads of the Past* (2 vols., Washington, D.C., 1952–53).

Sturt, G., *The Wheelwright's Shop* (Cambridge, 1934).

CHAPTER 16

SEA AND RIVER TRANSPORTATION

OCEANGOING VESSELS

The Heritage of the Fifteenth Century

THE EVOLUTION that took place in shipbuilding between 1500 and 1700 actually began during the fifteenth century; this was the period of the great voyages of discovery and of genuine navigation on the open sea. Various types of ships were in use; they were rapidly modified, and led up to the sixteenth-century models that were to continue to evolve, but at a slower and more steady pace.

In order to understand the origin of this evolution, it is necessary to review briefly the conditions of navigation during the fifteenth century.

Knowledge of the winds and currents

Until 1400, and with a few exceptions that we shall discover in the Mediterranean, navigation depended on favorable winds. No evidence exists to show that ships with more than one mast were to be found in northern Europe. Thus the square sail, suspended from a horizontal crossarm, remained the sole method of propulsion, and this type of sail did not permit beating to windward.

In the Mediterranean, ships with triangular lateen sails suspended from oblique yardarms had long been in existence; the use of this sail plan permitted beating to windward (already an ancient practice), and led to the use of two masts.

In the north, ships with more than one mast did not appear until around 1435, when the three-master arrived on the scene, as the result of a Mediterranean influence. The fact is that at this period Portuguese ships were revolutionizing maritime practices. In 1420 (and even as early as 1417) Prince Henry the Navigator was collecting and coordinating the equipment needed for major voyages — ships, astronomical observations, charts — and urging his sailors onto new routes, far from the coasts.

Figures 119 and 121 show how the navigation of this period led to the adoption of new types of ships.

Figure 119 shows how the winds and currents easily carried ships to Madeira and the Canaries. For the return trip, the ships set out in a northwesterly direction, in order to avoid confronting the wind blowing directly opposite to their route.

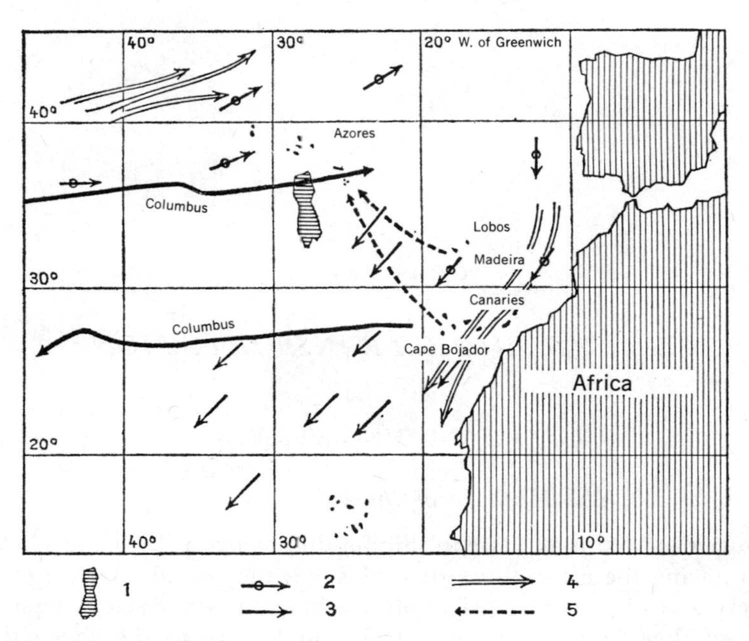

FIG. 119 Discovery of the Azores.
1. Hypothetical continent of Antilia. 2. Variable winds. 3. Prevailing winds. 4. Currents. 5. Route toward the Azores (1427).

Sailing with wind abeam was still possible for the ships of the period, which were thus able to reach the Azores by 1431 and perhaps even slightly earlier. North of the Azores the west winds naturally carried the ships back to the European coasts, without any need for tacking.

During the sixty-year period between 1432 and 1492, the knowledge of the wind pattern between the 30th and 40th parallels north was able to be solidly established. It was easy for Christopher Columbus to outline a program for his voyage, meeting only following winds on his return voyage, thanks to the knowledge that had been acquired by the Portuguese sailors and that Columbus had certainly learned during his sojourns in Portugal.

The situation was much different in the case of voyages of discovery southward along the African coasts. At first the sailors were frightened by the voyage beyond Cape Bojador; the most sinister rumors were spread about the dangers and impossibilities of navigation beyond this cape. It required all the authority of Prince Henry and the devotion of his equerry, Gil Eannes, to progress beyond this point (1434).

Difficulties were eliminated by the square sail, which sufficed to push the ship southward, and it was on a ship with square sails that Gil Eannes passed Cape Bojador (Figure 120). Once fear had been conquered, these southerly voyages became longer; Río de Oro was reached in 1444, and by the time Henry the Navigator died in 1460, Sierra Leone had been reached.

A problem of maneuvering ships then arose (Figure 121); off the coast of Africa, south of the equator, the ships had both wind and current against them.

← Fig. 120 Portuguese bark with square sails (1434).

Fig. 121 Routes to the South Atlantic. 1. A–B. Boundary of Spanish possessions (west) and Portuguese possessions (east). 2. Prevailing winds. 3. Variable winds. 4. Currents. 5. Christopher Columbus (1492). 6. Bartholomeu Dias (1487). 7. Cabral (1500).

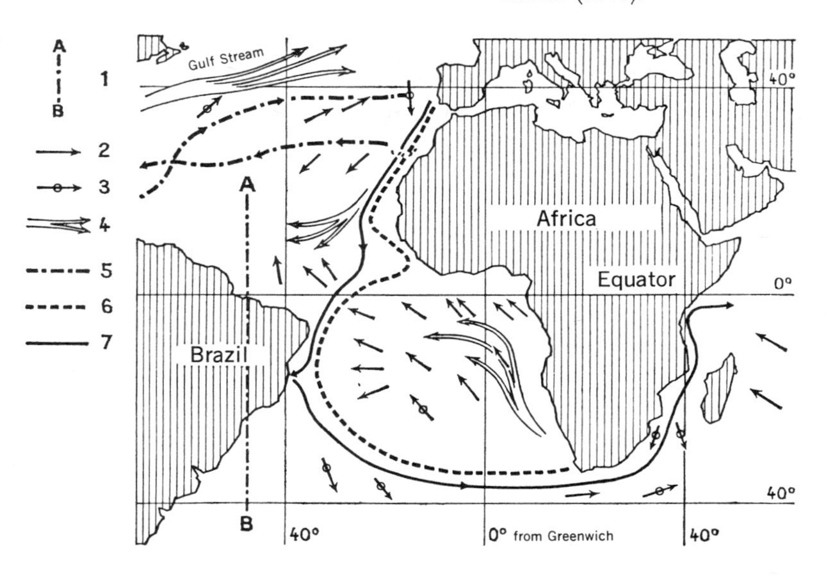

The appearance of the caravels

This fact explains the new use of the caravels. Our knowledge of the origin of the caravels is very vague. They were definitely utilized by fishermen in southern Spain, and also by the "Sarrasin" navies, the pirates who were the ancestors of the *xebecs* of later centuries. The adaptation of the lateen sail to an oceangoing ship must have come about quite naturally.

Around the middle of the fifteenth century, the use of caravels increased; we find two-masted (Figure 122) and three-masted caravels (Figure 123, taken from a sketch by Guilleux La Roérie). This divided-sail plan permitted an increase in sail so that they could be better distributed, by placing the center of the sail at a point along the length of the ship such that the action of the wind balanced the effect of drifting; thus the ship was able to hold a course beating up to the windward without difficulty and without the use of the rudder. In the fifteenth century the caravel was the instrument of the Portuguese discoveries; this ship made it possible to overcome the problem of the head wind and to

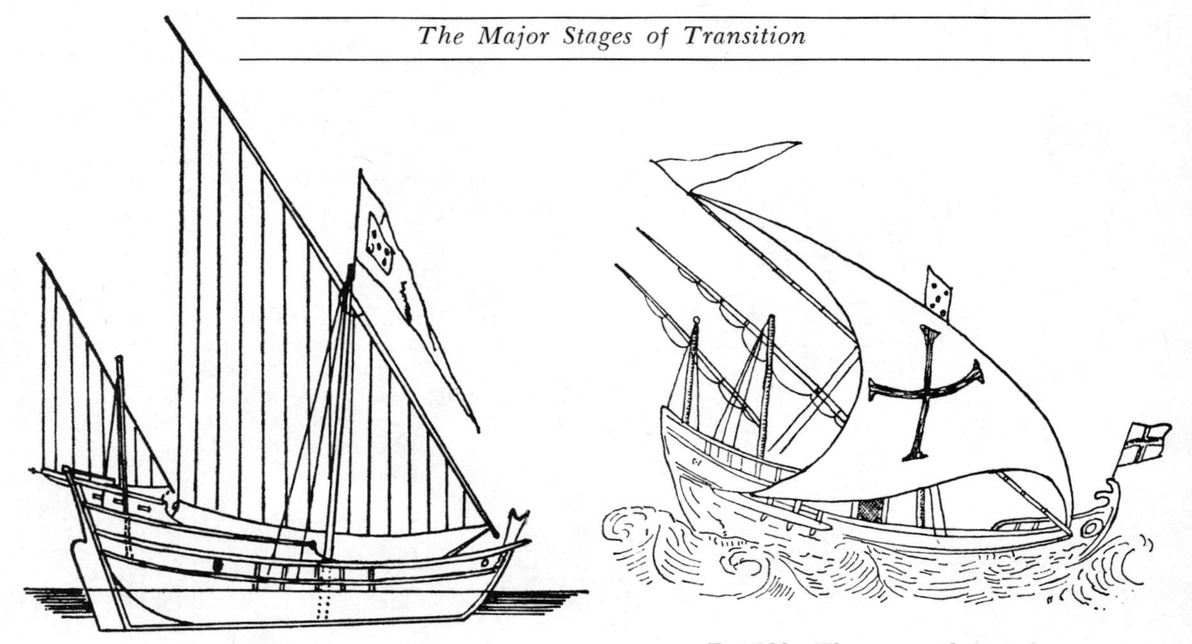

FIG. 122 Two-masted caravel.

FIG. 123 Three-masted caravel (from Guilleux La Roérie).

return by tacking to windward. The sailor, launched on new and distant routes, and already frightened by his distance from land, could not agree to depart on the adventure unless he were certain that he would be able to return, even with the wind against him.

After 1460, King John II energetically pursued the activity begun by Prince Henry, and the southerly voyages continued, this time with the goal of rounding the tip of Africa and reaching India, the essential goal of the Portuguese voyages.

With two caravels and a ship laden with provisions, Bartholomeu Dias tried to round the southern tip of Africa in the summer of 1486, but even with the caravels he was not able to reach South Africa. Heading directly southeast, the caravels were able to beat to windward, but the current pushed them north, and ultimately no gain in latitude was obtained. In 1487 Dias decided to head southwest, with the hope of meeting favorable winds that would permit him to turn eastward; this happened around the 40th parallel south. He thus became the first man to reach the route of the Indian Ocean to the East.

It must be noted that actually the caravel gave only a limited advantage in the navigation of discovery: beating to windward made it possible to reassure the sailors that they would return. Practically speaking, this advantage disappeared if the currents flowed in the same direction as the wind and canceled out the possibility of beating to windward; this is the case in the Atlantic.

There has been much discussion of the period 1487–1497, about which few definite facts are known. The Portuguese claim credit for having filled in the gaps in the knowledge of the currents and winds of the South Atlantic, a necessary prelude to further operations, and in fact during this period they did make a very sound study of all the possibilities of navigation. This study of the winds, conducted more as an oceanographic study than as a voyage of discovery, definitely appears to have been necessary in order to permit Vasco da Gama's ships to depart in 1498 with almost certain forecasts of the voyage.

Modifications in rigging While Bartholomeu Dias accompanied Vasco da Gama, the latter brought no caravels with him; these ships, which had been used for discovery, were not large enough for a trip to India, which required provisions, weapons, and a large crew.

Thanks to the experience with the lateen sail and multiple masts, the vessel (*nave, náo, nau*) had been modified during the fifteenth century. It now had three masts; the mainmast carried the large square sail, the foremast a second square sail, and the aft mast a lateen sail that was most convenient for maneuvering and tacking. This sail plan could be better balanced, and thus made a definite improvement in the ship's movement; it permitted the vessel to sail off the wind and in head winds, and even to begin to beat up to windward. While the *náo* was still inferior to the caravel when it came to tacking, it was larger and carried more personnel, supplies, and artillery.

The new route discovered by Dias with his exploration caravels utilized nothing but following winds (Figure 121); the same was true of the return journey, since all that was needed was to travel northward along the coast of Africa, where winds and currents carried the ships to the Canary Islands.

An English ship of the period is known to have been rigged with three masts and three sails, a type that was to continue to develop during the sixteenth century (Figure 124).

FIG. 124 English ship with square sails.

As for the caravels themselves, they underwent modifications in the course of the fifteenth century. The caravel of the Armadas (Figure 125) was larger and had four masts; one mast carried a square sail, which facilitated sailing with a following wind or off the wind. Thus it was an adaptation of the caravel for the routes recognized as being favorable to square riggings. Here we have the typical *náo,* with square sails, which formed first the squadron of Vasco da Gama (1498) and then that of Pedro Cabral. The latter, who pushed still farther westward, reached Brazil, and returned eastward in order to reach the Indian Ocean.

The dual evolution of the caravel toward the adoption of the square sail (to speak only of its sail) and of the ship (*náo*) with three masts and a lateen sail aft confirms the convergence of these models toward a single type that was to be the typical sixteenth-century ship. We find the confirmation of all these navigational procedures, developing with the ships themselves, in the fleet collected by Columbus for his first voyage (1492). Columbus could have and must have known the pattern of the winds in the two zones that would permit him to travel westward and then return without difficulties.

FIG. 125 The caravel of the Armada.

His ships

Thus it was with full knowledge of the possibilities for use offered by the various types of ships that he got together his small squadron. The admiral's ship, the *Santa María,* carried square sails — for the reason, it has been claimed, that it was a "rotunda" caravel (the Armada ship mentioned above). Actually, it was a cargo ship that had formerly been used to carry merchandise, chartered for the occasion, and one of its passengers was the owner, Juan de La Cosa, Columbus's patron. Thus it was a Nordic type of vessel closely related to the Hansa ships.

In contrast, Martín Pinzón's slightly smaller ships were clearly caravels. The reconstructions that have been made of both types are highly controversial, but it can be admitted that they are adequate representations of models of the period

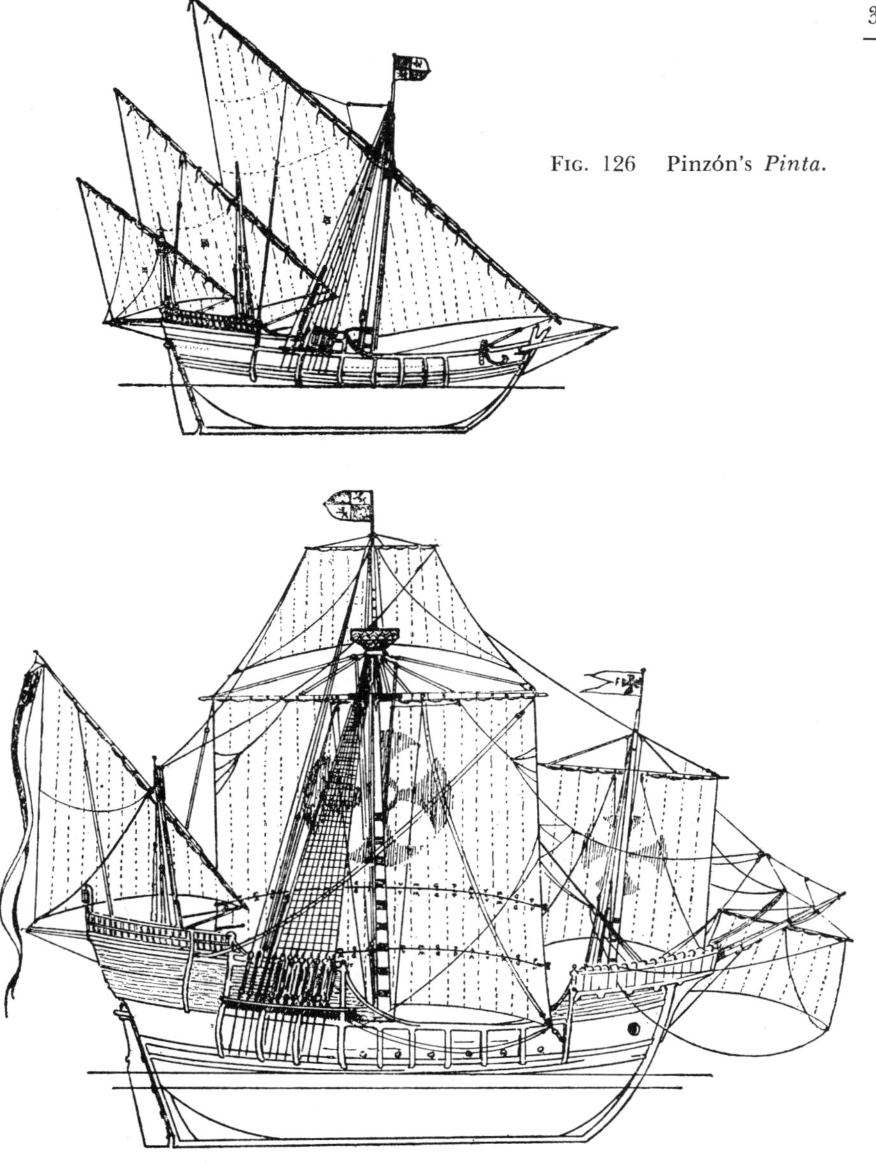

FIG. 126 Pinzón's *Pinta*.

FIG. 127 The *Santa María* of Christopher Columbus.

(Figures 126 and 127). Their sizes cannot have been very different from those given by various authors:

	Santa María	*Pinta*	*Niña*
Length on the deck	$73\frac{3}{4}$ feet	$65\frac{2}{3}$ feet	$65\frac{2}{3}$ feet
Overall length	$86\frac{1}{4}$ feet	$82\frac{2}{3}$ feet	79 feet
Beam	$26\frac{1}{4}$ to $27\frac{1}{2}$ feet	$24\frac{1}{4}$ feet	24 feet
Tonnage	200	110	105
Displacement	233	160	150

(A discussion of tonnage and displacement will be found at the end of this chapter.)

The flagship *Santa Maria* is believed to have had a total sail area of 4,951 square feet; this surface corresponded to a relation of 27 with the midship section of the vessel (Figure 127). Until the nineteenth century, the areas of sail were generally considered in relation to the midship section; the area of sail corresponded to the propelling power given by the wind, while the midship section furnished, in very approximate fashion, resistance to the progress of the moving ship.

The *Santa María,* a ship with round lines, was especially suited to a following wind; it was natural that Columbus should have chosen such a ship, the largest in his fleet, well suited to the route he intended to follow (steady east wind), and equipped to benefit on the return trip from the west winds; thus he traveled constantly with a following wind.

The *Niña* (Figure 128, left) was a caravel of a very different type from the *Santa María,* and therefore Columbus revamped her during his first stop, substituting a square sail for one of her lateen sails (Figure 128, right); this indicates that he had every intention of traveling always with a following wind. A similar transformation was drawn on the chart of Juan de La Cosa (1599 to 1620) (Figure 129).

In 1492 the *Santa María* carried only a fore-topsail above the mainsail; a square sail (the spritsail) under the bowsprit was utilized until sometime during

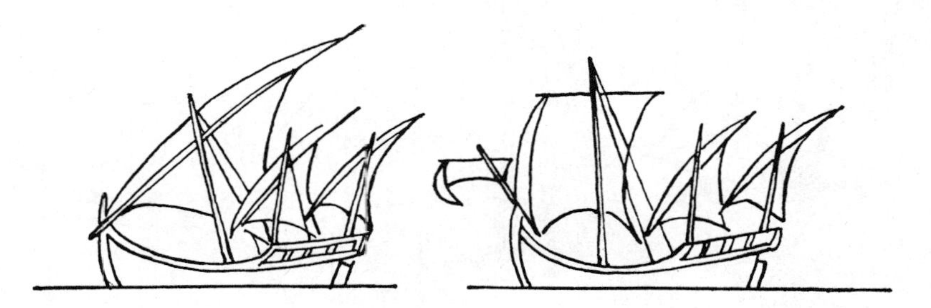

FIG. 128 The transformation of the *Niña.*

FIG. 129 Ships drawn on Juan de La Cosa's chart (beginning of the sixteenth century).

the eighteenth century. The sail was better distributed, and increased in area. In 1498, on the ships of Vasco da Gama, a topsail appeared on the foremast.

In contrast, Pinzón the elder retained the lateen sails of the *Pinta,* which ship, being slightly longer, and having finer lines than the *Santa María,* had greater freedom of movement. The mate's logbook (or rather the summary of it given by Las Casas) clearly indicates that the *Pinta* was a faster ship than the others; she preceded the others, returned, and had greater freedom of movement. This enraged Columbus, who suspected even Pinzón of desertion and betrayal.* Actually, Pinzón may have been able to reach several islands Columbus did not reach, and to bring back more information (and booty) than the admiral himself.

Thus the evolution of ships during the fifteenth century came to a close in such a way as to indicate that the ship with several masts, square sails on the mainmast and foremast, and lateen sail on the aft mast, which was able to make better use of the wind than the single-mast ship of 1400, was to take the leading role. It was also to acquire finer lines in order to regain the advantages of the finer ships like the caravelles, which could even beat up to windward.

As for the dimensions, the ships are very obviously not uniform; ships of between 400 and 600 tons can be found, like the 500-tonner reconstructed in the Science Museum of London (Figure 124). But the bulk of commercial traffic continued to be carried on in smaller ships of approximately 200 tons (around 300 displacement tons).

The lines of the ships At the time of the appearance of the three-master, the typical northern ship was clinker-built; that is, the thin planking overlapped like the shingles of a roof (Figure 130*a*), in a continuation of the practice begun on the ancient Nordic, Viking, and Saxon ships. The Mediterranean, on the other hand, retained the carvel-built type (Figure 130*b*). The lines of the bottom were crude, and were similar at both ends; the planking met the sternpost or curved over the bulging line of the hull. Above the upper wale, a transom held the rear portion of the aft castle, which was built on the hull and was still poorly incorporated into the structure of the bottom — a survival of the early medieval platform-type fore- and aft-castles, which were placed on, and were quasi-independent of, the hull.

This type of box was to survive until the end of the seventeenth century in Dutch ships (store ships) with crude clinker-built planking.

* P. Gille, "Navires lourds et navires rapides," in *Ve Colloque de l'histoire du navire* (Lisbon, 1960).

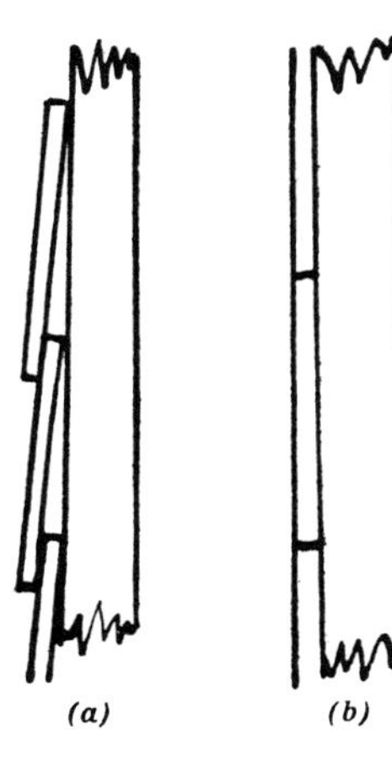

FIG. 130 Clinker- (*a*) and carvel-type (*b*) planking.

The triangular forecastle was placed far forward on the stem; it was to form an overhanging platform higher than the aft-castle until the end of the fifteenth century, and was used to hold the fighting men.

Sixteenth-Century Warships

The introduction of the cannon made it necessary to modify the warship, which in the course of the sixteenth century became increasingly different from the merchant vessels.

Cannons and the evolution of ships

The early cannons were light, and were placed on ships (in the tops, on the castles) only in time of war. When they became heavier they had to be placed on a solid deck, in the space between the castles. An exact date — 1501 (which, however, may not be precise, but is probably close) — is given for the first installation of gunports cut in the hull itself in order to place the cannons as low as possible. A native of Brest, whose name was Decharges, is believed to have been the innovator of this new arrangement.

In England, the *Regent* and the *Sovereign,* both constructed before 1500 (1487), had cannons on the castles; the *Mary grâce à Dieu* (1514) had cannons that fired through gunports cut in the hull. No detailed drawing exists to permit us to reconstruct the *Cordelière,* built at Morlaix and probably similar to the *Regent,* with which it was sunk in the famous battle before Brest in 1512.

The *Regent* and the *Sovereign* were 1,000- and 800-tonners; the *Regent* was a four-master, the first vessel to have two aft masts with a lateen sail, a mizzenmast and a bonaventure mizzen. The *Sovereign* was a three-master.

In the first quarter of the sixteenth century the separate topmast was introduced, which could be hauled down in bad weather in order to reduce the strain on the masting. At about the same time, in England, carvel-type planking was introduced. This technique, which permitted the use of thicker planks for the strakes of the planking, giving greater strength to the hull, is of Mediterranean origin. It was the technique used by the ancient Greeks and Phoenicians. A watertight joint was obtained by caulking the jointed strakes of the planking. This system very soon came into general use. Clinker-built hulls may have been too light and weak to bear the weight of the artillery; moreover, it is easier to cut gunports in a carvel-built than in a clinker-built hull.

Artillery then became heavier, and the variety of calibers increased — cannon, demicannon, culverins, and so on — which were later to correspond to the calibers of 32, 24, 18, 12, 9, 6, and 3 (pounds). Muzzle-loading bronze cannons remained almost unmodified in the following centuries.

In 1514 the *Mary grâce à Dieu* carried 186 cannons, including 122 light iron cannons with removable loading chamber (the breech), and four bronze culverins. In 1540 the ship was remodeled, and then carried only 122 pieces of artillery, 19 of which were bronze, the remainder iron.

In 1836 the Museums of London were given early iron breech-loading guns and very well-cast bronze cannons from the *Mary Rose,* which sank before Portsmouth in 1545. This mixed artillery is fairly characteristic of a period of transition that occurred around the middle of the sixteenth century.

Medium and light tonnage

Those of Brueghel's pictures that were painted between 1550 and 1569 reveal with unusual accuracy what the ships of this period were like; this is particularly true of his well-known painting "The Fall of Icarus" (Brussels Museum).

While in the second half of the sixteenth century we witness a marked increase in tonnage of the ships carrying heavy artillery, it must be noted that a number of ships of medium and light tonnage continued to exist. Among these was Francis Drake's famous *Golden Hind,* an explorer and pirate ship, which between 1577 and 1581 circumnavigated the globe, rounding Cape Horn and traveling up the west coast of the Americas. This ship, of which modern reconstructions show the likely characteristics, probably had the following dimensions:

Overall length	75 feet
Length of waterline	60 feet
Length of keel	47 feet
Beam	19 feet
Depth of hold	9 to 10 feet

Her tonnage was probably only 100 register tons, corresponding to a displacement tonnage of 150 tons.

It was with this ship and a crew of 88 men (reduced to 59 on the return voyage) that Drake crossed the Pacific in 89 days without restocking either fresh food or water. The drawing of a reconstruction (Figure 131) shows the hull with square stern and gunports; the fore- and main-masts have topsails and topgallants. Its small size shows the considerable differences that can be found among the various types of ships of this period, depending on their use: large warships, merchant ships, coasting vessels. Large ships reached 1,000 to 1,100 tons, while coasting ships could be as small as 100 tons (the *Golden Hind*) or even 40 and 50 tons.

The Great Armada

A major event, namely, the campaign of the Great Armada (1588), gives us a great deal of knowledge about this wide range of tonnage figures, as well as about the formation of a great combat fleet at the end of the sixteenth century.

The Spanish plan of attack against England was to become master of the English Channel and land in Great Britain with a large military force. This invading army was to consist partly of the troops brought from Spain by the Armada and partly of troops from the Spanish garrisons in the Netherlands.

The equipment utilized for this purpose was to be tremendous; the first plan (Santa Cruz) provided for:

		Tonnage	*Average tonnage per vessel*
Large ships	150	77,250	515
Transports (*Urcas*)	40	8,000	200
Small units	320	25,500	80
Totals	510	110,750	795

Plate. 29. Sailing vessel seen from behind; accompanying it is an oar-powered galliot. Engraving from a drawing by Peter Brueghel the Elder. *Photo Bulloz.*

FIG. 131 Francis Drake's *Golden Hind* (1577–1581) (from a reconstruction).

According to the practices of the period, 20 sailors per 100 tons would have been needed; this number had to be reduced to 15 because of the lack of experienced personnel, for a total of 16,612 sailors. A squadron of Spanish, Neapolitan, and Sicilian galleys was to add 8,000 oarsmen to this number; in addition, six galleasses included 7,200 sailors and 1,800 oarsmen. The fleet was to carry 55,000 men and 1,200 horses; the artillery included 4,300 officers and cannoneers. Navy and army combined made a total of 94,200 mouths to feed and 22,000 quintals of powder, 10,000 quintals of shot, 16,000 harquebuses, 5,800 muskets, 10,000 pikes, and so on.

These figures terrified King Philip II, and he reduced them to the figures provided by the program of the Duke of Medina Sidonia: six squadrons of ships, one squadron of transports (*Urcas*) and one squadron of light vessels. Tonnages varied from 70 to 1,050 tons; the total program called for 130 ships with a total tonnage of 57,868 tons, 2,431 cannons, and 29,305 sailors, oarsmen, and soldiers. Opposing them were 197 English vessels with approximately 16,000 men; the tonnage varied from 40 to 1,100 tons.

The armament of the two fleets was different. A combat at San Juan de Ulua had shown the English (under Sir John Hawkins) the advantage of long, small-caliber cannons with greater range, with which they could decimate the crews of their adversaries and tear up the rigging. The proportions of the types of cannons differed as follows:

	Short cannons	*Culverins and sakers*
on 124 Spanish ships	43.5%	56.5%
on 172 English ships	5 %	95 %

The English artillery, which was better handled by better-trained gunners, was able to carry out long-distance harassment of the Spanish fleet, whose crews, weakened by the long ocean voyage and by illness, were exhausted. The difficulties of preparing a disembarkation fleet in Flanders prevented the Spanish garrisons of the Netherlands from putting out to sea. The ill-fated return of the Armada, and its agonizing calvary around the English coasts, are famous.

The English ships

What the statistics show about this tremendous undertaking is that there were large ships on both sides. On the English side were:

the *Triumph*	1,100 tons
the *White Bear*	1,000 "
the *Ark Royal*	800 "
the *Revenge*	500 "

The *Ark Royal* carried 32 cannons and 12 small guns, the 32 cannons on the gun deck, the light guns in the tops. A diagram (Figure 132) based on the Westminster tapestry shows the silhouette of the *Ark Royal,* with its four masts and already well-developed sail plan: topsails and topgallants on the mainmast and foremast, as well as a bonaventure mizzenmast with lateen sail, a mizzenmast topsail above the lateen sail. It also shows the large two-story aftercastle, which was not yet completely incorporated into the structure of the hull; the forecastle is in this case not so high, but still projects quite far out over the stem.

We find different figures given for the Spanish ships by the Spanish and English sources. The *Santa Ana* (1,200 tons) was no larger than the *Triumph.* The captured *San Salvador,* listed by the Spaniards as being 958 tons, was rated at 600 tons by the English. This results from the difference in the units of measure, and shows how difficult it is to interpret early technical documents.

Characteristics of ships at the end of the sixteenth century

The end of the sixteenth century was marked by a refining of ships: the length of the keel increased from 2.3 to 3 and sometimes even 3.3 times the beam of the ship, which corresponds, for the length at waterline, to 3.5 and 4 times the beam. The depth of the hold was approximately half the beam.

The plans for certain ships of this period are still extant. A manuscript from the hand of the English shipbuilder Matthew Baker shows (Figure 133) how a ship was then outlined: a longitudinal cross section, the midship section, fore and aft sections, and the square stern. Such a drawing made it possible to outline the principal elements and to raise the components in dry dock; a continuous framework supported the intermediate ribs.

The outline of the midship frame was obtained by drawing a series of semicircles (Figure 134) whose centers lay on a single straight line. Upon superficial examination, the appearance of these plans could lead us to believe that scientific rules for the layout of ships were in use by the end of the sixteenth century. In reality, however, these were simply very primitive procedures used by the

FIG. 132 The *Ark Royal* (1588) (from a tapestry in Westminster).

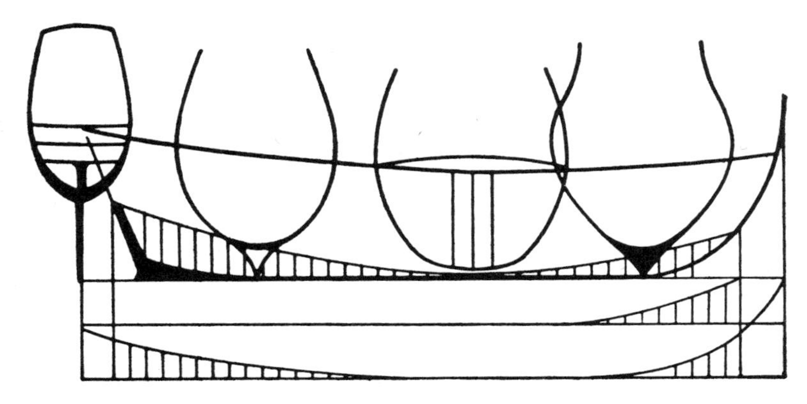

FIG. 133 Ship's plan (from Matthew Baker, end of the sixteenth century).

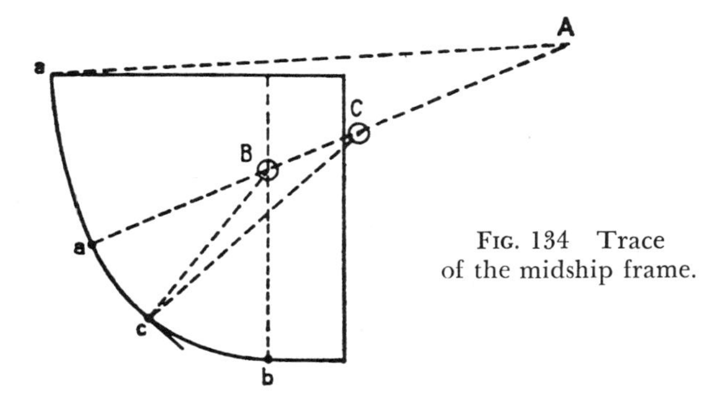

FIG. 134 Trace of the midship frame.

carpenters; their scientific knowledge went no further than the use of semicircles and straight lines. No calculations were made to justfy the diagrams, which were the product of nothing more than very elementary principles of draftsmanship.

The Development of Ships in the Seventeenth Century

The beginning of the seventeenth century witnessed a continued evolution in shipbuilding, without any major modifications.

The masts

The two aft masts with lateen sails, the mizzen and the bonaventure mizzen, continued in use. In the England of 1618, half of the royal fleet consisted of four-masters. Topgallants began to be installed on the mizzen in 1611, and by 1618 all the English ships had them. But the development of the lateen sail on the mizzenmast made possible the elimination of the bonaventure mizzen, and after 1640 there were no four-masters left in the English fleet.

The period between 1611 and 1618 saw the appearance of the vertical bowsprit mast, which carried a square fore-topsail. This mast was poorly secured, and was always a source of concern during high winds; nevertheless, it continued to exist until 1740.

At the beginning of Queen Elizabeth's reign the topmasts (foremast and mainmast) were stationary. Hawkins, who became Treasurer of the Navy in 1578 and continued in that office through the time of the Armada, ordered them to be made movable, which permitted them to be hauled down in bad weather. Beginning in 1600, removable topgallant masts were installed in the same manner, which made it possible to increase the sail plan by developing it upward.

The seventeenth century has left us a rich collection of documents that permits us to follow in detail the evolution of the ship. A discussion of the great differences between the organizers of naval construction in England and France will be of interest at this point.

Construction in England and France

In England a regular, permanent administration had already been established, and the names of well-known shipbuilders began to appear, for example Matthew Baker and the Pett family, who from father to son worked for the Admiralty.

In France the religious wars prevented the government from undertaking programs of organization and construction. Shipyards existed, but the navy did not yet have well-established, permanent services. Richelieu, who wished to make an effort to build up a fleet, conducted a search for a carpenter to whom he could entrust the construction of a large ship. He sent for Ch. Morieu, of Dieppe, who built the famous *Couronne* at La Roche-Bernard, at the mouth of the Vilaine; this spot was chosen because it was located near forests belonging to the Duke de Rohan, who placed them at the disposal of the shipbuilders.

The sizes of ships then began to increase markedly. The *Sovereign of the Sea,* built by Peter Pett in 1637, can be compared with the *Couronne* of the same period (1638). See page 377.

The sail was as yet little developed; the masting was quite poorly secured, and was fragile. It was in the areas of improvement of the masting and development of the sail that an evolution was to take place at the end of the seventeenth and during the eighteenth centuries.

	Sovereign	*Couronne*
Length at keel	127 feet	
Length at waterline	172½ feet	166¼ feet
Beam	46½ feet	49 feet
Draft	19.2 feet	18¼ feet
Tonnage	1,522 tons	
Displacement	2,270 tons	2,180 tons
Area of sail		1,798 sq. yards
Relation of sail to midship frame		24.7

When we consider the silhouette of the *Couronne* (Figure 135), we see that the aftercastle has become better incorporated into the hull, that the forecastle has become smaller and the cutwater is still in use; during the seventeenth century it decreased in size, to become a knee bearing the head with its ornamental prow figure.

The unfortunate period of the Fronde during the minority of Louis XIV (1648–1653) left naval construction in France in a sad plight, while the work of the English shipbuilders progressed steadily. The square stern found on all the English ships built at the beginning of the seventeenth century continued to exist for a long time on the Dutch vessels and those of the other continental navies, but in England it was gradually replaced by the round stern, in which the planking curved round to meet the sternpost in normal fashion. We find an example of this arrangement in England by 1617. The *Sovereign of the Sea* (1657) still had a square stern, but by 1654 and 1656 all the English ships were being constructed with round sterns.

The plans of the ships The English shipwrights of the second half of the seventeenth century left more detailed and complete drawings of their ships, as well as memorandums. The contemporaries of Samuel Pepys included Sir Anthony Deane with his *Doctrine of Naval Architecture* (1670); he gives mathematical details and plans for the sail of the six rates of ships that then composed the British Navy. In particular, it contains the plan of the lines of a third-rate ship with eighteen cross sections seven feet apart; this is the oldest such plan preserved in England. It does not give the waterline, which was not yet utilized in defining the lines of ships. Another manuscript gives the plans, cross sections, and decks of a first-rate vessel of 1680 *(Dummers Draught of the Body)*.

The method of construction indicated remained unchanged during the seventeenth century; however, we find the outline of crosswind bracing on the low gun deck. The idea of using inclined (but not intersecting) braces for wind bracing had already been suggested by Matthew Baker during Queen Elizabeth's reign.

Deane's drawings show the last applications of the long cutwater, which after 1670 became shorter and straighter to form the knee that supported the head.

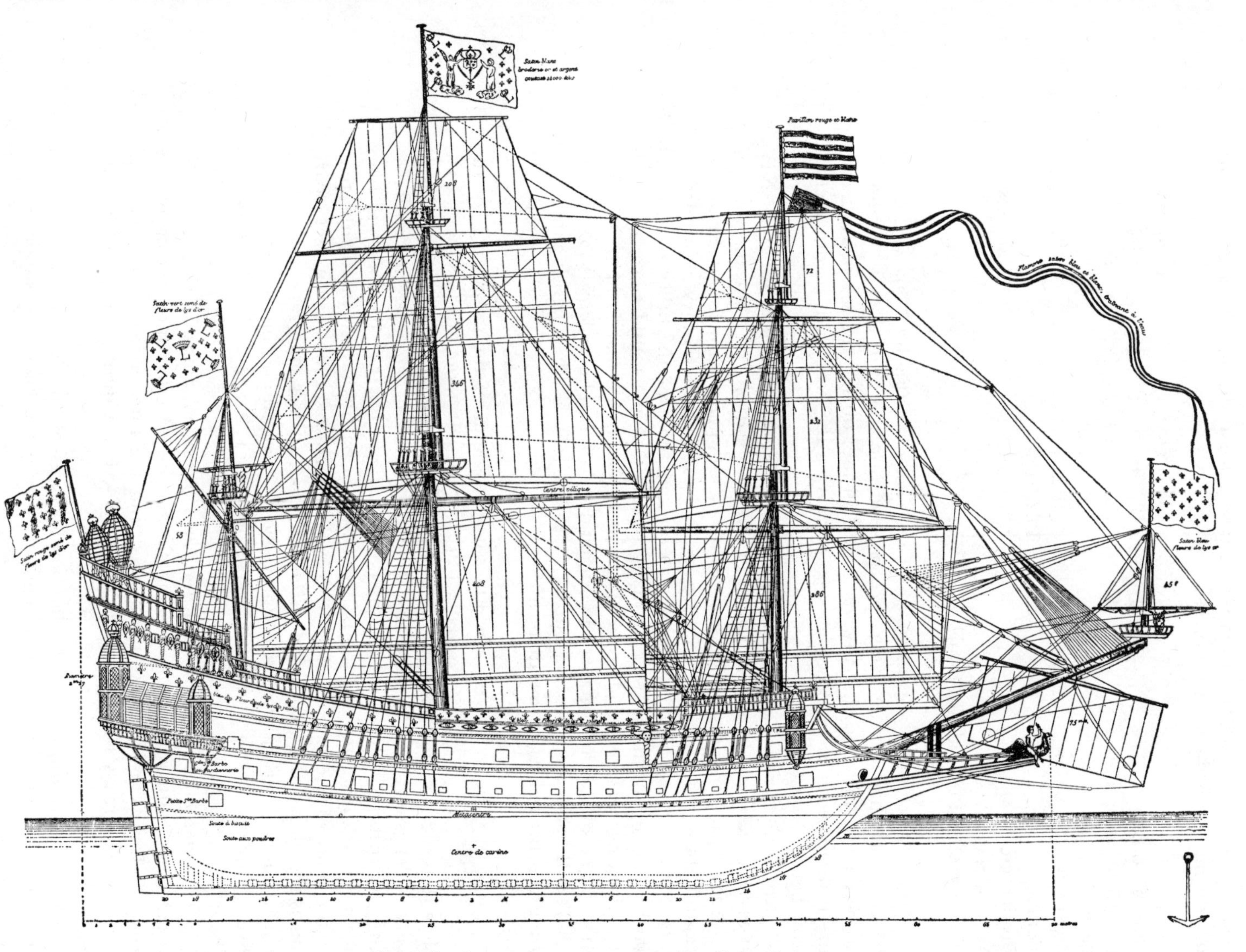

FIG. 135 Approximate reconstruction of the *Couronne*.

Shipbuilding efforts in France at the end of the seventeenth century

In France, first Colbert and then his son, the Marquis de Seignelay, attempted to revive a navy that had been abandoned since the time of Mazarin. Their chief concern was to increase the number of shipyards, and find shipbuilders for them. They sent for foreign shipbuilders – Italians like Biaggio Pangallo ("Maître Blaise") who worked at Toulon, and Dutchmen like Rodolphus, who rendered good service at Rochefort. But in the second half of the seventeenth century the evolution of ships in France must be considered from two very distinct points of view.

Colbert, whose inclinations are well known, tried to find the best plans and lines of ships; he wanted to compare the works of the French shipbuilders in various ports – Brest, Rochefort, Toulon – but also those of the foreigners. Like Seignelay, he wished to codify the various models, standardized in conformity with the best plans.

A financial drive served as support for the technical activity: the naval budget, which had fallen to 312,000 livres in 1656, was raised by Colbert to 2,000,000 livres for ships and 600,000 livres for galleys, in 1661.

In 1664, Brest was still only a village, where the experienced shipbuilder Hubac was working alone. Meanwhile, the English fleet now included 109 vessels, 21 fireships, armed with 4,192 cannons and staffed by 21,000 sailors. Holland was arming 103 large ships with 4,800 cannons and 22,000 sailors.

Between 1660 and 1690 Colbert created and developed Rochefort, despite the difficulties of establishing dry docks and buildings on the muddy banks of the stream. A regulation of 1674 fixed the working hours at 7:00 A.M. to 6:00 P.M. in winter, and 5:00 A.M. to 8:00 P.M. in summer, with a break from 11:00 A.M. to noon (in summer this was shortened to a half hour) for a "snack"; the salaries ranged from 15 to 18 sous per day. On the other hand, there were a good many holidays: France had up to 100, while England and Holland had only 52.

Types of ships

In order to classify the types of vessels, a "Construction Council" (1670–1671) was created; it discussed the sizes to be adopted, and passed the Regulations of 1673-1674.

For example, a three-decker was to be 148 feet from stem to stern (waterline); its beam was to be 38 feet at its widest part; the depth was to be 22 feet two inches. The relationship of length to beam was thus 3.894.

The result was the following classification:

First Rate: three decks – 1,400 to 1,500 register tons – 80 cannons
(2,100 to 2,250 displacement tons)
Second Rate: three decks – 1,100 to 1,200 register tons – 64 cannons
Third Rate: two decks – 800 to 900 register tons – 50 cannons
Fourth Rate: two decks – 500 to 600 register tons – 40 cannons
Fifth Rate: two decks – 300 register tons – 30 cannons

These vessels were narrower and deeper than the *Couronne*.

Discussion continued, however, on the various models to be adopted: when called upon to give their opinion Duquesne and Tourville supported respectively the Brest models (built by Hubac father and son) and the Rochefort models Maître Blaise, builder). Still others preferred the Toulon ships (Maître Coulomb, builder).

An attempt was also made to determine what reasons led the English and Dutch shipbuilders to adopt different models. Colbert du Terron (1618–1684) mentions (Figure 136) the most notable differences: the English ships were deeper and had a triangular line, while the Dutch ships were flat-bottomed, which permitted them to depart from the small backwaters of the Zuyder Zee.

The ordinance of 1689 indicates an increase in tonnages; the first-rate ship has a tonnage ranging from 1,600 to 2,200 register tons, that is, a displacement that could surpass 3,000 displacement tons. This was to be the model for ships of the line, with 100 cannons to be made of bronze (cast-iron cannons were adopted for other vessels).

Essays on mathematical theories

Another very important aspect of shipbuilding in this period was the introduction of mathematical considerations and procedures, which were to attract the attention and collaboration of scientists on the techniques of shipbuilding.

The question posed to Archimedes on the equilibrium of floating bodies had led this Greek thinker to the study of regular solids, for which he was able to obtain the center of thrust. But while the "principle" had been remembered for centuries, the ideas on stability that followed from the studies of Archimedes had remained unknown. His work became known only through a Latin version incorrectly attributed to Nicolò Tartaglia (between 1545–1555) — therefore, not until the sixteenth century.

In his famous *Hydrographie* (first edition in 1643, followed by two others in 1667 and 1679), Father Fournier traced the ribs of the ship with circles and straight lines, and obtained a calculation of volume at bottom.

Studies multiplied rapidly at the end of the seventeenth century. It must not be forgotten that the scientists of the period — Huygens, Newton, Leibniz — were then creating a general movement of application of the new mathematics. The Reverend Paul Hoste, S.J., a naval chaplain and an enthusiast of mathematics, published three major works in succession: *Traité de mathématiques* (1692), *Art des armées navales* (1696), and *Théorie de la construction des vaisseaux* (1697). In the last work he attempted demonstrations on stability and rolling, and discovered, not without astonishment, that the center of gravity of ships is above the center of hydrostatic thrust. He tried to demonstrate why the capsizing of the ship was not a consequence of this fact, and gave an erroneous explanation, thanks to an expedient (Figure 137): he supposes that the thrust of each half of the ship acts on its own side, and thus a sort of balance is maintained. The explanation is childish, but testifies to a spirit of research and a desire to establish a scientific demonstration.

In his *Manoeuvre des vaisseaux* (Paris, 1689), Renau d'Éliçagaray again took up the problem of the calculation of the bottom, and launched into mathematical considerations on the action of the sails and tacking. This led to polemics in which he found himself the butt of criticism by Huygens, L'Hospital, and Jean

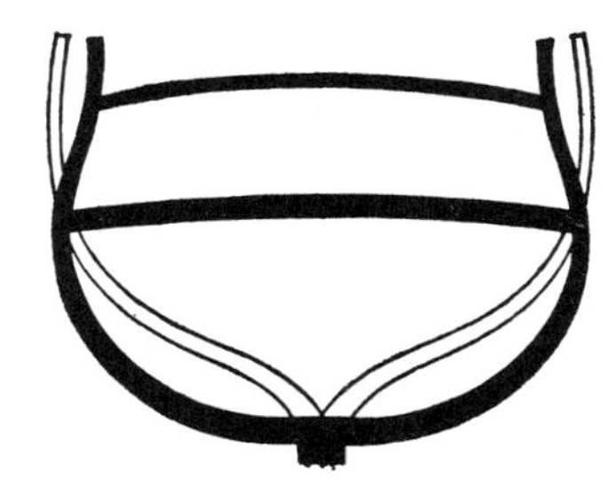

FIG. 136 Comparison of the midship frames of English and Dutch ships.

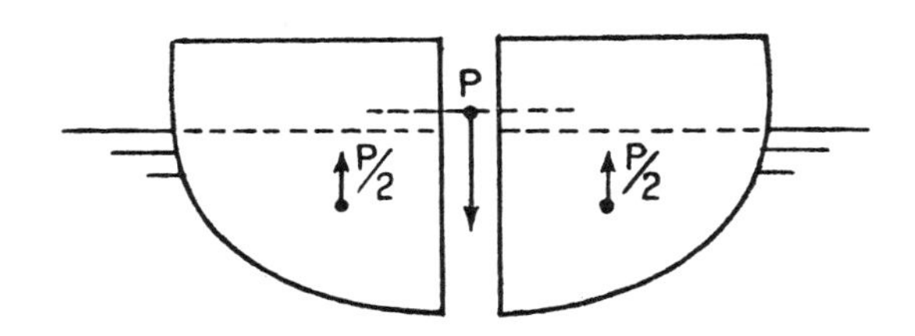

FIG. 137 Theory of stability (from Father Hoste).

Bernoulli; publications of these critiques and the answers made to them were exchanged, and captured the attention of scientific circles.

We shall see that exact solutions for most of the problems of hydrodynamics were to be given fifty years later by Father Pierre Bouguer, but it must be noted that the scientific spirit was already appearing in these domains at the end of the seventeenth century, and the attention of the scholarly world was directed to the numerous questions raised by the construction and utilization of ships.

Details of the structure of the hulls

In 1670 Anthony Deane suggested the use of iron for binding the stringers to the ribs; until then this had been done with large wooden right-angle knees. This innovation, which was unfavorably greeted by Samuel Pepys, was well regarded by King Charles and the Duke of York, and tried out on the *Royal James.* Less than one year after its completion, this ship was burned in a battle, and thus no conclusions could be drawn from this experiment. The use of iron was not envisaged again in England until the time of the Napoleonic wars, and then only because of the difficulty in obtaining the lumber necessary for the extensive needs of the navy.

During the same year Deane tried lining the bottom of a fourth-rate vessel, the *Phoenix,* with sheets of rolled lead. The planking was first coated with a mixture consisting of hair, tar, sulfur and tallow; this was then covered with thin strips of wood. The purpose of this combined protection was to defend the planking against the shipworms, stains, algae, and shellfish that attach themselves to the hull, considerably impeding the ship's progress. The attempt at lead lining was made in third-rate vessels and smaller vessels that sometimes went for ten years without scraping of the hull.

But the use of iron nails and bolts to attach the lead brought about electrolytic reactions that quickly destroyed the iron. This system was soon abandoned, but the idea was taken up again at the end of the eighteenth century.

Operation of the rudder

During the sixteenth and seventeenth centuries the rudder was operated in rather curious fashion, by a vertical lever pivoting at the level of the deck in a plane transversal to the axis of the vessel (Figure 138). This lever, known as the whipstaff, which increased the helmsman's power by the relation $\frac{L}{l}$ acted on the end of the tiller, which also acted as a lever in relation to the distance of the hydrodynamic effort on the afterpiece of the rudder. By the product of the relations of the two levers, it was possible, using one or two men, to balance the pressure of the water on the rudder.

The tiller was a square bar; for a ship of 133 feet it had to be 23 feet long, nine inches thick at the end joined to the rudder, and six inches at the other end. This free end had an iron peg that fitted into an iron ring in the lower portion of the whipstaff. For the same vessel, the length of the whipstaff was equal to approximately one-third the beam of the vessel, that is, nine feet.

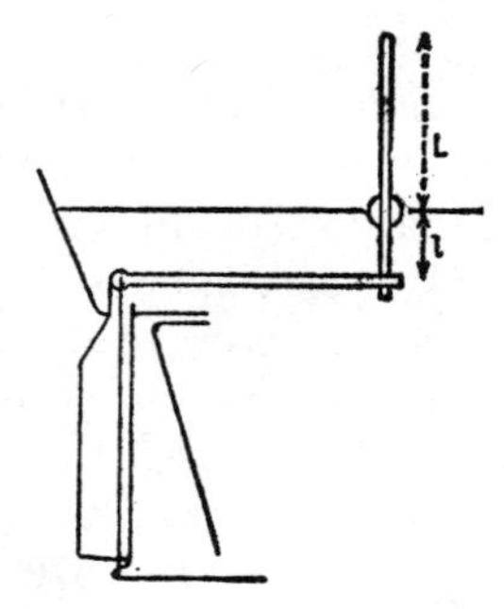

FIG. 138 Operation of the rudder by the whipstaff.

This theoretically ingenious arrangement had the disadvantage of permitting only a small angular movement of the tiller, five degrees on each side. Under these conditions the activity of the rudder was very limited, and could serve only to rectify minor deviations from the ship's course. To change course it was necessary to shift the sail plan's center of action forward or backward in order to make the ship sail into the wind or out of the wind's eye. Generally a change in the balance of the sail could be obtained by working on the lateen sail of the mizzenmast, which was easy to maneuver.

In the first decade of the eighteenth century, however, the whipstaff was replaced by a tiller rope rolled around a drum, which acted on the tiller; a wheel with handles made it possible to turn the drum, roll up the tiller rope, and maneuver the tiller. In this way a larger angle of swing and more effective action of the rudder could be obtained.

Sail

In the second half of the seventeenth century, major modifications were made in the sail plan by the gradual adoption of staysails and jibs. The triangular staysails were placed between the masts in the longitudinal plane of the ship; the jibs were placed between the foremast and the bowsprit. This gave a greater area of sail in the longitudinal plane. The jib made it possible to change the position of the center of the sail more easily than did the fore topgallant on the bowsprit, with its fragile vertical mast; between 1720 and 1740 the latter was completely replaced by the jib.

In 1588 Walter Raleigh mentioned studding sails as a recent invention, but

Buckingham's description of the attack on the island of Re in 1627 indicates that they were not yet used as a matter of course on the English vessels (the vessels of Dunkerque in the service of France were equipped with them). In 1690 they were being regularly used both on the topsails and on the lower sails of the mainmast and foremast. Thus they date from the adoption of three masts on vessels. They consisted of a rectangular strip of canvas laced to the boltrope of the sail; the studding sail's length was one quarter that of the sail. Below the studding sail a second narrow strip of canvas (the English drabbler) could be added, in order to make better use of slight breezes (Figure 139). Depending on the strength of the wind, the studding sails were added or removed. In order to do so, it was necessary to lower the yard to the deck, and then hoist it back up again.

The result was frequent maneuvering of the yard, which had to be guided by a parrel ("necklace") around the mast. This collar (Figure 140) consisted of wooden balls strung on ropes, in strings separated by thin pieces of wood; when the sails were large, the parrel collar had a large number of these balls, thus distributing the strain on the mast more evenly. Flemish and Dutch drawings and paintings of this period, which are often very faithful in their depiction of details, distinctly show these parrels; they appear very clearly in Brueghel's "Fall of Icarus."

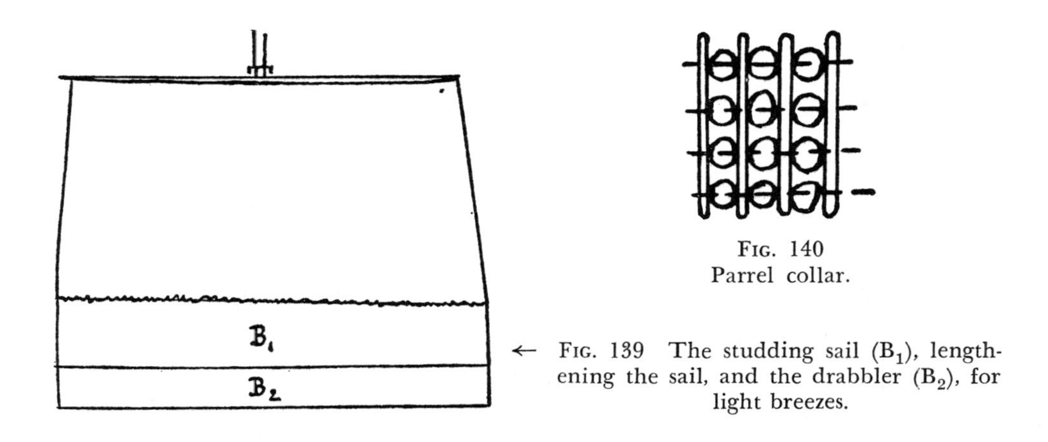

FIG. 140
Parrel collar.

← FIG. 139 The studding sail (B_1), lengthening the sail, and the drabbler (B_2), for light breezes.

Reefs began to appear in the middle of the seventeenth century; they made it possible to reduce the area of the sail by pulling it from the top up to the yard. A painting (1672) by Van de Velde of the Battle of Solebay gives a very detailed representation of a reef line on the upper portion of the lower sails and the topsails. Double and triple reefs were then quickly introduced; by the end of the eighteenth century there were four. Maneuvering the sails became quite different: instead of lacing and unlacing the studding sails on the deck, the sailors had to climb up the mast and, standing on the yards, pull up the sail and furl it by attaching it with the reefs. It was no longer necessary to haul down and hoist the yards, which could now remain fixed in position. The parrels became less important, and the yard was finally fixed and suspended by a permanent chain.

Plate 30. Sixteenth-century ship. Detail from "The Fall of Icarus" by Peter Brueghel, 1558. Musées Royaux des Beaux-Arts de Belgique, Brussels. *Photo Musées Royaux des Beaux-Arts de Belgique.*

THE GALLEYS

A history of naval technology would be incomplete without some discussion of the galleys. In particular, it should be noted that the sixteenth and seventeenth centuries witnessed the golden age and decline of these boats.

During antiquity and the Middle Ages the galley had been the principal warship: the Athenian triremes, the Byzantine dromonds, and the Venetian galleys formed the military fleets, while sailing ships were exclusively merchant vessels.

The galley's progress was strictly independent of the winds; since it had its own propulsion, the oarsmen, the galley could follow the route it desired. It could also maneuver during combat, carry combatants on its platforms, and attack enemy galleys with its ram, by utilizing the shock of its own mass. For centuries the sailing vessel remained defenseless against the oar-powered ships that could attack it in great numbers and ultimately overpower it. As late as the fifteenth century, during the Turkish siege of Constantinople (May 1452), several large sailing ships came bringing ammunition and some combatants. They were suddenly becalmed a short distance from the port, and were attacked and nearly sunk by a number of small Turkish oar-driven ships; they were saved only by a sudden wind that pushed them into port, without which they would have been lost.

For long centuries, then, this special characteristic of independence from the wind made the galleys the only instrument of naval warfare.

The history of the galley beginning from its origins will not be discussed here; it is our intention merely to discuss the last phase of its development, that is, the sixteenth-century galley and its decline during the seventeenth century. It had by then become a much larger ship than the galleys of antiquity; the Athenian trireme's displacement could not have been more than ninety tons, that of the Roman *liburnes* with two rows of oarsmen barely sixty tons.

The oar and the oarsmen

Their small size permitted them to utilize quite short (approximately 14 feet), lightweight (approximately 44 pounds) oars, each maneuvered by a single oarsman. At the beginning of the Middle Ages the galleys began to increase in size, probably under the influence of a strategy that required placing more combatants on them. (The Athenian triremes carried few combatants—some ten soldiers.) But while the oar-driven vessel became larger, we have no documentary evidence showing the successive steps in this development. Between the ninth and thirteenth centuries larger Venetian galleys* seem to appear like the fruit of a kind of sudden mutation.

In order to explain this mutation, we must refer to Figure 141, which indicates the position of the oarsmen. On the ancient galley, only about one-third of the total length of the fourteen-foot-long oars was inboard, with the results that the oar did not leave a great deal of space above the bench on which the oarsman sat, and there was only a slight displacement of the handle for each stroke of the oar.

* G. Casoni, *Venezia e le, sue lagune* (Venice, 1847).

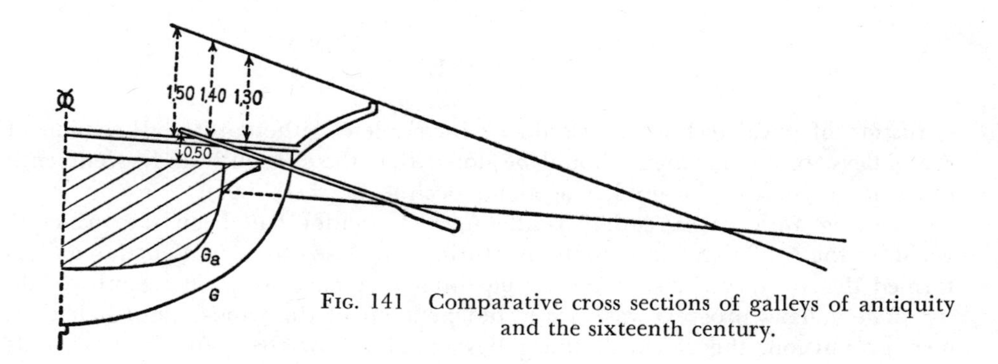

FIG. 141 Comparative cross sections of galleys of antiquity and the sixteenth century.

For both of these reasons, there could be only one oarsman; he remained seated, and worked his oar by bending first forward, then back, without moving from his bench. When the ship was larger (line "G" of Figure 141), the tholepin of the oar was higher above the water; the oar was longer and heavier, and could no longer be worked by a single oarsman. The inboard portion of the oar was longer, and rose higher (see figure) above the deck. The oarsman had to work in a standing position, for since the circle described by the end of the oar was now larger, it was impossible for a man in a sitting position to describe this circle with his hand. The oarsman stood and bent forward with his entire body, then fell back into a seated position on the bench behind him, as the blade of the oar came out of the water.

Figure 141 shows that the length of the handle of the oar and its height above the deck permitted several oarsmen to be placed on the same oar — a practice that was made absolutely necessary by the increased weight of the oar. There was room, for example, for two or three oarsmen side by side; if five oarsmen were desired for a still-larger oar, three could be placed on one side and two facing each other, the former acting by pushing, the latter by pulling.

Thus we witness the appearance of oars operated by several oarsmen. According to Giovanni Casoni, in the second half of the fourteenth century galleys of the following dimensions were in existence:

Keel	$66\frac{2}{3}$ feet
Beam at bottom of the hold	$10\frac{1}{4}$ feet
Beam at three feet above the keel	$18\frac{2}{3}$ feet
Beam at midships above the keel	28 feet
Depth	11 feet
Length at bottom	97 feet
Overall length	$102\frac{3}{4}$ feet

The largest galleys may have measured, according to a document of 1318 mentioned by the same Mr. Casoni, 133 feet in length and 33 feet in breadth, with 100 oarsmen, four or five to each oar, when there should have been more than 150 oarsmen. Perhaps these were professional oarsmen whose number was supplemented, when it was necessary to work all the oars, by a number of the soldiers on board. The sailors and soldiers on board plus the oarsmen may thus have totaled 150, in contrast to the 200 carried on a Greek galley. By the sixteenth century the galley carried a total of 250 men, which justified the increase in size (Figure 142).

The seventeenth-century galleys

This sixteenth-century galley, which underwent a final enlargement in the seventeenth century, is the type of which we have the most detailed knowledge, for it is depicted in numerous drawings and documents.

The typical galleys of this period were the ordinary galleys, the *patronnes* and *capitanes;* the latter carried commanders of squadrons and fleets. The *reale,* a luxurious vessel reserved for the commander of galleys, was a state vessel that was utilized only on special occasions.

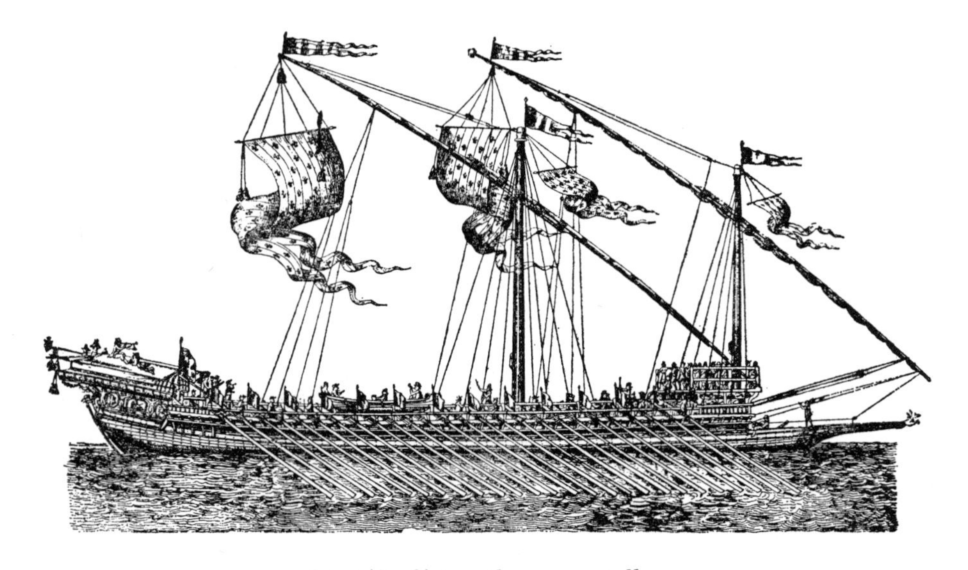

FIG. 142 Sixteenth-century galley.

The ordinary galley is represented by a very detailed reconstruction made in the nineteenth century by the engineer Garnier of the galley *Ferme* (*Association technique et martime,* 1909) (Figure 143). Its dimensions were:

Length	153 feet
Beam	19 feet
Depth (top of the keel to the deck beams)	7⅔ feet
Draft	7 to 7¼ feet
Disaplacement	260 to 280 tons

The hull was surmounted by a rectangular platform that held 26 benches on one side and 25 on the other, the place of the missing bench on the left being

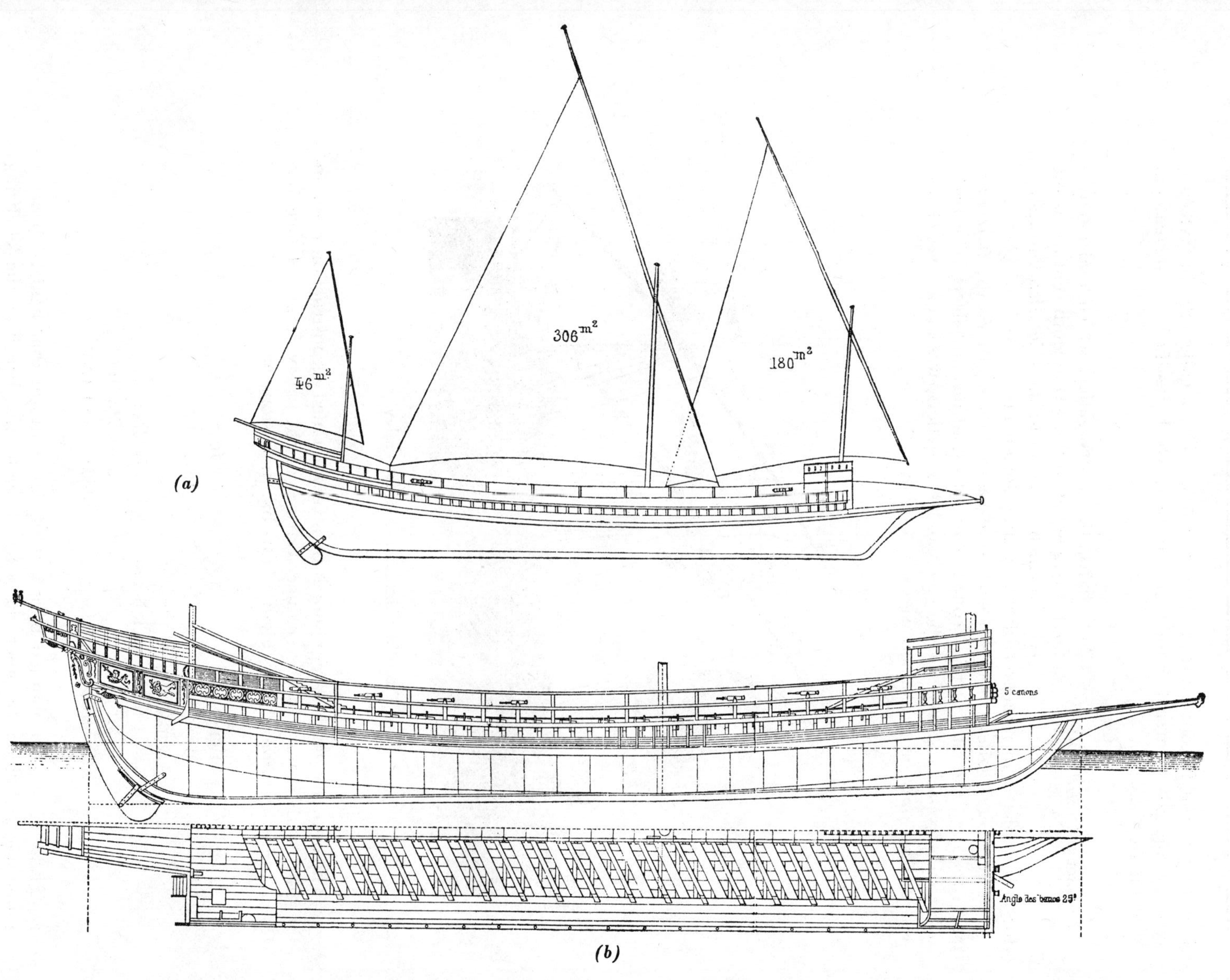

FIG. 143 The seventeenth-century galley *Ferme*, as portrayed by Garnier: (*a*) the sails; (*b*) plan and cross section in elevation.

occupied by the clay hearth used for cooking and heating food. In this regard it should be mentioned that the galley terminology was a very specialized one; the officer corps, consisting solely of officers belonging to the upper nobility, was a closed group, and its very exclusive spirit is revealed in the terms used on board the galleys, and even in the components of construction. This very special language, which was different from that of ships, disappeared almost completely when the galleys were eliminated in 1748.

Each of the 51 oars was operated by 5 oarsmen; the "gang" thus consisted of 255 oarsmen in the *zenzile,* or ordinary galley. As for the soldiers, sailors, and so on, their number varied depending on the nature of the expeditions undertaken by the galleys.

As for armament, only the front of the ship had space for artillery; five cannons, firing in the ship's axis, were placed in a parallel line on sliders that took the recoil. In the axle was a large cannon, the bow gun; it was generally 36-caliber, but could be as much as 48-caliber. It was flanked by two smaller guns (8- and 6-caliber).

On the edges of the platform were smaller guns, in most cases two-pound swivel guns, mounted on pivoting crutches. These were not considered artillery; they were almost portable weapons, more like rampart guns than cannons.

The soldiers on board the galley had blunderbusses or harquebuses, as well as pikes and sabers for boarding expeditions.

The galleys of Lepanto

As has been mentioned in connection with the fleets of sailing vessels in the time of the Armada, it is instructive to note the characteristics of the Christian fleet that won the victory of Lepanto (1571), as regards number of vessels, cannons, and crew. This fleet consisted of:

Galleys	207	(including 105 Venetian, 31 Spanish)
Ships	30	(10 Venetian, 20 Spanish)
Galleasses	6	(Venetian)

The galleys carried 5 cannons, the ships 27 (the Spanish ships had 20), the galleasses 18. The crew consisted of 43,500 oarsmen, 12,920 sailors, and 28,000 soldiers, for a total of 84,420 men (of which 41,000 were Venetians; 15,900, Spaniards).

These figures indicate that the models were highly standardized; the Venetian, Spanish, papal, and miscellaneous galleys were practically identical in size, armament, and crews. The galley had become an almost standardized type at the peak of its development, in contrast to the variety of types mentioned in connection with the English squadrons that in 1588 attacked the Spanish Armada. This was the end of the evolution of the oar-driven ship and the beginning of that of sailing ships.

The Turkish fleet was composed of 271 ships, 208 galleys, and 63 smaller ships. The Turkish ships were less uniform in size, and were in general smaller and sat lower in the water; they were dominated by the Venetian and Spanish galleys, whose redoubtable Spanish infantry, well trained and armed with harquebuses, overwhelmed the Turks with deadly fire.

This last action in the life of the galleys is also noted as one of the great battles of history, since it ended in the destruction of the Turkish fleet: 80 ships sunk, 140 captured, 30,000 dead, 8,000 prisoners taken, 10,000 Christian slaves liberated. The allies, in contrast, lost 12 galleys (8 Venetian), and counted 7,600 dead (4,800 Venetians).

But this was the last victorious appearance of galleys on a battlefield. Their cannons were insufficient in number, and could be aimed only by turning the galley; they fired only once before the boarding of the enemy vessel took place. The sailing ship was developing during the seventeenth century; as we have seen, its artillery was increased. The high-decked vessel crushed the galley, which was too low in the water. In 1684 a Spanish squadron of 36 galleys attacked the *Bon* before Porto Ferrajo on the island of Elba; the ship sank several galleys and damaged the rest, which had to retreat.

Galleasses

An attempt had been made to increase the offensive power of the galley by giving it more, and more powerful, artillery. This led to the birth, in the sixteenth century, of the galleass, constructed by the Venetian shipbuilder Bressano, who was inspired by the size of the "merchant galley," whose use will be described later.

A deck like the deck of a ship had to be installed above the oarsmen, and on it guns were emplaced. The galleasses were gradually equipped with 18 cannons of 4, 6, 12, 36, and 48 calibers, and 30 two-pound pivoting swivel guns. But the necessary stability and displacement were acquired only by considerably increasing the galley's size:

Length	170½ feet
Beam	34 feet
Draft	14 feet
Displacement	1,000 to 1,100 tons

The result was a ship similar to the sailing vessels, with a large crew: 450 oarsmen (50 oars, 9 men to each oar), 150 sailors, and 120 soldiers were required to man it.

The major defect of these galleys was that they traveled too slowly when under oar power; a too often neglected comparison between the ancient and modern galleys and the galleasses, namely, that of the driving power of the rowers, is suggestive, and shows the great inferiority of the galleasse.

If we relate the driving power (man power) of the oar-driven boats, determined by the number of oarsmen, to the strength of the hull, characterized very approximately but satisfactorily by the surface of the midship section, we find the following:

	Number of oarsmen	*Midship section*	*Men per sq. yard*
Greek trireme	177	5½ yards	35.4
Seventeenth-century galley	255	11 yards	25.5
Galleass	450	49 yards	10

The Athenian trireme was able to attain an average speed of 5.2 knots; the galley of the sixteenth and seventeenth centuries attained only 4.5 knots, the galleass 2 to 2.5 knots.

The advantage of oar propulsion became illusory; the galleass could be little more than a floating fort, powerfully armed, but remaining almost motionless, without being able to maneuver. At Lepanto the first line of the allied fleet consisted of six Venetian galleasses that helped to disorganize the Turkish fleet passing before them, but the actual fierce battle was fought behind the galleasses, between galleys of similar dimensions that were able to maneuver in combat.

Construction of the galleys

The galley was constructed in dry dock, beginning with the body, which consisted of the keel, the ribs, and longitudinal reinforcements. The curved ribs were numerous (142 in an area of 154 feet) and consequently small (a little more than 3 inches). The ribs were joined at the top by a long, strong piece of wood acting as a wale, on the inside by six longitudinal pieces that joined them inside the hull. The keel and the ribs were of oakwood, the six longitudinal pieces inside the hull of pine. Along the sides the bottom was closed by 73 strong deck beams (two per yard) 9¾ inches long and 3 inches thick.

The hull was strongly reinforced longitudinally, which was necessitated by its resistance to flection, for it is noticeable in drawings of galleys that they were very flat, the relationship of length to height of hold being much less than on the sailing ships.

The planking, which closed the bottom from the wale to the keel, was attached to the ribs. Up front was a ram, a kind of spur 19½ feet long, made of pine, and attached to the stem by a supporting brace; planking 30 feet long attached it from above to the deck, while on the side two pieces of elmwood joined it to the ribs and to the planking of the forward portion.

On the bridge a gallery *(corsia)* 3 feet wide that overlooked the benches permitted movement from fore to aft, for the maneuvering of the yards and sails, or for the personnel in charge of the galley slaves. (The terms peculiar to galley construction are in large part of Italian origin.)

On the hull was a large rectangle *(talar, palamante)* that held the benches of the rowing gang (see Figure 143); thus all the benches and oars were the same length. The outboard projection of the oars was therefore greater at the ends than in the center of the galley. The galley measured approximately 19½ feet beam amidships; the frame was 26¼ feet, overhanging the hull at the midship section by three feet. Along the longitudinal side of the frame were the pegs (now called tholepins) or supports for the oars; they were lodged in iron sockets mounted in the *apostis,* and the oar was held in place by a hemp rope *(estrope).* The rubbing parts of the *apostis* and the oar were fitted with pieces of hard holm oakwood, which wore little on each other and were easily replaced.

The construction of a galley required three months' work with a crew of 70.

The rudder

The arrangement of the rudder is seen in drawings of galleys. The head was equipped with a tenon; the tiller had a mortise that fitted over the tenon. The tiller was made of elmwood, and was approximately 6½ feet long and 9¾ inches wide at the joint of the tenon. Two ropes pulled it to right and left for maneuvering, which was done by helmsmen standing at the head of the tiller.

Ballast

The galley was relatively narrow, and would have lacked stability under sail if its stability had not been increased by placing ballast at the bottom of the hold. In addition to two months' food supply, the ballast consisted of:

500 baskets of stones — approximately 15 tons
620 stone balls of 36 lbs. — approximately 11 tons
144 stone balls of 8 lbs. — approximately ½ ton
120 stone balls of 6 lbs. — approximately ½ ton

The total ballast was approximately 27 tons, or 1/10 of the displacement.

Benches and oars

The benches were made of pine; there were 26 on the right side and 25 on the left side, where the small "kitchen" was located. (The *patronnes,* which were slightly larger, had 28 and 29 benches, the *reale* 30 and 31.) Onto these benches fell the galley slaves when, having plunged the oar into the water from a standing position, they sat down in order to bear down on the oar, whose blade rested on the water.

The benches were 8 feet long, 6⅔ inches wide, and 5½ inches thick; the round upper face was covered with worn woolen and cowhide covers, "to prevent the oarsman from being chafed when he fell back." The benches were placed obliquely, as is indicated by Figure 143.

The distance between the tholepins (which was also the distance between the benches) was 4 feet. A footboard (*pédague*) opposite the bench served as a footrest for the oarsman when he rose; it was 5½ inches wide and 3 inches thick, and its upper face was round so that the bare foot of the oarsman would not be injured. Between the benches, lower down, was the small step on which rested the oarsman's chained foot.

Starting in the sixteenth century the oarsmen were kept constantly chained, those on the right side being chained by the left foot, those on the left side by the right leg. Four of them had to sit two by two, feet facing, in a space of 4 feet; the fifth was able to sleep on the bench.

The oars (Figure 144) were made of beechwood, which was flexible and strong; they were all of the same length — 38¾ feet for ordinary galleys, 45⅓ feet for the *reales* and *patronnes.* About 26¼ feet of the oar was outboard; its blade was 9¾ feet long and 7 inches wide at the end, narrowing to 3½ inches toward the *tiers,* the portion between the tholepin and the blade. Inboard was the *fiol,* the thickest part of the oar (the diameter diminished from 6 inches to 3 inches as it approached the blade), and the handgrip.

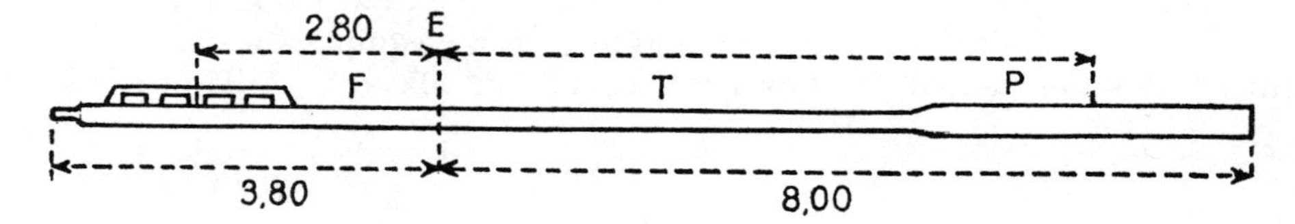

FIG. 144 Galley oar (seventeenth century).

The heavier weight of the *fiol* was not sufficient to balance the weight of the outboard portion of the oar, and lead plates were therefore embedded in the *fiol* to make it heavier.

The end of the *fiol* could be narrowed to form a handgrip, but for four of the oarsmen, on an area of approximately 6½ feet, the diameter of the oar was too great to be grasped with the hand. To these 6½ feet was fastened a piece called the *manille;* it was flat and made of wood, and it had as many openings as there were oarsmen on this section of the oar.

Masts and sails

The medieval galleys had only a single mast; in the sixteenth century they almost always had two, the mainmast (*arbre de mestre*) and foremast (*arbre de trinquet*). They were square-headed, round from the head down to the deck, and octagonal belowdecks, and carried the yards to which the sails were bent.

The Construction Council assembled at Marseille in September 1691 by the bailly de Noailles, lieutenant general of the royal galleys, whose members included squadron chiefs, galley captains, and the shipbuilders J.-B. Chabert, Louis Chabert, and Hubacq, fixed the sizes and proportions, as well as all the details of construction, masts, and rigging.

The mainmast was 5 feet forward, the foremast 52½ feet forward, of midship; the masts were parallel and vertical. The lateen yards were composed of two partially overlapping pieces (the *penne* and the *quart*); they were 123 feet long and had a diameter of 14 inches at their widest part. The two sections were held together by four *ligadures,* tarred ropes that gripped the beveled sections.

The size of the sails depended on the force of the wind. When a slight breeze was blowing, a small spar (the *espigon*) was lashed to the end of the yard; it was 21 feet in the case of the mainmast yard, 16 feet for the yard on the foremast, and 6¼ inches in diameter, decreasing to 4 inches at the tip. The total surface area of the sail of an ordinary galley was evaluated at 5,576 square feet, with its center located at 10¼ feet in advance of midship and 35 feet above the waterline.

The largest sail, the one carried by the *espigon,* was called the *le grand marabout.* However, depending on the force of the wind, smaller sails might be used. The surface of the *marabout* was about 80 percent of that of the *grand marabout;* the *misaine* (or *méjane*) approximately 75 percent, the *voilette* or *boufette* 36 percent. The *polacron* was the smallest sail, used in heavy weather. The foremast yard might carry either the large *trinquet* or the small *trinquet,* similar to the *maraboutin* and the *boufette,* respectively.

Lastly, there was a square sail, the *tréou,* whose area was about the same as that of the *boufette,* and which could be bent on a square yard 54 feet long, composed of two spars with a diameter of 9¾ inches at the center, 3½ inches at the ends. This sail was used with a strong following wind.

The maneuvering of the sails required lowering the yards to the *corsia,* bending the desired sail, and then hoisting the yards back into position.

Navigating by oar

The regulations of the sixteenth and seventeenth centuries indicated four different rowing maneuvers, depending on the widest angle described by the oar; in none of these methods did the oarsman remain seated. The widest arc described by the handle was that of "rowing by touching the bench"; in this case the oarsmen bent forward, brought the handle down to the bench in front of him, then fell back into a sitting position on his bench, pulling the oar with him and leaning backward.

It has been calculated that 26 complete strokes of the oar, with return to the starting point, could be performed in one minute, given maximum effort by the oarsmen. For prolonged rowing this was cut to 22 strokes.

Pierre-Alexander-Laurent Forfait estimates that in a calm sea the best-equipped galley could make no more than 4.5 knots during the first hour, 2.5 to 1.5 knots thereafter. The average maximum speed of 4.5 knots, which corresponds to the figures given earlier for the number of oarsmen on the galleys and galleasses, required efforts of from 11 to 15½ pounds per oarsman – which, though possible, demanded a total displacement of the human body in less than three seconds, and quickly brought the crew of oarsmen to the point of exhaustion.

Navigation by sail

It must also be taken into consideration that, contrary to the common belief, the use of the oars was limited. Rowing was indispensable in calms and for maneuvering during battle, but it could not be continued for any length of time. In contrast, the use of the wind was very advantageous for the galleys. For a midship frame of 107⅔ square feet, the sail surface of 5,382 square feet gave a relationship of 50, while on vessels of this period the relationship was only 27.

These relationships indicate that the galley, when very heavily equipped with sail, could utilize all available wind power, which explains the variety of sails that could be adapted to every case.

Given the characteristics which have just been mentioned for the ordinary galleys of the seventeenth century, their progress with a following wind must have been better than that of the best frigates of the end of the eighteenth century. It has been estimated that the speed of the frigate was half that of the wind pushing it; galleys equipped with sail must have reached a speed 10 to 15 percent higher. Thus it was easy for them to attain speeds of more than 10 to 12 knots under sail, that is, a speed considerably greater than that which could be obtained by the gang of galley slaves. In addition, the lateen sail gave them the power to beat to windward; they easily kept close to the wind at less than five quarters (five times 11°15′); with a slight breeze and a little help from their oars they could come a quarter closer to the wind's eye, that is, follow a course that formed an angle of 45° with the current of air blowing the breeze.

Despite their qualities as sailing ships, and the possibility of maneuvering with their oars, the galleys had serious defects. We have already mentioned their lack of cannons, which made them powerless before high-decked vessels; their freeboard prevented them from confronting heavy seas. In the Mediterranean, where one was seldom out of view of land, they quickly sought shelter in case of a heavy wind. They appeared in the ocean only on rare occasions, and then played only a very limited role. For all these reasons, by the seventeenth century the galleys were declining, and were gradually abandoned.

Comparing the figures given on the fleet that fought at Lepanto, we note that in December 1676, one hundred years after Lepanto, the number of oarsmen in France had dropped to 4,710 men. It is true that Colbert stimulated the zeal of the courts in order to send the largest possible number of condemned men to the galleys. But it was no longer worth the effort. The edict of 1748 signed the death warrant of the galley. The last galleys, useless and disarmed, were found in 1797 by the French troops upon their entrance to Venice.

The merchant galley

Although the galley, with its large gang of rowers and the provisions it required, was not capable of carrying much in the way of passengers and merchandise, it could, however, be easily adapted to commerce.

We have already noted that the galley was as much, if not more, than a sailing ship as it was a rowing ship, and its propulsion by oars might be only occasional. The crew of oarsmen, reduced to a secondary role that could be decreased even further in a commercial vessel, could be cut to a smaller number of oarsmen. Once it was no longer a question of warfare but merely a matter of temporarily compensating for the absence of wind (since we have seen that the galley can beat up to windward if the wind is contrary) it mattered little that the galley was slower; as soon as there was a wind, from whatever direction, the oar was no longer indispensable.

Gaining on the number of oarsmen, the provisions, and the ballast, which could be replaced by merchandise, the warship could be transformed into a commercial vessel. And so the merchant galley appeared. It quickly increased in size, as the following example shows: overall length, 150 feet; beam, 21 feet. This was no longer the narrow galley of yore: its beam had increased, and the relationship of length to beam had diminished from 7.5 to 6.6. Since the galley now had a higher freeboard in relation to the increase in beam, the oars were now 29½ feet long, and only two or three men were placed on each oar, which indicates that less speed was required.

These mixed galleys made it possible for Venice to carry on trade during the Middle Ages and the fifteenth century with Syria and Egypt, with the islands, and to transport numerous pilgrims to the Holy Land.

For heavy transport and commercial relations with the northern countries, which meant confronting the ocean, Venice had several sailing vessels; in the Lepanto fleet we find ten, which were capable of carrying heavy artillery. But Venice retained her military fleet of galleys, which she continued to regard and employ as her principal weapon. Even when it was a question of escorting and protecting a convoy in the ocean, it was always a squadron of galleys that was sent to do the job.

The experiment with the galleass mentioned earlier was a development of the "heavy galleys" for trade, with a deck that supported a more powerful artillery. As we have noted, from the military point of view the experimental use of this hybrid ship-galley was a failure.

The ordinary galley therefore continued to be the principal Venetian weapon of combat, and the Venetians were very reluctant to adopt sailing warships. Not until the end of the seventeenth century (1696) was Venice obliged to recognize that the galley was out of date. Even then, however, she still obliged the commander of the fleet to sail and exercise his command on a galley.

Attempts were made, however, to construct sailing ships; discussions were held on the qualities of the various models tried, which is not surprising given that in the France of Colbert, as we have seen, discussions were held on the subject of the best lines for ships.

But the decline of Venice was practically sealed after the Treaty of Passarovitz (1718); the Venetian military navy had lost its reason for existing. The shipyard ceased its activity in 1718, and vessels under construction at the time of the Peace of Passarovitz rested in dry dock for fifty-five years. French troops, upon their entrance into Venice in 1797, found only a phantom navy.

THE USE OF WOOD IN NAVAL CONSTRUCTION

The supply of building material, wood, rope, and so on, was always a major problem for those nations whose fleets were a major element of their power. In the writings of the ancient authors we find numerous references to the Athenian preoccupation with procuring wood for construction. Greece first turned to her forests, which were rapidly exhausted. Plato remarks that in his day there were no trees left that could furnish large beams like those still existing in some houses. Athens forbade the exportation of wood and wax (for caulking), and tried to import, even from distant lands, the materials needed for construction.

In order to create her last fleet, besieged Carthage, her communications with the hinterland broken, had to demolish the houses and use their beams and doors; since she had no more hemp, rope had to be made from the hair of the Carthaginian women. When they were organizing their ports and shipyards near Naples, the Romans found wood in abundance in the neighboring forests. Venice obtained her wood from Dalmatia, and the inhabitants of this area, whether rightly or wrongly, accused Venice of having ruined their country's forests by leaving the ground bare.

Lumber for masts

Thus it is not surprising to find the same problems in England and France, between the sixteenth and seventeenth centuries, when an effort was being made to construct fleets. In England, the supply of lumber was for centuries a subject of serious concern to the British Navy. The English forests easily furnished oak for the hulls, but the masts could not be found in England. The lumber used in masts has to have qualities of length and elasticity that are furnished only by large pines, and these trees grow in climates that have a long winter and a short summer. Thus England had to turn to New England, Canada, and the Ukraine for topmasts, to Norway for smaller species.

The Scotch pine, which is a fir, has fibers of coarse texture that lack sufficient elasticity. The presence of resin in the conifers gives them the required qualities, and the resinous content of the wood depends on the climate. The best is the *Pinus silvestris* from the Baltic region, but large trees of this species are found almost exclusively in North America. The Baltic forests, long exploited, furnish few trees of a diameter greater than twenty-seven inches, whereas the American forest can supply trees of diameters in excess of three feet.

It is interesting to note the sizes of masts and yards used in various eighteenth-century ships:

	1ST RATE (120 cannons)		3RD RATE (74 cannons)		28-CANNON FRIGATE	
	Diameter	Length	Diameter	Length	Diameter	Length
	(in feet)		(in feet)		(in feet)	
Mainmast	$3\frac{1}{3}$	$119\frac{3}{4}$	3	$108\frac{1}{4}$	$1\frac{2}{3}$	72
Main yard	2	105	2	$98\frac{1}{2}$	1	$62\frac{1}{3}$
Main topsail mast	$1\frac{3}{4}$	69	$1\frac{1}{2}$	$65\frac{2}{3}$	$1\frac{1}{3}$	42

The difficulties of obtaining lumber for masts drew the attention of the explorers to the trees of the newly discovered lands. A map of 1502 mentions, in connection with Newfoundland, that "here there is much lumber for masts." The propaganda for the colonization of Virginia stressed that the lumber from America could replace the Baltic wood.

There existed a broad band of "white pine" forests in the northern portion of the new continent. The first American cutting probably dates from 1623. The arrival at Deptford of a cargo from New England drew attention to the great size of the imported lumber. The first Dutch war completely cut off shipments from the Baltic. Beginning in 1652, all the masts for the British Navy came from Portsmouth (New Hampshire) or Falmouth (the modern Portland), Maine. Naturally, trade in tar, hemp, and also oak could easily be added to the lumber trade.

Thus the British Navy was able to satisfy its preference for masts made from a single piece of wood, even for large masts. In contrast, France and Spain, who obtained their supplies from the Baltic, where only small trees were found, had to build their thick masts by jointing smaller pieces.

Measures were taken in England to reserve for the navy all imported trees whose diameter was greater than twenty-four inches. The French and Spanish navies, which had much thinner trees at their disposal, combined them (see Figure 145 for the various types of joints) by banding them with iron hoops in sufficient number to hold them together. The distance between the hoops was generally equal to the diameter of the mast.

Lumber for the hull For the hull itself, the most suitable and highly valued wood was oak. As regards oak, every country was generally able to supply its own needs. *Quercus robur* was the type most frequently employed; it grows from Sicily to Norway, and from Ireland to the

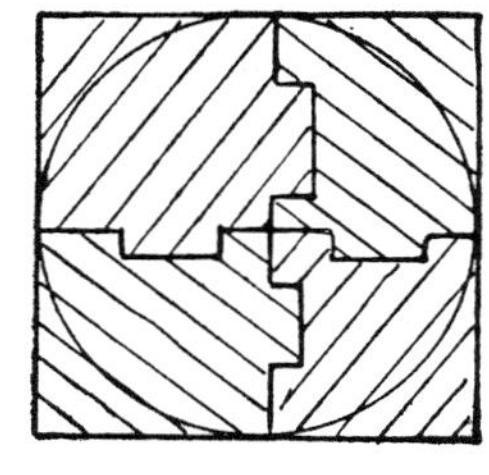
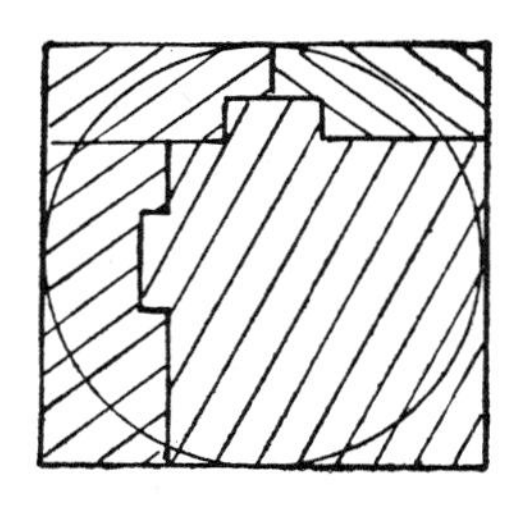

FIG. 145
Jointed masts.

Urals, but its quality varies depending on the soil and the climate. The American white oak (*Quercus alba*) was susceptible to attack by "dry rot," which was capable of quickly destroying the strongest pieces. The susceptibility of the wood to the attack of the small mushrooms that form this rot is determined by the time when the trees are cut down and the presence of the sap in the wood.

Moreover, the trees chosen for naval construction had to be either straight pieces or curved pieces – the latter for the ribs or knees for joining the deck beams to the ribs. Whence the practice, both in England and France, of going into the forests on foot to choose the trees, the builder noting the use of each tree by the pieces which he could obtain from it (Figure 146). The planking required straight pieces 39⅓ feet long and 27⅔ inches thick. At the beginning of the nineteenth century the size of the wooden vessels was limited by the maximum thickness of the pieces of lumber that went into them.

FIG. 146 Choice of trees for construction.

In the eighteenth century a vessel carrying 74 cannons, the type that formed the bulk of the military navies, required 3,000 loads of lumber at 50 cubic feet per load, an amount that could be supplied by total deforestation of 60 acres of oak trees. In England the heavy call for lumber for the navy was equaled by the needs of industry (the blast furnace) and house construction. The latter, which was particularly vigorous during the reconstruction of London after the Great Fire, dealt the *coup de grâce* to the English forests.

In 1677 Parliament voted funds for the construction of thirty vessels, intended for the struggle against the fleet being readied by Colbert. Phineas Pett and Anthony Deane, the foremost shipbuilders, who were searching the country for lumber, had to admit that large pieces for first- and second-rate vessels were impossible to obtain. The thirty vessels were finally constructed, but the haste with which they were built meant cutting down the trees in the wrong season; they were not dry, and remained full of sap, and in a few years they were destroyed by dry rot. The trees had to be replanted, and England had to wait the one hundred years necessary to have adequate lumber.

Protection of the forests

An Act of 1668 provided for the enclosure and replanting with oaks of 11,000 acres; in 1698, 2,000 acres were set aside in the New Forest to be replanted with oaks, and an additional 200 acres per year for 20 years.

In France, arrangements were made to ensure a supply of lumber for the navy. An ordinance of Philippe le Long in 1318 served as the basis for the ordinances of 1376, 1402, and 1515, which indicate that the king then had only two "masters of works" for choosing lumber in Champagne, Paris, and Normandy, and two in Languedoc. It was forbidden for anyone "possessing forests, thickets, and full-grown trees suitable for use of the navy within five *lieues* of the sea or rivers, to sell, cut down, or damage them, except with the express permission of His Majesty and after they have been inspected by the Commissioners of the Navy and carpenters employed in its service."

An ordinance of August 1669, and the Council's decrees of September 7 and September 28, 1694, testify to the same concerns. A choice was made of the trees, which were then hammered – that is, on a strip from which the bark had been removed a brand was made with an iron marked with an anchor and the fleur de lys.

The engineer-shipbuilders were entrusted with these operations in the eighteenth century; their reports listed the owner of the forest (and possibly ecclesiastical property, property in mortmain, or royal forests). The trees had to be cut down during the period of the waning moon, between November 1 and March 15.

It was realized that it was preferable to submerge the trees in water in order to remove the sap from them; the water gradually replaced the sap. The trees were then dried, attention being given to the circulation of air around them.

In this way huge stockpiles of lumber were built up that represented a period of as much as ten years between felling and utilization in construction. Lumber was thus obtained that was not attacked by rot, and the vessels constructed with it had an extraordinary longevity, sometimes even surpassing one hundred years.

PORTS

During most of their existence ships must make stops or stay for periods of time in ports, where they must find a shelter. The role of the port is thus of primary importance for navigation, and from time immemorial the sea-dwelling peoples have tried to create such shelters for ships. The Phoenicians, Greeks, and Romans carried out numerous projects for this purpose, and the modern world has had to solve the same problems they faced.

Two stages are to be distinguished in the period 1500 to 1700: the sixteenth century, when wars and political dissensions impeded the execution of major port programs, and the seventeenth century, when continuous efforts made possible the achievement of remarkable accomplishments.

Silting

The sixteenth century utilized natural shelters that were well known and had long been frequented, such as the ports of Marseille and La Rochelle, and the estuaries of rivers and streams. But in the latter, deposits of sand and mud, made unstable by storms and currents, transformed the natural water levels by decreasing their depth, and succeeded in filling up the ports or preventing access to them.

Harfleur, on the mouth of the Lézarde in the Seine estuary, was totally silted up, and became useless. In contrast, an unusual tide created a pass through a coastal strip bordering a swampy lagoon; the water accumulated behind it found a sudden outlet in an opening created through the strip of land. Once an entrance had been accidentally created, the lagoon became a shelter that was later to be the first basin of Le Havre.

In northern France, the mouths of the Somme, Liane, and Aa (to mention only a few) were utilized, but they were frequently blocked by silting; canals had to be dug, and canals already in existence had to be maintained.

When the Spanish Armada arrived in the English Channel, the departure of the debarkation fleet assembled at the mouths of various rivers in the Netherlands was imperiled by their silting. Canals and passages had to be dug and other work done, which hindered the fleet's concentration and departure.

The rias

In Brittany and on the Spanish coast (the Basque country of the Asturias and Galicia), the peculiar nature of the rias offered secure shelters. The rias were valleys dug by streams in early geological ages, and later invaded by the sea, whose level had risen. In this way the valley, where a shallow stream flowed, became a long, deep-water estuary, which was naturally well sheltered, since it extended deep into the coastline (Figure 147).

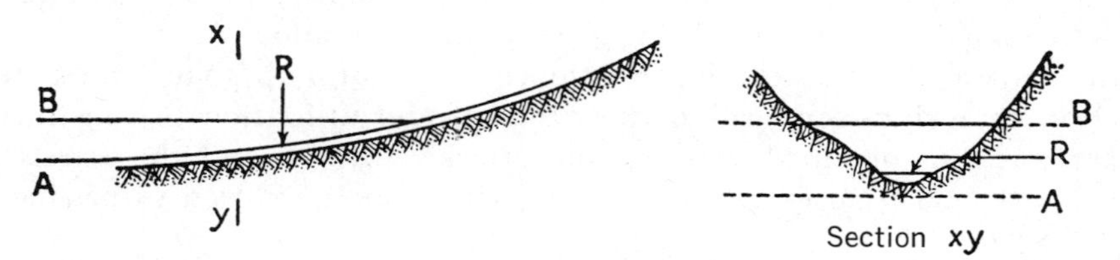

FIG. 147 Formation of the rias.

Such was the case, for example, with the port of Pasajes; and also the port of Brest, where the Penfeld offers a deep, narrow port whose entrance is well defended by a cliff. The Penfeld, a genuine ria, extends between steep schist banks for 1¾ miles between the Avant Garde and the Arrière Garde; it is 328 feet wide, and its depth ranges from 32 to 42 feet.

Another example was Morlaix, which was defended by the castle constructed on an island at the entrance of the stream (the Taureau), and by many creeks and secondary ports formed by the rise of the sea level in the coasts cut into by erosion. In the sixteenth century the ria of Morlaix formed an important port. Downriver were shipyards, and it was here, at Le Dourdu, that the famous *Cordelière* was built.

In 1542 the citizens of Morlaix, who were often attacked by the English and were anxious to be sheltered from the latter's incursions, constructed the first fortress on the rocky cliff in the estuary; this was the castle of Le Taureau, where the city maintained a garrison. Louis XIV confiscated the castle in 1660; in 1680 Vauban installed in it a low battery in vaulted casemates, armed with large cannons. The ancient keep, which crumbled in 1609 and was rebuilt in 1614, dominated the platform built on the casemates; it was equipped with light guns, including old eight-sided culverins bearing the coat of arms of Brittany, surrounded by the cordelière (knotted rope).

However, the streams gradually silted up the rias, and dredging therefore had to be done to maintain the depth. However, this was simply a matter of removing mud, not of digging into hard rock, and the machines of that time, which as we shall see had little power, were sufficient for this work.

The constantly active mind of Leonardo da Vinci had envisaged devices for cutting into rock from water. They remained on the drawing board, and no attempt was made to translate them into reality: only after metals came into greater use were machines constructed that made it possible to straighten and deepen basins and passages in rock.

Brouage

In the seventeenth century the uninterrupted and more energetic activity of Richelieu, later repeated under Louis XIV, was translated into major achievements. While Richelieu's attention was drawn to the advantages of Brest, the creation of Brouage was his personal achievement; he needed a port near La Rochelle that could serve as a base of operations against the latter.

Brouage was situated in the center of a gulf (now filled in) approximately six miles wide. At the end of the seventeenth century only a harbor 437 yards wide remained, while the channel we see today is only 54 yards wide; thus the filling process was quite rapid. However, in the process of their formation the new lands left a basin sheltered by the emergence of neighboring deposits. As the new lands formed, salt marshes were created; exports of salt from the region are mentioned by the seventh century, and they increased during the Middle Ages.

In 1555 Jacques de Pons founded a city he called Iacopolis, on the site of the present-day Brouage. This city was constructed on lands formed by the dumping of ballast: the sailing vessels that came in search of salt arrived with their holds full of sand, stone, and pebbles, which they dumped on the spot, being careful not to block the basin. On the small hillock thus formed, the first fortifications were constructed of planks, masts, and earth. By the middle of the sixteenth century, Brouage had become a beautiful, well-protected port, "the best port in France," it was said, "especially for large ships . . . ; the fleet of ships harbored there sometimes is a wondrous sight: ships from England, Scotland, Flanders, Germany . . ." (J. Vige, *Brouage* [1960]).

Brouage survived the vicissitudes of the religious wars, and the king entrusted its government to François d'Espinay Saint-Luc, who was succeeded by his son Thimoleon. The latter ordered the construction of a church and a convent; in 1627 he ceded the town to Richelieu, who decided to make it into a large port and a stronghold, and had ships constructed and cannons cast. The fortifications were completely rebuilt by d'Argincourt, Richelieu's favorite engineer (1630–1640). But the silting of the port continued. Vauban visited it in 1685. The incessant cleaning no longer sufficed to keep the port clear, and the city declined; by 1765 it was reduced to forty inhabitants.

Typically, a natural cove forms the port in front of the city, which is a strong fortress surrounded by salt marshes (Figure 148).

The difficulties of maintaining the harbor of Brouage led Colbert and Seignelay to turn their efforts to Rochefort, located on a stream quite far from the sea, so that it was protected from attack by a fleet, and Brest, where the Penfeld offered a secure shelter and the depths of its ria. Arsenals and shipyards were therefore installed in both ports.

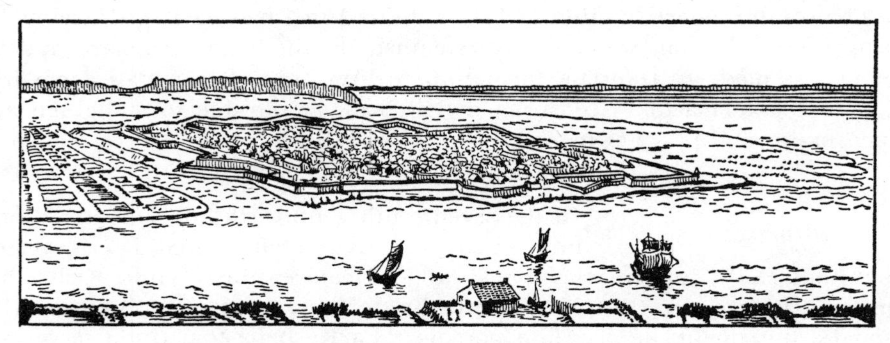

FIG. 148 General view of Brouage, seventeenth century.

Dunkerque

In the north, the major effort was concentrated on Dunkerque, whose organization was due to the very important work of Vauban. Here it was on the alluvial lands, won in part from the sea, as in Holland (Moëres), that the basins and channels were established.

It was necessary, in particular, to secure the access channel to the basin, which crossed the *estran* (the area included between the limits of the low tides and the high-water point of the spring tides). The channel, formerly an outlet for small coastal streams, was now dredged and made regular; however, it could be upset by the movements of the tide and the currents, and had to be protected along both banks by jetties capable of resisting the attacks of the sea (Figure 149). These projects were noted and justly described with admiration by seventeenth- and eighteenth-century technicians, particularly Bernard Forest de Bélidor.

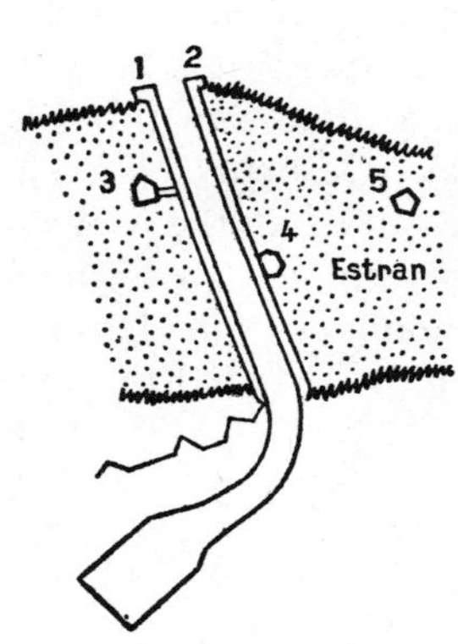

FIG. 149 Dunkerque channel, laid out by Vauban: 1–2, Jetties. 3, Fort Risban. 4, Fort Blanc. 5, Château-Gaillard.

Dunkerque, which was the subject of a dispute between the Spanish and the French, and was occupied in 1656 by the English, was repurchased by Louis XIV in 1662; he decided to make it one of the greatest strongholds in Europe, and a major port between the English Channel and the North Sea. Vauban drew up the general project and outlined the program of works.

A channel 50 fathoms (320 feet) wide was dug between two jetties; it was bounded by Fort Risban with its 42 cannons to the west, on the east Fort Blanc, both constructed of masonry; closer to the city, to the east, lay the wooden Château-Gaillard, and behind the west jetty the masoned battery of Revers. The channel gave access to the basin; it was capable of holding a squadron of 40 vessels, kept afloat by a large lock of 138 feet constructed in 1686 according to Vauban's plans. The fortification system surrounded the city with a triple wall, covered ways, and moats in which multiple locks permitted the use of the water for the defense of the stronghold.

Louis XIV put 30,000 men to work on the project. At four o'clock in the morning a cannon shot was the signal for the departure of 10,000 armed men; they marched in battle formation to a place near the construction site, where they laid down their arms and picked up their tools. At nine o'clock another cannon shot interrupted their work; they picked up their weapons and returned to the camp. Two other teams, each consisting of 10,000 men, shared the period from nine o'clock in the morning until eight o'clock in the evening.

Construction of jetties

At Dunkerque, it was decided to construct jetties on each side of a narrow channel that could be deepened by a continuous series of flushing locks. At first these jetties were constructed with fascines. The terrain was first leveled as much as possible, at low tide and with the power shovel. Next came the digging of ditches three feet wide and of the same depth, which were filled with carefully compressed clay. In this way the land was protected from the effects of the movement of the water.

Fascines were then laid, one against the other, to a depth of one foot; they were fixed in place with stakes three feet long, which projected at the top. Between the projections were entwined "plaits" made of three rods fifteen to sixteen feet wide, whose purpose was to hold the fascines very tightly together. Layers of fascines were built up in this manner. Then the surface of the embankment thus formed was covered with beams four inches square, forming a grillwork of open squares four feet long (inch of 27 millimeters, foot of 325 millimeters); these were held in place by piles pounded into the angles. The spaces were filled with large stones wedged and pressed together with blows from wooden clubs.

Experience with such jetties had shown that at Ostend mussels and other shellfish attached themselves to the surface and helped to protect the works against the undertow; it was therefore forbidden to remove them.

These jetties, however, could not help but be precarious and temporary in nature. Some twenty years after the construction of these early projects, Vauban directed jetties to be constructed of wooden coffers. The embankment of fascines was leveled to a depth of three feet below the level of high water at neap tide, which made it possible to work constantly no matter what the level of the tide; thus the base of the embankment, which had already settled and been filled with

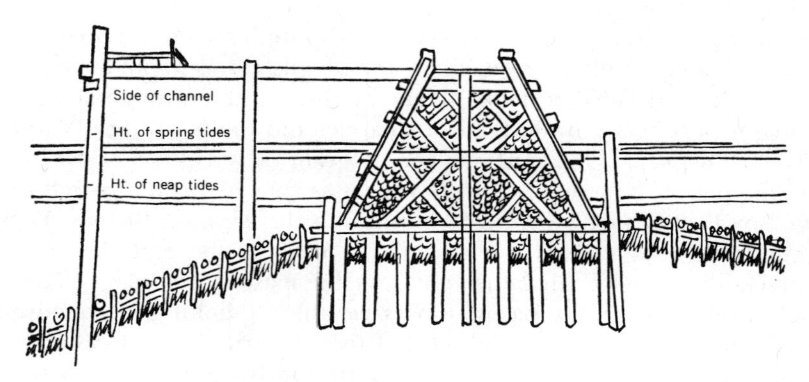

Fig. 150 Wooden coffers of the jetties at Dunkerque.

sand, was preserved. On this foundation were raised wooden frames (Figure 150) eight feet apart, joined by a planking that formed a coffering. In this coffering the masons placed stones and hard rocks; the empty spaces between them were carefully filled with chips of stone.

In order to extend the jetties into the sea beyond the fascine embankments, piles were laid, also to a level of three feet below the level of high water at neap tide; the space between the ground and the top of the piles was filled with stones, enlarging this foundation into a gradual slope on each side. Wooden trusses were laid, then planking, and finally came the filling of the coffering. The jetties were thus laid for more than a thousand fathoms, and formed a solid protection for the channel.

But the extensive use of wood put the project at the mercy of destructive animals, and in the eighteenth century the construction of such projects was improved by a generalized use of masonry with solid facings of dressed stone.

Docks and basins In the early stages of naval activity, the only method available for repairing ships, or simply scraping and repainting their hulls, was to haul them up on land or to place them on an inclined slip. With small, shallow, light ships, as for example the galleys of antiquity, hauling could be done easily. For heavier ships with greater draft, slips with winches were required; it was possible, using several capstans with a group of men working on their bars, to haul large ships out of the water. This maneuver was delicate and dangerous, for a "walking back" of the capstans knocked over the men harnessed to the bars.

When ships became larger, they had to be hauled into a basin from which the water could be drained. This was the dry dock, which appeared in many ports at the end of the seventeenth century.

In a tidal port, the construction of the dry dock and its use were relatively easy; all that was necessary was to keep the apron side above the level of low tide. In this way the basin could be constructed dry during low tides. A vessel could enter it at high tide; then the water was permitted to drain out at low tide; folding wooden miter gates closed off the entrance and kept the basin dry when the tide rose again.

The basins of Rochefort were constructed in this way, but with two sections placed end to end; the entrance opened into the deepest one, from which the ship passed into the second, more shallow section (Figure 151). The first section was able to hold first-rate vessels, while the second section took second- and third-rate vessels. The difference in level between the two was seven feet; it was believed that the upper section would be less disturbed by the water that flowed into the basins and would naturally drain into the deepest portion.

Despite this arrangement, and despite the pumping machine (a description of which is given later), the apron of the deep basin was at too low a level in relation to the low tide at neap-tide period, and therefore the basin could not be emptied by the natural drainage of the water. In 1720 it was necessary to fill the bottom with a mass of masonry in order to raise the level of the apron.

At Rochefort the basin had been constructed in an opening made in alluvial lands. At the depth at which the masonry for the bottom of the basin was laid, the builders discovered a firmer, watertight terrain, on which the masonry of the apron was laid.

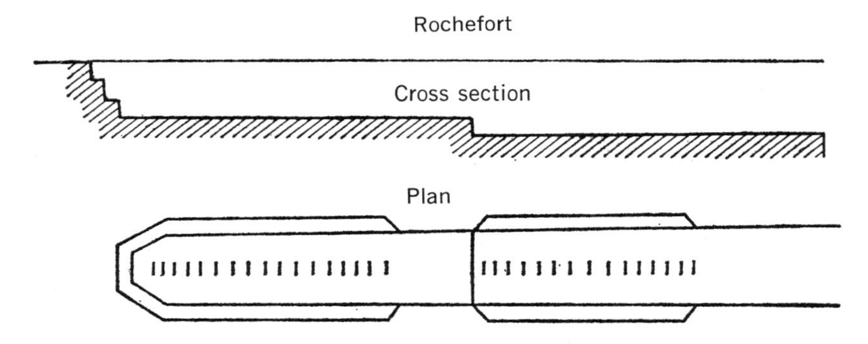

Fig. 151 Plan and longitudinal cross section of the dry dock at Rochefort.

At Brest a different type of construction was required. We have already discussed the nature of the Penfeld and its character as a ria. Secondary rias flowed into the ria of the Penfeld — like the latter, channels that had been dug by the erosion of the small tributaries of the Penfeld. From the highest point of Brest (205 feet) valleys radiated in the direction of west-south-west, toward the stratas of mica schist that constitute the terrain. The first, on the side of Brest (left bank of the Penfeld), is the valley of Villeneuve, which opens out near the entrance to the shipyard.

This channel, extending under the level of the sea, offered a location where a dry dock could be established without the rock-clearing activity that would have been very difficult in the seventeenth century. By filling in the rock with masonry, a fairly deep basin could be obtained without any need for preliminary terracing.

This very advantage (since economy could be effected in the masonry by keeping the bottom as low as possible) led to defects that hindered the develop-

Plate 31. The entrance to the port of Brest, at the end of the seventeenth century. Engraving in the Cabinet des Estampes of the Bibliothèque Nationale. *Photo Bibliothèque Nationale.*

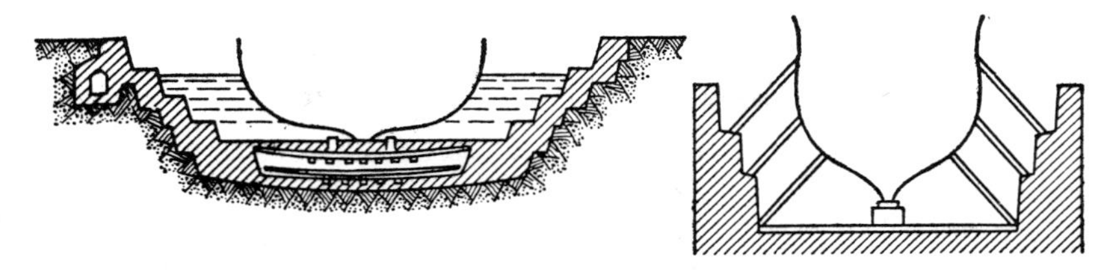

FIG. 152 Dry-dock basins: (left), Rochefort; (right), Brest.

ment of the basin. Comparing cross sections, on the same scale, of the basins of Brest and Rochefort (Figure 152), that of Rochefort was too wide, while with the utilization of the basin of Brest came the realization that it was much too deep and that the disadvantages found at Rochefort were still more of a problem at Brest.

The defect of this basin was that its platform was at approximately the low-water level of the strong equinoxial tides; therefore the water that filtered in, down the slopes of the surrounding terrain, from springs, rain, or the seawater that filtered in through the gate, could not be evacuated by natural drainage at low tide except at the equinox; the use of a machine operated by several horses was needed in order to keep the basin dry, which was very burdensome.

The builders would have avoided these difficulties by raising the level of the section 6 inches above low-water level during neap tide; they would still have had a margin of 17 feet of water for the beaching of ships, for the tide is usually 20 feet at Brest.

It was recognized that the length of the basins between the bottom (upper edge) and the gate had to be 190 feet in order to take first-rate vessels; the basin of Brest with its 240 feet was felt to be too long, necessitating an excessively high expenditure for construction and, when in use, too much pumping. As for the width, it was estimated that 48 feet were needed for the lock, to which must be added the recession of the banks, that is, 3 to 5 feet; this meant that at the level of the terrepleins the basin would be 78 feet wide.

Mediterranean ports and basins: Marseille

In the Mediterranean the problem was very different, both for the ports and for the dry docks, for the level of the water always remained approximately the same. The natural cove of the Lacydon, at Marseille, had served as a port since the earliest days of antiquity; its waters were shallow, but their depth remained constant. All that was required by way of improvement was to construct wharves where small and medium ships could unload (large ships remained out in the center of the basin, where the water was deeper).

The first dock was constructed at the beginning of the sixteenth century, the need for one having been felt by the end of the fifteenth century (ships came in to the beach, where disorder reigned). This dock was completed in 1512; it consisted of a wall 13 feet high above the water, 6½ feet under the water, faced with stones. What seems more extraordinary is its extreme narrowness: squeezed between the sea and the houses, it was only 5 feet wide. Thus it was a simple passageway, by means of which the merchandise brought in by the ships was unloaded.

However, maintenance of the port was one of the most serious problems of the municipal authorities. Being a sunken area in the relief of the land, it was a receptacle for mud, and for garbage and dirt dragged in by the streams, whence its continuous tendency to become silted up. As early as the fourteenth century (1323) the construction of a dredging machine, a bucket wheel for the cleaning of the port, was planned. These machines were perfected, and by the seventeenth century were found in almost every port; they had a scoop at the end of a long shaft (Figure 153). This was the power shovel, which remained in use during the eighteenth and even the nineteenth centuries.

This still crude device, which was used for cleaning and even digging basins, was limited in its effectiveness by the lack of power utilized. It had two drums operated by two or three men — all that was then available as usable power (Figure 153). However, we find attempts being made to use horse-operated treadmills installed on pontoons. Men and animals were the only means of production of power for these devices.

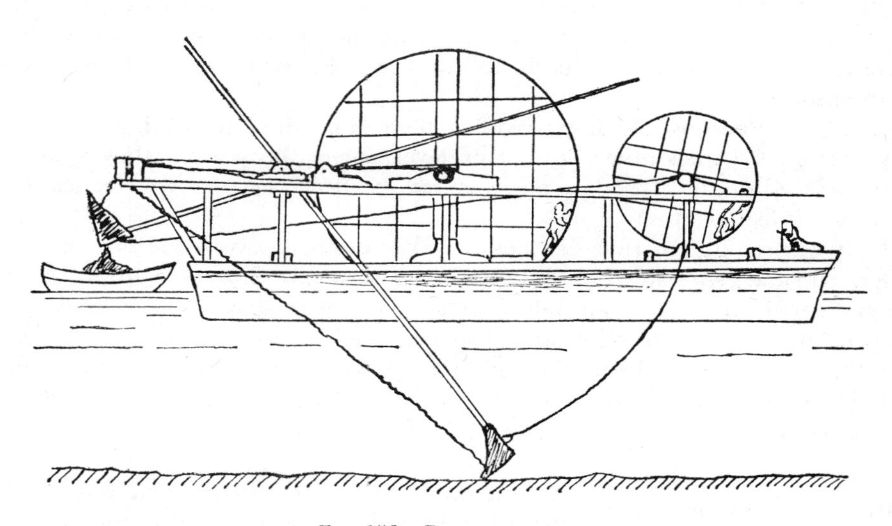

FIG. 153 Power scoop.

Toulon

In the case of many Mediterranean ports, the absence of a natural sheltered basin imposed the creation of an artificial shelter. It was then necessary to construct protecting jetties or moles. This technique was ancient; the Romans had successfully employed it at Ostia, where jetties protected the entrance, and at Centumcelloe (Civitavecchia), where from his villa Pliny the Younger could see the barges bringing up the rock fill and throwing it into the sea, and the jetties gradually emerging.

This was the type of work that gradually built up the port of Toulon. The first port establishment was simply the unprotected coast on which the boats

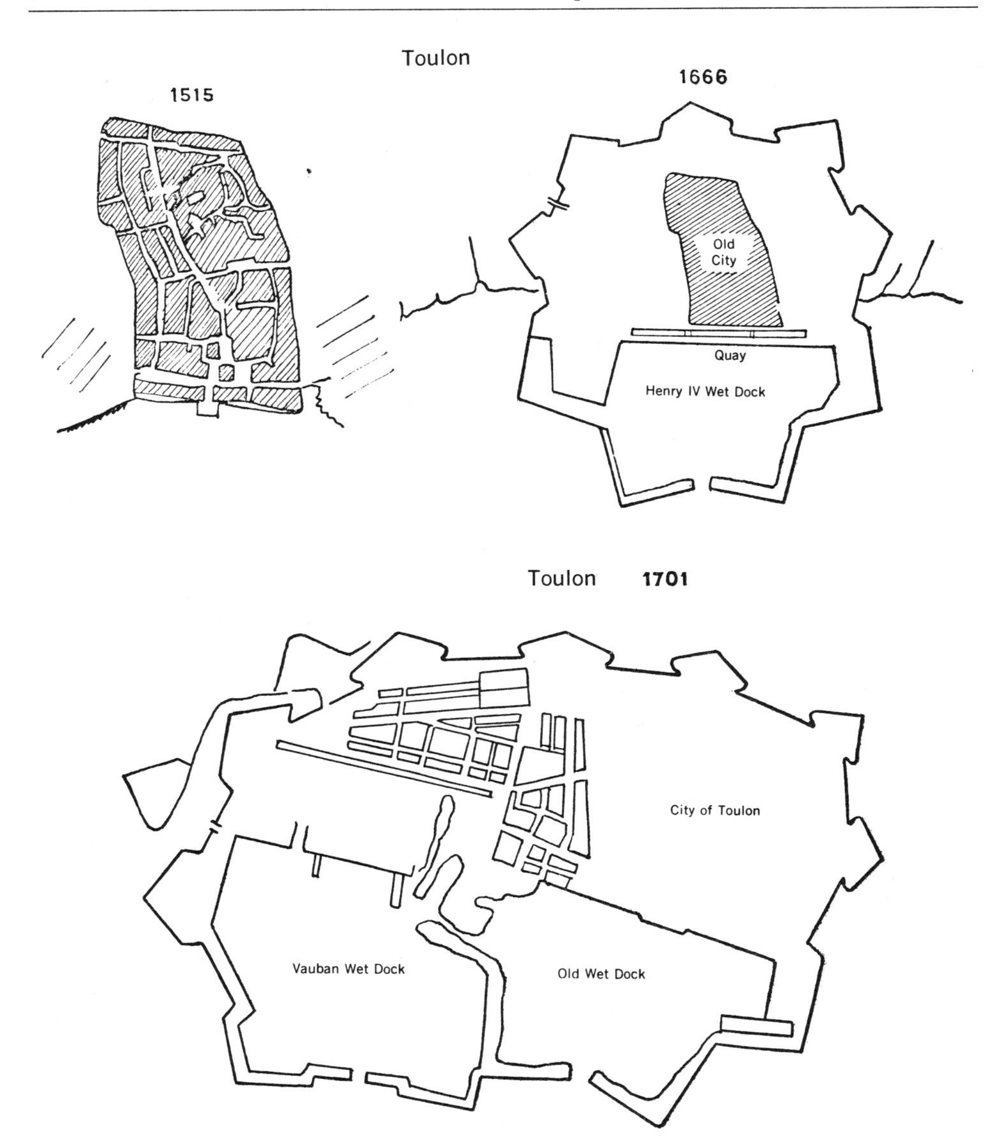

FIG. 154 Successive enlargements of Toulon, 1515 to 1701.

were beached (Figure 154). It is true that the port, being situated at the bottom of a roadstead, was already somewhat sheltered. But in order to have a port that would offer shelter in all weather, it had to be protected. In the middle of the sixteenth century the agglomeration was only a village with old, ruined ramparts. A new city wall was completed in 1596, but not until 1599 did a decree of the Cour des Comptes of Provence order the setting aside of land needed for the establishment of a shipyard.

In 1595 Henry IV had ceded to the inhabitants of Toulon any land they might reclaim from the sea by constructing the wall of the port. Not until 1604, when they had obtained the necessary funds, was the work begun; the port (Figure 154) was completed in 1609. Henri IV then had the royal galleys transferred from Marseille to Toulon. Land for the shipyard had been set aside on the west side of the port, but the "royal garage" for galleys became inadequate, and various projects for enlarging it were developed between 1660 and 1676. These included the major project of 1676, presented by Pierre Puget, for a new harbor to the east of the original, with a group of structures that would have been magnificent but which, when examined in Paris, remained in the portfolios of the Ministry.

In 1678 Vauban was consulted; he adopted Puget's idea of a second port, which he located to the west of the original one. In 1681–1682 he drew up a complete new plan, several features of which he modified in 1701 (Figure 154). The port of Toulon thus comprised two contiguous basins, that of Henri IV and that of Vauban; this group was an object of general admiration throughout the eighteenth century.

Construction of dry docks

The Mediterranean dry docks were limited by the difficulties of construction and utilization in a tideless body of water, a fact which explains why nothing but galley dry docks existed in the seventeenth century. Their size was limited; for the largest galley a length of 35 fathoms (224 feet), a width of 44 feet in the upper portion, and a depth of 10 feet measured above the planking of the platform sufficed; at the bottom, the width could be reduced to 26 feet.

In order to construct this dry dock, all that was necessary was to dig out an area to a depth of 13 feet below the highest level of the sea, which left 3 feet for the mass of the apron, and 10 feet for the water needed for the ship's entrance into the galley. With these shallow depths, pumps were able to keep the basin dry. But the absence of tides prevented the use of basins sufficiently deep to take vessels.

For walls of piers, as for the larger and deeper dry-dock basins, the methods were the same: creation of an enclosed area in which it was posisble, with the limited methods of pumping, to work in dry conditions. If firm ground (tufa) was discovered, as in the case of the basins of Rochefort, the foundations were laid on it. If the ground was sufficiently solid, pilings were built. Pile-driving procedures were unchanged. For longer and heavier pilings the mechanical methods available in that period were used (Figure 155): a gin placed on a pontoon. The ram was hoisted by a windlass operated by a squirrel-cage drum. The pilings were sawed level, even under water. These methods had long been in existence; the notebooks of Villard de Honnecourt had already depicted several examples.

The bottom of the basin was given a masonry lining of cemented stones to a depth of 2½ feet. Pieces of wood (crossbars) were first laid down; they were 9 to 10 inches square, 3½ feet apart, and extended under the ledges. The intervening spaces were then filled with masonry. On top of this was laid a flooring 2 inches thick, caulked, tarred, and coated with pitch, on which the men could work when the form was dry. The strengthening side walls were constructed of a masonry formed of bricks and small stones, and were lined with a facing

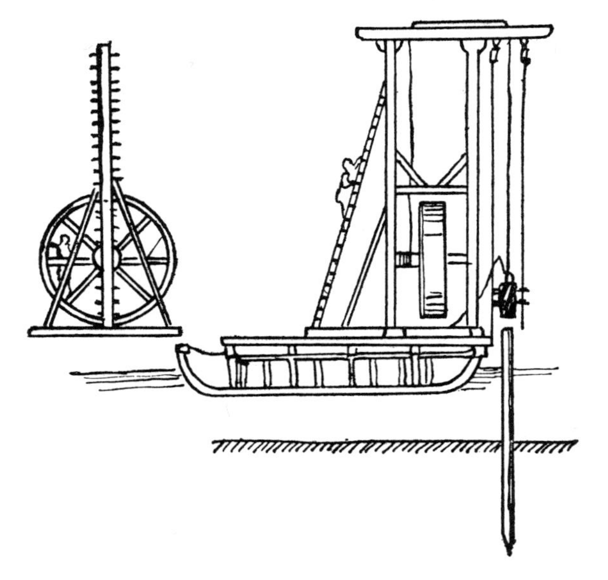

Fig. 155
Pile driver.

of stretchers and headers. On the upper portion, which was 2½ feet wide, a paving was laid; along the edge, stones were fixed in place with iron clamps sealed in lead.

The construction work for the dry-dock forms of Marseille was done between two lines of pile planks; the foundation was of spalled stone or bricks laid in carefully regulated courses in a bed of cement.

Construction of jetties

By the seventeenth century the very ancient technique of constructing jetties was already very well known and established on logical principles; earlier projects had led to observation of the movements of the sea and the manner of making these projects resistant to the shock of the waves.

It was known that precautions had to be taken to establish such projects on a fundation of recent alluvial deposits or mud; it was necessary to dig through the excessively soft ground in order to reach a sufficiently firm bottom. Into the opening thus dug, stones were thrown — but not stones of just any size. The action of the sea had been observed; it was known that large stones are less easily displaced than small stones, and also that the agitation of the sea diminished with increasing depth. It was estimated that at twelve or fifteen feet down, the sea is only slightly agitated, and that at twenty-four feet even the smallest stones are no longer moved by the water.

The stones were therefore divided into three sizes. The smallest were used at the bottom; at twelve feet, medium-sized stones were used, and the largest were placed in the most exposed area, that is, around eight or nine feet. In addition, the large stones were attached to each other and were laid manually by divers, the small side facing forward, the end wedged in place by the upper layers. Moreover, when the stones were being thrown into the sea, the largest were placed preferably on the outside of the pile, medium stones inside; the relationship of base to height of the slope was planned at two to one. The empty spaces were filled with quarry debris of pebbles and gravel, in the proportion of one boatload of gravel for two of stones.

In this way the foundation was built up of stones to a level of three feet below lowest low-water level, the stones being laid as carefully as possible at this level, while between three and six feet below it the spaces between the stones were carefully blocked up. This masonry was then covered with a mortar composed of equal amounts of sand, gravel, and well-crushed limestone and pozzolana; this was left to rest for twenty-four hours and then emptied with buckets equipped with valves so that the water would not dilute the binding agent. Pebbles were buried in this mass of mortar, and several layers of mortar and pebbles were superimposed, the last layer being carefully leveled to receive the masonry of the superstructure of the jetty. For this, ordinary masonry was used to form the core, while the outside was covered with large stones laid as headers and stretchers, using a cement mortar and iron clamps sealed in lead.

In addition to the precaution of dredging the bottom if it was uneven, it was customary to let the rocks settle, generally for one year, before constructing the superstructure; during this time the activity of the sea and the shock of the waves made "the stones fall into the mutual position which was most satisfactory" (Bélidor) (Figure 156).

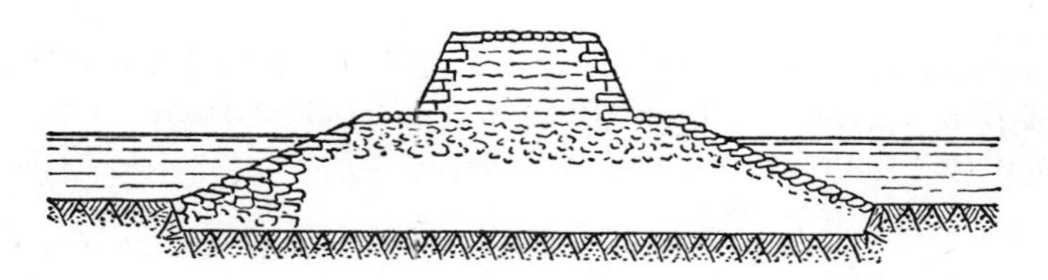

FIG. 156 Protecting dike of a roadstead.

By permitting the infrastructure of stones to settle, excessive dislocations of the masonry jetty were avoided; if the work had to bear a defensive rampart, casemated shelters for cannons, storerooms, and so on — that is, major structures, frequently of an architectural nature — these structures had to be protected against any subsequent movement of their base. In this case the mole, constructed of stones with an upper portion of a concrete made with pebbles and mortar, was leveled to 18 inches below low-water level; the mole was then surmounted with a grille of "longitudinal beams and crossbeams" 10 to 12 inches square, forming compartments that left an empty space of 30 inches at maximum. These beams were carefully pegged together, and were assembled in rafts of from 8 to 10 fathoms (51 to 64 feet) long. They were then floated out, weighted with rocks, and sunk. The compartments were filled with a mortar of pozzolana or, if this was not available, "Tournay cinder." The edge of the grille was recessed 15 feet from the edge of the rock, in order to avoid putting the weight too close to the embankment.

The walls were constructed 2 feet in recession from the longitudinal beam at the edge of the grill. But before the superstructure was raised, the grille was weighted with stones, and left to settle for several months under the effect of the weight and the sea. The entrance to the port, between the ends of the moles, had to be large enough to permit the passage of the largest ships, but sufficiently narrow to be closed with one or several chains; a distance of 16 fathoms (102 feet) was felt to be suitable.

This technique was uniformly employed in the Mediterranean area (at Civitavecchia, Nice, et cetera) in the seventeenth century and, having proved satisfactory, continued to be utilized throughout the eighteenth century.

Draining of the dry dock; pumping machines

The shallowness of the dry-dock forms at Marseille made it possible to use pumps operated only by men. The drainage pumps were of the customary type (Figure 157): a pipe through which passed a chain with cups that rose vertically and pulled up the water, except for the water that leaked out because the cups were not watertight. To decrease the leakage, cups shaped like small basins, with leather joints, were used. Two shackled convicts turned the cranks: after one hour of work they were replaced by other convicts (Figure 158).

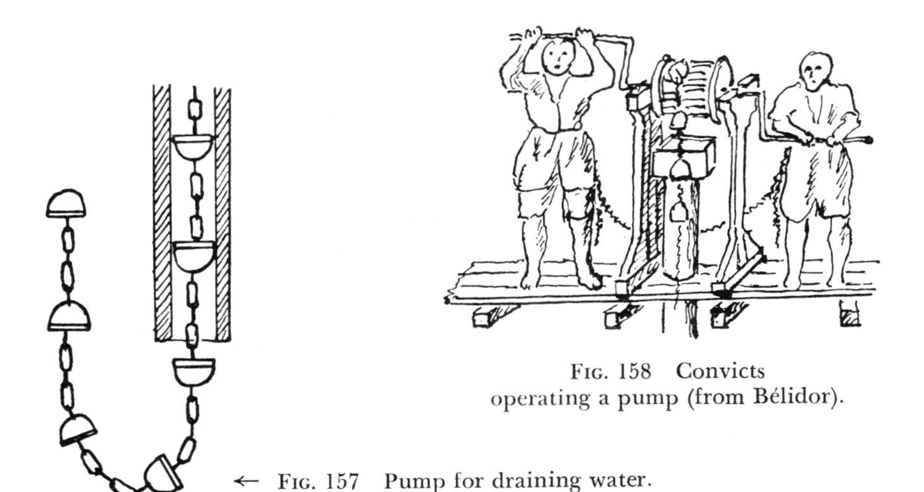

FIG. 158 Convicts operating a pump (from Bélidor).

← FIG. 157 Pump for draining water.

This installation, which was not very effective, sufficed for shallow dry docks; it would have been powerless for large establishments similar to those found in ocean ports. In the latter case, if the dry dock had been situated above the low-water level at neap tide, the forms could have been drained at low tide. But the additional water filtering through the earth or flowing along the surface, and leakage through the gates with an excessively low apron, required pumps for insufficiently low tides. This was the case for Rochefort, as we have noted, and Brest, whose basin was excessively deep.

In this case the system of a treadmill with four horses harnessed to bars turning a vertical axle was used. The installation of Rochefort (1722) included a system of wheels and lanterns with three bucket chains, the buckets being wooden boxes (Figure 159) whose sides were held together by iron knees. A hole

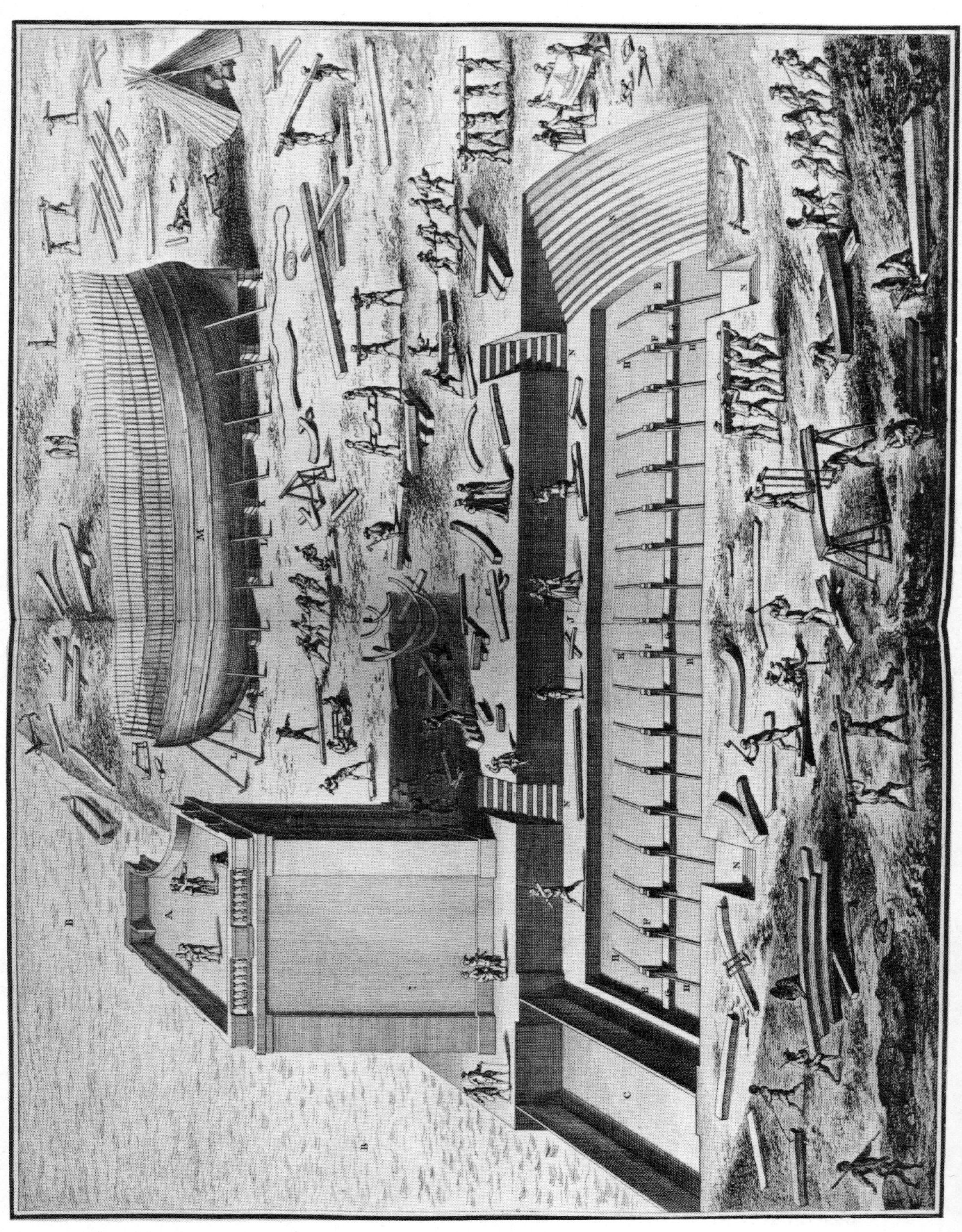

Plate 32. Shipyard at Rochefort, eighteenth century. From collection of illustrations for the *Encyclopédie*. Conservatoire des Arts et Métiers. *Photo by the Conservatoire.*

in the upper face permitted the water to enter, and then to flow out when the bucket was turned over by an octagonal wheel at the top of its path.

On the vertical axle of the mechanism a sprocket wheel pulls three lanterns, while each of the latter in turn has a small wheel having on the same axle the octagonal wheel that holds the chains (Figure 160).

The dimensions of this device were as follows:

Sprocket wheel	diameter 3 feet	48 teeth
Lanterns	" 15 inches	16 teeth
Small wheel	" $2\frac{1}{2}$ feet	32 teeth
Octagonal wheel	" 2 feet	

Each of the three chains had 30 buckets of half a cubic foot, forming a chain of 10 fathoms (64 feet). The water was brought up from a depth of 24 feet at the rate of 1,296 cubic feet per hour.

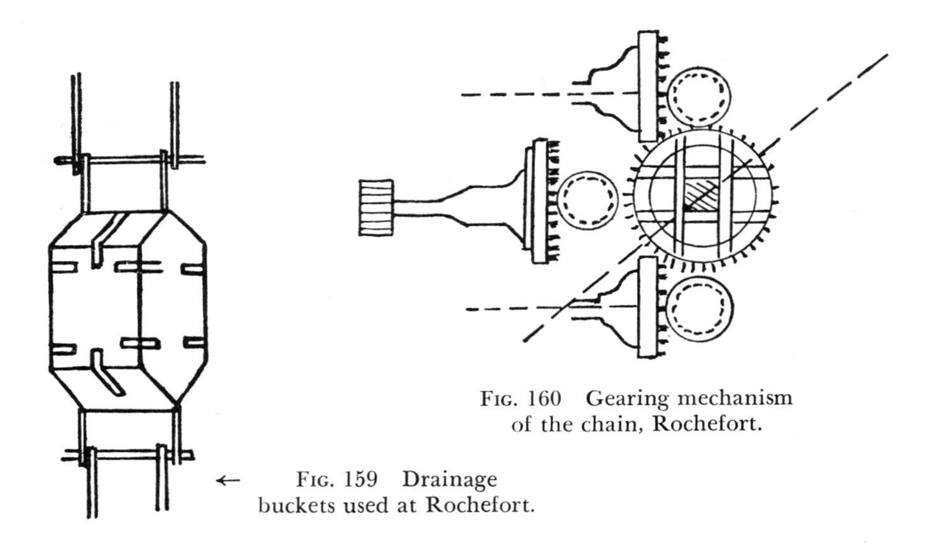

Fig. 160 Gearing mechanism of the chain, Rochefort.

← Fig. 159 Drainage buckets used at Rochefort.

The horses, harnessed to bars 15 feet long, turned at the rate of 72 revolutions per hour, covering a circumference of 15 5/7 fathoms, for a total of 1,131 fathoms per hour. Upon analyzing this machine, Bélidor found it inadequate, since the horses made only 1,131 revolutions per hour ($2\frac{1}{3}$ feet per second), when they could have made 1,800 ($3\frac{1}{4}$ feet), and the mechanical output was very low.

The information just given shows a power in water raised of 22 kilogram-meters/second, that is, approximately 30/100 horsepower, and with a harnessed horse less than one-tenth horsepower. Bélidor felt that with a better arrangement two horses would have been sufficient; we can, however, deduce that even in his time the losses of output in the wooden gears, which were all that existed at this period, were not completely understood. It seems that the fatigue of the horses was such as to require changing the teams every hour.

The same calculation applied to the hand pump of Marseille gives the following:

Diameter of the pipe	4¾ inches (.12 meters)
Radius of the cranks	2 feet (.65 meters)
Radius of the winding drum for the chain	1 foot (.325 meters)
Height for which water was raised	9 feet (2.885 meters)
Pressure per man on the cranks	9 kg. (20 lbs.)
Speed of ascension of the chain	1 foot per second
Water raised per second	4½ quarts (4.32 liters)

This corresponds to an available power of 2.885 × 4.32 × 12.46 kilogram-meters, that is, 1/6 hp, or 1/12 for each man. This indicates an output of the engine, which has no gearing, superior to that of Rochefort. But with a horsepower of 1/6, the pump nevertheless remained an extremely weak engine; the Rochefort pump with its 30/100 hp. and its four horses was not very powerful.

Thus the absence, at the beginning of the eighteenth century, of methods for producing power in sufficient quantities limited the size of the dry-dock forms, and especially their depth. As in the Mediterranean ports it was necessary to limit the dry-dock engines utilized by the galleys with shallow draft. It was impossible to envisage dry-dock forms for large ships at Toulon or Marseille.

However, mention should be made of an original installation at Portsmouth shipyard (1690–1704) (Figures 161 and 162). A canal that linked the upper basin with the sea by means of a series of gates supplied a powerful current of water acting on a paddle wheel. By means of cranks the wheel operated piston pumps and gave a power greater than that of the horse-driven treadmills; in any case it required less work and fewer expenditures, and could function in any low tide by the difference of levels between the basins and the sea.

LIGHTHOUSES

As soon as ocean navigation developed, it was necessary to give ships at sea during the night a means of navigating difficult passages and entering port.

A lighted fire at Cape Sigeion may have been in existence as early as the ninth century B.C. Another fire was later established at the approach to the Piraeus. The tower on the island of Pharos, regarded as one of the seven wonders of the world, was constructed at the beginning of the third century by the Cnidian Sostrates; it is supposed to have measured one thousand cubits, which seems incredible.

The Tour d'Ordre at Boulogne

The Romans installed signal fires at Ostia, Pozzuoli, and Ravenna; they also established one at Boulogne in Gaul, to which the Dover fire corresponded. Every night the fire in the Boulogne tower (later known as the *Tour d'Ordre*) lighted the route (insofar as its weak candlepower could do so) for the numerous

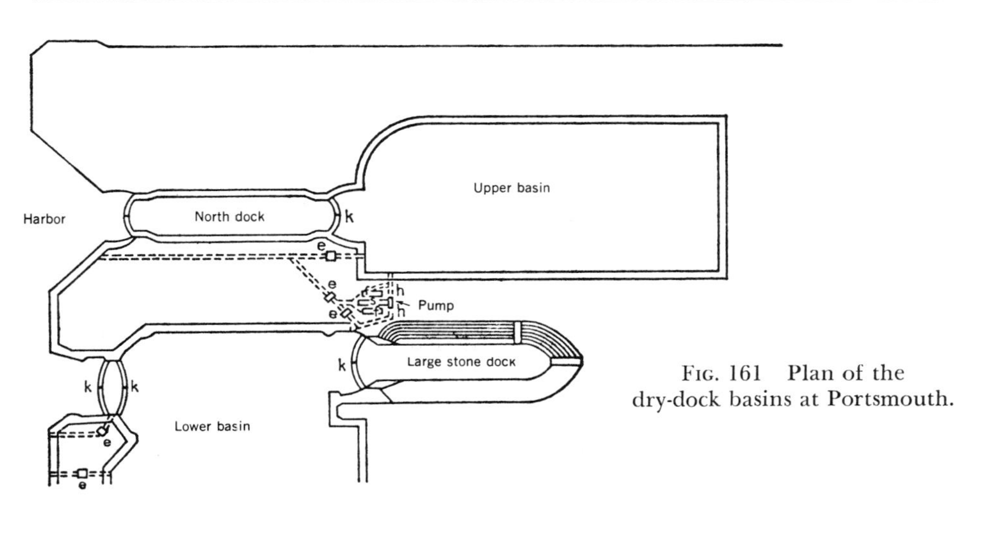

FIG. 161 Plan of the dry-dock basins at Portsmouth.

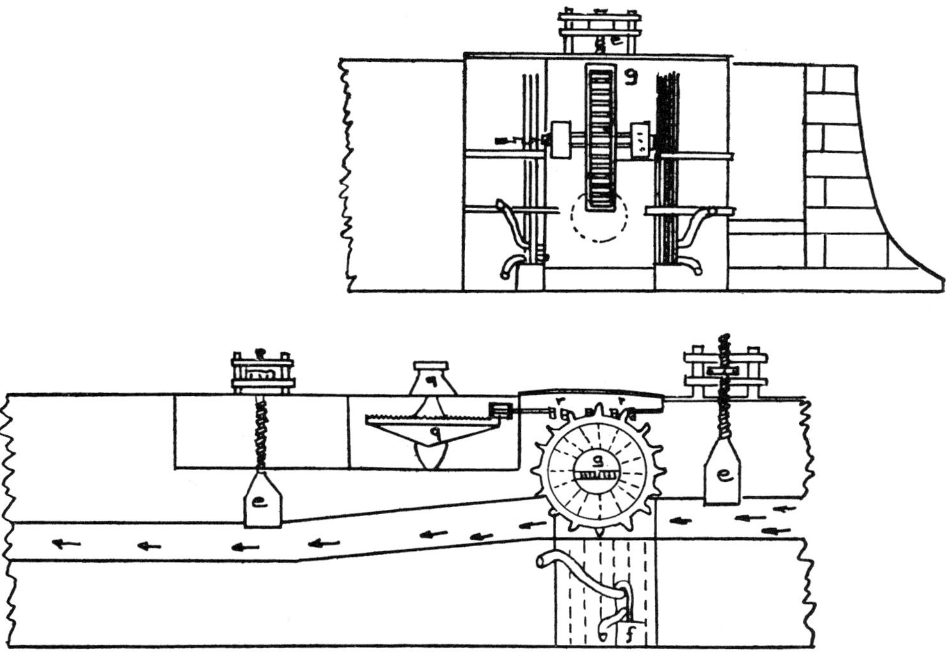

FIG. 162 Diagram (plan and elevation) of the pumping engines at Portsmouth.

ships crossing the channel. It had been established by Caligula to celebrate his supposed campaign in Britain:

"In commemoration of his victory he built a very high tower, in which fires shone like those of a lighthouse, in order to regulate the path of ships during the night" (Suetonius, XLVI).

In 811 this tower, which appears to have been maintained during the Roman domination, and which was probably abandoned at the time of the barbarian invasions, was restored to its role as a lighthouse by Charlemagne, on the eve of the expedition he was preparing against the Norse pirates. It was preserved, for the same purpose, throughout the Middle Ages.

During the English occupation (1544 to 1599) the *Tour d'Ordre* was surrounded by brick ramparts that formed a fortification wall with redans and bastions, a genuine citadel equipped with artillery. This redoubtable fortress resisted the assaults of armies, but the rock was gradually eaten away by the sea. After several attacks by very violent seas, between 1640 and 1645, the fort and the tower crumbled two or three times.

Several sketches of the period permit us more or less to reconstruct the edifice: an octagonal pyramid with 12 stepped stories, each recessed by a depth of 1½ feet. Every face of the first story is supposed to have been 24 feet, that is, 192 in perimeter and 64 in diameter. The tower is believed to have been approximately 200 feet high, which when added to the 100 feet of the rock cliff must have given the light quite a wide range.

Three vaulted chambers one above the other, with doors opening to the south, and a staircase to reach them, were still visible at the beginning of the seventeenth century. The construction consisted of alternating courses of yellow stones from the cliffs and red bricks; the masonry was bonded by a very hard cement composed of limestone, sea sand, and crushed bricks. Thus it clearly appears to have been built in the Roman technique.

In the Middle Ages and the Renaissance the tower-lighthouses were incorporated into the architecture of fortresses and churches, as at Dover, where the old Roman lighthouse tower became a bell tower. The Constance Tower at Aigues-Mortes has a thirteenth-century turret with a lantern; a coarse iron grillwork cage surrounds the light. At La Rochelle the fifteenth-century lantern tower has a bell tower; on the side a turret houses a staircase that leads up to the lantern.

Isolated watchtowers are found on the coasts of Corsica, southern Italy, and Cyprus; in the fifteenth and sixteenth centuries they served as guardhouses, refuges, and lighthouses. The fires were maintained by watchmen.

The lighthouse of Cordouan

On the cliff of Cordouan, monks had installed a signal fire during the reign of Louis the Pious, to guide ships through the difficult passages in the Gironde estuary. It was restored and fortified in the fourteenth century (1362–1376) by the Black Prince, son of King Edward III, who at that time owned all Guyenne; during this period navigation between Bordeaux and England became particularly active.

The lighthouse formed part of an agglomeration that included a chapel and a small village; the village was inhabited by fishermen, the tower by a hermit. The first of these hermits – Geoffroy de Lesparre – had to light and maintain a wood fire on the platform, and received two silver gros for each boat laden with wine.

This service, which was performed fairly regularly, was continued until 1584.

Plate 33. Tour d'Ordre. Roman pharos at Boulogne, France. From the contemporary picture of the British seige of Boulogne, 1544.

The tower was then rebuilt by the engineer-architect Louis de Foix, who around 1579 had straightened the path of the Adour for the benefit of the port of Bayonne and had cooperated on the construction work at the Escurial. He brought to Cordouan the sumptuousness of the construction of the palace on which he had worked in Spain: pilasters, pediments with rich sculptures, an inner chapel, and the "king's chamber" decorated with sculptures and medallions.

The tower, built at the expense of the Governor of Guyenne and completed in 1610, threw the light only 65½ feet above the highest water level. In the eighteenth century its range was increased by the elevation of the tower and by the improvement in the source of light.

The Eddystone lighthouse

In England, the entrance to the port of Plymouth is marked by the Eddystone Lighthouse. Four towers were built in succession on this rock.

In the summer of 1696 a handful of men disembarked on the reef and began to drill holes in the sea-washed rock. This was the beginning of work on the first Eddystone lighthouse. Its builder, Henry Winstanley, an eccentric genius, attached to the rock a pillar of solid masonry 14 feet in diameter and only 12 feet high; this core was topped by the wooden structure of a tower 80 feet high, whose upper portion (Figure 163) earned for it the name of "pagoda." Winstanley quickly realized that the shock of the waves shook the wooden structure, whose weight was not sufficient to resist it; he raised the core of masonry eight feet. Five years later the builder was directing reinforcement works, when a sudden blast of wind carried away the structure along with its unfortunate builder and the watchmen.

Shortly after this catastrophe, a ship (the *Winchelsea*) seeking to enter Plymouth was shipwrecked on the reef. From these dramas the two obvious conclusions were drawn: the light was necessary, and it had to be very sturdy. In 1709 the new lighthouse was constructed by Rudyerd. In the eighteenth century it was destroyed by fire, and was later rebuilt twice.

Lights and lanterns

As for the light itself, it can safely be said that between the Wonder of Alexandria and the eighteenth century, no notable improvement had been made: the wood-burning (later coal-burning) brazier was still in use. However, an attempt was made to improve its visibility and to eliminate the smoke from the fire that occasionally hid the light.

The cage of Aigues-Mortes had strips of horn that protected the fire from the wind and created a draft that removed the smoke. Glass was then substituted for horn; the first lantern at Cordouan was of stone (Figure 164) with glass windows in its faces. Inside the lantern the masonry was calcined by the heat of the fire, and had to be replaced. The hearth was lowered beneath its original level, but the visibility was thereby lessened. In 1720 the hearth was returned to its original position, and a windowed lantern tower with an iron framework was installed (Figure 165).

The fire had originally been fed with oakwood; the new fireplace burned coal. The stove held 225 pounds of coal, which was lighted at sunset; the fire

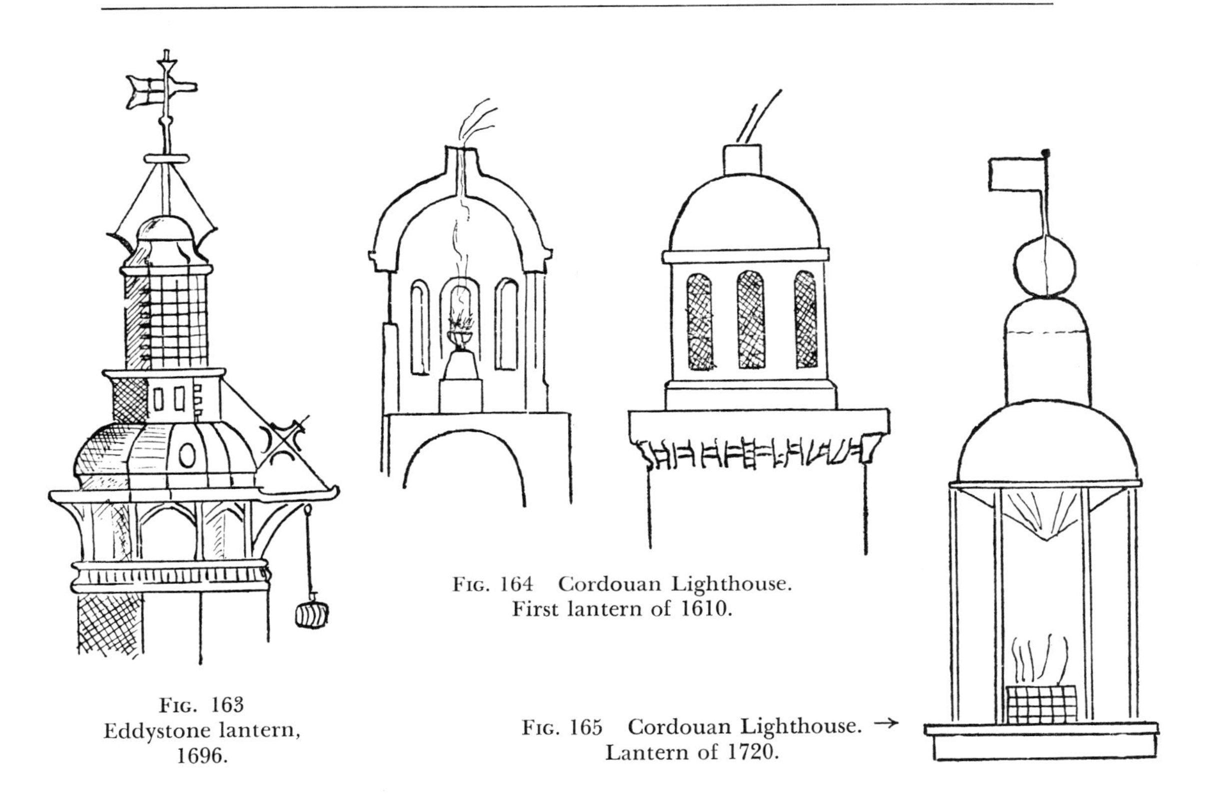

Fig. 163 Eddystone lantern, 1696.

Fig. 164 Cordouan Lighthouse. First lantern of 1610.

Fig. 165 Cordouan Lighthouse. Lantern of 1720. →

burned all night. The wood stove was small but gave a large flame; however, the fire burned for only three hours, at the end of which time the stove had to be reloaded. The visibility was only about two or three miles.

The same arrangements were also in use in England; there was, for example, the lighthouse on Agnes Island in the Scilly Isles, constructed in 1680, which although no longer used has been preserved in its original form. It functioned until the nineteenth century, the only modification being the addition of a ventilator.

The only major improvement was made in the first three structures at Eddystone, namely, the use of candles, the coal-fire system being no longer applicable to it. It was these new arrangements, as well as the later adoption of oil lamps, whose development was to be continued in the eighteenth century.

NAVIGATION ON STREAMS AND CANALS

From earliest ages sea navigation had made it possible to link the continents and the islands; it was an indispensable means of communication. On the continents themselves, road transportation offered travelers and merchandise the possibility of circulating freely, but we have seen the problems encountered by overland transportation: roads that were difficult and burdensome to construct, the tractive power required of teams.

Thus navigation by stream had long been used to facilitate the movements of travelers and merchandise. Strabo indicates *(Geography,* Book I) that navigation was active on the Loire, Rhône, Seine, and Garonne rivers. But their irregularity made travel on them very slow.

The Rhône and the Loire

The example of the Rhône is typical. At its lowest low-water level the river is only 1⅓ feet deep; 250 days of the year it reaches or surpasses 5¼ feet; only during 140 days does it reach 6½ feet. If the ship is of a certain tonnage and of considerable draft, in order to navigate it must wait until the water is sufficiently deep. Formerly it required between 28 and 30 days to make the journey between Arles and Lyon; in winter this was increased to two months.

The ship was also halted during floodtime, when the current was too rapid and the ship was unable to breast it. At any time of year the Rhône has a certain speed, and breasting it requires considerable tractive power, which was supplied by teams of horses and oxen. Teams of 30 or 40 horses towing a string of six boats carrying 300 or 400 tons were still to be seen in fairly recent times.

Until the nineteenth century the Loire was an important channel of communication, but its irregular flow permitted use only by boats of small draft, flat-bottomed barges with square sails that utilized the wind as a method of propulsion. The river carried heavy traffic: wood from the Forez, agricultural produce from the Charollais and the Bourbonnais, grains from Beauce, wines, and on the return trip products from America. Travelers also made use of the river, for example itinerant salesmen and pilgrims. François I traveled by boat from Blois to Amboise; in the seventeenth century Mme. de Sévigné descended it, her carriage traveling by raft.

Shifting sandbanks impeded its passage and changed the bed of the river; at very low water or in perfect calm it was necessary to stop and camp on the riverbank. Ten days and perhaps more ("six days or six months," as the old saying went) were required to travel from Orléans to Nantes.

Economic advantages of the waterways

But by comparison with the roads the waterway offered notable advantages: boats carried heavier loads than carriages and wagons, whose capacity was limited by the poor condition of the roads. From the viewpoint of economy, transportation by water is cheaper; a comparison can be made by comparing the tractive power of the wagon and the ship.

We have already indicated the effort required to pull a wagon; on a well-paved surface (rare at that period, since good paving was limited to the major highways around Versailles) 44 pounds of power were needed to pull one ton. On a canal this dropped to 1.1 pounds, on a sufficiently deep and wide river to ¾ or ½ pound. In addition, the weight of the vehicle itself must be taken into consideration; in a wagon this was almost equal to the load.

On a stream, and better yet on a canal, the shipbuilder, who need not be concerned with the stresses inflicted by waves on a seagoing vessel, could construct a very light ship; Flemish barges weighed only 15 percent of the load they carried. Thus the tractive power was exercised in a more useful way, the dead weight being reduced to a minimum. Let us consider, for example, the barge of 280 tons, whose dead weight is 42 tons; the total weight to be pulled (322 tons) then requires ¾ of 322 = 241.5 pounds; pulling the 280 tons of merchandise will require only two horses, instead of the hundreds that would have been needed to pull the same weight along the highway. Thus it is very easy to explain the high regard in which water transport was held and the ease with which its slowness and irregularities were overlooked.

Regularization of the waterways

We have already mentioned the efforts made to improve the streams and those of their tributaries that led to more remote areas. The Eure had already been improved under Charles VII. Active companies were exploiting the tributaries of the Rhône and the Ouvèze, and the Ardèche, whose flow was perhaps less torrential than it has now become as a result of the deforestation of its basin.

The construction of locks also improved navigation by making the flow of water more regular, reducing the speed of the current, and maintaining the level of the water. Their invention appears to have originated in the region around Venice, after which it was very soon (fifteenth century) utilized in the Netherlands. The first gates must have been constructed of wood, on pilings, with wooden aprons (Figure 166). When the poor support of the wooden side walls (*bajoyers*) was noticed, they were replaced with masonry walls, which continued to be established on a wooden apron linked with and supported by the pilings.

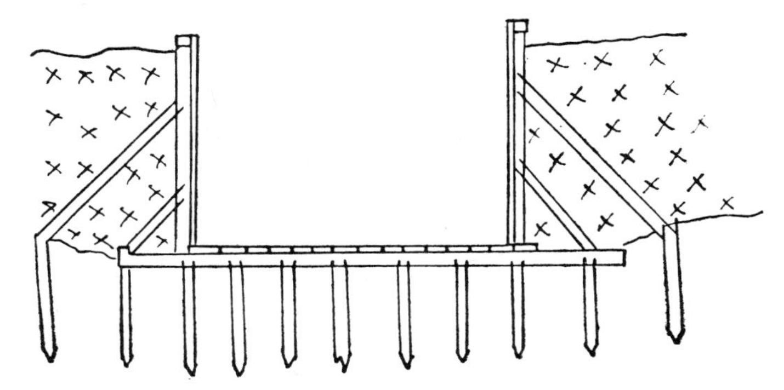

FIG. 166 Wooden lock on pilings, with wooden apron (fifteenth century).

In 1500–1501, after Louis XII's campaign against Ludovico Moro, Leonardo da Vinci settled in Florence, where he studied the improvement of the Arno with a view to making it navigable between Pisa and Florence. He is supposed to have established communication between the two canals of Milan by means of locks with turning gates, and when he settled in France he was inevitably consulted on the subject of possible projects for the Loire.

At this time the governing authorities in Brittany were preoccupied with regularizing the course of the Vilaine, in order to link the entire country with the sea by way of Redon. An engineer who had attempted similar work on the Mayenne without great success was called in. The work was begun around 1540, but it conflicted with the needs of the riverbank dwellers. An Italian named

Bartolozo and several Dutchmen had cooperated on the construction of wooden locks, but these did not hold back the water. An Angevin followed them, but he too had to give up the project. Another Italian engineer constructed two masonry dams, but they were carried away with the first floods.

After several additional attempts an association was formed, and in 1575 was authorized by Henri III to construct ten locks modeled after the lock of Mons (which was not demolished until after 1840), on the basis of a thirty-year license. Navigation was assured after 1585, the effective date of the diversion of the Vilaine.

The construction of canals

The ideas thus spread abroad concerning the advantages of navigation by canal inspired numerous projects. Not until the reign of Henri IV did people begin to benefit from these advantages, following the example of the Dutch and Flemings, who linked their cities by this method, as for example Brussels-Antwerp and Ostend-Bruges.

Sully conceived various projects, including that of the junction of the Seine and the Loire by means of the Loing; this was begun shortly before the king's death, abandoned, then completed by Richelieu. This canal, built by Bouterone and Guyon, originated in the area of Briare, passed Châtillon-Coligny and Montargis, and joined up with the Loing, which pours into the Seine at Moret. This was the first canal with two slopes, crossing the ridge that separated the basins of the two rivers. To feed it, the builders counted on seven *étangs* (ponds) situated at the point where the water divided; in time of drought, however, there was not enough water for regular use.

Given the importance of this Loire-Seine junction, it had to be made more reliable. Louis XIV granted to his brother, the Duke of Orléans, a license to build for his benefit another canal that branched off from the Loire around Orléans and joined the first canal near Montargis. To avoid low water in the Loing during periods of drought, the regent ordered the construction of a new canal along the Loing, from Cépoi near Montargis to Moret. This project, which was completed in 1724, definitively ensured the joining of the Loire and the Seine.

In 1626 Elisabeth-Eugénie, Regent of the Netherlands, ordered the digging of the canal that links the Rhine with the Meuse, in order to transport German merchandise to Brabant —which diverted this traffic from the Dutch waterway. Spinola and the Count of Berg, who were directing the work, protected it against the attacks of the Prince of Orange by twenty-four well-armed redoubts defending the construction sites. The Prince of Orange did attack them on several occasions, but was not able to prevent the completion of the canal.

Construction of the Canal du Midi

The most important project of the seventeenth century was the consummate achievement of the Languedoc Canal (also known as the Canal du Midi), which linked the Mediterranean with the ocean by way of the Garonne River. The idea was already old when Riquet (Pierre-Paul Riquet de Bonrepos) was able, by means of leveling done on his own land, to outline the project. It was a question of crossing the summit of Naurouze in the Lauraguais, 100 fathoms (640 feet) above the Mediterranean, and reaching the Garonne at a point situated, at its average level, 31 fathoms (199 feet) above this ridge. The plan, carefully prepared in order to determine the proper levels for the various sections, resulted in the establishment of a stretch of 96,315 fathoms between Naurouze and the *étang* of Thau, which gave access via Sète to the Mediterranean, and 29,366 fathoms between Naurouze and the Garonne, for a total of 125,681 fathoms or 50½ *lieues* (a *lieue* being 2,500 fathoms or 4,870 meters).

To compensate for the difference of level, it was necessary to construct 74 locks on the Mediterranean slope and 26 on the side toward Toulouse, for a total of 100 locks each of which had an 8-foot drop.

Riquet immediately encountered serious difficulties, created by the people who owned lands through which the canal had to pass. Colbert intervened to obtain from Louis XIV an edict (October 1666) by which the king took over all expropriations and indemnities and "erected as a fief the said canal of communication." The path having been thus paved, the work, which was officially begun on July 29, 1666, by the laying of the first stone at Sète, was completed in May 1681, after fifteen years of work. Riquet was not present at its completion; he died in 1680, leaving the profits to "Messrs. de Bonrepros and the Count de Caraman, his worthy sons." (The income had been assigned by the king in perpetuity to Riquet and his descendants.)

The cost of the canal was 14 million livres. Colbert succeeded in obtaining half of this amount from the king and half from the States of Languedoc, but Ricquet himself had to lay out considerable sums, and his entire family fortune, for the project. After stormy debates, the license for exploitation was given to Riquet and his family, with rights of establishment of stores, construction of boats, and so on.

The completed work aroused much admiration, and justly so, for in addition to Riquet's activity it had been necessary to construct numerous major works: the locks, of which the most famous were the octuple lock of Fonséranne and the Malpas Canal.

At Fonséranne, near Béziers, eight elliptical lock chambers (diagram Figure 167) were put into operation for an area of 156 fathoms (approximately 984 feet), forming a gigantic staircase and providing a change in level of 70 feet, 8

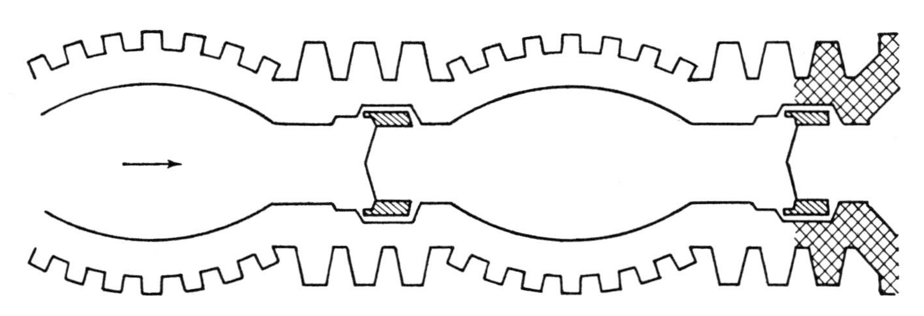

FIG. 167 Elliptical lock chambers of the Fonséranne locks on the Languedoc Canal.

feet 3 inches per chamber. At Malpas, instead of going around a projecting piece of land, it was felt to be more advantageous to dig a tunnel 180 yards long, which had to be supported internally to avoid the crumbling of the rock. Within ten years the supporting framework was attacked by rot; the vault had to be masoned for 80 fathoms (508 feet), and its two ends opened out in the form of a trench. The tunnel was as wide as the canal, with the addition of a towpath four feet wide.

The depth of the canal (cross section Figure 168) was 6 feet; the width at the water surface was 64 feet, at bottom 32 feet. To cross declivities it was necessary to construct bridges that supported the canal on piers and a certain number of round arches. In this case, the cross section of the canal is slightly narrower (Figure 169).

One very important project was the scheme for the summit-water supply.

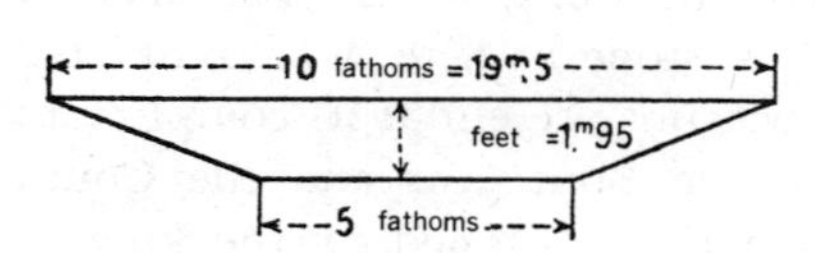

FIG. 168 Standard section on the Languedoc Canal.

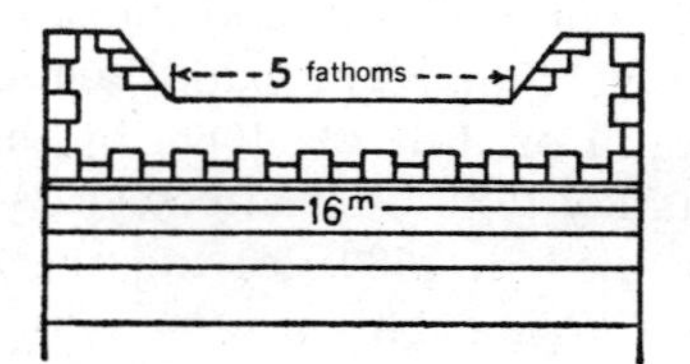

FIG. 169 Cross section of canal bridge.

Riquet created the Naurouze Reservoir, a large octagonal basin 1,279 feet by 958 feet, into which flowed several streams, notably Saint-Fariol (or Ferréol). At the same time it served as a dividing level at which ships passed through one lock at the entrance and another at the exit.

Riquet had planned for several streams to flow into the canal at various points along its course, with catch basins in these sections to catch the overflow. Twenty years after it was put into operation, the disadvantages of this system became apparent: it brought alluvial deposits into the canal, and especially into the Naurouze basin, which was almost completely silted up. There was much criticism, also, of the passage of boats into this basin; because of its large size the wind agitated the water, so that it became dangerous for the boats using it.

The idea of "hazards of navigation" on such a small body of water may seem odd, but it must not be forgotten that precisely because of the lack of agitation of the water, light boats loaded to the maximum, and which therefore had a very low freeboard, were utilized. The smallest waves could submerge the gunwale and bring dangerous quantities of water onto the ship.

At the beginning of the eighteenth century, Vauban reviewed the project, and created dikes and aqueducts over the canal in order to protect it from water carrying too much extraneous material. He gave up the idea of cleaning out the Naurouze Reservoir, and replaced it for purposes of navigation by a level of ordinary size around the basin. The volume of water in this level, which was now the dividing level, was sufficient to compensate for the flow in the locks on each slope.

The water supply, however, continued to be furnished by the same basin (Saint-Ferréol); a dam 105 feet high blocked a reservoir covering an area of approximately 130 acres, to which the water was supplied by large copper valves in a feeder channel after it had crossed a distance of 14 *lieues,* starting from the Montagne Noire (the Saint-Ferréol basin is 44 miles from the canal as the crow flies). The reservoir was able to hold 250 million cubic feet of water. Thus Riquet's work was finally brought to completion, thanks to this terminal achievement by the great engineer Vauban.

The lock of Boesinghe The most interesting projects of the seventeenth century include one that is original to the Flemish canalizations: the lock of Boesinghe, constructed in 1643 at Furnes on the Ypres Canal, which joins the Yser Canal around Dixmude and Nieuport (the Dixmude-Ypres line of combat from 1914 to 1918). Bélidor called it the most beautiful lock chamber he had ever seen.

This lock was 128 feet long and 20 feet wide, and had a drop of 20 feet, although the rule was that a lock chamber should drop no more than eight to ten feet. The then famous engineer Dubié achieved the drop with a single lock chamber. The chamber is sturdily constructed of masonry; the apron is held by a solid wooden grille resting on a masoned foundation.

What made the project special, however, was the fact that it was a model of the lock with side ponds. The level between Hetsas and Ypres was to be the dividing point between the canal leading toward the Yser and a branch to the south of Ypres. An attempt was made to extend the canal southward, but the construction of the locks proved impossible; in this shifting soil of almost liquid alluvions, the block formed by a lock constituted a kind of floater that was shifted by the hydraulic pressure. In addition, the canal ended in a cul-de-sac at Ypres. At this high point, supplying the canal with water could be difficult in the dry season. The surrounding plain is engorged with water, the water level being very close to the soil, but it is low, and the water was needed at the level of the section of the canal that crossed a rise in the ground.

Dubié succeeded in saving two-thirds of the water flowing from the lock downstream at each locking operation. This high point formed a rise in the ground, at which point the ingenious builder established two basins at different levels. To empty the lock, instead of sending its entire content of water downstream (see diagram, Figure 170), the water of the upper third (1) was directed to one of the side ponds through an aqueduct equipped with gates (*A*); the second third was sent through a second aqueduct (*A′*) into a slightly lower basin. Only the last portion was drained into the downstream section of the canal. In

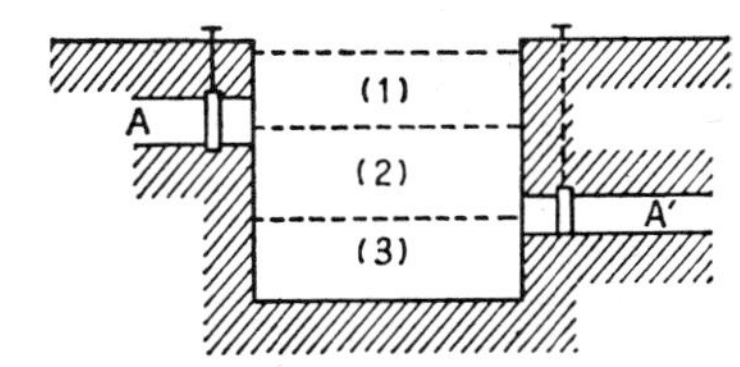

FIG. 170 Lock with side basins.

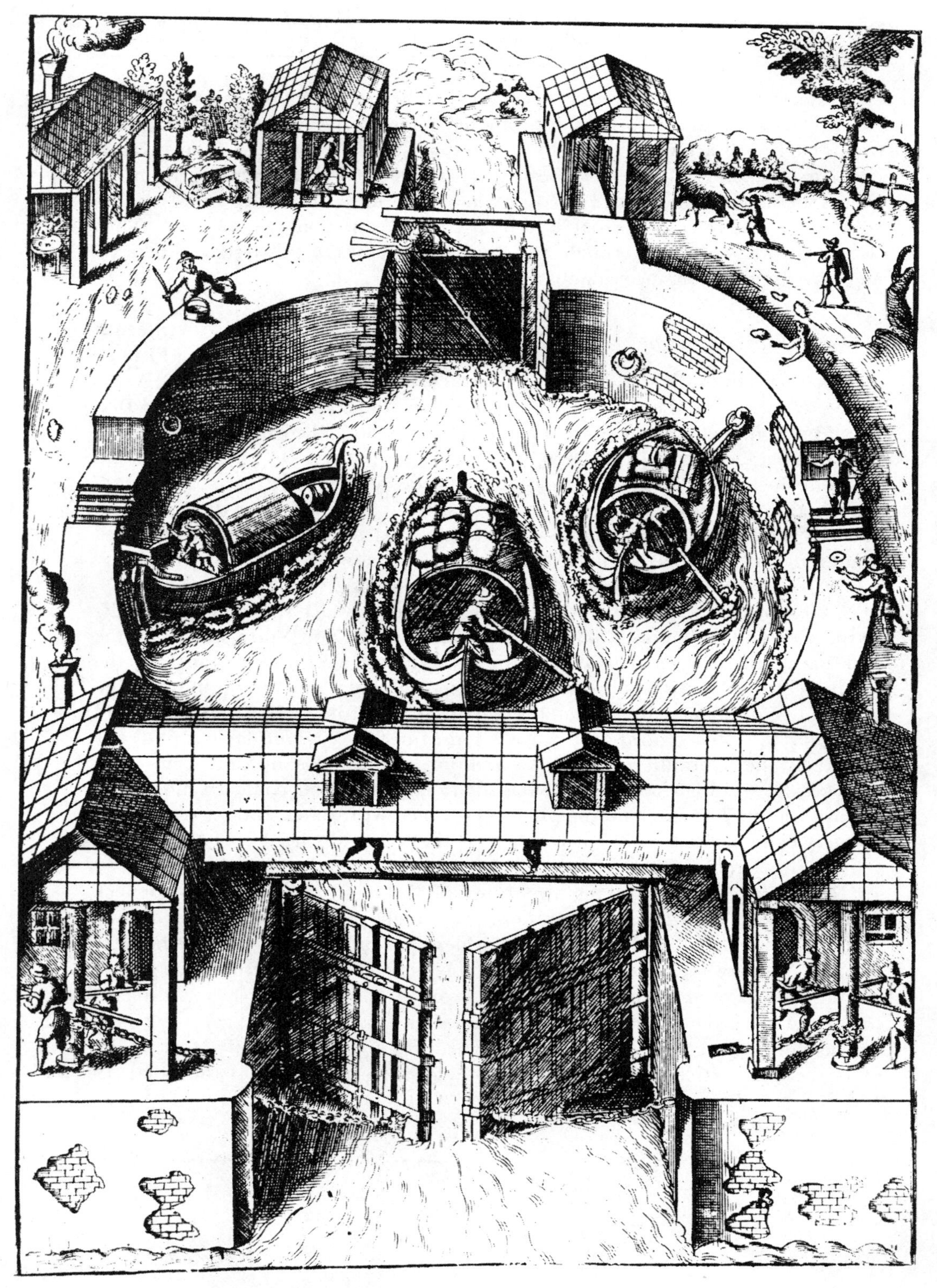

Plate 34. A seventeenth-century lock. Engraving taken from Zonca, *Novo teatro di machine et edificii,* Padua, 1656. *Photo Conservatoire des Arts et Métiers.*

order to fill the lock again, the two-thirds economized was by simple gravity turned back into the lock, the second third into the bottom part, the first third into the middle section. To complete the filling operation, all that was necessary was to supply the water for the upper third. Thus each passage required only one-third the volume of water necessary for the entire operation.

While the canal received its greatest development in the eighteenth and nineteenth centuries, the engineers of both the beginning and end of the seventeenth century already demonstrated a consummate knowledge in the field of hydraulics.

APPENDIX

MEASUREMENT AND TONNAGE OF SHIPS

Reference has been made on several occasions in Chapter 16 to the sizes of ships by allusion to their "tonnage." Much confusion surrounds this word; the words "tonnage" and "ton" and their corresponding values are often used or interpreted incorrectly. The *displacement* of the ship, which according to the principle of Archimedes is equal to the total weight of the floating vessel, is too often confused with "tonnage" figures, which refer to the ship's capacity.

The notion of displacement is a modern conception. In order to know the volume of water displaced (buoyancy), it would have been necessary to measure the volume of a body that has a complicated shape, an operation shipbuilders did not learn how to perform until the seventeenth and eighteenth centuries. In contrast, the notion of the weight of the merchandise that could be stowed in the hold was accessible to everyone; this was what determined the economical utilization of the ship and the possibilities of chartering it.

This idea must have been current in antiquity; the fleets that brought wheat from Pontus Euxinus and wine and oil across the Mediterranean are known to us through the shipwrecks they left on the coasts and through the writings of the ancient authors, who give statistics on these ships. From the latter we can deduce that the average ship crossing the Hellespont could carry a load of approximately 250 tons.

Wheat was transported loose or in bags, while at that time wine was transported in terra-cotta containers (amphoras) of farily uniform content (20 liters [21 quarts] in the Greek amphora, 26 liters [27½ quarts] in the Italian amphora).

It is possible to determine the internal capacity of a ship by the number of "jugs" it carried: thus the "10,000-jug ship," at 26 liters to each amphora, would have carried 260 tons of wine.

During antiquity and the early Middle Ages, only terra-cotta amphoras were in existence, and their capacity could be used as a common unit of measure. Thus it is not surprising to find in very early English measures "wine jars" of one-half a cubic foot, giving 80 jars per ton.

The jar being equal to:

$\frac{1}{2}$ cubic foot $= \dfrac{28,316}{2}$ cubic decimeters [dm³] $=$ 14.15 kg. of wine

80 jars give

$$80 \times 14.15 = 1,132 \text{ dm}^3$$

This is the English freight ton or shipping ton of 40 cubic feet.

In the Middle Ages the terra-cotta amphora was replaced by the wooden barrel, and thus the amphora was replaced by the wooden hogshead as a measure of capacity. From then on, it was these hogsheads that were to be used for measuring the tonnage (internal capacity) of a ship. How did they come to be imposed as a unit of measure in England and France?

As we have already noted, from early antiquity the wine trade had been an important element in world economy, and it continued to be very active, beginning in the Middle Ages, between England and Bordeaux, where the Plantagenets obtained the wine needed in England. Thus it was the hogshead (French *bordelaise*) that became the model for measurement of capacity. Another reason was that a cargo in hogsheads is a typical cargo for ships; properly stowed casks completely fill the hold of the ship and constitute exactly the weight that gives the ship its normal floating line.

Thus the first English measurement of tonnage includes a unit, the *tun* of two pipes, or ton of 252 gallons (see Figure 171 for the size of this 126-gallon pipe). What is the metric value of this ton? The 252 gallons are wine gallons of 3.785 liters, and thus the ton represents 252 × 3.785 = 954 kg.

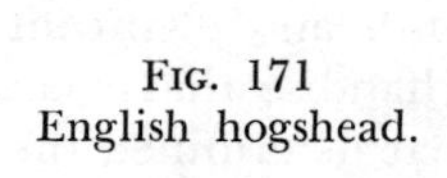

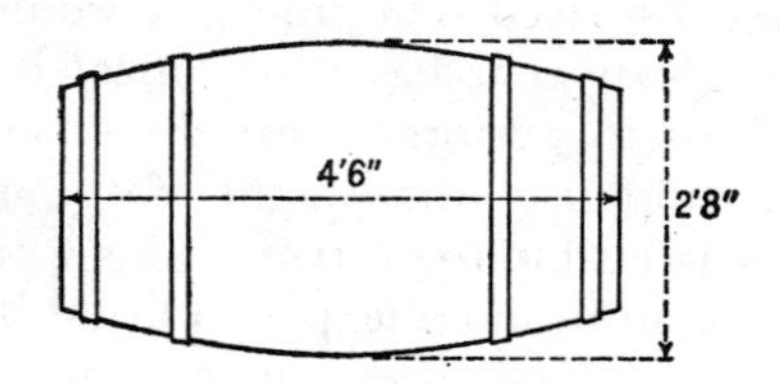

FIG. 171
English hogshead.

The American gallon is equal to 231 cubic inches, while the English gallon is ten pounds avoirdupois weight), or 4.546 (Winchester gallon). The *bordelaise,* which equals one-quarter of the French ton, is thus 244.75 liters, and the English half-pipe is 238.5 liters; these two measurements are very similar, and seem to be related to the same hogshead of the Bordeaux origin.

The ancient jars were piled vertically one against the other, in two or three layers, a position in which they have been found in the numerous shipwrecks along the Mediterranean coasts. However, the best method of loading them is to place their axes in horizontal position, which makes it possible to shift them simply by rolling them.

The shape of a ship's hold is irregular, not geometrical; experience was required in order to determine the number of hogsheads that could be loaded onto a ship. In 1582 Matthew Baker was asked to make this experiment. He found that one ship, the *Ascension,* could hold 320 pipes; at two pipes per ton, the tonnage was therefore 160 tons burden.

The problem then was to determine capacity without measuring each ship individually. Assuming that at this period the shapes of ships were similar and their dimensions proportional, all that was needed was to reduce the volume of the hold to the product of three dimensions: length, beam, and the depth of the hold.

Using the dimensions of the *Ascension* — keel 54 feet, beam (inside planking) 24 feet, hold 12 feet deep — the product is 15,552; to obtain 160 tons, the tonnage determined by the experiment, it sufficed to divide the product by 97⅕. Baker adopted the divisor of 100 for the rule of tonnage (this rule was slightly modified in 1588).

A second measurement is also used in England: the ton and tonnage obtained by adding ⅓ of the first measurement, which would represent the actual load that can be placed on the ship.

Are all tonnages comparable? If the measurement of the dimensions were done in the same way, the results would be comparable, but actually (Figure 172) an understanding must be reached on the manner of measurement.

When the ship is in the shipyard, the length of the keel can easily be measured.

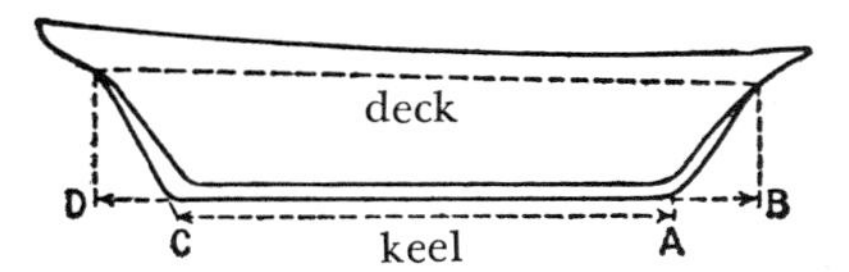

FIG. 172
Dimensions of ships.

But the ship is raked fore and aft (represented by *A, B — C, D* in Figure 172). For the period under discussion we have no plans giving the waterline, which it would, moreover, be difficult to mark on the hull. Measuring the length is done either by increasing the length of the keel by a conventional value (the English length for tonnage) or by taking the length of the gun deck, which will be close to that of the waterline.

Forty-six years after Baker's experiment, the builders of the *River* (Thames Builders) requested permission to use 94 instead of 100 as the divisor, and to measure the beam outside the planking; as it was a question of selling ships according to tonnage, they would thus have obtained an increase of 20 percent. However, the king had the last word, and made Matthew Baker's rule law.

The tonnages measured in other countries did not have to coincide with the English system of measurement. The Spaniards did not have the same reasons for utilizing the Bordeaux hogshead, which was not an element of their transportation, as their point of reference. They chose figures relating rather to the weight of the cargo, and in the case of a warship increased this by 20 percent.

Thus in the time of the Great Armada the English and Spanish tonnages did not have the same significance. Depending on the various systems of measurement, in 1587, under the Baker rule, the *Dreadnought* would have been a 360-tonner; in 1588, a 400-tonner, according to the modified rule. By Spanish measurements it would have been a 478-tonner, or 574 tons if considered as a warship.

In the seventeenth century the regulations became somewhat more precise. In the first half of the century Father Fournier clearly seems to hesitate in his calculations in the *Hydrographie universelle* of 1643, whose publication followed close on the construction of the *Couronne*. He feels "that it is morally impossible to be able to determine precisely, and give a general rule by which it is possible to learn, the weight which can be carried by a vessel" (Book XVIII, Chapter 15). It is nevertheless true that a 300-ton ship is defined as one which "in addition to the weight of its body can carry 300 tons of water." He admits that the full hold would be the volume of 600 tons of water and the displacement of 635 tons. By applying several calculations to the famous *Couronne,* he finds for it a tonnage of 1,152 tons.

The methods of measurement were clarified in England by the B.O.M. (Builder's Old Measurement) and in France by the Ordinance of 1681. The English B.O.M., which was modified in 1770 (Act. 13, Geo. III, c. 74), gives the formula,

$$\frac{(L - \frac{3}{5}B) \times B \times \frac{1}{2} B}{94}$$

L being the length of the keel, and B the beam, the depth is accepted as being half the beam.

In France the ordinance of 1681 fixed the measurement (shipping) ton at 42 cubic feet. The linear foot (royal foot) of 324.8 millimeters gives a cubic foot of 34.277 cubic centimeters; in metric measure, 42 cubic feet is equivalent to 1.44 cubic meters. This is the old French measurement ton, theoretically corresponding to an actually transported burden of 2,000 pounds, or 979 kilograms in a space of 1.44 cubic meters for four *bordelaises.*

How can these measurements of capacity be related to the displacement (weight in tons) of the ship?

The theoretically parallelepipedic hull of a ship would weigh one-third of the displacement; thus two-thirds of this displacement would be available for the cargo (Vial du Clairbois). At the rate of 72 pounds per cubic foot of seawater, the hull will have a volume of 28 percent for 2,000 pounds ($28 \times 72 = 2{,}016$). Thus: volume of 2,000 pounds of water = 28 cubic feet; volume of cargo according to the Ordinance of 1681: $28 + \frac{1}{2}\,28 = 42$ cubic feet. But the ship is round, the hull contains posts and other encumbrances; thus the theoretical burden is not actual, and one-third additional space must be added (experiment with the *Adventure* in England in 1626):

Actual load: $42 + \frac{1}{3}\,42 = 56$ cubic feet

The cargo of the ton (2,000 pounds) of water requires the double of the exact volume of the liquid. The density of the cargo is .5 when the latter consists of hogsheads of wine.

If, on the other hand, we add to the weight of the hull itself the rigging, equipment, and accessories, the total weight of the hull would then be (as Father Fournier had already noted) the same as that of the cargo. Thus:

P total displacement = $\frac{1}{2}$ P hull, rigging, etc.
plus: $\frac{1}{2}$ P cargo.

The internal volume of the hold available for the cargo will be equal to the volume of the hull. If the hull displaces P tons of water (28 percent seawater) we will have P tons of actual displacement by $\frac{P}{2}$ tons of cargo in a volume of hold equal to 56 cubic feet by shipping tons.

With this information we are able to link the tonnage calculated according to the Ordinance of 1681 with the displacement, for vessels of the same category: For one measurement ton, the volume of the hold defined by the measurement is 42 cubic feet, which corresponds to $\frac{42}{56}$ of a ton (2,000 pounds); the displacement will be double this weight.

$$\frac{42}{56} = 3/4 \text{ ton of 2,000 pounds; } 2 \times .75 = 1.50 \text{ displacement tons.}$$

Thus:

1. Weight of cargo in tons: 3/4 of the measurement in ordinance tons;
2. Displacement = 1 1/2 times the measurement in ordinance tons, or twice the weight of cargo.

The figures furnished by the formulas of measurement served to fix charter prices or to determine the subsidies given to the ships depending on their sizes.

In sixteenth-century England a subsidy was given to ships of more than 100 tons; the shipbuilder therefore had an advantage in having his ship measured to a tonnage slightly higher than 100 tons, but from the viewpoint of port duties it was to his advantage if the measurement gave a lower figure. Thus, calculating the measurement could be a very important matter, and the measurer, the master of its determination, could play a considerable role.

Connected with the work of measuring was the task of verifying that the ship taking on a cargo met certain conditions of safety at sea; it should not be overladen. The organization of official safety checks began in ports of assembly for the Crusades and trips to the Holy Land, which necessitated the formation of genuine convoys. Inspectors at Marseille (1253 to 1255) and Genoa (statutes of 1330) examined each ship leaving for Palestine, and supervised the exact observation of the rules — especially as to the space reserved for the passengers, provisions, and so on. According to Augustin Jal: "Very wise laws, passed as much in the interest of the sailors as in that of the merchants, regulated the shape of the ships and their proportions, depending on the voyages they were to make and the cargos they were to carry. The *Capitulare Nauticum* of Venice (1255) contains strict provisions."

According to the statute of December 10, 1339, every hull leaving a Genoese port had to be banded with three strips of iron attached by the *ferratores;* if this law was not obeyed, a stiff penalty was imposed (statute of March 17, 1340). Ancona (Maritime Statute of 1397) forbade the overloading of ships under penalty of a fine, and provided that the cargo must not submerge the irons. This was the beginning of the application of rules of freeboard.

Similarly, when Spain was assembling the fleets which were to find the wealth of America, strict checks were instituted (A. Thomazi).

"Before its departure, every ship was inspected three times by *Visitadores* answerable to the judges [of the *Contratación*].

"If the cargo was too heavy [the commissioner] ordered part of it removed.

In order to determine this maximum with precision, an edict of 1618 provided for placing an iron ring at each edge of the hull in such a way that it touched the water when the cargo had reached the safety limit.

"The *visitadores* had the task of measuring the ships. This complicated operation was done by measuring five dimensions, expressed in cubits of 55 cm. (more precisely, 557 to 559 mm.); using these numbers, there were three methods for obtaining the measurement in tons and pipes. A standard cubit was deposited at the *Casa de Contratación.*"

Detailed precautions were taken to prevent ships that had been loaded and checked in the port from going outside to take on additional cargo before their departure, in this way evading inspection.

In France until the eighteenth century (until the nineteenth in England), the calculation of the measurement was applied not only to commercial vessels but also to warships.

In the sixteenth century and at the beginning of the seventeenth it was difficult to calculate the displacement of a ship, for the summary plans drawn up by the builders did not include a plan of the contours, which would have made such calculations possible. Thus the only easily verifiable elements were the measurements that could be made on the ship itself when it was in the water; the tonnage could then be deduced from these measurements by means of an empirical formula.

Approximate Calculations of Tonnage

The formula of the English Builder's Old Measurement reduced the calculation of the tonnage to the use of two linear elements, length and beam. If during the period considered in this chapter the dimensions of the ships remained more or less proportional, it can be accepted that the length is also proportional to the beam; the formulas of tonnage can then be replaced by a function of a single linear element, the beam, which leads to an expression of the form.

$$\text{Tonnage} = \frac{B^3}{A}$$

This was accepted by certain builders at the beginning of the eighteenth century, who accepted (Collomb) A = 4, the measurements being still the tonnage in measurement tons (Ordinance of 1681), the dimension retained in feet. This was a recognition that the ship measuring one ton (obviously a theoretical ship) would have a beam of four feet and that all the other ships can be measured by their resemblance to this theoretical one-ton ship.

This rule is not invariable, but can give sufficiently close results; it makes it possible to compare not only French ships with each other but also with foreign ships. It must be specified that in order to obtain the results in French (ordinance) measurement, the beam must be taken in royal French feet (325 millimeters).

This method can be applied (and verified) to the *Couronne,* which is the first ship described in detail and with sufficiently precise dimensions, namely:

Length at waterline	54 m.	(177 feet)
Beam	14.94 m.	(49 feet)
Draught	5.4 m.	(17¾ feet)

Our approximate formula will give (with 4 feet = 1.30 m. beam):

$$\text{Tonnage} = \frac{14.94^3}{1.3} = (11.5)^3 = 1{,}520 \text{ tons}$$

Father Fournier estimated the tonnage of the *Couronne,* which he then considered as its weight in cargo, as 1,125 tons (ton of 2,000 pounds).

Following what we have just said, the weight in cargo would be ¾ of the ordinance tonnage:

$$.75 \times 1{,}520 = 1{,}140 \text{ tons [ton of 2,000 pounds]}$$

a figure almost identical to the one given by Father Fournier. The displacement of the ship would be twice the weight in cargo = 2,250 or 2,280 (or 1½ times the tonnage).

This can be checked in another way: if we take the three dimensions — length, beam, draft — the parallelepiped corresponding to a volume of

$$54 \times 14.94 \times 5.4 = 4{,}356.5, \text{ in seawater weight } 4{,}460.$$

The relationship with the displacement calculated above constitutes the coefficient of filling (the English block coefficient).

$$\delta = \frac{2{,}250}{4{,}460} = .504$$

which is approximately exact for the contours of the hull of the *Couronne.*

We may thus conclude that this approximate formula for the entire period of the sixteenth, seventeenth, and eighteenth centuries furnished a very precise method of comparing various ships, whether French or foreign, by means of a single dimension — their beam.

BIBLIOGRAPHY

ABEL, WESTCOTT, *The Shipwright's Trade* (Cambridge, 1948).
ALBION, ROBERT G., *Naval and Maritime History. An Annotated Bibliography* (3rd ed., Mystic, Conn., 1963).
BOWEN, J. P., *British Lighthouses* (London, 1947).
CLOWES, G. S. D., *Sailing Ships, Their History and Development.* Part I, *Historical Notes* (London, 1958).
LANDSTRÖM, BJÖRN, *The Ship* (Garden City, N.Y., 1961).
LANE, FREDERICK C., *Venetian Ships and Shipbuilders of the Renaissance* (Baltimore, 1934).
MALLETT, MICHAEL EDWARD, *The Florentine Galleys in the Fifteenth Century* (Oxford, 1967).
PAYNE, ROBERT, *The Canal Builders* (New York, 1959).
ROBINSON, G., *Elizabethan Ships* (London, 1956).
SYKES, P., *History of Exploration* (London, 1950).
TAYLOR, EVA G. R., *The Haven-Finding Art: A History of Navigation from Odysseus to Captain Cook* (London, 1956).

Section Four

The Production of Power

CHAPTER 17

FROM THE TRADITIONAL METHODS TO STEAM

THE TRADITIONAL MACHINES

THE Middle Ages witnessed the gradual substitution of water mills for man- and animal-power; slightly later, the windmill also made a useful and effective contribution to the production of power. All these methods remained in use during the sixteenth and seventeenth centuries; man still contributed to the production of power in certain cases, for example movable engines, cranes, and dredgers, in which he operated the squirrel-cage treadmill. These machines, however, required only a relatively low power.

Water- and wind-mills

The hydraulic machine appeared in two forms. The Nordic mill, with vertical axle of rotation, turned the millstone in the upper portion directly; in the lower portion a small horizontal wheel with buckets was turned by the water. The engine produced very little power, but the slowly turning millstones had the same speed of rotation as the hydraulic wheel, which eliminated the need for an intermediate gear.

These mills, which made it possible to fill only limited needs (for example, to grind the wheat needed by a single family), were much in vogue until the end of the Middle Ages; in France they were still in use in the valley of the Garonne around 1588, and they have survived into modern times in the Orkney Islands, Norway, Romania, and Lebanon. This mill was a remote precursor of the hydraulic turbine.

We find the oldest description of the water mills in Vitruvius. The axle of the wheel is horizontal, and operates the vertical shaft of the mill by a multiplying gear train; the millstones of the Roman mills usually made five revolutions for every one revolution made by the hydraulic wheel. The overshot wheel, whose troughs collected the water coming from an overhead source, was preferred to the undershot wheel; the weight of the water in the troughs caused the driving wheel to revolve. In all the early industrial establishments hydraulic power acquired the preponderant role in the production of power, and retained it until the eighteenth century.

The windmill, on the other hand, played an important part in the plains areas, and after 1600 became the customary driving element of the drainage system in the Netherlands.

Both windmill and water mill required a method of transmission: the toothed wheel. For centuries this was made exclusively of wood. Being easy to construct and repair, its meshing could be gradually perfected. But transmission by wooden teeth always gave a low output.

However, the need for methods of producing energy was limited to small amounts of power, for which the mills sufficed. Until the eighteenth century the mills remained the principal source of energy. The water mill was even combined with the first steam engines used for pumping: the water lifted by the steam engine (Savery's, Papin's projects) was used to turn the wheel of a water mill.

In order to understand the relative importance of the various engines for power production, it will be helpful to compute the power of each and compare their possible utilization.

Man- and animal-power A man turning a handle with a radius of one foot at the rate of one revolution every two seconds can exercise a power of 7 kilograms (15 pounds) continuously, which corresponds to a power per second of

$$7 \times \frac{.7}{2} \times \pi = 7.7 \text{ kilogram-meters/sec., or approximately } 1/10 \text{ hp.}$$

On a treadmill 26¼ feet in diameter, a horse is able to supply an effort of 45 kg. at a speed of .90 meters/second, that is,

$$45 \times .9 = 40.5 \text{ kilogram-meters/second.}$$

But man and animal are able to maintain this pace for only eight hours. Thus in one day they produce a total output of approximately:

Man: $7.7 \times 60 \times 60 \times 8 = 220{,}000$ kgm.
Horse: $40.5 \times 60 \times 60 \times 8 = 1{,}200{,}000$ kgm.

A one-hp. machine (i.e., 75 kgm/sec), working continuously would in 24 hours supply

$$75 \times 3{,}600 \times 24 = 5{,}180{,}000 \text{ kgm.}$$

Functioning for 16 hours (work in a factory) it would supply

$$75 \times 3{,}600 \times 16 = 3{,}500{,}000 \text{ kgm. (approx.)}$$

A comparison of these figures clearly shows how the early machines, whose power when evaluated in terms of horsepower seems feeble, did nevertheless supply a relatively large amount of power in comparison with that of the man- and animal-driven machines.

Power of the mills What actually was the power of the wind- and water-mills? The hydraulic wheel, even with its somewhat improved gears, supplied at most a power of 10 hp., but its average was 5 hp. This power, when operating continuously 24 hours a day, could supply approximately 26 million kgm. per day, that is, the equivalent of the amount that would have been supplied by 23 or 24 horses with all the maintenance they require (food, shelter, and so on.)

This comparison demonstrates the advantages of the hydraulic engine. Thus even in the eighteenth century, when the steam engine appeared, the power produced by the water mills continued to be widely used.

The windmill, which was not highly valued in England, played a major role in the Netherlands. Since the fourteenth century it had been employed for draining polders, and it remained the principal source of power for the pumps until modern times. Around 1600, windmills were utilized in the Netherlands for sawing wood, fulling, grinding grain and tobacco, and for driving the cylinders of paper mills.

By the end of the seventeenth century, the banks of the Zoon, a stream that crosses the largest industrial area of Holland, had 900 windmills (including those used for pumping out polders); other mills, also used for industrial purposes, were located around Leyden, Rotterdam, and Dordrecht. In 1900, 2,000 such mills were still in existence.

The use of engines that lessened the use of human labor always aroused opposition; artistans' guilds protested in 1581, claiming that they would cause unemployment.

The studies of John Smeaton (1759), confirmed by modern measurements of a pumping mill constructed in 1648, show that while the power applied to the axle of the shafts might be 40 hp., on the driven shaft it was only 15.6 hp. (39 percent); thus 61 percent was lost in transmission. This was improved slightly in the eighteenth century by developing the use of metal parts, but in reality the windmill remained essentially a wooden structure, and this limited it to a maximum of 50 hp., a large part (almost $2/3$) of which was lost in transmission.

The use of wood had one advantage: it softened the shocks and vibrations in the gears better than could have been done by cast-iron parts cut and fitted in an age of little precision. But by the eighteenth century the windmill had reached the limit of its possible development, and was very gradually to give way to the steam engine, which could supply much greater power at lower cost.

On the other hand, the windmill made it possible to satisfy the power needs of very small establishments that required little power. Numerous windmills could be found that supplied amounts ranging from only $1/3$ to 3 hp. In the eighteenth century few establishments required large amounts of power. Large water-supply installations were still unusual. These installations and the figures concerning them are mentioned in the chapter on water distribution, notably the first machine at Marly, with its two horse-driven treadmills, which was supposed to yield 600 cubic meters (785 cubic yards) of water and yielded little more than 300 cubic meters. Thus, for an elevation of 62$1/3$ feet,

$$19 \times 300 \times 1{,}000 \text{ kgm.} = 5{,}700{,}000 \text{ kgm.}$$

were required daily.

If the total output of gearing and pump was .4, the work of the treadmill had to be 13,500,000 kgm., representing approximately eleven to twelve horses working for eight hours, which thus required three treadmills, each with four horses. Each team worked for eight hours, which necessitated large expenditures for su-

Plate 35. Hydraulic undershot wheel operating a sharpening wheel, a wheat mill, and a hammer for crushing. Engraving taken from Zonca, *Novo teatro di machine et edificii* (Padua, 1656). Conservatoire des Arts et Métiers. *Photo by the Conservatoire.*

Plate 36. Undershot wheel for operating the bellows of a furnace. Engraving taken from Böckler, *Theatrum machinarum* (Nuremberg, 1662). *Photo Conservatoire des Arts et Métiers.*

pervision and maintenance of the animals. Thus it is easy to understand why this arrangement was soon replaced by hydraulic machines.

The hydraulic machine at Marly consisted of 14 wheels supplying a total of 700 hp., that is, 50 hp. per wheel, which seems to agree with the data given above for the waterwheels. But the actual power obtained was only 150 hp., corresponding to an output of only .215. This is not surprising when we consider the series of connecting rods, transmissions over great distances, and so on.

From this description of the installation at Marly, we can realize what enormous machinery was needed, and at what incredible price, in order to obtain an actual power (in water raised) of only 150 hp.

When the development of industry in the eighteenth century, and especially its continuation in the nineteenth century, required new and more powerful installations for the production of power, the water mills, as conceived in the traditional manner, were incapable of satisfying the need for power. It was the development of the steam engine that made it possible to obtain more power at lower cost.

FIRST STEPS TOWARD THE STEAM ENGINE

Since ancient times it had been readily observed that water, when heated in a closed vessel and changed into steam, can send a blast through a narrow opening at great speed. However, it required many centuries of experimentation to utilize the power that can be furnished by steam.

The pumping installations described by Agricola (1546) were no different from those known in antiquity. By the middle of the seventeenth century they proved to be insufficient for the pumping out of large mines. Not until 150 years later were the possibilities of steam put to use.

In 1575 the first printed translation of the *Pneumatics* of Hero of Alexandria, by Frederick Commandine, appeared in England. This book, which had a wide distribution, mentioned several uses for steam for the purpose of turning rollers operated by the action of a blast of steam; these were aeolipiles, and their direct use could produce no useful results. But Hero's book called the attention of the inventors to the resources that could be obtained from steam.

Unachieved ideas

In 1601 Gianbattista della Porta described a device in which the pressure of steam was used to raise a column of water; moreover, the condensation of the steam created a vacuum that sucked up the water. This was the very principle that was to be utilized by Savery almost one hundred years later; however, it had already been pointed out by Hero, who did not realize the power produced by heating water.

In 1615 Salomon de Caus, engineer and architect of the King of France, published in Frankfurt a work entitled *Les raisons des forces mouvantes, avec diverses machines tant utiles que plaisantes* (The principles of moving forces, with various machines as useful as they are agreeable). According to his demonstration of his "theorem," water rises with the help "of a fire higher than its level" (Figure 174). A tube rests on the bottom of a heated container that contains water; the boiling of the water creates steam pressure, which causes the water to rise through the tube. When all the water has been removed, it can easily be replaced in the cooled vessel, into which it is sucked by the vacuum.

A different system was described by Giovanni Branca in 1629 (*Le machine diverse del Signor Giovanni Branca, cittadino romano, ingegniero, architetto della Sta. Casa di Loretto, Roma* MDCXXIX) (The various machines of *Signor* Giovanni Branca, Roman citizen, engineer, architect of the Sta. Casa di Loretto, Rome 1629). A bronze boiler kettle heated to a high degree (Figure 173) emits a blast

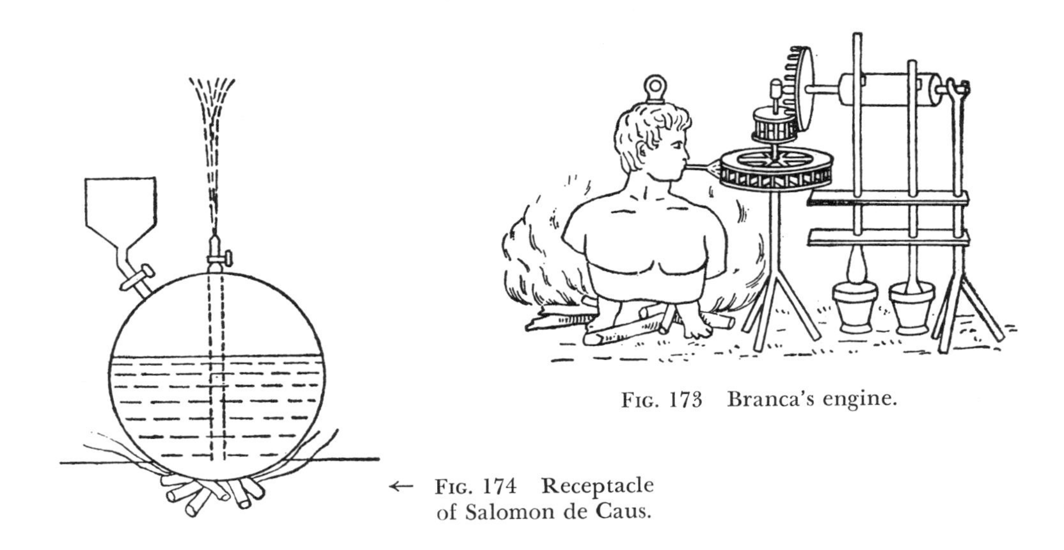

FIG. 173 Branca's engine.

← FIG. 174 Receptacle of Salomon de Caus.

of steam that strikes the blades of a wheel with sufficient strength to turn the wheel so that it can be used for "a great number of uses." This was the first example of a steam turbine with an impulse wheel. Its practical utilization was not to be discovered until 250 years later.

Worcester and condensation

By 1650, actual experiments, rather than simply ideas on the use of steam, were beginning to appear. This was the beginning of a period of experimentation that was to continue for approximately fifty years. Depending on the historians and their nationalities, the actual invention and construction of the steam engine is attributed to one inventor or another. In France, Papin is in most cases regarded as the sole inventor, while in England, Savery is considered to be the actual builder of the first steam engine.

In 1663 the second Marquis of Worcester, Edward Somerset, published a collection of descriptions of his inventions. It is written in a more or less obscure manner, but seems to contain one of the first devices imagined by the author: "Some one hundred names and specimens of inventions already created by me." Between 1630 and 1645, he is supposed to have installed a water-lifting machine at Vauxhall, near London, after having first constructed an experimental device at Raglan Castle. According to Thurston, the principle of this device may have been similar to those of della Porta and Salomon de Caus. Hollow spaces are supposed to have been found in the walls of the keep of Raglan Castle in which the vessels that formed the device were placed; this is in agreement with the texts of a patent of 1663.

Thurston deduces from this that the Marquis of Worcester is the first person to have constructed a steam engine. Dickinson, who is less affirmative, feels that the contemporary testimony is too vague and that it is possible neither to affirm nor to deny the existence of this hypothetical machine. In any event, Worcester may not have been the inventor of the principle already described by Porta, but he may perhaps have made an early attempt to apply it.

The inventors who succeeded him are better known; they have left less questionable drawings and proofs, and it is possible to follow the sequence of the successive attempts that were to lead to the achievements of the eighteenth centurn. At that time two principles were to be applied simultaneously: the idea described by della Porta and S. de Caus, which may or may not have been put into concrete form by Worcester; and the principle that utilized atmospheric pressure and that appeared in the middle of the seventeenth century. The first idea was used by Savery, the second by Papin, who then abandoned it and adopted the first system, while attempting to improve it. Shortly thereafter the atmospheric-pressure system outlined by Papin was adopted by Newcomen and expertly perfected by James Watt at the end of the eighteenth century.

Guericke and atmospheric pressure

In 1641 the fountain builders of Florence knew that a suction pump in a pipe could not lift water more than thirty-two feet and that above this height the pipe remained empty. Below it the water filled the pipe, which fact was explained by the saying "Nature abhors a void." When consulted on this newly observed phenomenon, Galileo answered that the "abhorence of the void" ended at thirty-two feet; this answer did not satisfy him, but he did not develop his reflections further, and in any event he died shortly thereafter.

It was Evangelista Torricelli and Viviani who discovered the cause of the phenomenon: the pressure of the atmosphere, which was counterbalanced by the height of water observed. Using a tube closed at one end and filled with mercury, Torricelli in 1643 verified that the height of the column of mercury corresponded to the relationship of the densities of water and mercury, that is, 28 inches. The phenomenon of atmospheric pressure, which was verified by Pascal in several experiments and again at Le Puy de Dôme in 1647, aroused general curiosity.

Various experiments were made by Otto von Guericke, a physicist and burgomeister of Magdeburg; he invented the pneumatic pump, and in this way was able to demonstrate the tremendous pressures exerted by air on surfaces. This opened the door to the use of atmospheric pressure in machines.

In 1654, when his official functions called him to Ratisbonne, von Guericke performed a remarkable experiment before the Prince of Auerberg (Figure 175). A piston placed in a cylinder open at the top and closed at the bottom was retained in the upper portion of the cylinder by means of a rope passing over a pulley and held by twenty people. A vessel from which the air had been pumped

Plate 37. First depiction of the principle of the steam turbine. Beginning of the seventeenth century. Engraving taken from Bronca, *Le machine* (Rome, 1629).
Photo Conservatoire des Arts et Métiers.

out by means of the pneumatic pump was connected to the bottom of the cylinder; when a tap communicating with the cylinder was opened, the air in the cylinder expanded, and the pressure of the outside air on the piston, becoming greater than that in the cylinder, pushed the piston up, despite the efforts of the twenty people to hold it in place.

This was an early experiment with the cylinder and piston-driving system, acting under the effect of atmospheric pressure, but a vacuum had to be created in the cylinder, below the piston. Abbot Jean Hautefeuille conceived the idea of placing a small quantity of powder in the bottom of the cylinder; when it was ignited, the powder heated the air and expelled it from the cylinder.

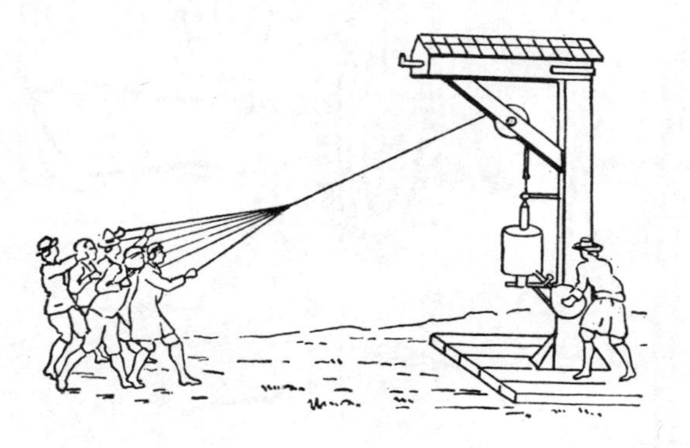

FIG. 175 Otto von Guericke's piston cylinder.

The powder machine of Huygens

This was the system that Huygens presented in 1673 in a memorandum to the Académie des Sciences (Figure 176); two small, flexible leather flaps in the upper portion of the cylinder permitted the escape of hot gases when the powder was ignited. When a vacuum was created inside the cylinder owing to the cooling of the gases, the flaps flattened and prevented the outside air from entering. The piston dropped as a result of the difference of pressure on the two faces, and was able to lift a heavy weight.

The device that Huygens tried out in the Royal Library, with Papin's help and collaboration, consisted of a vertical cylinder one foot in diameter and four feet high. This cylinder was constructed of strips of iron soldered with tin, but since no reaming system existed that would permit corrections to be made to the inside of the cylinder, it was coated on the inside with plaster. Approximately four grams (one drachma) of powder were used to heat the gases.

The device was obviously very imperfect. The porous plaster crumbled, the leather flaps functioned poorly; half of the "air" produced by the powder remained under the piston and thereby lessened to a certain extent the length of

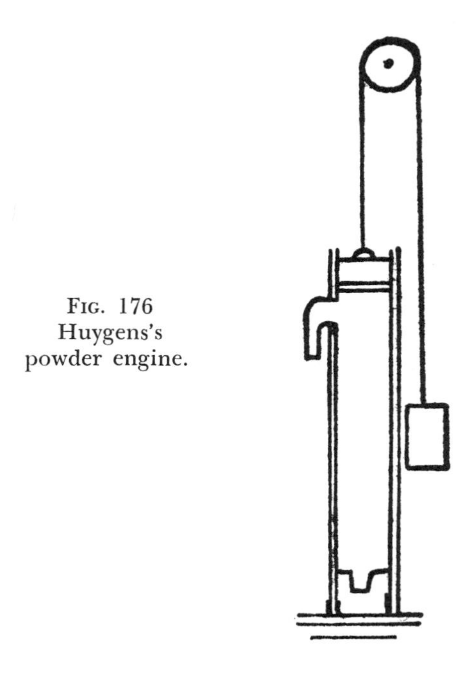

FIG. 176
Huygens's
powder engine.

its drop. How this experimental device could have been transformed into a machine with a useful (and systematic) function is not apparent. Huygens appears to have forgotten the idea for the utilization of steam that he had mentioned in his proposals of 1666.

These experiments of Huygens were performed with the help of an assistant, Denis Papin (1647–1712). It was this experiment that inspired Papin with the original idea of the invention with which his name is connected. Papin's collaboration with Huygens lasted for two years (1673–1675); he published a memorandum on this subject, under the title *Nouvelles expériences du vide avec la description des machines qui servent à le faire* (New experiences with the vacuum, with the description of the machines used to produce it) (Paris, 1675).

At this time there was no thought of machines to produce power, but rather of a miscellany of experiments using the pneumatic machine that he had perfected; these experiments included the conservation of perishable foodstuffs and the impregnation of wood or plaster with the help of glue. What appears especially evident from his notes is Papin's ingenuity in solving problems of detail in his devices. His book was well received, and a summary of it was published in the *Journal des Sçavans* on January 2, 1675.

This was also the period during which Huygens invented the spiral spring in order to apply to watches the principle of the balance wheel, which he had already applied to clocks (January 20, 1675). It was a resounding success, although various individuals both in England and in France attempted to claim credit for it. Lord Brouncker, president of the Royal Society, asked Huygens to send him one of these marvelous clocks, and Huygens decided to utilize his young assistant as messenger to take the instrument to London. On July 17, 1675, he sent Papin to England with a letter of recommendation:

"He has been with me for two years, assisting me in all kinds of experiments. And you have seen, I believe, a small treatise which he has had printed regarding experiments with vacuums, and which also contains his method for assembling these machines, which is ingenious and very successful in its application. He wanted to see what is being done in your country, and even has the intention of settling down there, if he should find an opportunity."

Papin's digester

By sending Papin to Lord Brouncker and H. Boyle, Huygens put him in contact with the milieux in which the arts and sciences were at least as much in honor as in France and in which Papin could hope to make his way.

It was in the course of Papin's twelve-year sojourn in England that he invented his "digester," a simple closed cylinder in which he placed meat and bones to be heated in the water brought to a high temperature under pressure. This device would be of no direct interest to the history of the steam engine but for the fact that it had a safety valve, in the common form of a weight suspended from a lever that prevented a cover placed on the opening from rising (Figure 177). Papin knew the danger of excessive pressures exerted on the inside of a vessel full of water and heated to a high temperature. He had first evaluated the temperature, which he estimated in relation to the internal pressure, by the time of evaporation of a drop of water poured on the cover of the digester. A pendulum marking the seconds gave this time:

4 seconds for 5 atmospheres of pressure
3 seconds for 9 atmospheres of pressure

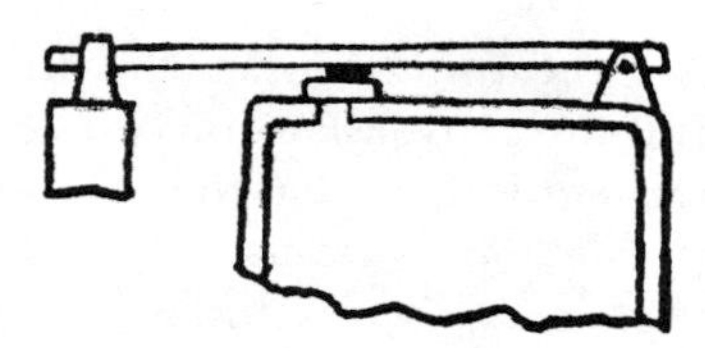

FIG. 177
Papin's safety valve.

His apparatus consisted of a cast-iron cylinder 6 inches in diameter and 18 inches wide. The polished upper portion was attached to the cylinder by means of a leather joint. Papin succeeded in raising the temperature of the water up to the point at which tin melts (210 degrees), thereby attaining a pressure of 270 pounds per square inch. While this device was extremely successful (without becoming a source of income for its inventor), it is only one episode in Papin's life, and is not one of the links in the history of the steam engine.

Though it won for him a nomination as a Fellow of the Royal Society, Papin does not appear to have continued his study of the steam engine before leaving England (1687).

THE ACHIEVEMENTS OF PAPIN AND SAVERY

The exploitation of the mines and the production of power

Around the end of the seventeenth century an increasingly urgent problem was confronting those engaged in the exploitation of the English mines. The consumption of the combustible mineral was increasing, and the mine shafts were descending further into the earth, reaching areas into which large quantities of water filtered, impeding or even paralyzing the miners' work. What was needed was to operate machines capable of pumping out the water.

The only available source with sufficient power was that of the hydraulic wheels, for the power which could be supplied by men and horses was, as we have seen, inadequate and excessively expensive. The geographical location of the mine usually placed them far from a source of water power; it would have been necessary either to install a method for producing power on the site of the mines (which the steam engine was to make possible in the eighteenth century) or to transport the power supplied by a distant waterfall (which was to be accomplished by electricity two centuries later).

It is true that the water could have been taken from the neighborhood of a waterfall, then sent to another point; but this would have required aqueducts over long distances, tunnels, and so on — structures it was impossible to build because of their great cost. Mechanical transmission, which was also costly (witness the machine at Marly), could be used only for short distances, and even then with great loss of mechanical output. Over a great distance all the power would have been lost through friction.

This problem appealed to Papin's fertile mind; he envisaged a connection through the vacuum between the place where the power was produced and the place where it was utilized. Pneumatic pumps operated by hydraulic wheels of the classic variety in use at this period would suck the air into a piping system that would transmit the drop in pressure over a distance; the latter would operate the mine pumps through the disparity between this pressure and the atmospheric pressure. Denis Papin had made a study for Boyle of the perfecting of the pneumatic machine, and the sequence of his ideas can be seen.

But the methods then available could not produce airtight pipes and joints. Losses of vacuum through the entrance of air into the pipes resulted in a loss of output as great as, if not greater than, the loss through mechanical transmission. We do not know whether Papin attempted to apply his invention; in any event, he must have quickly realized that it could not be utilized (1685).

Papin at Marburg

Until 1687 Papin remained in London, where he published several articles in the *Philosophical Transactions* of the Royal Society. But at that time he saw no future possibilities for him in England, and so he accepted the offer of the Landgrave and Elector of Hesse-Kassel, Karl Augustus, of a chair of mathematics at the University of Marburg. This position was perhaps not much more advantageous than the one he had already achieved in London, but Prince Karl Augustus had a reputation as an enlightened individual capable of taking an interest in the mechanical arts and therefore in original inventions. Moreover, the revocation of the Edict of Nantes (1685) had closed the doors of France to Denis Papin, who was a Protestant, and a part of his family (as well as a number of their French co-religionists) were welcomed to Germany. The Landgrave of Hesse-Kassel ordered the issuance of a French-language proclamation inviting "All those who would like to settle in our states in order to exercise or establish manufactures which do not yet exist here, and other arts, works, and crafts useful and necessary, whatever they may be."

At Marburg, then, Papin found emoluments (150 florins per year), advantages in kind, and the hypothetical fees that might be brought in by pupils. The prince also gave him several subsidies that helped him to build the machines he invented. In addition, he was able to benefit from the *Acta Eruditorum* published in Leipzig in order to write and publish information on his inventions; he was able to find in Leibniz a support that was often useful to him, while maintaining contact with Christiaan Huygens, who had returned to Leyden.

Despite the difficult situation in which he found himself in 1690, he continued to pursue the most varied inventions; he collaborated on a diving engine, and in particular went back to the idea of the steam-operated cylinder-and-piston driver.

The first idea of using the force of expansion of steam for projects more useful than Hero's toys, and more practical than the ideas of della Porta and de Caus, seems to have been expressed by Huygens. In 1666 the latter had submitted to Colbert a program of various experiments that called for "examining the power of cannon powder by enclosing a small quantity of it in a very thick iron or copper box. – Examining in the same way the power of water rarefied by fire."

It is true that this is somewhat vague, and of these proposals Huygens seems to have retained only the gunpowder engine.

Gunpowder engine and steam engine

This engine and the experiments performed with it have been mentioned earlier. Papin remembered these experiments, but does not appear to have envisaged resuming them in England. But around 1688, sensing possible assistance from the Landgrave of Hesse, he took up the idea again. The prudent Landgrave, who was probably careful with his money, consented to finance a small experimental device, "from which the principles necessary for the construction of very large machines could then be deduced" (*Acta Eruditorum,* September 1688).

The cylinder was decreased to 5 inches in diameter and 16 inches in height. Papin replaced the pieces of leather by genuine valves; we have no information about the construction of the cylinder, except that it was to be of bronze.

The results obtained were slightly better than those of Huygens, but one-fifth of the air still remained under the piston, which made utilization of the method by "very large machines" illusory. Papin determined one of the causes of the failure: the internal surface of the cylinder was imperfectly polished, and the first step needed was to have very straight pipes well polished on the inside. He undoubtedly continued to reflect on methods of perfecting the machine, but not until 1690 did he publish (in the *Acta Eruditorum*) the idea of the atmospheric steam engine, under the title, *"Nova Methodus ad vires motrices validissimas levi pretio comparandas"* (New method of obtaining very great power at a low price). Was this a repetition of the early idea of Huygens or an original conception? Papin does not appear to have owed this inspiration to Huygens's article, of which he perhaps knew nothing. We are therefore justified in crediting him with the idea.

The device was modified in order to permit the use of steam. The bottom was a thin sheet of metal, so that the fire underneath could easily heat the water contained in the cylinder. The piston was raised by steam pressure (which was then called "rarefaction" by fire). When the piston had reached the top of its course, the fire was banked, condensation occurred, and the piston dropped under atmospheric pressure.

Papin then envisaged a *use* for the device by repeatedly removing and then replacing the fire under the bottom of the cylinder. He did not construct a machine functioning along these principles on a large scale, but probably used a portion of the material already utilized in the earlier experiments (1688). To lessen the disadvantages of the poor fit of the piston in the cylinder, which created a harmful play, he put a little water on the piston, hoping to prevent water from entering when condensation occurred; this is indicative of a fear of finding a cushion of air under the piston that would prevent it from dropping completely (air that could not have been eliminated and that would have accumulated in increasing quantity in the cylinder).

Papin compared the activity of the powder engine with that of the steam engine:

"Approximately one-fifth of its total capacity of water always remained in the pipe. Whence two disadvantages: First, only half of the desired effect is obtained, and a weight of only 150 pounds is raised to the height of one foot, instead of the 300 pounds that should have been raised had the pipe been completely empty; second, as the piston drops, the force which pushes it from top to bottom gradually diminishes . . .

". . . it would be much more convenient to have a driving power that would be constant from beginning to end. With this in mind we made several attempts to obtain a perfect vacuum with the help of cannon powder; for in this way, since there would no longer be any air to resist the piston, the entire upper atmospheric column would push the piston to the bottom of the pipe with a uniform power. But until now all attempts have been fruitless, and after the extinguishing of the ignited powder approximately one-fifth of the air always remained in the pipe. I therefore tried to achieve the same result by another means, and since a small quantity of water, by a property which is natural to this liquid, when reduced to steam by the action of the heat acquires an elastic power similar to that of air, and later returns to liquid state by cooling, without retaining the slightest appearance of its elastic strength, I was led to believe that it would be possible to construct machines in which the water, by means of a moderate heat, and without major expense, would produce the perfect vacuum which could not be obtained with the help of cannon powder."

Placing water in his cylinder (Figure 178) to a depth of 3 or 4 times (6, 7, to 9 millimeters), he pushed the piston down until water came out through the hole *C* drilled in the piston. *C* was closed with the rod *M;* thus there was no longer any air under the piston, and it was possible to obtain first vaporization and then condensation of the steam, and therefore to cause the piston *B* to rise and fall.

Papin's experimental pipe, which was approximately 2 inches in diameter, could, he believed, lift 60 pounds; the pipe itself weighed only 5 ounces. He believed that with cylinders weighing 40 pounds he would be able to lift a weight of 2,000 pounds for 4 feet; the action could be performed in one minute. He thus had a glimpse of the enormous power that could be obtained with this engine.

However, these seemingly high figures should not give rise to any illusions about the power that could be supplied by such a device. Papin envisaged only a momentary action of the piston; the power of the device was thus reduced to

$$\frac{979 \times 1{,}299}{60} = \text{approximately } 20 \text{ kgm./sec.}$$

which is less than one-third hp.

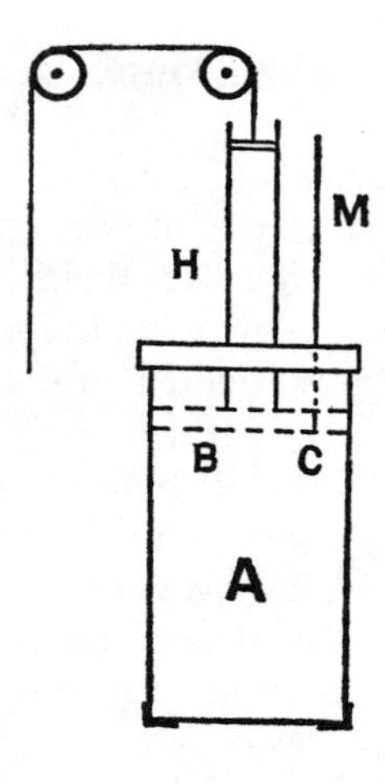

FIG. 178
Papin's cylinder.

He conceived of a pipe 2 feet in diameter that could lift 8,000 pounds to a height of 4 feet per minute; in addition, he thought of coupling together several pipes, under each of which the fire would be placed in succession, which corresponds to 68 kgm./sec., or almost one hp. By utilizing several "tubes," the power would be multiplied in corresponding proportion.

First experiment with an atmospheric engine

After the Peace of Ryswick, the Landgrave decided to construct various installations in his domain, with particular attention being paid to the project of bringing water from the Fulda to the top of a tower. Papin suggested the use of his machine, but Karl Augustus, undoubtedly finding that it was not sufficiently perfected, gave preference to the classic system of hydraulic wheel and pumps.

The disappointed Papin turned his attention to new applications of his idea, for example a carriage, for which he may have made a model, and a ship; he does not describe them in the *Acta Eruditorum.* He appears to have changed his orientation, and to have adopted another principle for his steam engines; perhaps the system of the cylinder with movable fire seemed difficult to achieve. He must, however, have thought of a system of several cylinders under each of which the fire would have been placed in succession. But in this case the mechanical connections of the pistons became more complicated, and the continual shifting of the fire was difficult and would have required human labor.

Thus he progressed to a simpler system, which consisted of producing steam in a heated vessel and causing the steam to pass into another vessel filled with cold water. The water was pushed back into the ascending conduit; then the tap was closed, and the vacuum produced by condensation permitted another quantity of water to be sucked into the second vessel through the supply pipe. Papin thought of pumping the water to a height of 70 feet; in this way he eliminated the mechanical intermediaries, but ended with the system Savery was just then perfecting in England.

Damage had been done to his experimental device, and the now wary Landgrave was tardy in contributing the monetary assistance requested; thus Papin's new achievements failed. They were to be perfected several years later, but by then Savery's machine had definitely made a place for itself in England (Savery's patent is dated July 25, 1698, the time when Papin was already being forced to suspend his work). His position in Marburg having become for various reasons difficult, in 1696 Papin went to Kassel.

Thomas Savery

Thomas Savery, son of a Devonshire family, began his working career in the mines; he then became a captain in the merchant marine, a skilled engineer, and in addition an inventor. His patent of 1696, No. 347, for a mill for grinding, polishing, and other uses including that of propelling ships, had been proposed to the Admiralty in connection with the latter purpose, but it had been greeted with disdain. Like many Englishmen, Savery was greatly concerned with the problem of pumping water out of the mines, and he was acquainted with Papin's activity. But the system discussed by Papin in 1690 in the *Acta* was vigorously criticized, especially by Robert Hooke: the slowness of the movements of the piston, which could make only one oscillation per minute, and the necessity of shifting the fire back and forth, were obvious obstacles to its practical use.

Hooke's criticisms turned Savery against the use of Papin's cylinder-piston system; he adopted the principle of suction of the water by the vacuum produced by condensation, then compression by steam pressure. This, however, reduced the steam engine to the exclusive role of a water-raising machine; it was also a return to the principle of Hero, della Porta, and Salomon de Caus.

Savery, who was anxious to obtain a patent, wished to affirm the originality of his conception, and in 1702 he described the circumstances which had suggested the engine to him *(The Miner's Friend: Description of an engine for raising water by fire, and the manner of placing it in the mines, with an explanation of the various usages to which it may be applicable, and an answer to the objections raised against it,* by Thomas Savery, London, 1702).

According to his story, having drunk a bottle of wine at an inn, he threw the empty bottle into the fire in the fireplace. As he asked for a small basin of water to wash his hands, he saw the steam escaping from the neck of the bottle, which was heated by the fire. The idea then occurred to him to pull the bottle out of the fire with one of his leather gloves, and to turn it upside down in the water in the basin (Figure 179). The water immediately rose into the bottle and filled it, the condensation of the steam having created a vacuum in the internal atmosphere. This phenomenon, which Savery was not the first to observe, is clearly being claimed as a new observation by Savery in order to justify the originality of his conception and ward off the prior claims of della Porta, de Caus, and even Papin.

FIG. 179
Savery's bottle.

Savery's atmospheric machine

In this way he succeeded in obtaining patent No. 356 of July 25, 1698, the object of which is defined as follows: "Raising of water and operation of machinery, by means of the moving of fire, which can be utilized for draining mines, supplying cities with water, and operating all kinds of machines wherever there is neither water nor steady winds" (Alphabetical Index, March 2, 1617–Oct. 1, 1852)

On June 14, 1699, Savery presented to the Royal Society in London a small model of the machine patented, and the *Philosophical Transactions* for that year contained a picture of the model presented (Figure 180).

At the end of this meeting, the society decided that the device was interesting but that it could be improved upon, and, remembering its Fellow Papin, charged one of the members to write to Marburg to ask him to come to London and follow the experiments.

At the same time Savery was named a Fellow of the Royal Society. Being

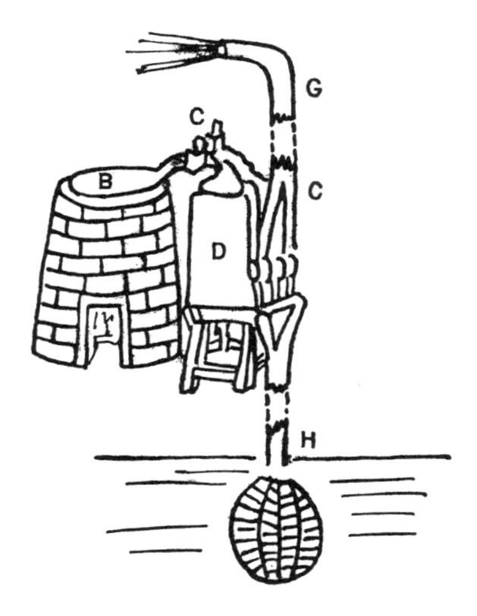

FIG. 180 First drawing of Savery's engine.

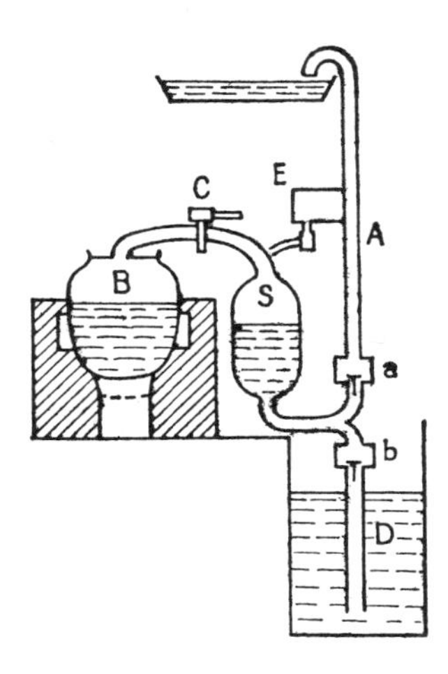

FIG. 181 Diagram and operating principle of Savery's engine.

more aware than Papin of the possible results, Savery immediately pursued the possibility of exploiting his fire engine commercially, and succeeded.

The principle of the machine was as follows (Figure 181):

The steam supplied by the boiler *B* passes through the tap *C* into the vessel *S*. Pushing the water contained in *S*, it causes it to rise through the valve *a*, which then opens into the pipe *A*, where it rises to the higher level. When *S* is empty, by closing *C* the steam is able to condense and create a vacuum in *S*; the water is then sucked up by the valve *b* and the pipe *D* and fills *S*. During this period the valve *a*, returned to its place, prevents the column of water *A* from falling. When *S* is filled with water, the operation is begun again.

The machine was able to lift four times the contents of the vessel *S* (52 gallons of water) to a height of 55 feet (17.95 meters) per minute.

Thus there were two successive phases: suction and compression. In order to have a continuous flow, two *"S"* vessels could be combined, one being filled while the other was being emptied. This is the method depicted in the *Philosophical Transactions* of 1699.

The power produced by the device is indicated by the following figures:

Work done in one minute: $4.54 \times 54 \times 17.9 = 4{,}389$ kgm.

Horsepower (75 per second): $1/75 \times 60 = \frac{1}{4}\ 500 \times 4{,}389$

that is, slightly less than one hp., two hp. if two containers were used.

Savery had overestimated the efficiency and power of his device; he thought it would be capable of lifting water to a height of 500 or 1,000 feet if he were able to have "vessels" that were sufficiently strong, but he actually envisaged as possible a height of 60, 70, or 80 feet *(The Miner's Friend)*. Actually, it was barely possible to surpass 40 feet, which was the limit of pressure for the devices it was then possible to construct. An engine tried out at York with a height of 300 feet fell apart, "the heat being so great that the soldering melted and the machine opened at the seams" (Désaguliers).

In addition, it must be noted that the direct contact of the steam with the cold water caused a rapid condensation and that the very high consumption of steam made the engine economically inefficient. In order to install it in the mines (Figure 182, diagram of an installation), it was necessary to put the engine less than twenty-four feet from the level of the water to be pumped out; the presence of this engine, which always aroused a fear of explosion, did not engender much confidence on the part of the miners. Despite the urgent need for methods of pumping, the fact had to be faced that there was little future for the use of Savery's machine in the mines; it had some success in open-air installations. But its development quickly ended in the eighteenth century when Newcomen's machine was used.

Papin's use of the piston

While Savery was successfully carrying out his projects, Papin was in Kassel. He was aware of Savery's writings and the creation of his engine; Leibniz, visiting London in 1705, had sent him detailed information about, and drawings of, the engine. Papin sent Leibniz's letter and sketches to the Landgrave, who, being very much interested, urged Papin to resume his work and – more important for its success – allocated subsidies to him to construct a new engine.

Papin did not slavishly copy Savery's device; he was able to benefit from his own experiments and reflections, and to conceive a machine that eliminated the greatest defects of Savery's invention. He called his new machine the "Elector's Engine," in honor of the prince who had helped and supported him, and published a description of it in 1707, in Latin and French, in a treatise entitled "New method for raising water by the power of fire, explained by M. Denys Papin, medical doctor, professor of mathematics at Marburg, conselor to HRH of Hesse and Fellow of the Royal Society of London. At Kassel, by Jacob Estienne, Court Librarian, and Johann Gaspard Voguel, printer, M.D.C.C.VII."

Remembering the good contacts he had had at London, Papin addressed his treatise to the Royal Society; his letter reflects the ideas and tendencies of the day, as well as the author's desire to obtain an approval from which he could profit for the exploitation of his invention.

Papin's work contains the following detailed description of his engine (Figure 183).

"Here is the manner in which the engine is presently constructed. *A* is a large copper vessel which I call a retort, because of its similarity to the instrument of that name used in chemistry; it is 20 to 21 inches (540 mm.) in diameter at its widest point, and is 26 inches high. It must be enclosed in a brick furnace . . .

". . . at its very top this retort must have a curved pipe *ABB,* to which the tap *E* is soldered, by which the passage is opened for the steam . . .

". . . the steam, heated in the vessel *A,* erupts violently into the pipe *ABB* as soon as the spigot *E* is opened, and pushes the piston *F* in the pump *D* . . .

"The vessel *D,* which acts as a pump, is 20 inches in diameter, and its piston *F* travels through an area 16 inches high. Thus it is easy to calculate that at each operation this piston is able to remove 200 pounds of water from the pump *D.*

"We are also able to calculate that if the pipe *G* is made so that at its narrowest place it is 8 inches in diameter, and that the opening *G* (through

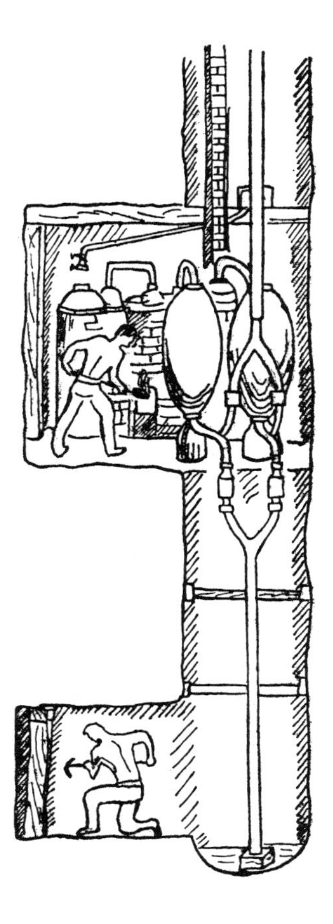

Fig. 182
Savery's engine
installed in a mine.

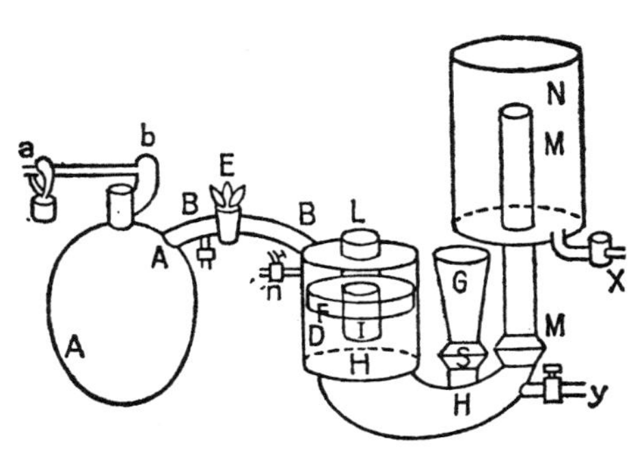

Fig. 183 Papin's
engine of 1707.

which the water must return in order to fill the pump *D*) is 8 inches higher than the spigot *n* (by which the water must leave as soon as there is a sufficient quantity of water in *D*), we may, as I said, demonstrate by calculation that this will suffice to permit the pump *D* to be filled with 200 pounds of water in less than one second, the said pipe *G* communicating with the large curved pipe *HH* which has its principal opening in the pump *D*.

"The piston *F* is a hollow metal cylinder, tightly stoppered lest the water entering it make it too heavy, for it should float upon the water so that it will always rise to the top of the pump when the latter fills up. In this piston, note the pipe *I*, open at the top and closed at the bottom, and passing through the center of the said piston, where it is carefully soldered both to the bottom and to the cover. This pipe serves to receive red-hot irons, which are introduced through the opening *L* and which remain suspended from the top of the pump; their purpose is to increase the power of the steam which enters the pipe *ABB* when the tap E is opened . . .

". . . to the small end of the pipe *HH* is soldered the pipe *MM*, which fits partly into the cylindrical vessel *N*: this vessel should be 3 feet high and 23 inches in diameter so that in a height of 1 foot it contains 200 pounds of water, and in all 600 pounds. Thus when it is filled with water to a height of 2 feet, the air in it will be reduced to occupying no more than one-third of the space which it ordinarily occupies, and will support only 16 feet of water, in addition to the ordinary height, which is 32 feet."

Papin thought that the tank could be filled in one second and emptied in the same amount of time, which would have given 30 pulsations per minute. By virtue of all the maneuvers to be done with the piping, it was possible to evaluate at 10 seconds the time that would actually have been required for one pulsation; moreover, if water is raised, it is necessary to figure on the minimum height, that is, 16 feet, or 5.20 meters.

Improvements and actual power of Papin's machine

The machine as described by Papin would thus have a power/sec. of $\frac{200 \text{ pounds} \times 5.2}{10}$, approximately 50 kgm./sec. or ⅔ hp.

But Papin, figuring on an average of 64 to 16 feet, that is, 40 feet, and on 30 pulsations per minute, overestimated it at $\frac{200 \times 13}{2}$, approximately 650 kgm./s. or 8.65 hp, which he was able to equate with 50 men.

By increasing the dimensions of the machine and the pressure of the steam, he envisaged the possibility of obtaining a power 500 times greater than that of one man, while one man would have been needed to operate the engine.

While the use of heated irons seems rather impractical, it should nevertheless be regarded as a credit to Papin; it represents a kind of superheating, an idea which was not to be adopted until much later. The air chamber was not an invention of Papin, as he himself admitted, but was due to one Mautsch of Nuremberg, who had had in mind the regularization of the blast from the fire pumps. Papin adopted it in order to have water under pressure without a tank placed at a great height, which could be more convenient in case the engine was being used to propel a boat.

The floating piston is particularly interesting, to prevent the steam arriving in the pump from coming into contact with the cold water. In addition, Papin's arrangement of the closed cylinder eliminated the entrance of air which he had found so undesirable in his first device, the cylinder and the piston being subject to air pressure.

How did the Royal Society greet Papin's letter? The *Transactions* did not publish either the report or Papin's letter. In the twenty years that had passed since his departure from London, Papin must have lost many of his former contacts. Now a stranger in England, he had just introduced a competitor for the system already put into operation by Savery, who was also a Fellow of the Royal Society. A number of the Fellows had been following Savery's progress, and knew of his success. Many people undoubtedly felt it improper that a foreigner should question the tried and proved system of an Englishman. There could be no favorable answer to Papin's proposal.

In Germany, the machine built with the Elector's subsidies aroused curiosity; people came to see the water rise, but the machine appeared to be of no practical use in a country whose industrial development was far behind that of England. The Landgrave wearied of subsidizing a machine that was of no immediate use, and Papin's invention fell victim to lack of interest (1707). All that could now be done was to cease all further expenditure and take the machine apart, probably to save a small quantity of the metal. Despite Papin's continued correspondence with Leibniz, who still willingly supported him, Papin did not find the money he would have needed in order to build a new machine (beginning of the year 1707). In order to live he had to put other inventions to use, and his fertile brain did not fail to find others — chemical products, steam engines for hurling grenades, and a ship, which he again thought of equipping with a steam engine.

The idea of the steamboat

For a long time Papin had been dreaming of utilizing steam engines on boats. In a report in the *Acta Eruditorum* of August 1690, he explained at length his ideas on the subject:

". . . However, I shall now list the variety of ways in which a propelling force of this type would be preferable for the movement of a ship. First, ordinary oarsmen overburden the ship with their weight, and lessen its capacity for movement. Second, they occupy a large space, and consequently are a great hindrance on the vessel. Third, it is not always possible to find the number of men needed. Fourth, whether working at sea or resting in port, oarsmen must be fed, which is no mean increase in expense. Our pipes, in contrast, would burden . . . the vessel with only a very light weight; they would occupy little space; they could be procured in sufficient quantity if there existed a factory to make them; lastly, these tubes would require wood only while in operation, and would require no expenditure in port. But since ordinary oars would be less easy to operate with tubes of this type, it would be necessary to utilize paddle wheels of the type I recall seeing on the engine constructed in London on the order of His Most Serene Highness, Palatine Prince Rupert. It was operated by horses with the help of oars of this type, and it left the royal launch very far behind, although the latter had 16 oarsmen. There is no doubt that our tubes could give a rotating movement to oars attached to an axle, if the shafts of the pistons were equipped with teeth which would gear into toothed wheels attached to the axle of the oars. Three or four tubes would have to be adapted to the same axle, so that its movement could continue without interruption. While one piston was at the bottom of its tube, and consequently was unable to cause the axle to turn until the steam had raised it to the top of the tube, at that moment the locking device on another piston could be removed, and this piston would by descending continue the movement of the axle. Another piston would then be pushed in the same fashion . . .

". . . Moreover, a single furnace and a small fire would be sufficient to raise all the pistons one after the other.

"But someone may object that the teeth of the shafts geared into the teeth of the wheels will exercise movements in the opposite direction on the axle when they rise and fall . . .

"Every mechanician is well acquainted with a method by which toothed wheels are fixed to an axle; when moved in one direction they pull the axle with them, when moved in the other direction they communicate no movement and permit the axle to obey the opposite rotation. Thus the principal difficulty is to find a factory where large tubes can easily be forged, as has been described in detail in the *Acta Eruditorum* for the month of September 1688."

"Oar-wheels"

At the time, Papin believed that his steam cylinder could be put into use immediately; he failed to realize the necessity of perfecting it, with all the details and additions that this always entails. This document, which is so remarkable from every point of view, nevertheless contains his numerous ideas and problems regarding the difficulties to be solved.

Paddle wheels are not an invention of Papin; they had been known and suggested long before. Papin mentions that they had already been tried on Prince Rupert's boat, where they were operated by a horse-driven treadmill. References to such wheels had already been found in the authors of the ancients; the Spaniards claimed for Blasco de Ganay the priority for the utilization of the steam engine to drive a 200-ton boat with a paddle wheel, but the document they advance in support of this claim appears highly dubious (Thurston).

What Papin is clearly advocating is the use of multiple cylinders that undoubtedly would have been placed in vertical position and that would have been heated by the movable fire beneath them; also, the axle of the wheels would have been operated by racks provided with click-and-ratchet systems, an idea which we find again in the essays of the Marquis de Jouffroy at the end of the eighteenth century.

However, Papin did not attempt actually to construct the project, which inevitably would have been extremely difficult. He does not appear to have returned to the idea until 1704, and then only because he was again thinking about how the "oar-wheels" should be constructed, and was devoting all his attention to the question.

He now constructed a boat in order to experiment with this system of propulsion. As yet the boat had no engine, and until 1707 Papin did not think of adapting his atmospheric cylinder to it — that idea was now in the background of his thoughts. During the first part of 1707 these thoughts were full of bitterness: the "Elector's Engine" had been taken apart, and Papin was unsuccessfully searching for the means that would permit him to build another one. He wrote to the Royal Society in London, but received no answer. His difficult nature had made him many enemies in Kassel. He nevertheless believed that he had perfected the steam engine in a form superior to that of Savery, and felt it was capable of being applied very extensively. He thought, too, that he would find more fertile terrain for the development of his invention in England, where he had once been happy and highly regarded, and where industrial activity was making rapid progress, and he therefore prepared to cross the Channel. In order to simplify his preparations and reduce the costs, he decided to utilize the boat he had constructed in 1704 and prudently put into storage.

The story of this period of Papin's life has too often been romanticized and distorted in story and picture. Papin is depicted as installing the engine on his boat, making a successful experimental run, and embarking with his family in an attempt to reach England.

His last uncompleted projects

The examination of the texts must lead us to a very different view. What was Papin's boat like? We do not know exactly, but we know that it was used for the experiments with the "oar-wheel," and that until then it had not been equipped with a steam engine. We know little about its size; it has been said (Ch. Cabanes) that it was quite large, since the ship was able to carry a 4,000-pound cargo. However, this is barely two tons, which indicates that actually it was a fairly small boat.

On the other hand, we know that in the theory of steam propulsion which he proposed in 1707, Papin had abandoned his earlier idea (1690) of atmospheric-pressure cylinders with gears, racks, and click-and-ratchet systems. The engine he wished to employ this time was the "Elector's Engine," which supplied only a fountain of water; in order to transform this power, the system then known and used was the utilization of the jet of water to turn a bucket wheel (Figure 184) acted upon by the water coming out of the compressed air chamber.

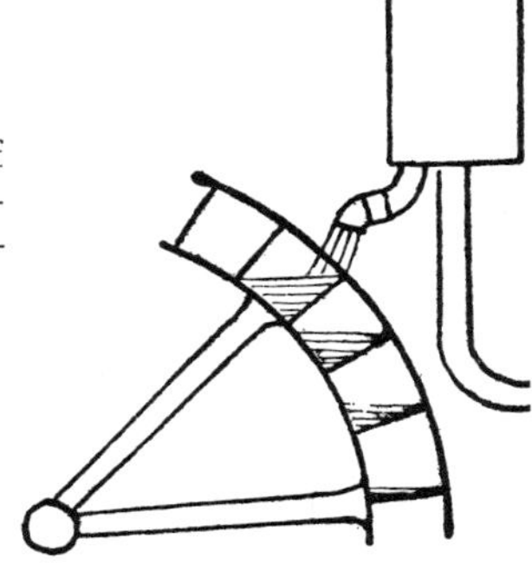

FIG. 184 Combination of the steam engine (for raising water) and the hydraulic wheel.

Was the boat sufficiently large to hold such an engine, the boiler, the cylinder, and the furnace with its brickwork? It is doubtful. In what terms does Papin speak of it? On July 7, 1707, he wrote to Leibniz declaring his intention of "retiring to England" and his desire to put "his new device to the test in a seaport like London, where it can be given enough depth to apply the new invention, in which thanks to fire one or two men will be made capable of producing greater effect than several hundred oarsmen."

This sentence clearly indicates that it was in England that he intended to construct a larger boat and install a steam engine in it. At that moment he did not have the engine: the first one had been taken apart, and he did not have the money to build another one.

He adds, "It is my intention to make the trip on the same boat which I have had the honor to mention to you before, and we shall see that with this model it will be easy to construct others to which the fire engine can very easily be applied." This clearly proves that the fire engine was not and could not be applied to his small boat. Thus it was a model boat, moved by an "oar-wheel," but without an engine, which Papin had at his disposal in 1707.

Perhaps he had saved a few mechanical parts and pipes from the destruction of the engine, which the Elector permitted him to keep; the prince had undoubtedly allowed Papin to take and use the boat. In his letter to Leibniz, Papin stresses the consent given by His Highness, "with circumstances which demonstrate that he is still, as he always has been, much more kind to me than I deserve." Does this not point to an abandonment of equipment to Papin, rather than just a simple permission to return to England?

Papin was particularly concerned with the difficulties he might encounter on the part of the powerful boatmen's "Ghilde" on the Weser, and again requested help from Leibniz to avoid them. Having received no answer by August 1st, he again wrote to Leibniz:

"The experiment with my boat is completed, and it succeeded in the manner which I had hoped; the power of the stream's current was so weak in comparison with the *power of my oars* that it was difficult to realize that it traveled more quickly downstream than it did upstream. My Lord was kind enough to show his satisfaction at seeing such a good result, and I am convinced that if God permits me to arrive safely in London, and to construct vessels *of this type sufficiently deep* to permit the use of the fire engine to move the oars, I am convinced, I say, that we *shall be able* to produce effects which *will appear* incredible to those who have not seen them."

These expressions are clearly directed at the type of construction of the boat and the wheels; it is not a question of the steam engine – since the latter is envisaged only for a future time – but is still the hope of a future, not confidence in an engine whose operation Papin would not have failed to boast of if he had made even a simple experiment with it.

Thus we are clearly to believe that contrary to what has been so often repeated, Papin departed on a small boat without an engine, its sole cargo (and sole cargo possible) being his family and possessions, and perhaps a few parts from the "Elector's Engine." Consequently, the too widely accepted fiction of the destruction of the engine by the boatmen must be completely altered, along with the elimination of the idea that the latter saw in the new engine a possible competitor which they hastened to eliminate.

What actually happened (and it was already sufficiently cruel for Papin) was a quarrel for rights and privileges of navigation on the Weser, which Papin had feared from the moment of his decision to go by water to Bremen and which brought about disaster: the boat was pulled to the riverbank (which proves that it was not heavy) and broken up. No mention is made, however, in the various corerspondence on this unfortunate adventure, of the supposed destruction of an engine of any kind.

Leaving his family (or abandoned by them in the face of this failure that ended a last hope), Papin departed alone for England. At the beginning of 1708 we find him living in London; on February 11 the Royal Society received from him a report on the steamboat. Having seen the importance acquired by Savery's machine, he had to demonstrate the superiority of his own, and for this purpose he suggested a comparative test:

"I offer . . . to build a machine constructed like that of Kassel, and equipped in such a way as to be applied to the propulsion of ships [it had been realized that Savery's machine was ill suited for this use], simultaneous with 'another machine made by Savery's method' . . .

"In the event that I succeed, and only in this event, I humbly request that my expenses, my time, and my difficulties be reimbursed to me, and I estimate the whole at 15 pounds sterling."

The report was read, as was required for every communication by a Fellow of the society, but Savery's solidly founded situation pleaded in favor of his engine against the hypothetical advantages of Papin's device. The latter's proposal was not accepted, and apparently was not even discussed or answered.

Now an old man, in an environment that had changed greatly since his departure from London, an old man who had brought from Hesse only the reputation of an unfortunate creature ruined by failures, Papin and his ideas were given no consideration, and he was accorded only a disdainful pity. From then on, despite his repeated requests addressed to the members of the Royal Society, he could arouse no interest in any quarter. Condemned to live miserably on tiny subsidies ("alms" might be a better word), he disappeared around 1712 or 1714.

The story of Papin's life carries us into the first years of the eighteenth century, but it is part of the seventeenth century, ending with the final failure of his ideas. He introduced the idea of the atmospheric steam-and-piston engine, only to abandon it thereafter; he sought to perfect an engine on the same principle as that of Savery, without succeeding in imposing his system.

In the same way, Savery's system was replaced by Newcomen's engine born of the cylinder-piston system that was developed fully in the eighteenth century; Papin played no role in its practical utilization and ultimate success. Papin's life and work, although they ended at the beginning of the eighteenth century, belong, not to the period of development of power by steam engines, but rather to the age of trial and error and fruitless experiments.

BIBLIOGRAPHY

CABANES, CHARLES, *Denys Papin, inventeur et philosophe cosmopolite* (Paris, 1935).

DICKINSON, H. W., *A Short History of the Steam Engine* (Cambridge, 1939).

JENKINS, RHYS, "Savery, Newcomen and the Early History of the Steam Engine," Newcomen Society *Transactions,* 3 (1922–24), 113–130.

KERKER, MILTON, "Science and the Steam Engine," *Technology and Culture,* 2 (1961), 381–390.

NEEDHAM, JOSEPH, "The Pre-Natal History of the Steam Engine," Newcomen Society *Transactions,* 35 (1962–63), 3–58.

RETI, LADISLAO, "A Postscript to the Filarete Discussion: On Horizontal Waterwheels and Smelter Blowers in the Writings of Leonardo da Vinci and Juanelo Turriano," *Technology and Culture,* 6 (1965), 428–441.

SKELTON, C. P., *British Windmills and Watermills* (London, 1947).

STOWERS, A., "Observation on the History of Waterpower," Newcomen Society *Transactions,* 30 (1955–57), 239–256.

VOWLES, H. P. and M. W., *The Quest for Power* (London, 1931).

WAILES, REX, *The English Windmill* (London, 1954).

WILSON, P. N., *Watermills with Horizontal Wheels,* Watermills Committee Booklets No. 7 (London, 1960).

Section Five

Military Techniques

CHAPTER 18

FORTIFICATIONS

THE PERIOD between 1500 and 1700 witnessed the creation of modern fortifications, of the type that were to be defended against attack by cannons and to answer the assailant with his own firearms. This type of fortification, the work of architects and engineers from Sanmicheli to Vauban, was characterized by the adaptation of the bastioned profile to the defense of cities. The polygonal trace had been known long before, we may say since ancient times, for city walls and castle walls, with curtains and towers that permitted the defenders to counterattack an enemy approaching the walls.

Beginning in the fifteenth century, the use of the cannon, with a greater force of projection of balls, and the bullets of the portable weapons, which could travel over a greater distance, profoundly modified conditions of attack and defense. The old fortifications were more or less adapted to firearms; these weapons, and particularly the early cannons, which were quite light, could be installed on the top or in the individual stories of the towers and fired through embrasures cut in the masonry. Heavier cannons, however, could not be placed on the towers or behind the walls.

The bastion

Originally the bastion formed a ground-level rampart rather than a flanker. It consisted of a circular terreplein constructed around and in front of the towers of castles and cities; its purpose was to hold the guns whose weight and bulk now made it impossible to shelter them inside the masonry structures.

Later this rampart replaced the walls, which had become useless as the principal element of fortification. The necessity for flankers for infantry and artillery fire that could beat the flanks of the parapets and the approaches to neighboring bastions then became apparent.

In order that all its faces would be covered by the defenders' guns, the bastioned front (Figure 185) with the general direction of AA_1 was formed by a polygonal line $EDABC\ C_1B_1A_1$; AB, A_1B_1 being the faces of the bastion, whose

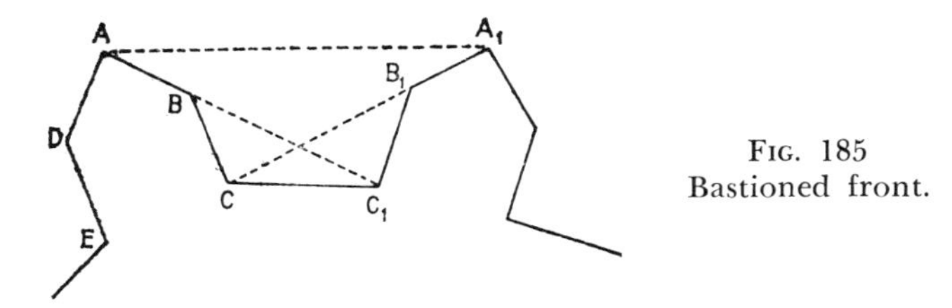

Fig. 185
Bastioned front.

flanks BC and B_1C_1 beat the approaches. The guns on the flank B_1C_1 were able to cover assailants who might arrive at BA; similarly, the curtain CC_1 covered flanks CB and C_1B_1.

The dimensions of the various components of the polygon were determined by the range of the weapons, particularly the guns of the defense. The perfecting of the harquebuses and muskets very quickly increased their effective range from 270 yards to 550 yards; thus the distance $A\ A_1$, between the projections of the bastioned profile, increased from 383 yards in the sixteenth century to 875 yards in the nineteenth century.

Complementary works

A weak spot, in front of the curtains, was reinforced (Figure 186) by a secondary outwork built before the walls: the half-moons, whose outer contour was in turn flanked by the faces of the bastion.

When the enemy reached the ditches of the works in preparation for his assault on the walls, he could ensconce himself in the dead space of the walls, at their base. He would be raked by the guns of the caponiers, masonry outworks constructed under the escarpment that could conceal the guns until the last minute and then enfilade the base of the walls. These caponiers were sheltered from the attackers' artillery fire and hidden by the orillons (Figure 187).

The dimensions of the various components of the fortification were determined by the number and size of the weapons assigned to each face of the polygon. Each of the flanks CB, C_1B_1 had to be able to support two cannons; the banquette where the infantry was to be installed was also determined in their trace.

The composition of the components of a fortification was thus the work of an engineer, but at the time of the Renaissance it was also the work of architects and artists, and we find the greatest names in architecture appearing among those of builders of fortifications, an activity in which they played the role of innovators.

Mention has already been made of city plans established in terms of their defense and the layout of their fortified works, for example those of Antwerp (1540) and Palma Nova (1595). Even before the beginning of the sixteenth century, however, architects were working on military projects.

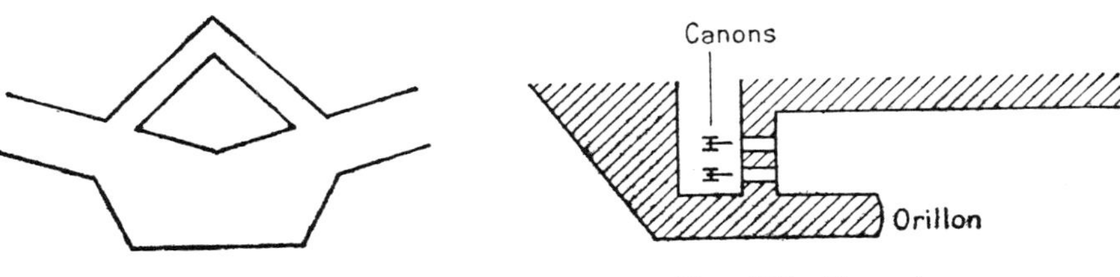

Fig. 186 Half-moon.

Fig. 187 Caponier.

The builders

Filippo Brunelleschi (1375–1446), architect and sculptor, constructed a fortress in Milan, that of Vico Pisano, the two citadels of Pisa, and the fortifications of the port of Pesaro. Vasari said of him that "if every state possessed a man of his like, there would no longer be any need of weapons for defense."

Michelozzo di Bartolommeo (1396–1472), creator of the Palazzo Riccardi at Florence and the water-supply system of Assisi, drew up the plans for the citadel of Perugia. Around 1502, Leonardo da Vinci drew up fortification plans that are still extant. Bramante (Donato d'Agnolo) (1444–1514), who was also a military engineer, succeeded by means of his restrained ornamentation in making the fortress of Civitavecchia into a remarkable work of art.

The San Gallos were a family of civil and military architects. Juliano da San Gallo (1445–1516) fortified the Castellina in 1478, but was unable to defend it against the King of Naples; he also constructed the fortifications of Borgo. His brother Antonio remodeled those of the Castel Sant' Angelo, and worked for the papacy on the ramparts of Nepi, Civita Castellana, and Montefiascone. His son Antonio (II), architect of Saint Peter's (1485–1546), renewed the fortifications of Civitavecchia, and drew up the plans for the citadel of Ancona; his work can also be seen at Florence, Ascoli, Parma, Piacenza, and the Borgo. After 1527 he took refuge with Pope Clement VII at Orvieto, where he constructed a famous monumental well.

Fra Giocondo (died 1515) cooperated on the fortifications of Verona, which were later to make Sanmicheli famous.

In 1527 Albrecht Dürer (1471–1528) published in Nuremberg his work *Instruction on the Fortification of Cities, Castles, and Towns.* He sought to protect the ditches, which he made very wide, by low casemates, an idea that had already been expressed by Machiavelli; he recommended a distance of one mile as the open zone before the ramparts, to be covered by a culverin. His projects included large angle towers leveled to the ground and equipped with a low casemated battery and an open battery on the platform, with two flanking guns and nine guns on the rounded face. In his plans these towers form an enormous block of masonry 295 feet by 146 feet, and 146 feet tall. Behind the escarpment a gallery shelters cannons that are able to hit the ditch. He invented the principle of discharging vaults that relieve the escarpment, and ingenious arrangements of masonry courses inclined in order to resist overturning under the thrust of the mass. He was able to apply his theories to the construction of the fortifications of Nuremberg, but it meant abandoning the idea of the towers, which would have involved considerable financial effort.

The universal genius Michelangelo was a great builder of fortifications. After the sack of Rome the Italian cities surrendered one by one to Charles V (Charles I of Spain). Florence decided to defend herself against Pope Clement VII. On January 10, 1529, Michelangelo was selected to direct the fortification work. He ordered, and carried out, the destruction of every house outside the walls for a distance of one mile, in line with Dürer's idea.

After a visit to Florence, Machiavelli wrote:

"It seems to the captain that the San Nicolo gate as well as the entire suburb (left bank of the Arno) as far as the San Miniato gate cannot be held, given that the entire area is overlooked by the mountain, that there is no hope of defending it in any way, and, what is worse, that it is impossible to fortify it."

Nicolò Machiavelli's theory, which he expressed in 1521, was that a detached outwork in front of a stronghold could not be defended, being "constantly exposed to the fury of the artillery"; as an example, he mentioned the bastions constructed by the Genoese on the hills surrounding their city. "The taking of these bastions [by Louis XII], which were conquered in several days, brought about the loss of the city itself."

But Michelangelo did not hesitate to fortify the hill of San Miniato, from which the enemy would have been able to bombard Florence. On April 5 he was named *Governatore et procuratore generale sopra le forticazioni* for one year, at a salary of one gold florin per day. In April and May he went to Livorno, in June to Pisa. In July he went to Ferrara to investigate the new fortifications organized by Duke Alfonso d'Este.

It is said that Vauban greatly admired Michelangelo's work at San Miniato. Attacked on October 29, and bombarded on the 30th, it was the target of 150 cannon balls. Treason delivered the city into the hands of the papal armies and Charles V (Charles I of Spain). But the Pope, being in need of Michelangelo's services, became reconciled with him.

Back in Rome, Michelangelo disagreed with Antonio San Gallo on the subject of the fortification of the Borgo; however, he continued to work on the ramparts and gates of Rome, particularly the Porte Pia, which he constructed in 1559 (Pontificate of Pope Pius IV).

Sanmicheli

The architect whose original creations and breadth of conception and work seem to surpass all other, even the most fertile, architects, clearly appears to have been Michele Sanmicheli (Verona, 1484–1559), who is better known in Italy, where his reputation retains uncontested prestige, than in France. His fellow citizens raised a statue to him that bears this legend: *"Grande nella architettura civile et religiosa, massimo nella militare."*

In addition to constructing a large number of palaces and churches, he may be regarded as one of the early founders of modern fortifications, which defense against cannons had made a necessity. He was one of the first masters of the bastioned fortification, which one hundred years later was to be developed to its highest form by Vauban. Sanmicheli is the perfect example of the union of artist and engineer in one individual; he followed in the footsteps of his illustrious Romans predecessors, including Vitruvius, *Praefectus Fabrorum* of Caesar's armies, that is, chief engineer of the Roman legions and author of the famous architectural work that was utilized for centuries.

According to Vasari, Sanmicheli was employed by Clement VII, together with Antonio San Gallo, on the fortifications of Parma and Piacenza; at Verona he built a greatly admired bastion, which led the Venetians to entrust him with the fortifications of Legnago and Porto and those of Orzi Nuovo.

Sforza called him from Venice (the Venetians granted him a leave of absence of only three months) to inspect the fortresses in the Milan area. Back in Venice, he inspected the strongholds of La Chiusa, Friuli, Bergamo, Peschiera, and Dalmatia. Since he was not able to cope single-handedly with all this work, he had the construction of the defenses of Sebenico entrusted to his nephew Giovanni Girolamo, who had fortified Zara. He hastily repaired the defenses of the islands threatened by the Turks — Corfu, Cyprus, and Crete. The Turks succeeded in taking the latter island only after a two-year siege (May 1, 1667–September 17, 1669).

After the completion of these projects he returned to Venice, where he was entrusted with the fortification of the Lido — a difficult project, for the land is swampy and shifting. In order to obtain a solid base in this uncertain ground, he constructed a cofferdam with jointed piles, and pumped out the water. Then, assembling all the stevedores of Venice, in a single day he emplaced a series of large blocks that packed the ground by compressing it. On this more solid base Sanmicheli built his fortress, which on the outside presented enormous bossages of very hard Istrian stone. According to Vasari, "You would take this work for a carved rock so huge are the blocks of which it is formed and so perfectly are they fitted together."

There was no shortage of jealous people and detractors to criticize the fortress and claim that it was not solid, that being built on a swamp it would crumble at the mere sound of cannon shots. The Venetian Senate decided to test it. The largest cannons were brought out from the arsenal, were loaded up to their muzzles, and were all fired at the same moment. "You would have thought the end of the world had come; the fortress resembled a volcano. . . . Not a single sign of a crack appeared in the structure."

Sanmicheli renovated and improved the fortifications of Murano. His fame was universal, as was that of his nephew Giovanni Girolamo; Charles V (Charles I of Spain) and François I unsuccessfully made attractive offers to them in an attempt to obtain their collaboration. But Sanmicheli preferred to work at Verona, where he constructed the fortifications that were the first major examples of bastioned polygonal traces.

At the end of the fifteenth century Machiavelli recognized the weakness of the Italian defensive works facing the invasion of Charles VIII (1494): "poorly built embrasures, excessively light battlements." In France, in contrast, the battlements were thick and strong, the embrasures "wide on the inside, diminishing toward the center of the wall, then becoming larger toward the outside . . .; the portcullises in the form of a grille, far superior to our solid ones. . . ."

The Italian campaigns came as a shock to the Italian engineers, who brought a more enlightened knowledge to their construction. At Verona, between 1525 and 1530, Sanmicheli adopted traces that were not to be introduced into France until after 1550, the polygonal bastion with projecting angle, and with efficient crossfire. He rectified the ramparts of the Scaligeri, and enlarged the wall of Verona (Figure 188), adding new bastions (the "Spanish," the "Maddalena," and so on).

The Maddalena bastion (Figure 189), dated 1527, is laid out along new lines, with flat faces, slightly raised platforms, covering casemates with cannons, two-story flankers. The Spanish bastion (Figure 190) dominates the Adige from its sharply pointed salient; it has cannon vaults to shelter the flanking guns. A vaulted gallery runs under the escarpment, giving access to the ditch.

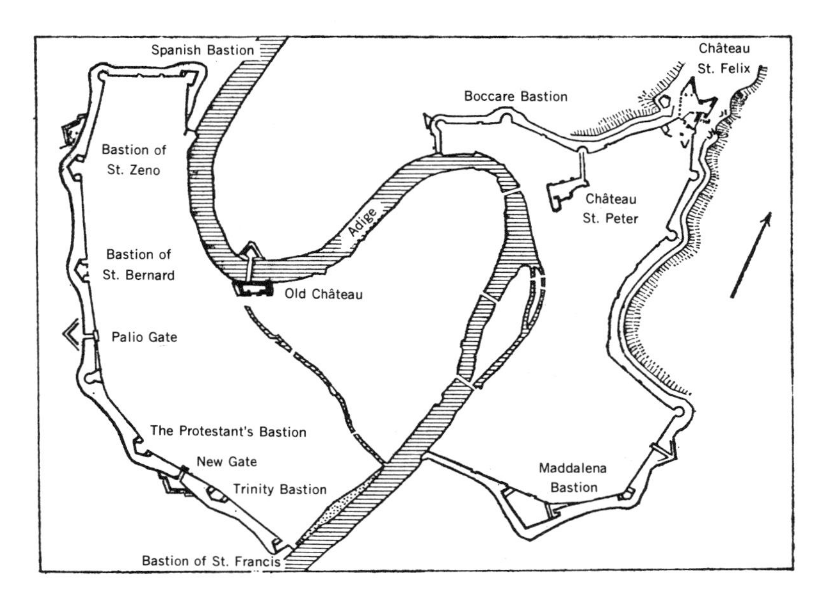

FIG. 188 Fortifications of Verona.

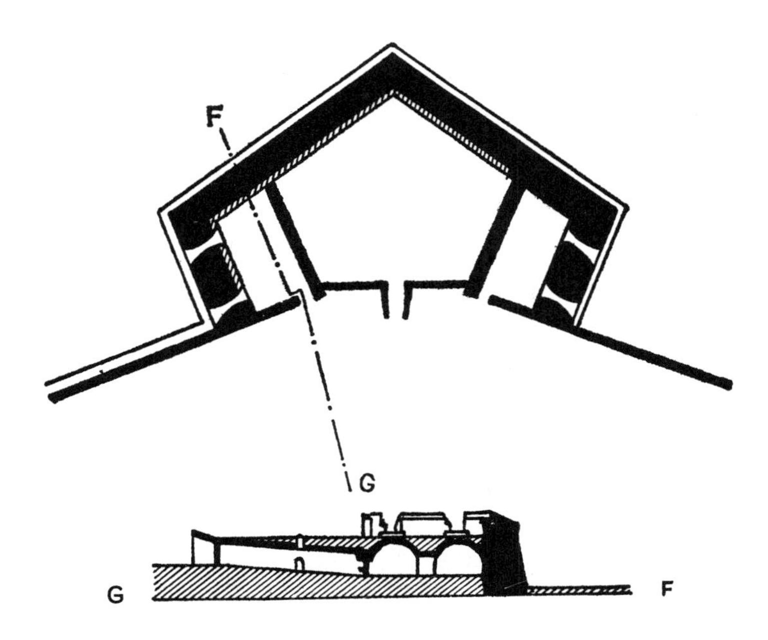

FIG. 189 Verona, Maddalena bastion.
Plan and cross section following *F–G*.

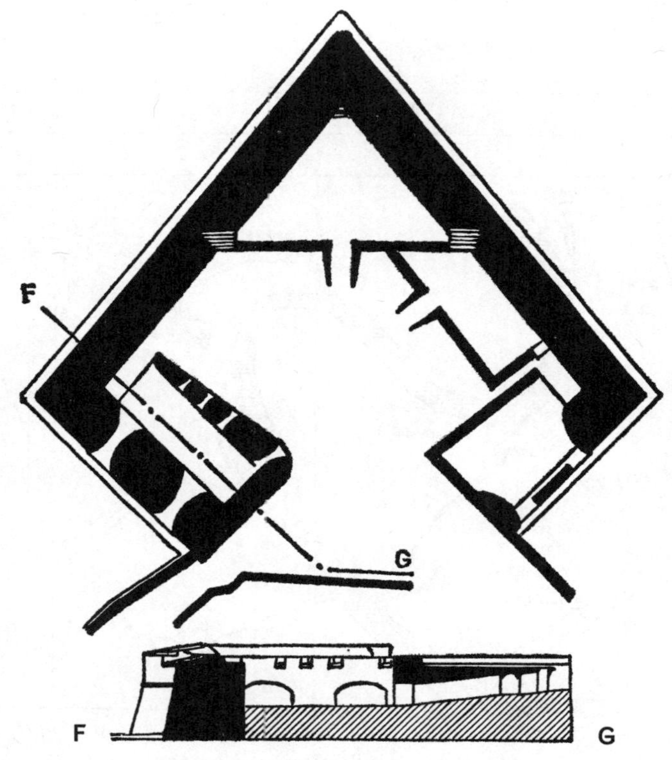

FIG. 190 Verona, Spanish bastion.
Plan and cross section following *F–G*.

The gates

While Sanmicheli's talents as an engineer are apparent in the arrangements of the ramparts, his architectural genius is revealed in the city gates of Verona, which are genuine masterpieces.

The Porta Nuova, on the southern face of the fortifications, opens onto the road that leads from Mantua to Legnago. It is a combination of a gate and a cavalier, which is a completely new idea, and forms a rectangular mass in the center of a curtain between the Trinity and Reformers' bastions. The ornamentation consists of semiengaged columns, with architrave, frieze, metopes, and triglyphs in imitation of the ornamentation of ancient monuments. The interior façade is constructed of specially chosen materials: soft limestones for the bossages, hard stones for the entablature and the capitals, red brick for the filling. The external façade, which was exposed to the enemy fire, is built completely of hard stones, limestones and marble. The dates of 1535, 1537, and 1540, are still visible on the subfoundation, interior façade and the metopes. The work is redolent of equilibrium and power; Giorgio Vasari declared that "a more grandiose or more comprehensive work had never been seen." The Porta del Palio, which is similar in conception, was even more carefully planned and constructed. It is understandable why the Veronese called their great builder *"massimo nella architettura militare."*

Vauban

The Marquis de Vauban (Sébastien Le Prestre) is justly regarded as one of the greatest military engineers, and the frequently repeated reference to works *"à la* Vauban" can lend credence to the belief that he was the inventor of these defensive systems.

In the evolution of the techniques of fortification, the role of Sanmicheli was that of an innovator of ramparts and bastions adapted to artillery. The role of Vauban is apparent especially in the application of the principles adopted in the sixteenth century, and in the adoption of the logical, definitive forms of the defensive works.

Vauban, whose logical and thoughtful mind cannot be too highly admired, completed the process of reducing the ramparts to their proper role, lowering the tall structures, decreasing the dead areas, developing low casements and counterscarp coffers. He had traveled in Italy, and was acquainted with the defensive works of Florence and Verona; from his studies he drew conclusions useful in providing maximum effectiveness in the defense of strongholds. He therefore remodeled many older systems, while constructing new ones. The novelty of his ideas lies in his effort to accomplish the defense of an ensemble forming a continuous frontier, which led him to build and rebuild an impressive number of citadels and city fortifications.

Being familiar with the defects of the older fortifications, he organized attacks on cities; he created the method of approach by means of parallels, temporary fortifications that protected the assailant until, having reached the walls, and a breach having been opened by the artillery, he was able to launch the assault with the fewest risks and losses.

Vauban did not fail in a single siege. He rebuilt the walls of the cities he took, and thanks to his work they became impregnable. The range of his activity as an engineer arouses our amazement. In 1655 he was licensed as the royal engineer. After the Peace of the Pyrenees he worked on the frontier of northern France. When hostilities were renewed in 1667, he directed the sieges of Tournai, Douai, Lille, and Dole. His work at Lille established his reputation; Colbert and Louvois (François Le Tellier) assigned him the task of transforming the strongholds.

In 1673 he directed the siege of Maastricht, in 1674 those of Besançon and Dole. In 1678 he became Commissioner General of Fortifications, and after the peace of Nimègue he created a belt of fortresses around France, from Dunkerque to the eastern Pyrenees. He took Luxembourg in 1683, and remodeled its defenses; Mons (1691), Namur (1692), and Charleroi (1693) followed. In 1694 he defended Brest against the English by a remarkable organization of the troops operating on the coast, basing the defense on the forts and fixed defenses. He repaired three hundred old strongholds, constructed thirty-three new ones, and directed fifty-three sieges. His maritime projects have been discussed elsewhere in this book.

The model of the Vauban defensive system and a number of his fortresses survived into modern times, and did not have to be remodeled until the end of the nineteenth century, when the power of the artillery was considerably increased.

There is another aspect of his work that is too often ignored, but it is of prime importance, for it sheds much light on Vauban's ideas and judicious spirit of research. He was continually traveling about to supervise projects, and lived on construction sites. He became interested in the use of the labor supply; given the tremendous amount of work he had to accomplish, he was forced to attempt to improve the work output and to perform it with a minimum of workers. This led him to search for the most economical methods and the most suitable tools, to calculate the work performed by the teams of workers, and to organize them in rational fashion. In this domain he was a somewhat misunderstood precursor of modern research, begun in the closing years of the nineteenth century and continuing into the twentieth century, on the utilization and economy of labor.

BIBLIOGRAPHY

CROIX, HORST DE LA, "The Literature on Fortification in Renaissance Italy," *Technology and Culture,* 4 (1963), 30–50.

OMAN, SIR CHARLES, *A History of the Art of War, the Middle Ages from the 4th to the 14th Century* (London, 1905).

O'NEIL, B. H. ST. J., *Castles and Cannon: A Study of Early Artillery Fortifications in England* (New York, 1960).

SHELBY, LONNIE R., *John Rogers: Tudor Military Engineer* (Oxford, 1967).

TOY, S., *A History of Fortification from 3000 B.C. to A.D. 1700* (London, 1955).

CHAPTER 19

WEAPONRY 1500–1700

LAND GUNS

In another chapter of this work we have seen that beginning in the fifteenth century a major evolution can be observed in the construction of cannons: the cast bronze cannon replaced the cannon composed of iron sections assembled and bound together with rings of the same metal. Both techniques continued in existence during the sixteenth century.

Iron and bronze cannons

Large bombards of forged iron, with a rear section that screws onto the gun, can be seen in a number of museums, for example those of Ghent and Edinburgh. The Basel Museum also possesses various specimens, and guns of smaller calibers can be viewed in Brussels (Porte de Hal) and Paris (Musée de l'Armée).

Their construction technique is simple (Figure 191). In the cannons of Ghent and Basel, for example, the iron rods are jointed and held in place by iron rings forming a continuous envelope; the Ghent cannon weighs 32,120 pounds, which is almost identical with the weight of bronze cannons of the same period. This cannon was capable of hurling a 748-pound stone ball.

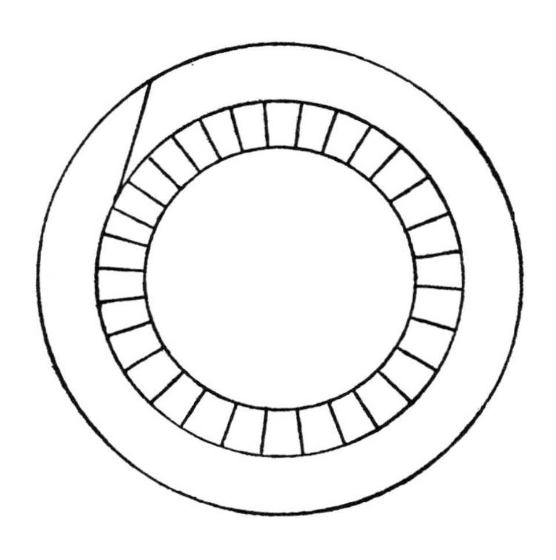

Fig. 191
Cross section of
a large iron cannon.

We possess very precise information about the early bronze guns, and certain specimens of the cannons themselves and their stone balls are still extant. They made a sensational appearance in 1453 at the siege of Constantinople by Mohammed II, and proved to be of decisive effectiveness in the attack on the old fortifications, which until then had resisted many assaults by armies equipped only with battering rams and catapults. Balls hurled by these enormous cannons and left where they fell have been measured by their contemporaries and by modern experts (Pears, *History of the Siege of Constantinople);* their circumference has been given as 98 inches, and their weight as 1,200 pounds, which corresponds to a caliber of approximately 75 centimeters.

The making of bronze cannons

Contemporary narratives describe the technique used in making one of these large bronze cannons. G. Schlumberger has reproduced the notes of Aristobulus of Imbros, who gives a firsthand account of them in his *History of Mohammed II,* from 1450 to 1463.

Three months were required for the construction of one cannon. The mold was made of thoroughly kneaded clay bound with flax and hemp. A cylinder formed the bore; it was enveloped by an outer mold, the space left between the two determining the thickness of the gun. All the evidence indicates that the mold was placed in a vertical position. The outer mold was wrapped with iron circles and pieces of wood, so that it would not burst under the pressure of the molten metal. The mold was supported by an embankment of stones and soil piled up around it. A small basin at the top, 29¼ inches high, contained the riser, where dross collected and the shrinkage hole formed.

Two furnaces built on the side, and operated constantly for three days and three nights, smelted the materials that in liquid form were to be poured into the mold through pottery pipes. These materials were mixed with charcoal to maintain the purity of the bronze alloy.

Thus this was an already quite advanced technique of smelting bronze; the decomposition and oxidation of tin in the open air was understood, as was the necessity of covering the molten metal with charcoal in order to maintain the proportions of the components, and the usefulness of the riser in obtaining a flawless upper section in the cast gun.

Thanks to this information we are able to reconstruct with almost absolute certainty (Figure 192) the arrangements of the furnaces and the mold. The latter,

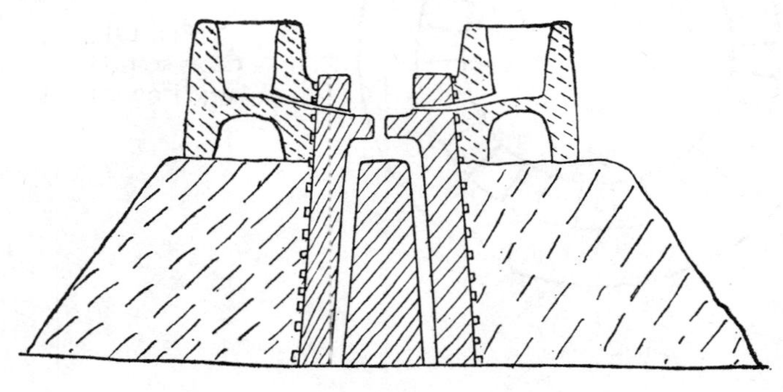

FIG. 192 Casting the cannons of Mohammed II.

which was constructed on the ground and supported by an embankment, was surmounted by the opening left for the riser. Since the smelting vat for the bronze had to be on a higher level, it could thus be placed with the furnaces on the outer embankment. The earthen tubes naturally led the metal by gravity into the furnace. It is hardly to be supposed that the liquid metal could have been led from the base of the furnace by a bottom casting. It was therefore necessary to have a sufficient height of liquid metal (approximately thirteen feet), which created a strong pressure inside the mold and necessitated banding it with iron bands, as Aristobulus notes.

The techniques for making molds and smelting metal were known by the middle of the fifteenth century; we need only recall that the first large equestrian statue to reappear in Europe since the end of antiquity was that of the condottierre Erasmo Gattamelata, completed by Donatello in 1453, that is, at the very period when the Hungarian Urban was casting the cannon of Mohammed II.

Once the metal had cooled, all that was needed was to remove the earthen embankment and the mold in order to uncover the cannon, which could then be scraped and polished. The gun thus obtained, according to the most reliable information, must have weighed around 15 tons.

Lacking trunnions and gun carriage, the cannon was dragged from Adrianople to Mohammed's camp. All the historians have described this operation as being extraordinary: it was dragged by thirty pairs of oxen, accompanied by 450 men, laborers and carpenters who were to prepare, level, and strengthen the road. This suffices to show that at that time there existed neither roads nor wagons that would permit the transportation of such weights.

The use of large cannons

As for the emplacement of the gun, the available information indicates that it rested on the ground and was chocked up with stones. After each shot its aim undoubtedly had to be rectified and the cannon again wedged in place. Aiming and loading therefore required much time, and the historians tell us that the gun fired seven shots during the day, one during the night.

While this cannon has left numerous traces in history, only a few of its balls remain in existence to indicate its exact dimensions. But other guns dating from the same period can apparently be recognized in cannons that are still in existence. There is, for example, the bronze cannon that rests on the lawn of the Tower of London. This beautiful gun was presented to Queen Victoria in 1867 by Sultan Abd-el-Aziz; it was part of a battery of 42 identical cannons commanding the entrance to the Dardanelles. It consists of a barrel almost 10 feet long; the caliber of its bore is 24⅓ inches, and consequently it hurled stone balls of 671 pounds (See Figure 193). The rear section, which was 3¼ feet long,

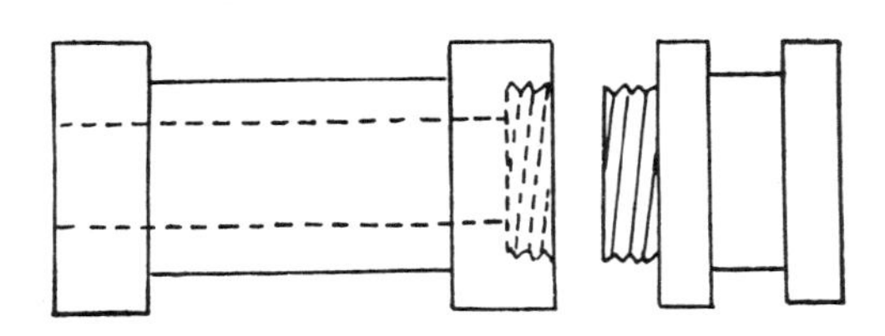

FIG. 193
Bronze Turkish cannon.

screwed onto the body of the cannon, in a manner similar to that of the Edinburgh cannon. The total weight of the gun is 18 tons. It is decorated, and among other motifs the date 855 A.H. (that is, A.D. 1463) can be distinguished.

From this information we can conclude that ten years after the taking of Constantinople, which was marked by the use of cast bronze cannons weighing fifteen tons, the construction technique of these large engines had been continued and even perfected; they have survived to the present day, after centuries of use. The last time they fired on a fleet trying to enter the Dardanelles was in 1807, and their target was a British squadron.

The embellishments with which the fifteenth-century casters decorated these guns make them works of art, and this artistic character was maintained in the casting of bronze cannons until the eighteenth century. Very often the casters were themselves artists. In his letter (if indeed it is by him) to Ludovico Moro, Leonardo da Vinci does not fail to mention that he is able to make cannons and even statues. Moreover, in his drawings we find prettily ornamented cannons (Figure 194, Villardi Collection, Louvre Museum), which must date from between 1480 and 1490.

Improvements

But these large cannons, which as we have seen were difficult to transport and in most cases lacked gun carriages, could not be commonly utilized by armies. The military authorities were faced with the problem of developing lighter weapons that would be easy

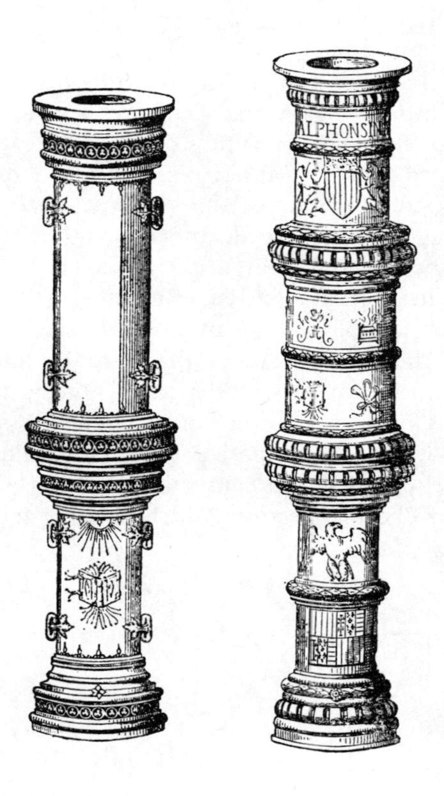

FIG. 194
Leonardo da Vinci's cannons.

to pull so that they could follow the armies' movements. The stone balls could no longer be conveniently used or transported. References have been made to the reforms of the Bureau brothers, Jean (died 1463), and Gaspard (died 1469), and the improvements they made in Charles VII's artillery; it seems certain that these perfections consisted in a more generalized use of metal shot (iron or cast iron) and the adoption of gun carriages that were more convenient and easier to move.

Shortly before 1480, various improvements were introduced that were to give artillery a modern appearance. The first of these was the improvement of cannon powder by "corning." The irregular powder, mixed with "unclassified" grains and dust, which had been in use until then produced very different results in the same cannon — an explosion that was either rapid or slow burning. By making the shape of the grains more regular and "classifying" them, after 1480 it was possible to obtain powders that gave uniform results.

From the viewpoint of aiming and transporting the gun, the adoption of trunnions permitted the organization of a gun carriage on which the gun could turn. Formerly the gun, lacking trunnions, had to be held by a cradle above which a jointed arrow made it possible to aim the gun. An arc attached to the arrow permitted the cradle to be immobilized at the desired angle of fire. This was already a great improvement over earlier guns, whose aim was uncertain; the gun found at Bouvignes and preserved in Brussels (in the Musée de la Porte de Hal) has no aiming device whatsoever, and must date from the beginning of of the fifteenth century. It is still very crude; its gun carriage has solid wheels, and the cradle, although it already has an aiming mechanism, is very rudimentary (Plate 38).

Thereafter, improvements seem to have followed in rapid succession; witness the cannons of Charles the Bold, preserved in the museum at La Neuveville (Canton of Bern) since 1476. After the defeat of Charles at Morat on June 22, 1476, the 90-man militia of La Neuveville, which was part of the confederate army, received as its share of the booty seven cannons and three bombards. These guns, which have been carefully preserved by the citizens of La Neuveville, reveal what the cannons of a Burgundian army in the second half of the fifteenth century were like (Figure 195). They are made of forged iron, are relatively light, and are mounted on gun carriages with spoked wheels. The Burgundian artillery may also have had several bronze cannons.

The adoption of trunnions led immediately to the gun carriage, composed of two cheeks between which the cannon could be depressed and elevated. The cheeks were held in place by means of transoms, and the breech was supported and immobilized at a fixed angle of fire, first by wedges and later by a screw.

Very soon thereafter the invention of reaming made it possible to obtain a cylindrical bore in cannons. The artillery of Charles VIII benefited from all these improvements, and showed its superiority in the Italian campaigns. In the same period Galiot, who was in command of the artillery under both Louis XI and Charles VII, is supposed to have invented "dishing," that is, the inclination of the spokes of the wheels toward the outside; this ensured greater stability in the carriages, which aroused Machiavelli's admiration when he saw the vehicles of the French artillery. At Ravenna in 1512, Paul de Bensérade, who

a

b

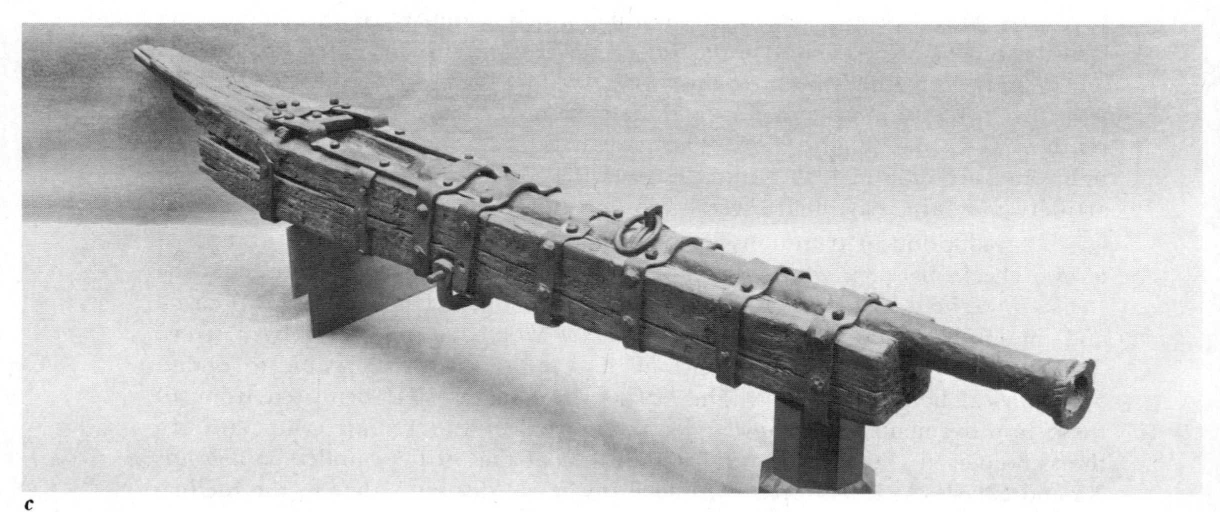

c

Plate 38.

a. Fifteenth-century culverin. Musée de la Porte de Hal, Brussels. *Copyright A.C.L., Brussels.*

b. Veuglaire. Fifteenth century. Musée de la Porte de Hal, Brussels. *Copyright A.C.L., Brussels.*

c. Veuglaire. Sixteenth century. Rijksmuseum, Amsterdam. *Copyright Foto-Commissie.*

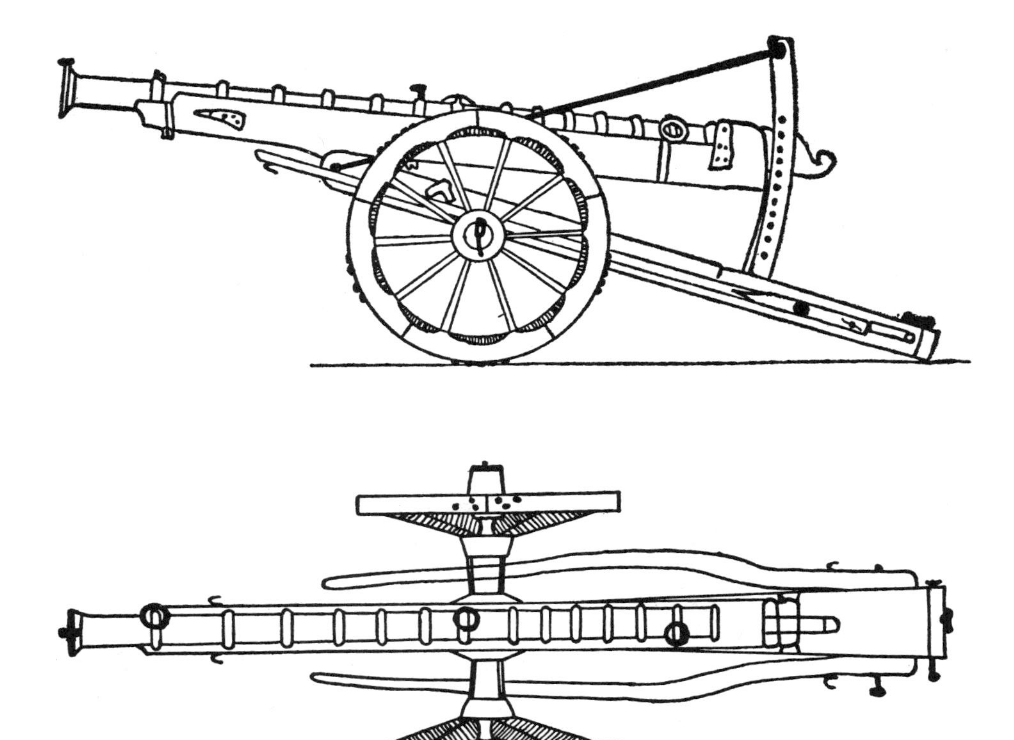

FIG. 195 Cannons of Charles the Bold (Museum of La Neuveville).

was commanding the artillery, was killed. He was replaced by Galiot's nephew, Galiot de Genouillac (1465–1546), who increased the importance of the artillery. At Marignano in 1515 the line was composed of seventy-two large cannons, each dragged by thirty-five horses. At Pavia, the thirty cannons of François I would have ensured victory if the king's poorly planned charge had not hidden and paralyzed them.

Stabilization of calibers

Between 1521 and 1530 Charles V (Charles I of Spain) subjected the calibers to fixed rules, placed their production under supervision, and ordered a program of experimentation. For the first time the type of each cannon was determined by the weight of its projectile; the great variety of calibers then in use was reduced to 40, 24, 12, 6½, and 3.

Under François I the French guns were again limited in six sizes which d'Estrées, artillery commander from 1550 to 1569, standardized, more or less permanently, as follows:

	WEIGHT OF CANNON	CALIBER		WEIGHT OF PROJECTILE
Cannon	5,400 pounds	6 inches	3 *lignes*	33 pounds
Large culverin	3,800 "	4 "	4 "	16 "
Bastard culverin	1,970 "	3 "	10 "	8 "
Medium culverin	870 "	2 "	9 "	3 "
Falcon	750 "	2 "	4 "	2 "
Falconet	450 "	1 "	10 "	1 "

These are the calibers known as "the six calibers of France." They were pulled by:

21 horses for the cannon
17 horses for the large culverin
11 horses for the bastard culverin
7 horses for the medium culverin
4 horses for the falcon and falconet

The teams were harnessed in shafts, and the horses pulled in a single file. Each type of gun was mounted on its own gun carriage; all the gun carriages, however, were of the two-wheeled variety, and had no limbers. The principal parts of the gun carriage were the two cheeks and the four supporting transoms. The wheels had twelve spokes, and the fellies consisted of six pieces; iron fittings bound the various parts of the wheels. The large 33-pounder was supported on a special carriage.

During the major disturbances occasioned by the religious wars, everyone had guns cast according to his own ideas, departing from the standard types. In 1572 Charles IX attempted to reestablish the "six calibers of France," but this was achieved only under Henri IV.

Around this time iron rings were placed on the French guns as an aid to maneuvering them – a practice that had already been common in Germany for two centuries. In the second half of the sixteenth century the bomb mortar appeared in Germany; it was used for the first time at the siege of Lamotte, in Lorraine, in 1534. An attempt was also made during this period to obtain a more standardized composition for bronze (91 percent copper, 9 percent tin).

The artillery of Henri IV and Louis XIII remained unchanged; not until the time of Louis XIV do we witness a new series of campaign guns (1698), under slightly different names:

French cannon	33-pound ball
Spanish demicannon	24-pound ball
French demicannon (culverin)	16-pound ball
Spanish quarter-cannon	12-pound ball
French quarter-cannon	8-pound ball
Falcon and Falconet	2- and 1-pound balls

During this time short guns (8 and 12) were introduced into artillery nomenclature.

NAVAL ARTILLERY

The technique of cannon production inevitably evolved in the same manner for both naval and land artillery. But the installation and use of artillery on ships differed from practices on land.

While several early dates marking the use of land guns have been found, the first definite appearance of cannons on a battlefield is claimed for Crécy in 1346. Guns were utilized on ships several years earlier, at the battles of Arnemuiden in 1338 and L'Écluse in 1340, thus preceding the Battle of Crécy by several years.

The fifteenth century witnessed the beginning of improvements in land guns and the development of lighter weapons, which inevitably benefited naval artillery. But the evolution of the latter was influenced by another factor peculiar to its use on board ship.

Small-caliber iron cannons

The first cannons, which were of the iron variety, were placed on a wooden cradle. It was the sailing ships in particular that were able to carry these cannons; in the sailing vessels the entire space along the decks and on the castles was available, and here the combatants were placed. The guns made possible more effective firing, at greater range than was possible with the arbalests.

Use of these cannons, which began as small-caliber guns, required that they be equipped with a pivoting form (Figure 196*A*) held in the framework of the

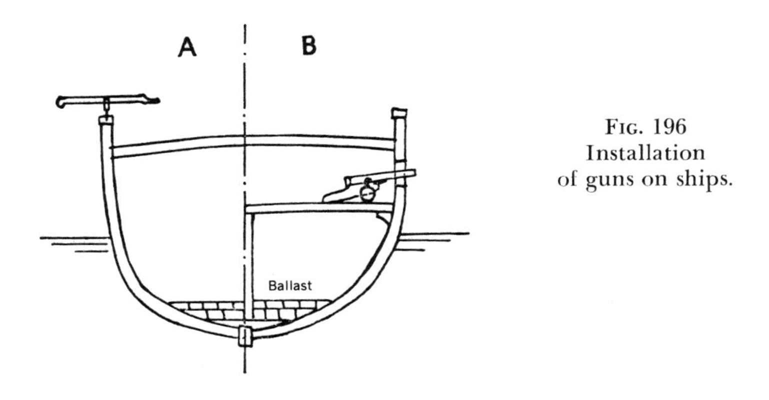

FIG. 196
Installation
of guns on ships.

bulwarks (specimens can be seen in a number of museums). These small-caliber guns were used especially to attack the personnel on the deck and castles of an enemy ship, rather than the ship's hull, which was capable of resisting the shock of such projectiles. The tops of a ship could be equipped with a great number (as many as 100 and even 200) of cannons; this, however, was a menace to the crew and the rigging.

The use of higher-caliber cannons capable of attacking the hulls of ships, demolishing their tops, and even sinking them, was more difficult; it would have required placing the cannons on the deck or the castles. This raised the problem of the stability of the ship: a sailing vessel, of the type then used as a merchant ship, was stable only on condition that it carried a cargo that uniformly filled its hold. Its stability was threatened when the cargo (artillery) was placed in the tops; it was then necessary to compensate for this weight by placing a ballast at the bottom of the hold. The plan of the sail-powered warship, an offshoot of the merchant vessel, was developed around the end of the fifteenth century; it was able to support the weight of the artillery, counterbalanced by ballast.

In the course of the following centuries the galleys were adapted, with greater difficulty, for the use of the cannons; their sides, which were occupied by the oars and oarsmen, could carry only a few very light guns, while the aft portion was occupied by a platform and a small quantity of equipment indispensable to the officer staff. Only the narrow foreportion was equipped with cannons that could fire in the axis of the ship. The role of the galley diminished in importance as the use of artillery developed on sailing vessels.

The gunports In order to retain sufficient stability, it was useful to place the cannons as low as possible, and to avoid overloading the tops and fore- and aft-castles, whose decks, moreover, were not strong enough to support the weight of the cannons and the shock of firing (Figure 196*A*).

It is generally accepted that around 1500 a man named Descharges, of Brest, may have had the idea of placing the cannons on a lower deck and opening gunports for the artillery. The guns, placed on an orlop deck or in the castles on the deck, were quite low, and thus were sheltered by the walls and bulwark from the firing of the light guns (Figure 196*B*).

The period around 1500 thus marks a major turning point in the use of cannon on ships. This was also the period when bronze cannons were fitted with trunnions, which had been invented shortly before (1480), for use on ships. But as always happens, the new invention continued to be used simultaneously with the old machines: iron cannons assembled by forging continued to be used along with the new bronze cannons.

Specimens of both types have been found on the same ship, the *Mary-Rose,* which capsized before Portsmouth in 1545, and whose cannons were saved. These included forged iron cannons, specimens of which are housed at the Tower of London and the Maritime Service Museum, Whitehall, and bronze cannons (Figure 197). The latter weighed 3,000 pounds apiece; it is believed there were five of them.

Bronze and cast iron When ships constructed prior to 1500 were remodeled, their artillery was transformed by replacing a part of the iron cannons by bronze cannons. The *Harry-Grâce-à-Dieu,* which originally (1514) had 186 iron cannons, was reequipped with 122 breech-loading iron cannons and 11 bronze cannons; thus the 11 bronze guns replaced 64 iron cannons, which were lighter.

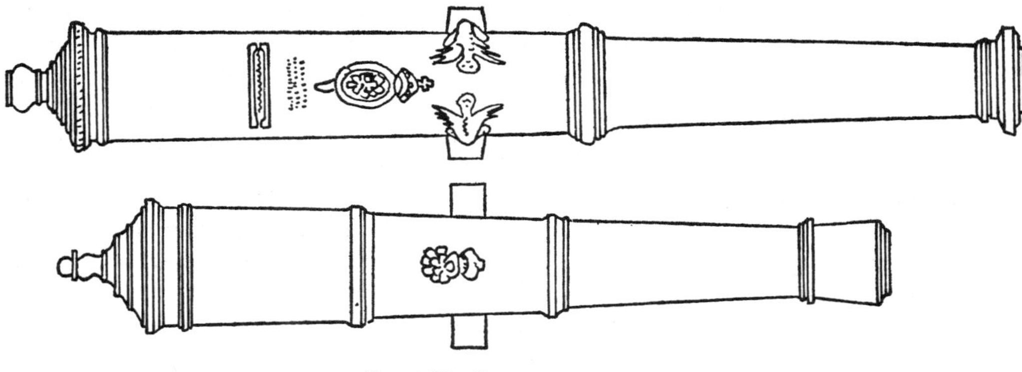

FIG. 197 Bronze guns from the *Mary Rose*.

The gun carriages (Figure 198) were modified, and acquired the form they were to retain practically unchanged until the nineteenth century: the cannon pivoted between two sturdy wooden cheeks mounted on wheels.

All these bronze guns were covered with decorations and inscriptions, and were genuine works of art. Their defect was that their materials (which were imported into France) and their modeling, which required experienced artists and metal casters, made them very expensive.

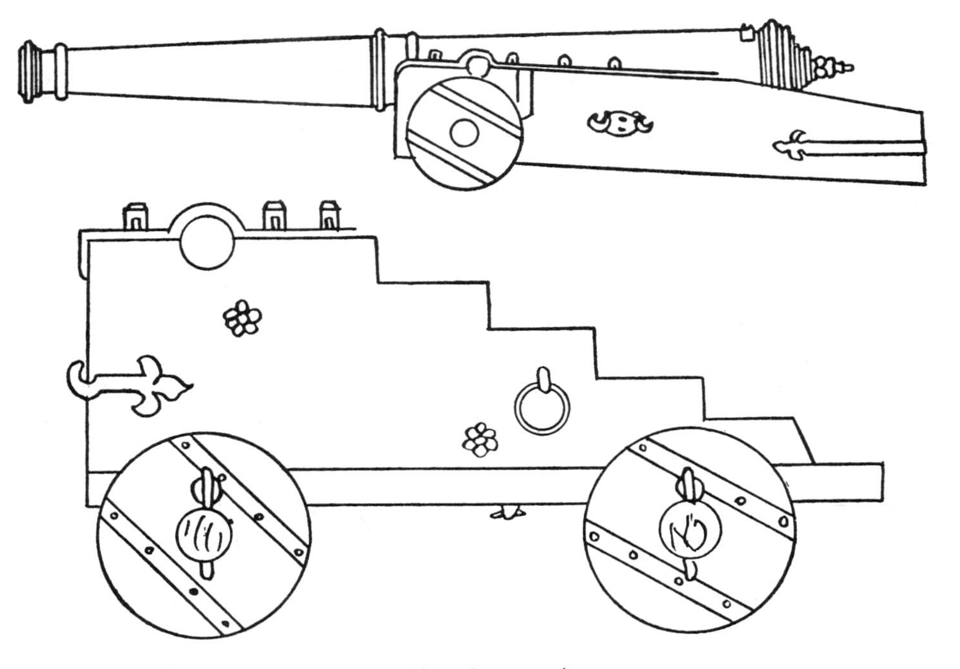

FIG. 198 Gun carriages on wheels (sixteenth century).

In France, Richelieu concentrated his efforts to build up a fleet on the creation of foundries in the shipyards. He was obliged, however, to purchase materials (copper and tin) abroad. Not until the advent of Colbert was the making of cannons organized rationally. For the most part, cast iron then replaced bronze. Cast iron seems to have been in use for naval artillery before this period; at Lepanto (1571) the Venetian galleys had large 36-caliber cast-iron guns, the bow gun being placed in the ship's axis.

Colbert ordered the development of the foundry of Saint-Gervais-sur-Isère, and actively promoted the technique of casting (1678); in a few years most ships' guns were being made of cast iron. Around 1685 the stores of the arsenals included 1,052 bronze cannons and 4,183 cast-iron cannons.

The documents of this period must be read with an understanding of the terms used, and the error that has sometimes been made by the historians must be avoided. The old cannon made by assembling forged iron guns had by then disappeared completely, and thus it did not have a special name (iron cannon). The bronze cannon, in contrast, was known as a *"fonte verte"* (green casting), and more often simply as a *"fonte."* The cast-iron cannon, in contrast, was often called an "iron cannon," whence the possible confusion of the cannon then called the "cast" cannon (that is, the cannon made with molten bronze) with the "iron cannon," which was no longer the old cannon made by forging but a cast-iron cannon.

These cast-iron cannons, which were now widely employed, no longer bore the abundant sculptured ornamentation of the bronze cannons, but simply metal rings and a few very simplified markings. In this way economies in production were realized.

The ordinance of April 15, 1689, provided for the distribution of the types of cannons on the ships:

"All first-rate vessels will be armed with cast [that is, bronze] cannons, without admixture of any iron [that is, cast iron] gun."

Second-rate vessels, if commanded by a general officer, "will also have only cast [bronze] cannons, and if commanded by a squadron chief or a captain two-thirds of their cannons will be cast [bronze] and one-third will be iron."

For the other rates, the proportion of bronze cannons descended to the fifth rate (three-quarters cast-iron cannons). "Frigates and all other vessels will carry only iron cannon."

The same ordinance provided for bronze metal casters to work at Rochefort and Toulon, established the calibers of guns at 36, 24, 18, 12, 8, 6, and 4, and provided for the composition of the material from old cannons, bells, Swedish copper, et cetera, to which was added one-tenth yellow copper scrap and one-twentieth fine tin.

Iron cannons were to be undecorated, and there were only five calibers: 18, 12, 8, 6, and 4. The proportions, weight, length, and diameter were to be fixed by law.

Since there were no known methods of analyzing the materials utilized, it was to be expected that the smelting of the metal created irregularities. The phenomenon of liquation, which had been observed but was unexplained because of lack of knowledge of the chemistry of metals, gave the castings a lack of homogeneity in the composition of the metal, which weakened the strength of the cast objects.

The only tests of the strength of the cannons were the examination of the inside of the bore and the firing of the gun, using a powder charge that was two-thirds of the weight of the ball. Iron cannons were examined visually and then tested by firing, with a powder charge that varied with the caliber of the gun:

18-caliber:	2/3 of weight of ball
12-caliber:	3/4 of weight of ball
8-, 6-, and 4-:	6 1/2 pounds, i.e. the weight of ball

The metal casters

By the sixteenth century cast-bronze artillery had thus achieved a remarkable degree of technical perfection, combined with an artistic appearance indicative of founders who were masters of their craft; they are known through history itself, and through their signatures, which they engraved on their molds and which appear on their guns.

These guns were obviously in demand in all countries for the arming of vessels and ports; the guns supplied by certain workshops are met with almost everywhere. The Dutch, who were at war with the Spaniards, supplied the adversaries of the Spanish fleets with cannons that are still in existence in the old batteries of the Moslem countries of Africa. Some of these guns can be seen at Sfax; the famous "Consulaire," 16 1/2 feet long and of Venetian origin (1542), was brought from Algiers to Brest and installed on the terrepleins of the arsenal. At Rabat and Safi they formed the defense against attacks by fleets seeking to flush out the pirates' haunts. The cannons of Safi (from Dahr et Bahar) bear inscriptions giving the names of the casters: "Arent Van der Put, of Rotterdam in 1619"; "Willem wegen waert, of The Hague in 1625," with Arabic invocations to Allah to protect the emir of the Believers.

The origin of these cannons, which is already significant from the political point of view, proves the existence of a genuine industry of talented metal casters whose works were highly valued articles of export.

Underwater investigations have turned up equally remarkable works dating from the time of Louis XIV. A cannon found in wrecked vessels lying off Le Croisic (Figure 199), and thoroughly described in the magazine *Neptunia* (No. 42), can be dated from 1670 and attributed to Master Jean Baube of Narbonne, who was called to Toulon to cast twelve 24's destined for the *Royal-Louis* and the *Royal-Dauphin;* here, again, the caster signed his work. It is one of these guns

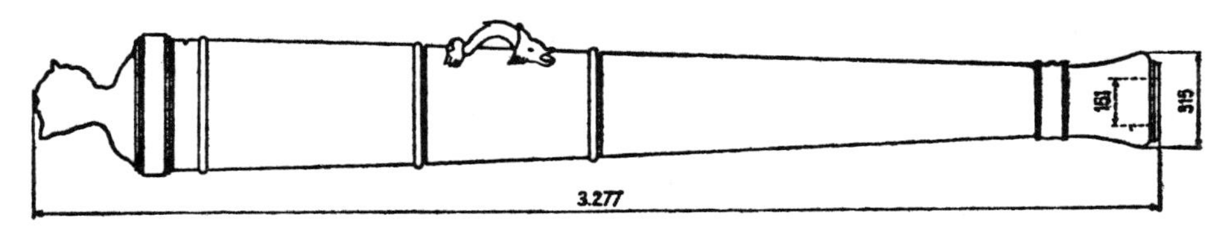

FIG. 199 Bronze cannon (period of Louis XIV).

Plate 39. Bronze cannon by Hans Fussli III (1616–1684). Schweiz Landesmuseum, Zürich. *Photo by the museum.*

thus brought to light that reveals the name of the caster; he bore the title of "Caster-General of Naval Artillery of France at Toulon." The medallion showed the profile of Louis XIV, the inscription in the name of the Count of Vermandois, with abundant decoration; the breech button was fashioned in the shape of a fawn's head. The gun weighs 4,704 pounds (Figure 199). The cannon, which probably comes from a vessel sunk in 1759, belonged to the *Soleil-Royal,* an 80-tonner, whose (completely bronze) artillery included thirty 36-calibers, thirty-two 24-calibers, and eighteen 8-calibers. (Six 36-caliber guns were found in 1762).

Thus it appears that bronze guns were in use on ships for more than a century, while cast-iron guns were introduced with increasing frequency during the eighteenth century.

HAND WEAPONS

While cannons were replacing the trebuchets and ballistas, portable weapons were adopting the use of gunpowder. The first portable firearms appeared around 1424, and were gradually brought to complete development during the fifteenth and sixteenth centuries.

The harquebus and its development

The first "fire rod" or "hand cannon" was simply a barrel held by the gunner with both hands; an assistant ignited the powder through a touchhole in the upper side of the barrel, which at first had no stock or support.

By attaching the iron barrel to a primitive grip that the user rested on his shoulder (Figure 200), his left hand was left free to ignite the powder; thus the portable weapon now required only one man for its operation. Improvements were next made in the grip (better to resist the recoil of the weapon) and the igniting of the powder (to make firing more reliable and rapid, without upsetting the weapon's aim).

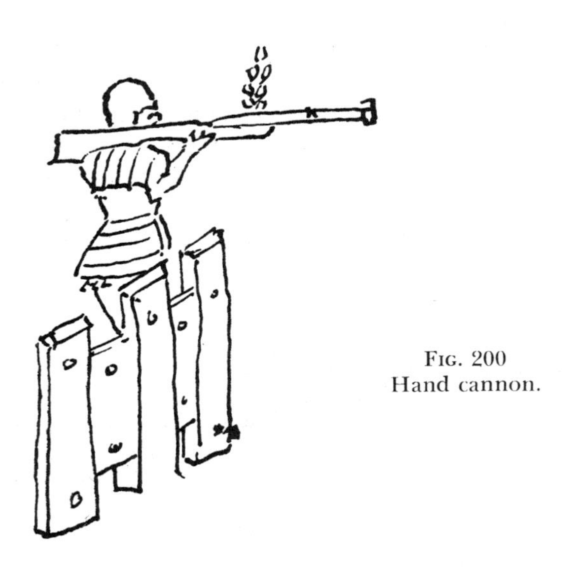

Fig. 200
Hand cannon.

The first hand weapons were heavy, and their weight, combined with their recoil, made them difficult to handle. A kind of hook was placed underneath the barrel, to serve as a locking device when it was supported on a wall preparatory to use. This *"arquebuse à croc"* completely replaced hand cannons; it sometimes weighed as much as 50 to 60 pounds, was 5 to 6 feet long, and could be used for little else except as a rampart weapon. An attempt was made to make it lighter so that the infantry could be equipped with it, but it still required a support, either stationary, like a parapet, or mobile, for example a portable crutch, which was utilized for a long period.

The process of igniting the powder was improved by the use of a "match" (a smoldering cord made from strands of twisted tow and prepared by boiling it in vinegar or the lees of wine) held in a serpentine; it was lowered into the firing pan, which contained a small quantity of powder and was placed in front of the priming hole (Figure 201). The position of the priming hole then had to be changed from the top of the cannon to the side. This was the matchlock harquebus, which in the hands of the Spaniards turned aside François I's ill-planned charges at Pavia in 1525. The matchlock developed into the musket, which remained the weapon of the French infantry until the reign of Louis XIII. It had serious disadvantages: the user had to keep a constantly lighted match with him, or a means of relighting it; for each shot, he had to regulate its length so that the end caught in the serpentine would fall exactly into the firing pan; when he was ready to fire he had to blow on it to revive the fire. These operations were impossible for a horseman who was using one hand to hold his horse.

Because of their deadly power, firearms were in use by the infantry by the beginning of the sixteenth century; in France, the first establishment for their manufacture was organized in 1516 at Saint-Étienne.

The *arquebuse à croc,* with its crutch fixed in the ground to support its weight, and its irregular and inaccurate fire, was still an unperfected weapon. The rain prevented the match from taking fire, and if the charge went off at all it caused violent backfire. The barrel was more or less carefully forged from iron, but the weapon often burst. In any case, even with a well-built and skillfully handled weapon, only one shot could be fired every eight or ten minutes.

The harquebus was gradually made lighter in weight by shortening the barrel; the process of igniting the powder was improved by a more ingenious and reliable device. The musket began to replace the harquebus toward the end of the reign of Henri II; the balls fired by the musket weighed one ounce and two gros (slightly more than 38 grams). In the musket the sulfured match was replaced by the steel wheel lock.

Around 1517 the Germans invented the wheel lock, which consisted of two

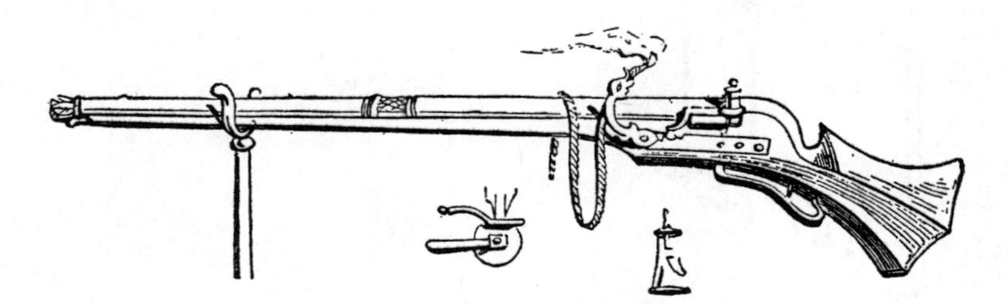

Fig. 201 Matchlock harquebus.

parts. A steel-toothed wheel was wound up by a key to tighten a spring; a pull of the trigger brought the revolving wheel in contact with a lump of pyrites (later flint), the sparks produced ignited the power in the firing pan (Figure 202).

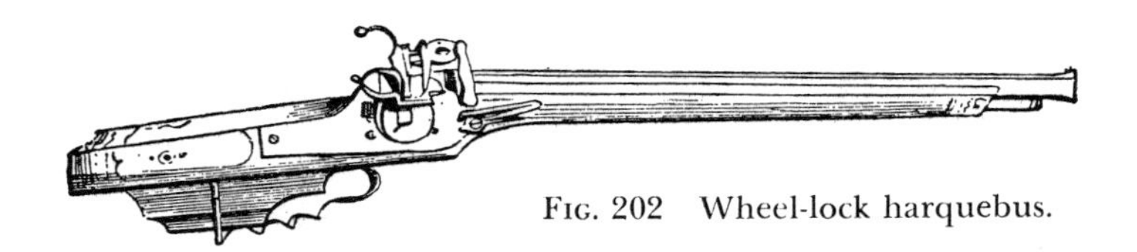

FIG. 202 Wheel-lock harquebus.

The flintlock

The Spanish then introduced a major perfection, whose use continued into the nineteenth century: the flintlock. The wheel had been eliminated, and the sparks were produced by the shock of the flintstone clamped between the jaws of the cock being exercised on the steel rim of the pan.

The Spanish flintlock had an outside spring that rested on the arms of the cock; later the parts were covered, with the exception of the cock, which remained exposed. This invention was not utilized until around 1630; prior to this time the wheel lock, although a complicated device that frequently failed to fire, continued to be used, often with the addition of a serpentine holding a wick; when it was thus equipped with two ignition devices, the weapon was more reliable. The flintlock gun was not fully adopted in France as a weapon of warfare until 1670.

Attempts were then made to create rifled weapons, carbines only three feet long, with longer range and more accurate fire. But the difficulties of perfecting the weapon caused it to be abandoned (it was taken up again in the nineteenth century). The ordinary weapons, muskets and later flintlocks, remained in general use; their defect, as we have already noted, was that they fired very slowly. The weapon first had to be loaded, that is, a fixed quantity of powder had to be poured into it. This was done at first with a powder horn whose beak-shaped neck measured the charge (Figure 203); a valve at the base of the

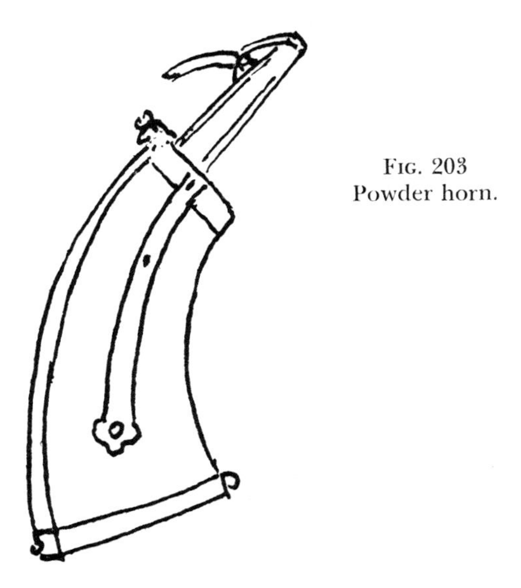

FIG. 203
Powder horn.

neck and a lid at the end permitted it to be filled and then emptied into the tube of the musket.

The priming powder, which was finer, was contained in a small leather or metal powder flash that had no measuring device; the quantity of powder being poured into the pan could be measured by sight. Another bag contained the shot, which was rammed into the barrel with a rod.

To decrease the time required for loading the gun, cartridges were invented; they contained a fixed charge, and were carried on shoulder straps. The envelope was simply torn open with the teeth and the powder was poured into the barrel, a small quantity being reserved for the priming to be put into the priming pan. In this way it was possible to eliminate the entire apparatus of powder horns and powder flasks, and at the same time to control the charge better, and increase the speed of loading.

Despite these improvements, firing was still slow, and when the infantrymen had fired they were defenseless until their weapons had been reloaded. For this reason numerous pike companies, without firearms, were maintained; these pikes were maneuvered with skill and strength to cover the temporarily disarmed riflemen against an unexpected attack. Before the charge that decided the victory of Cérisoles in 1544, Montluc gave detailed instructions to his pikemen on the use of their pikes against the maneuvers of the enemy pikemen.

In order to equip all soldiers with firearms, the latter had to be made equivalent to the pikes. At first a kind of spear was used; it was one foot long and had a wooden handle of the same length, which could be fitted into the barrel of the gun. The gun was transformed into a sidearm, but with the barrel of the gun thus blocked, neither loading nor firing could be done.

At the end of the seventeenth century Vauban invented or brought about the adoption of the bayonet with a hollow socket that left the barrel of the gun open and permitted firing with the bayonet in position. It is not very likely that Vauban himself was the inventor; it was more probably the invention of smithies or locksmiths of Bayonne (whence its name), who were then renowned for metalwork of all kinds; in any case, Vauban's merit was that he understood the tremendous advantage in adopting it.

Equipping every soldier with a firearm that was at the same time a side arm was the means of transforming the army's firepower. Louvois immediately declared his support of the idea; Louis XIV is said to have been opposed to it, supported by many generals who remained greatly attached to the organization of companies of pikesmen. The old method persisted for twenty-one years, and gave way only after the defeats of the Spanish War of Succession. Thereafter the French infantry made great use of the bayonet, in particular at Steenkerque and Denain. By the end of the seventeenth century, armament included the flintlock gun with its bayonet, which, with several improvements made during the eighteenth century, survived into the nineteenth century.

In 1605 the cavalry ceased to carry the lance, which for centuries had been its most characteristic weapon; it was replaced by the short harequebus or the pistol.

The pistol The pistol did not appear until the second half of the sixteenth century. Actually, it was a small version of the harquebus, lighter and easier to handle, which a man on a horse could use with one hand. It was originally the weapon of the *Reiters,* the German mercenary cavalrymen who made their first appearance at the battle of Renty (1554), where their mode of combat greatly impressed observers. They charged in large squadrons, halting within pistol range before the enemy infantry; the entire first row fired simultaneously, then parted to right and left to reveal the second row, and retired to the rear to reload their weapons.

Actually, the pistol was simply a reduced version of the weapon used by the infantry, and thus it had the same components. The pistol used by the Reiters was a wheel-lock model; the grip had an oval knob, and the barrel was short. The parts of the wheel-lock, at first uncovered and on the outside of the pistol, were later enclosed in a drum. The grip formed an angle of 45 degrees with the barrel, then straightened; in the model in use during the reign of Henry IV, barrel and grip formed a straight line.

In ten years this small-scale model of the harquebus became a common weapon; ordinances were passed forbidding its use for purposes other than the outfitting of the army, without success. By the second half of the sixteenth century it was the most deadly weapon in existence, creating a number of victims of political murders; during the religious wars, the pistol killed as many people as the musket.

The Italian and German arms manufacturers who produced pistols rivaled each other in the richness of the wood, ornamentation, and encrustrations they employed. Double-barreled pistols were produced, the barrels being placed on top of the other; each barrel was equipped with its own firing mechanism (Figure 204). The double-barreled pistol was an attempt to remedy the slowness of loading, for a shot fired by the pistol left its user momentarily disarmed, as had been the case with the harquebus. Men acquired the habit of carrying several pistols, hung on the saddlebow or slipped into the belt, breeches, or boots.

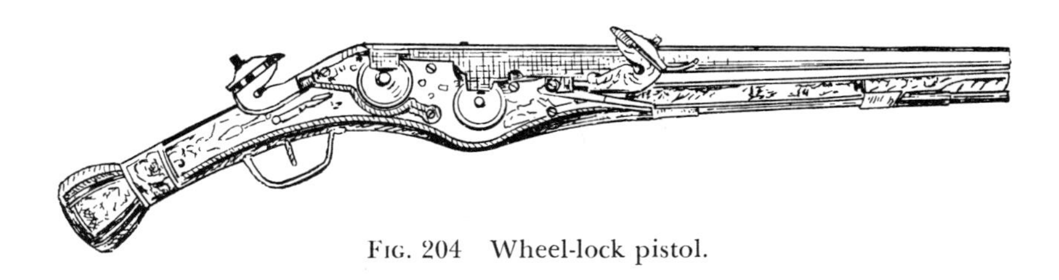

FIG. 204 Wheel-lock pistol.

The development of the pistol naturally followed the evolution of the infantry weapons. The process of firing was simplified, and the weapon became lighted once the wheel had been replaced by the cock holding the flint.

BIBLIOGRAPHY

CARMAN, W. Y., *A History of Firearms* (London, 1955).
CIPOLLO, CARLO M., *Guns, Sails, and Empires: Technological Innovation and the Early Phases of European Expansion, 1400–1700* (New York, 1965).
FFOULKES, CHARLES, *The Gun Founders of England* (Cambridge, 1937).
HALL, A. RUPERT, *Ballistics in the Seventeenth Century* (Cambridge, 1952).
HAYWARD, J. F., *The Art of the Gunmaker* (2 vols., London, 1962–63).
HIME, W. L., *The Origin of Artillery* (London, 1915).
NORMAN, A. V. B., and POTTINGER, DON, *A History of War and Weapons, 449–1600. English Warfare from the Anglo-Saxons to Cromwell* (New York, 1966).

Section Six

Construction and Building

CHAPTER 20

URBAN PLANNING AND THE DEVELOPMENT OF CITIES

IN THE FIRST PART of this volume we have seen that by the fifteenth century various authors and architects had outlined the principles of street layout and that groupings of streets and public squares had been created during the sixteenth century.

Political changes were now bringing about transformations in municipal administration. The Middle Ages had witnessed the formation of groups, and battles between the princely families, causing the concentration of their followers in neighborhoods, and the formation of suburbs. Both city and suburb were organized spontaneously and without planning, for purposes of defense against the neighbors rather than for the establishment of relations and methods of communication.

The remains of the past

The heritage of the past survived in the cities, which were still small and squeezed for defensive reasons within a ring of walls on a rise in the terrain. A reminder of antiquity could also be seen occasionally in the rectangular trace of the old, reconstructed walls of Roman cities.

This plan can still be seen in "old cities" like Boulogne-sur-Mer, where it testifies to a long history and the passage of Julius Caesar; it is also found at Aosta (Figure 205) and Spalato (Split), where the city was built up within the wall of the palace of Diocletian, a genuine fortress city.

With the coming of independence and a growing consciousness of their usefulness, the cities began to create centers of municipal life, with the city square as the core. However, the medieval period still lacked commercial traffic and the need for major transportation; the roads and bridges of the Roman period were abandoned, if not destroyed. It was the Renaissance that gradually gave a new face to the cities. The city of Venice gave this movement a special

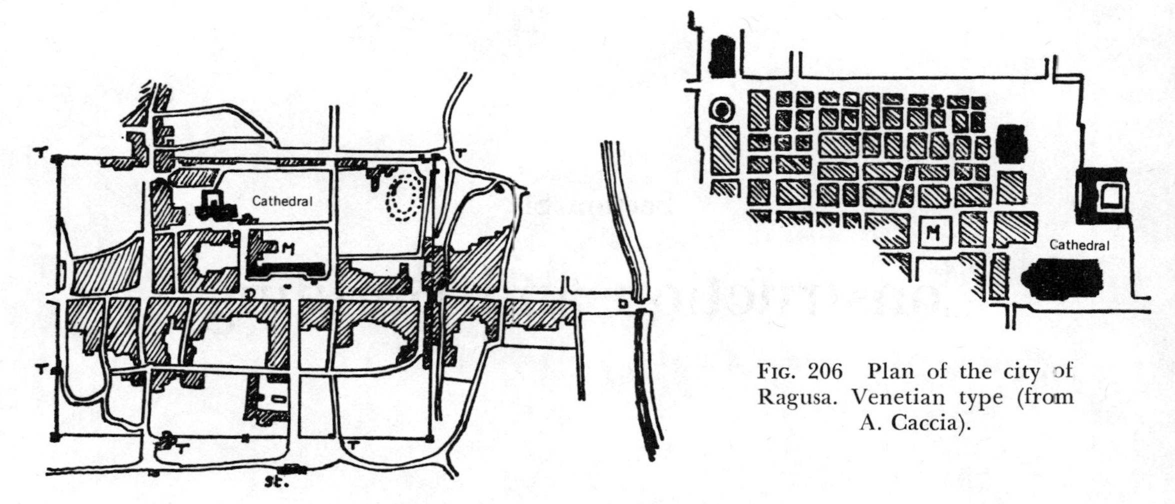

Fig. 205 Plan of the city of Aosta, showing the original layout of the Roman camp (from A. Caccia).

Fig. 206 Plan of the city of Ragusa. Venetian type (from A. Caccia).

character, thanks to a genuine municipal administration that very soon established rules concerning the canals, bridges, and even the houses. The Venetians applied these principles on both sides of the Adriatic, on the one side at Sebenico (Sibenik), Zara, and Ragusa (Dubrovnik) (Figure 206), at Forli and Pesaro on the other side.

A regular pattern of streets laid out in straight lines at right angles was the general rule; L. B. Alberti's ideas on curved streets twisting through the city appear to have remained a theoretical conception, with the exception of several secondary roads. The symmetry and scope of urban-planning conceptions reached their peak during the pontificates of Popes Julius II, Leo X, and Paul III.

The remodeling of the cities consisted of more majestic lines for the major arteries, and public squares created in order to leave space for large movements of traffic.

Influence of the artillery In the case of new cities, however, innovations in weaponry created new requirements of layout.

Prior to the use of artillery, the city's defenders, who were equipped with short-range weapons, were protected by the walls. The approaches to the city had to be cleared for only a short distance, and the suburbs created by the development of the city could press very close to the walls, almost at their foot.

The modifications brought about by the use of powder artillery profoundly influenced the defensive system. The walls became lower, but the firing of cannons required a system of ditches and glacis, which came to occupy an unusually large area of terrain; this is clearly evident in the diagrams of Vitry-le-François (Figure 207) and Palma Nuova (Figure 208).

There were two possible plans for city streets: the right-angle layout, which was a repetition of the old Roman system, and the circular or polygonal plan, which gave minimal length of walls in relation to the area enclosed; the latter plan led to the radial-concentric system of urban planning.

The layout of a new city could provide only for a fixed area established on

FIG. 207
Vitry-le-François in 1554.

FIG. 208
Palma Nuova in 1553.

the basis of a fixed and almost constant number of inhabitants. Later, enlargements had to be added outside the wall, along the roads that led out of the city, or (as in most cases) without preliminary planning, almost always haphazardly and chaotically.

The sudden development of certain cities meant additional city walls had to be built to include new districts, and posed urgent and continually recurring problems of municipal administration. The development of Paris and its administrative organization is a striking example.

The development of Paris

The wall constructed around Paris, beginning in 1190, by Philippe Auguste included an area of 253 hectares (625 acres), distributed over both banks of the Seine. In the thirteenth century the major avenues of Paris were only 16 to 29½ feet wide, and only the main roads were paved; most of the streets were foul-smelling mud roads. Refuse was trampled underfoot; the houses had no sewers, and the only disposal system was the "window method" or the common stream. The small brooks that carried the natural flow of water down to the Seine served as open-air sewers for several centuries.

By the end of the thirteenth century Paris covered an area of 350 hectares (865 acres) crisscrossed by 300 streets. Until 1382 the streets were closed off with chains for security reasons; the chains were eliminated for a short period, but were put back into use between 1407 and 1436. The city wall begun by Étienne Marcel (died 1358) was completed by Charles V; it included the districts that had been created outside Philippe Auguste's wall.

The great plague of 1348 led to the proclamation of the first regulation (1350) concerning municipal hygiene; it was renewed several times (1388, 1506, 1531, 1577), but was not effectively applied until the seventeenth century.

Charles V's wall was crenelated and flanked with square towers. The use of artillery began to be taken into account for its traces, with low *chemins de ronde,* and double moats. The protected area was increased to 439 hectares (1,085 acres). The Seine was barred with chains.

Around the middle of the sixteenth century the appearance of the houses, which had remained unchanged between the thirteenth and fifteenth centuries, changed with the advent of the Renaissance style; mullioned windows were replaced by windows with sashes, and squares of transparent glass replaced thick stained-glass windows inserted in lead. The façades rather than the gables of the houses now faced the street; they were crowned with high, slate-covered roofs.

Catherine de' Medici (1550) imported from Italy the carriages that were to introduce new requirements in street layout. However, they did not come into widespread use until fifty years later.

A new city wall was begun under Charles IX, and was completed during the reign of Louis XIII; it brought new districts to the west of the city within the walls (1633–1636). Charles V's wall on the right bank was then bastioned. The trace for the left bank was planned but was never finished. After the restoration of order, Henri IV ordered the resumption of work to improve the city; he repaired ruins, paving, and drains according to a methodical plan.

By 1636 the number of streets had surpassed 400, and the growth of Paris was becoming a burden, making it difficult to supply the city with food, and creating fear of insurrectional movements by such a dense population. Successive edicts (1627, 1633, 1638) renewed the prohibitions of an earlier edict of 1548, forbidding construction beyond the limits of the suburbs under pain of fine and confiscation. These edicts, which were renewed on numerous occasions during the eighteenth century, were powerless to prevent urban development, and the boundaries had to be continually pushed back, which imposed new limits on building.

Attention was also given to measures of hygiene. The drive of Civil Governor d'Aubray and Chief of Police La Reynie had its effect: the city was cleared of garbage. In 1667 streets began to be opened up, and the number of paved streets increased. The houses were equipped with sewerage ditches; regular services of road repair, fire protection, and public transportation (fiacres and five-sous carriages) were formed. The construction of barracks provided lodging for the troops (particularly the Musketeers), who until then had been lodged with the inhabitants. The effects of these new arrangements were manifested in the elimination of epidemics: there were no epidemics between 1693 and 1832.

Street lighting also made noteworthy progress. The citizens were first requested to light a lantern in front of each house between November 1 and January 31 (1318, 1345, 1504, 1551), and later to maintain lights at the street-corners (1558, 1594, 1639). Lighting became a public service in 1667; 5,500 lanterns were then distributed throughout the 900 streets of the city, where 20,500 houses existed. By 1630 the population of Paris had reached the 560,000 mark.

Italy

City-planning projects did as much as architectural masterpieces to place Rome and Italy in the front rank in the history of art.

The boundless energy and desire for grandeur that inspired popes and princes were revealed not only in the decoration of palaces, in painting and in sculpture, but also in urban planning units that inspired the architects and artists of every country. It cannot be said, however, that a rational study of urban planning, in the sense in which we understand the term today, always governed the layout of public squares and roads.

The squares were often constructed with public use in mind, as for example vegetable markets (Piazze delle Erbe) and fruit markets. Other squares were created to open up an empty space in front of the town hall. But in no city was there an establishment of a general network of streets, carefully planned organization of public squares and communications routes, or improvement of sanitation.

Sometimes the fierce struggles between noble families or parties (for example the Guelphs and Ghibellines) influenced city layouts or the plans of buildings in unusual ways. The well-known Florentine example of the Palazzo Comunale was begun by Arnolfo di Lapo at the end of the thirteenth century. It was called by Vasari *"magnifico e grande."* The layout of both palace and square are reminiscent of the passion and hatreds of the parties. The Ghibellines, temporarily masters of the city, were chased out after the death of Frederick II, and the houses of the great Farinata family were razed. The Guelphs, now masters of Florence, refused Arnolfo permission to build part of his edifice on the site formerly occupied by these houses. The result, according to Vasari, was the "ambiguous and irregular" outline of the Palace. Forty years later an ephemeral tyrant swallowed up the neighboring quarter and surrounded it with walls. The Palazzo Vecchio and the square still retain this erratic, forbidding appearance, the result of their violent history.

Fires and revolution succeeded each other in this area, whose appearance can be properly judged only by keeping this history in mind, including Savonarola's agony on May 23, 1498, which must be imagined within the context of the great stern walls of the Palazzo.

Venice

The very unusual character of Venice merits some discussion of the development of this extraordinary city.

The small islands formed in the lagoons by the alluvial deposits of streams had served as refuges for people from the neighboring mainland fleeing from the barbarian invasions. These movements formed the localities of Torcello, Malamocco, and Chioggia, and then the groupings of Rialto and Olivolo which later formed Venice itself.

At first, light wooden houses constructed on the still shifting mud flats sheltered the inhabitants, who lived from the products of their fishing and farming. Being obliged to use boats for fishing and communication, they quite naturally became a sea people.

In front of the Ducal Palace, rebuilt after the fire of 976, stood the first Piazza San Marco, a narrow, rustic public square; gardens, vineyards, and stables occupied large areas right in the heart of the city. But once the Republic became an important commercial center, Venice grew in size and beauty. Dorsoduro and Spinalonga were settled, and Rialto supplanted Torcello. By the end of the eleventh century, a Norman adversary of the Venetians, William of Apulia, was praising the wealth of the city, and recognizing "that no other race in the world was more valorous in naval warfare, more knowledgeable in the art of handling ships on the sea."

The enlargement of the Piazza San Marco in the mid-twelfth century provided a suitable foil for the beauty of the reconstructed church. In 1156 the Rio Batario, which bounded the square, was filled in; in 1172 the church of San Geminiano disappeared, and the enlarged piazza was paved and lined with buildings.

The palace of the Doges was magnificently rebuilt after the riot of 976, and continued to be modified during the fourteenth and fifteenth centuries. It now dominated the Piazetta, looking much as it does today. The early wooden houses of the city were replaced by stone buildings. In order to support their foundations on this alluvial ground it was necessary to increase the piles buried in the layer of mud. On the piles a wooden superstructure was laid that supported the foundations of the walls and even formed vast platforms on which the entire building rested.

Since the twelfth century the two banks of the Grand Canal had been linked by the wooden Rialto bridge, near the principal market (*merceria*). With the exception of the market, the unpaved streets of the city probably resembled sloughs in which pigs wandered about freely. Inside the city there still existed vacant areas covered with meadows, vineyards, and clumps of trees. Numerous horses were still in use.

A decree had to be passed (1342) forbidding the riding of horses in the market, in order to prevent accidents; it also required riders to carry bells on their saddles. The government began to impose measures for cleanliness and hygiene, and to embellish the city. The great aristocratic families utilized their wealth for the construction of sumptuous palaces. By 1367 there were more than two hundred patrician palaces.

These Venetian palaces were not citadels; civil strife was exceptional in Venice, and there was no need to barricade oneself in a fortress. The houses opened freely onto the canals, rivaling each other in luxury and elegance.

Wealth continued to increase in the fourteenth and fifteenth centuries. Polychrome painting inspired by the East enriched the Venetian Gothic architecture. The palaces built in the fifteenth century turned the Grand Canal into a wondrous highway, "the most beautiful street in the world, I believe," in the words of Philippe de Commynes.

By the end of the fifteenth century the wealth of the shops in the *merceria,* the San Paolo market, which in any other Italian city would have seemed miraculous, and the Rialto bridge lined with shops, contributed to the wealth of the island of Rialto.

The Seigniory paid careful attention to the condition of the city. It ordered the lighting of lanterns, at state expense, in the narrow and unsafe streets. Projects were carried out to deepen or fill canals, remove piles of garbage from the streets, and furnish drinking water to the city. The night watchmen charged with policing the city had to see to the execution of all the laws. A capitulary fixing the assignments is filled with picturesque details; it was particularly forbidden to throw garbage into the canals.

Much effort was also devoted to diverting the mouths of the rivers away from the lagoon, for the material which they carried was capable of filling it up. Other projects included reinforcing the *lidi* which protected the lagoon, and removing silt from the passages that provided communication with the sea, the chief of these being Saint Erasmus, Malamocco, Lido, and Chioggia, which are still in existence.

According to Marino Sanudo, a well-known Venetian author and traveler, the city is supposed to have had between 180,000 and 190,000 inhabitants at the end of the sixteenth century, a figure which is probably exaggerated; better established testimony claims, with greater probability, 110,000 in 1509 and 191,000 in 1540.

In the sixteenth century, despite the loss of her Eastern empire and the onset of decadence, Venice seemed to be remaining at the peak of her greatness. The Piazza San Marco, although already a subject of great admiration, was still encumbered with trees, vineyards, shops, and piles of garbage. In 1601 the Seigniory had it cleaned, and erected a series of buildings on it. Between 1495 and 1517 the *Procuratie vecchie* were built; in 1506 the library and the façade of the mint replaced the shops on the piazzetta, where butchers and fish merchants had their stalls.

In 1582 the old San Marco hospital was demolished and the *Procuratie nuove* were constructed on the site. Thus the public squares acquired their present appearance in the course of the sixteenth century. The various sections of the city were embellished with new buildings; in 1591 the old wooden Rialto bridge was replaced by Antonio da Ponte's bridge, after a contest which was open to the most famous architects of the day. Among them was Palladio, whose project (the conception of which was very inferior to that of da Ponte) was rejected.

Every bit of green disappeared to make room for sumptuous buildings; the earth itself could no longer be seen. The horse disappeared, to the exclusive benefit of the gondola, which had become the customary method of transportation. At the end of the sixteenth century there were more than 10,000 gondolas in Venice, some of which were so luxurious that in 1567 and again in 1584 the Senate was forced to restrain their excesses.

The seventeenth century marked the end of the colonial grandeur of Venice and the last efforts at embellishment of the city. In its still sumptuous decor, unique in the world, which it has left for our admiration, the Venetian state died, being no longer concerned with maintaining its former power, but preserving only the brilliance and frivolity that attracted visitors to its luxury and pleasure.

Rome

The successive remodelings, and even vicissitudes of Rome have caused the plan of the Eternal City to change in such a way that at least a brief description of it must be given.

Within the great wall of Aurelian (275–276), thirty-seven and a half miles long, which was supposed to protect Rome from the barbarian incursions, the population may have reached one million inhabitants; some estimates have claimed two million, which seems highly exaggerated.

The barbarian invasions that dismembered the empire, and the transfer of the imperial house to Constantinople, gradually reduced Rome to ruins, and the departure of the papacy for Avignon reduced its importance to a minimum. It has been estimated that at the return of the popes to Rome there remained approximately 30,000 inhabitants, living among abandoned, crumbling, ruined palaces and ancient temples.

Pope Nicolas V, who reigned from 1447 to 1455, undertook the restoration of the fourteenth-century ruins. A one-time resident of Florence, Nicolas conceived grandoise projects, and sent for one of the best architects of Florence, Leone Battista Alberti (1404–1472), a theoretican and universal artist comparable to Leonardo da Vinci.

On one side of the Tiber, Rome was to be transformed by immense road-building works; on the other side the city was to be reconstructed as a new city with palaces and straight avenues. Plans were drawn up simultaneously for the Vatican Palace and St. Peter's Basilica. The jubilee-year celebrations of 1450 brought a flood of people to the Vatican, and the gifts that poured in furnished the means to begin these immense projects. But the death of Nicolas V halted them, and Alberti departed for Rimini, to work for the glory of the Malatestas.

The successors of Nicolas V were not able to complete this barely begun work, but they did continue to build palaces, taking the materials from the ruins of the ancient buildings; blocks taken from the Colosseum can be found in many of these edifices. However, in 1475 Pope Sixtus IV announced his determination to do everything for the beautification and the improvement of the sanitation of this city that was the "capital of the world and seat of the Prince of the Apostles." The Jubilee of 1475, which was as glorious as that of 1450, brought in a flood of gold. Sixtus IV was able to pull down a number of feudal towers and abolish various slums.

The plans of Nicolas V, which were adopted by Sixtus IV, were still no more than projects when Alexander VI (Rodrigo Borgia) died in 1503. However, Rome owed to him the cutting through of the Borgo Nuovo (the former Borgo Alessandrino) in 1494; it was followed by the opening of the Borgo Angelico under Popes Julius II and Leo X, which contributed to the improvement of sanitation in the district between Saint Peter's and the Tiber.

Julius II wished to rebuild his capital as well as the basilica and his palace. An inscription of 1512 defines his great projects: "After . . . freeing Italy, he beautified the city of Rome . . . by opening new roads. . . ."

On both banks of the Tiber, a great straightened avenue and a new road parallel to the river led toward the Vatican: the Lungara led to the Trastevere, while the other road was known as the Via Giulia.

But at his death Bramante, the great architect employed by Julius II, left Rome in ruins; he had conducted large-scale demolitions in order to clear the necessary spaces and accumulate materials, and his victims included old basilicas and ancient monuments. He was criticized, praised, and nicknamed *"ruinante,"* the creator of ruins. In addition, during the direction of this audacious master there was a shortage of funds needed to complete an excessively ambitious program imprudently undertaken all at once.

Pope Pius IV of the Medici family (pontificate 1559–1565) resumed the monumental undertaking. The Capitol was transformed according to the plans of Michelangelo, who built the magnificent stairways himself. The three majestic palaces were constructed on his plans, but work continued on them until the seventeenth century.

The reconstruction program was resumed and hurried forward by Pope Sixtus V (pontificate 1585–1590); he demolished ancient and medieval monuments, but also built a number of great palaces. For forty years the seventeenth century witnessed the activity of Bernini (1598–1680). A universal artist, whose taste was more widely appreciated by his own age than by later ages, Bernini can in large measure be credited with the present appearance of Rome.

Bernini's works included the leveling of St. Peter's Square, for which he razed a palace built by Raphael, and various houses, and built the double colonnade that forms a majestic frame for the square and the atrium of the basilica (Figure 209). Several palaces built according to his plans were completed in the eighteenth century. The Trevi Fountain was constructed in 1733 by Nicolas Salvi (1699–1751) according to Bernini's design. The largest staircase in Rome, that of the Trinità dei Monti, with its winding flights of stairs whose movement is similar to that of Saint Peter's portico, was completed in 1725, in the spirit of Bernini.

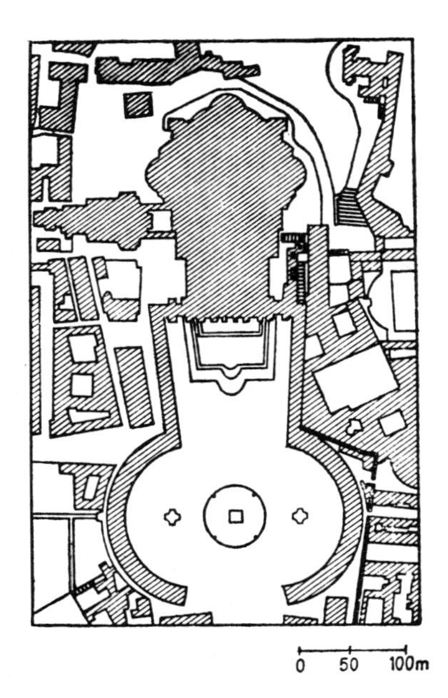

FIG. 209
Saint Peter's,
Rome.

Architectural units

At the beginning of the seventeenth century, urban planning began to become a more exact technique; in France and in Rome it was oriented toward the creation of public squares surrounded by groups of buildings with identical façades. Behind this decorative system the plan of the houses had to be adapted to the façade, and the requirements of use and human habitation were unable to influence the arrangement of the façades, whose style and window and door arrangements were arbitrarily imposed.

In Paris, the Place Royale (now the Place des Vosges) is the oldest monumental square created in this way. It was Henri IV's intention to give this square, which was built on empty lots and space created by demolition of old buildings (for example, the Hôtel des Tourelles), a useful role, and to make it a "promenade for the inhabitants, who are closely crowded together in their houses"; he decided that all the pavilions bordering on the square "will be built in the same symmetry for the decoration of the city" (1605).

In 1607 he ceded to Achille du Harlay, first president of the Parlement, the land included between the palace and the Pont-Neuf, which had just been completed. Harlay was to construct a triangular lot whose houses would have uniform façades.

These ensembles aroused admiration and encouraged other examples. Clément Métezeau (1581–1652), a collaborator of his elder brother Louis, was called by Charles de Gonzague et de Clèves to create the new city of Charleville. Clément II, who came of a famous family of builders, was the grandson of an architect (Clément I) and son of Thibault (1533–1587), successor to his father and brother of Louis (1559–1615). The latter was charged with numerous commissions by Henri IV, at the Louvre and the Place Royale, where he replaced Androuet du Cerceau. He laid out his city like a vast checkerboard of oblongs instead of squares, in imitation of the plan of Vitry-le-François, with a palace occupying an entire side of one oblong. The other sides were occupied by buildings inspired, needless to say, by the façades of the Place Royale in Paris.

The architectural unity was extended to include the entire city. The streets and squares were lined with symmetrical buildings in the style of the Place Ducale, except that these were two-story buildings, and the shops had round windows. The same materials — stone, brick, slate — were used everywhere. It was a total plan, not only of the layout of the streets and squares, but of all the buildings, which were made uniform throughout the entire city. It was complemented by strong fortifications — so strong that Richelieu became concerned, and finally seized them in the name of the king. The cardinal commissioned the same engineer, after the failure of an Italian engineer, to construct the dike of La Rochelle in five months (October 1627 to March 1628).

Versailles

Versailles was not only a castle with immense gardens and outbuildings, but a city whose plan is of major interest. The development of the city itself as a result of the installation of the court at the castle brought about the achievement of a genuine urban planning project.

Louis XIII envisaged only the institution of fairs and markets to attract inhabitants through small-scale but constant commercial activity; the clearing of lands on the side toward Paris was the beginning of this project. In 1664 Louis XIV's practice of keeping several favorites by his side led to the construction of the first large residences. Their number increased rapidly, and in 1671 Louis XIV decided to create a genuine city. The terms of his resolution of May 22 laid down the principles of its establishment:

"His Majesty, being particularly partial to the village of Versailles, and hoping to make it as flourishing and much frequented a village as possible, has decided to make gifts of land to every person who desires to build beyond the pump of the said Versailles as far as the farm of Clagny . . . so that the said lands may be enjoyed by each of the individuals to whom the said lands will be delivered over in full ownership to them, to the charge of them, their heirs and assigns, to maintain the buildings in the state and in the same symmetry in which they will be built and raised."

Thus he simultaneously imposed a style and a uniform type of construction. In addition, the layout of the major avenues was fixed by Le Vau. Mansart replaced the buildings already on the site by the Royal Stables (the Great Stable on the north side of the Avenue de Paris, the Small Stable on the south side), at a cost of three million livres. Then came the Great Common, on the site of the hovels of the old village. The next additions were the Venery and the castle of Clagny (Figure 210).

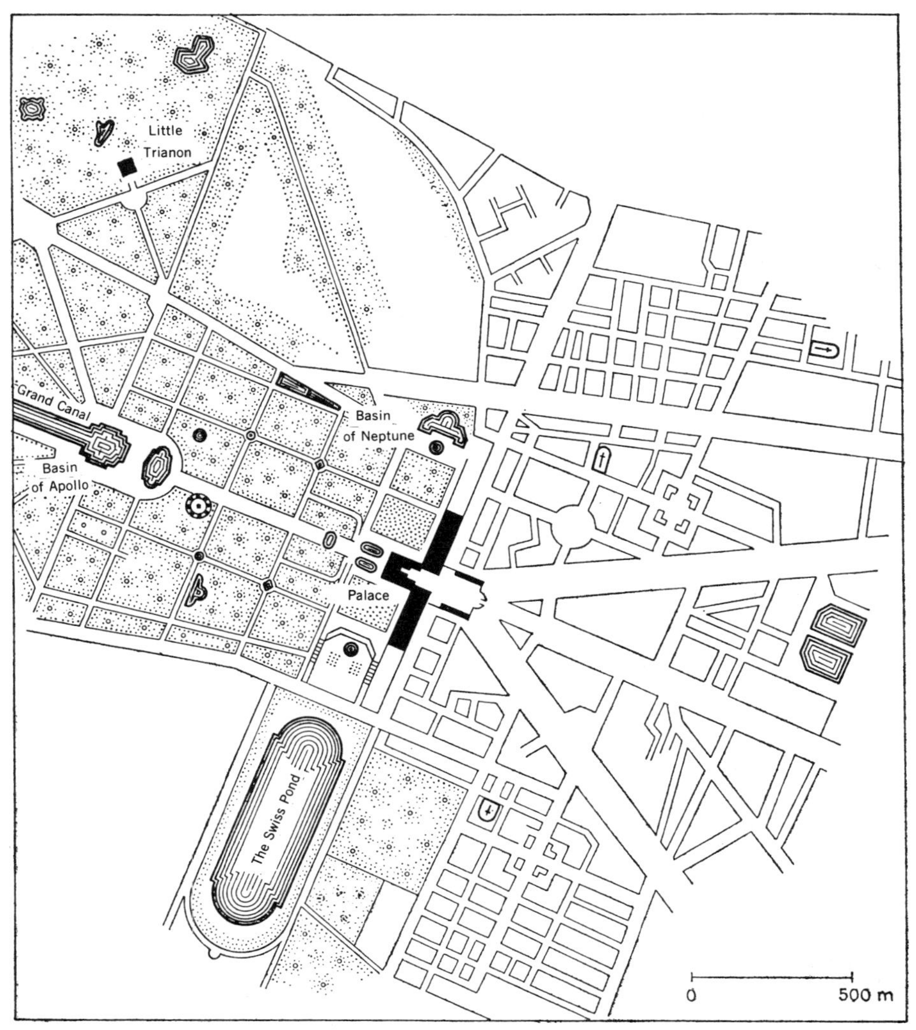

FIG. 210 Plan of the castle and grounds at Versailles.

Versailles, which was to have 30,000 inhabitants in 1715, with its gardens, regular streets, plantings of elms, homes of seigneurs, shops and inns, was thus created from nothing. The buildings were of uniform height and color, of stone and brick, and were covered with single-story slate roofs. Their height was determined by the rule that the roofs of the houses must be no higher than the level of the marble court. This gave the city a uniformity of height that inevitably brought more light to the streets by reducing the number of stories. But it also confirmed both the position of the castle on the height and its domination over the city built at its feet, and the supremacy of the monarchy, whose palace dominated every other building.

We have seen that the most ambitious plans of beautification, for example those of Rome, failed for lack of means. Let us review the extraordinary amounts of money and labor devoted to the construction of the castle, gardens, and surroundings of Versailles. A single director presided over the works: for thirty years, beginning in 1683, Jules Hardouin-Mansart was to be the executor of Louis XIV's projects. These projects were to require an incredible amount of labor. An army of masons, levelers, and laborers moved the earth, dried up the swamps, and cooperated on the construction of the palace. In Dangeau's notes we find two references to the personnel utilized: 22,000 men and 600 horses in August 1684; 36,000 men in May 1685.

As early as 1678 Mme. de Sévigné related that the death rate among the laborers was prodigiously high, and that at night "wagonloads of corpses" were carried away. This mortality rate is explained by lack of understanding of the microbian effects of malaria, which was made more active by leveling operations in the mud and the overturning of the earth.

On May 6, 1682, Louis XIV installed himself at Versailles, and turned his energies to accelerating the work. Colbert's reservations about these colossal undertakings, and his vain efforts to reduce them, date from this period. The wars, which required additional expenditures and the return of workers and soldiers to the armies, completely halted the work. While the buildings and the major portion of the enterprise remained intact, it was necessary to sacrifice a part of the decorations begun, and even to destroy almost all the marvels of goldsmiths' work that decorated the building in order to recover the precious metals.

The burning and reconstruction of London

The destruction of a large area of a city by fire immediately raises the problem of planning its reconstruction. This was the case in London in the seventeenth century.

Between September 2 and September 6, 1666, an exceptionally violent fire destroyed a major portion (approximately three-fifths) of the city of London. The destroyed area included 13,000 houses and 87 churches—a small-sized city from which the inhabitants fled in the face of this unexpected calamity, in which they lost everything.

All Europe was moved by the catastrophe. England's enemies declared that London had been struck by "the hand of God." But Louis XIV, then at war with England, while admitting that this catastrophe was a stroke of good luck for his cause, prohibited all rejoicing over the misfortune, because it was "a deplorable event," and proposed to send food and other necessities to the stricken city. The Dutch, in contrast, did not show a similar compassion; they represented the fire as a punishment from Heaven, and shrewdly spread abroad the rumor that England was ruined.

However, the country reacted promptly, and the city of London was reconstructed in a relatively short time. Certain special aspects of the causes of the fire, the reconstruction, and the influence of economic conditions on urban planning merit discussion.

The fire, which at first was not very large (it occurred in a baker's house), spread rapidly because of a violent east wind, which carried it to a pile of hay in a neighboring courtyard. The sparks and flaming debris immediately spread across the narrow streets, and the great amount of wood used in the walls and roofs, which had been dried out by the hot summer, quickly caught fire.

At this period there were few available methods of combating fire. There was no water-supply system capable of supplying water in large quantities—only buckets that had to be filled at the fountains, or the river if it was not too far away, and heavy sprayers that threw very small quantities of water on the façades and roof beams.

The most effective method consisted of creating an empty space in the path of the fire by pulling down the houses; this was done with hooks attached to ropes or poles, which when hung on the roofs and pulled vigorously caused the façades to fall and the houses to collapse. The fire, which could not cross the empty space in front of it, was restricted and died for lack of new fuel.

But under the circumstances the lord mayor at first hesitated to take such radical steps, fearing complaints from the owners and renters of the houses which would have had to be destroyed—although these houses were destined to disappear on the first day, a fact that was not foreseen in the early hours of the disaster.

Moreover, there was general confusion. The inhabitants, fleeing in carriages with what they were able to carry, crowding the streets in their attempt to reach the Thames and the boats moored on its banks, hindered the organization of the struggle against the fire. Not until four days later, when the wind died and no longer pushed the flames onward, did the fire die of its own accord.

The adoption of the principles of reconstruction aroused lively discussions. What could be done in this vast, practically flattened zone from which rose only an occasional portion of a wall and a few partially destroyed buildings?

The example of the achievements of the great urban planning projects in Paris impressed various people in London. The layout of cities like Vitry-le-François, which had inspired Métezeau at Charleville, could also serve as an example. Was it not necessary to profit by the presence of this vast leveled area, to be reconstructed according to a carefully laid-out plan? There was lively discussion of Christopher Wren's plan, which was conceived in this spirit (Figure 211). There were three opposing conceptions: Wren's idea, a return to the old plan to satisfy landowners who did not wish to be dispossessed, and a third conception, inspired more by the second idea than the first, which was limited to a few enlargements, open spaces, and straightening of secondary streets.

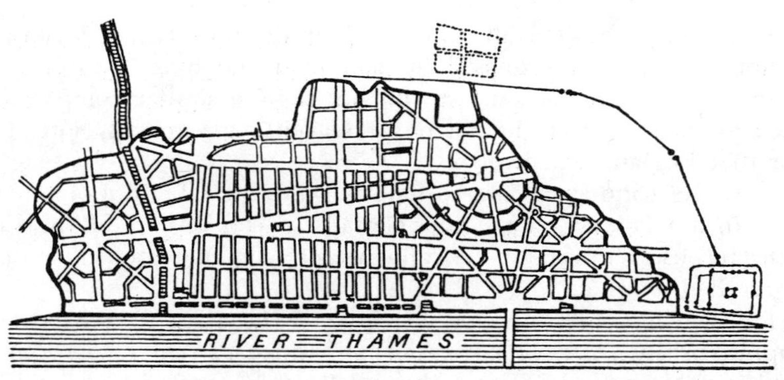

FIG. 211 Wren's project for the reconstruction of London.

Wren's partisans and descendants maintained that an opportunity to make London a modern city had been missed, that the plan so broadly conceived was sacrificed to the egoism and lack of understanding of the inhabitants. The solution adopted, the third, perhaps made a more rapid reconstruction possible.

The problem was to supply the construction sites with the unusual quantity of materials needed. Materials salvaged during the leveling operation were utilized, but naturally they were insufficient, given the nature of most of the old buildings – the wood and daub walls had burned and disappeared. Though it was necessary to open and develop quarries in England itself, the builders were able to profit by the proximity of the Thames to the major construction site, importing from abroad certain materials, such as the beautiful stones destined for the public buildings. The construction of ordinary houses, which was the subject of regulations designed to secure them against the danger of fire, required less costly materials. The builders profited by the possibility of establishing brickworks on the outskirts of London to obtain sufficiently strong materials more economical than stone, which involved troublesome carting services or expenditures for its transportation by water.

Brick walls were substituted for the old wooden ones. However, a broader series of regulations were initiated which were not limited to the architectural regularity of the façades but covered all kinds of ordinances. These were included in a series of proclamations by the king and the lord mayor, and in an act of reconstruction. The latter fixed the heights of the houses (Figure 212) in relation to the width of the streets, and the principal dimensions and thicknesses of the walls.

The concerted action of great architects like Christopher Wren (1632–1723), scientists like Robert Hooke (1635–1703), Secretary of the Royal Society, and commissioners named by the king or chosen by the city, so impressed world opinion that two and one-half centuries later, when Tokyo was ravaged by earthquake and fire, the Japanese came to London in search of ideas and examples to help them rebuild.

Streets and public services From the Renaissance to the seventeenth century, the art of urban planning was devoted, as we have seen, to the creation of several large roads and especially architectural ensembles

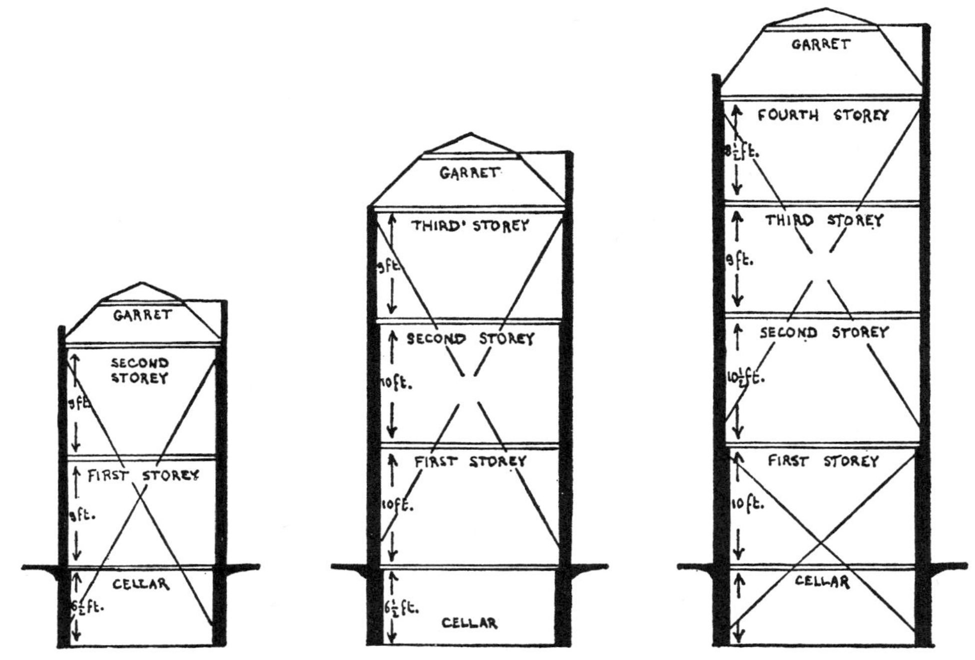

Fig. 212 Regulation of building plans (London).

that framed the large public squares. But the old streets continued to give access to the inhabited quarters behind these squares and major roads, and the traffic network still took the form of a tangle of often very narrow streets.

The cities, being tightly compressed within a belt of ramparts, were unable to open up empty spaces sufficient to provide wide passages and give air and sunshine to the houses. Astonishing figures have been uncovered concerning the width of certain streets (two feet four inches at Genoa and Venice); moreover, the streets were lined with houses of four and even more stories.

Several old streets of Paris have survived as lanes or dead ends *(impasses)* in the modern quarters. For example, the *impasse* Sourdis, which is still intact, was considered average for the centuries we are discussing. It is ten feet wide, and is lined with buildings ranging from one to three stories in height; it is made still more narrow by the guardrails that protected the façades of the houses from damage by carriage axles (Figure 213). In the days when the horse was still a mode of transportation, these guardrails also served as mounting blocks for riders.

Where paving existed, it sloped down from both sides into the middle to carry water into the center of the street. If two pedestrians met, courtesy required that the better dressed of the two be permitted to walk on the inside, so that he would not have to splash about in the river.

There were no sidewalks; a great innovation was created with the building of the Pont-Neuf (1578–1604), with its absence of shops and its sidewalks on which pedestrians could walk and enjoy a view of the river. For this reason it became a place for conversation and demonstrations by boatmen.

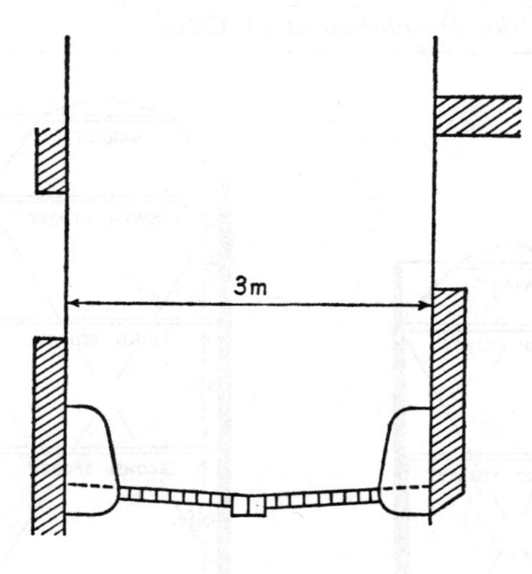

FIG. 213 Impasse Sourdis, Paris.

The carriage did not appear until the middle of the sixteenth century, and its use spread slowly; it was difficult for carriages to circulate in such streets. We see the carriage of Henri IV traveling with difficulty along a narrow street; his bodyguards are obliged to detour to meet him at the place where the street opens out. This makes it easy for his assassin, Ravaillac, with one foot on a guardrail and the other on one wheel of the carriage, which has been stopped by an obstacle, to wound his victim mortally.

Old streets and lanes have survived in London as in Paris. After the fire of London, the city fathers saw the difficulty and even impossibility of realizing Wren's plan, which would have created an ensemble better adapted to traffic, whence the reconstruction on narrow, old streets that still exist.

It was only in newly created cities — Versailles, Richelieu, Charleville — that a unified plan applied to virgin land made it possible to lay out the streets in a more normal and hygienic manner.

Paving

Even the most summary histories do not fail to relate the story of how Philippe Auguste looked out of his palace on the Cité in 1184, saw carriages bogging down in fetid mud, and decided to have a portion of the city paved.

The first paved street was the Rue de la Barillerie (now the Boulevard du Palais, and undoubtedly much wider than it was then). Its paving consisted of large square paving stones 4½ feet long and almost 10 inches thick, sealed with mortar. These heavy stones, which were difficult to handle, were later replaced by blocks 20 to 24 inches long and 6¼ to 8 inches thick, and still later by cubes whose sizes were fixed by a series of regulations:

1415 — cubes of 6⅓ to 7⅓ inches
1567 — cubes of 7⅓ to 8½ inches
1720 — cubes of 8½ to 9½ inches

But this expensive paving was adopted only for streets that carried heavy traffic; until the nineteenth century a great many Parisian streets consisted only of packed earth, without any strong facing. Paving was also adopted for the major roads in the vicinity of Paris that were utilized by the king for his journeys, and particularly the road from Versailles to Marly and Saint-Germain; it eliminated mud and dust and gave the carriages a smoother rolling surface.

Debris and garbage

Though liquids thrown out of the houses could flow down the central gutter of the street in order to reach a stream or sewer and thence the river, this was not the case for household refuse, which obstructed and poisoned the streets with pestilential wastes.

It very soon became necessary for the inhabitants of a street or district to group together in order to rent tipcarts, remove the garbage, and transport it outside the city. In 1508 a tax was established for street maintenance and removal of garbage. In 1539 an ordinance of François I made it obligatory to use baskets for the garbage, instead of throwing it pell-mell into the street. It also fixed the sizes of the wagons used to transport it (six feet long and two feet high), and ordered that they be closed up "so tightly that neither garbage nor refuse can escape" and operated at regular hours for the removal of rubbish. A similar ruling was not applied to other cities until the end of the eighteenth century.

The problem of the garbage dumps then arose: they had to be outside the city, but sufficiently close to keep transportation charges at a minimum. If the city had a fortified wall, the garbage was dumped not far from the gates. But dumps on the outskirts of a large city became very large, and created mounds that grew larger each day. In the fourteenth century two buttes, or artificial rises, were formed between the present Avenue de l'Opéra and the Rue Richelieu by the dumping of refuse and rubbish from demolished buildings. The first, situated at approximately the junction of the Avenue de l'Opéra, the Rue Thérèse, and the Rue des Pyramides, was named the Butte Saint-Roch. Formed by demolitions and earth fill from the construction of Étienne Marcel's wall, it became a place of execution, and served as a base for Joan of Arc when she was directing her fruitless attack on the nearby Saint-Honoré gate on September 8, 1429 (the attacking artillery was located on the butte). It grew still higher in 1536, thanks to fill from the new fortification works. In 1615 it was partially leveled in preparation for the construction of houses, but not until the opening of new streets in the nineteenth century was it completely leveled.

The second butte, called the Butte des Moulins, at the crossing of the Rue Sainte-Anne and the Rue des Petits-Champs, was begun in 1536, also as the result of the work we have just indicated. On this butte were built mills, which disappeared in 1688 when entrepreneurs leveled the butte to build a new district (1668–1677).

On the left bank, fill also formed an obstacle to the cutting through of the Rue Saint-Guillaume, an old fourteenth-century roadway, and forced it to form a bend that later became the Rue Perronet. A mill was constructed on this rubbish heap, whence the name of Rue de la Butte (1530).

Drains

Removing rubbish and debris posed an often difficult problem for the large agglomerations, and the disposal of rainwater and liquid wastes also created a dilemma on occasion.

The natural slope of the land permitted the center gutters of the streets to carry the waste water to a drain, which was normally a brook or nearby stream. The contours of the terrain and natural slopes were thus put to use to remove waste and rainwater. For many centuries the water was carried to a stream, by open ditches, some of which had once been small brooks themselves. A Parisian example is the old Ménilmontant stream, which in the sixteenth century was transformed into a drain, starting from the ditches of the Temple and ending at the modern Place de l'Alma. Its profile was straightened in the seventeenth century, but it retained its character as a ditch dug in the ground. Not until the eighteenth century was its transformed into a canal paved with paving stones; it was not covered until the end of the eighteenth century.

Another example, this time in Dijon, was the regularization of the Suzon, carried out by Hugues Sambin, a pupil of Michelangelo. Originally intended to supply water, it was paved, springs were drained into it, and it could be used for pumping water. But in the city, where it was vaulted, it collected rubbish. It had to be cleaned out in 1512 and 1559, and in 1599 rubbish was still flowing into it.

Not until the nineteenth century did rationally conceived and executed projects succeed in improving the sanitation of a certain number of large cities. In the Paris of pre-Haussmann days there were only ninety-three miles of drains, some of which were still open and foul smelling, to carry water to the Seine.

The lack of drainage and the sanitation difficulties weighed heavily on the sanitary condition of the castle of Versailles, which had no sewers. The accumulation of garbage for months at a time and the draining of urine into the corridors made the atmosphere unbearable, and obliged the court to spend the summer in another residence so that the necessary washing and cleaning could be done.

BIBLIOGRAPHY

DAWSON, PHILIP, and WARNER, SAM B., JR., "A Selection of Works Relating to the History of Cities," in Oscar Handlin and John Burchard (eds.), *The Historian and the City* (Cambridge, Mass., 1963).

HAMBERT, PER G., "Vitruvius, Fra Giocondo and the City Plan of Naples: A Commentary on Some Principles of Ancient Urbanism and Their Rediscovery in the Renaissance," *Acta Archaeologica,* 26 (1965), 105–125.

MUMFORD, LEWIS, *The City in History: Its Origins, Its Transformations, and Its Prospects* (New York, 1961).

MUNDY, JOHN H., and RIESENBERG, PETER, *The Medieval Town* (Princeton, 1958).

PIRENNE, HENRI, *Medieval Cities* (2nd ed., Princeton, 1939).

REDDAWAY, T. F., *The Rebuilding of London* (London, 1940).

SCHEVILL, FERDINAND, *History of Florence* (New York, 1936).

STEPHENSON, CARL, *Borough and Town: A Study of Urban Origins in England* (Cambridge, Mass., 1933).

CHAPTER 21

HYDRAULIC WORKS AND WATER-SUPPLY SYSTEMS

SINCE THE most remote ages of history, water, an indispensable element in human existence, has been an object of concern in every country. The Romans created the largest water-supply systems, and left to posterity achievements that were utilized for centuries after their disappearance. These accomplishments have been described in the first volume of this work. But before discussing the period of the Renaissance and the sixteenth and seventeenth centuries, which benefited from the water-supply systems left by the Romans, a brief summary of the Roman activity will be useful.

The Roman aqueducts The methods of water supply available to the Romans are later found in use, unchanged from the technical point of view, in the Middle Ages and the Renaissance, namely, the use of natural slopes that permitted the water to flow. This in turn required the use of aqueducts and occasionally siphons in order to cross valleys. When they had to raise water by artificial methods, the Romans had at their disposal hydraulic wheels, "tympanums," the Archimedean screw, the noria, and finally the piston pump, which if it was not a Roman invention was at least frequently used by them.

As for the quantities of water needed by a large city like Rome, they were considerable, even for the million inhabitants assumed to be the total population of this large agglomeration in the imperial period.

Frontinus, who was the supervisor of Rome's waterworks around A.D. 96–98, indicates that until about 312 B.C. the Romans were satisfied with the water from the Tiber, wells, and a few springs. But with the increase in population, the water of the Tiber became polluted, and the wells and springs became inadequate; it was necessary to construct aqueducts to carry water from the neighboring hills, where the supply was fairly abundant.

By around A.D. 97, 880 miles of aqueducts were supplying a quantity of water that has been estimated at between 961,380 and 1,242,600 cubic yards per day. Of this quantity, three-tenths was consumed outside the city, seven-tenths in Rome itself. For an estimated population of one million inhabitants, this gives a per-capita figure of between 153 and 171 gallons daily. This figure seems enormous, but it must be reduced to a lower amount for periods of drought, and also during the frequent repairs to the aqueducts, whose maintenance required a labor force of between 250 and 450 men. In addition, the water flowed freely in a large number of fountains, and a great portion of it was thus lost. Of the amount distributed in the city, 17 percent was consumed "in the name of Caesar," and 44 percent by public establishments (military camps, public buildings, public theatres, and 591 fountains).

The consumption of water was increased by the construction of new aqueducts in the second and third centuries; the last Roman aqueduct was constructed in the fifth century.

. . . restored during the Renaissance

The destruction wrought by the barbarian invasions reduced the water supply to nothing and the inhabitants to an infinitesimal number. When the popes decided to rebuild a mighty Rome, their first concern was to restore the ancient aqueducts, whose water brought back life, health, and joy to villas and gardens.

The only aqueduct in existence in the Renaissance was the Aqua Virgo, which had been restored by Nicolas V. Sixtus IV put the Aqua Martia and the Aqua Claudia (which he rechristened Aqua Felice) back into operation; at the entrance to the Borgo Felice the clear, fresh water from the Alban hills supplied a fountain built by Fontana and ornamented by Sixtus V with a colossal "Moses" by Prospero Brasciano, a caricature of Michelangelo's "Moses." In this new district of the city built by Sixtus V, and in the rest of the city, water was now flowing from twenty-seven fountains.

Following the repair of the Aqua Claudia, other aqueducts were repaired in the seventeenth century. The Aqua Trajana, which was put back into use by Pope Paul V, brought water from the top of the Janiculum to the Aqua Paola fountain. A number of fountains in imaginative shapes came to ornament the great public squares; their elegance can still be admired.

The villas that began to proliferate in the sixteenth century formed a belt of greenery and flower beds with playing fountains around the city. A French engineer of the last quarter of the seventeenth century recorded his admiration of these "genuine earthly paradises and enchanted places."

Methods of pumping

Many cities lacked the advantage of being able to harness, at a relatively short distance, water that could reach the city by gravity. In these cases the water had to be pumped from wells or streams.

Pumps, which had been known to the Romans, were in general use during the Renaissance and succeeding centuries. In his *De re metallica* (1541), Agricola describes various types of pumps in detail; they are operated manually, by a horse-driven treadmill, or by hydraulic wheels when it is possible to install them on a waterfall or a stream whose flow is sufficiently rapid. In addition to piston pumps, he describes a pump with leather-and-horsehair balls; an endless chain to which the balls were attached passed through wooden pipes banded with iron (Figure 214). This arrangement has survived into modern times, and in similar form, in domestic and garden devices.

Agricola also describes the functioning of norias, operated by a hand crank, with a demultiplication of $1/36$. Troughs capable of holding almost two quarts of water were attached to the chain by means of leather thongs; the chain wound around drums that were equipped with hooks. The inspiration for this device

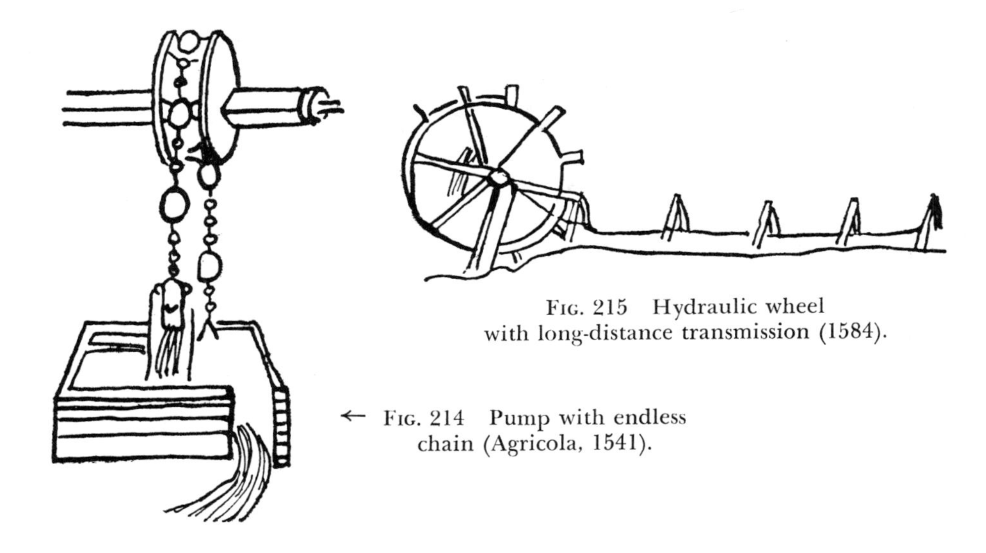

FIG. 215 Hydraulic wheel with long-distance transmission (1584).

← FIG. 214 Pump with endless chain (Agricola, 1541).

derived from Vitruvius. Agricola mentions more powerful machines with larger troughs, which were operated by hydraulic power.

In the first book of his *Instruments méchaniques et mathématiques* (1584), J. Errard of Bois-le-Duc desribed methods of pumping that were similar but could also be activated by a hydraulic wheel acting at a distance (Figure 215) by means of a rigid shaft supported by a series of balance beams. A windmill, which, however, is simply a reproduction of the numerous devices utilized in Holland, can also be used for this purpose.

Ramelli's work *Dell' Artificiose Machine,* published in 1588, contains a drawing of a pump with eccentric rotor and sliding pallets. Given the rather crude methods of construction available at this period, it was difficult to construct this device satisfactorily. But it is interesting to note that the principle of this rotating pump, which was revived at a later period, was already known at the end of the sixteenth century.

Piping systems

The water carried by an aqueduct actually formed a small stream, and the channel had to be made absolutely watertight. The Romans were very familiar with the composition of linings and cements made with mixtures of lime and materials like the various types of pozzolana, ensuring a hardening and a strength comparable to those of modern artificial cements.

For underground pipes, the problem was less extensive but more difficult. It had been solved millennia ago by means of fitted sections of terra-cotta pipe. Specimens of ancient Greek piping have been found; they range from 17½ inches to 23½ inches in length, with a diameter of between 6 and 8 inches and a thickness of ⅓ to ⅔ of an inch, depending on the diameter.

The Romans used both terra-cotta and lead for their pipes. Vitruvius describes them as follows:

"If you wish to pipe water at less expense, you will use terra-cotta pipes,

Plate 40. Pump with eccentric rotor. Engraving taken from Ramelli, *Le diverse e artificiose machine.*
Photo New York Public Library.

which should be at least $1\frac{1}{2}$ inches thick, and narrow at one end so that they can be fitted into each other. The ends are joined together with lime blended with oil.

". . . . Terra-cotta pipes have the advantage of being easy to repair when necessary, and the water is much better than when piped through lead pipes, in which there forms a substance called white lead, which is considered to be very dangerous for the human body. . . ."

The Roman pipes were shaped either on the potter's wheel or on a slightly conical core. The latter, very crude procedure was still in use in the Middle Ages; it produced a pipe that had a fairly regular, smooth interior, while the outside was very rough, since it was carelessly worked with the hand. The end destined to receive the narrow portion of the next section of pipe was widened with a tool (Figure 216).

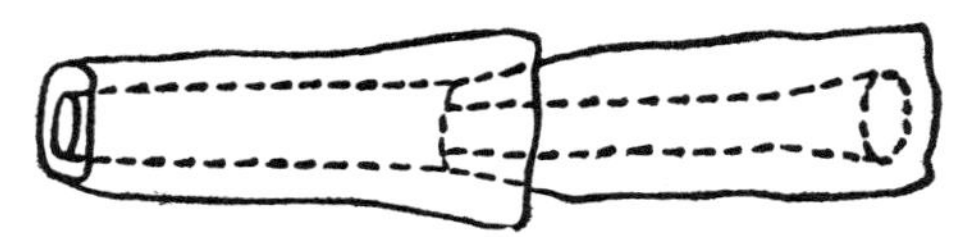

FIG. 216 Terra-cotta pipe sections.

Pipes made on the wheel were more smoothly shaped. In this connection it should be mentioned that while the potters' wheels of antiquity were hand-operated, those of the sixteenth century were foot-operated, which freed the operator's hands and made better workmanship possible.

In the Middle Ages and the Renaissance, as in Roman times, lead pipes were made by soldering together sheets of lead shaped on a core. The Roman pipes were soldered by applying molten lead along the slit between the edges of the sheet. Later, the lips were chamfered and soldered with tin and a hot iron.

These procedures were still being applied in the seventeenth century. The pipes were joined with a sleeve cast from lead or tin solder, which was then worked by hammering. Specimens of these pipes dating from the seventeenth century are still in existence; their diameters vary greatly, ranging from $1\frac{1}{2}$ to $6\frac{2}{3}$ inches. The tin solder of this period which has been found varies in composition:

Lead – 62.5% to 55.5%
Tin – 37.5% to 44.5%

Cast-iron pipes. When the charcoal-burning blast furnaces began (in the sixteenth century) to function on a regular basis, they supplied a cast iron that was sufficiently fluid to be cast in thin sheets. It was now possible to make cast-iron pipes; however, they were very rarely used prior to the seventeenth century. According to Bélidor, the first French cast-iron pipes may date from 1672. Very crude specimens with conical joints are supposed to have been in use in fifteenth-century Germany. Longitudinal ridges on the outer surface clearly indicate the joint of the mold box in which they were horizontally cast.

A pipe of this type, with an interior diameter of $1\frac{1}{2}$ inches and walls $\frac{1}{2}$ inch thick, was utilized from 1661 to 1875 to carry water under a pressure of ten atmospheres. Pipes of the same period with diameters of 2, 4½, and 5⅔ inches, and 5¾ feet in length, have been found; all of them have conical joints. These are the only cast-iron pipes known prior to those of Versailles.

Wooden pipes. In the great piping systems of the capitals and large cities, lead, terra-cotta, masonry, and later cast-iron pipes were generally used. The use of wooden pipes was rare and even exceptional. In small cities and establishments, in contrast, wooden pipes were widely used from the fifteenth to the seventeenth centuries; they were less expensive, and could be made on the spot by craftsmen.

These pipes were obtained by boring tree trunks, on which the bark was often left, since it helped to protect the wood. Oak, elm, or softwoods like alder were utilized for this purpose. The boring was crudely done by hand with a series of augers of increasing diameters, the last one having the shape of a semicircular spoon, reaming the hole to the desired diameter. The augers were short, which necessitated boring from both ends (Figure 217:1).

In joining the pipes, one end could be whittled down and the corresponding end of the pipe to be fitted enlarged into a cone (Figure 217:3 and 5). The sections could also be joined with biconical iron rings (Figure 217:4 and 6). To ensure watertightness, the joints were coated with moss or oakum, and sometimes with a mastic made of mutton fat or crushed brick kneaded together. When there was a hole or a crack to be blocked, wooden wedges were used with oakum and mastic (Figure 217:2).

The diameter of these bored pipes would not have been sufficient to carry a heavy flow of water. For this purpose a channel was cut in a tree trunk, and was then covered with planks pegged along the edges. However, joining such large pipes posed problems of watertightness that it was difficult to solve. Parts of a piping system formed by joining square pieces have been found (Figure 218).

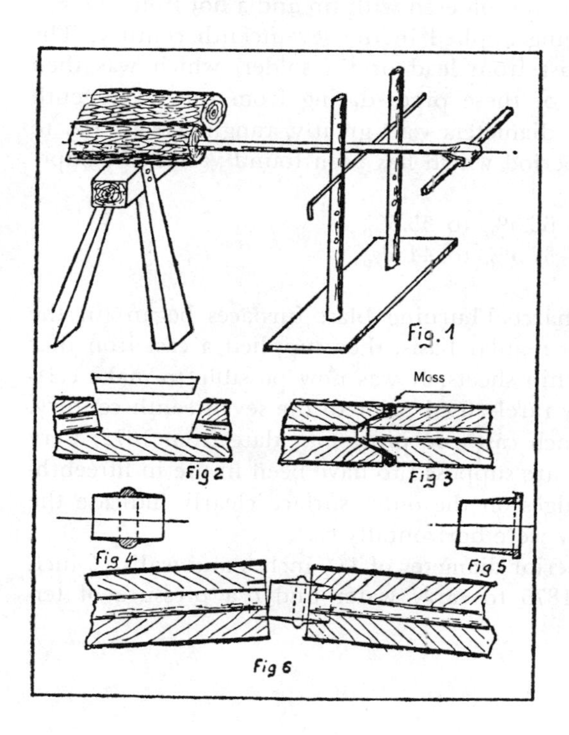

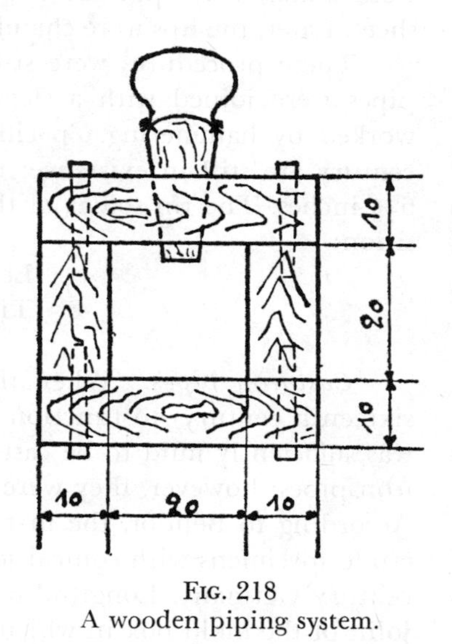

Fig. 218
A wooden piping system.

← Fig. 217
Making wooden pipes.

Hand-boring of pipes was long and required much labor. According to the *Encyclopédie,* in one day a worker could drill a hole 2 inches in diameter in 38⅓ feet of elm or alder pipe; if he was working with oak, this figure dropped to 6⅓ feet. An attempt was therefore made to mechanize this work by a hydraulically operated machine for boring (Figure 219). The drill, which was supported

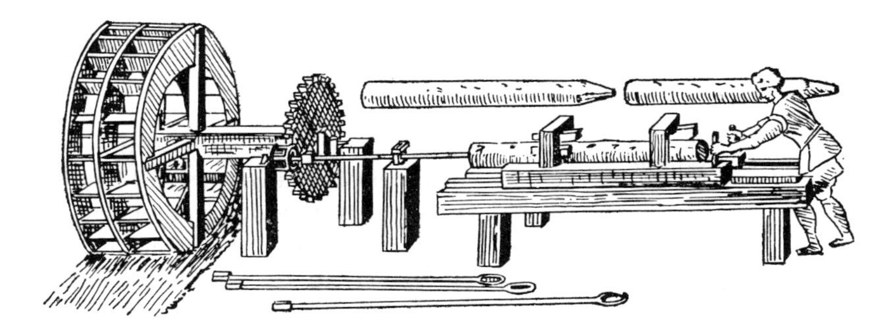

FIG. 219 Machine for boring wooden pipes.

so that it would not bend, penetrated the piece of wood, which moved on sliders. Such machines were still being built by Périer at the end of the eighteenth century.

To open a branch connection in such pipes, all that was needed was to drill a hole and insert a lead pipe with a conical bronze faucet joint at its end.

While they were easy to make, wooden pipes had the great disadvantages of rotting quickly and of being able to sustain only relatively low pressures. The very ease with which branch connections could be opened in them encouraged dwellers along the riverbanks to tap (sometimes secretly) the pipe, which when tapped in various places gave an inadequate supply at its "official" outlet, and often dried up completely.

Measurement of water. The Romans measured water in *quinarii,* sectional units of connections to a common basin on the same level. In the Liège region a similar measuring procedure was used: the *xhancion,* the discharge of an opening one-quarter of a (Liège) inch (equal to slightly more than 8 millimeters) under a pressure of almost 6 inches, which gave 1⅓ gallons per minute, or 1,903 gallons in 24 hours.

Elsewhere, and particularly in Paris, measurements were at first much less precise. The taps were "about the size of a pea, a vetch grain, or the metal tip of a lace" – vague dimensions that were clarified by attaching a ring of the diameter of the tap granted to the customer's certificate. At the end of the sixteenth century the diameter was fixed in inches and lignes. This was the origin of the unit known as *le pouce fontainier,* an opening of one square inch.

In 1626 waterworks or reservoirs began to be installed in Paris and the users had to be linked up with them. This system standardized the pressures in the consumers' pipes under these conditions, the water corresponding to a *pouce fontainier* amounted to a discharge of 5,112 gallons.

FIG. 220 Water catchment.

FIG. 221 The aqueduct of Arcueil.

The water-supply system in Paris

Roman Lutetia had two water-supply systems: the Chaillot system (third century) which daily carried 654 cubic yards of water from Passy, through terra-cotta pipes, to public baths that are believed to have been on the site of the present Royal Palace, and the Arcueil aqueduct (fourth century) which carried 2,224 cubic yards of water from the springs of Rungis through an open concrete channel 9½ miles long. This channel was constructed of a very hard concrete made of "limestone, flint, and gravel." The inside was lined with a watertight coating of two layers of cement (coarse for the first layer, fine for the top layer) made with fragments of tile.

Thus Lutetia had a supply of 2,878 cubic yards of water daily. These two aqueducts were destroyed during the Viking sieges of Paris. Sometime before the tenth century, the monks of Saint-Laurent and Saint-Martin-des-Champs constructed the aqueducts of Pré-Saint-Gervais and Belleville, which were fed by tapping springs. These taps were small galleries made of small stones, without mortar, and laid on the impermeable layer of clay beneath the water level. The water passed through the openings between the stones and flowed into the gallery, which was lined with paving stones coated with a waterproof layer of clay in order to prevent infiltration of surface water. The water was collected in a common basin at the Pré-Saint-Gervais, which was the starting point of a pipe six miles long. The pipe, originally of terra-cotta, was later replaced by lead pipes 5¼ inches in diameter.

The four-mile-long Belleville aqueduct was larger; it was a vaulted gallery 4 feet wide and 6 feet high.

These two aqueducts combined supplied Paris until the sixteenth century with the meager quantity of 458 cubic yards of water per day (262 cubic yards in periods of drought). By the end of the Middle Ages this water was being distributed to the public fountains, where individuals and water sellers came to obtain their supplies. Given that medieval Paris had more than 200,000 inhabitants, this meant that only one or two quarts daily per capita were available; however, this figure does not include the water that could be supplied by numerous wells located throughout the city.

In 1598 Henri IV ordered the installation of a fountain at the Palais de Justice, to be supplied from the Pré-Saint-Gervais aqueduct. He also made a special effort to reduce the numerous taps made by individuals, which had increased despite Charles VI's edict of 1392.

By the beginning of the seventeenth century, the Louvre and the Tuileries were consuming half the water supplied by the two aqueducts. In order to supply the Luxembourg Palace and the neighboring districts, Marie de Medici ordered the restoration of the old Roman aqueduct of Arcueil, a concrete channel 13⅔ inches wide and 23½ inches high, surmounted by a vaulted gallery 3 feet wide and 5¾ feet high (Figure 221). The old channel sufficed to carry the 745 cubic yards of water supplied each day by Rungis.

Between 1603 and 1608 Henri IV ordered the construction of the "Samaritaine" pump, which carried 915 cubic yards of water daily from the Seine to the Louvre and the Tuileries. The water was distributed to various individuals and to the public fountains. In the fourteenth century only monasteries, the Royal Palace, and six public fountains were being supplied; by 1500, there were twelve fountains in Paris and five outside the walls, and nineteen seigneurs and religious establishments were also being supplied. By the beginning of the seventeenth century the number of fountains had increased to twenty-two, that of the "consumers" to forty. Thus the need for the installation of pumps—the "Samaritaine" under Henry IV, and the Notre-Dame bridge pump in 1670.

In 1608 the Samaritaine added 915 cubic yards of water to the 450 cubic yards being supplied by the aqueducts; in 1623 the restoration of the Arcueil aqueduct raised the volume available to 1,962 cubic yards. In 1670 the Pont Notre-Dame pumps added 2,616 cubic yards, but the aqueducts, which were in poor condition, were now supplying almost nothing.

In the seventeenth century the role of the pumps became of primary importance in the Parisian water-supply system. They were able to ensure a maximum total discharge of 3,532 cubic yards of Seine River water. If we exclude from the available water supply that tapped by the palace, a maximum of slightly more than three quarts daily per capita was all that could be supplied to the 500,000 inhabitants of Paris.

The pumps

The Samaritaine and Pont Notre-Dame pumps were similar in operation (Figure 222). The Samaritaine was operated by the current of the river, through the intermediary of a wheel 17 feet in diameter, with 8 blades 19 feet by 4¼ feet. At each end of its axle there was a crank 22¼ inches long that operated a connecting rod acting on balance beams 21 feet long. These beams operated two pumps 9 inches in diameter, with a displacement of 38 inches, which were submerged in the water. The water, compressed through a pipe 6 inches in diameter, rose to a tank 75½ feet above the pumps, and from there to the distributing pipes.

The power wheel could be raised by jacks to the proper level, depending on the variations in the level of the Seine, by varying the length of the connecting rods linking the cranks with the balance beams.

Supplying moderate-sized cities: Rennes

While it was difficult to supply water to large agglomerations like Paris, it was absolutely necessary that the king, who resided there, deal with the problem and carry out the projects required. In smaller cities the problem was equally difficult, while fewer means were available for its solution. It would be

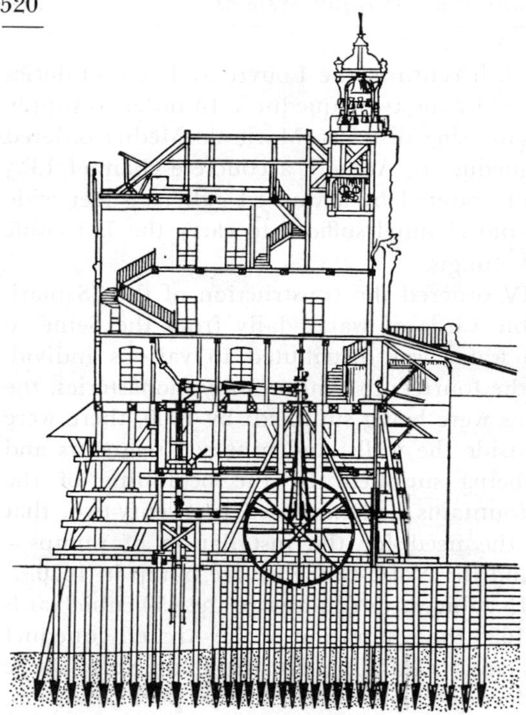

Fig. 222
The Samaritaine pump.

interesting to study the various operations carried out to supply average-sized cities with water. We shall choose only one example, that of Rennes, which demonstrates the difficulties peculiar to, and the inadequacies of, such undertakings.

We must keep in mind that Rennes was built on a rise bounded on the northwest by the course of the small stream called the Ille, on the south by swampy meadowlands traversed by the Vilaine. Being separated by these valleys from the nearby hills, the only water the city had was the rainwater, more or less polluted by the waste water of the city, which collected in crevices in the impermeable, schistose ground.

In 1443 it was suggested that drinking water be brought in from a certain distance. Beyond the village of Saint-Grégoire, slightly more than three miles to the north, were springs that issued from geologically different terrain: a basin of calcareous faluns, whose water could rise after it had been filtered. An abundant spring with good water, the Blanchart, was discovered, and it was decided to bring its water to Rennes through elm and alderwood pipes, at a cost of three sous per foot.

On October 9, 1444, the water from the Blanchart fountain was brought to the junction of another spring, the Vivier, beyond Saint-Grégoire. Then, following along the valley of the Ille, the water finally reached the Cartage fountain in the center of the city.

Even before the work was completed, those in charge realized that the supply from these two springs was insufficient and that they would be obliged to supplement it with water from other springs closer to the city (1505). These springs were led into tanks that were made as watertight as possible with a clay compound covered with cement; in 1523 the tanks were vaulted with a stone-covered brick vault.

In 1519 the first two springs reached the Place du Cartage through a piping system of terra-cotta rather than wooden, pipes. These springs had been supplying the fountain since 1510, but each day the poor condition of the wooden pipes decreased the amount of water by a certain amount, and on February 15, 1595, the flow of water stopped completely.

Other springs near the city walls — Tour-le-Bart, Porte-au-Foulon — were sought, while at the same time an attempt was made to utilize the original piping system by causing it to discharge at various places along its course. In 1604 the piping system was reconstructed with "good Dutch pipes" purchased in Rouen, that is, terra-cotta pipes brought by water to Rouen.

At the same time, the water of the Tour-le-Bart spring was carried to another square (Saint-Germain). Here the water from the fountain flowed into a basin, which quickly became a cloaca where the butchers washed their meat and the housewives came to wash clothes and household utensils. However, this water, which was obtained from the surface layer of the schistose terrain, was an unhealthy mixture of rain and waste water. In 1654 the municipal authorities decided that it was better to abolish the fountain and install another one farther north. In short, an unhealthy water supply from land close to the city on the one hand, and inadequate piping and springs on the other, were unable to supply a city that in the seventeenth century had 25,000 to 30,000 inhabitants.

Other springs (La Marre, Le Vivier, and so on) were sought in the Saint-Grégoire region, and the Cartage fountain was replaced by another fountain in a better location in order to facilitate the flow of the water. But other factors already mentioned tended to lessen the quantity of water it carried. One Sieur La Touche-Cornulier was the owner of various lands crossed by the Saint-Grégoire piping system; as compensation he requested, and obtained, a tap for the use of his household. The petitioner, being a *"Général des finances,"* was in a position to obtain a favorable decision, which was granted on condition that he would "not let the pipe flow constantly, and not exceed one twelfth of the supply from the pipe."

Lead pipes gradually replaced terra-cotta pipes, just as the latter had replaced the wooden pipes. A report of 1716 indicates that by then 506 feet of metal piping were in use.

The fire of 1720 caused such disruption of the streets under which the supply pipes ran that the water service was practically halted. Every landowner near the route of the pipe then felt justified in making use of the water, which no longer reached the city, and the old system fell into complete ruin. In 1727, Gabriel (Jacques III, 1667–1742), who was in charge of the general reconstruction of the city, drew up a new program for the reorganization of the piping system and the installation of fountains; he estimated that the water from the slope of Quincé could supply "nine *pouces fontainiers,"* and that by combining the other sources of water 15 *pouces fontainiers* (that is, 1,080 *muids* of about 72½ gallons in 24 hours) could be obtained.

The plan provided for pipes of 3 inches and 1½ inches. This would have yielded a supply of 15 × 25¼ cubic yards in 24 hours, for a total of 76,500 gallons, or approximately 2½ gallons for each inhabitant. Water for the low-lying districts on the left bank of the Vilaine was to be supplied by a spring from the slope of Guynes. In actual fact, however, in 1765 the pipes were supplying barely 26,475 gallons, or slightly more than 3 quarts per capita, a figure similar to the one given for Paris. The supply was even further reduced by numerous taps *à la* Lord Cornulier.

These deficiencies were met by digging wells at various points in the city; ten or eleven such wells are known, but the water they supplied was of dubious quality because, as we have already noted, it came mainly from the surface of the schistose ground rather than from deep springs. Reconstruction and attempts to replace the old pipes, hindered by financial arguments, and the inadequacy of funds, deprived the city of a well-organized, dependable water supply until the nineteenth century.

We should add that the still primitive science of chemistry and the absence of knowledge of microbiology permitted only a very inadequate check of the water. Some of the sources utilized, which would today be classified as dangerous, were then considered safe for drinking; waters to which dissolved carbonic acid gave a slight taste were defined as "very pleasant." Thus the spread of fearful epidemics in the cities is easily explained.

The water supply of Versailles

The modern system of water distribution through metal pipes under high pressure originated at Versailles. This was a difficult project: the castle was situated at a higher level than most of the neighboring areas, with the exception of the plateaus to the south and west, which lacked water. It was necessary to create artificial ponds (*étangs*) that would collect the surface water, or to search out water at a distance in streams or at higher levels, or to pump nearby sources with enormous engines. The series of efforts and the scope of the projects demonstrate the method by which Versailles was supplied and the need for the considerable quantities of water for the gardens, water fountains, and so on was met.

In order to establish the dates, let us first note that it was the festival given in 1661 by Fouquet at his castle of Vaux that aroused the king's jealousy, brought about Fouquet's downfall, and led Louis XIV to begin the great water-supply project of Versailles. Vaux had 1,200 fountains and cascades, and the creation of the gardens and the castle had required the services of 18,000 workers. For the gardens of Versailles, Louis XIV employed the men who had created Vaux: Mansard, Le Vau, and Le Nôtre. He had 1,400 fountains and cascades installed in Versailles, of which only 600 are still in existence.

Even in the time of Louis XIII, a pump had supplied the castle with water from the *étang* of Clagny. In 1663 Le Vau had a building constructed for a new pump. The pump, constructed by one Jolly, consisted of four lift-and-force pumps operated by two horse-driven treadmills (Figure 223), and it pumped the water to the tank on the upper floor of the tower. This method could supply 785 cubic yards daily.

Reservoirs were constructed in 1666 and 1667; the three reservoirs of Glaise

FIG. 223
Clagny pump works:
the horse-driven
treadmill.

(6,540 cubic yards) collected water from Clagny. In 1668 three windmills were added to lift the water stage by stage to the Glaise reservoirs. A reverse windmill was added, to bring the water from the basin of Apollon back to the Clagny *étang* through four cast-iron pipes 6 inches in diameter.

In 1672 the Clagny *étang,* which had become inadequate, was fed by the drainage of surface water from Le Chesnay and Glatigny. In 1674 a "great pump" operated by two horse-driven treadmills supplied a constant discharge of 3,819 cubic yards. By 1680 the pumps were supplying all the needs of Versailles. When, later, they were stopped, the *étang* continued to pour its water by gravity into the head of the Grand Canal.

Water from the Bièvre

Four windmills pumped water from the Le Val *étang* in successive stages to the plateau of Satory; cast-iron piping led it from there to a reservoir. It crossed the low-lying sheet of water known as the *"pièce d'eau des Suisses"* by means of a cast-iron siphon; a pipe (which was actually two 8-inch pipes) carried it to the castle. In 1668 the Launay mill was pumping the water from the Bièvre to Satory through a cast-iron pipe.

Water from the Loire (1674)

The famous engineer Riquet planned to bring water from the Loire by means of an aqueduct, but the Abbot Picard caused the project to be abandoned through his demonstration, by means of a level with an eyepiece (which he invented around this time), that the planned water tap was below the castle of Versailles.

Upper and lower étangs

With more precise methods of leveling, the planners were able to discover that the plateau to the southwest of Versailles (vicinity of Trappes) drained the streams coming from Satory, while the streams flowing from the Saclay plateau, situated 65½ feet higher than the courtyard at Versailles, could reach the castle (1675–1685).

The piping was accomplished either with pipes and siphons or over an aqueduct in the Buc Valley.

The machine at Marly On October 7, 1678, a twenty-five-year-old Liègeois named Arnold de Ville (born at Huy in 1653) suggested the idea of pumping water from the Seine to supply both Marly and Versailles. The idea was tried on a small scale at the mill of Palfour near the castle of Saint-Germain, and then the main project was begun. Seven years were required for the construction of the machine, at a cost of 3,859,583 livres (equivalent to approximately four billion francs in 1959), which included the cost of the Palfour experiment. The plan called for a discharge of 7,848 cubic yards; in actual fact it never surpassed 3,924 cubic yards. Most of this water was utilized at the castle of Marly; the surplus went to Versailles.

The machine of Marly, whose fame was due to its size, had other particularly interesting features.

The driving power was supplied by 14 large paddle wheels 39 feet in diameter and arranged in three rows, beginning from the upriver side (see Figure 224), with 7 wheels in the first row, 6 in the second, and 1 in the third.

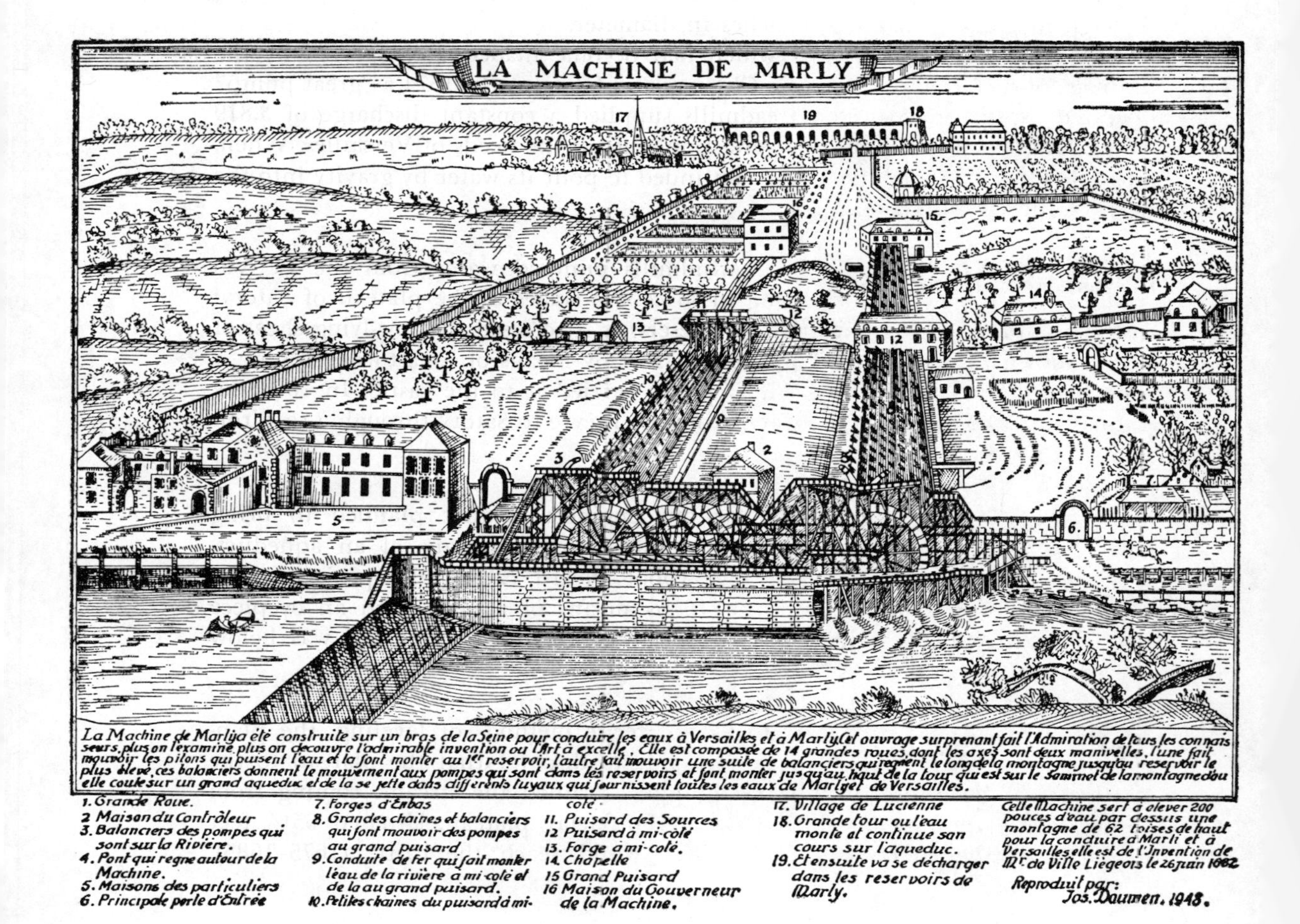

Fig. 224 The machine of Marly.

The water was pumped to the top of the plateau by three groups of pumps. The lift-and-force pumps at the level of the river pumped the water to a first level 150 feet high; here a second group of pumps took over and raised the water 175 feet; a third group then raised it 177 feet, for a total height of 502 feet, over a distance of almost a mile.

At this period there was no possibility of direct pumping to such a height. The current methods of machining cylindrical pump chambers and pistons did not permit precision reaming and turning; not until the end of the eighteenth century did machine tools capable of precision machining come into existence. In the seventeenth century watertightness had to be ensured by means of pieces of leather, especially under strong pressures, which with direct pumping would have reached 33 pounds at Marly. In addition, the irregular movements of the water in the pipes, produced by the functioning of alternating pumps, would have caused knocking, and a long length of pipe would have dangerously increased this knocking.

Consequently the pumping operation had to be broken up and pumps installed at three different levels. This posed the problem of the long-distance transmission of the power produced by the hydraulic wheels. A shaft system has been described earlier (Errard of Bar-le-Duc, 1584). At Marly, this transmission was accomplished by means of suspended chains acting on the balance beams attached to the pumps (Figure 225). These were arranged as follows:

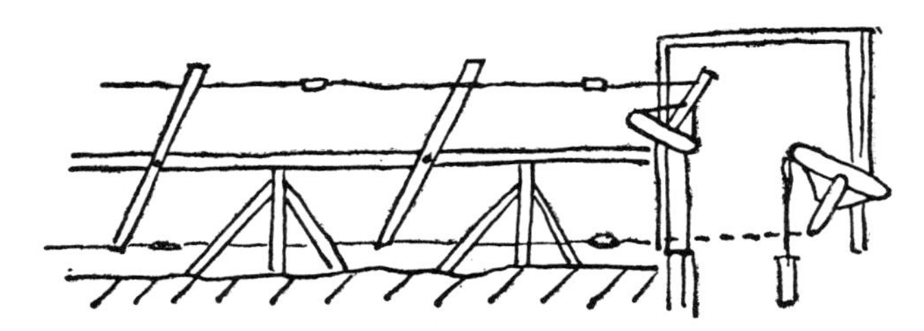

FIG. 225 Remote-control regulator of the pumps of the machine of Marly.

At the level of the Seine there were 64 lift-and force pumps; in addition, 8 supply pumps were planned that would feed a basin raised to the level of the pumps, which could thus be reprimed.

At the next level were 79 lift-and-force pumps; these were supplemented with 14 pumps that took water from neighboring springs.

At the third level were 78 lift-and-force pumps, and 16 pumps used for raising spring water.

This gave:	64 + 79 + 78 =	221 lift-and-force pumps
	16 + 14 =	30 pumps for the springs
		8 feed pumps
for a total of:		259 pumps

Arnold de Ville hired as his assistant a carpenter from Liège, named Rennequin Sualem, who had built a machine at Modave, around 1668, for Marshal de Marchin; the latter may have recommended him to Louis XIV. Sualem, with the assistance of his brother Paul, built the entire structure.

The chain method of control had already been utilized around 1600 by a Swede named Pohlen; however, it seems to have been only a variant of the one described by Errard of Bar-le-Duc, functioning with rigid rods. Thus the machine of Marly was not a new invention, strictly speaking, but a project of unusual scope for pumping water to a height unusual for that period.

While the lumber needed was obtained in France, the builders appear to have gone to Liège to obtain the metal joining parts and also the cast-iron pipes. On the whole, the project was well planned, taking into consideration the techniques of the period. The machine was built on piles sunk in the bed of the Seine and protected by wooden bumpers.

The cost of installation already mentioned can be completed by information on the salaries paid to Arnold de Ville and the Sualem brothers. Between 1680 and 1695 Arnold de Ville received 238,600 livres, while the Sualem brothers received 41,095 livres each, for a total of 82,190 livres. This represents a total of 320,790 livres paid for the supervision of the work, or approximately 9 percent of the total cost of the project. A permanent team of 60 workers, directed by a supervisor, was needed for the supervision and maintenance of the machine.

Theoretically the machine of Marly was supposed to supply 7,848 cubic yards of water per day, but it seems never to have surpassed 3,924 cubic yards. The lack of precision of the metal parts caused an infernal noise in the operation of the wheels and the transmissions of the pumps, which could be heard at a great distance. It does not appear to have aroused vigorous protest on the part of the inhabitants of the area; perhaps the very sound of the machine, by proclaiming its exceptional importance, made it an object of even greater admiration.

In comparison with the Samaritaine and Pont Notre-Dame machines, the machine of Marly was no different in principle. However, the engine was much larger in size. While the Samaritaine pump furnished an actual power of 8 to 10 horsepower (steam), that of Marly furnished 150. Marly was also characterized by an improved arrangement of the paddles on the wheels. There were only 8 paddles on the Samaritaine wheel, distributed around the circumference at a distance of $\frac{3.14 \times 17}{8} = 6\frac{2}{3}$ feet, whereas at Marly there were 48, distributed around a circumference of 39 feet, that is, at intervals of $\frac{3.14 \times 39}{48} = 2\frac{1}{2}$ feet. Thus the movement must have been more regular.

As for the mechanical output of the installation, the power supplied by the 14 wheels has been evaluated at 700 horsepower, for an effective power of 15 horsepower, which gave an output of only 21.4 percent.

The water supply from the Eure

Despite all the money and effort expended on the hydraulic system of Versailles, in 1685 it was discovered that the water brought to the castle would be inadequate. Colbert had limited the moneys spent at Versailles, insofar as it lay in his power to do so. But his disappearance in 1683 gave Louvois the freedom to envisage a still larger project; the diversion of the Eure. In 1684 he appointed La Hire to make a study of this project, with a tap at Pontgouin and a leap across the valley at Maintenon.

Vauban, aided by Mesgrigny, suggested supplying the water through ditches and using a siphon to cross the valley. Louvois countered with the idea proposed by various members of the Académie des Sciences, which called for a large viaduct. Louvois argued against the siphon on the grounds of the cost of the cast-iron piping; using 9 miles of pipe 1 foot in diameter, the cost would have been 525,000 livres.

The great viaduct was therefore chosen; it was to carry an aqueduct 7 feet wide with 3 feet of water. Philippe de La Hire did the leveling in 1684 and 1685; Vauban drew up the plans and prepared the projects with detailed specifications that later served as models for those of the Bridge and Road Authority.

The work was begun in 1685, and was far advanced by 1688, when it was halted by the war (League of Augsburg). The approximately 30,000 men working on the construction site under the command of the Marshal d'Uxelles, decimated by malaria, were sent to the army.

By the time of the Peace of Ryswick, the bad financial condition of France and the disappearance of Louvois halted the project. It would have included 50 miles of piping, the embankments for which were between 49 and 66 feet high in some sections, and two valley crossings. The Larris Valley crossing (3,280 feet) was to be accomplished by means of an aqueduct bridge; in 1688 this idea was replaced by that of a cast-iron siphon. The crossing of the Eure Valley at Maintenon (16,400 feet) was to be done with a three-story aqueduct, whose maximum height was to be 239½ feet; only the first story was constructed, and in 1688 the two upper stories were replaced by a siphon supported on the first story. The entire project was to supply Versailles with more than 130,800 cubic yards of water at a cost of between 10 and 11 million livres (approximately 10 billion francs in 1959). But if we compare the cost of the machine of Marly and the amount of water that could be supplied by each installation, excluding the expenses for builders and maintenance workers required at Marly, the Eure aqueduct must be considered vastly more economical.

Utilization of water

If we were to take into account only the population of Versailles (city and castle), the need for such a large water-supply system could not be explained. We have given a figure of 30,000 inhabitants. The court included approximately 20,000 people, but of this number 9,000 soldiers were lodged in the city, 5,000 servants in the outbuildings, and 1,000 *"gentilshommes"* of minor importance, who were attempting to work their way into the court circle, in the city. The castle itself was home to 1,000 seigneurs and 4,000 servants, for a total of only 5,000 people. Thus a total population of 35,000 people had to be supplied with water. At three quarts, the per capita rate of the cities of that period, which we have indicated for Paris and moderate-sized cities, or even at the rate of four gallons that was sometimes envisaged, 1,654 cubic yards of water per day would have been needed — a very low figure in relation to the amount actually supplied.

The reason for the discrepancy is that the principal goal was to supply the fountains and basins in the gardens with "the living, bubbling water, the pleasant murmur of fountains that accompanies and enchants the promenader in the gardens of Italy" (A. Peraté, *Versailles*). The network of piping that fed the gardens was an imitation of the Italian system. Le Brun, Le Vau, and Le Nôtre, who constructed the palace and organized the gardens, were assisted by François and Pierre Francine or Francini, of Italian origin and natives of that very Florence whose Medici gardens were vivified by water.

Versailles was a rival of Chantilly, where a closer and more abundant supply of water offered greater possibilities. A miracle of determination, talent, and expense was required at Versailles to achieve this creation in a plain of stagnant water and swamps. This is what Saint-Simon mercilessly criticized:

"The marvels of the fountain are dried up . . . despite provision for those seas of reservoirs that cost so many millions to build and pipe over the shifting sand and the mud. . . . This failing became the ruination of the infantry. He [Louvois] conceived the idea of diverting the Eure River between Chartres and Maintenon and of making it come to Versailles. Who can count the cost in gold and men over several years . . . ? The war finally interrupted the work in 1688, and it has not been resumed since then; there remain only shapeless monuments, which will immortalize this cruel folly."

Leveling instruments

Between the Roman period and the seventeenth century, leveling devices were neither modified nor perfected: they were rulers held in a horizontal position by a weight or a plumb line, with sighting done through small holes.

In 1675 Abbot Jean Picard invented the level with eyepiece, which permitted much more precise measurement of level (Figure 226). The windows were equipped with cocoon-silk threads crossed at right angles. A plumb bob passing through a tube made it possible to regulate the horizontal position of the eyepiece.

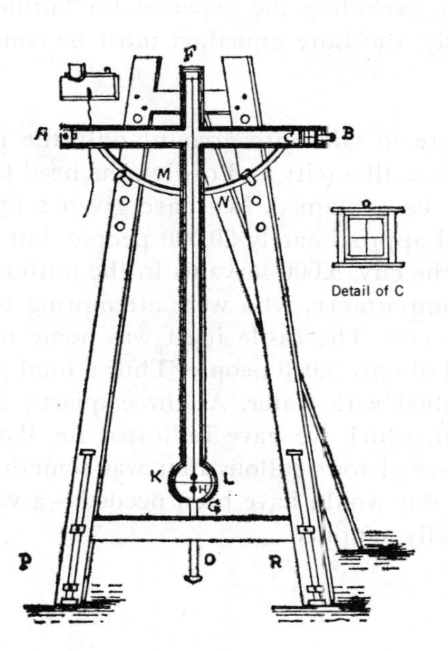

FIG. 226
Picard's level
with eyepiece.

This was the device that proved that the water of the Loire could not reach Versailles through gravity and that in contrast it was possible to supply the castle with water from Bièvre.

Picard's leveling operation on the *étang* of Trappes made it possible to control the slope, which was only eight inches per mile.

Pipes

At Vaux, and in the early work at Versailles, lead pipes were utilized. After 1668 only cast iron was used for the large pipes, except for elbows and small, narrow branches, which continued to be made of lead.

In the seventeenth century lead pipes were still being made by rolling sheets of lead around a core and brazing the lips with tin. They were 4 and 8 inches and 1 foot in diameter, and were joined by means of rings of soldering alloy.

The first large-scale use of cast-iron pipes occurred at Versailles; more than 24 miles were laid. The first sections laid carried the water from the Bièvre. The pipes were utilized as they came from the foundry, with crude thick flanges and bolt holes; no mechanical methods of machining cast iron existed at that period.

The joints were sealed with a diamond-shaped lead ring; but because of imperfections in casting, the action of tightening the bolts did not compress the ring perfectly. The watertightness of these joints was mediocre at best, and failed under strong pressures. This is what made the construction of siphons difficult; as we have seen, siphons were replaced by aqueducts, or their difference of height was lessened by installing them on an aqueduct half the normal height, as at Maintenon.

These cast-iron pipes used in the first installations are 6, 13, and 19 inches in diameter. When they were dug up in 1939, they proved to have undergone little change since their emplacement in 1685. Their general dimensions were:

INTERNAL DIAMETER	THICKNESS	LENGTH	FLANGES	NUMBER AND DIAMETER OF THE BOLTS
6 inches	2/3 in.	3 1/2 feet	1 in.	Four 1 in.
13 inches	4/5 in.	"	1 2/5 in.	Six 1 2/5 in.
19 inches	1 in.	"	1 4/5 in.	Eight 1 4/5 in.

They were cast in mold boxes laid in horizontal position.

The first pipes had been made in France, but those used in the machine of Marly came from Huy, probably from one of the foundries belonging to Wynaud de Ville, the father of Arnold de Ville. Their diameters ranged from 6 to 6 3/4 inches; the thickness of the metal varied from 8 lignes (18 millimeters) to 10 lignes (22.5 millimeters), and their uniform length was 6 feet. Apparently they were acknowledged to be stronger than the pipes made by the French foundries, and it seems that stronger pipes were needed to resist the high pressures and knocking of the Marly pumps.

Faucets

There was no check on the quantities of water utilized, since all the water brought to Versailles was intended for the use of the king. On the other hand, a shortage of water was, as we have seen, a constant fear. This led to economy measures and to an attempt to prevent the water from flowing constantly, as was the case with small individual distributing systems, where the flow was limited only by the small diameter of the pipe.

The result was the widespread use, at Versailles, of faucets on supply pipes. Specimens that have been found show that these faucets were of the type well known since the days of the Romans: a cone with a hole drilled through it and a handle on top, which was inserted in the supply pipe. When the cone was turned so that the hole faced in the direction of the flow of water, the water was able to flow out of the supply pipe. A lead mass soldered around the supply pipe supported the faucet in the pipe.

These faucets were carefully made, the turning of the cone of the faucet pipe and the reaming of the body being done with precision. The nature of the alloy that went into them was as follows:

	Body	*Cone*
Tin	4.80%	3.98%
Lead	22.10%	16.24%
Zinc	2.05%	14.73%
Copper	71.05%	66.05%

The complexity of this mixture was probably due to the reuse of material whose composition was irregular, and to the desire to obtain a sufficiently fusible metal (thanks to the presence of tin). The use of lead is characteristic, for it is on lead that we now depend for the quality of parts subject to slight rubbing.

Despite the impossibility of determining by analysis the composition of the alloys and of regulating the percentages of the constitutent elements, the experience of the eighteenth-century technicians led them to rational solutions that are still being used today.

BIBLIOGRAPHY

Bélidor, Bernard, *Architecture hydraulique* (Paris, 1782).

Keller, A. G., "Renaissance Waterworks and Hydromechanics," *Endeavour,* 25 (Sept. 1966), 141–45.

Kirby, R. S., and Laurson, P. G., *Early Years of Modern Civil Engineering* (New Haven, 1932).

Reddaway, T. F., *The Rebuilding of London* (London, 1940).

Robins, F. W., *The Story of Water Supply* (London, 1946).

Rouse, Hunter, and Ince, Simon, *History of Hydraulics* (New York, 1957).

Shapiro, Sheldon, "The Origin of the Suction Pump," *Technology and Culture,* 5 (1964), 566–574.

CHAPTER 22

CONSTRUCTION TECHNIQUES

THE CONSTRUCTION OF BUILDINGS

THE STUDY of technology can be broadened to include construction in a very wide range of building types and sizes. The building of rural dwellings plays a major economic role by virtue of their number. The construction of a city dwelling can be very different from that of its rural counterparts, while the construction of large edifices — castles, palaces, cathedrals — is essentially different in nature from both of these.

In all these buildings, however, we find similar concerns on the part of the builders: a search for the most economical materials, a desire to employ the fewest possible workers, the need to adapt the work to a clearly defined goal that conditions its dimensions, and the obligation to protect it from the ravages of weathering and to make it permanent.

The rural house

At the bottom of the scale, the rural house is obviously the simplest type of construction, and the type in which the search for economy is pushed to the maximum. It thus consists only of elements found *in situ:* clay for the walls, a wooden framework when wood is abundant in the vicinity, stones found nearby, for example in the fields, and thatch, which is supplied by the cultivated fields after the harvest has been brought in.

Thus the labor involved in making the pisé for the walls, whether of dried bricks or of clay compressed between planking, is very inexpensive. The clay is kneaded on the spot; it is mixed with straw, which creates a kind of reinforcement and prevents cracking and crumbling. This method, which is still used in rural areas, dates from remote antiquity; the Bible depicts the enslaved Israelites making bricks mixed with straw, one of the most serious subjects of dissension with their Egyptian masters being the question of whether the working of the straw was included in the agreement for brickmaking.

Few changes have appeared over the centuries in the methods of constructing rural dwellings; the predominant consideration is the necessity of finding materials that are easy to work and that are available in the immediate vicinity of the construction.

The urban dwelling

The problem is similar, but already slightly different, in the cities; the materials are sought close by in order to reduce the cost of transportation, but the limited space within the city walls makes more advanced methods of construction necessary, in particular the construction of houses with several stories, which require floors and a supporting framework.

In the older European cities numerous wooden houses can still be seen. Generally they do not antedate the end of the fourteenth century, and the most recent date from the seventeenth. This type of construction became more rare during the seventeenth century, since the economic growth made it possible to pay masons to build stone walls; moreover, lumber had become expensive in the fifteenth century, and the disappearance of the advantage of low cost that had formerly characterized this material no longer permitted it to compete with stone construction.

The use of wood

Of course, houses in which wood was widely utilized existed long before the fourteenth century. The chronicle of Lambert, parish priest of Ardres, makes mention of the palace of Arnoul II, Count of Ardres (1099), which by its magnificence contrasted strongly with the wooden houses existing throughout Flanders. It had been built by a master carpenter, Louis de Bourbourg, who was responsible for its ingenious and complex plan.

Mud walls, widely used in the north (they have now been replaced by brick), were formed of laths nailed to the wooden framework and covered with clay. The workers who executed this type of facing were known to the Middle Ages as *plakeurs.* Again for the sake of economy, the laths were made by splitting staves of old barrels with a hatchet.

The forests were extensively utilized for construction purposes. In the fourteenth century and at the beginning of the fifteenth century, the framework of a building consisted of large beams resting on a masonry base (*solin*) and extending the entire length of the two stories. The corner and other posts, vertical planks, horizontal stringers, small vertical planks, and the inclined struts that braced the framework, were all jointed to each other.

Later, when the supply of large trees had been exhausted, the carpenters were obliged to use smaller planks, and this led them to superimpose independent stories. This technique made it possible to corbel the upper floors and thus gain more space.

Corbeling

The local authorities had to regulate the exaggerated corbels that, especially in narrow streets, decreased the space between the houses. In Rouen at the end of the fifteenth century, in 1516 at Amiens, measures were proclaimed to limit these "great and excessive" projections that were "very dangerous to the public welfare." The magistracy ordered that the overhang of the second-story corbeling should be no greater than one foot, that of the second a half foot, and that the total overhang for a three-story house could be no more than one and one-half feet, a half-foot projection for each story.

Many of these old houses still exist. Like certain stone buildings, their walls reveal a noticeable bulge, visible at the level of the first floor. Various explanations of this bulging have been advanced. Some authorities, for example A.

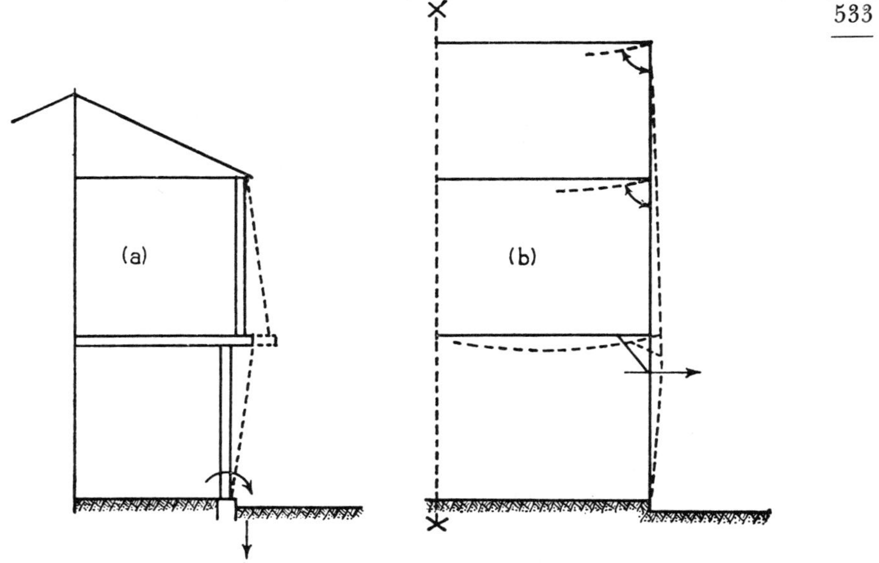

FIG. 227 Bulging of façades
(from Choisy (*a*) and Viollet-le-Duc (*b*)).

Choisy, attribute it to a settling of the ground under the street in front of the house, created by the ground-compressing effect of water and traffic (Figure 227*a*) Viollet-le-Duc gives a more plausible explanation *(Histoire d'une maison*). The beams of the flooring warp under the heavy burden of their flagstones (Figure 227*b*), and a rotating movement of their wall supports exerts a horizontal thrust, pushing on the outer wall. In the upper stories the wall, now inclined, forms an angle with the flooring of each story that corresponds exactly to the rotation of the support. The façade thus rests flat, inclined as a unit, on the bulge of the ground floor.

Generalization of the use of masonry

In some cases, and especially in Burgundy, the houses have stone gables that support the ends of the horizontal planks in the wooden framework. Thus these houses, having two masonry walls and two walls of wooden planking, do not warp like houses that have three or four wooden walls. Masonry dividing walls between houses, which acted as fire stops, were made obligatory in certain areas of reconstruction (London; Rennes after 1720).

In the twelfth century stone façades began to be sumptuously decorated; the cutting of the stones and the columns with sculptured capitals seen in certain façades at Cluny testify to a highly perfected workmanship. In the large agglomerations whose economic life was vigorous, the middle class had sufficient financial means at its disposal to bear the expense of a luxurious and costly façade.

Elsewhere, in small lanes and in less frequented streets, the old and inexpensive techniques and materials continued to be used. The house constructed with materials within range of the poorest was modestly priced.

The same situation occurred when a catastrophic fire razed a large area of a major city. Such was the case in London and Rennes (1720), both of which examples are instructive from many points of view. After a few experiments, the authorities finally adopted a plan of reconstruction, enlarging and straightening the streets and imposing uniform façades. The bases of the latter were built of hard stones (for example granite), with arcades; their upper stories were of a softer stone that was easy to cut.

The reconstruction of Rennes

Statistics concerning the areas destroyed reveal the nature of the houses constructed in the days when the city was enclosed within the ancient walls (the first wall, whose trace dated from the Gallo-Roman period):

Area ravaged by the fire	approx. 7½ hectares (18½ acres)
Number of houses destroyed	850
Families without shelter	2,400

This area is obviously small by comparison with the areas devastated by the Great Fire of London and especially by the great fires of the twentieth century, for example that of Smyrna (1922), which ravaged hundreds of acres. But the case of Rennes is more typical as regards destruction of medieval houses.

If we count on an average of four persons per family, we arrive at a figure of 10,000 persons for the number of inhabitants in the burned area. The houses were tall wooden structures with at least four stories. The number of inhabitants per acre averaged out at 540, a high figure, since in the most densely populated districts of Paris we find, in 1800, the following figures: Hôtel de Ville, 281, Temple, 346. The Marais district of modern Paris (the area delimited by the quays, the boulevards Henri IV and Beaumarchais, and the rues du Temple, Filles-du-Calvaire, Turbigo, du Renard, and Beaubourg), which is considered to be overpopulated, now has 252 inhabitants per acre, living in houses most of which are more than one hundred years old. In the nineteenth century some of the old quarters were rebuilt, and the inhabitants moved to new areas (Passy, Auteuil, and Montmartre, for example).

As a matter of comparison, we may mention the figures accepted in the twentieth century for additions to major cities:

Low density	81 inhabitants per acre
Moderate density	91 inhabitants per acre
High density	103 inhabitants per acre

In the reconstruction of the devastated area of Rennes, the new houses had only two or three stories, with larger apartments and open areas (streets and courtyards). As for the construction techniques, the façades had to be of stone; however, the use of wood was permitted for the façades opening onto courtyards.

This transformation of the type of construction used in private houses brought about an economic evolution that should be discussed, since it is closely linked with the construction technique.

Wood was the original building material, as we have already mentioned; it was supplied by the forests, which were at that time still close to the city. It was accompanied by materials obtained nearby: clay for filling in the studs and mud walls, and even stone. For the latter material the builders utilized the products of numerous quarries opened near the city; they supplied a very soft stone that was easy to cut but had little strength and split quickly under the effect of water. (This was pre-Cambrian schist rock from the phyllite stratum of Saint-Lô [Algonkian]). In the reconstruction of the city, greater use was made of a harder stone, a very strong red Cambrian schist with irregular grains, which was located at a certain distance from the city but whose quarries bordered on the banks of the Vilaine.

This experience demonstrated the usefulness and cheapness of transportation by water in comparison to land transport. The result was the appearance of a desire to assure ease of navigation on streams. On December 11, 1721, the Estates of Brittany, seeking to contribute to the reestablishment of the city of Rennes, appropriated the sum of 300,000 livres for this purpose; it was approved by decree of the Council on August 3, 1725. This sum included a subsidy to "make the Villaine River navigable all year round, from Rennes as far as Messac" – that is, on the stretch that included the *cahot* quarries the (name by which this red schist was known until modern times).

A demographic consequence should also be mentioned. Houses constructed cheaply, in many cases by the inhabitants themselves, were regarded as being of little value; an evaluation of losses caused by a cataclysm ravaging a city included the value of the furnishings and the terrain, while that of the buildings was considered negligible.

In reconstructed areas, however, the new houses built by paid state workers, masons and carpenters, acquired a value that later exceeded that of the furnishings. The larger space offered in the new dwellings contributed to this increase in value. Thus the former inhabitants who had been driven from their ruined and very ordinary houses could not return to the new houses, which were beyond their means. They fell back to the outlying districts, where low-priced houses were being built, often with the private means of their inhabitants.

This was typical of the reconstruction of Rennes, in which the former inhabitants of the destroyed areas were replaced by a new and wealthier class of rich bourgeois, government employees, parliamentary councilors, and other affluent people, who were able to pay for more expensive housing. In this particular case, ownership of a house by a single owner was replaced by ownership of individual apartments, which for centuries was almost unique in France, and which was a consequence of this economic phenomenon.

In short, a great variety of construction techniques were in existence during the entire period under discussion. They were either crude and inexpensive techniques that made use of local materials, or more complex techniques that utilized choice materials (dressed stones) whose price was higher, not only because of their quality but also because of the cost of transportation.

The "Mansard" roof

This period witnessed the appearance of a new type of roof. Roofs were still being made of wooden trusses covered with tiles, slates, or thatch, with a single steep slope. François

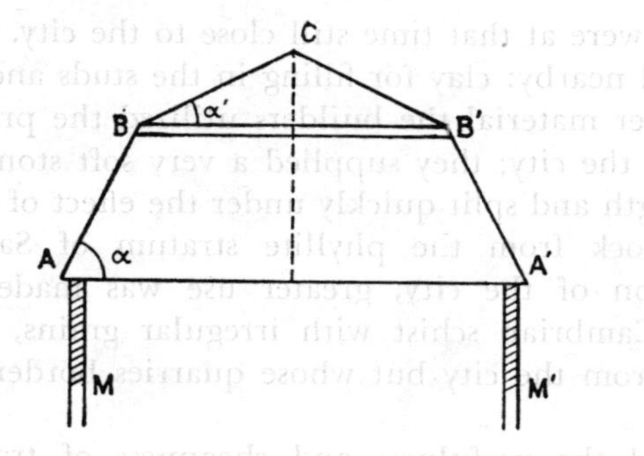

FIG. 228
Mansard roof.

Mansart is credited with the so-called "Mansard" double-pitched roof (Figure 228), which had a larger space between its slopes (*ABCB′A′*). The advantage of this style was that a smaller story with almost vertical walls could be constructed above the building walls *M, M′* that was almost identical with the lower stories, and the attic could be utilized as living space.

The Mansard roof, which has been very frequently used since then, lends itself very well to the plan generally imposed by the modern rules of urban planning, which fix a limit defined by a circle above a vertical line. (See Chapter 20, "Urban Planning and the Development of Cities."

It should perhaps be pointed out that this arrangement, which is advantageous because of its improved and more economical utilization of habitable upper stories, is an intuitive application of mechanics; it is the solution to a problem of statics. The form *ABCB′A′* is in a state of equilibrium under the stresses created by the weight of the roof surfaces *AB, BC,* and so on. Although it is an instable equilibrium, by the very fact that it exists the joints of the truss (tie beam *BB′*, joints of the angles *A, B, C*) can be greatly reduced.

If p and p' are the weights of the roof surfaces *AB, BC,* the equilibrium is achieved for

$$\text{tg}\alpha = \frac{p + 2p'}{p} \times \text{tg}\alpha'$$

and if $p = p'$:

$$\text{tg}\alpha = 3 \ \text{tg}\alpha'$$

At the time when Mansart introduced the use of this type of roof (1650), the work of neither Huygens nor, in particular, Maupertuis and Hamilton had yet been done. Thus it was, we believe, the idea of making better use of the attic space in houses that led him to the adoption of a system whose mathematical justification he did not foresee.

Venice

Between the Middle Ages and the eighteenth century, economic conditions rather than technical development led to the construction of buildings that are remarkable from both the artistic and architectural points of view. Venice is a significant example.

The primitive houses on the small islands in the lagoon that sheltered the peoples chased by the barbarians from the destroyed cities (Heraclea, Altinum) must have been made of branches and rushes, probably coated with daub. But by the eleventh century the new agglomerations (Torcello, Murano, the Rialto) were undergoing an economic development that made it possible to construct more solid houses. Stone appeared in the eleventh century; brought by boat from the coast, it had been removed without difficulty from the ruins of the ancient abandoned cities.

The construction of new cities in the lagoons necessitated consolidating the ground by means of banks of fascines, between which the current created by the tide flowed out through canals. The tide in the Adriatic reaches three feet in the north, and in the lagoons of Venice it creates strong currents that are very useful for cleaning out the canals. Under stone buildings of a certain height, which weighed more than the primitive huts, foundations were especially necessary; walls everywhere rested on piling, the piles being linked by a wooden grillwork. Beginning in the fourteenth century, pilings were installed not only under the walls but under the entire surface of the building, which rested on a vast platform.

By 1367 more than two hundred patrician palaces, representing a value of three million ducats, were already in existence. The technique of their construction contributed no innovations, aside from the richness of the work — as for example the famous Ca d'Oro, with engraved columns and arches and painted and gilded cornices.

The Venetian state bore the expense of lighting the narrow streets, and contributed to the vast and sumptuous constructions of public buildings: the Procuratie vecchie and the Procuratie nuove (1495–1517), the Library on the Piazzetta (1536) by Sansovino, the Fabbriche Vecchie e nuove, and the Ducal Palace, which was remodeled and rebuilt on several occasions. The great palaces belonging to the Venetian aristocracy were too numerous to be counted, and there were a great many churches.

In all these buildings the techniques used were similar, but the cost of the work, because of the high cost of the materials and expenses of an artistic nature, was in direct relation to the extensive financial means at the disposal of the Venetians between the twelfth and the seventeenth centuries.

The same observations can be applied to all the great, sumptuous dwellings constructed in other countries. The techniques used for the walls, roofs, and flooring presented no unusual features. We might mention that walls constructed with panels framed in cut stone but covered externally with brick appeared at the end of the sixteenth and beginning of the seventeenth century (for example Place des Vosges): this was the "Louis XIII style."

The possibility of obtaining various kinds of materials led to a variety of types of buildings. When the house was constructed with economical materials obtained locally, the result was a uniform model that persisted for centuries, as for example the Norman houses with open studwork and clay fillings. In contrast, the possibilities offered by wealth made it possible to develop a great variety of models, the only guiding rules being the desire to make a display of wealth or to make the dwelling proof against attack.

Palaces

While Venice displayed her sumptuous buildings, which, resplendent with sculptures, paintings, and gold, opened onto the canals, the political life of Florence created dwellings that could withstand assaults. From the stern Palazzo Vecchio of Arnolfo di Cambo (1298–1314), with its few small windows on the ground floor and its walls constructed from top to bottom with regular courses of hard stones, and the Podestà Palace (the Bargello, begun in 1255), down to the palaces of the fifteenth century, the construction technique used was that of the fortress. The great Ghibelline families had to be sheltered from the Guelph riots and the violence of groups like the "Ciompi" and the "Piagnoni," whom Savonarola was inciting against the rich:

"Let them be declared rebels and their wealth confiscated . . . let them the [*gonfalonieri*] first sack the houses of the *signori;* a quarter of the booty will belong to the *gonfalonieri,* and the rest is for the companies. . . . This is what you must do, comrades!"

The ground floors of the fifteenth-century Strozzi (1483, Benedetto da Maiano) and Riccardi palaces (the Medici) were of large rusticated stones, and the Strozzi's very small windows were equipped with heavy iron grilles. Thus the palaces could be closed and barricaded. The austere stories of the Pitti Palace, a strange and severe building begun by Brunelleschi and built between 1440 and 1465, seem to rise from a heap of just-quarried, rusticated stones.

The volume of these enormous blocks, if it had corresponded to their exterior face, would have been extremely heavy, and transporting them from the quarry to the construction site would have required excessive expenditures. This appearance, however, is only a false front; the structure behind it consists of an enormous mass of bricks. Brick was abundant; the clay deposits used in its production were located close by. These deposits were created by accumulations of fine sand that permitted the precipitation of a very slow current of water. The necessary combination of conditions was found in the great alluvial plains of the lower reaches of rivers (the Po, the Arno, and even the Tiber), where the greatly reduced slope of the terrain causes the water to flow very slowly. The outer surface of the buildings was faced with very thin slabs of stone whose outer surface formed bossages.

In this country where it is cheap and widely used by the Italian masons, who are masters in its utilization, brick often formed the core of the structures on which stone and even marble facings gave the appearance of a homogeneous construction. The façades of the buildings themselves then acquired an artificial character, without any relation to the members of the internal structure. Typical examples are still in existence, such as the façade of San Lorenzo in Florence with its crude, bare, undecorated bricks. In other buildings, too, for example Santa Croce and Santa Maria Novella (both in Florence), the facing on the façade is only an artificial decoration, sometimes executed several centuries after the completion of the building itself.

Palaces and castles required no special construction techniques; only the cost of the construction differentiates them from the ordinary houses. Surviving examples include the palace of Jacques-Cœur at Bourges (1443–1453), and the castle of Vaux-le-Vicomte, built for Fouquet over a five-year period (1656–1661), which employed 18,000 workers and cost the equivalent of 10 billion francs (1959). Fouquet chose such eminent architects and artists as Le Vau, Le Brun, and Le Nôtre, and jealousy led Louis XIV to surpass Fouquet's masterpiece with the Palace of Versailles, for which the king employed the same artists.

Churches

Technical problems were more important and their solutions more remarkable in the construction of large edifices, particularly churches, whose conception, plans and elevations, and utilization of material determine the results.

Being large edifices built to hold large numbers of people, they had to be illuminated and roofed. It was the relationship between the openings in the walls, and the resistance of the walls to the weights and thrusts of the roofing, that led the architects and builders to devise rational, daring, or exaggerated solutions, an analysis of which reveals the evolution of the construction techniques.

The simplest solution for the large covered building (Figure 229*A*) is that of a nave consisting of two walls and covered with a wooden roof. If the nave is flanked by side aisles, their roofing, which is lower than that of the nave, leaves the upper portion of the walls free, and windows can be opened in this area. The lower windows of the side aisles and those in the nave walls permit the light to enter. As for the roof, it was often left visible; its large beams, tie beams, and kingposts could be painted or sculpted. Sometimes a ceiling that incorporated the tie beams of the trusses was decorated with paneling and paintings. This roof, which has been discussed earlier, is the type that is so frequently seen in both northern (Florence, Rome) and southern (Naples and Sicily) Italy.

If the walls of the nave are too high, they may lack stability; in this case a barrel-vaulted (or more often a groin-vaulted) side aisle helps to support the wall by locking its foot. Then there is no difficulty in opening windows in the wall of the side aisle (Figure 229*B*), especially when a groined vault frees the internal wall, leaving all the space needed for the windows. In the case of a nave and vaulted side aisles that do not leave the upper parts of the structure open, windows are cut only in the side-aisle wall (Figure 230).

This is the model that was used for centuries for the roofing of naves. Beautiful examples have survived, even in the East (for example, the church of Qalb-Luzeh, Figure 231).

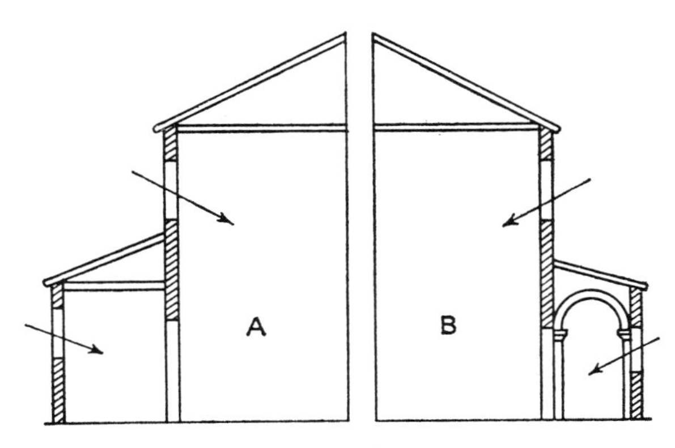

Fig. 229 Lighting of naves.

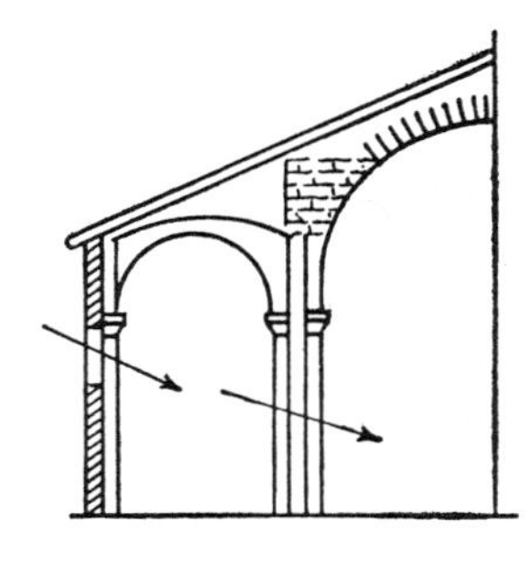

Fig. 230 Lighting of vaulted churches.

Fig. 231 Church of Qalb-Luzeh (from Vogüé).

But wooden roofing inevitably presented serious disadvantages, the first being the danger of fire. In the immense destruction caused by the ravages of the Norsemen in the ninth and tenth centuries, the fires that devoured wooden-roofed edifices are mentioned everywhere.

Other accidents – more rare, it is true – led to the replacement of wooden roofs by stone vaults. The wooden roof of the abbey church of Beaulieu near Loches, founded in 1004 by Foulques Nerra, was carried off by the wind on the eve of the ceremony of its consecration in 1012. It was quickly repaired, but this accident drove home the idea that stone vaults offered better security, and after 1050 a vault replaced the roof truss.

When the Normans had been tamed and stabilized by the Treaty of Saint-Clair-sur-Epte, other pillagers came to set fire to churches and abbeys. On several occasions the Hungarians ravaged the country as far as Burgundy (Tournus, 937). Upon beginning the reconstruction of these lamentable ruins, the builders must immediately have realized that masonry would offer greater resistance to fires and even to inclement weather. This led, not to the adoption, but rather to the generalization of already well-known systems of vaulting, many examples of which had been left by Roman and Eastern architecture.

The construction of the vaults

But new problems faced the builders in the construction of their vaults. In order to construct a vault it was necessary to accumulate the materials on centerings, which was a costly practice when the vaults were very high. It must have been realized, also, that the vault exerted a thrust on the walls that tended to overthrow them.

The builders experimented for a long time to find satisfactory solutions. It was only the repeated occurrence of fires that encouraged them gradually to abandon the wooden roof truss, which was easier to build, for the masonry vault. Builders working in the period after 1000 lacked accurate theoretical information on the stresses exercised by vaults on their supports and on what dimensions should be adopted for the latter.

Scientists of the period regarded arithmetic as a science best suited to solve theoretical problems, rather than as a body of knowledge applicable to concrete calculations; they studied the hidden properties of numbers rather than their combinations. Neither at the end of the Middle Ages nor for several centuries to come did mathematicians have the means of resolving the problems of static mechanics, thrusts, resolution of forces, or moments of overturning thrust. Knowledge of the Greek scientists of antiquity, transmitted by the Arabs of Spain and Sicily, did not appear until the twelfth century. Leonardo da Vinci learned of the traditions of the school of Jordanus of Nemore, which were beginning to enlighten Simon Stevin and Galileo, by way of Blaise of Parma. Not until the seventeenth century was precise scientific knowledge that could be used in this field established.

It is incorrect to attribute to the builders of the Romanesque and later the Gothic churches of the golden age of this architecture (thirteenth-fourteenth centuries) knowledge they did not possess. In the absence of notes and calculations, it has sometimes been rather hastily accepted that they *must* have possessed elements of calculation needed for the construction of such daring projects, that this knowledge, and even that of the strength of materials, was kept secret, and that even modern builders have not discovered their secret.

Superficial examination of Villard de Honnecourt's notebooks has also led certain historians to view them as proof of an already highly evolved technique. The geometrical graphs of these sketches, however, are only indications of proportions of human faces (Figure 232), or of regular polygons established on the principal points of the face (Figure 233, the star). In actual construction studies (Figure 234), the distance between the supports of a portico can be doubled by eliminating one column, done by keying a depressed arch beginning from a center O. The cutting of the stones is reduced (Figure 235) to simple proportions of the sides of triangles.

During this period (thirteenth century), according to Villard's manuscript, tracing the vaults was simply a matter of tracing arcs from given centers (Figure 236); the same radii were moreover utilized to trace a round vault with the center O_1, and pointed vaults with the centers OO_1 or O_1 O_3 or O_4, thus creating arches in tierce point, quarter point, and so on.

But in this notebook we find no trace of the mechanical knowledge that would have been so important in the creation of very stable vaults resting on

FIG. 232

FIG. 234

FIG. 233

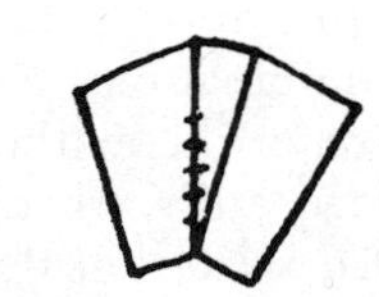

FIG. 235

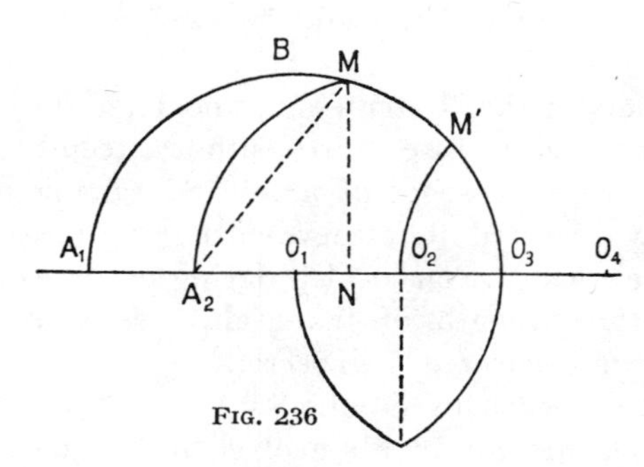

FIG. 236

FIG. 232 Proportions of faces (from Villard de Honnecourt).

FIG. 233 Face inscribed in a pentagon (Villard de Honnecourt).

FIG. 234 Elimination of a support. Keying of the arch (from Villard de Honnecourt).

FIG. 235 Diagram of voussoirs (Villard de Honnecourt).

FIG. 236 Graph of tierce-point and quarter-point vaults (Villard de Honnecourt).

walls or supported on logically proportioned abutments and flying buttresses. How then was it possible to build these structures, which became increasingly remarkable with the passage of time between the eleventh and thirteenth centuries?

Mystique of numbers and forms

Long before then, profiles and traces had been determined by the use of a very simplified geometry. Since antiquity men had believed in a certain mystique of forms, in equilateral triangles and hexagons, in the Egyptian triangle (sides 3, 4, 5) which has the advantage of supplying an easy and rapid method of raising a perpendicular (the ancient cord with regularly spaced knots). With this triangle (Figure 237, shaded portion) the lengths of the sides gave the graph of the Sassanid vaults (Firuz Abad): the opening $BB' = 2 \times 4$, the two arches of the vault traced from the top of the triangles by means of its sides: arcs BC and $B'C'$ from centers B and B', arc $C'DC$ from the center O.

The general layout of the vaults and arches of Santa Sophia (sixth century) is based, in the main lines of the walls, the cupola, and the quarter-circle vaults, on the repeated use of equilateral triangles.

This was a reflection of the Pythagorean mystique of geometrical forms, the persistent influence of Platonism, which reduced the essence of all elements to the regular solids, and which had survived through the ages.

But between the eleventh and thirteenth centuries, all that remained was a very elementary use of a few simple geometric figures. There exists a thirteenth-century geometrical drawing of the graph of the façade of a great cathedral (Figure 238), composed of very simple rectangles and triangles. These proportions were considered to be the only ones capable of giving a harmonious appearance, and for centuries builders continued to lay out works of art in a similar fashion (Figure 239, Saint-Denis gate, by the famous architect François Blondel, which we shall discuss in detail). During the Renaissance and Classical periods the same method was used for doors and windows (Figure 240) and decorative objects (Figure 241).

The mystique of numbers supplemented that of forms; the properties of numbers astonish us, and hint at a mysterious character. The Bible was the basis for a mystical inspiration that produced the Cabala. But in the cloisters this deviation of the use of numbers was viewed unfavorably, and arithmetic seemed suspect; Abélard declared it to be dangerous and baneful.

However, the numbers 3, 7, and 12, and, for large dimensions, relationships in the round figures 10 and 20 or 30, 100, were commonly used.

What we are seeing here are no more than very simple arithmetical procedures; nowhere in Villard's notebooks does the idea of problems of stability and equilibrium of weights appear. Not until much later, with Stevin (Simon de Bruges) (1548–1620), Gilles Personnier de Roberval (1602–1675), Isaac Newton (1643–1727), and Pierre Varignon (1644–1722), was a clear idea formed of forces and their composition. The construction of large edifices was the product solely of experimentation and trial-and-error methods, not of secret scientific knowledge.

Moreover, when we pass from theoretical knowledge to the crafts, which worked with matter, we note the existence of a feeling of contempt for those who did the manual labor. Here, as in the Platonic system, science (or what was then considered science) meant speculation about ideal forms, numbers, and the Aristotelian theories on movement, that is, an abstruse mechanics that was closer to metaphysics than to static mechanics. The scientists argued at length about movement and impetus before ideas on weights and forces were clarified.

The masonry of antiquity Yet daring and impressive buildings were constructed, and it will be useful to examine them in order to learn how and by what theoretical or practical means their construction was accomplished.

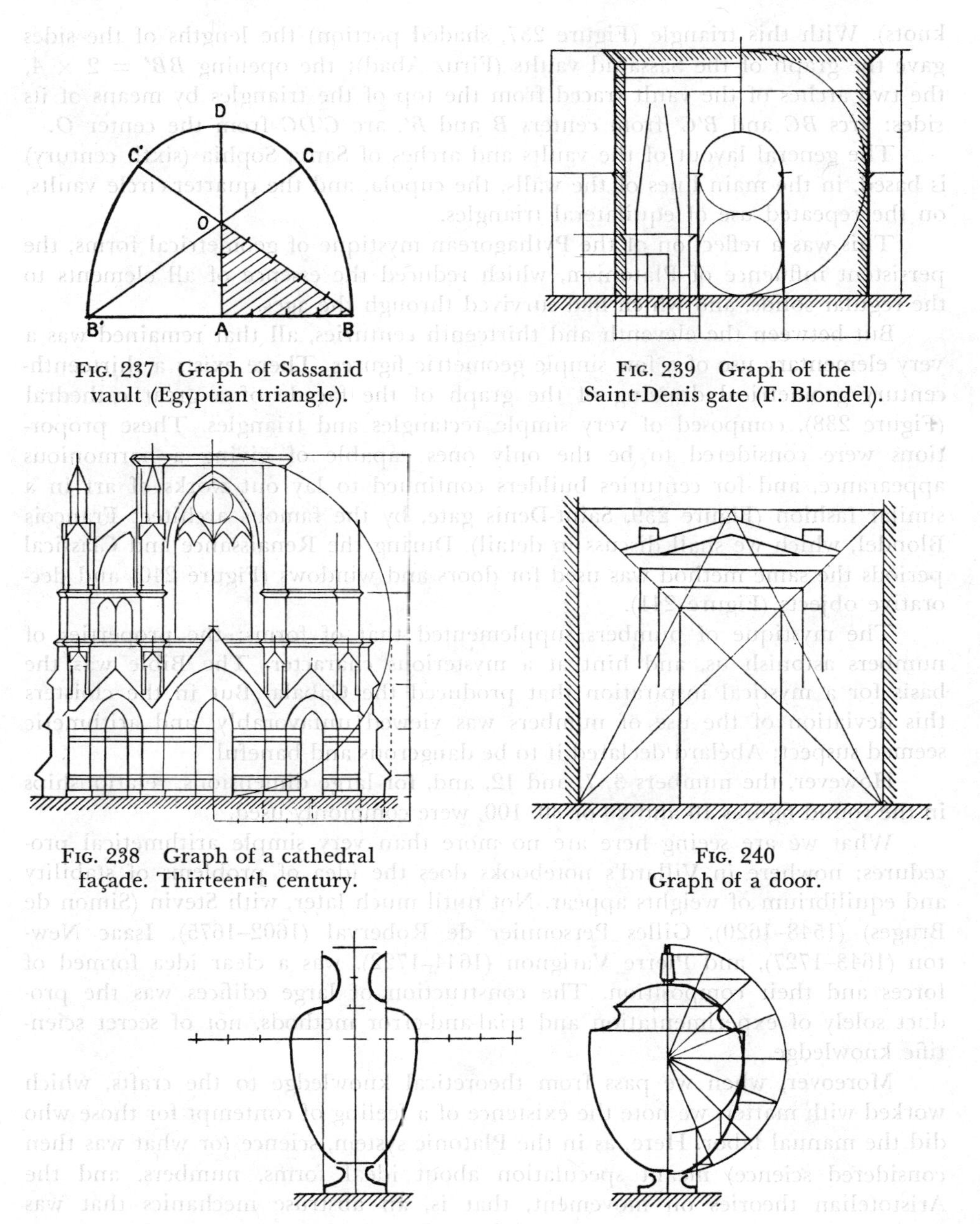

FIG. 237 Graph of Sassanid vault (Egyptian triangle).

FIG. 239 Graph of the Saint-Denis gate (F. Blondel).

FIG. 238 Graph of a cathedral façade. Thirteenth century.

FIG. 240 Graph of a door.

FIG. 241 Graphs of ornamental vases.

The dominant factor in the evolution of construction techniques was that of the masonry vault, including the methods available to the builders and the manner in which they employed them.

Surviving examples dating from Roman antiquity were admired and studied by architects in every century down to the Renaissance, whose greatest masterpieces were inspired by its creations. Agrippa's Pantheon is one of the most important works from this point of view, and various characteristics of this remarkable structure should be noted. The vault is built of masonry; a cement composed of limestone and pozzolana produces a concrete that forms a block strongly resistant to internal stresses.

The vault rises above, and is a continuation of, its vertical supporting walls. It is formed of courses of concrete, small stones, and mortar, laid in horizontal layers. Since it has neither radiating courses nor voussoirs reminiscent of the shape of the dressed stones found in the Middle East (Syria and other countries), the emplacement of these concrete courses could be done with unskilled labor. However, a precision-shaped, sufficiently rigid support was required. The genius of the Romans was that they reconciled the rigidity of the centering with great lightness and a minimum of expense. The Roman structure had a massive vault poured into a framework embedded in the mass that incorporated it.

The use of dried brick, which was light and strong, permitted the construction of a framework of arches forming a latticework, with a layer of flat bricks under it. Sometimes this continuous envelope of flat bricks stiffened by a kind of brick ribbing formed a coffering and centering; it formed as it were a kind of curved tiling. The concrete was poured on top of, and engulfed, the brick framework. When a layer of mortar had been poured and had hardened, it formed a ring that remained fixed without additional support and that enclosed the framework coffering incorporated into it; the solid concrete helped to fix the rest of the brick centering and to hold it firmly before it in turn was engulfed.

This curved "tiling" forming a vault is strong, and its use has continued in Italy and especially in Rome; the modern Italian builders are still constructing curved ceilings with the help of bricks laid flat. But the ancient Romans were satisfied to use them as supports for the construction of the concrete mass. Judging by the current practices of the Italian workers, we can claim that the Romans executed the work without the use of centering, proceeding in successive layers of triangular or fan-shaped cells (Figure 242). The small fans seem to carpet the vault, when they are not hidden by decoration; this is the case, for example, in the vault of Diocletian's mausoleum at Spalato (Split), which has been transformed into a cathedral. Each flat brick is supported on two adjoining bricks that are already sealed, and it is fixed in place by quick-drying plaster, which makes it possible to work without a support. When it is completely covered with masonry, the vault becomes a block that rests on the walls without exerting a horizontal thrust.

The vault of the Pantheon in Rome (first century B.C.), a gigantic half-circle, rests on the walls, which form a circular drum. The walls were made

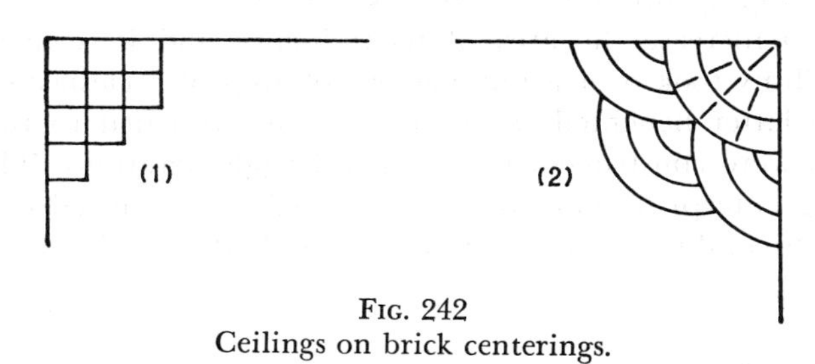

FIG. 242
Ceilings on brick centerings.

lighter by means of deep niches, and provided there was no cracking in the vault they were not obliged to resist any horizontal thrust.

Lateral supports thus became superfluous. However, in edifices with a complex plan the walls of adjoining chambers could serve to buttress the adjacent vaults or cupolas. This balanced grouping of buttressed vaults could be utilized in cases where cracks occurred in the curved vaults, or as additional strengthening for unsupported vaults.

The same desire to build without the use of scaffolding or temporary wooden centering, especially in countries where trees are rare, appears in the construction of large Eastern edifices. The construction of great vaults (for example the palace of Ctesiphon of Khosrau I, third century, Figure 243) is particularly remarkable. The light, thin vault rests on supports that are corbeled out, thus eliminating the need for supports. It is formed of superimposed rings; the lower ring was constructed without any centering, of light elements supported by the already completed adjoining element. Here, again, these light brick elements can be laid one after the other, since the joint formed by quick-drying plaster is sufficient to hold them in place. The first ring could be placed on the vertically constructed façade, which then supported it; this ring then acted as a centering for the laying of the second, and so on in sequence.

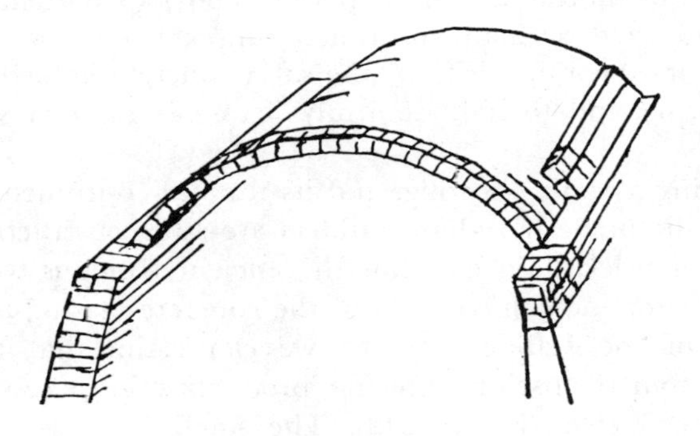

FIG. 243
Vault at Ctesiphon.

The influence of Cluny Completely different methods were available to the Burgundian builders, and particularly those of Cluny, whose extensive building program called for a number of small edifices to be constructed as economically as possible. This period (tenth century and first half of the eleventh) was that of the great abbots – Mayeul, Odilon – who achieved both the expansion of the famous abbey and the Christianization of a country that was still far removed from Christian practices. The monks formed a group that was capable of coordinated and powerful action, and whose power was developing prodigiously at this period.

In order to increase the area of its influence and its evangelizing action, the abbey of Cluny took over a number of holdings of ecclesiastical domains. In most cases the latter benefited from the revenues attached to them; numerous small churches were constructed in tiny rural agglomerations. The direction of such a far-flung enterprise inevitably necessitated an organized center, obviously at Cluny, which combined its own resources with the local resources, including

Plate 41. Romanesque house at Cluny (Seine-et-Loire). *Photo Archives photographiques d'Art et d'Histoire.*

the rural labor force, which was at its peak outside the periods of heavy agricultural work (plowing, harvesting, threshing).

Simplicity of construction was a necessity, as was the use of materials that could easily be found *in situ,* whence the opportunity to construct vaults of crude, undressed materials incorporated in and bonded with mortar, and dressed on the inside with a smooth facing to conceal the irregularities of the masonry.

However, this method of construction was not new; many examples of it appear in Sassanid architecture (Figure 244, Fars). The supporting wall was corbeled out, up to the level of *a;* this could be done easily and without centering, the masonry forming a block that needed no support.

The circular vault

If we construct a sufficiently pointed, very highly pitched arch, the courses can be corbeled out up to the top, thus eliminating scaffolding. This method had been in use since remote antiquity (for example the Treasury of Atreus at Mycenae), and reappeared in French Catalunya (Catalogne). But during the period under discussion, the vaulting technique generally began with the circular vault, which required a centering. The first courses could be corbeled out, but after that it was necessary to position the masonry mass of the vault, buttressed on its two supporting walls. This necessitated a centering to support it during construction.

This procedure continued to be employed in vaults of dressed stone (Figure 245). The wall *M* carries the corbeled courses *C* to *C′* of the vault, whose weight exerts no thrust except at the base. Above, the voussoirs *V* rest on the last horizontal course *C′*; the wall area *M′* buttresses the vault, with its horizontal courses resting on the last course of the corbeled area *MC′*.

There is a breaking point (*N–N′*, Figure 246) in the vault, from which the thrusts (*f*) are exerted on the corbeled supporting courses, and an overturning thrust (f_1) on the wall; the latter was to be buttressed by an abutment *B*. The latter thus appears as soon as we substitute the masonry vault for the wooden roof. The problem now arose of determining the size of the lateral supports, which was to bring about the evolution not only of the abutments but later of the flying buttresses.

Since the builders did not have the necessary scientific knowledge, they had to experiment and even outline explanations and semiempirical, semireasoned procedures. No trace of their work remains, either in manuscripts or in the engineers' notebooks. We can assume that the obligatory use of a centering may have led to the formulation of practical rules that guided the process of determining the size of the architectural members. The habit of utilizing the hexagon in traces may provide the key to this mystery.

Considering the round vault *ACMDB* (Figure 247), let us position two beams, *AC* and *DB,* whose length, being equal to the radius, can therefore be determined without difficulty; these two beams can be held at the top by a third, *CD,* of equal length. This centering *ACDB* is made rigid by wedging it against the lower portions of the vault, *AC* and *BD,* which have already been constructed with the corbeling method. This centering requires the knowledge of only one length, that of the semiopening of the vault (*AB,* that is, the radius of the semicircle *AMB*); in addition, it can be directly and easily laid, and

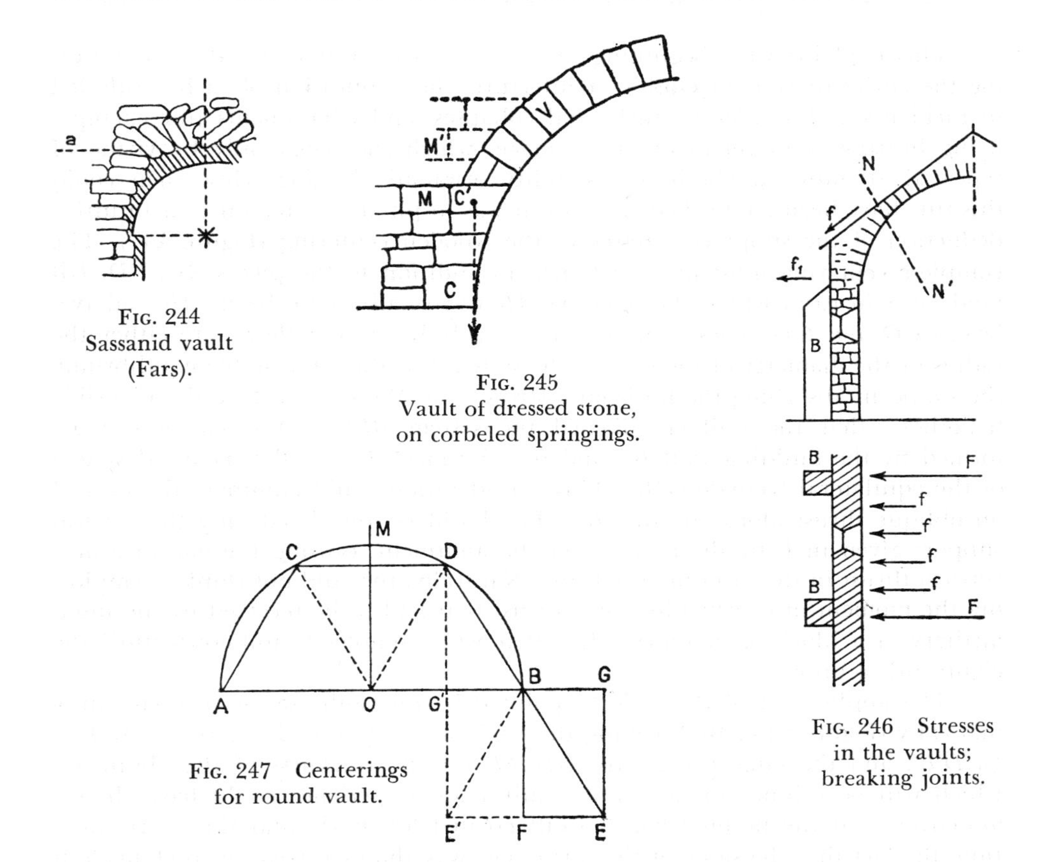

Fig. 244 Sassanid vault (Fars).

Fig. 245 Vault of dressed stone, on corbeled springings.

Fig. 246 Stresses in the vaults; breaking joints.

Fig. 247 Centerings for round vault.

immediately verified. On this centering the voussoirs completing the arch *AC,* and those of the vault *CMD,* can easily be positioned.

In addition, the thickness of the abutment must be calculated. Much later, in the seventeenth century, we discover the rule that confirms the suppositions just made and gives the rule that must date from a much earlier age and that establishes the determination of the sizes for the centers of the vaults. In his famous *Traité d'architecture* (1675), François Blondel (1617–1686, Lord of Les Croisettes, diplomat and architect) gives the practical rule for determining the thickness of the piers of the vaults: he divides the round vault into three equal parts, which is the same thing as constructing the hexagon *ACDB;* he extends the side *DB* with a line *BE,* equal in length, and takes *FE* as the thickness of the pier.

Thus the origin of this rule may be dated from the period when the round arch was adopted in vaults, and it is quite logical to see in it a recollection of the wooden centering supporting the construction.

This rule is obviously not exact, and in Book II of his *Science des ingénieurs* (1729) Bélidor correctly remarks that it ignores the *weight* of the vault and the *height* of the pier. But these two ideas, force and its direction, and the moment of overturning, were unknown to those who established it.

This explains why, beginning in the eleventh century and thereafter during the entire lifetime of Gothic architecture, the application of such a rule led to incidents and accidents, cracks and collapses, and experiments (for example flying buttresses) in repairing the damage, which succeeded only by virtue of repeated attempts to obtain the stability originally lacking. How then could this rule have been established? Perhaps it was inspired by a reasoned or intuitive deduction of the supposed stresses in the wooden centering (Figure 247). The complete centering must have included, in addition to the pieces *AC, CD, DB* (and even in cases where the opening *AB* was small), a tie beam *AB* and two beams *CO* and *OD,* whose lengths could easily be traced: they were either the radius or the diameter of the circle. The system was thus in a stable equilibrium, the tie beam absorbing the horizontal stress in *B*. We can advance the plausible hypothesis that the builders likened the system *BFE,* a right-angle triangle formed by the building wall *BF* and the abutment *FE,* to the balanced system of the equilateral triangle *ODB*. This consideration could eliminate the idea of an oblique thrust along the line *BE*. The builders considered only the vertical support given in *E* by the masonry of the abutment, causing the entrance of a vertical thrust in the sustaining polygon. Naturally, they did not think of employing the moment of overturning, or a thrust that had to be received by the intermediary of a block of masonry, ideas that were completely unknown until the eighteenth century.

The application of this rule to pointed barrel vaults seems to occur automatically (Figure 248). By breaking the circle at the point *M*, we retain $MD' = \frac{1}{2} D'B'$, and the same procedure gives *FE* for the thickness of the abutment, which will be thinner for a similar vault span. Experience could have shown, moreover, that the pointed barrel vault created less horizontal thrust. In addition, the fact that the slope of the springings was almost vertical in *B′D′* made it easier to corbel out the lower portion of the vault, and reduced the importance of the centering needed for the construction of the upper portion that formed the vault.

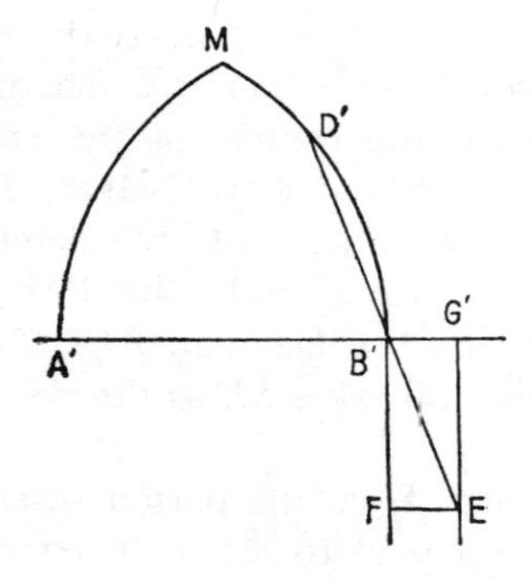

Fig. 248
Graph for abutment (Blondel).

The evolution of vaults, supports, and braces

It is easy to grasp the development of the technique of masonry vaults, their supports and braces, from this foundation, which resulted from the use of the round vault.

The round barrel vault (Figure 246) can be buttressed by the abutments *BB,* but outside these supports, between which there is a window, the vault can exert thrusts *ff*. In order for the thrusts to be combined in isolated thrusts *F, F,* corresponding to the abutments, the masonry of the walls and the vault must form as it were solid blocks, that is, the masonry must be able to distribute the stresses *f, f* to the distant points *F, F*. Modern builders do not attribute to stone masonry composed with lime or even cement the property of being able to perform this distributing function. Earlier builders could have been led to the same conclusion when cracks occurred in the masonry.

Two improvements were gradually made in the emplacement of vaults on their supports. Early examples have survived, for example in the church of Farges-les-Mâcon, where the simple vault rests on the wall and there is no separation between the two. The vaults of the side aisle buttress the springing of the vault, whose weight falls on arches and through the arches onto sturdy columns. This is the primitive model, crude, simple, and heavy. The only light in the nave comes through the windows of the side aisles; the apse, a half dome, is more brightly lighted by several windows cut in the circular wall, and contrasts with the darkness of the nave — a light that could seem symbolic to the worshipers in the shadowy nave.

Numerous edifices were undoubtedly built in this fashion during this early period; as examples, we can cite the first vaults of Blanot, those of Chapaize, those of the first great church of Cluny, where a barrel vault replaced the wooden roof (dedicated in 981) during the rule of Odilon (994–1049). While some of these churches have survived, others (Chapaize, Blanot) fell, apparently as a result of the thrust exercised on the walls, which they pushed apart.

The church builders of the period 1030–1050 were led, by intuition and reflection, to a major modification of the emplacement of the vault on the walls, to prevent the supports and buttresses from being excessively thick. If we consider (Figure 247) the vault (axis *OM*) placed straight on the wall, the rule discussed earlier leads to a wall and an abutment *B*. If we move the vault forward of the wall on isolated supports (Figure 249), the application of the same rule produces an abutment *C′*, with the possibility of making the wall between the isolated supports lighter. The pressure of the vault, however, must be organized on a support, an arcade in front of the wall, distributing the weight and the thrusts on the interior pillars that parallel the abutments. This distribution is complemented by large transverse ribs which divide the vault into compartments, and which are also in line with the supporting elements. The wall between these active supports can be reduced to the thickness of a simple filling.

This new architectonic arrangement had pilasters in front of the wall, arches from which the vault springs, and diagonal ribs in line with the pilasters (Figure 250). This is the system of distribution of stresses, whose logical sequence was the equilibrium of the so-called "Gothic" or "French" vaults.

The compartmentalization of the vault by the transverse ribs also made it easier to construct the segments of the vault between the ribs. Once these ribs, which require only light, narrow centering, have been constructed, it is possible to construct the triangular masonry cells *P, P* by the corbeling method; then,

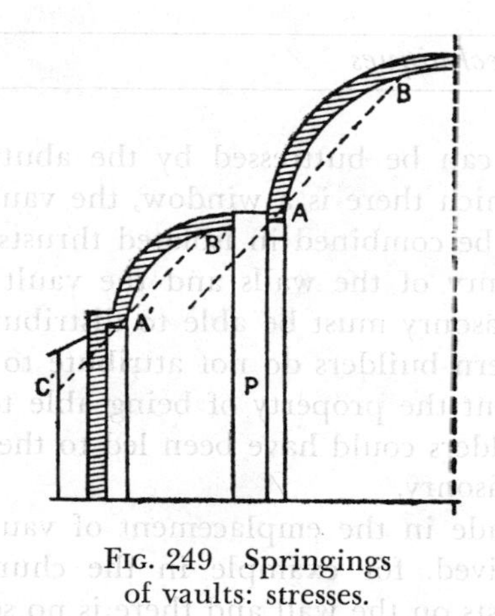

FIG. 249 Springings of vaults: stresses.

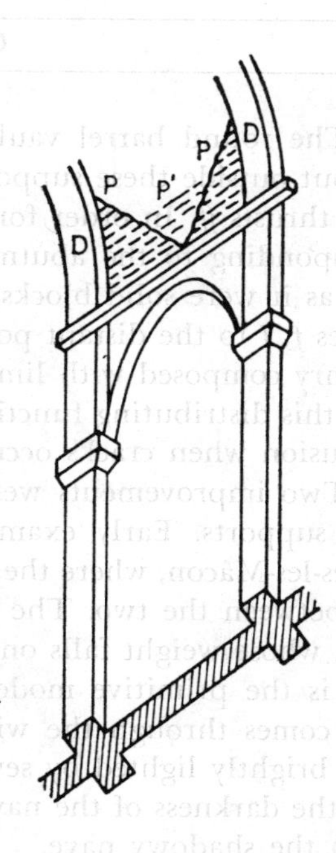

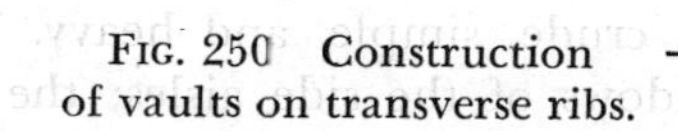

FIG. 250 Construction → of vaults on transverse ribs.

after their mortar has hardened, the cell *P′*. Proceeding by a series of triangles, it is possible to raise the springings of the vault higher, without scaffolding or centering, and to build the upper portion using only a light centering.

The vault springs from a mass firmly fixed to the wall, which retains the projecting portions of the corbeling (see Figures 244 and 245). Simultaneously, the pointed barrel, substituted for the round barrel, contributed the advantage of reducing the stresses and permitting the corbeled portion to be raised higher.

The Gothic vault This is the origin of the style generally known as "Gothic" (some authorities call it "French"), which must be dated from the very beginning of the twelfth century. We have earlier discussed the construction method used in the Romanesque vault and the heaviness of the corbeled masonry supports at the springing of the vault.

During this period knowledgeable builders were able to envisage easier methods of construction designed to make the vaults lighter, and thus at the same time reduce the difficulties caused by the thrust of excessively heavy vaults. This was the beginning of the vault with diagonal ribs (ogives).

According to A. Choisy, they may have appeared around 1100. The date of 1110 has been given for the bottom room of the bell tower of Louvres; the ogives of Morienval date perhaps from 1120–1130, followed closely by those of Saint-Pierre-de-Montmartre and Saint-Denis (before 1150). The ogives were placed, experimentally, over the small span of the compartments in the deambulatory vaults of Morienval (Figure 251).

But the idea of the ogive arch appears to have grown out of a modification of the groined vault. As we have already seen, at the beginning of the twelfth century no convenient geometric method existed for cutting stones that would form the strong ribs of a vault; moreover, even with an adequate geometrical graph it would have been necessary to determine the cutting of each stone indi-

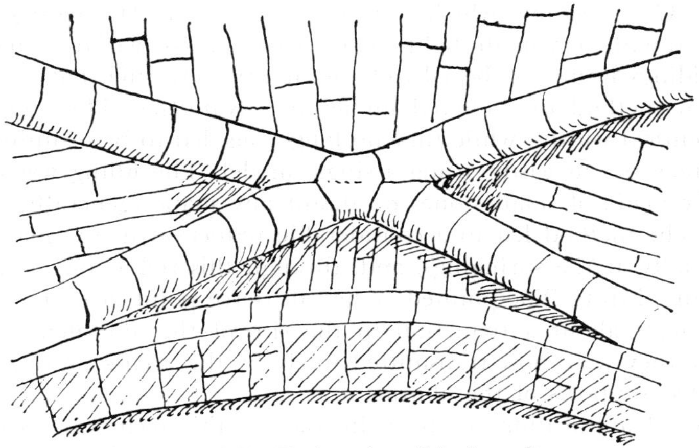

FIG. 251 Early ogives (Morienval).

vidually, depending on its position between the springing of the arch and the keystone.

The groined vaults constructed according to the Roman techniques with a filling of small stones and mortar did not even have to be raised on geometrically established centerings, which would have represented the intersections of cylinders (or of tori, in the case of the trapezoidal cells of the deambulatories).

In many cases it was possible to utilize the classic corbeled-out method for the springings, finishing with keystones closing the vaults with the help of a simplified support. The visible inner face of the masonry was then given a coating so that it would have a smoother appearance.

The idea has been suggested (and seems justified) that in certain cases the builders made use of a scaffolding holding sand or earth, on which the vault was modeled by hand. A first layer of mortar and filling could be applied on this model, which gave the desired form for the vault after the masonry had hardened. The sand could easily be removed to reveal the internal form. This would explain the existence of shapes of groined vaults composed of surfaces with multiple curves, for which it would have been practically impossible to make a wooden centering.

In vaults whose coating has not been renewed, ribs formed of very large stones, crudely embedded in order to avoid cracks in the mortar along the rib, are plainly visible. It is the replacement of these crude vault ribs by an arch of dressed stones that constitutes the great transformation of the Romanesque vaults into Gothic vaults.

It is clearly evident in Figure 251 that the ogive is formed by a series of identical stones that forms a thick ring. This ring later acquired a much lighter appearance: two tori back to back, then increasingly deeper profiles forming bundles of ribs.

This discovery profoundly transformed the technique of the vault. Instead of proceeding by corbeled-out springing courses, the diagonal ribs and the arches were first positioned on light centerings, since the arches carried only their own weight. The framework thus formed received the filling of the sections of the vault. These sections, which had to cover only small triangular cells, could be constructed with very small, light elements laid almost without support — very small scaffoldings hung on the ribs of the arches were sufficient.

The new method thus offered numerous advantages. The voussoirs could be cut to identical widths, while they no longer needed to be of uniform length; they could now be cut easily, using a single model. The filling was easy to lay, and could be made of small stones of almost any shape (generally small parallelepipeds). The method led to a considerable reduction in the masonry of the vaults, especially at the springings, and, through this reduction of weight, to a decrease in the thrusts. The pointed arches, in tierce point or even more sharply pointed, brought about a reduction in the size of the abutments. Lastly, the thrusts, since they were now transmitted solely by the arches, were exerted only on the isolated supports of the abutments.

The logical application of the principles of this technique permits us to follow the evolution of the Gothic style. In countries with a supply of stone that could easily be cut, the thickness of the vault could be decreased, and in some instances (for example vaults of Anjou) the vault could simultaneously be raised in the form of a dome, with arches that were simply very light ribs. This is what has led certain authors (Corroyer) to believe that they were derived from the cupolas of the southwest (Périgueux, Angoulême).

From the simple square plan with its two main arches, the builders soon progressed to a multiplicity of ribs in sexpartite vaults; this made it possible to make the filling cells still lighter, and facilitated their construction without centering. This system inspired the vaults of both Anjou and the Île-de-France, and reached its maximum development in England with the fan vaults of the English cathedrals.

Supports, abutments, and flying buttresses

Without going into the architectural details of the great edifices constructed before the fourteenth century, it will be useful to discuss the basic nature of the methods of buttressing, and the experiments made by these builders who had no accurate methods for making calculations.

The principal idea that dominated architecture beginning in the twelfth century was to open the walls of the nave to the light, and to increase the size of the windows as much as possible. The Cistercian opposition to this tendency, as well as the Cluniac solutions, are well known. But this opposition did not succeed in stopping the trend toward increasing height for the vaults, in the desire to open larger windows that would admit more light.

The Île-de-France chose this path, and this impetus was strengthened in other areas after the death of St. Bernard in 1153. It can be characterized by the relationship of the height under the vaults to the width of the principal nave, a relationship constantly increasing (at Fontenay it is slightly less than two, at Pontigny two).

The church of Pontigny, begun in 1150, had certainly not been vaulted by the time of St. Bernard's death in 1153; it was to be the first Cistercian church with ogive vaults, preceding that of Châalis by a very few years. In any case it shows that the Île-de-France's example of vaults with ogive arches was quickly imitated everywhere, an indication that this system presented undeniable advantages from the builder's point of view.

The relationship of height to width gradually increased. At Laon (1155–1160) it was 2.22; at Notre-Dame de Paris (1163) 2.90; at Amiens (1220) 2.89; at Beauvais (1247) 3.3. At the beginning of the twelfth century the Cluny system gave 2.57 at Paray-le-Monial and 2.56 at Cluny.

But this exaggerated height of the walls with windows above the side aisles produced the unfortunate consequences already mentioned: spreading of the supports, cracking, and collapsing of the vaults. Such was the case of Cluny in 1125. This is undeniable proof that the builders' technique was at fault.

All the evidence shows that the builders experimented by raising vaults that would exert less thrust on the side walls. The very original solution of covering the nave, between the façade and the transept, with vaults with transversal axes that buttressed each other was applied on a grand scale at Tournus (Figure 252); it was imitated only once or twice in smaller projects (for example Mont Saint-Vincent).

FIG. 252
Vault of Tournus
(narthex).

Vézelay had been rebuilt after a major catastrophe — a fire that occurred during a pilgrimage and which is said to have taken a thousand lives. This seems to indicate that it was covered with a wooden roof. An attempt was made to cover the nave with groin-vaulted compartments between transverse arches, which left open wall spaces for windows. The same arrangement is found at Anzy-le-Duc (first quarter of the twelfth century).

But this construction was too heavy, and the poorly evaluated thrust against the abutments proved to be too strong. At a later date it was balanced by abutments that imitated the older abutments of the choir; the latter had themselves been installed after the termination of the construction (they must have had to be added when problems appeared in the vaults). Choisy thought that the structure had been shored up at first with wooden beams, which were later replaced by a stone flying buttress; he regarded this as the origin and first example of flying buttresses. But holders for the iron tie beams have been discovered in the nave; thus it appears that the thrust of the diagonal ribs was at first balanced by a tie beam that prevented the springings of the arches from spreading apart.

Empirical experiments All these attempts reveal the uncertainties and hesitations of the builders. As we have already noted, they lacked all notion of the importance of stresses and their directions. They were also unaware (and were to remain so until the nineteenth century) of the effects of compression: distortions, both permanent and elastic, under weights. But the builders of the pillars and flying buttresses did realize the inequalities of distribution of stresses in the horizontal sections, probably when they saw cracks and openings in joints. This taught them that in a single section there was pressure and compression.

The builders did not hesitate, especially in the twelfth- and thirteenth-century cathedrals, to cut openings for circulation in the galleries in the areas under tension, or to place springings in an overhang, having found that they involved neither reactions nor compressions. But in no document do we find a trace of an attempted explanation based on knowledge of a direction or on intensity of stresses. Only a graph dating from the seventeenth century, but which surely originated five or six centuries earlier, reveals a method that must have been applied during that period.

Thus there was a continuous series of attempts to buttress vaults, beginning with (Figure 253) side-aisle quarter-circle vaults that contained the thrusts of the barrel vault over the central nave. The inadequacy of the pillar as a defense against overturning must then have become evident, the supposed rule used for determining the thickness of the pillars immediately revealing (Figure 249) the need for abutments (*B′A′C′*) to receive the thrust.

The quarter cylinder was suitable for the support of a continuous vault; if the stresses were distributed in line with a certain number of points, localized at the diagonal ribs, the builders could simply buttress the springings of the arches and those of the ogives. This they did, for example (Figure 254) by placing a solid wall, or a wall made lighter by an arch, above the diagonal ribs of the side aisle but hidden under its roof. The idea was to ensure the support of the springing of the arches, with the thought that it was sufficient to place a support at the point indicated by the customary diagram. Cracks probably occurred in the compressed arches, the builders having failed to realize the existence of permanent or elastic distortions under the effect of pressures.

Thus it became absolutely necessary to have a support placed higher, above the roofs of the side aisles (Figure 255, Gonesse, left portion of drawing). This was an important modification in the external aspect of churches; the initial design (Figure 255, right side) originally provided for no constructional member outside the roofs, and left all the windows completely open to the outside, as they appear, for example, in the apse of Pontigny.

The first flying buttress projecting from the roof gives the impression of having been added after (often quite long after) the completion of the construction. This is understandable, since the stresses must have sometimes occurred gradually, by slow distortion of the masonry structures.

This support must then have proved inadequate (Figure 256), and so it was necessary to add a second flying buttress (*A′* to the first (*A*), which supported the springing of the arches, and to increase the projection of the buttress (*B-B′*) to resist the thrusts.

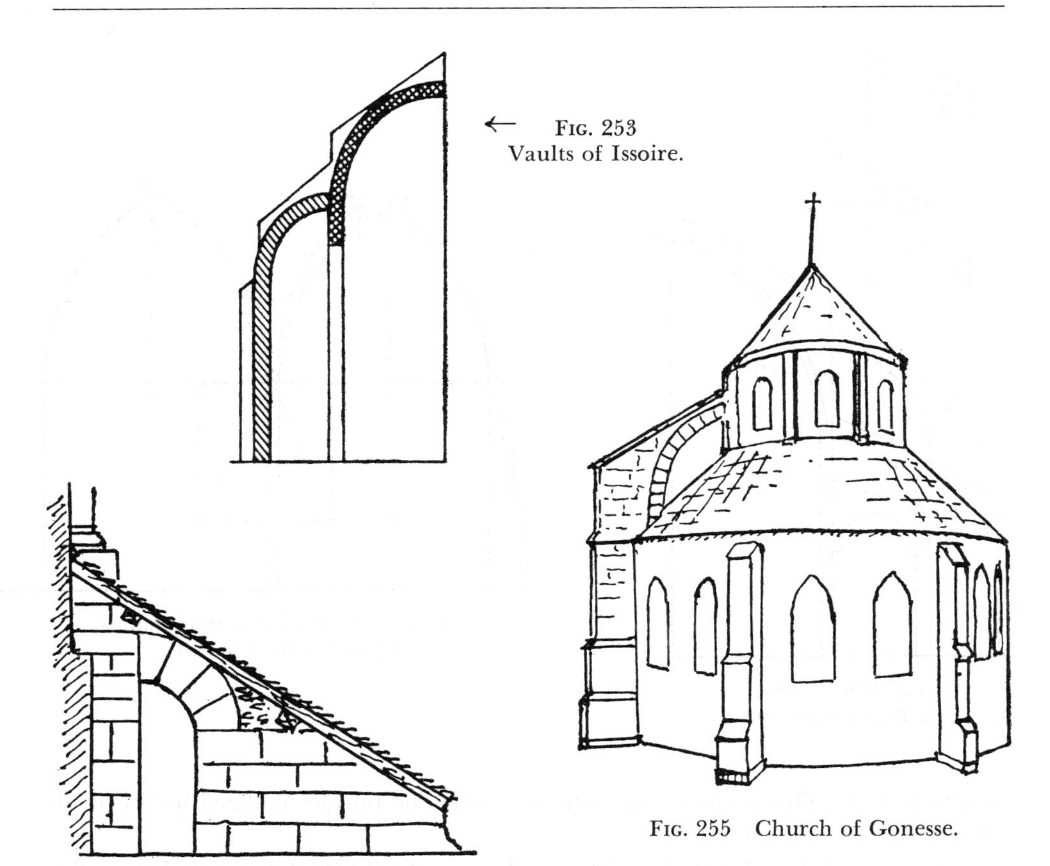

← Fig. 253 Vaults of Issoire.

Fig. 254 Early flying buttresses, concealed under roof.

Fig. 255 Church of Gonesse.

During this period, which was that of the great masterpieces – Paris, Chartres, Amiens, Beauvais – the construction of flying buttresses and the reinforcement of the abutments appear in all cases to have been posterior to the cathedral itself. Eug. Lefèvre-Pontalis recognizes (*L'origine des arcs-boutants,* Paris, 1921) that the observed breaking point may have been the upper portion of the corbeled support of the lower part of the vaults.

The builders also came to realize that once they had been buttressed, the internal pillars of the naves were no longer subject to the effects of overturning and could be made lighter. Instead of using massive pillars that were cruciform or had a column at each corner, it was now possible to adopt thinner pillars that opened up a view and no longer isolated the nave from the side aisles; the internal appearance of the edifice was that of an ensemble open to the eye as far as the exterior walls.

It was also realized that symmetrical oblique stresses eliminated their thrusts (Figure 257), as is the case in the refectory of Saint-Martin-des-Champs. The diameter of the columns was decreased, and in fact they appeared exag-

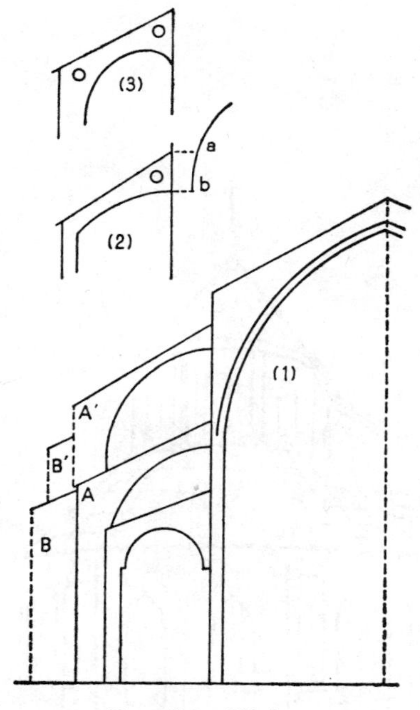

FIG. 256 Flights of flying buttresses.

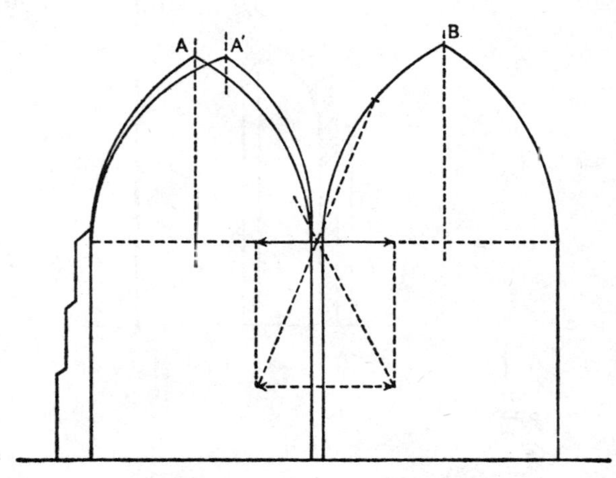

FIG. 257 Saint-Martin-des-Champs: equilibrium of thrusts.

geratedly thin. This is clearly an indication that the builder realized he was dealing only with a vertical thrust.

Again at Saint-Martin-des-Champs, there is an unusual trace in the arches whose top *A′* is related to the axis of the columns; perhaps this was an attempt to add additional security by decreasing the horizontal thrusts of the internal arches, which, however, had to be balanced from one vault to the next. Is this merely a search for perspective? In any case it indicates a line of reasoning that is already highly developed, although lacking in a profound knowledge of static mechanics and the strengths of materials. Only later, in the fourteenth and fifteenth centuries, did experience finally prove that since the flying buttress was an absolute necessity it would have to be constructed at the same time as the vault.

In the twelfth and thirteenth centuries, however, it was empiricism that clearly seemed to dominate, and the effect of its domination appears in insufficiently understood thrusts in a structure like that of Beauvais, in which the abutments required for the support of multiple flying buttresses succeeded in hiding the development of the windows from the exterior.

When it was finally recognized that the flying buttress should be part of the initial plan, it became possible to give the buttress a more logical shape (Figure 256[2]) and to spread it over the springing of the arches (*ab*) in order to hold the latter rigid. It was even possible to improve the design by curves (Figure 256[3]). whose graceful line nevertheless heralds the approaching period of decadence.

The Renaissance and the sixteenth and seventeenth centuries modified their techniques, as we shall see, by the addition of methods for precision-tracing the shape of the dressed stones. The problem that had faced the Middle Ages, a simple, crude technique of masonry, was replaced by the problem of cutting the stones, which it now became possible to solve.

By a rather strange twist, the light ogive arches, which were logical in relation to the medieval practices, disappeared, and the groined vaults that had never been successfully applied to main naves during the Middle Ages (except in almost isolated cases such as Vézelay and Anzy-le-Duc) now reappeared, this time in dressed systems. These were either groined vaults or vaults with intersecting diagonal ribs in the principal barrel of the nave, which made it possible to cut large windows. Geometrical graphs for cutting stones no longer frightened the builders; they had been made possible by geometrical tracing methods used, for example, for the vault of Saint-Sulpice.

The vaults were perhaps heavier, but their supports were built with large dimensions, and the flying buttresses acquired new forms; they present (Figure 258, Saint-Nicolas-du-Chardonnet, 1656–1709) a concavity that opens out widely at the top, supporting the wall at a great height, and an arch freeing the lower portion in order to distribute the stress on the abutments without crushing the vaults of the side aisles. A similar arrangement (Figure 259, dome of Les Invalides, 1676) holds the drum on which the thrust of the dome is supported.

Development of techniques of masonry

At first sight it appears that after the construction of the great Oriental, Greek, Roman, and lastly Byzantine monuments, the builders had nothing more to learn and had only to follow techniques that had been tried and tested for millennia.

But Europe, having been overrun by the barbarians, had only slowly and sporadically resumed contact with the East. Economic disorganization and ravages and destruction in the West had reduced the available materials for and possibilities of creating major works.

The tenth century had replaced the Roman masonry, which was often a mass with an external facing or which was characterized by perfectly flush chains of bricks forming continuous binding courses, by crude masonry walls. In order to make these walls resistant to cracking, the builders used courses of flat, obliquely laid stones (Figure 260), which played, on a less decorative level, the role of the Roman brick tying course. This was the *opus spicatum,* consisting of either a single row of flat, obliquely laid stones or of two rows laid in reverse order. The latter was the "herringbone" pattern, and it is found practically everywhere beginning in 970 at Tournus (chevet, 987), at Saint-Clement-sur-Guye, and at Saint-Gildas-de-Rhuys.

Later, with the gradual establishment of political and economic order, wealth increased and the financial means available made it possible to devote more labor to the task and to build with more refined masonry. Still crudely cut stones (Figure 261) were used in 1003 at Saint-Bénigne in Dijon and in the narthex of Tournus. Stonecutting was then introduced more widely, first in angular tying courses (the cells were still made of small-stone masonry) after 1050, and finally for general construction (after 1090). Southern France, how-

Plate 42. Vault of the church of Saint-Sulpice, Paris. End of the seventeenth, beginning of the eighteenth century. *Photo Atlas Photo.*

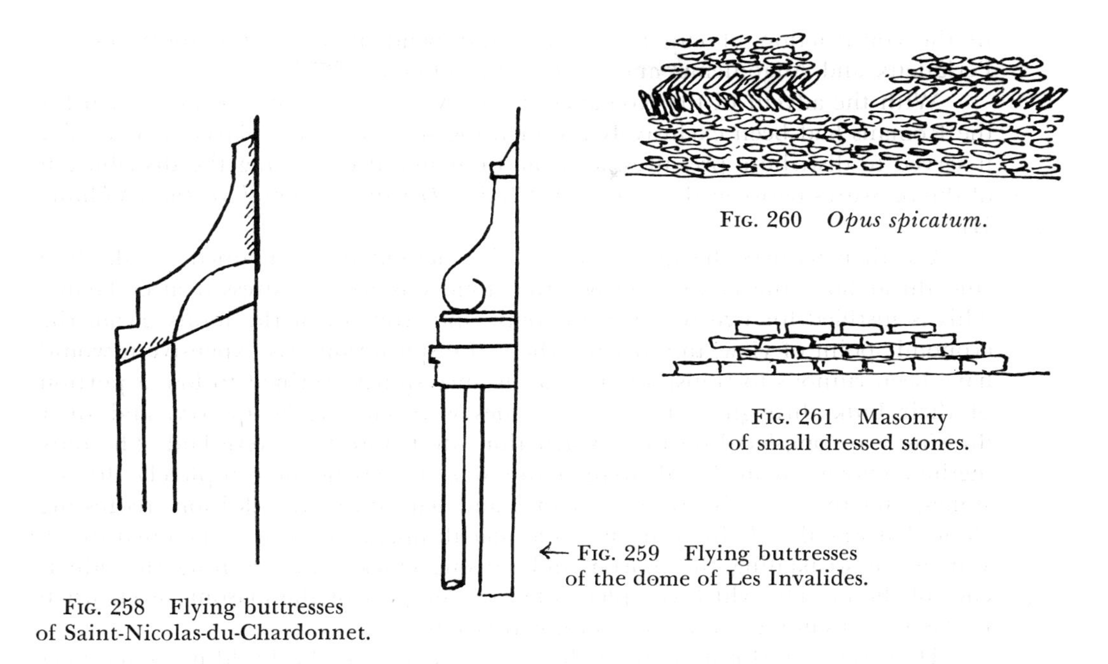

FIG. 258 Flying buttresses of Saint-Nicolas-du-Chardonnet.

← FIG. 259 Flying buttresses of the dome of Les Invalides.

FIG. 260 *Opus spicatum.*

FIG. 261 Masonry of small dressed stones.

ever, was influenced by factors that had not played a role in the countries invaded by the barbarians.

Stonecutting

The Greek technique had always utilized cut stones, which are found almost everywhere. It was characterized by a geometrical cutting of the stone, which practice is explained by the preservation of the knowledge gained by Hellenic science and technology. The builders of Santa Sophia, Anthemius of Tralles and Isidorus of Miletus, were natives of Asia Minor, and were famous as mathematicians and geometers. Geometrical cutting reduced the thickness of the joints, making the use of mortars unnecessary. This method of assembling stones, which was unbroken in the Greek technique, required precision cutting complemented by careful dressing of the faces through rubbing and wearing of the stones against each other. The right-angle joint distributed the pressures over all the surfaces in contact; thanks to precision cutting, the stresses were completely normal at the joints, and no slippage need be feared, except in the event of seismic shocks. The builders often complemented the support of these unbonded stones with cramps or metal tenons.

This technique, which was practiced in the East, spread by way of Sicily and the Mediterranean to the southern countries and the French Midi: to Spain, for example at Narranco in the Asturias (where the barrel vaults, however, are constructed of small stones) and San Julian de los Campos (Oviedo), to southern France (Alet [Aude]), and sometimes even to the north, as at Soissons (whose masonry dates from 850), which was probably constructed by a southern builder. In the construction of the Palatine chapel at Aix, Charlemagne had the assistance of architects from Asia Minor (Syria?). The technique was then continued in the construction of cupolas, arches, and pendentives in the southwest, at Périgueux and Angoulême, and Loches (Saint-Ours).

With the advent of the Romanesque age of dressed stone, mortar, which for the Romans had been exclusively a cementing material, now played a new role: it served to distribute pressures, as a plastic material equalizing the distribution of the pressures between the stones (A. Choisy, *Histoire de l'architecture,* Volume II).

We then witness the imposition of the method that remained an absolute rule throughout the Gothic period: the stone was never redressed after laying. This is justified for two reasons: economy, and strength of the joints. From the economic point of view, in a period when transportation was expensive it would have been ruinous to transport from a distance stones destined to lose a portion of their bulk through cutting. The stones were cut at the quarry, and their dressing was completed on the construction site before they were laid. The only method known to medieval architecture was the laying of completely dressed stones, and this was also true of facings and sculptures. In addition, redressing stone that was already in position was a difficult operation; it had the great disadvantage of weakening the mortar and causing cracks, thus ruining the adherence of the mortar, which complemented the support of the masonry even when it was not absolutely necessary (tensional stresses).

However, transformations in the very structure of the buildings sometimes necessitated the demolition of certain parts and on-site cutting of certain elements. This was the case in the cathedrals of Sens and Paris. In the conception and construction of the edifice, the windows that lighted the upper portion of the nave were already quite large. But when they were compared with those of more recent cathedrals, they were judged to be inadequate. It was therefore decided to redress the masonry and to cut taller windows, by developing them either from the bottom (Paris) or the upper portion (Sens). In the latter case reconstruction of the vaults became necessary. This must have caused the appearance of certain cracks, which entailed the construction of new flying buttresses; the latter probably would not have been necessary in a structure left intact. This thirteenth-century remodeling produced the large flying buttresses of Notre-Dame of Paris, which leap unsupported across the side aisles and which probably are not replacements of primitive flying buttresses, as Viollet-le-Duc believed; they appear to have been put in place to support the upper portion of the vaults weakened by the window work. Such recutting of masonry already in place could produce only unfortunate effects and dangerous cracks.

The use of costly and fragile tooth-shaped stones was forbidden. The builders of antiquity had on occasions given their arches "stepped" extradoses. This practice was not accepted in the Romanesque period: "parallel" extradosing, the method always used, was easier to cut and economized on stone. In Gothic construction the stones were very carefully cut. It was necessary to ensure the the equal transmission of pressures that were often very great, and the beds and joints were cut as carefully as the faces.

The cutting equipment was also modified. The tools of the Roman period had smooth-edged blades. After the twelfth century stonecutters employed the bushhammer. As for percussion cutting with the bushhammer, which crushed and cracked the stone, it was (correctly) considered dangerous, and was used only for cutting granite (western France).

Metal clamps, which had been unknown in the Roman period, were introduced into Gothic structures when cracks, indications of a shifting of masonry were found. All the stones forming leveling courses were joined by iron clamps sealed in lead (an example is seen in Notre-Dame of Paris).

During the Renaissance, brick, always used in Italy, was introduced into France, but the Italian method discussed earlier in this chapter was not used in France. In Italy the brick masonry was built up, and the casings of bays, façades, and so on, were added later. In France no decoration was independent of the structure; the materials, even when they were different from the mass of the masonry, were incorporated into and built up with it.

The practice of redressing is a Roman tradition that was continued only in Italy. This method, which made possible a decorative appearance that did not correspond to the joints of the dressing of the stones, was slow to reach France. The Gothic tradition survived through the reign of Louis XII and the early years of the reign of François I. Ribbed construction of vaults is found until the middle of the sixteenth century. The only novelty consisted in the use of bricks for the filling cells.

Not until the time of Philippe Delorme (born Lyon 1518, died 1577) do we find complex vaults and the use of corresponding dressing and tracing of joints. This was a period in which the development of geometry facilitated the making of graphs for stonecutting. It was Delorme who, upon his return from Rome in 1546, geometrically studied the graph of the volutes of the Ionic capital, a study he published in 1567, preceding the work of Palladio (1570), to whom the invention of this method is sometimes attributed.

The appearance of stereotomy

The stonecutters had long been using similar procedures for the sometimes very complicated tracing of skewed arches, pendentives, and squinches. The ancients had possessed geometrical knowledge that helped them in preparing these graphs; this knowledge had been preserved in the East. But the medieval builders had lost or forgotten it, and had to build these surfaces of small-stone masonry as best they could. This again indicates how risky it is to attribute to the medieval builders knowledge they are supposed to have kept secret and which was then lost, so that modern builders are supposedly incapable of reproducing the medieval work, particularly the squinches.

These tracing methods, which were only approximate, were the object of reflection and study. But very remarkable projects had already been executed in sumptuously constructed buildings, for example the Hôtel Pincé at Angers (1523–1530), the castles in the Loire Valley, and others.

The development of the mathematical sciences in the seventeenth century was to provide precise geometrical solutions. Desargues (1598–1662) had propounded theories whose utilization, by virtue of their unusual presentation, appeared difficult. Desargues, however, was able to apply them to the cutting of stones. In 1643 Abraham Bosse (1602–1676) published his book *La Pratique du Trait, à preuves, de M. Desargues, Lyonnais, pour la coupe des Pierres en*

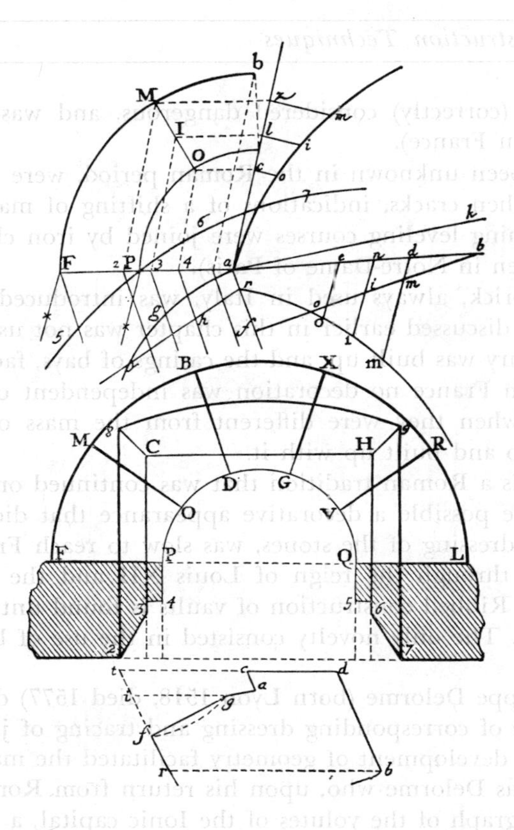

FIG. 262
Graphs.
From Desargues.

l'Architecture (The practice of tracing, with demonstrations by M. Desargues of Lyon, for the cutting of stones in architecture) (Figure 262). He borrowed from Desargues the elements of projective geometry that were only later completely developed (by Poncelet).

Stereotomy, which had once again become an exact science, now made an effective contribution to the work of the stonecutters, and made it possible for them to return, after the long interruption, to the traditions of the stonecutters of antiquity.

The technique of cupolas Because of its important role in several works of architecture, we believe it will be useful to devote a paragraph to the technique of cupolas, a discovery of the conceptions of their authors and the technique they used.

We have already discussed the Pantheon in Rome and the method used in the construction of its cupola. For centuries, and particularly during the Renaissance, this cupola was a subject of admiration and study for the Italian builders (Figure 263).

The internal diameter of the room is 142 feet, and the circular wall that supports the vault is 17 feet thick. The distance from the floor to the top of the vault is 145 feet; thus the diameter is approximately equal to the total internal height. The springing of the vault begins 74 feet from the ground, or at approximately the halfway point.

The great church of Santa Sophia in Constantinople, reconstructed by Theo-

FIG. 263 Pantheon, Rome.

dosius in 415, was destroyed by fire in 502, and reconstructed by Justinian on a grandiose scale. Completed in 539, it was shaken by several earthquakes (555 and 557), and on May 7, 558, half of the cupola collapsed. The building was constructed on a masonry platform 20 feet high; the pillars were built of calcareous stones jointed with iron clamps. Justinian had the cupola rebuilt immediately by a nephew of Isidorus of Miletus. Its height was increased in order to decrease the thrusts, and the main arches were reinforced. Later, under Murad II (1572–1596), enormous abutment walls were added, which give the outside its heavy, pudgy appearance.

The cupola, whose diameter is 107 feet, rests on four pendentives. The thrusts are contained by quarter-circle vaults on the east, and by arches buttressed by abutments (Figure 264) on the north and south.

In succeeding centuries we find other examples of cupolas, but they are

Fig. 264 Santa Sophia.

Fig. 265 Angoulême.

much smaller. The palatine chapel of Aix, constructed at the end of the eighth century, consists of an octagonal hall whose diameter is only 47½ feet. The octagonal shape corresponds to flat walls, in which windows can be cut more easily than is the case in a cylindrical wall. There is no window in the cupola, and the difficulties of tracing are therefore reduced to a minimum. Around it, over an area of 21 feet, a 16-sided polygon combines with the central octagon. The same number of diagonal arches distribute the thrusts of the cupola onto the wall.

The cupola was introduced, in very crude form, into structures in central

France, specifically in Burgundy, beginning in 1050. The church of Avenas demonstrates this very primitive system of squinches constructed of small arches of crudely cut stones, with small vaults of coarse masonry. Above this is the section of wall that creates the octagonal form at the base of the cupola. We find the same system at Blanot, where we also note that it requires neither a precise plan nor complicated stonecutting. The masonry was masked with a coating, to provide a smooth surface for a decoration of simple or stylized forms.

Churches like those of Périgueux (Saint-Front) or Angoulême (Figure 265) are much more highly developed from the viewpoint of stereotomy; the cutting of the stones shows traces of Eastern influence via the Mediterranean and Sicily.

At Saint-Front, the pendentives are dressed normally to form a spherical surface; this is the solution used in Santa Sophia for the transition from the square plan to the spherical cupola (end of the eleventh and beginning of the twelfth centuries).

The spherical cupolas of Fontevrault, farther north, are similar in solution, but have no pendentives; the spherical surface rests directly on diagonal arches.

Limiting our discussion to the masterpieces whose history is known, we should mention several cupolas of later centuries, which were and still are an object of universal admiration. There is, for example, the cupola of Santa Maria del Fiore at Florence, which was constructed on rational principles by Brunelleschi, and of which Michelangelo said, "It is difficult to build as well; it is impossible to do better."

Brunelleschi

It must be remembered that artists like Brunelleschi went to Rome to learn from the great works of antiquity; between 1404 and 1415, in company with Donatello, he enjoyed the sight of the grandiose monuments; *"pareva fuor di se,"* said Vasari.

On his return to Florence, Brunelleschi conceived the idea of the cupola, and in 1417 raised the drum, as far as the *oculi* that still light it, on four gigantic pillars. The lantern that caps the upper opening of the cupola was not completed until 1461, seventeen years after the architect's death.

However, this plan and its daring conception were vigorously criticized: both the church wardens and Lorenzo Ghiberti, the creator of the famous Baptistery doors, felt that it was impossible to build a vault without a keystone.

Ideas on the distribution of stresses, incomplete as they were at that time, could differ widely (see Figure 266). When we consider a vertical cross section along the axis, the two sections of arc have no support; there is no keystone in the vault, and thus the vault cannot stand. This was Ghiberti's thesis.

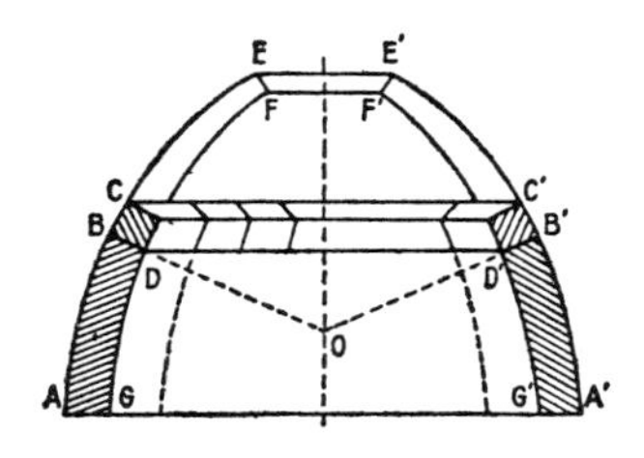

Fig. 266
Equilibrium of a cupola.

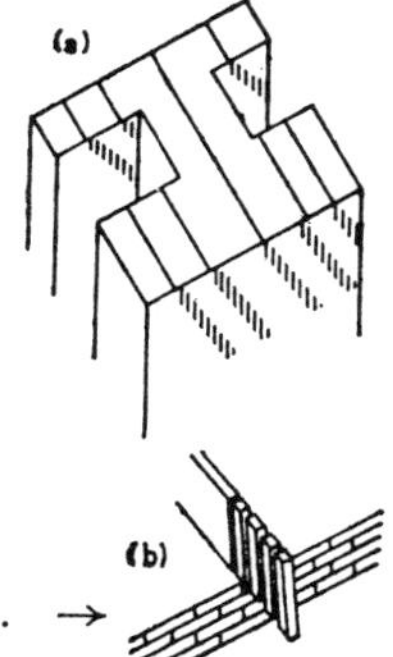

Fig. 267 Brick construction (Florence). →

If on the contrary, says Brunelleschi, we consider the cupola as a series of rings constructed on top of each other, the fact that each element is compressed by its own weight makes the dome a stable structure.

The technique of the construction clearly demonstrates the architect's line of thinking. The cupola is double; there are two surfaces that fit into each other and are joined by longitudinal ribs, the whole forming a light, rigid framework.

In this country, where brick construction is the general practice, the cupola of Florence naturally followed this rule. The bricks are arranged so that helicoidal leveling courses made of vertically placed bricks break the horizontal or slightly inclined courses at regular intervals; the dome is relatively pointed.

It seems that in this way Bruenelleschi introduced a system of vertical leveling courses in order to avoid breaks and cracks in the mass of the vault. Choisy saw in this an expedient for tracing the outlines of the vault on stretched wires or carpenter's lines, but this is a very complicated manner of obtaining a simple graph, and it should be regarded instead as a constructive system established in line with a deliberate, rational conception.

Thus the Florentines were astonished to witness the triumph of Brunelleschi's conception and the raising of the cupola in successive rings, without scaffolding, abutments, or flying buttresses. The cupola itself was completed in 1425 and was later surmounted by the lantern, whose weight, according to its architect's theory, only increased the stability of his vault.

The cathedral of Florence was barely completed when Pope Nicolas V, profiting from the resources supplied by the jubilee of 1450, decided to build a new Saint Peter's Basilica that would surpass the Florentines' monument. The plans, which were revised several times, were adopted by Michelangelo in 1547. But its diameter was smaller than the Florentine cupola's 141 feet (comparable to the cupolas of Santa Sophia and the Pantheon), and its construction, although one century later than that of Florence, is indicative of no technical progress. The less pointed profile did not lend itself as well to construction without supports; thus the vault had to be constructed on centerings. The centerings supported only the longitudinal ribs, so that they would not be too heavily weighted; the cells were added later, without centerings and without weighing on the trusses. This type of construction was not as well bonded as that of Brunelleschi.

These great projects exercised a profound influence on seventeenth-century construction; the churches of the Sorbonne (Lemercier) and Val-de-Grâce (François Mansart), dating from the time of Louis XIII, are derived from them. The dome of Les Invalides, the best work of the school of J. -H. Mansart, dates from the reign of Louis XIV. It rests on a cylindrical drum with external buttressing systems along its entire length. Saint Paul's of London, a work of Sir Christopher Wren (1700), is almost a contemporary of the dome of Les Invalides. One of the last applications of vaulted architecture in the eighteenth century is Soufflot's Panthéon (1755).

Arabic construction

While the Arabic building techniques produced remarkable structures, they contributed no technical innovations. The Arabs utilized the methods of the local builders, which had been in existence long before the Moslem, Persian, Syrian, Coptic, and Western invasions (for example into Spain).

However, their special characteristic (which is especially that of Arabic art) is the use of the geometrical linear graphs that give the decorations their highly original appearance. The use of a variety of polygons and their combinations is the determining factor in these graphs (Figure 268). The law of combination is simple: the sum of the angles around a point is always equal to four right angles (A. Gayet, *L'art arabe*).

FIG. 268 Arabic graphs.

The primitive combinations used by the Copts were limited to triangles, squares, diamonds, and hexagons, whose combination produced four right angles. The Arabic algebraists combined polygons of various shapes; they gradually realized that it was possible to combine two octagons and a square, since when calculated in right angles this sum of angles represents

$$2 \times 1\tfrac{1}{4} + 1 = 4;$$

two hexagons and two equilateral triangles give

$$2 \times 1\tfrac{1}{3} + 2 \times \tfrac{2}{3} = 4;$$

while two dodecagons and one equilateral triangle give

$$2 \times 1\tfrac{2}{3} + \tfrac{2}{3} = 4$$

and so on.

They also discovered the possibility of combining three different polygons: a hexagon, two squares, and a triangle give

$$1\tfrac{1}{3} + 2 + \tfrac{2}{3} = 4$$

The Arabs attributed a mystical meaning to these combinations. The figures

Fig. 269 Arabic squinches and pendentives (plan and view).

of the even numbers reflect calm and serious feelings, the odd numbers, disturbance and uncertainty; whence combinations giving mixed impressions. Interlacings are a superposition of elements of polygons in an attempt to create an impression, whether of immutability, mysterious disturbances, or concern.

In short, the framework of polygonal decoration is based on four elements: the network, the grouping of the basic polygons, tangential circumferences, and derived figures. The same spirit rules the strange stalactites that offer various solutions of squinches and pendentives (Figure 269), also derived from geometrical constructions, for the blocking of angles.

The same methods were applied to furniture and architecture; the construction results from combinations of prisms with triangular, square, or pentagonal cross sections. All this construction reveals a perfect knowledge of stonecutting.

The horseshoe arch

Another characteristic aspect of Moslem art is the use of the horseshoe arch, whether round or pointed at the top.

There has been much discussion of the origin of the horseshoe arch. Choisy saw in it a development of the round vault (Figure 270); the recessed piers, leaving brackets on which the wooden centering could be supported, were hidden by a filling, and this expedient was later deliberately adopted.

Actually, however, the study of various arches, even those of early periods, contradicts this theory. It seems that the springings of the arch were constructed by the corbeling method, but sloped out rather than in (Figure 271). Thus the corbeling could not fall in. In addition, the slopes of the two sides of the arch acted as buttresses for the establishment of a wooden centering. There was no need for brackets bearing the centering, as in the early French Romanesque; on the centering were placed the voussoirs of the arch, which were either of dressed stones or small-stone masonry.

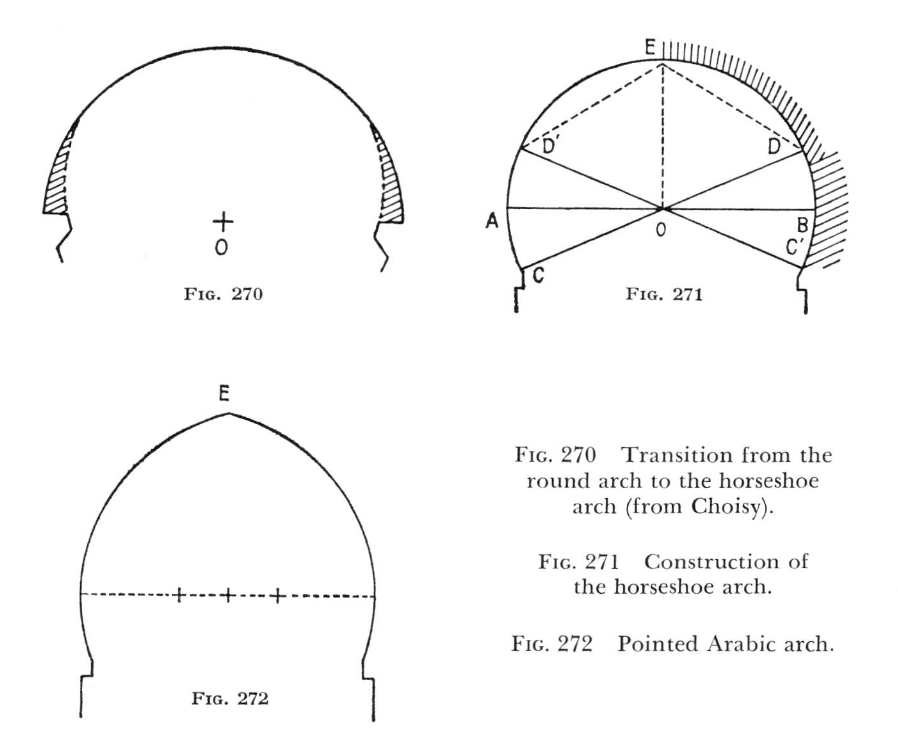

FIG. 270 Transition from the round arch to the horseshoe arch (from Choisy).

FIG. 271 Construction of the horseshoe arch.

FIG. 272 Pointed Arabic arch.

It is easy to follow this technique in buildings constructed in line with this conception, whether of large stones joined with masonry and then covered with decoration covering, or of dressed stones. The joints vary in appearance, revealing the method of construction. Those of the corbeled portion are thinner, the soft mortar having been crushed during laying; in contrast, the joints are thicker in the portion laid on the centering. A similar tracing, with arches beginning from the two centers (Figure 272), gave the pointed horseshoe arch.

The innumerable examples include the monumental gate of Bab Aguenaou in Marrakesh, the oldest of the Almohad gates, constructed under Abdu-el-Mumin (1130–1163). It is constructed of blocks from the Gueliz, which is only three or four miles away, and is a powerful specimen of Maghrebin art (Figure 273). Rabat has the gate of the Casbah of the Oudaias (reign of Yakoub-el-Mansur, 1185–1194) whose beautiful ocher-colored stones rival those of Bab Aguenaou.

The same technique can be followed, unchanged, in the gates of the Alhambra at Granada, constructed for the most part under Jusuf I (1333–1354) and Mohammad V (1354–1391). Details on the two faces of the Wine Gate can be plainly seen, and it is easy to distinguish between the corbeled portions and those constructed on a centering. In the Gate of Justice, brick was used in the same manner, and was then given a faïence decoration.

The construction of the minarets was simple: massive walls whose stones are dressed and sculptured, using the same methods of decoration, which gives them a pronounced and striking appearance of sumptuousness. One example is the

FIG. 273 Monumental gate of Bab Aguenaou, Marrakesh.

Koutoubya of Marrakesh (1153), which rises 221 feet on a square base of 41 feet. A sister to the Hassan Tower of Rabat (1195), it is 53 feet long, but is an uncompleted 144 feet tall. It was constructed with materials transported by boat from quarries situated seven miles from Wadi Akreuch at the junction of the Bou Regreg.

Turkish structures A few words should be said about the techniques of the great Turkish architects. To characterize them, we need only recall that they were inspired by the Byzantine buildings, which they had before their eyes.

From the green mosque of Nicaea (fourteenth century), which had only one large cupola in a modest masonry cube, they progressed to projects that were already large: the mosque of Andrinopoulos (Eski Cami, 1350), the Green Mosque (Yesil Çami, 1470) of Bursa, and the mosque of Mehmed Fetih, constructed by Greek builders during the reign of Mehmed II.

Mention should be made, in particular, of the numerous achievements of the great Moslem architect Sinan (1489–1578), who worked under Suleiman, Selim II, and Murad III, and who was the creator of 132 buildings, including numerous mosques: the Sehzade (1543–1548), the Suleiman mosque (1550–1566), the diameter of whose cupola is less than that of Santa Sophia (85 feet) but which is 20 feet taller and is better balanced. Sinan was wonderfully skilled at drawing from his model a logical audacity, and tracing its lines in a geometrical spirit; he constructed in dressed masonry, which was colder and less sumptuous than the Byzantine works, but marked with a severe grandeur. His masterpiece was the Selim mosque in Adrianople.

All these structures adopted the same general trace, the cupola being supported on the main arches or buttressed by either two (as at Santa Sophia) or four (as at Şehzadé Çami) quarter-circle vaults.

These are all applications of geometrical traces whose equilibrium seems to result from intuitive, successful proportions rather than from a method of calculation that is unknown to us and that could not be justified by a still-nonexistent knowledge of stresses and their directions.

VARIOUS ASPECTS OF THE PRIVATE DWELLING

The fireplace

The climate of certain regions makes heating a necessity, and therefore fireplaces have always been an important element of buildings.

In single-story buildings the Romans utilized hypocausts, in which the heat from a fire circulated under the stones of the flooring, which rested on small brick piers, while fumes and smoke escaped through an isolated masonry chimney placed at a distance from the buildings. This system, which was used especially in the Roman baths, was still in use in the ninth century, at Saint-Gall in Switzerland. But we also find at Saint-Gall a large stone, acting as a hearth, in the center of the room to be heated, and a hole in the ceiling above it to permit the smoke to escape. This was almost the system of the ancient huts, with the opening in the center of the roof.

Primitive as this fireplace may have been, it was still in existence in England in the sixteenth and seventeenth centuries. The first addition to it was an open chimneypot that protected the opening; next came a structure inside the house, over the hearth, consisting of four pillars supporting on four lintels a hollow masonry pyramid which opened into the chimneypot. This fireplace (sometimes called the "Saracen" fireplace), which was open on all sides, survived in Bresse into modern times.

Multistoried houses in compact rows replaced the single-story houses when space became limited, as for example within city ramparts. In Rome itself there were areas where single-story, sumptuous houses (of the type to be seen in Pompeii) predominated, and overpopulated areas with multistoried houses. Here the heating system with an opening in the roof to permit the smoke to escape was not applicable.

In France, after the eleventh century, the use of fireplaces placed against a wall began to spread. The back, the core of the fireplace, and the bottom of the hearth were constructed of tiles laid flat or on edge, to withstand the heat.

Beginning in the fifteenth century, the use of cast iron made it possible to line the back of the fireplace with decorative sheets of cast iron. Above the hearth a lintel or mantel, resting on brackets or jambs, supported the hood, which guided the fumes and smoke up the chimney. The flues ended at the roof in a stone or terra-cotta miter that protected the opening.

Chimneys dating from the thirteenth century (for example at Fougères, Ille-et-Vilaine) are still in existence. They are still completely open on all sides, and the hood is supported on pillars.

The lintels and hoods, which began as simple structures, were later developed with architectural motifs, arches, and sculptured panels, forming decorated vertical coffers.

For heating large rooms, twin (twelfth century, at Laon and Mont-Saint-Michel) and triple (Fontenay Abbey, thirteenth century; palace of Poitiers,

1385) fireplaces, back to back, were used. Some fireplaces were constructed of wood or of masonry faced with wooden paneling. In the fifteenth and sixteenth centuries they were made of plaster, on a wooden framework.

The hearths were large so that they could burn large pieces of wood. Coal, which was beginning to be used industrially (glassworks and lime kilns, Douai, 1350), was occasionally used, beginning in the fourteenth century. Coal burning required grills; these were also utilized for small logs. Movable ovens (braziers) were frequently employed in houses where there were no permanent fireplaces. The brazier, which had been used in Roman days, was the customary method used for heating multistoried houses.

Windows

The window, which is used to light and supply air to buildings, belongs by virtue of its form and size in the domain of architecture. But the methods used in its construction lie within the realm of technology.

Until the middle of the fifteenth century, wooden shutters were used in private houses to close the spaces between the mullions, preventing the entrance of light except through small slits.

The windows of large buildings such as churches had stained-glass windows. Because of their weight the early glass windows of private houses could only be held within a rigid sash frame above windows that were closed only with shutters. Moreover, the price of windows prevented them from being widely used.

In many sixteenth-century houses this arrangement was complemented by sashes holding oiled, translucent canvas. Then paper, which was not as strong but was more transparent, gradually replaced the canvas.

During the first half of the sixteenth century stained-glass windows were still rare. After 1550 they began to become more common, but their cost still limited their use. During the reign of Louis XIII sashes filled with paper were utilized, along with window glass, even in the royal residences.

The era of the glassed window began in the second half of the seventeenth century. Clear glass replaced stained glass. The small pieces of glass, which were all that could be obtained, led to the construction of windows in small squares, a practice that remained in use almost unchanged until the middle of the nineteenth century.

The technique of closings for windows was therefore determined by the techniques of glassmaking. But the technique that has left the greatest mark on the history of art is, of course, that of the stained-glass window, whose use was widespread and indispensable for closing the window spaces of large buildings, such as the large bays of churches.

Antiquity had used strips of either mica or translucent stones such as alabaster, which could easily be cut into thin sheets, to close openings that had to admit light. These were rigid windows, not movable panels; the weight and fragility of the strips of stone necessitated their being firmly fixed in the frames.

Glass had also been known since the most remote ages of antiquity, but its systematic use, especially for the windows of large buildings, developed only in Byzantium, Rome, and Merovingian Europe. References have been made to stained glass at Saint John Lateran, Saint Paul's Outside the Walls, Saint Peter's in Rome, and Santa Sophia in Constantinople, and the basilicas mentioned by Gregory of Tours and Fortunatus.

A distinction must be made between painted glass and tinted glass. The latter was made with the use of metallic oxides incorporated into the mass of the elements and fused together. This procedure had already been used in Egypt for counterfeiting precious stones; it was later applied to thin strips which by a combination of small pieces could form translucent mosaics, an advantageous replacement for mica. The number of small pieces was consequently very large – up to 350 to 400 per square yard of window.

Painted glass was colored only on the surface, and could thus be covered with figures or drawings. The colors were then permanently fixed by firing. The art of window painting, whose resources were greater and more interesting than those of tinted glass, was not known when the oldest Christian basilicas of the East and Europe were being constructed; it does not antedate the ninth century. Thus tinted glass was the point of departure for stained glass, and its early uses and gradual progress were achieved by means of tinted glass.

Stained-glass windows, probably very crude models, have been mentioned as early as the ninth century, for very narrow and small windows. Not until the twelfth century are large stained-glass windows definitely attested. There are, for example, the windows of Saint-Denis, whose execution between 1140 and 1144 monopolized all of the Abbot Suger's attention. They are said to have been very expensive; the glaziers are supposed to have used sapphires to color the glass. Subsequent remodelings have modified them, and the present windows probably contain only a few pieces from the early windows. The effigy of Suger kneeling may be the oldest known donor portrait.

Clear glass was also utilized, especially in the Cistercian churches, under the influence of St. Bernard, who had vigorously criticized the sumptuousness of the lighting and the dazzling color of the stained-glass windows, particularly those of Suger at Saint-Denis.

Absolutely clear glass was unknown during the Middle Ages; the glass used in this period always had a greenish, yellowish, or bluish cast. But the glassmakers obtained decorative effects from this diversity.

After the twelfth century, while much use was made of relatively clear windows, such windows in public buildings were brightened with escutcheons and emblems made from pieces of colored glass. The pieces were assembled with fairly wide strips of lead; the panels thus formed were fastened with bars and small rods of wrought iron, so that when they were raised in vertical position they would not bend under their own weight.

The ironwork, at first simple – more or less equal squares (twelfth century) – gradually became more complex; by the thirteenth century it was forming medallions, portions of circles, diamonds, and so on (Figure 274). The iron skeleton progressed to a point where it became an element of architectural decoration. But its use decreased when the bays were broken up by stone mullions, leaving smaller openings.

There were also developments in the designs; large figures were introduced into the windows of the great bays. The ironwork was then simplified into diagonal bars, since it was no longer necessary to forge fifteen or twenty (and

FIG. 274
Ironwork of
stained-glass
window.
Thirteenth century.

often more) medallions of various shapes in order to assemble the windows. The glazier's work was also simplified, since he was able to paint large surfaces with human figures more quickly.

After the thirteenth century there was widespread use of grisaille, a kind of pearlized glaze that gave a sheen to all the tones.

Many of these early windows disappeared in the eighteenth century, for example at Notre-Dame of Paris, Rheims (1739–1768), and Chartres (1757 and 1770–1788); they were removed in order to admit more light, either for the purpose of showing off modern sculptures or (particularly) to permit the faithful to read the books that were now available to everyone.

The art of window-making for public buildings was greatly developed during the fourteenth century; kings had them installed in their palaces, the great seigneurs in their dwellings, the rich bourgeois in their homes. In the fifteenth century methods of production were improved, and new colors enriched the palette of the glass painter. Flashed glass, invented in the fourteenth century and greatly perfected in the fifteenth, permitted an increase in the intensity of the tones. The sheets of glass were blown in two colors, one thick and colorless, the other colored and thin. By removing areas of the thin colored layer with a graver, cutting wheel, or emery, the glassmakers obtained very elaborate designs that could be enhanced by painting them with silver stain or with certain enamels.

The methods made available by the glassmaking technique led to the production of stained-glass windows in which the decorative elements and human figures spread over increasingly large sheets of glass. The making of the leading was also modified during this period. The lead extractor, a machine that replaced the plane for shaping the leading, turned out leads that were very easy to handle: long flexible ribbons that easily followed the contours. The work of the glazier was thus facilitated, but the strength of the window was diminished; the thick, broad, coarser leading of early windows was a greater guarantee of strength.

At the beginning of the seventeenth century diamonds began to be used in glass cutting; until then it had been done with an iron heated in the fire. According to the ancient precepts of Theophilus Presbyter (*Diversarum artium schedula*), "Push the iron along the line you would like to cut, and the crack will follow." The cutting was then made more regular with the grazing iron. The diamond was safer and easier to use, but the grazing iron gave the cut small, sharp, jagged teeth that bit into and became firmly embedded in the lead, making the joint more rigid. The thickness of the glass was decreased; its strength was also diminished by virtue of the larger sizes of the pieces.

The stained-glass window now entered a period of decline, caused by the fact that in this period the windows of private houses, palaces, and public buildings were in general being equipped with clear windows.

The making of window glass

The glass industry was flourishing in Constantinople. Byzantine glaziers settled in Venice, where their art underwent considerable development in the thirteenth century. A Venetian named Nicolò brought the art to Bohemia in the sixteenth century.

In France, where the glassmaker's art had been known since the Gallo-Roman period, the first glassmaking establishment (for bottles and various purposes) is believed to have been created in 1290 near La Capelle (Aisne). Numerous other glassmaking establishments appeared in the course of the following century. But until the sixteenth century progress was slow, despite the privileges granted to the glaziers. Windowpanes were difficult to obtain; their translucency was uneven, and impurities in the material gave them a greenish cast.

In order to develop glass production, Colbert imported a number of Venetian glassmakers. A mirror-blowing factory was established at Tourlaville in 1665. In 1688, Abraham Thévart conceived the idea of pouring glass, and installed his first shop in the Rue de Reuilly. This is the shop that was later transferred to Saint-Gobain; the area around this forest supplied siliceous sand and wood for heating the furnaces.

Venice and Austria retained the glassblowing process, which survived into the nineteenth century, even for large windows, without notable modification. One of the exhibits at the Vienna Exhibition in 1845 was a mirror three feet by seven feet which had been made by blowing.

The pouring process replaced older methods. At first the softened, vitreous paste was drawn out with pliers. With this method small sheets of glass could be obtained, but they were quite thick (several millimeters). It was this type of small sheet that was known to the early authors (particularly of the Carolingian period) as window glass.

Then came the blowing-iron method (Venice). The blowing iron is an iron tube that flares out at one end and is inserted in a wooden handle at the other. An assistant took a small quantity of the semifluid material on the iron, turned it over repeatedly on the slab of the furnace, and then handed the blowing iron to the master glassmaker. After he had blown it slightly, the glassmaker turned it rapidly, obtaining a flat disk of fairly even thickness (Figure 275),

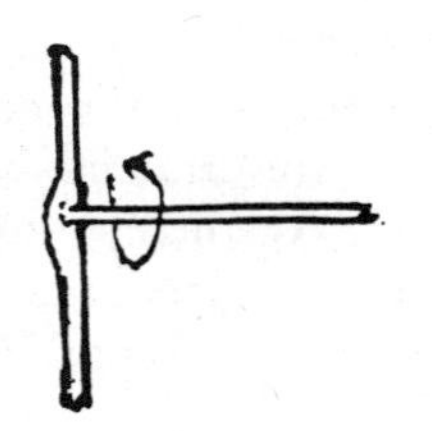

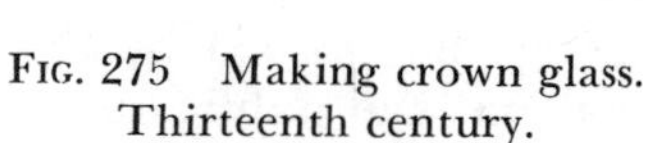

Fig. 275 Making crown glass. Thirteenth century.

Fig. 276 Steps in making window glass. Seventeenth century.

except at the point where the disk was attached to the end of the blowing iron. When the disk was detached, the result was a sheet with a bulge in it. Cutting it produced a rather uneven sheet, pieces from which were suitable for use in the most beautiful stained glass of the thirteenth century: its very unevenness helped to create plays of light, preventing the illumination from being too harsh.

For the less irregular window glass, which began to be used, as we have seen, in the seventeenth century, a cylinder was blown (Figure 276). The blown material at the end of the blowing iron was swung back and forth; under the effect of gravity this movement drew out the bubble of glass until it became a kind of large bottle with a cylindrical center section. A hole was drilled in the bottom, and the opening was enlarged with a plank placed on a support; the piece of glass was separated from the blowing iron with a cold iron. The top was then detached by rolling a thread of hot glass around it, then touching the thread with a rod of cold iron, which produced the break. The cylinder thus obtained was softened in the oven, and split diametrically by chilling the line of the break with a cold iron. The cylinder was placed on a heated metal table, and unrolled. The result was a very even, flat window glass.

As for poured mirrors, the glass material was poured out on a heated table, and was spread evenly with a heavy metal cylinder, giving a uniformly thick sheet.

All these procedures continued in use without major changes until the eighteenth century. Not until the nineteenth did mechanization make the glassmaker's work (especially blowing) less difficult, and production of bottles and other objects more economical.

Mosaic pavements and ceramic tiles

In the most primitive single-story houses, beaten earth has always sufficed as a flooring, and in the rural areas many examples of this practice are still to be found. In the houses of the rich, large buildings, and palaces, cut and polished stone and marble pavements were universally used. The large, thick stones were heavy to handle and transport; by breaking them up it was possible to make them thinner and obtain paved floors by combining squares, sometimes in a variety of colors.

The mosaic is composed of small cubes assembled and fixed in a cement. These cubes can be of stone, terra-cotta, or opaque enamel (called *smalto* in Italy) colored while in the paste state. It is a characteristically Roman technique that was continued in Byzantium and Venice, where it was practiced universally, using exactly the same technique.

Mosaic works abounded in the Roman edifices, whether public buildings, palaces, or homes of the wealthy; they served as paving and also as wall decoration. The pavements depicted either designs, large compositions with human figures, or borders both decorative and narrative. The famous mosaic in the Naples museum represents (according to the most plausible conjectures) the Battle of Arbela (Issus), but it must be a copy of a Greek picture translated into mosaic. It was discovered in Pompeii (first century A.D.).

After the fourteenth century the mosaic became the principal decoration of religious edifices; we need only mention the marvels of Ravenna, Venice, Byzantium, and Sicily, whose technique remained unchanged and whose study belongs within the realm of art history. Mosaic was a specialty of Italian artists, but it is also found in France as early as the Carolingian period. In the twelfth century the red marble was replaced by terra-cotta; and pozzolana cement, which ensured an excellent binding of the components, was used. The mosaic was then replaced by encrustations of stones and terra-cotta. In the course of the fifteenth century these expensive procedures were replaced by ordinary marble and terra-cotta paving.

Mosaics depicting human figures, in Biblical and secular scenes, were vigorously criticized. St. Bernard was angered by the pavements with historical scenes, and scandalized by the sight of the faithful walking on the faces of angels and saints.

These developments served as the basis for the creation, in the thirteenth century, of the varnished tile (or "leaded tile," as it was called in the Middle Ages, by virtue of the fact that its transparent glaze was obtaining by the firing of a lead salt). The use of enameled tiles became widespread throughout Burgundy, beginning in the middle of the thirteenth century, and the relaxation of the Cistercian rule permitted the use of these tiles, for example at Fontenay, in a thousand decorative combinations (Figure 277). They varied in size from three to six inches. In some tiles a design was incised in the malleable clay and covered with a black, yellow, or brown enamel. Others were decorated with a technique that was not peculiar to Burgundy but was known in Champagne and in the Île-de-France: the design was first stamped in the soft clay, and was then filled with colored, friable enamel that stood out in a lighter color against the dark background. The enamel of the designs was generally less resistant than

FIG. 277
Enameled floor
tile of Fontenay.

Plate 43. The Battle of Arbela (Issus). Roman mosaic from Pompei, second-first century B.C. Naples Museum. *Photo New York Public Library.*

the clay, and, being more sensitive to the rubbing of feet on the pavement, was more quickly worn down.

The same procedures were used in public architecture; between the thirteenth and seventeenth centuries pavements of encrusted terra-cotta were extremely common in public architecture.

As for the flooring of houses, in many cases it was a simple planking; it is very probable that decorative combinations in the flooring were an innovation of the Renaissance. In ordinary houses the floor was formed of ordinary planks laid side by side. In the houses of the rich and in palaces, the paving on the ground floor was reproduced in the upper floors: the space between the beams was filled in, and a layer of thin mortar was put down to form a base on which paving stones or tiles could be laid.

Between the fourteenth and sixteenth centuries painted tiles were sometimes commissioned from internationally known potters; the great seigneurs also patronized the shops near their castles, whose tradition has been continued down to modern times by the royal porcelain establishments.

The production of both *smaltos* and tiles was identical to that of glass (fusion in the crucible) or a kiln-fired pottery. Upon analysis, the *smaltos* produced in Italy prove to be of a composition similar to the following:

Sand (silica)	1,300
Minium	600
Potassium nitrate	60
Lime fluoride	300
Sodium carbonate	400

The mixture was colored by the addition of metallic oxides: manganese for violet, cobalt for blue, uranium for yellow, copper for green and red, chrome for green, iridium for black. The *smaltos* for gold backgrounds were made from a thin gold sheet appliquéd on a glass and covered with fused *smaltos.*

The slabs of *smalto* were divided with a cutter into small pieces. The technique of laying them was always the same: after he had prepared the solid surface that was to receive the mosaic, the mosaicist coated it with cement (sometimes with soft plaster, if it was a mural decoration), and encrusted the cubes in it.

The cements of antiquity had been composed either of two parts of crushed mortar and one part of travertine lime, or two parts of pozzolana and two parts of aged hydrated lime. More modern varieties of cement consisted of:

	First Layer	*Second Layer*
Pozzolana	10½	8½
Crushed brick	4½	3
Slaked lime	8½	10½
Water	1½	3

In the first layer, three-fifths of the pozzolana and lime was in the form of lumps between two and three millimeters thick; the remainder was in powder form. In the second layer, the lumps were reduced to one millimeter. When the first layer was destined to compensate for irregularities in the masonry (as at Saint John Lateran, thirteenth century), it could be as much as 2⅓ inches thick. On very flat masonry surface, the builders were satisfied with a half inch for the first layer and one-third of an inch for the second; sometimes a single layer was sufficient. The same types of cement were used for pavements.

For decorative mosaics on vaults and walls, the sixteenth century utilized a different technique. Muziano of Brescia (1528–1592) was the inventor of the "oil" method, which he utilized in certain mosaics of Saint Peter's in Rome. The cubes were held in place by a mastic which included:

Powdered travertine	60	parts
White lime (of travertine)	25	"
Raw linseed oil	10	"
Sediment of cooked linseed oil	6	"

The advantage of this binding was that it remained malleable for two, three, or four days in summer, and one week in winter, which made the work easier and corrections possible. Lime cement remained malleable only for several hours.

Mural painting The study and history of painting belong particularly within the domain of the history of art. However, the use of painting in the decoration of buildings includes methods that a history of technology cannot completely ignore.

The art of mural painting was well known in Roman antiquity, a period which has left magnificently decorated walls, for example at Pompeii and at the Palatine in Rome. But much older paintings can be seen in the Egyptian monuments. In order to trace their techniques, it will be useful to mention briefly the various procedures employed since antiquity.

For mural painting, the wall was always given a special treatment (the *tectorium* of the Romans). According to Pliny, it had to receive three layers of lime and pozzolana, and two layers of lime and marble stucco. Each successive layer had to be applied on the still-damp underlying layer, in order to ensure binding.

Naturally these rules were not absolute, and we find layers of differing nature and number under the various paintings. The total thickness varied, depending on the nature of the supporting wall, from as little as one-twelfth or one-fifth of an inch to as much as two and three-quarter or three inches. In order to ensure adherence to the wall, iron nails or cement pegs were sometimes used.

The methods used by the ancients included three types: distemper, fresco, and encaustic.

Distemper, which had been known to the Egyptians, consists of binding the colors with a substance that fixes them: egg (white and yolk, or yolk alone), fig juice, milk, gum, and so on. The painting is applied on a preparation of chalk or plaster mixed with glue.

Fresco is done on the freshly coated wall, whose lime component combines with, and fixes, the mineral matter of the painting. But the lime would decompose the animal and vegetable organic matter, and therefore only a limited number of mineral materials can be used, which reduces the painter's freedom of expression. Upon contact with carbonic acid contained in the air, the calcium hydrate forms a transparent film of carbonate that can resist even bad weather and washing. This method of painting the old temples *(Lanuvium),* which had been utilized in the Aegean and was quickly adapted in Italy, aroused Pliny's admiration.

Encaustic made it possible to obtain brilliant effects. It was first employed in Egypt, where the temperature facilitated the use of the melted wax that surrounded the colored pigments. It was applied on a layer of white wax impregnated while hot in the coating of the walls. Corrections could be made with a hot iron or by an overlay of additional wax. This method permitted the use of all types of coloring matter, both vegetable and mineral; the painting was altered neither by sun nor by heat. However, it was the slowest and most complicated method; fresco on plaster was less costly and easier to execute on large surfaces. The encaustic method was applied only to small areas.

These procedures spread through Roman Gaul; Gregory of Tours speaks of the paintings of Saint-Martin, and perhaps even exaggerates the use of rich, vivid colors. A law of Charlemagne ordered that paintings and gilding be limited to the choirs of churches; the nave had to remain bare and austere.

After the troubled period of the tenth century, churches again began to be decorated (999, Châlons-sur-Marne). During the eleventh century the old methods were renewed, and the twelfth century witnessed a splendid development in this field.

A magnificent testimony to the activity of this period survives in the building known as the Monks' Chapel, near Cluny. The very well-preserved paintings have been carefully studied, and the original methods (1105 to 1108) have been discovered. Around 1930, Fernand Mercier made a thorough investigation and study, in which he lists the basic points of the techniques:

1. The foundation of the picture was very carefully prepared in five layers:
   1. Ocher-tinted whitewash, applied directly to the stone;
   2. A layer of coarse mortar;
   3. Whitewash;
   4. Blue-tinted wash;
   5. Very fine mortar, one-fifth or one-sixth of an inch thick, containing an oil.
2. The drawing was executed with brush and red ocher.

3. The painting was executed with eight colors: blue, yellow ocher, red ocher, cinnabar, cupreous green, ceruse blue, lampblack, and brown (a mixture of white and black). Violet did not exist as a separate color, but was obtained by applying a layer of blue on a cinnabar ground.

The medium shades were laid by a distemper procedure in which the color was mixed with glue. The presence of oil in the preparation of the wall indicates that the paint was to be laid dry. Then the light and dark shades were added; here the colors were mixed with wax, which relates this procedure to encaustic.

This technique, which was very similar to encaustic painting, was inspired by Italian methods that St. Hugues, abbot of Cluny, may have seen at Monte

Plate 44. Frescoes in the Chapelle des Moines, France, twelfth Century.
a. Apse of the chapel.
b. Detail of Christ enthroned in a mandorla. *Photos Combier, Macon*

Cassino in 1083, as well as by the Byzantine character of Italian works of the period.

This is the art that produced the magnificent paintings of the great church of Cluny (Cluny III), which like the paintings of Saint-Denis (1137–1144) aroused the vehement protests of St. Bernard. The paintings of Saint-Savin, and those of the chapel of Le Liget (now almost completely disappeared), mark the end of the twelfth century.

Fresco painting was employed by the early painters, Giovanni Cimabue, Giotto di Bondone, P. Uccello, Masolino da Panicale, and Masaccio, down to but excluding A. Castagno. Distemper was replaced by oil painting in the fourteenth century in northern France and Flanders, but the fresco continued to be utilized by the greatest painters, for example by Raphael in the Vatican *Stanze,* and by Michelangelo. In France, where the fresco was introduced in the sixteenth century by Francesco Primaticcio and Rosso (Giovanni Batista de'Rossi) it was later to cover the walls of the palaces of Fontainebleau, the Louvre, Versailles, and Marly.

BIBLIOGRAPHY

ADAMS, HENRY, *Mont-St.-Michel and Chartres* (Boston, 1904; paperbound, Boston, n.d.).

BLUNT, ANTHONY, *Art and Architecture in France, 1500–1700* (London, 1953).

BRANNER, ROBERT, "Villard de Honnecourt, Archimedes, and Chartres," *Journal of the Society of Architectural Historians,* 19 (1960), 91–96.

BRIGGS, MARTIN S., *A Short History of the Building Crafts* (Oxford, 1925).

COWAN, HENRY J., *An Historical Outline of Architectural Science* (London, 1966).

EVANS, JOAN, *Art in Medieval France, 987–1498* (London and New York, 1948).

FITCHEN, JOHN, *The Construction of Gothic Cathedrals* (Oxford, 1961).

LAVEDAN, PIERRE, *French Architecture* (London, 1956).

LOWRY, B., *Renaissance Architecture* (London, 1962).

MALE, ÉMILE, *Religious Art from the Twelfth to the Eighteenth Century* (New York, 1949).

MOREY, CHARLES R., *Medieval Art* (New York, 1942).

MURRAY, PETER, *The Architecture of the Italian Renaissance* (New York, 1963).

PANOFSKY, ERWIN, (*ed.*), *Abbott Suger and the Cathedral Church of St. Denis* (Princeton, 1946).

——, *Gothic Architecture and Scholasticism* (New York, 1957).

SCAGLIA, GUSTINA, "Drawings of Brunelleschi's Mechanical Inventions for the Construction of the Cupola," *Marsyas: Studies in the History of Art,* 10 (1961), 45–68.

SHELBY, LONNIE R., "Medieval Masons' Tools: The Level and the Plumb Rule," *Technology and Culture,* 2 (1961), 127–130.

——, "Medieval Masons' Tools II: Compass and Square," *Technology and Culture,* 6 (1965), 236–248.

——, "Setting Out the Keystones of Pointed Arches: A Note on Medieval *Baugeometrie,*" *Technology and Culture,* 10 (1969).

SIMSON, OTTO VON, *The Gothic Cathedral* (New York, 1956).

Section Seven

Techniques of Expression

CHAPTER 23

TECHNIQUES OF THE DECORATIVE ARTS

THE HISTORY of the decorative arts is very often confused with that of their techniques, each discovery by their practitioners having brought about a good or bad stylistic evolution. A history of technology should thus complement the history of the various arts discussed. This history has in fact been only partially written; each historian of art is in general satisfied to incorporate into his works a chapter devoted to the procedures of the artists, thus permitting the reader to follow a stylistic evolution regarded as the only evolution worthy of interest.

Understanding the techniques and procedures utilized requires a certain amount of elementary knowledge that we have briefly supplied, deliberately simplifying it so as to give only the general principles of production.

Although we have included it in the volume devoted chiefly to the seventeenth century, this chapter will go well beyond this chronological limit. The most logical ploy seemed to be the individual review of the ceramic arts, the arts of wood, metal, and fabric. We have deliberately omitted from this chapter the art of book production, which requires extensive treatment.

THE THERMAL ARTS

The term "thermal arts" has on occasion been disputed; some authorities prefer the term "clay" arts, which refers to the raw material rather than to the methods that make it possible to transform the material. Heat, on the other hand, has been employed in numerous techniques that have no connection with the thermal arts as such. Whichever term one chooses to use, both include glass objects and stained-glass windows, mosaics, enamels, and ceramics. All these will be studied individually.

Glass

The glassmaker's art is the simplest and also the oldest of the thermal arts; it was already widely practiced in Egypt. The chief properties of glass are its translucency and impermeability. To make glass, the craftsman utilizes only one raw material: sand (silica), which, when fused at high temperature, produces glass. Actually silica is practically never employed in its pure state. In the twelfth century Theophilus Presbyter advised a formula of one part silica to two parts beechwood ash for the composition of glass. The chemists clarified these empirical data by saying that in order to lower the fusing point of silica one adds to the sand "a flux, that is, a basic oxide which lowers the fusing point. The principal fluxes are soda (carbonate, sulfate, nitrate), potassium, calcium (chalk, limestone . . . , etc." (J. Barrelet, page 5).

Blowing

In fusible state, silica combined with oxides has a special property: its malleability, which explains the various phases of production.

The silica, fused in the furnace in clay pots, is removed from the furnace and laid on a sheet of marble. The molten material, called the "parison," is "collected" with a hollow iron tube between six and six-and-a-half feet long, one end of which is wooden, the other, pointed end (the jaw) of iron. The assistant grasps the parison, which has been slightly cooled on the marble, and hands it to the glassmaker, who works seated on a bench with armrests on which he can rest his blowing iron. He blows into the tube to give the molten material the desired shape. Rotating the material permits him to obtain circular, cylindrical, or other shapes; blowing into wooden or metal forms makes trial-and-error unnecessary when it is a question of reproducing several objects of the same shape. Irregularities and burrs are cut out with the iron or with chisels. If the parison has become too cool and lost its malleability, it can always be reheated.

Here, in brief, are the various stages that have been used in the production of glass objects since the invention of the blowing iron, which seems to have been known to the Egyptians. A more detailed description of the method will be found in the chapter on construction techniques (Part Two, Section One); cf. Figure 13 of Chapter 10 "The Extraction of Chemical Products."

Decoration

The decoration of glass objects is extremely varied; it is done sometimes with heat, sometimes with a cold method.

"Cold" decoration is obtained with the help of wheel-operated molds and glass cutters, and tips. The Romans were perfectly acquainted with this type of decoration. Not until very late in the nineteenth century was the practice of engraving glass with hydrofluoric acid developed; the parts that were to be left plain were protected by a varnish.

"Heat" decoration is more varied; the play of forms constitutes its essential element. In its molten state glass sticks to glass, whence the great ease with which a number of additional elements have been added: rims, threads, vase handles, and so on. In the Merovingian period small decorations were also heat-pressed on solidified glass by means of matrices, probably of terra-cotta; drops of enamel, both colored and clear, were also added while hot to shaped pieces.

Color is another important element of the decoration; objects are colored by the addition of metallic oxides to the material. Not all the colors are deliberately produced; some result from the very nature of the material used—metallic oxides existing in natural state in the sand. The use of colors was known to the Egyptians. If the glass is not tinted before the production of the object, the completed object can be partially enameled, as in the case of the glasses of the sixteenth, seventeenth, and eighteenth centuries, and ornamented with letters, portraits, and coats of arms. A white enamel obtained with tin oxide was sometimes utilized to produce partial or total opacity; the Venetian filigrees of the sixteenth century (the so-called "*a latticinio*" glasses) made this technique famous. In Picardy a similar process that consisted of coating finished glasses with white enamel came into use in the Merovingian period.

The gilding of glass was well known to the Romans; in the twelfth century Theophilus Presbyter tells us that "the Greeks" gilded their glasses with gold leaf, attaching it with the help of a hot flux.

As for flashed glass, it was obtained as follows: a second layer of glass, of the same or a different color, was applied in fused state on a finished object. The crackled glass so dear to the sixteenth-century Venetians was simply plunged while hot into cold water (J. Barrelet, page 158). The "*verre églomisé*" which was very popular in the sixteenth, seventeenth, and eighteenth centuries, involved painting *under* glass; its name comes from one Glomi, an eighteenth-century practitioner of this art.

The great malleability of glass permitted it to be drawn and shaped into small human figures, often reinforced with iron. Beginning in the seventeenth century the production of these figurines, crèches, and animals, which were sometimes extraordinary, became a specialty of the city of Nevers.

Windows, crystals, and mirrors

The composition of window glass differed slightly from that of ordinary glass; generally soda or potassium was added to the silica. Artificial crystals, which should be very carefully distinguished from rock crystals, are lead crystals.

The fused material was blown into cylinders, which were placed on the marble slab, split lengthwise, and flattened. It was also possible to blow the molten material into the form of a sphere, which was then flattened into a disk by swinging it in a horizontal plane; this made the glass thicker in the center than at the edges.

The mirrors most frequently utilized in the Middle Ages were of polished steel. However, Barrelet notes (page 42) that "mirrors of glass lined with metal were known throughout the Roman Empire, in Gaul as in Egypt. They were cut from balls of blown glass, which gave them a curved shape, a method that is still employed today to obtain certain eyeglass lenses. Into the concave portion was poured a thin layer of molten lead; the glass itself was so thin that it did not shatter in this operation. These mirrors were small, and were inserted in a framework. They gave a very clear image."

In the fifteenth century the Venetians replaced the lead with an amalgam of tin and mercury, which continued to be used until the nineteenth century. These mirrors were first blown in the form of a cylinder. At the end of the seventeenth century one Perrot, an Orléans glassmaker, discovered a method of pouring glass instead of blowing it; mirrors thus cast on a marble slab were much larger than their predecessors. The Royal Mirror Factory installed in the castle of Saint-Gobain in 1685 contributed greatly to spread the use of this method.

Stained-Glass Windows

The stained-glass window has often been defined as a "mosaic" of translucent glass. Actually, it would be better to say that it is a series of pieces of translucent glass tinted in the mass or painted with vitrifiable colors, and juxtaposed by means of lead rods, or held by plaster or cement.

The glass utilized is that of "window glass." In the twelfth century stained glass was made from potassium obtained from ferns; soda, in the form of sea salt, later replaced potassium, at least after the thirteenth century.

Color was obtained from metallic oxides (oxides, bioxides, and cuprous oxides) added to the molten glass. In the thirteenth century blue was produced from copper or cobalt, red from copper, green from manganese, and yellow from iron. These colors varied in the course of the centuries; in the fifteenth century blue was obtained from manganese oxides, gray from nickel. Changing the length of the firing period produced colors of varying intensities. Once tinted, the parison was blown following the methods already described for the sphere and the cylinder.

Once it has been cooled, the glass must be cut with a hot iron, according to the composition to be executed: the cartoon. This cartoon can be prepared on parchment, canvas, or metal. Theophilus Presbyter advised composing it on wooden tables and outlining the glass, panel by panel, on these tables, with lead and tin; all the information concerning design and color was written on the table, and could be read through the transparent glass.

Assembling the pieces of glass was done by means of lead rods grooved on each side to hold the glass. (In some Arabic specimens of this art, the pieces were held in place by means of plaster.) The rods of lead were then soldered to each other. The completed window was supported by iron rods that divided it into geometrical panels and gave it a rigidity the leading alone could not provide. The evolution of architecture and the increasing size of the windows posed increasingly difficult problems for the glaziers.

In modern times a new technique called "*verre éclaté*" ("splintered glass") produces effects of iridescence and unlimited plays of colors. The "splinters" are produced by means of special hammers, and cement replaces lead for assembling the pieces.

Numerous improvements were very quickly made in the early practices. To depict human flesh (faces, hands) and folds of garments, the glaziers conceived the idea of coating the finished glass with a mixture of copper filings, iron dross, crushed glass, and resin; this was applied with a coarse brush, and the pieces of glass were returned to the kiln. This was the process used in making the "*grisailles*" so dear to the glaziers of the Middle Ages.

In the fourteenth century a new color was created. Called "silver stain," and made from silver chloride and ocher, it too was applied with a brush and refired. The watered and warm yellow tones of garments were "painted" in this manner; so too were flesh tones at the end of the fifteenth century, thanks to sanguine mixed with a flux.

Flashed glass was also frequently utilized, particularly for the red tones, which, being very deep, would have made the windows too dark; clear glass was "flashed" with a thin film of red glass. Flashed glass could be made of several layers. In the fifteenth and sixteenth century specimens, plays of light were obtained by cutting into the outer layer with a glass cutter, thus causing the underlying layer or layers to appear.

Grisaille, silver stain, and sanguine were not, strictly speaking, painting on glass; the color penetrated in depth, but we cannot actually speak of tinting in the mass. "Painting on glass," while derived from the same principle, consists of painting on already hardened glass with colors mixed with fluxes. This method, which was perfected in the sixteenth century, signaled an unmistakable degeneration of the stained-glass window.

Mosaics

The "mosaic" is a combination of materials, whether homogeneous or diversified. There exist mosaics of stone, marble, and other materials; marquetry is a wood mosaic. The glass mosaic is composed of pieces of glass tinted in the mass and cut with the hammer into small cubes of uniform size; these cubes are assembled in accordance with a design, and are attached to the wall by means of one or several layers of mortar. Colored canvas was sometimes placed on the wall; occasionally it was discovered under the mortar. The cubes can be assembled in the workshop and glued face down onto the cartoon, then put in place on the wall; after emplacement, the cartoon is peeled off with water.

The gold-ground mosaics with their rich reflections received special treatment. Theophilus Presbyter devotes considerable discussion to the "Greek" manner, which consisted of sheets of glass "as thick as a finger; they cut them with a hot iron into small square pieces, and cover them on one side first with gold leaf and then with a very shiny layer of glass . . . they then assemble them on an iron table covered with lime or ashes, and fire them." The result was a sheet of gold between two layers of glass.

Enamelwork

Philostrates notes, "It is said that the barbarians near the ocean spread colors on burning bronze, where they become as hard as on stone, and the design they depict is preserved." Who these "barbarians" were, we do not know; on the other hand, it seems almost certain that the Egyptians and Etruscans knew the art of enameling, which was being practiced at the beginning of the Christian era by the "barbarians" from the East.

What exactly is enameling? Molinier's excellent definition is still valid: "Enameling is the art of applying on metal a molten flux or glass tinted various colors by means of metallic oxides." These enamels are translucent (that is, they permit the underlying foundation to be seen) or opaque, depending on whether or not tin oxide is used in their composition.

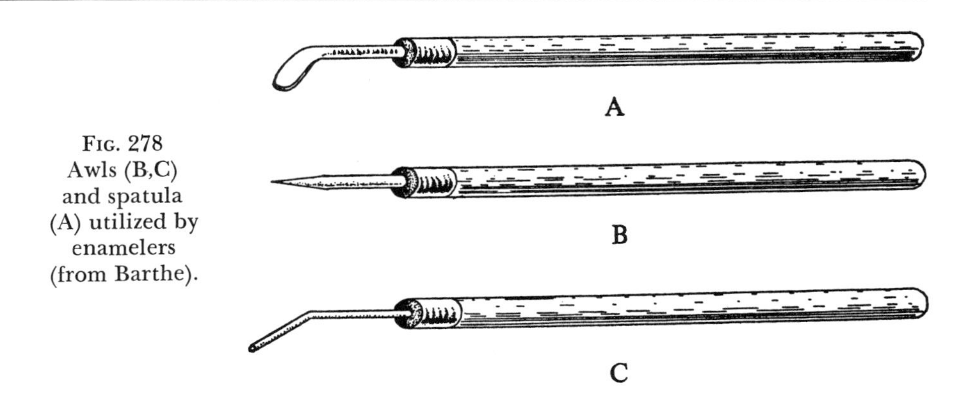

FIG. 278
Awls (B,C)
and spatula
(A) utilized by
enamelers
(from Barthe).

This simple definition suffices to distinguish enameling from, for example, cloisonné goldsmith's work, which has sometimes been confused with enameling. In the case of cloisonné goldsmith's work, the craftsman works with cold material, and inserts pieces of colored and cut glass into metal compartments.

The preparation of enamels is the same as that of glass. The kilns must be constructed so that they can be regulated at various temperatures, and during each firing the temperature must remain constant. Their construction was always a subject of considerable concern to the ancient enamelworkers. The enamels were supplied in colored cakes; they had to be broken and pulverized in a mill, then crushed, washed, and dried.

The metals most frequently utilized as foundations are gold, silver, copper, and, more rarely, iron and tinplate. The metals are first worked as for goldsmithing. Silver must be free of the alloys that favor its preservation but diminish its brilliance. Red copper is preferred to yellow, which contains zinc.

Cloisonné enamels

Cloisonné enamel, for which the Byzantine enamelers were particularly renowned, is an enamel whose colors are separated by small wire compartments placed in upright position on the sheet of metal that acts as a support. The preparation of the sheet to be enameled can be done in three ways.

In the first case, if the entire support is to be enameled, the artist bends its edges up to make a cavity resembling the bottom of a box, and then fixes his tiny metal wires forming the compartments to the support, either by soldering or gluing them, for example with gum arabic.

In the second case, if it is a question of a partial enameling, the portion to be enameled is hollowed out in the metal base, and the compartments are arranged as in the first method. This method was utilized by the Byzantine enamelers, who must have cut out wooden or metal wedges giving the general design of the pieces to be reproduced. All the objects made in this way with the same wedges will obviously have the same dimensions.

Lastly, it is possible to solder onto one sheet acting as a support a second sheet that has first been perforated and that gives the general outline of the design to be reproduced. (Figure 279).

Once the metal had been prepared and the partitions fixed in place, the

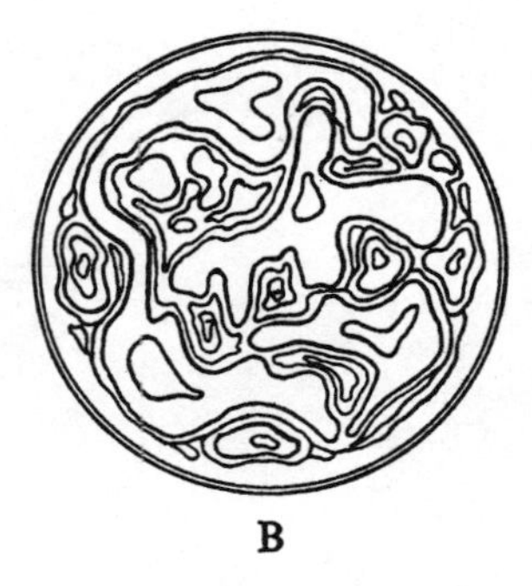

FIG. 279 Compartments arranged on a plate in preparation for *cloisonné* enameling (From Barthe). A: Incised Patten. B: Arrangement of the compartments.

enameler lays his colors in powder form in each small cavity, and then fires the object (Figure 280). The next step is polishing.

Cloisonné enamels are generally executed on gold, more rarely on silver. In both cases, because of the beauty of the metals and their reflection, the enamels are translucent; opaque enamels are utilized on the other metals. The cloisonné enamels are generally small in size.

The so-called "*plique*" (opplithe, applique) enamels, which are frequently mentioned in inventories of the fourteenth and fifteenth centuries, appear to have been, as C. Enlart has demonstrated, simply cloisonné enamels that, after having been practically abandoned in the thirteenth century, become fashionable again in the fourteenth. The word "*plique*" may come from the Latin "*plicatura*," and is supposed to refer to the bending (*pliage*) of the partitions.

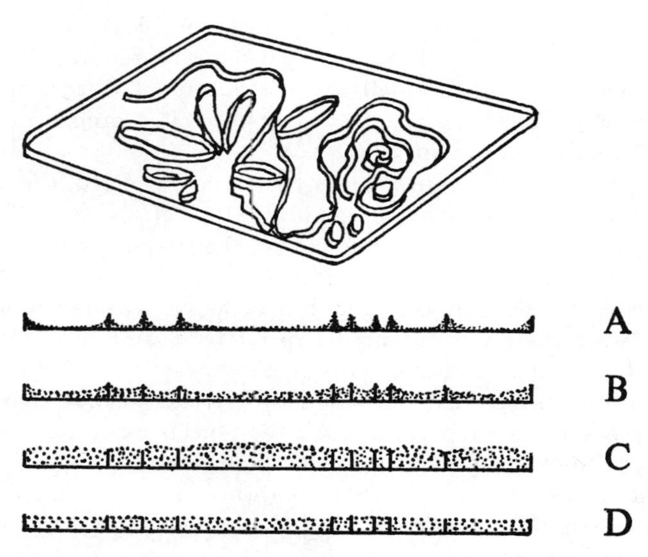

FIG. 280 Compartments arranged on a bas-relief to be enameled, seen in plan and in profile. Profiles A,B,C, and D show the arrangement of the enamels after the various firings (from Barthe).

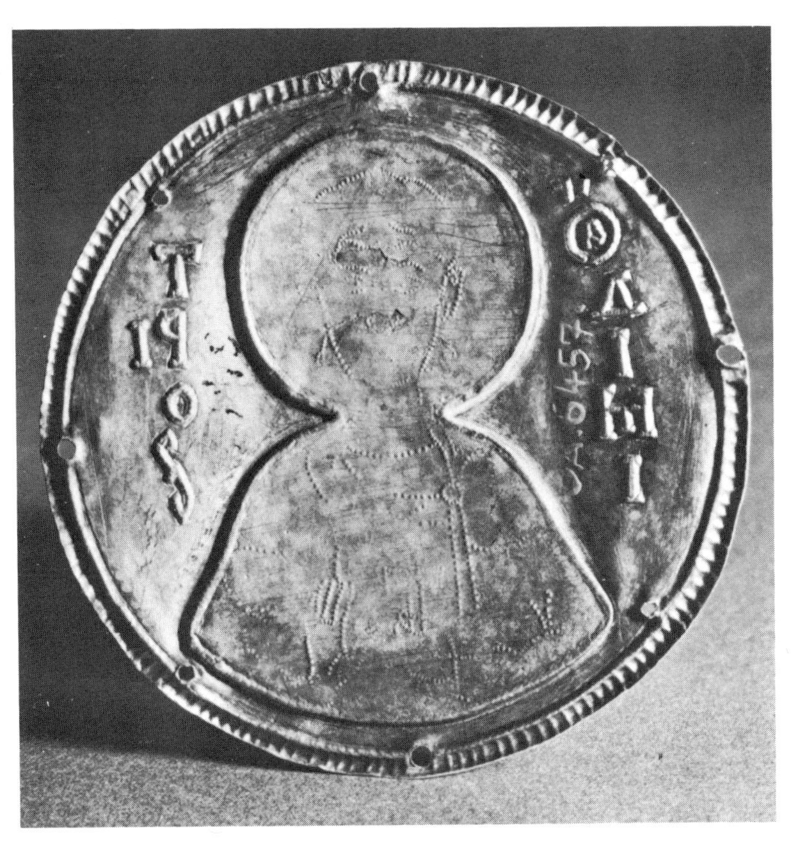

Plate 45. Medallion of Demetrios. Byzantine cloisonné enamel, tenth-eleventh century, obverse and reverse. Musée de Cluny. *Photo by the museum.*

Cloisonnés à jour

The "cloisonnés *à jour*" enamels are enamels of a special type; they are mentioned by Cellini, who explains that they must be fired in a kind of box, the enamels being carefully isolated from the bottom of the box by means of soil or other material. After firing, the soil is removed. The result is "cloisonnés without a support" or "cloissonés *à jour.*" The enamels are translucent.

The *émaux de résille* so fashionable at the beginning of this century are of the same type: the support of translucent cloisonnés enamels is mechanically abraded after the firing. The term *"résille"* designates partitions that hold the enamels like the holes of a net. Molinier has incorrectly applied this name to a very different type, namely, cloisonnés embedded in a flux. A completed motif, generally small, in cloissoné enamel on gold is embedded in a colorless flux, in much the same way as objects dug up in the course of archaeological excavations are embedded in a small block of plastic material in order to protect them from the air.

Champlevé enamels

Champlevé enamelwork was done everywhere in the West in the Middle Ages; the early Limoges, Limousin, Rhenish, and English enamels were all of the champlevé type. The method consists of hollowing out cavities in a metal support, which are then filled with enamel, fired, and polished. Thus the technique is very similar to the cloisonné method. The metal most frequently utilized was copper, a poor metal that can be "champlevé" without damage; the enamels are opaque. The enameled objects can be quite large (the plaque of Geoffroy Plantagenêt preserved at Le Mans measures two feet by one foot). This method can be used for enameling convex and concave surfaces, which is practically impossible with the cloisonné method.

"Mixed" enamels are enamels that are both cloisonnés and champlevés. The major portion of these enamels is generally champlevé, while certain delicate details (eyes of the figures, borders of small crosses, and so on) are cloisonnés. Numerous examples of mixed enamels are known, notably certain Limousin and Limoges enamels of the twelfth century.

Other varieties

Basse-taille (or translucent) enameling, so prized in the fourteenth century, is a variant of champlevé enamel. For ordinary champlevé enamels there was no need to devote special attention to the cavity that was to receive the enamel, since it was covered with opaque enamel. But if translucent enamel was to be used, background effects became possible, and the reflections differed, depending on the colors used and the depth of the champlevé. These are the *basse-taille* enamels — small "enameled bas-reliefs," as Emil Molinier so well defines them.

In later periods it was possible to attack the support with acid, as in the case of simple engravings, and modern procedures of stamping made possible the industrial production of small enameled plaques; their technique is very remote from that of genuine champlevé, and yet they are derived from the latter.

Enamels in the round came into fashion in France at the court of Charles VI. One or several translucent or opaque layers of enamel are simply applied on gold statuettes, and fired. Enamels *"sur fond guilloché"* are produced on gold, silver, or copper. Their name comes from a lathe-operated cutting tool, the "guilloche," which reproduced symmetrical designs on the support to be enameled. Enamels *"sur fond flinqué"* are applied on a support (generally gold) pricked with tiny points, which have sometimes been compared to small honeycombs. The *"guilloché" and "flinque"* techniques of champlevé enamelworking were widely used by snuffbox manufacturers at the end of the eighteenth century.

Painted enamels are to champlevé enamelwork what *grand feu* pottery is to *petit feu* pottery. The decoration is applied on an enamel that has been already fired. In other words, a thin copper sheet is given a coat of enamel, while its reverse side is covered with a flux (the *contre-émail*) to prevent the sheet from becoming cloudy during firing. Once the enamel support has been applied and fired, the colors are laid on it, and the object is refired as many times as necessary. The range of colors is very extensive.

The early painted enamels were made in slightly different fashion. On the first layer of enamel the artist outlined his figures, generally with brown enamel. The piece was fired, and then the figures themselves were enameled in various colors, but this time in translucent enamel, thus emphasizing the reliefs.

Grisaille enamels, a favorite with the Limoges enamelers, are similar to flashed and engraved glass and to slipped ceramics. A sheet of copper is completely coated with white enamel, and this first layer is covered with a layer of black enamel that is then partially removed with the help of a pointed stylet, so that the underlying white layer appears in selected areas and the design stands out in white on a black ground. The painted enamels, executed on copper, are generally opaque, with the exceptions mentioned above. Sometimes certain elements of the decoration were heightened with translucent enamels; for vivid splashes of color, the artist placed under his translucent enamels tiny sheets of gold or silver whose brilliance was intended to reflect the light. Tiny drops of translucent enamel laid on in this way sometimes acquired the appearance of small precious stones.

Nielli

Nielli, which are very often confused with enamels, were to be found in large numbers in Germany in the twelfth century and in Italy in the fifteenth and sixteenth centuries. Actually, they are simply silver sulfides. Cellini advises smelting one ounce of silver, two ounces of copper, and three ounces of lead; these molten metals are combined in a bottle that is half filled with sulfur, and are cooled. The bottle is then broken. If the material is not sufficiently homogeneous, borax is added.

The gold or silver sheet to be worked is engraved, boiled, and cooled; it is then reheated, and the crushed and diluted niello is spread on it like an enamel. The piece is cooled, after which it is polished with a steel burnisher.

Ceramics

The term "ceramics" comes from the Greek word *keramos,* meaning clay, while the word pottery comes from the Latin *potere,* to drink. The two meanings are complementary, since the essential concern of the early potters was to make clay vessels capable of holding liquids and beverages.

When kneaded with water, clay becomes very malleable, and hardens in drying. Originally the modeling was done by hand, but at a very early stage the invention of the potter's wheel (Figure 281) made possible a variety of shapes, while the invention of molds permitted the repetition of the same models. When dried in the sun, these pottery objects remained permeable and porous. To correct this porosity, potters created the various types of ceramic products we are about to define.

Glazed pottery

In order to make the early pottery waterproof, it was covered with a vitreous, transparent or opaque, colored or clear coating; as in glassmaking, the colors were obtained by means of metallic oxides. De Groote proposes for these vitreous coatings the name "glaze," which he defines as "transparent, clear or colored compositions covering a permeable clay body, or added by second firing at low temperature on an already vitrified ceramic product." This term "glaze" in the ceramic art thus corresponds to that of "flux" employed in enamelworking.

These glazes vary greatly. The lusters are silico-alkaline glazes used to cover shaped and fired clay; the black lusterware, in which black lusters were laid on red earthenware, was the most famous of the Greek ceramic wares. The "varnishes" (the term is incorrect) are transparent glazes, both clear and colored, with a lead base. "Varnished" pottery was very common in the Middle Ages.

The clay pottery could be given a special decoration before the application of the glaze; this was the case for pottery encrusted with clay of various colors, and also for "slipped" pottery. The latter was prepared as follows: a clay of a given color was completely or partially covered with another clay of a different color; slight incisions were made in the top layer, permitting the underlying clay to show through and thus giving a two-tone decoration; a transparent glaze was then applied.

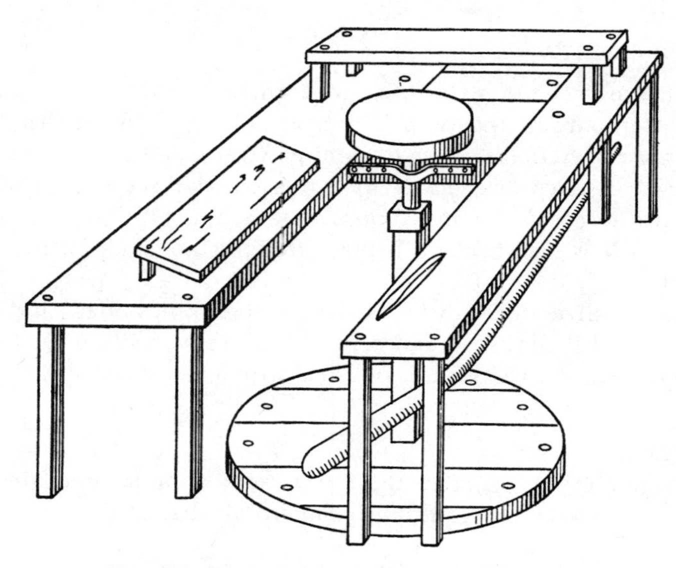

FIG. 281 Potter's wheel (from Piccolpasso, *Arte del Vasaio*).

Plate 46. Potter's kiln, sixteenth century. Engraving taken from E. Piccolpasso, *Arte del Vasaio. Photo Conservatoire des Arts et Metiers.*

Faïence

"Enameled" clay does not necessarily have for the ceramicists the same meaning that it has for enamelers. If enamel is, according to our definition, "a glass colored with metallic oxides and heat-applied to a base," then pottery objects covered with a glaze are enameled potteries. For the ceramicists, however, the term "enamel" is more particularly reserved to the "opaque vitreous coatings laid on any ceramic body" (de Groote). Opacity, which is due to the presence of tin salts, would thus be the criterion used to distinguish "enamels" from glazes.

The term "faïence" owes its name to the ceramic products of the city of Faenza in Italy, where they were produced between the fourteenth and eighteenth centuries. This term too lends itself to confusion. De Groote wishes it to be applied to "ceramic products with porous walls, covered with a vitreous coating." Thus it seems that for this author the terms "faïence" and "enameled terra-cottas" are not completely synonymous. For our part, we would be tempted to reverse the terms of these definitions, so that all terra-cottas covered with any kind of a vitreous coating would be considered enameled, while the term "faience" would be applied only to those terra-cottas that are covered with an opaque enamel, as for example the terra-cottas produced at Faenza. In any event, no one hesitates when it is a question of defining the "faïences of Bernard Palissy," or *grand-feu* and *petit-feu* faïences.

We have mentioned Palissy's "faiences" as a separate category for the reason that these products are as it were intermediate between "varnished" potteries with transparent lead glaze and opaque faiences; only a small quantity of tin salts went into the composition of the famous potter's ceramics.

Grand-feu faïences

Genuine faïence, for example the Italian majolica, is completely covered on both sides with a stannous white enamel that acts as a background; the colors, made from metallic oxides, are applied on the dried, unfired enamel. Only then is the object fired, at quite high temperature. The colors, which are generally dark and deep, are few in number, because of the high temperature of firing.

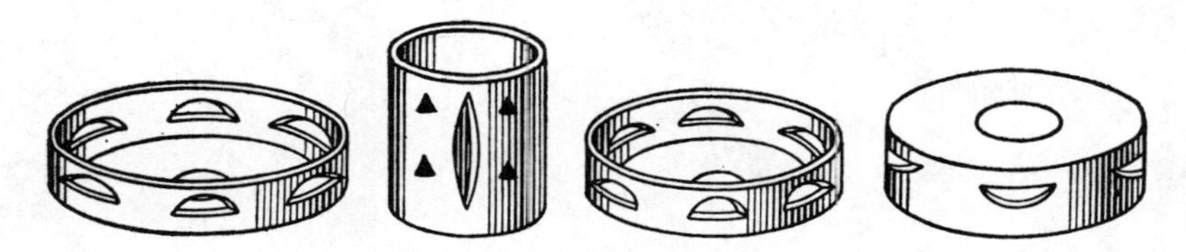

FIG. 282 Containers used in the firing of pieces of majolica (from Piccolpasso, *Arte del Vasaio*).

The methods utilized by the Italian ceramic makers are well known, thanks to a manuscript left by one of their number, Cipriano Piccolpasso, who in the sixteenth century wrote for the benefit of his colleagues *L'Arte del Vasaïo* (The Potter's Art). This manuscript describes the various phases of production: preparation of the clay, kneading, turning, molding, first firing, enameling, preparation and application of the colors, firing in the kiln, and cooling. The metallic lusters of certain types of faience (Hispano-Moorish faiences, faiences from Deruta and Gubbio) are due to the presence of copper, silver, and gold salts.

The early faiences of Lyon and Nevers were *grand-feu* faïences.

Petit-feu faïences

The decoration of *grand-feu* faïences is applied on unfired enamel, that of the *petit-feu* faïences on fired enamel. The object may be refired several times, whence the possibility of a much greater range of colors, and also of retouching. *Petit feu* faiences, which appeared at the end of the seventeenth century, underwent considerable development, especially in France, in the eighteenth century.

"Faïences fines"

The inaccurate term "*faïence fine*" (which designates earthenware washed over with a white slip and covered with a clear glaze) emphasizes (if emphasis is needed) the confusion in the terms employed, even in early times, in ceramics. Alexandre Brogniart (1770–1847; became director of the Sèvres establishment in 1800) defined *faïence fine* as a "white-clay, opaque pottery whose texture is fine, dense, and sonorous, and which is covered with a lead crystalline glaze." The material varies in composition; that of the English *faïences fines* contains clay and crushed flint. A mixture that includes kaolin produces opaque porcelains. Like stoneware, the *faïences fines* are intermediate between potteries and porcelains.

Stoneware

Brogniart carefully distinguishes between two types of stoneware: ordinary ceramic stoneware and fine stoneware. He defines the ordinary stonewares as "thick, very hard, sonorous, opaque potteries with a fine grain." The clay is given a glaze or vitreous coating, which is fired at the same temperature as that which causes the vitrification of the object. Recently a more general definition has been suggested: "The term 'stoneware' can be applied to any ceramic product which after the final firing presents a vitrified broken edge, but remains opaque when thin" (P. Fouquet). Glazing of ordinary stoneware can be accomplished by throwing sea salt on the objects during firing.

The clay used in fine sandstone is more complex than that of the ordinary stoneware; kaolin and flint often enter into its composition. Fine stoneware is very similar to *faïences fines.*

Porcelains

Fouquet defines porcelain as "any ceramic product that after final firing presents a vitrified broken edge and a certain translucency when thin." This definition clearly marks the dividing line between porcelain and stoneware: translucency when thin.

Soft porcelain

For centuries the Western pottery makers tried to imitate the beautiful porcelain brought back from the Far East by travelers and businessmen. Their efforts resulted, in the sixteenth century in Italy and in the seventeenth and eighteen centuries in France and Germany, in the creation of a type of ceramic product that was as closely related as possible to the hard porcelain of the East: soft porcelains.

The composition of the clay used in these porcelains could vary, and many porcelain makers died refusing to reveal their secrets. The clay generally contains "a white, calcareous diorite-sand that is not very plastic, then a subtle mixture of sand, sodium, sea salt, and niter, made homogeneous by the presence of a binding element" (G. Fontaine). After the first firing the objects are given a transparent lead glaze. The colors, which are made from metallic oxides, penetrate this glaze, which during firing combines with the flux added to the colors and permits perfect adherence. Gold – "gold powder obtained by the trituration of gold leaf" (Brongniart) – was utilized without a flux.

Hard porcelains The discoveries of the German ceramists at the beginning of the eighteenth century, and notably those of Böttger in 1709, made it possible to perfect a new type of porcelain: the hard porcelains made from kaolin, a nonfusible white clay. The clay used in the hard porcelains included among its principal components:

1. Kaolin (which constitutes the refractory base);
2. Quartz, a very pure silica which combines with the other components to form a silicate which acts as a binding agent;
3. Felspar, which acts as a flux and gives transparency.

After a first firing at 700 degrees or 800 degrees C., the object is covered with a very hard glaze, or vitreous coating, often made from felspar, and is fired at approximately 1,250 degrees to 1,280 degrees C. The decoration is then painted on the glaze, and the object is refired.

The hard Chinese porcelains were also made from kaolin (the name "kaolin" comes from the city of Kingtechen) and felspar.

Biscuit Biscuit (or bisque) is undecorated porcelain. The Sèvres bisques are particularly famous under the form of statuettes or groupings that in some cases are interesting models of sculptures. Depending on the composition of its clay, bisque is either soft or hard porcelain.

THE WOODWORKING ARTS

Wood Various types of wood are utilized in the woodworking arts. The soft woods – chestnut and particularly poplar – are preferred by sculptors. Fine wood – fruitwoods of varied colors, such as cherrywood and pearwood – are used for veneers and rustic furniture. The resinous wood – fir, pine, and larchwood – are sometimes used in ordinary furniture. The hardwoods – oak, beech, walnut, and ash – which are harder to work but are stronger, are used for first-class work. Then there are the precious woods – mahogany, amboyna, and such island woods as rosewood, kingwood, violetwood, and cayenne – which add their warmth to veneers.

The choice of material is of primary importance. The lumber should be chosen neither from the sapwood (the young portion of the tree, under the bark) nor too close to the woody center. Artists are able to turn even defects in the wood to good use; gnarls (pathological excrescences of various trees) and knots in the grain (in portions taken from the intersection of the large branches and the trunk) often decorate fine cabinetwork.

In any case, it is important that the wood be dried for a long time and that once worked it does not shrink. Today the drying process is accelerated by various machines. In the nineteenth century it was still being done naturally; the pieces of lumber, carefully piled up but isolated from each other and protected from rain, were left to dry for several years in the craftsman's shop. They were then sawed up before being sent to the workshop. Sawing is done with the grain, that is, parallel with the fibers; when done slightly on the bias, it produces a more varied effect in veneers. End-grain lumber is sawed perpendicular to the grain.

Many craftsmen have utilized and continue to utilize wood: sculptors, cabinetmakers, carpenters, manufacturers of small articles and inlaid work, carriage makers. We shall briefly describe the techniques involved in the arts of furniture making.

Carpentry

The carpenter's equipment, which is now relatively mechanized, was formerly quite simple: a workbench, various types of saws (the first mechanical saws date from the very end of the eighteenth century), hammers, planes and jointers, brace bits, and various chisels.

The principal problem facing carpenters has always been that of joining. The oldest and also the strongest of these joints is the "tenon-and-mortise" joint: the tenon, the male element, is cut in one of the parts to be joined, and the female element (the mortise) is hollowed out in the other part; the whole ensemble is strengthened by rigid pegs. In the dovetail joint, both parts are carved in the shape of a bird's tail. This joint, which is found as early as the thirteenth century, has been gradually perfected; for example, the hidden dovetail joint is partly invisible.

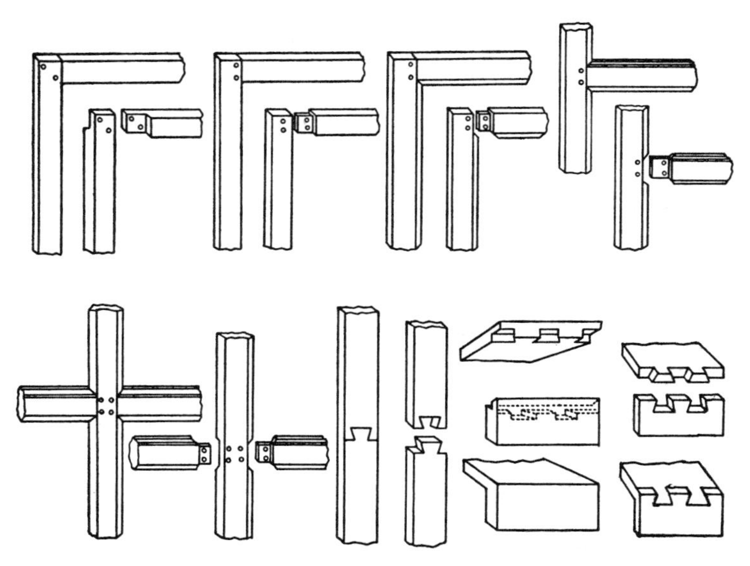

Fig. 283 Various types of joints used in carpentry work (illustration from the *Encyclopédie,* Vol. VII).

The grooved joint is very simple, and can be modified in an infinite number of ways. Roubo's *L'Art du menuisier,* published in 1772, and the *Encyclopédie* describe at length a number of other joints: keyed, forked, and so on – the combinations seem to be infinite. Glue was in principle forbidden; an article of cabinetwork was supposed to be built so that it could be taken apart and reassembled.

Moldings were cut from a solid piece of wood, with chisels and planes; nowadays they are machine made. Such items as furniture feet and small decorative columns were always shaped on the lathe.

A piece of furniture could be simply polished and waxed; it could also be painted, varnished, gilded, silvered, lacquered, or inlaid.

Painted and varnished furniture

There is much painted furniture in existence, and certain regions are very partial to this type. Watin's *L'art du peintre doreur, vernisseur,* published in 1774, gives numerous instructions for painting on wood, using water, glue, oil (linseed, walnut, poppyseed), turpentine, or varnish. The oil and glue methods were utilized for such purposes as paneling and certain types of furniture; the varnish method was also used in certain types of furniture, panels of carriages, and the like.

Varnish was made from resin. Watin writes that "the art of making varnish consists of dissolving one or several resins in a fluid, or incorporating a fluid into resins melted over an open fire in such a way that they are unable to return to their former consistency." The liquid serves only as a binding agent, and evaporates when the substance is laid on the furniture, leaving the varnish "with its own transparency." The liquids prescribed for dissolving resins are turpentine, spirits of alcohol (light varnish) and oil (oily varnish); the resins utilized are either genuine resins (including copal), gums, or bitumens.

Gilding

The methods used for gilding wood in the seventeenth and eighteenth centuries were so-called "leaf" methods, in which very thin gold leaf was applied on the parts to be gilded.

Fig. 284 Various phases in the making of a sofa. (illustration from the *Encyclopédie,* Vol. VII).

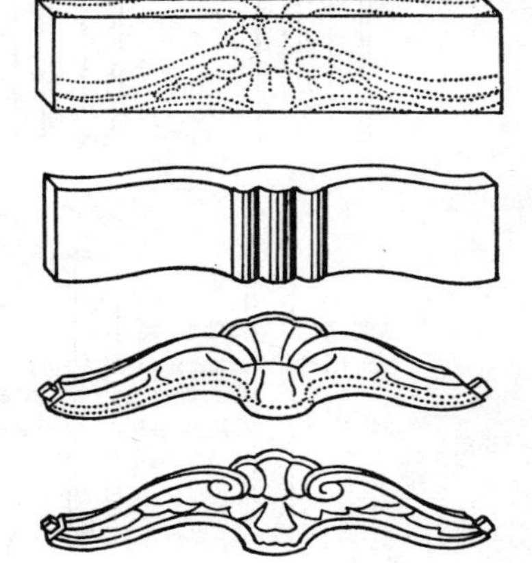

Watin carefully distinguishes several methods of making the gold lead adhere to the surface. Distemper gilding includes a number of operations that can be summarized as follows. The wood is first given a preparatory sizing so that the gold will "stick"; this dressing, which is applied in several layers, must be "stripped and redressed" to restore to the sculpture all its delicacy of line. The finished dressing is "yellowed" with glue mixed with yellow ocher in order to "fill the pieces which the gold cannot enter." The next operation consists of "laying the support" of the gold. This support is composed of "Armenian bole, a small quantity of sanguine, a very small quantity of graphite, and a few drops of olive oil," the whole moistened with glue. The gold leaf is applied to this support with cold water, and is burnished or "matted."

"Grecian" gilding is quite simillar to the foregoing, but Watin notes that the unpolished areas are reserved and treated later.

"Oil gilding" is done on a dressing made from white lead, yellow ocher, and litharge, moistened not with glue but with oil. The support is then laid on, and the gold leaf is applied and polished.

Silvering is done in the same way, by laying silver leaf on a preparatory base.

The gilding methods used in the nineteenth century did not have the quality of leaf gilding. "Varnish gilding" was prepared by pulverizing gold leaf and blending this gold powder with a quick-dry varnish; this mixture was laid on the dressing. A still more mediocre method consisted of rubbing the object with a paste that contained a certain proportion of gold chloride.

Lacquers

In the seventeenth and eighteenth century lacquered objects were generally imported from the Far East. They were either "painted" or sculpted.

Sculptured lacquers (Coromandel lacquers, for example) are composed of a thick paste made with filasse, boiled paper, and eggshells. This paste is crushed in camelia oil, and can then be colored. Several layers of this mixture are spread on a wooden support covered with a very thin piece of cloth, to a thickness of as much as one-third to two-thirds of an inch. A design is outlined on it, and is engraved with the graver.

The "painted" lacquers are varnishes prepared with a resin (or, more precisely, a latex) obtained from certain trees that grow in the Far East (China, Japan, Indo-China). This resin is moistened with oil to which ferrous sulfate and rice vinegar have been added. The colors are made from natural substances: black from animal substances and oil of tea, yellow from pork gall and oil, red from cochineal, and so on. The wood is coated with a preparatory layer of emery powder, vermilion, gamboge, and oxgall. After polishing, between six and eighteen layers of lacquer are laid on with a brush; each layer must be perfectly dry before the next layer can be laid on. On the subject of lacquers, an amusing but vague *Mémoire sur le vernis de Chine,* written by Father d'Incarville, was repeated by Watin.

Western artists made every effort to either reuse or copy the Oriental lacquers. The Louvre, for example, possesses a Louis XV commode decorated with Coromandel lacquers that still bear the signature of a famous lacquer maker;

Carel, who stamped the commode, merely cut out and installed an imported screen. Watin devoted an entire chapter to "the manner of imitating Chinese lacquers." The varnish was made with either "carob" or lacquer gum, and spirits of alcohol.

A completed chair or bed was sent to the tapestry maker, who upholstered it with horsehair straps (springs did not come into existence until the nineteenth century) and rich fabrics, and outlined its contours with braid.

Marquetry

The origin of marquetry must be sought in the encrusted work made fashionable by the Italians during the Renaissance, traces of which are sometimes found in certain rustic furnishings. Small grooves are cut in the wood (generally a softwood is used, and pieces of wood of contrasting colors are glued in these grooves. Small pieces of polychrome marble are sometimes used instead of wood. The Italian word *intarsia,* which originally designated only this type of work, is today used to mean inlaid work in general. The first French articles imitating *intarsia* were decorated with tiny tin wires incrusted in and glued onto an often precious type of wood.

Shell and copper inlay work, which made the name of A.-C. Boule famous during the reign of Louis XIV, is also a form of encrustation, but it is applied, not to a portion, but to the entire panel being inlaid. Following a given design, the cabinetmaker carefully cuts out the elements of the ornamentation – flowers, human figures, and so on, from a sheet of copper or tin. These elements are then heat-inlaid in a thin sheet of shell that acts as the base; this is called *"première partie"* work. When shell is inlaid in a copper base the work is known as *"contre-partie."* A careful worker can simultaneously prepare the decoration for two similar articles of furniture that are to be executed in *"première-partie"* and *"contre-partie"* respectively. The copper or tin elements can also be inlaid in horn that has been dyed red or blue. The shell itself can be beige or red, and the shell, tin, or copper can be complemented by hard stones (for example, small pieces of lapis lazuli).

In genuine marquetry, the wooden support disappears completely under a "veneer" of varicolored precious woods, sometimes accompanied by small pieces of mother-of-pearl or bone. The panels of veneer are coated with glue and heat-applied by means of an "iron that heats the glue" and a veneering hammer. The

FIG. 285 Marquetry and bronze work. Detail from a Louis XV desk (Musée de Versailles).

pieces of veneer are very thin (one millimeter at most); some are left in their natural state, others are dyed with natural coloring matter. If the design is complex, the carefully drawn panels of veneer are cut out and assembled, possibly by gluing them on the wrong side onto a sheet of paper. Once the veneer has been glued in place, it is polished and often varnished.

The last step in making inlaid articles of furniture is additional ornamentation with pieces of marble and strips of copper and bronze gilded with mercury, varnished, or silvered.

THE ARTS OF METALWORKING

The Precious Metals

Gold

Gold is the principal metal, the metal sought and undoubtedly utilized before all the others; the goldsmith (*Aurifaber*) is essentially "the man who works gold."

Gold, which can be recognized by its untarnishable shine, is extraordinarily malleable; the gold leaf used by gilders is sometimes as thin as a cigarette paper. Gold has a density of 19.26; when hammered, its density increases to 19.36. It is easy to work, but has one defect: its lack of strength, whence the numerous alloys intended both to increase its strength and to lower the cost of objects made from it. From the earliest ages of human history, governing authorities and guilds strictly regulated the composition of these alloys; certain markings (in France the first such markings date from the thirteenth century) were stamped on objects solely as a guarantee of the quality of the metal.

In modern alloys, gold is said to be of first grade when it consists of 920 parts of gold to 80 of copper; of second grade when the proportion is 840 to 160; of third grade when it is 750 to 250. There is another method of designating these alloys, namely, by carat, which is the weight of pure gold representing $1/24$th of the total mass of an alloy. Pure gold is 24 carats. As for the weight of the gold, in old inventories we frequently find it expressed in *marcs;* the ordinary *marc* weighed 244.75 grams, and four *marcs* were thus equivalent to slightly less than one kilogram. In certain cities the weight of the *marc* was different; for example, the Troyes *marc* weighed 260.05 grams.

The goldsmith possesses only a small number of tools, frequently of his own making: hammers of all kinds, stakes (round-headed anvils) and two-beaked anvils. The very frequently utilized swage is described by L. Lanel as being "made of a steel rod bent twice at right angles, in opposite directions; its center portion is approximately 14 inches long. One of the bent ends is tightly clamped in a vise; the object is threaded on the rod, and the other bent end, which is of appropriate length and shape, is placed under and touches the part to be embossed. When the swage is struck with the hammer, it bends; then, because of its elasticity it immediately springs back to a point beyond its original position, striking a blow on the metal the force of which is in proportion to the blow it received. A bump appears, which indicates to the workman the exact location of the tip of the tool. Guiding the object with his left hand, he continues to hammer with his right on

Plate 47. Grilles of the Golden Gates, Place Stanislas, Nancy. Eighteenth century. *Photo Georges Viollon-Rapho.*

the swage, carrying on his work without hesitation." The goldsmith's worktable is a workbench curved so as to leave a concave portion for the seated workers; it is placed on a flooring equipped with movable wooden hurdles to catch the metal droppings.

Silver

Silver has a density of 10.414; when the metal is beaten, this rises to 10.510. Its malleability is astonishing; when drawn out on old or modern drawing benches, one gram of silver can furnish more than eight feet of wire. Its strength is equally surprising: one silver wire one-twelfth of an inch in diameter can support 187 pounds.

Silver, like gold, is most often found in the state of alloy. A first-grade alloy is 920/1,000, second-grade 850/1,000, and third-grade 750/1,000. The old method of expressing these alloys was by "denier"; pure silver was 12 deniers.

Metalworking

By utilizing the properties of metal, and notably its malleability, the worker gives a piece of gold or silver the desired shape through the almost exclusive use of the hammer (embossing, recessing, planing). The skill of the goldsmith, both ancient and modern, lies essentially in his ability to handle the hammer, which should not strike the same spot twice in succession. This skill often depends on dexterity. Lanel writes that "we have known capable planers to create, with their hammers and anvils, a vase out of a five-franc silver piece, while preserving intact on the neck of the vase the inscription on the edge of the coin." When hammered, gold and silver lose a portion of their malleability, and they must be heated to regain a certain elasticity. If the metal is not hammered it can be "taken in the mass": the smith draws the objects to be reproduced on a block of silver, and then cuts it out with cutting instruments, following the general outline of the object.

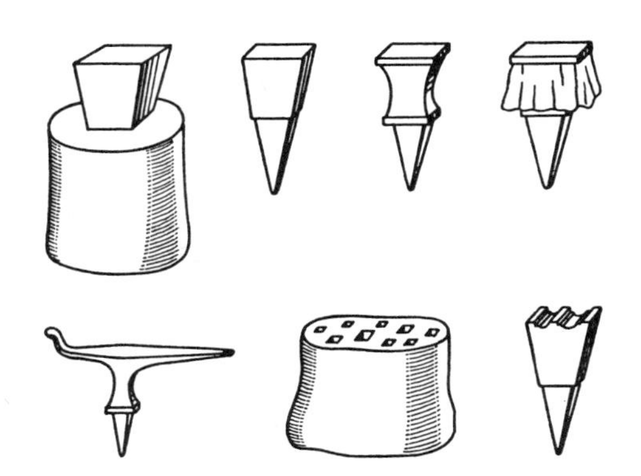

FIG. 286 Various types of anvils used by goldsmiths (illustration from the *Encyclopédie,* Vol. VIII).

If the smith works on the wrong side, the work is said to be embossed; numerous gold and silver figures were treated in this manner in the Middle Ages. Theophilus Presbyter recommended filling the hollows with a mixture of crushed brick and wax to strengthen the figures.

Certain parts of an object, such as beaks and handles, can be shaped separately and then soldered; in this case, the quality of the metal used for the additions is inferior to that of the body of the object, and their fusing point is lower. Excessive solderings were very soon forbidden because of the danger that they would lower the intrinsic value of the object.

To simplify their work — to "standardize" it, as we would say today — goldsmiths soon learned to make wooden or metal forms or cores, on which they shaped the sheet of metal by hammering. Eighteenth-century gold boxes owed their shape to the "embossing dies," described in the *Encyclopédie.*

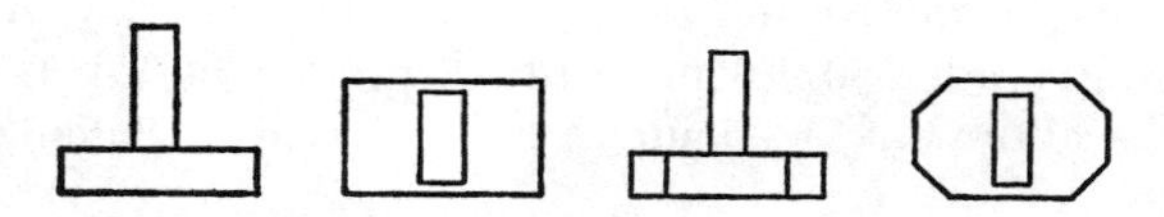

FIG. 287 Models for snuffboxes
(illustration from the *Encyclopédie,* Vol. VIII).

The lathe could not be utilized as easily in goldsmith's work as in ceramics. However, eighteenth-century smiths conceived the idea of equipping certain complex lathes with bronze models, plates for example, which guided the movement of a cutting tool; all the plates turned in this way had exactly the same shape as the models. The *guilloche,* which we mentioned in connection with enamelwork, was lathe-operated. A description of these lathes can be found in Chapter 12, "Industrial Mechanization."

All the goldsmiths' methods described so far were cold-working methods. A number of early pieces of goldsmiths' work were not cold-hammered but smelted, notably spoons and forks, which at the end of the eighteenth century began to be struck with the balance press, like medals.

Objects cast in a mold are more rare, but we are acquainted with a fair number of ewers, sauceboats, and even plates executed in the seventeenth and eighteenth centuries by the casting methods described in connection with bronzes.

Decoration

The completed object, whether cast or hammered, could be simply polished; this brought out all the brilliance of the silver. Before polishing, it could also be decorated, for example with pearls, filigree work or wires, engraved or incised work, niello, enamel, or gilding.

Silver (or gilt silver) wires for filigree work were drawn separately and soldered or riveted to the object. The filigree work of the eleventh, twelfth, and thirteenth centuries would merit a separate study: there were filigrees done with one or several wires, filigrees with varying degrees of complexity, and imitation filigree. Simple or multiple rings were also added, being held in place with rivets.

Engraved ornamentation was executed with cutting tools, which attacked the metal to a certain depth and "raised" it. In contrast, chasing, which as its name indicates was done with chasing tools, eliminated imperfections and emphasized the forms without attacking the metal.

Gilding was done with mercury. The carefully prepared object, scoured with acid and rubbed, was coated with an amalgam of one part of gold and two parts of mercury; when fired, the gilding adhered to the metal and the mercury volatilized. "Distemper" gilding was of mediocre quality; the scoured object was dipped in a bath of gold chloride and potassium. Modern electrogilding gives satisfactory results.

The last step consisted of polishing the objects with the wheel, emery powder, or a burnisher with an agate tip.

Other Metals

Tin

Tin is a low-density metal (7.29) whose particular characteristic is a very low fusing point (232 degrees); therefore it is most often utilized in the form of a solder. Tin objects are cast in molds; they can then be stamped or engraved and polished.

Tin oxide is utilized, as we have seen, in glass- and enamel-working. Very few medieval tin objects have survived into modern times. The great nineteenth-century collections have made the works of Briot and the German sixteenth-century casters fashionable.

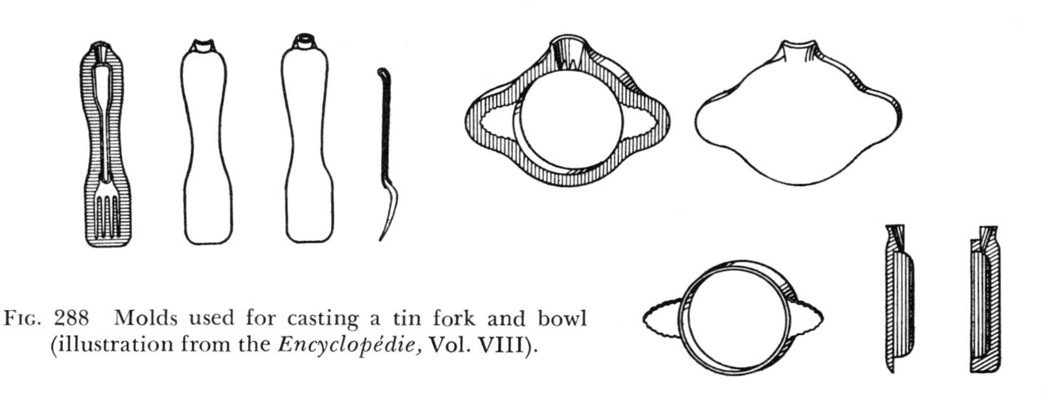

Fig. 288 Molds used for casting a tin fork and bowl (illustration from the *Encyclopédie,* Vol. VIII).

Iron

The density of iron is 7.86; it has a high fusing point (1,500 degrees C.). There is no need for a lengthy description of the method of working iron: it is a matter of common knowledge that when heated red-hot on the anvil, iron takes the shape imposed on it by the hammer, and can be welded to iron; it can be gilded, silver-plated, enameled, and bronzed. There are innumerable artistic uses for iron, the most justly famous being those connected with architecture (balconies, stairways, grilles), which were specialties of the seventeenth and eighteenth centuries.

Lead

Lead, which is very malleable, has a higher density than tin (11.37); its point of fusion — 327 degrees — is also higher than that of tin, but still relatively low. Like tin, it has few uses outside of castings, which may be either small or monumental. Many famous fountains, for example at Versailles, are made of lead; it is impervious to attack by the mineral salts contained in water from streams. Lead oxides are utilized in ceramic and enamelwork.

Copper

Although known as a "poor metal," copper has very often been utilized by the smiths. A number of medieval shrines and reliquaries are of embossed copper, without any further decoration. Enameled and gilded copper made the reputation of the Limousin, Rhenish, and Limoges goldsmiths.

The very high fusing point of copper (1,084 degrees) makes it a difficult metal to cast or mold; this is done by alloying. The famous "copperware" is the product of an alloy of copper and zinc.

Gilt copper was obtained by the "molded-gold" gilding process, in which mercury was used. It could also be simply varnished. Copper was silver-plated by a similar mercury process; it can also be done with silver leaf, distemper, and by electroplating.

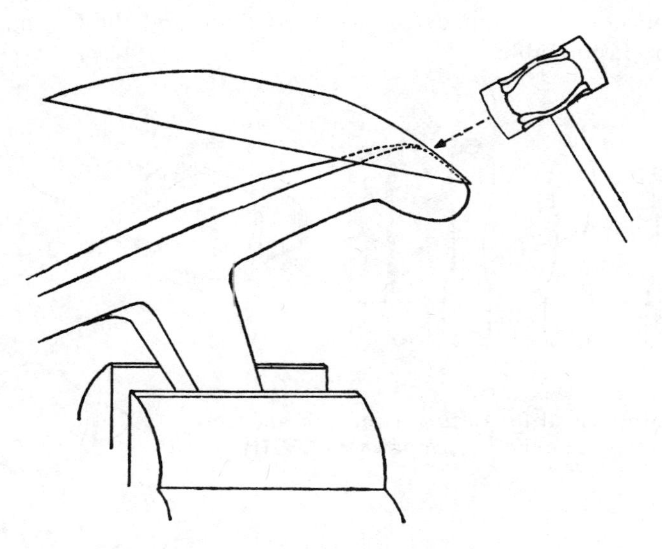

FIG. 289 Planing with a hammer. Preparation of a copper plate before enameling (from Barthe).

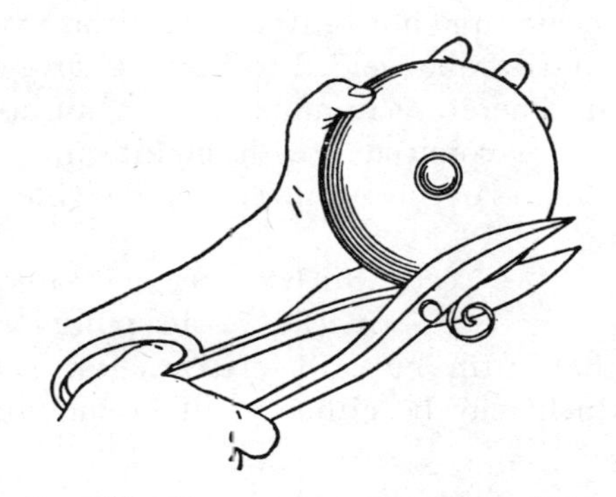

FIG. 290 Preparation of a copper dish before enameling (from Barthe).

Copper plays a part in the composition ot numerous alloys, including electroplate, nickel silver, and silver plate. Electroplate, discovered in the eighteenth century, consists of copper coated with silver by rolling; as its name indicates, the plating is silver fixed to copper by soldering. Nickel silver is an alloy of copper, zinc, and nickel. Silver plate, which has been popularized by silverware for table use, is a nineteenth-century invention. The first silver-plate items were of brass silver-plated by the electroplating method; the brass was later replaced by nickel silver.

Casting

The casting methods described below are valid for bronze and for the other metals.

Bronze is an alloy of copper and tin, in varying proportions; lead and zinc are sometimes added to the original alloy. The alloy used by the Italians in the fifteenth and sixteenth centuries generally consisted of between 88 percent and 90 percent copper. Some bronzes of antiquity contained 70 percent copper and 23 percent tin, to which lead and zinc were added. In the seventeenth century the Kellers used an alloy of 90 percent copper, 2 percent tin, 1 percent lead, and 7 percent zinc.

Cire-perdue *casting*

Until the beginning of the nineteenth century the "lost wax" (*cire-perdue*) process was the method used for casting works of art; examples are the "Gattamelata" and "Colleone" statues, and Girardon's "Louis XIV."

The sculptor began with a sketch, from which he made a scale model in wood, plaster, clay, or wax in the case of small objects. The finished model was molded of plaster in order to obtain a shell in two parts, front and back. Certain delicate decorations could be molded separately.

For large castings, a core was then constructed with iron rods, clay, and crushed brick stuffing. This core, which was smaller than the original, was placed in the shell, which was then sealed; wax was now poured between the shell and the core, so that when the core was removed from the shell it was covered with a layer of wax. Any adjustments were made at this point by the sculptor himself.

The wax-covered core was now placed in a jacket made from plaster, iron dross, sometimes tin and cement (seventeenth–eighteenth centuries). Channels, gates, and vents were drilled for the passage of air and wax, and the mold was then placed in the furnace. The wax slowly flowed out through the channels, and the molten metal, which had been separately prepared, was poured in through the channels. The bronze thus replaced the wax, which was "lost."

When the metal cooled and the jacket was opened, the metal was reworked, engraved, redressed, and given a patina using various methods (oil, acid, varnish, lampblack). The metal might also be gilded with mercury or simply polished.

For small cast bronze or silver objects, there was obviously no need to make a core; the shell was completely filled with wax, and the finished bronze was a solid piece.

Plate 48. Casting of the equestrian statue of Louis XIV, raised in 1699 on the Place Louis-le-Grand (Place Vendôme). Collection of illustrations in the *Encyclopédie. Photo Conservatoire des Arts et Métiers.*

Sand-casting

Sand-casting for bronze works of art is still being practiced. From the plaster model the worker makes a hollow model, also of plaster, called a "lost shell"; this is used to obtain a model that resembles the original. This model is placed in an iron frame and laid on a bed of sand that is to reproduce the original in sunk relief. The core is laid in the sand mold, jointed and firmly lashed, and the channels and vents prepared. The bronzemaker then proceeds to fill the space between the core and the sand mold with molten bronze.

Large statues are cast in several parts and then assembled. Engraving, patination, and gilding are done as for "lost-wax" casting.

FABRICS

The raw material and its preparation

The term "textile material" is applied to any fibrous material that can be spun and woven. The fibers utilized in "artistic" fabrics are classified into major categories, depending on their origin:

fibers of vegetable origin, extracted from seeds (cotton, kapok), stems or roots (flax, hemp, jute, nettle), fruit (coconut);

fibers of animal origin: wool, silk, cashmere (goats' hair), camel's hair, alpaca, vicuña, cat, rabbit, horsehair;

fibers of mineral and chemical origin, whose use has been popularized by modern industry. We shall not discuss them here in detail.

Animal and vegetable fibers were the only ones utilized in the arts until modern times. Flax, hemp, and wool were used in antiquity; silk became known to the West, by way of the Arabs, in the sixth and seventh centuries.

Formerly, the extraction of fibers was done manually. Today, at least in the West, it has been mechanized; for example, cotton ginning (separation of the fibers from the seeds), once done by hand, is now done by the cotton gin. Flax is soaked, a process known as "retting." Silk is unwound from a cocoon, after the worm that inhabits it for about two weeks has been killed before it can become a butterfly. The cocoon is made of fibrous materials secreted by the worm after it has gorged itself on mulberry leaves. The largest cocoons can yield as much as 3,000 to 5,000 feet.

Discontinuous fibers (for example wool, as contrasted with the continuous thread of a fiber such as silk) must be carded, spun, and dyed before they can be sent to the weaver. The spinning wheel is one of the first "mechanical" tools for weaving, both commercial and domestic; the first spinning wheels date from the sixteenth century, and were a replacement for distaffs and spindles. Various inventions of the eighteenth, nineteenth, and twentieth centuries have made possible almost complete industrialization of the spinning process.

Weaving

Weaving is the interlacing of two separate sets of threads known respectively as "weft" and "warp." The weaving is done on a loom, which can be either simple or complex.

The preliminary operations of weaving include spooling and warping, that is, the processes of winding the warp threads around a bobbin known as the "beam," and coating them with a glue that disappears in the finishing of the completed fabric.

In very simple weaving, for example that of canvas, the interlacing of the threads is simple: the weft thread is passed alternately over and under the warp threads. In order to make this possible, the warp threads on the beam are divided into odd-numbered and even-numbered threads. On the loom these threads pass through mails (eyelets) in the center of heddles (pieces of wood separated by hemp or steel wires); in the case of canvas, the loom has two heddles. These heddles, which are regulated by a cam, alternately rise and fall, separating at one time the even-numbered, the next time the odd-numbered, threads, and permitting the passage of the shuttles that hold the weft thread.

In the case of more complex weaves, all that need be done is to increase the number of heddles on the loom; however, cam-type looms can hold only eight heddles, and thus permit eight different combinations of weft and warp threads. The medieval weavers utilized levers that separated the warp threads manually by pulling on ropes linked to the mails through which these threads passed; the difficulty and slowness of the work is easily imagined. After the beginning of the seventeenth century weaving became gradually mechanized, particularly at Lyon, thanks to Claude Dangon, the protégé of Henri IV and Laffemas. Dangon succeeded in perfecting a loom that made possible considerable variety in the warp threads with the help of perforated cards; the threads, which were linked to needles, separated only when the needles pulling them came into position opposite one of the holes in the card. Falcon's discovery is at the origin of the "selecting box" mounted above the loom, which made it poossible to utilize up to 32 heddles. Starting with Dangon's and Falcon's discoveries, and utilizing Vaucanson's findings, at the very beginning of the nineteenth century Jacquard perfected the loom that still bears his name. The principle of separation of the warp threads remained the same (that is, perforated cards), but the combinations were theoretically unlimited. Still in use, the "Jacquard loom" was greatly perfected, and served as the basis for perfecting the modern automatic loom.

The warp threads were not the only threads that posed problems for the engineers. The passage of the weft threads rolled on spools inside shuttles must also be done automatically when the warp threads separate. On ordinary looms the passage of the shuttles from left to right and right to left is done by a horizontal batten with a picking-stick; it is operated by a leather thong activated by a lever, which in turn is operated by a cam shaft and a spring. The shaft that throws the shuttle is also regulated by a camshaft. Looms with several shuttles produce a great variety of weft effects. Automatic looms permit changing the empty spools without stopping the operation; some of these looms may have bobbins with a sufficient amount of thread, rather than shuttles.

In using the looms, an infinite variety in the effects of weft and warp is possible, and fabrics produced by them have astonishing variety and sumptuousness.

The weave

The interlacing of the threads of warp and weft is called the weave. This weave can be defined in the form of a relationship, which is expressed graphically by a design. For this purpose a paper marked off in squares is used; the vertical lines represent the warp threads, the horizontal lines the weft threads, the squares the crossing point of the threads. If the warp thread passes over the weft thread, the square is colored; if the warp thread passes under the weft thread the square is left white, and this alternating pattern of the first squares is repeated throughout. The simplest relationship is that of the canvas, which comprises a continuous series of alternating one-to-one white and colored squares. For the expression of complex relationships, several colors may be used. The patterns are divided into the basic weaves (canvas, twill, and satin) and their derivatives. Obviously the design does not take into account the materials used, and two fabrics that look very different may have the same weave.

Although it is a very imperfect description of the fabric, the design considerably facilitates its study. The old fabrics, which have been little or poorly studied, have often given rise to inadequate descriptions; even the terms employed vary with each author. To overcome these difficulties, an International Center of Fabric Studies was created several years ago in Lyon; it has already edited for its members a multilingual dictionary that gives precise definitions of the terms utilized, and suggests models for stylistic and technical analyses.

Fabrics woven or worked by hand

Embroidery consists of needlework on a fabric; it makes use of a great variety of special stitches. Appliqué work, for example, consists of "applying" a piece of fabric or precious material (pearls and the like) on cloth. Mechanical embroidery is used today even on the craft level.

Lace is an openwork fabric formed of a network of stitches; it is made with a needle, with bobbins and a small loom, or with a hook. Genuine lace did not appear until the Renaissance; its development produced the "bobbin" lace of Flanders and the needle lace of Italy and Alençon (seventeenth century).

Tapestry

Tapestry should not be confused with needlepoint, which was so popular in the last century. Needlepoint is done with a needle and colored wools or silks on a canvas. The strength of this version of tapestry has made it a popular covering for chair seats.

Genuine tapestry is executed on a loom rather than a canvas; its raw material is wool, but silk thread is found in the finest tapestries. Gold metallic thread (actually gilt silver) and silver thread are sometimes combined with the weft threads to produce special effects. The wool is rolled on shuttles, each color having an individual shuttle. The design is ordered from a painter (known as a cartoon painter), who draws a cartoon to scale; the colors are not reproduced on it but are simply indicated by numbers.

Tapestry makers were numerous in Paris in the thirteenth century, but the earliest surviving tapestries do not antedate the fourteenth century. Colors were few in the Middle Ages; by the end of the seventeenth century the Gobelins were utilizing approximately 120 colors, while the eighteenth century had an infinite variety (1,000 shades of some thirty colors). "Modern" tapestry makers have returned to the practice of utilizing only a small number of bright colors.

The tapestry loom is sometimes vertical (the high-warp loom), sometimes horizontal (low-warp loom). Each loom has two beams or rollers on which the warp threads, divided into odd and even threads, are stretched. The separation of the two "sheds" is done by hand by means of threads (heddles) (high-warp looms), or by foot by means of treadles (low-warp looms). Weaving consists of passing the shuttle of weft thread by hand between the sheds of warp threads.

In the use of the high-warp loom, the cartoon is placed behind the tapestry maker, who consults it in a mirror; he is aided by ink marks on the warp threads. The tapestry maker works in reverse; looking at his model in the mirror, he reproduces his cartoon on the right side.

The tapestry maker who uses the low-warp loom works in a seated position. The cartoon was first cut out and placed on the warp threads; it was thus reproduced in reverse. In the eighteenth century the tapestry makers conceived the idea of replacing the cartoon with a tracing, which reversed the design in such a way that it could be reproduced on the right side. In the same period Jacques de Vaucanson envisaged a balance loom that combined the advantages of both types of looms.

A different shuttle is utilized for each color; the shuttles are hung one after the other behind the loom, and can be reused if necessary in the course of the work. A comb is used to lock the filling in place as it is finished. The sewing of the slits between the color areas is not done until the tapestry has been com-

← Fig. 291 Working on a high-warp loom (illustration from the *Encyclopédie,* Vol. IX).

Fig. 292 Working on a low-warp loom (illustration from the *Encyclopédie, Vol. IX*).

pleted. On extremely narrow low-warp looms the borders are sewn on after completion of the weaving.

Carpets

Classification of carpets is based on their method of production, whether knotted by hand or woven mechanically.

Hand-knotted carpets are obviously the oldest; they are woven on high-warp and low-warp looms reminiscent of the tapestry looms, with their rollers and warp threads of cotton, hemp, or wool, stretched and separated into two sheds, as in tapestry work, by a glass rod. The wools of various colors are prepared on spindles or cut into pieces of the desired length. The worker, who sits facing the warp threads, works on the right side; he places his wool thread (weft) on the warp thread, knots it, and cuts it. This is the simplest type. In the Ghiordes (or Turkish) knot the pieces of yarn appear in pairs between two warp threads, while in the Sehna (or Persian) knot each piece appears singly between a pair of warp threads. After each horizontal row of knots the carpetmaker passes a weft thread between the sheds of warp threads and tightens the work with his comb.

In the case of Savonnerie rugs (produced by the seventeenth-century factory of that name near Paris) and their imitations, "the piece of wool is not cut after the knot is made. After the knotting, the wool is wound around a kind of round metal bar with a sharp cutting edge at one end, which the worker holds in horizontal position against the warp threads. After a certain number of knots the worker draws the shaft from left to right, which cuts the rings around it and creates a number of small wool tufts" (F. Windels). In the other cases, the threads are cut with scissors as soon as the knots have been made.

Machine-made carpets are woven on looms that bear more resemblance to weavers' and tapestry makers' looms. Like the latter, they have heddles (in this case they are called harnesses) to perform the selection of the warp threads. The Jacquard method also transformed the rug-making industry by systematizing the use of perforated cards. The most recent of these mechanical looms are of the Sehna type, and imitate hand-woven carpets.

IVORY AND PRECIOUS GEMS

Ivory

It is impossible, even in the most summary of discussions, not to make at least a brief reference to the art of ivory carving, which has created exceptional works of art in the course of the centuries. The ivory most often seen is that of the elephant; it can also be obtained from the hippopotamus, narwhal (the "unicorn" horn of the old treasures), and walrus. Strips of horn are often used on lesser objects to imitate the appearance of ivory. African ivories were greatly prized in antiquity and the Middle Ages. In the twelfth century the Germans made much use of walrus ivory.

The various parts of a single tusk, depending on their shape, were sawed separately by hand (by machine, in more modern times) in accordance with the use for which they were intended. The ivory was then worked with a tool or on the lathe into the desired shapes and objects. Thin sheets were cut by means of very fine saws.

Precious gems

Gems are crystallized mineral substances, and consequently their surface has flat faces and ridges. The density of these stones varies from approximately 2.5 (agate) to approximately 3.5 (diamond); their hardness is evaluated on the "Mohs scale," which indicates in decreasing order the power of each gem to scratch other gems. A diamond can be scratched only by another diamond; its index is ten on the scale, while soapstone has an index of one.

Gems are sought for their brilliance (the diamond) or their color — red (rubies and garnet), blue (sapphire and lapis lazuli), green (emerald and malachite), yellow (topaz), and violet (amethyst).

The cutting of gems

The cutting and polishing of gems requires a high degree of skill. Cutting is made possible by two of the characteristics of gems: *crystallization,* which gives gems of the same substance a characteristic geometrical shape (octahedral or tetrahedral for the diamond, for example), and the *wearing* of one stone by another, diamond being the only stone that can scratch or wear all the others.

In cutting a diamond, for example — the most delicate of all gem-cutting operations — the cutter must begin by choosing the most favorable "cleavage line" in the rough diamond (by virtue of their crystallization, crystals have a tendency to split along certain planes of least resistance). Once this plane has been chosen, the stone is fixed with a fusible cement on a support with a handle; it is cleaned with another, cut, diamond. With a third diamond the lapidary makes a notch in the crude diamond. Then a quick blow on the hammer with a special tool breaks it open.

The simplest and poorest cut is the "cabochon," the only one known prior to the fifteenth century. Faceted cutting is more elegant and skillful, whether it is the "rose" cut (a pyramid with a flat base) or the "brilliant" cut, which is done by dividing the octahedron of the diamond into 16 parts. The cutter removes $6/16$ to form the "table" (the flat face of the gem); the "table" is surrounded with a faceted crown $2/16$ in height. The next $5/16$ constitutes the pavilion, which ends in a point at the culet; the remaining $3/16$ is eliminated.

The number of facets varies; modern cutting tends to decrease the size of the table to show off the facets of the crown, but the essential operation remains practically the same.

The polishing of the various faces of the diamond is done with bronze wheels coated with diamond dust; the other gems are also polished on the wheel. In earlier ages the polishing shops were generally installed on the edges of streams, and the current was utilized as the driving power to turn the wheels.

For use in jewelry, the stones are mounted in settings of precious metal. Settings with individual "prongs" came into existence in the fourteenth and fifteenth centuries; prior to this time the setting completely surrounded the stone.

The weight of a diamond is expressed in carats. One carat weighs 200 milligrams (about 3 troy grams).

BIBLIOGRAPHY

ANGUS-BUTTERWORTH, L. M., *The Manufacture of Glass* (London, 1948).

BATES, K. F., *Enamelling, Principles and Practice* (Cleveland, 1951).

BEDINI, SILVIO A., "A Renaissance Lapidary Lathe," *Technology and Culture,* 6 (1965), 407–415.

BURGESS, E. M., "The Mail-Maker's Techniques," *Antiquaries Journal,* 33 (1953), 48–55, 193—202.

———, "A Reply to Cyril Stanley Smith on Mail Making Methods," *Technology and Culture,* 1 (1960), 151–155.

A Diderot Pictorial Encyclopedia of Trades and Industry, ed. by Charles C. Gillispie (2 vols., New York, 1959).

FRANK, EDGAR B., *Old French Ironwork: The Craftsman and His Art* (Cambridge, Mass., 1950).

GLOAG, J., *A Short Dictionary of Furniture* (London, 1952).

GOODMAN, W. L., *Woodwork* (Oxford, 1962).

HONEY, WILLIAM B., *European Ceramic Art from the End of the Middle Ages to About 1815* (New York, 1949).

LISTER, RAYMOND, *Decorative Cast Ironwork in Great Britain* (London, 1960).

SAVAGE, G., *Porcelain* (Harmondsworth, England, 1954).

SMITH, CYRIL STANLEY, "Methods of Making Chain Mail (14th to 18th Centuries): A Metallographic Note," *Technology and Culture,* 1 (1959–60), 60–67.

———, "A Jeweler's Shop (1533)," *Technology and Culture,* 8 (1967), 207–209.

THEOPHILUS, *De diuersis artibus,* transl. by C. R. Dodwell (London, 1961); also transl. and edited by John G. Hawthorne and Cyril Stanley Smith as *On Divers Arts: The Treatise of Theophilus* (Chicago, 1963).

TREUE, WILHELM; GOLDMANN, KARLHEINZ; KELLERMANN, RUDOLF; KLEMM, FRIEDRICH; SCHNEIDER, KARIN; STROMER, WOLFGANG VON; WISSNER, ADOLF; and ZIRNBAUER, HEINZ (eds.), *Das Hausbuch der Mendelschen Zwölfbruderstiftung zu Nürnberg: Deutsche Handwerkbilder des 15. und 16. Jahrhunderts* (2 vols., Munich, 1965).

WATERER, J. W., *Leather and Craftsmanship* (London, 1950).

WELSH, PETER C., "Woodworking Tools, 1600–1900," U.S. National Museum *Bulletin* 241 (Washington, 1966).

CHAPTER 24

PRINTING
ORIGINS AND EARLY DEVELOPMENT

THE TECHNICAL evolution that brought printing to the threshold of the Machine Age falls between two extremely active periods, both of which were characterized by an intense pressure of all the needs of the Western world on the graphic arts. Typography could have been invented when the alphabet spread through the Mediterranean world, but in fact was not invented until the fifteenth century, because at that particular moment of the Humanist revolution such a discovery was absolutely necessary. Similarly, on the eve of the Machine Age, in that century of the Enlightment which was sharpening men's awareness of everything connected with science, there was an urgent need to discover the technical possibilities of preservation and rapid multiplication of the basic elements of the book. Finally broad avenues were opened to humanity toward a kind of mechanistic culture that had been somewhat despised by earlier centuries and that, in light of the facts, may appear to certain thoughtful modern minds as the source of many delusions.

These two powerful periods of capital importance were the prehumanistic fifteenth century and the premechanistic eighteenth century. As for the intervening centuries, the fact is that they were simply a period of the perfection of simple techniques based on the crafts and the orientation of these techniques toward the tremendous achievements of modern times.

In short, the fifteenth century consolidated the age of book publishing, and the eighteenth century prefaced that of the newspaper.

The manuscript or "work of art"

The word "manuscript," although it has the general meaning of "piece written by hand," immediately calls to mind the magnificent creations of the Middle Ages, which form the foundation of our great libraries. The history of the manuscript is one of the most dazzling stories of Western civilization: the splendor of the covers and the graphic methods, the zeal, knowledge, and art of its creators produced genuine masterpieces. It must be realized that the wellsprings of this remarkable movement were the enthusiasm, feeling, and faith promoted by Christianity.

In contrast to the sober artistic creations of antiquity, the medieval manuscript presents a splendor and a surprising perfection of execution. With the faith of the believer it celebrates a Messianic hope, in an esthetic language that is still naïve but whose richness of expression is reminiscent of that of children and primitive artists. The Christian medieval period is completely bathed in a threefold art: music, which creates an atmosphere that is both religious and in a sense magical; the imagery of the stained-glass windows and sculptures in the cathedrals, in which the people learn the history of their faith; and the miniatures of the manuscripts that contain the prayers of the clergy.

The riotous imagination of the artist had never before or since reached such a pitch. Stimulated by the constant presence of the Bible and especially the Apocalypse, it produced amazing work, for example that eleventh-century *Apocalypse of Saint-Sever* which, in the words of Émile Mâle, exercises a veritable tyranny over the imagination.

As this mystical orgy was calmed, however, and as men's minds turned with greater willingness toward the everyday world, a secular literature was born that was to be the beneficiary of the expressive art of the religious manuscripts. As early as the thirteenth century, works written in the vernacular offered stories of knighthood, songs of the *trouvères,* and satirical narratives such as the *Roman de Renart* and *Lancelot du Lac.*

During the fourteenth and at the beginning of the fifteenth centuries there appeared numerous books that were later to be widely disseminated by printing: The *Treatise on the Properties of Things,* the *City of God,* the *Great Chronicles of France,* and innumerable sumptuous *Books of Hours.* Like their predecessors, they are characterized by the importance of their decoration, an importance that had gradually become a characteristic of book production. As early as the fourth century, St. Jerome had condemned "books of purple parchment written in gold and silver letters, masterpieces of writing rather than books."

For the honor and glory of God, the medieval miniaturists strained to the limits of their skill as decorative artists. But the copyists kept pace with them, and in increasingly numerous monasteries and workshops they lovingly worked to improve on their handwriting, achieving such perfection of line and regularity of hand that the first printers had only to copy (or have copied) their forms in order to obtain, painlessly and without experimentation, the raw material of their art: *the letter.*

However, there is no doubt that the letter or sign, whether produced by stamping or by patterns, had long been in use for minor printing procedures. These methods had been used since prehistoric times to "brand" objects, animals, and even slaves. The antagonism of the two primary materials used in printing techniques — wood, which was generally used for stamping, and metal, which in addition was utilized for the establishment of "patterns," that is, plates hollowed out in certain areas to permit the fixing of a sign on a support by means of ink — is already in evidence.

There is also the example of molded bronze sheets utilized by the Roman potters to mark their works; they contain a text in relief and in reverse, which is "printed" right side up in the soft clay of their pottery objects, thus forming a perfectly clear metallographic impression (see paragraph on the metallographic book).

BLOCK PRINTING

The First Appearance of Printing Techniques

Black-line wood engraving

The first efforts toward block prints were made in the Far East during the first centuries of the Christian era. The Chinese were the teachers of the West in everything connected with the graphic arts, and it is therefore proper to begin by searching through their subtle and poetic literature for traces of their graphic experiments. Numerous American authors have concentrated on attempting to solve this major problem; while their conclusions are hypothetical, a few precise facts have nevertheless been established by their research.

To begin with, it seems that by the third century of the Christian era the Chinese technicians were in possession of all the elements that were to make possible the discovery of block printing. They had paper, which according to Dr. Laufer was already being made with a pulp composed of waste silk, both raw and woven. They had ink, which in that period was made by boiling together glue, various aromatic substances, and vegetable matter combined with the lampblack obtained by calcination of pinewood. They possessed the technique of block printing in the form of a very specialized "sunk lithography." In certain places of pilgrimage, texts and pictures were engraved on steles in the order in which they were to be read, to aid the pilgrim. A sheet of paper could be placed on the stele, and pressed into the engraved elements; then the surface of the paper was coated with ink, so that when the paper dried and was removed from the stele, the areas that had been pressed into the incisions stood out in white against the dark ground.

Given these three requirements of paper, ink, and technical capability, one requirement for the triggering of the invention was still missing: the social context, that is, the intense pressure of a more widely diffused culture requiring the creation of an instrument of reproduction. During the sixth and seventh centuries we find in the literature vague traces of various attempts at block printing, but the irresistible need of discovery does not seem to have manifested itself prior to the seventh and eighth centuries. In any case, in the eighth and ninth centuries numerous engraved images, often accompanied by prayers, made their appearance, and in 770 this technique spread from China to Japan.

Moreover, we are certain that by the year 835 calendars were being printed by the same method; an edict of that date orders the cessation of their production. As for the book itself, it must have rapidly benefited from the immense possibilities offered by block printing. Carter feels that the first book ever printed was the *Diamond Sutra,* in 868. Around this same period, however, we find traces of several other printed books.

Block printing in the West

How did these secrets relative to the graphic arts reach the West? Two highways linked these distant areas, a land route and a sea route, and the theory that both were used can be accepted. The most important commercial center of western Asia in ancient times was the Samarkand market, and André Blum (*La route du papier,* Grenoble, 1946) suggests that paper from China could have been imported and sold there by the seventh century; from here, its use spread throughout all western Asia. The rest of the route, at least as far as paper is concerned, was by sea: by coastal trade as far as Spain, and also by the Crusaders of the seventh crusade, who were long prisoners in Syria, and who brought the secret to Italy and France around the middle of the thirteenth century.

The secret of block printing probably traveled by the overland route, since impressions on engraved wood are mentioned in Egypt and Persia around the tenth century. However, it is also possible that under the widespread pressure of need it was directly reinvented in the West, where woodworking had long been known.

Popular imagery and black-line wood engraving

We have seen that the clerics and the copyists' shops were creating sumptuous copies of pious books for an élite that was hungry for mystical knowledge. But the Church soon sought to complete the evangelization of the masses by popular teaching, a task that was entrusted to the religious orders and particularly to the mendicant orders. Since the very first efforts at evangelization in the West, the problem had always been one of penetrating ever deeper into the ranks of the masses, who had remained deeply attached to their old animistic and pagan cults. These impulsive, imaginative masses, so different from the Jewish people who had given them a new, essentially moralistic religion, needed an imagery in order to become more deeply attached to their God. Discovering the Biblical fairyland on the portals and in the windows of the cathedrals and churches was one thing; understanding its moral significance was another.

This moral development of the masses, who were full of faith but illiterate, could be done only through the picture, and especially through the colored picture accompanied by a short text or a prayer. It was eloquently complemented by the production of the "Poor Man's Bible," consisting of booklets of pious pictures with short explanations. All these works were designed in the shops of the image makers, colored, and completed with small accompanying hand-copied texts.

At this point there appeared the budding technique of block engraving, obtained by the method of black-line engraving, which brought to the image makers the assistance of its wide possibilities. We shall not discuss the timid earlier experiments, notably those of stone rubbings and printing on fabrics, for the work in which we are chiefly interested now must have the essential characteristic of *systematic reproduction of the same motif in a very large number of copies.* This characteristic does not seem to have appeared until the second half of the fourteenth century. It is possible, as has so often been declared, that the first manifestations of this new technique were counterfeits of the hand-drawn picture, which could be sold at the same price as the latter. But the course of progress cannot be halted, and soon the "mold-cutters' " guild took its place among the guilds that formed the great family of the graphic arts.

Another of these industries was also growing: that of the "*cartiers,*" or manufacturers of playing cards. This industry was very soon obliged to utilize

engraved wooden tablets to multiply their card plates. Here, again, the city of Lyon enjoyed a certain importance; its archives for 1444 mention one Jean du Boys as working in that city as a "cutter of molds for cards."

Xylography and metallography

Black-line engraving was characterized by the fact that the engraver cut into the wood or the metal to remove the white areas, leaving on the surface of the plate only the areas to be printed. The printing was then done by inking the relief areas (Figure 293).

Xylography, which was a black-line engraving on long-grain wood (Figure 294), permitted the inexpensive development of a religious imagery that filled the obvious need of the people to learn by picture and preserve the history of their faith (Figure 295).

For the same purpose metallography, or black-line engraving on a metal plate, was utilized. Wood had the disadvantage of being very sensitive to dampness and changes of temperature; in addition, over a long period of time it became worn out, which precluded numerous reprintings. Several books whose engravings were done on copperplates are still in existence; they generally date from the first half of the sixteenth century. The appearance of the pictures differs little from those obtained from the wooden block, with one exception: The material for the plate was strong and stubborn, and an ordinary knife was no longer enough to work it; it had to be engraved with a gouge. In order to

Plate 49. Workshop for the production of playing cards, Place Dauphine in Paris, around 1680. Musée Carnavalet. *Photo Giraudon.*

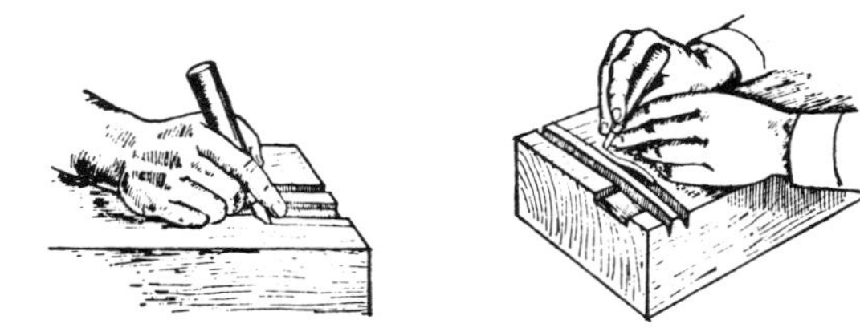

Fig. 293 Position of the hands in cutting a block for black-line wood engraving.

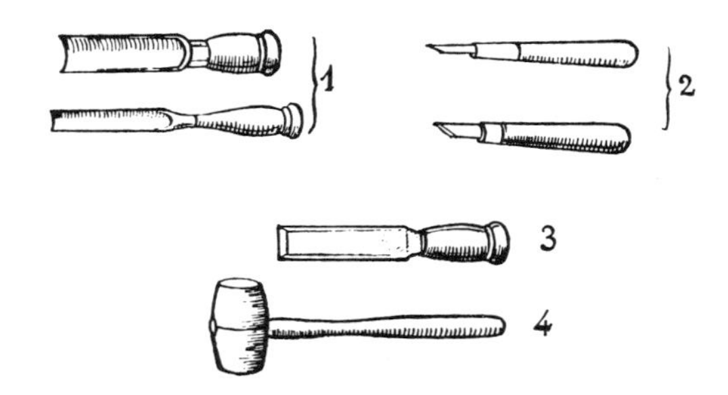

Fig. 294 Tools for carving long-grain wood:
1. Gouges 2. Penknives 3. Chisel 4. Mallet

Fig. 295 "Carrying of the Cross," woodcut. End of the fourteenth century.

Plate 50. I. Techniques of metallography (first half of the fifteenth century), reconstructed by the Imprimerie Nationale.

a. Matrix cut in a copper block with punches of letters struck side by side.

b. Relief block obtained by molding a lead alloy on the matrix.

c. Printing of the text on paper with the relief block.

II. Lyonese printing types of the fifteenth century, found in the bed of the Saône (Brassart collection).

d. Chamfered foot and heel.

e. and *f.* Types with round hole and slit, for two-color printing.

Bibliothèque Nationale. *Photos Bibliothèque Nationale.*

avoid working the entire plate, the engravers sometimes employed steel punches carved with points, stars, and other signs in relief. This made it possible to make the background lighter by means of a series of identical impressions. This "dotted" method was soon abandoned, for it gave a much harsher design than xylogravure, whose cut was much more flexible (Figure 296).

For a long time accompanying texts were added by hand to the prints thus obtained; these texts were sometimes quite large, especially in the *Biblia pauperum*. Then, in order to save time in adding these written letters to the picture, the engravers attempted to engrave them on the plate itself. The Protat woodcut, which has been claimed (perhaps somewhat arbitrarily) as a work of around 1380, shows letters engraved on a streamer; the "Saint Christopher" of 1423 has two lines of text under the picture, while the "Saint Sebastian" of 1437 contains 13 lines.

Thus the engravers slowly became accustomed to engraving increasingly large areas of lettering copied from that of the magnificent manuscripts. The time was approaching when they would be able to engrave all the pages of a small book (Figure 297).

Printing small books of images and individual prints was done as follows. The surface of the wood was inked with ink-soaked pads of leather. A sheet of paper dampened on the back was laid face down on the inked surface, and was rubbed with a fabric cushion to cause the ink to adhere to the paper.

But this method produced a very pronounced impression in which the relief

FIG. 296 A "dotted" metal engraving.

of the engravings penetrated deeply into the water-softened paper and appeared on the reverse side. Thus only one side of the paper could be printed. The sheet of paper, which consisted of two pages, was folded in half and joined to the preceding and following pages along its edges, producing an accordion-pleated sequence of pages. The Far East still utilizes this method for printing on its very thin, transparent sheets of paper.

The Raw Materials

It is very certain that in Western Europe, as earlier in the Far East, certain conditions had to coincide in order to permit block printing to develop without difficulties. The "social context" was present, as we have seen, along with the basic technical element: the methods of black-line engraving. Only two elements remained to be supplied: the support and the ink.

The support The Far East apparently began even before the Christian era to make paper pulp from silk wastes, both raw and worked, and bark, undoubtedly that of the mulberry tree. A certain Tsai Luen, Minister of Agriculture under Emperor Ho Tin, is supposed to have had the idea, around the second century A.D., of looking for old ropes and cast-off fishing nets whose hemp fibers could, after long boiling, form a fibrous pulp suitable for making paper. This was the starting point of a technique that was to revolutionize the world of graphic art. It is noteworthy that

FIG. 297 A *Biblia pauperum* engraved on wood (Bibliothèque Nationale).

the texture of the sheet of paper requires the presence of fibers in the basic preparation. All living material, whether vegetable or animal, which retains its fibers after the disappearance of actual life, can thus be used in the making of paper pulp.

At a very early date the Far Eastern papermakers replaced waste hemp with bark from small trees, which they washed, scraped, and crushed until it was reduced to a pulp by methods that remained practically unchanged until the nineteenth century.

The secret of papermaking appears to have reached the periphery of Europe around the middle of the eighth century, by way of the plains of Turkestan or the Samarkand route. By the twelfth century the Genoese and Venetian merchants, who were trading with the Arabs, brought back from their travels a new support for writing, which was quite fragile but which could be used for written documents whose life span was of short duration. However, it was not until the middle of the thirteenth century that the actual secret reached Western Europe.

The basic material used in the West was "rag pulp," made from waste canvas, linen, or hemp rags. Since these elements were rare, paper remained very expensive for a long time, while the book trade was beginning to develop. But the fourteenth century witnessed the general establishment of the use of discarded undergarments. The result was that the supply of waste rags became abundant, and the papermaking industry benefited from this situation to produce paper in much larger quantities and to offer it at lower prices. "Paper mills" were installed everywhere in the valleys that were collecting points for very pure, noncalcareous water, and also near major printing centers. Thus for Paris these mills were established in the regions of Saint-Cloud, Essonnes, and Troyes; for Lyon, in Auvergne, Beaujolais, and Dauphiné.

Papermaking was, and still is, done in two stages:

1. Preparation of the pulp. – After the rags had been sorted, they were left to rot slowly. Then they were torn up in beating troughs – tubs with hydraulically operated, spiked mallets. The pulp thus obtained was washed, and was compressed into thin bundles in preparation for the second operation.
2. Preparation of the sheet of paper. – The paper pulp was maintained in suspension in a large tub of water. A mold with a metal latticework was plunged into the tub; when quickly lifted out with a shaking motion, it raised a thin layer of pulp (Figure 298). The sheet thus formed was laid on a piece of felt; a number of sheets were piled one on top of the other (today one pile consists of one hundred sheets), and the water was pressed out of them under a very powerful press operated by an equally

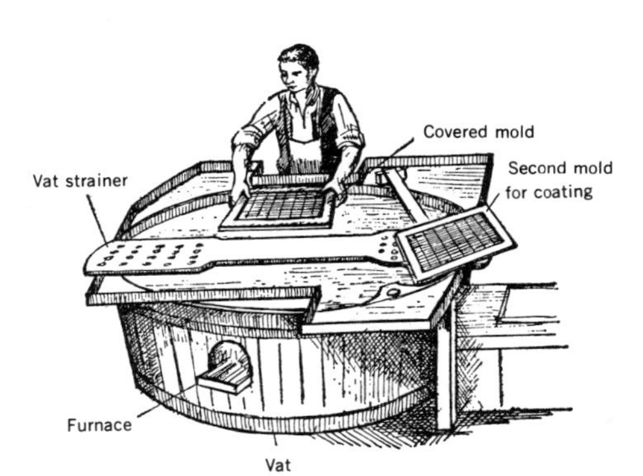

FIG. 298
Worker plunging
a mold into a tub.

powerful capstan. The sheets were then hung until completely dry in an air dryer which was generally located over the main halls of the mill.

Figures 299 and 300 are diagrams of the construction of a mold. It is a rectangular frame with a longitudinal, closely spaced network of brass wires crisscrossed by a network of much more widely spaced wires. A deckle placed on the edges of the form when it is dipped into the bath permits it to hold a larger quantity of water and pulp. The mark of the network of wires is clearly visible when a sheet of paper is held up to the light.

In the eighteenth century an attempt was made to eliminate the closely spaced network of wires; the marks it left on the paper were a hindrance in printing. Around 1750 the English printer John Baskerville conceived the idea of replacing the brass latticework by a very fine metallic canvas. The paper obtained by this method was christened vellum, because it bore a faint resemblance to the beautiful vellum sheets of earlier ages.

Watermarks

The papermakers very soon began marking their sheets with watermarks obtained by attaching brass wires to the latticework, which produced a faint marking representing a very schematized design or letters (Figure 301). In France the first watermarks appeared at Troyes. These figures, which were at first manufacturers' identifying marks, eventually came to indicate the "format" or size of the sheet, since each mill, which always used the same molds ("*formes*"), made paper only in the corresponding "*format.*"

Ink

Ink for printing had to be very different from the ink used in handwriting, for it had to adhere to the printing surfaces and not create blurs. The writing ink of the Greeks and Romans was composed of lampblack, glue, and water; in the twelfth century ferrous sulfate, nutgall, gum, and water were used. But these inks could not have properly adhered to the relief areas of the printing surfaces, and it was necessary to find another substance that could accomplish this purpose.

This substance had to possess a certain strength and sufficient fluidity to spread evenly on the surface of the wood block, while being thick enough not to flow around the features to be reproduced. It was made with lampblack, which produced a beautiful dark color; glue was added, and the mixture was macerated in oxgall. This composition was sufficient for xylographic printing, since it penetrated the wood to a certain extent. But when metal surfaces and pages of type had to be inked, it was found that the ink flowed over the impermeable metal, and thus a more oily ink was needed.

The new substance, which was black and shining, was made from lampblack, turpentine, and an oil mixture (probably walnut oil) reduced by heating to the consistency of varnish. By the seventeenth century the Dutch had achieved such superior proficiency, both in the art of printing and in the making of ink, that Sublet de Noyers, superintendent of buildings for King Louis XIII, requested French Ambassador Brasset to send him a few pressmen and compositors, one of whom must be able to make printer's ink.

The eighteenth-century printing establishments ordinarily made two kinds

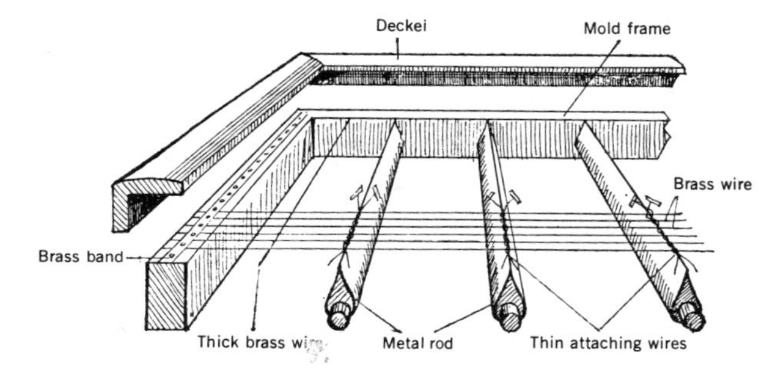

FIG. 299 Detailed view of the construction of a mold.

FIG. 301 Watermark (fourteenth century).

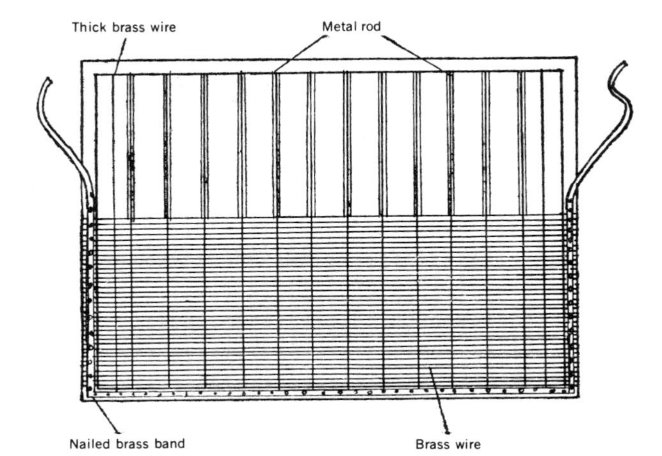

← FIG. 300 Construction of a mold.

of black ink, one very thick for use in hot weather, the other more fluid for winter use. These two products could when necessary be combined for use in certain types of work. By the end of the eighteenth century the most famous inks were those of Ibarra in Spain, Bodoni in Parma, and Cambridge University, which in 1700 offered to supply France with a certain quantity of ink in exchange for Greek characters known as "the king's Greeks."

Until the beginning of the nineteenth century every printing establishment made its own inks; the preparation was delicate and required a certain "knack." In 1818 Pierre Lorilleux Senior created the first industrial ink manufacturing plant in Paris.

Copperplate Engraving

After black-line wood engraving (fourteenth century), a new type of engraving became common in Western Europe around the middle of the fifteenth century. It involved a technique that was exactly the opposite of the black-line technique: intaglio engraving in metal, called "copperplate." The design was incised in the copperplate (Figures 302 and 303), and the plate was then coated with ink. The surface was wiped, and the sheet of paper firmly pressed on the plate so that it was impregnated with the ink at the bottom of the engraved lines.

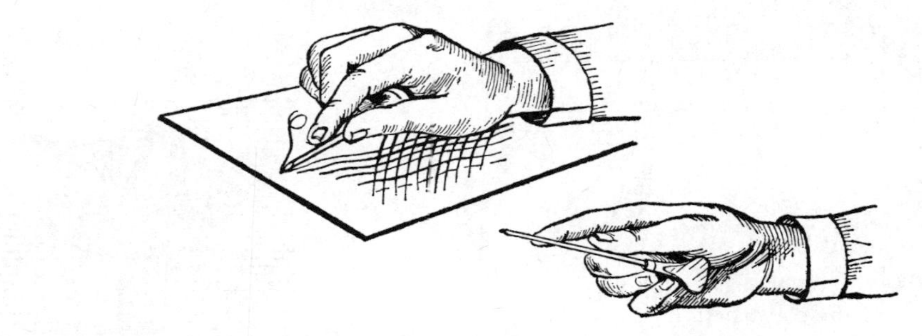

FIG. 302 Position of the hand in copperplate work.

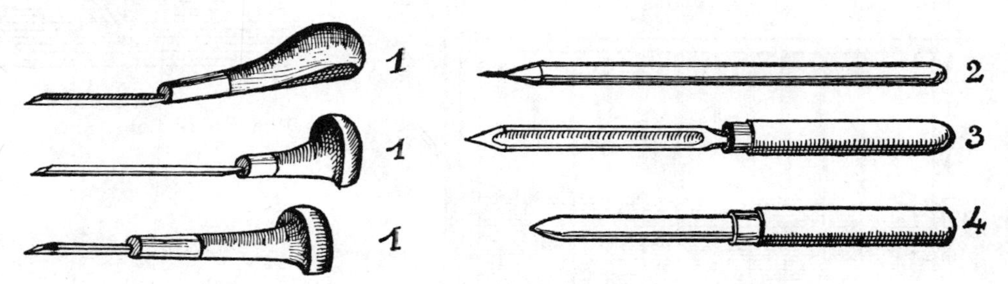

FIG. 303 Tools for copperplate engraving:
1. Gravers 2. Etching-needle 3. Scraper 4. Burnisher

It is generally accepted that copperplate engraving was derived from the niello technique, which utilized an incised silver plate into whose lines a black enamel was poured; to find out how far the engraving work had progressed at a given time, it was possible to pull a proof simply by filling the incised areas with lampblack or ink. In Berlin there exist prints obtained by this method in 1446, depicting the "Flagellation of Christ." However, it has been long recognized that the inventor of copperplate engraving was the goldsmith-niello-worker Maso Finiguerra, who utilized it for the first time at Florence in 1452 (or 1460). The National Library has an indirect proof of a "Coronation of the Virgin" believed to have been engraved by Finiguerra; its plate is preserved in the Bargello at Florence.

These experiments were to have major consequences, especially from the esthetic point of view. By bringing to the attention of all those who desired it the elements of a culture that until then had been limited to the clergy, the Renaissance, in addition to its need for the innumerable libraries that were to be established almost everywhere, created other obligations relative to the works of art, which also determined its esthetic system. These works of art became increasingly rich and technically perfect, thanks to the "finish" sought by the Renaissance sculptors and especially painters.

The art of the Renaissance, which had now become independent, filled the world with its numerous achievements, which became increasingly liberated from the obligations of the past and more and more an art of the personal "master-piece." This art, which, unlike that of the Middle Ages, was no longer an "art of faith" but an "art for the sake of art," could not have been created without the existence of methods of confrontation, comparison, and judgment. This meant that it had to be possible to reproduce the works of the masters as exactly as possible, and to bring them to the knowledge of a curious public, and even at times to give a small group of enlightened amateurs the sensation that they were the owners of these works.

The technique of the black-line wood engraving was too crude to accomplish this purpose, and was incapable of transcribing the subtle values of works of art, especially those of the painters. In contrast, the copperplate technique permitted extreme finesse of modeling thanks to its sharp prints. Thus every work of art, complicated or simple, heavy or light of line, could be reproduced — in a single color, to be sure — and the desired references could thus be made. Moreover, in contrast to black-line engraving, which spread a uniform film of ink on the paper, intaglio engraving, because of the varying depth of the incisions, produced a variety of inked areas that gave the print a relief ranging from the imperceptible line produced by the simple scratch on the metal to the compact mass produced by the deep incision.

The success of this method was immense, and surpassed all expectations, for it was not bound by the narrow and somewhat servile limits of a simple technique for reproducing works of art; the masters of painting — or at least a great many of them — utilized it directly to express themselves, thus transposing their genius into the domain of engraving: Botticelli, d'Andrea, Mantegna, Schongauer, Dürer (who is the creator of the first known etching), Callot, van Dyck, and especially Rembrandt, who brought the art of the etching, whose theme he transformed, to its peak.

In the seventeenth and eighteenth centuries this technique was the preferred instrument of the great portraitists as well as of so many minor masters of elegant pastoral scenes; it was unrivaled until the beginning of the nineteenth century, when other techniques began to appear, notably wood engraving across instead of with the grain, that is, engraved with the graver held at right angles to the fibers of the wood. This technique also permitted extreme delicacy of design on certain very hard woods that had a very close grain.

In the nineteenth and twentieth centuries, photography was to contribute the same virtues of rapid dissemination and comparison of works of art, that is, the possibility of studying and comparing, through the picture, the most widely separated or most unfavorably situated esthetic constructions.

The copperplate press

The press used for copperplate work was much simpler than the typographical press, which will be studied later. The engraved copperplate, with a sheet of damp paper laid on top of it, needed only to be subjected to a heavy pressure from above. Between two firmly braced side beams that supported the mechanism, two large wooden rollers tightly pressed a table that held the engraved plate and the sheet of paper, both protected by pieces of cloth. This was placed between the rollers. The action of the controlling wheel forced the copperplate to pass between the table, held by the lower roller, and the upper roller (Figures 304 and 305). In this way the paper absorbed the ink from the engraved lines.

FIG. 304 The copperplate press (from the *Encyclopédie*).

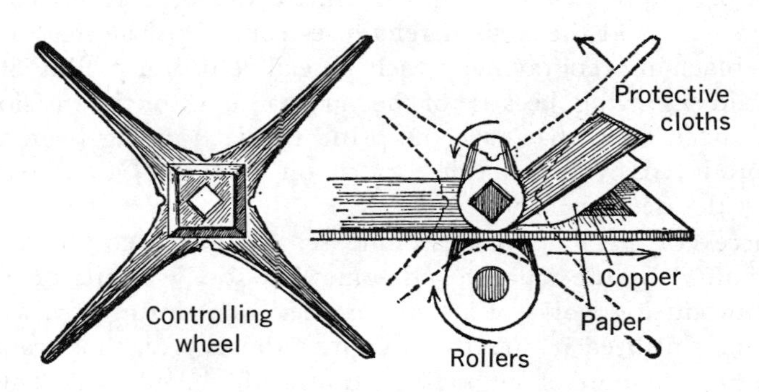

FIG. 305 Details of the copperplate press.

Considerable force was needed to turn the wheel; thus the pressman appears in a characteristic position in every picture of the press in operation, for example the illustrations of Abraham Bosse (1642) and Diderot's *Encyclopédie*. His legs are planted far apart, and with both hands he holds the top bar (the one facing him) of the wheel, while his foot rests on the low bar. In this way he moves the table forward until he is able to grasp with both hands the bar approaching him by virtue of the circular movement of the wheel (Figure 304).

Various methods of copperplate engraving

This technique of engraving on metal permitted such flexibility of work that it was possible to create several very different methods for expressing all its possibilities.

The first and most natural method, called engraving, was done as follows. The image to be reproduced was first drawn on the carefully planed copper, steel, or brass plate. The plate was then worked with gravers, sharp rods the engraver pushed before him with the palm of his hand and which dug series of parallel incisions crisscrossed by other, similar series of incisions either diagonal or at right angles to the first set. The depth, breadth, arrangement, and delicacy of the incisions made it possible to obtain all the details, modeling, and softness as well as power of the subject to be reproduced.

A second method, that of etching, utilized an acid rather than the graver as its active agent. The copperplate was coated with a warm varnish that was impervious to acid. On this ground the subject to be reproduced was outlined with needles, steel pens, small wheels, and gravers, leaving the metal exposed in the engraved areas. The plate was then placed in a bath of nitric acid. When the fine lines were adjudged to be sufficiently eaten away, the plate was washed and the lines were covered with a protective bitumen varnish; the acid process was repeated as often as possible until the engraving was completed.

The "soft-varnish" process was a variation of the etching. The varnish used was quite sticky, and clung unevenly to the tools and to anything pressed on it, causing an irregular line of engraving.

The design was generally traced on the plate with a metal instrument. The etching technique offered numerous possibilities. For the first time in the history of engraving, the "manual" work was transferred to a chemical agent, the acid. Centuries were to pass before these possibilities would be utilized, but in the nineteenth century, when the appearance of photography made possible the easy transfer of a subject directly onto a metal plate, this acid method ensured the automatic establishment of a picture in relief.

It was realized that acid had another advantage over the graver: a much freer and more cursive line, since it was traced, not by gravers, but by tools whose sole task was to cut into the varnish instead of the metal. The method even permitted the eating away of large areas and the obtaining of wash effects by direct applications of acid with a brush. The areas treated in this way had to be raised by very light hatching done with the graver, because the ink would not have clung to the smooth surfaces. This is why very few etchings were absolutely pure; generally the work of the acid was complemented by that of various engraving tools.

A third method, drypoint, was derived from, but was simpler than, engraving. It was done by means of a thin shaft of steel with a sharp point, which the engraver held in his hand like a pencil, and with which he scratched the metal. Printing was done in the same way as for the engraving. This method left thin lines of varying shades of blackness on the sheet of paper, quite similar to those of a pen drawing.

In the seventeenth century a new method was added: mezzotint, which was invented in 1643 by a Hessian lieutenant named Louis von Siegen, and was introduced around the middle of the century into England, with great success. Mezzotint utilized a metal plate granulated by means of a "cradle" with fine points that traced a series of parallel lines crisscrossing in all directions, thus creating a close network of points that created a unified, velvety tone in the print. The original grain was then cut or crushed in the areas that were to be white or light colored with a scraper and a "burnisher." The result was a play of tones ranging from very light to very dark, on a dark ground. This method was utilized particularly for depicting effects of light and dark.

The first experiments with a wash engraving, named aquatint, are attributed to one Johann Adam Schweikard of Nuremberg, at Florence in 1750. Leprince utilized the new method in France around 1760. It consisted of first covering a metal plate with a granular substance, either grains of resin or a layer of mastic dissolved in alcohol, which left a natural granulation on the surface of the plate when it dried. The surface was then attacked by applying acid with a brush, whereupon it became covered with a protruding network of metal grains unevenly distributed over the surface. The plate was then worked like an etching, the engraver covering with a varnish the parts of the plate which were sufficiently engraved.

Each of these methods had numerous variations.

EARLY PRINTED BOOKS

The xylographic book

The fifteenth century bore within itself an extraordinary desire for knowledge that transformed the Western world. What better instrument could there be for the development of this irresistible urge than the printed book? Utilizing the black-line engraving method, all the pages of a small book began to be engraved on wooden tablets. Once they had been engraved, the xylographic book could be produced with such rapidity that the first creators of the printed book, who jealously guarded the secret of their technique, were afraid of being accused of witchcraft.

The "practical book," as opposed to the medieval "work of art," had thus been born, and its success was to be tremendous; for the first time an instrument was available that made it possible to multiply easily and in their exact text the essential elements of thought, and particularly Christian thought. People have a tendency to see in the appearance of the printed book only a contribution to the ease of the clergy in studying, assembling, comparing, and transmitting the key texts of Western literature; further, they say, it permitted the unhindered development of textual criticism. The major determinant of the development of the book, however, was that it taught the people to read.

To be sure, the engraving of a xylographic book was a long and tedious process. Our curiosity led us to commission the engraving by a professional engraver of the plate of a small (five inches by seven inches) edition of Donatus, and the experiment was a fruitful lesson. The xylographic booklets that have survived are all executed in Gothic or similar characters, and such letters, which were tipped top and bottom with points that formed a small triangle that could easily be detached with the gouge, were relatively easy to engrave. Our experiment revealed that the text was worked not line by line, but space by space; then all that remained to be done was to detach the letters from each other.

Judging by the time spent by our engraver in preparing one page, it can be estimated that the professional engraver of the fifteenth century executed a page of this type in thirty or thirty-five hours. The work had to be done in shops that combined three techniques under the same roof: drawing, transfer, and engraving. Probably these shops also did other engraving work, particularly plates for playing cards.

Xylotyping and the idea of composition

It is not so long since the very mysterious dividing line that in the West separates xylography from typography was crossed in a single leap. It is said that in order to discover the secret of typography, which is basically a com-

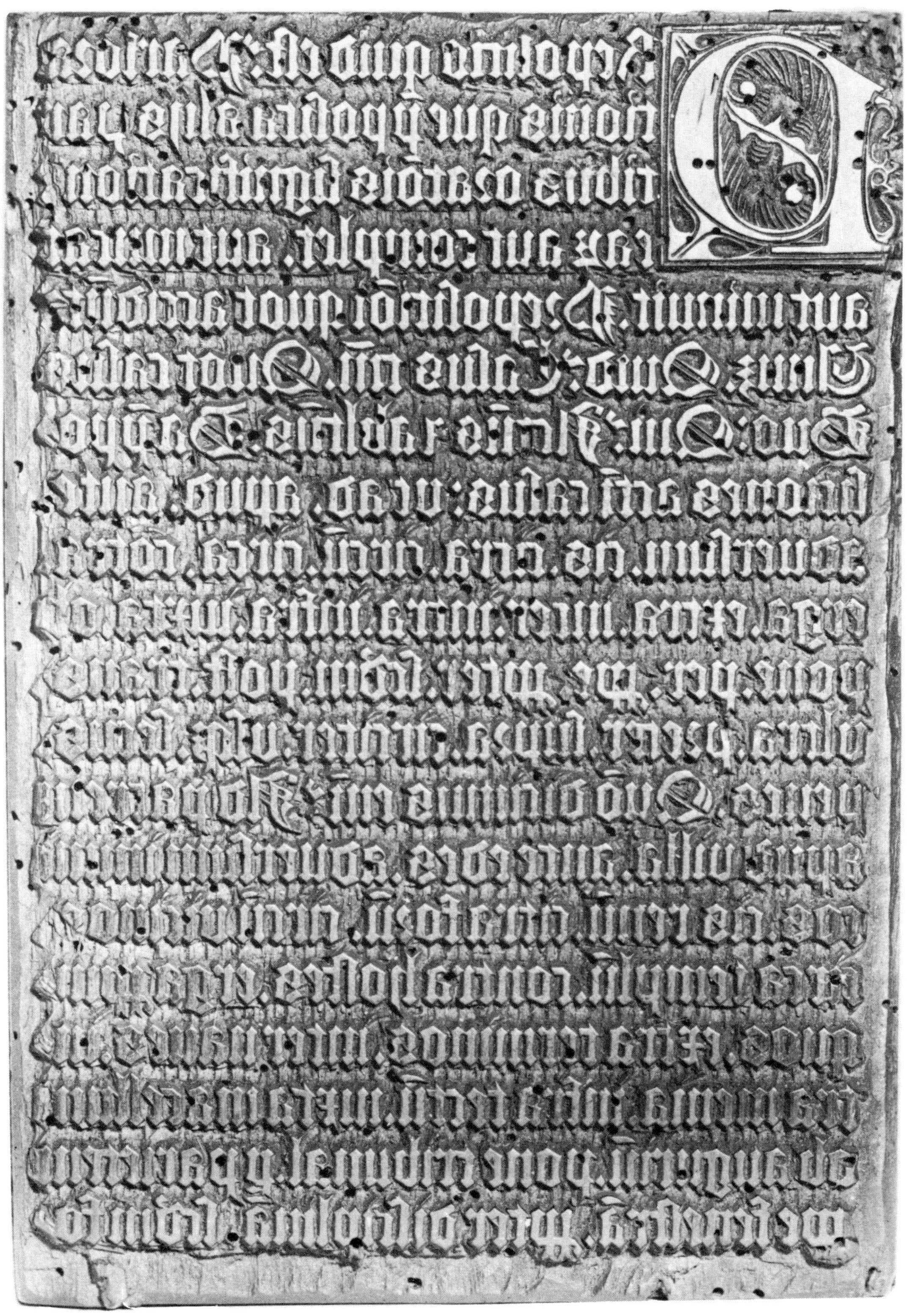

Plate 51. Xylographic block, fifteenth century. Bibliothèque Nationale.
Photo Bibliothèque Nationale

bination of small individual elements, all that was necessary was to cut the xylographic block into small parallelepipeds, each bearing one letter. But the problem is much more complex if we wish to take into account the numerous technical difficulties that argue against such an explanation. There is the difficulty of cutting out all these tiny wooden parallelepipeds so that they can be effectively reassembled in different order; there is the problem of preserving them so that they do not warp as a result of their fragile material, which is subject to hygrometric variations, or deteriorate and quickly break.

However, some consideration can be given to a well-established legend that claims that around 1423 or 1437, at Haarlem, one Laurens Janszoon is supposed to have utilized individual wooden letters that he cut from beechwood and assembled in order to print texts, thus utilizing the early version of typography invented by the Chinese around 1040. But given the small size of the signs of the Western alphabet, Janszoon does not seem to have been able to engrave the letters indispensable for the printing of a large book; it is our opinion that he composed large blocks of text by the xylotyping method and that he immediately encountered the difficulties inherent in the material utilized. It is nevertheless true that Janszoon could well have been the first man in the West to realize the "functional importance of composition" (that is, the assembling of prisms each of which bears one engraved letter) and that he must occupy a place of honor in the hierarchy of precursors. However, this was not yet genuine typography.

The metallographic book The woodworking arts had now contributed to the limit of their resources to the solution of the problem; they had given birth to the "mechanical book," and had driven home the idea of composition by means of separate letters. To continue further along the road that was opening so miraculously, new technicians had to be brought into the debate: the practitioners of the metalworking arts.

What degree of progress had been achieved by the metalworking arts by the beginning of the fifteenth century? It can be said that they were dominated by the technique of the punch. The punch had been used during the late Roman Empire for striking the bezels of rings used as seals. It later made possible the production of the matrix of certain seals, as well as the die with which coins and medals were struck. The binders also utilized the punch for embossing the boards and spines of their books. This technique was already highly advanced, since it was naturally supported by the goldsmiths' industry, which had reached a rare degree of perfection by the fifteenth century. As for the printing techniques, they were now widely used, since metal plates were occasionally engraved by the black-line method, in which the dark portions were made lighter by means of the punch.

The method of casting in molds had already been utilized in antiquity; the Gauls and the Romans had cast coins in terra-cotta molds. This technique, which had been abandoned in the Middle Ages, was adopted by the medalists of the thirteenth and fourteenth centuries, and the idea that their molds were prepared with engraved punches can safely be accepted. The metal casters (and particularly bell casters) of this same thirteenth century were also combining the use of the punch and the casting method to make hollow matrices of clay or sand into which they poured the metal, thus bringing out relief inscriptions on their cast objects. The makers of funerary monuments placed texts on tombstones in much the same manner.

While the xylographers were painfully engraving their first booklets, the metallographers, aided by their possibilities of stamping and casting in molds, were trying to find a definitive solution for the problem. What must be done in order to eliminate the long work of engraving printing blocks? — "Replace this engraved block by a molded block with the same properties."

This was not at all difficult if the technique and procedure just described were utilized. All that need be done was to create a hollow matrix by utilizing punches of letters struck side by side in the order of the text, and to cast in this matrix metallographic tables in relief that could be utilized exactly like the xylographic tablets. The saving in time was tremendous, but the result was poor: the molded block had major imperfections, owing especially to the lack of alignment of the letters, the depth of the imprints, and the distortion to which the striking of each letter subjected the adjoining, completed letter.

This procedure, which seems to have been utilized for the first time in Holland, if we are to believe a passage in the *Chronicles of Cologne,* must have produced only minor works: poorly printed booklets with many errors, which the *Memorial* of Saint Aubert's Abbey of Cambrai qualified in 1445 and 1451 as "doctrinal material thrown into a mold, worth nothing and completely wrong." From Holland this method naturally spread to the Rhineland provinces, where a few diligent metallographers, including Gutenberg, must have tried to perfect it. We are vaguely apprised of its existence at Strasbourg in 1439 by the deliberately obscure testimony of the witnesses in a lawsuit that placed Gutenberg in opposition to the family of one of his associates, who had just died; then in Avignon during the years 1444, 1445, and 1447, in connection with a strange method of "artificial writing" which a goldsmith from Prague (via the Rhineland provinces), Procopius Waldfoghel, was attempting to utilize and which, according to one of his partners, was "genuine, easy, and useful."

In an attempt to eliminate the disadvantages of the metallographic technique, the experimenters tried to assemble their punches before striking, in order to avoid distortions and poor quality in the type faces.

Johann Gutenberg

There is no question of Gutenberg's worthiness to bear the credit for being the inventor of typography: the very early attempts made in this direction by such men as Ulrich Zell, Mathias Palmieri, Johann Tritheim, Jacob Wimpheling, later Sebastian Munster, and many others proved useless, according to explicit statements by most of the chroniclers, technicians, and authors during the years that followed the invention.

The debate seems to be definitively closed by the statement of Johann Schoeffer, the son of Peter Schoeffer, a partner of Gutenberg and Fust, in the dedication of an edition of Livy which he printed (1505), that "the admirable art of typography was invented by the ingenious Johann Gutenberg in 1450 at Mainz." This is the same Johann Schoeffer who after 1509 systematically informed his readers that the *only* inventors of this art were his grandfather, Fust, and his father, Peter Schoeffer. However, we find a confirmation of the attribution of this discovery to Gutenberg in the simple existence of the lawsuit that opposed him to Fust in 1455, and which reveals that while Fust was the financier of the partnership, Gutenberg was completely in charge of technical operation.

Thus we are able to reconstruct the history of the invention, based on the evidence given in the Strasbourg lawsuit of 1439. In 1436 Gutenberg, who was keenly interested in everything concerning the budding arts of book printing, and very enthusiastic about the Dutch experiments in metallography, hired three partners: Johann Riffe, a wealthy merchant and the financial support of the company, and two assistants, Andreas Dritzehem and Andreas Heilman, as well as two technicians: a goldsmith named Hans Dünne who undoubtedly worked with Gutenberg on engraving the punches needed for the metallographic process, and a carpenter, Konrad Sahspach, who perfected the handpress. This implies that Gutenberg was working on two levels: improvement of the page-preparation technique, and creation of a printing device. This fact is moreover confirmed by the documents in the lawsuit, which state that Gutenberg had concealed some of his secrets from his partners and that he later promised, in exchange for a certain sum of money, to complete his revelations.

The typographical book

In his attempt to improve the metallographical process, Gutenberg was naturally led to utilize his punches directly by assembling them in lines and pages, "composing" them for use in small printing projects ("laboratory experiments," as we would call them today). But it is certain that the large typographical book, as remarkable as the manuscript work it was desired to multiply, could be achieved only by means of *types obtained in series by molding.* This has long been realized by the historiographers of printing, who declare that the essence of the typographical invention was the creation of *molded,* not engraved, type.

In order to obtain this result, however, it was first necessary to create a molding technique sufficiently precise to permit the casting of small types of good-quality metal. This was certainly what Gutenberg was doing during the years that passed between the accepted date for the invention of movable type (1440) and the appearance of the major works of the printer's art. It can, however, be accepted that it was Peter Schoeffer, the partner of Gutenberg and Fust after 1452, who must have invented the *hand mold,* which was to facilitate the obtaining of individual types. Gutenberg probably obtained his types by the sand-casting method (Figure 306).

The typographical book was the undeniable solution to the technical problem of the book (Figure 307). While there was sometimes opposition to this solution, as we shall see, it was aimed, not at its principle (which remained sacrosanct), but at the difficulty of handling and preserving the extremely "mobile" types. An invention of equal importance was the only one that could have competed with this solution, and it appeared much later: photography. The preeminent position of typography still weighs on the arts of the book, and we are only now beginning to envisage its elimination and replacement by radically different processes.

Fig. 306
The type founder
(after Jost Amman, 1568).

The power of the book

Before entering the printer's shop and looking at the various tools used there, particularly at its essential piece of equipment, the printing press, let us consider what typography could have represented at this period in the eyes of those who were beginning to be its beneficiaries. The invention was greeted by a burst of enthusiasm; as Wimpheling stresses, the pioneers of printing were compared by the Humanists of the day to Christ's apostles because, like the latter, they sowed the Word and preached the new gospel of truth and knowledge. The masters of this "divine" art, as it was called, assure us themselves that it was "the symbol of the golden age and the era of bliss."

Was not the significance of some of these judgments prophetic? Like Christ's apostles, these pioneers, dispersed throughout Europe, created roads of hunger paved with misery, failures, disappointments, and despair. During the second half of the fifteenth century, that is, the "incunabula" period, they were generally miserable itinerants, driven from city to city by their desire to bring the benefits of their activity to a wider circle of people. Being without resources, they were often forced to give up their liberty and use their few material belongings in order to obtain the raw materials needed for their work; sometimes despoiled and forced to cease their activity in one place, they went to other areas, always seeking the wealthy patron they needed. Many French cities welcomed the infant printing industry within their walls – an action of which they can now be justly proud. The archives of some of these cities, notably those of Lyon, depict these pioneers faced with overwhelming problems, often oppressed by misery to the point of being relieved of all taxes, while they were preparing sumptuous works; a document of 1503, for example, explicitly states that the tax collectors forgave a printer named Jean Syber his taxes "out of pity."

et liber: sed omnia & in omnibus xp̄s.
Induite vos ergo sicut electi dei sancti
et dilecti viscera misericordie: benigni-
tatem·humilitatem·modestiam·pati-
entiam·supportātes invicem & donan-
tes vobisimetipsis si quis adversus a-
liquē habet querelam: sicut & domin⁹
donavit vobis ita et vos. Super o-
mnia aūt hec caritatem habete qđ est
vinculū pfectionis: et pax xp̄i exultet
in cordibus vestris·in qua et vocati
estis ī uno corpore: & grati estote. Ver-
bum xp̄i habitet in vobis abūdātēr:
in omni sapientia docentes et cōmo-
nentes vosmetipsos psalmis ymnis &
canticis spiritalibȝ in gratia cantātes
in cordibus vestris deo. Omne qđcūqȝ
facitis in verbo aut in opere: omnia
in noīe dñi nostri ihesu xp̄i: gratias
agentes deo et patri p ipm. Mulieres
subdite estote viris: sicut oportet ī domino. Viri diligite uxores vestras: &
nolite amari esse ad illas. Filij obedi-
tē parentibus per omnia: hoc enim
placitum est in dño. Pres nolite ad
iracūdiam provocare filios vestros:
ut non pusillo animo fiant. Servi
obedite p omnia dñis carnalibus: nō
ad oculū servientes quasi hominibus
placentes: sed in simplicitate cordis ti-
mentes deū. Qđcunqȝ facitis·ex animo operamini·sicut dño & non hominibus: scientes qđ a domino accipietis
retributionē hereditatis. Dño xp̄o ser-
vite. Qui enim iniuriam facit recipiet
id quod inique gessit: et non est per-
sonarum acceptio apud deū.
Domini quod iustum est et equū
servis prestate: scientes qđ & vos
dñm habetis in celo. Oratōni instate
vigilātes in ea ī grātiarū actione: oran-
tes simul et pro nobis: ut de⁹ aperiat
nobis ostium sermonis ad loquēdū
mistēriū xp̄i: propter quod etiam vin-
ctus sum: ut manifestem illud ita ut o-
portet me loqui. In sapientia ābula-
te ad eos q foris sunt: temp⁹ redimen-
tes. Sermo vester semp in grātia sale sit
cōdit⁹: ut sciatis quomō oporteat vos
unicuiqȝ respōdere. Que circa me sunt
omīa vobis nota faciet tychic⁹ carissimus frater & fidelis minister et cōservu⁹
in dño: quē misi ad vos ad hoc ipm
ut cognoscat q circa vos sūt: & cōsolet
corda vrā cū onesimo carissimo & fideli
fratre q ex vobis ē: q omīa q hic agū-
tur nota facient vobis. Salutat vos
aristarc⁹ cōcaptiu⁹ me⁹: et marc⁹ cōso-
brin⁹ barnabe: de quo accepistis mā-
data. Si venerit ad vos: suscipite illū.
Et ihesus q dicit iustus: qui sūt ex cir-
cūcisione. Hij soli sunt adiutores mei
in regno dei: qui michi fuerunt solatio.
Salutat vos epafras qui ex vobis est
servus xp̄i ihesu: semper sollicitus pro
vobis in oratōnibus: ut stetis perfecti
et pleni in omni voluntate dei. Testi-
monium enim illi perhibeo: qđ habet
multū laborem pro vobis & pro hijs
qui sunt laodicie et q ierapoli. Salutat vos lucas medicus carissimus: et
demas. Salutate fratres qui sunt laodicie et nympham: et q in domo eius
est ecclesiam. Et cum lecta fuerit apud
vos epistola hec·facite ut et in laodi-
censium ecclesia legatur: & eam que la-
odicensium est vobis legatur. Et dici-
te archippo. Vide ministeriū qđ acce-
pisti in dño: ut illud impleas. Salu-
tatō mea: manu pauli. Memores estote vinclorȝ meorȝ. Gratia dñi ihesu vo-
biscū amē.

FIG. 307 The 42-line Bible printed by Gutenberg and Fust around 1455.

However, the work done in the printing shops of Europe was truly prodigious. Lucien Febvre and H. J. Martin (*L'apparition du livre,* Paris, 1958) have given us figures for the first half of the century that appear fantastic: 30,000 to 35,000 separate editions representing perhaps 15 to 20 million copies.

With the advent of the sixteenth century, all this changed: the difficult battle of the pioneers had been won, and the printer was master of his art, completely installed in his shop, very often a bookseller and Humanist as well as a printer. He was now to reap the harvest of the exhausting struggle waged by his predecessors in the fifteenth century. From their well-stocked shops, the printers and booksellers of the sixteenth century were to "bring out" (according to the authors we have just quoted) 150,000 or 200,000 various editions; if we accept the figure of 1,000 copies per edition as an average printing – which does not seem exaggerated – this represents a total of 150 to 200 million copies. These authors, moreover, are taking into consideration only editions of books; they ignore the ordinary printing jobs (posters, brochures, job work, minor printed matter, etc.) which represent a certain supplemental activity. It is not surprising, then, that succeeding centuries suffered from a decreasing demand for the book, which had been so abundantly distributed during the Renaissance.

THE EARLY TYPOGRAPHICAL SHOPS

The shops of the early printers were located in nooks, cellars, and tiny remote rooms, for the equipment needed was very limited and took up little space. The engraving that depicts Matthew Husz's shop in Lyon in the fifteenth century (the oldest such picture known) is very interesting from this point of view. It shows a box of type, a compositor and a copyholder, two chases, the press and its inking pads – and that is all. Of course the engraver of the picture concentrated in a small space the essential elements of the equipment of a printer's shop, but the few items not shown are of little importance.

What materials, equipment, and raw materials were needed to set up a printing shop (Figure 308)? Let us first distinguish between two separate operations:

the composition of the pages,* whose essential piece of equipment is the case in which the types are assembled;

the printing of the form, which is done on the printing press.

These two operations and their equipment were to determine the use of a complete series of various tools, as follows:

For the compositor:

The cases and half-cases in which the various types bearing letters and signs and blanks for spaces within the line were placed.

A table with shelves on which were placed blank leads for spacing, within easy reach of the compositor; the composed pages were laid on the shelves as they were completed.

* I.e., the type set and arranged for the page.

FIG. 308 The printing shop of Matthew Husz at Lyon. End of the fifteenth century. (From *Dance of Death.*)

The composing stick, held in the compositor's hand, in which the types were placed to form the lines of text.

The composing galley, which received the composed lines from the composing stick. It was generally placed in the upper-right portion of the case, an area that was seldom utilized during composition.

The copyholder, which held the sheets of the copy, and which fitted into the edge of the case, generally toward the top center.

Tongs, used to extract a piece of type from the page, for example in the correcting of a composition.

For the pressman:

The press itself, with all its components – the most important piece of equipment in the shop, by means of which the prepared and properly assembled pages were printed on the paper.

The imposing stone, a table on which the pages were assembled and readied for printing. In the fifteenth century this work was probably done directly on the press stone.

A chase, in which the pages were tightened into a rigid form by means of wooden wedges.

An inkpot, which was part of the press and placed behind it; it was used to spread the ink until a perfect inking surface was created.

Inking pads, used to spread the ink evenly on the pages during the printing.

A table near the press that held two stacks of paper, one of plain white sheets, the other the printed sheets coming off the press.

A tub of water in which the sheets of paper were dampened in order to soften them; a better and more even impression was thus obtained.

A stretcher with which the printed sheets were lifted and placed on ropes to dry.

With a few exceptions, this list of equipment and material could also be used as a description of a printing shop at the beginning of the nineteenth century, which is to say that the technique had evolved very little in four centuries. At most, as we shall later see, the press became lighter and was improved so that it could do a better job, while the material was refined and increased in quantity.

But outside large, well-known printing shops turning out books that do honor to the graphic art, right from the start there existed numerous itinerant printers who traveled between cities that did not yet have a printing industry, printing posters and doing other minor printing jobs, perhaps even small brochures. To be sure, their equipment was extremely limited: types, undoubtedly carried in small boxes which when joined together acted as cases, a composing stick, a short composing galley, tongs, a small chase, inking balls, black ink, and paper. They had no presses, but this bulky piece of equipment was not needed for printing posters and other small jobs. When tightened into a chase, the page could be printed in two different ways: either by placing the page on top of the type and rubbing it, or (as is done by children with their toy printing sets) placing the paper under the type face and applying pressure.

Whether this latter method was employed in large printing shops remains to be proved; it would in particular be necessary to study closely the difficult problem of "misplaced types," which we shall discuss later and which raises very puzzling questions.

The printing press in the fifteenth century

Since the press is the essential piece of equipment in the book-printing shop, we are going to study it from its early stages. We have already mentioned a lawsuit involving Gutenberg and his partners with the heirs of their deceased partner Dritzehem in 1439. We are fortunate in possessing the text of the minutes of the case; they refer to a complete series of technical operations relative to printing that had been in use by the group since 1436. Since the invention of typography appears to date from 1440, this undoubtedly refers to an earlier, apparently metallographical process, probably the method discussed earlier under this heading.

However, the text of the witnesses' testimony reveals that the partners were concealing one or two new pieces of equipment that can only have been the prototypes of a printing press — the first press ever to exist — constructed by the carpenter Konrad Sahspach.

There has been much discussion of the nature of this device; notably, people have regarded it as a mill for casting the characters. We fail to understand how there could exist any further hesitation after a consideration of Sahspach's testimony, which states that another partner (Andreas Heilman) told him, "Since Andreas Dritzehem is dead, and since you are the one who made the presses and are acquainted with the material, go, take away the parts of the press, separate them from each other and take them apart, and thus no one will be able to discover what it is."

This Strasbourg team must certainly have been using for quite some time the technique of the metallographical book, which had come from Holland, and were seeking not only to profit by it but also (and especially) to improve on it.

However, Gutenberg must have quickly realized that the problem of the book had now been raised in its entirety; that is, what was needed was not only an improvement in the method of preparing the block pages, but also a way of printing these pages mechanically and quickly. Printing was still being done by the crude rubbing method: the sheet of paper was laid on the inked block and was then rubbed on the wrong side to create the impression. Gutenberg felt it was more urgent to attack this problem first, and he decided to create a printing machine that would be capable of multiplying the power of production of the book. Thus he conceived the idea of his first press.

What was this press like? We do not know. Many authors, notably Francis Thibaudeau, Karl Faulmann, and Charles Mortet have suggested that it was crudely built, that it was something like a simple wooden press, a descendant of the oil, wine, and paper presses, and that only much later, at the beginning of the sixteenth century, was a movable carriage adapted to it, making it possible to remove the form so that it could more easily be inked and covered with the sheet of paper.

Not all the authorities are in agreement with this idea. It is a striking fact that inventions are always achieved as an integral unit; that is, the inventor, having considered all aspects of the problem, discovers the principle of the solution in its totality, later improvements being simply technical improvements. The printing press without a mobile carriage bearing the form, and without the chase holding the sheet of paper in its place, was not a very efficient machine. Gutenberg and his partners were thus obliged to build a press, which was perhaps crude in appearance but completely perfected as far as its functioning was concerned. Engravings depicting the old presses, including even the very first press known, and the fifteenth-century engraving representing Matthew Husz's shop in Lyon, indicate that the carriage of the form was movable, and clearly show the iron fittings on the joints of the chase carrying the sheet (Figure 308). Figure 309 shows a reconstruction of the fifteenth-century press.

The value of Gutenberg's invention becomes even more apparent when we consider that the rigidity of the platen (the plate that causes the sheet of paper to adhere to the face of the characters) obliged the type founder to make the faces of his types as perfect as possible, if the impression was to be exact. This was one of the reasons for the abandonment of the metallographic plate, which was never sufficiently flat, and the adoption of the typographical method, which by the use of individual letters made possible a greater striking flexibility of the pages.

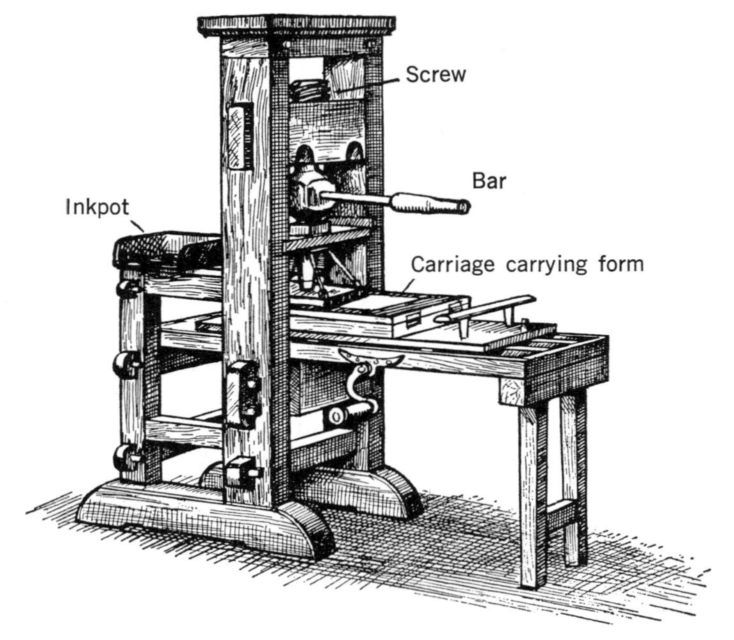

FIG. 309 Reconstruction of the fifteenth-century press (Musée lyonnais de l'Imprimerie et de la Banque).

However, there is one important point on which, to the best of our knowledge, little information is available. The presses that could have served as models for Gutenberg in the construction of his first press were undoubtedly the small binder's press and the papermaker's drying press. But these presses were built to obtain a slow and powerful pressure, and therefore their central screw had a single thread. In adapting these presses to the printing technique, Gutenberg encountered his first obstacle: in order sufficiently to raise and lower the platen over the form, it was necessary to turn the screw at least once, and perhaps several times, depending on the pitch of the threads on the core of the screw. For this purpose, the bar that turned the screw would also have had to make a complete turn, which was impossible because the framework of the machine was in the way. The head of the screw, which was cut in one piece with the screw and was generally quadrangular, therefore had to be drilled with four holes that one by one received the end of the bar, exactly like a capstan, so that the mechanism could make a complete turn and so that the platen touched the sheet of paper on the form. This was a tremendous loss of time, since the operation had to be repeated in order to raise the platen.

The solution to the problem lay in a transformation of the screw, which instead of having a single thread should have three and more probably four, all equal and parallel. In this way the threads could be steeply pitched; and the descent of the platen was made four times faster, which meant that with only a quarter-turn of the screw the same result was obtained. The bar could thus be permanently fixed to one of the four faces of the screwhead, since a quarter-turn sufficed to permit the platen to touch the sheet of paper.

Who was the first man to think of utilizing this screw with four threads? It is very difficult to say, but we believe that this device was in use at a very early period. Since by its very nature the printing industry tended to make increasingly rapid production necessary, the difficulties arising from the use of the single-thread screw must have been quickly solved.

But the four-thread screw had the corresponding vice of its virtues: while it did in fact permit the platen to descend four times quicker, it greatly reduced its striking power. This meant that the group of pages in the form could not be struck with a single impression on the sheet of paper. Two impressions were therefore made; that is, the movable carriage holding the form was pushed halfway under the platen; and the bar was turned once, permitting half of the sheet to be printed. The platen was then raised, and the carriage was pushed forward the rest of the way; a second turn of the bar printed the second half of the sheet.

This vigorous operation had one other disadvantage: the double blow shook the device and over a period of time caused the press to pivot on itself, thus contributing to its disintegration. To hold it in place it had to be braced, that is, fixed to the stringers of the ceiling by beams and, later, by decorated balusters.

The early types

"Misplaced types" are the name that has been given to strange marks that are sometimes seen in incunabula editions. They were produced as follows: a technical accident caused a type to be laid flat on a page of characters being printed. The pressman was unaware of it, and printed one sheet, which bears the mark of the side face of the type. An examination of this mark almost always reveals that the type has a chamfered foot and a heel, and occasionally a hole or slit. These "misplaced types" have been discovered on books printed in every Western European country in which typography originated – Germany, the Netherlands, Belgium, Italy, and France – which implies that this special technique was in general use in the making of types. Victor Scholderer ("The Shape of Early Type," *Gutenberg-Jahrbuch,* 1927) was the first to attempt to study these curious marks, and his still-unpublished notes (which he generously made available to the authors of this book) concerning his latest discoveries with regard to these misplaced types leads us to believe that certain books containing such marks could have been printed by the method of bringing the prepared form down onto the sheet of paper.

To the best of our knowledge only about twenty of these marks have been discovered to date. A recent discovery, appears in a book in the Musée lyonnais de l'Imprimerie (Figure 310).

Almost 250 varieties of types dating from the fifteenth and very beginning of the sixteenth centuries, were discovered in the bed of the Saône River at Lyon. They are divided between two collections, one of which belongs to the National Library, the other to M. Brassart of Montbrison. The morphology of the types corresponds exactly to that of the misplaced types; that is, almost all of them have a chamfered foot and a heel, and several of them have holes or slits.

The chamfered feet and the heels undoubtedly corresponded to a technique of sizing of the types, which were cast much longer than the type height needed, then broken on the bias to shorten them and carefully filed down until the exact type height was obtained. As for the openings on the sides, they could have been used to facilitate the process of registering two colors simultaneously on the sheet.

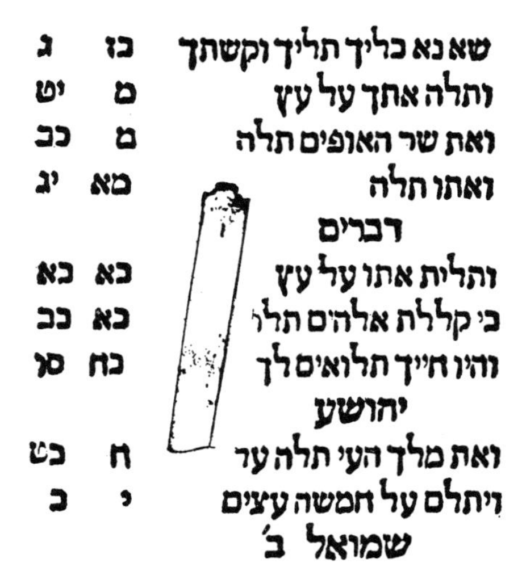
שא נא כליך תליך וקשתך כז ג
ותלה אתך על עץ מ יט
ואת שר האופים תלה מ כב
ואתו תלה מא יג
דברים
ותלית אתו על עץ כא כא
כי קללת אלהים תלו׳ כא כב
והיו חייך תלואים לך כח סו
יהושע
ואת מלך העי תלה עד ח כט
ויתלם על חמשה עצים י כ
שמואל ב׳

FIG. 310 A "misplaced type" in *Meïr Natib* (Concordance to the Bible) (Venice, D. Bomberg Tisri, 1523).

Two-color printing

The first printers almost certainly experienced great difficulty in achieving rapid two-color printing of books they were trying to produce in the spirit of the manuscript copies. With the crude methods at their disposal, it was very difficult for them to rubricate except by removing from the page the types which were to appear in red, inking them separately, and replacing them in the page (which had been separately inked in black), thus obtaining a two-color page in a single printing.

This was a lengthy process, but the more rapid method of printing in two impressions (first the black, then red) was extremely risky, for the exact registering of the two colors was made very difficult by the practical impossibility of always replacing the sheet at exactly the same place on the tympan of the press.

These openings in the fifteenth-century types probably served to hold together, by means of a wire, the types comprising the rubrication. How else can we explain the presence of such an opening, drilled after the type had been cast, and which increased the fragility of the body of the type? In addition we have noticed that in some of these drilled type pieces in the National Library collection, the hole, which normally opened on the side, ended under the foot of the type, which implies that these types were placed at the beginning or the end of the text of the rubrication and that the wire was introduced through the foot of the first type and came out through the foot of the last one. This method made it easier and faster to insert and remove the rubrication as a unit from the block of the page.

But by the beginning of the sixteenth century technical progress had made it possible to transform the making of the types. On the one hand their sizing was accomplished by the intermediary of a riser, which was removed once the type had been cast. On the other hand, the registering in two printings of the two colors of a single form was done by means of a press point, that is, two needles embedded in the form that pierced holes in the sheet of paper. These holes made it possible on the second printing to register the colors in exactly the right place, and the verso of the sheet in relation to the recto. The press of 1564, reconstructed by Jost Amman, shows the two press points of the form. Certain "reject pages" of books, dating from the end of the fifteenth century, prove that in certain shops these press points were already in use by then. Thus the types, now free of certain obligations, acquired their present form, and were equipped with the nick that permits the compositor to recognize their direction.

But there is another method of two-color printing that must have been utilized as soon as the registering of the sheets was achieved and that remained in use until the eighteenth century; Fertel describes its operation (*La science pratique de l'imprimerie,* Saint-Omer, 1723, p. 277). When the completed form had been placed on the press, and the frisket was in position, a trial run of the form was made directly on this frisket. Those parts of the text that were to be printed in red were then cut out of the frisket, and were glued on the tympan in the exact spots they were to occupy; this gave an extra pressure to the red parts and made it possible to print without dry impression of the black. The print was then made; only the portions corresponding to the openings of the frisket appeared on the paper, the rest of the form being printed and overprinted in red on the frisket. Care had to be taken that this constant overprinting did not ultimately cut through the parchment of the frisket. When this red printing was completed, the letters corresponding to the red portion were removed and replaced in the form by quadrats, and the black portion was printed normally.

The underlays could also be placed, not on the tympan, but under the letters in order to raise them. In certain printing shops that produced large quantities of highly rubricated pious books, types of two different type heights were used, those in red being clearly taller than the others; this eliminated the need of raising them by means of parchment underlays. The Musée Plantin at Antwerp possesses a collection of types of this kind.

The printing of music

It has long been thought that typographical musical notation was first practiced at the beginning of the sixteenth century by Ottaviano Petrucci, a printer of Fossombrone near Urbino. Actually, experiments had been made long before then.

In the early days of printing, to be sure, no one was concerned with solving the problem of the printing of music: a space was left in the text for the lines of music, which were then filled in by hand. This was the method used for the music in the *Missale lugdunensis,* printed at Lyon by Jean Neumeister in 1487.

The first typographical musical notation appears to date from 1473. In his edition of the *Collectorium super Magnificat* of Gerson, a printer of Esslingen named Konrad Fyner probably utilized full-face types in order to print very crude square notes without staves.

Since books were generally rubricated in this period, the music printers naturally came to utilize two colors in printing: red to denote the staves, black to place the notes on these staves simultaneous with the printing of the text. The first book printed in this way was the *Missale Romanum,* printed in 1476 at Rome by Ulrich Han; next came the *Missale Herbipolense,* printed at Würzburg in 1481 by J. Reyser, and another *Missale Romanum* printed at Rome in 1482 by S. Planck.

But an increasing number of small and even large books were being printed in a single color; it was therefore natural that ultimately they began to print the music together with the text, using the black-line wood engraving, as the illustrators were doing. In 1482 Ramos de Pareja printed the music for a *Tractatus de musica* on xylographic plates; he was followed by Hugo Rugerius at Bologna in 1487 and Pruss at Strasbourg in 1488. Music continued to be printed in this way for a long time, notably by Johann Froschauer, who printed a *Lilium musice plane* by this method at Augsburg in 1500.

With the music of the *Policronicum* of R. Higden, printed at Westminster in 1495, a new technique was born. Here for the first time we find a genuine typographical composition combining staves, notes, and stems of notes. Erhard Oeglin also utilized this process at Augsburg in 1507 for a small music sheet.

In France it was the Parisian prototypographer Ulrich Gering who, together with his partner Renholt, was the first to print music in two colors: the staves and bars of music in red, the notes in black, with the text of the *Psalter* of 1494. In the following year Jean Neumeister utilized two-color printing of music to present his *Missale uceniensis.* Then, in March 1497, another Lyonese printer named Michel Topié proceeded in the same manner to print the *Missale ad usum romane curie.* This method was used for a long time.

A talented type founder named Pierre Haultin then conceived the idea of casting each note with its fragment of stave on a single piece of type, thus composing lines of music, the stave of each note being matched to that of its neighbors. Another Parisian typographer, Pierre Atteignant, a son-in-law of Philippe Pigouchet, continued the experiment, and in 1528 began publishing an entire series of volumes of music with this method. A great number of printers followed his lead, notably the Ballards.

The first of the Ballard family, Robert, was the brother-in-law of the famous type founder Guillaume Le Bé, who was able to supply him with special kinds of type. On August 14, 1551, he obtained a royal license of nine years to print or have printed and exhibit for sale all kinds of music books, both instrumental and vocal, which might be printed by them. This license was regularly renewed for more than two hundred years, and thus it can be said that this dynasty of printers established a *de facto* monopoly over the entire French production of music until the approach of the Revolution.

We shall not devote much space to the copperplate method of engraving music, which was popularized by Plantin, for from the technical point of view it did not differ from the classical methods of engraving. Staves, notes, and texts were engraved on the copperplate, as was done for any other material.

THE INTERMEDIATE STAGE
(Sixteenth to Eighteenth Centuries)

The world of the book

Thus the adventure of the graphic arts began in the sixteenth century to change course. It was no longer the technicians who were to struggle to perfect a universal method of expression. The instrument was found in a form that was considered miraculous; it did have several defects, which were soon perceived, but at least it existed, and it was up to the technicians to ensure its continual efficiency, perfect it, and pave the way for major new inventions that were later able to furnish new solutions for the technical problem.

At the dawn of the sixteenth century, after the heroic age of the pioneers, a monumental new work had to be constructed. A spiritual civilization was in the process of being born that was to occupy the human mind for centuries and that had to be organized on all levels: the Civilization of the Book. The first task was to adapt the technique that was its basis to the immense needs of a human race that had suddenly become aware of science, art, and all the forms of the "encyclopedia"; the second was to establish consistent relationships, cultural as well as commercial, between a growing number of important authors and a public avid for everything which could fill its mind. The master printer was born of his difficult combat with matter; the master bookseller was to be born for the combat of the spirit, and from a utilitarian point of view was to dominate the intellectual life of this period. Many printers during these three centuries worked as booksellers or in close collaboration with booksellers.

Lucien Febvre and Henry-Jean Martin have entitled one chapter of their work "Le petit monde du livre" (The little world of the book). The word "little" seems to us to be, not a value judgment, but rather a way of expressing the fraternal nature of the enterprise. The dynamism of this little world of the book is visible in the simple movement of its history. The printer-bookseller formed its living center; he was, or became, a Humanist, and concentrated all the great minds of the arts and letters, and soon all the scientists, around his little shop.

Following Febvre and Martin, several leading lights can be distinguished among these printer-booksellers – enlightened and learned men who were the glory of this shining sixteenth century. There was, for example, Johann Amerbach, who in 1475 began to publish corrected editions of the writings of the Church Fathers from his tiny shop in Basel. His highly learned son, Boniface, became the proofreader and friend of Frobenius, the publisher of Erasmus. On the same level, Aldus Manutius, undoubtedly with the support of Pico della Mirandola, opened his famous shop in Venice, from which he published innumerable Latin and Greek works. In 1501 he had the first italic type engraved by Francesco Griffo; it was immediately copied by the Lyonese printers. Then there was Joss Bade, a Fleming, who published numerous editions printed by Trechsel in Lyon, then began printing in that city, and finally settled in Paris, where he published a quantity of works, notably with Guillaume Budé, before leaving his enterprise to his son-in-law, Robert Estienne.

Like parallel currents, shop followed shop in Paris and Lyon, all managed successfully and competently by printer-Humanists worthy of their predecessors.

In Paris there was the Estienne dynasty, which grouped around it all the great artists of the day. There were Simon de Colines and Vascosan; there was Claude Garamont, the eminent printer and type founder; there was Geoffroy Tory who in his *Champfleury* laid down the basic laws for the construction of the Latin letter. Lyon had Sébastien Gryphe, the imitator of the editions of Manutius, who launched so many books by Erasmus, Budé, Politien, Sadolet, and Rabelais; Olivier Arnoullet; Robert Granjon, who designed the famous French "official" script; François Juste, who printed the almanacs of Rabelais; Étienne Dolet, who began as a proofreader with Gryphe, then went into the publishing business himself in Lyon and later in Paris, and died on the pyre in the Place Maubert, surrounded by his condemned books. There was Rouillé, whose activity was boundless, and especially the de Tournes dynasty, similar to that of the Estienne family; together with the painter-engraver Bernard Salomon ("le petit Bernard"), the de Tournes family published many illustrated editions and made many strong friendships in the Protestant circles of Lyon. Finally, there is the somewhat isolated figure of Christopher Plantin, working in Antwerp, who made his mark on the art of the book by endowing it with a very special style and developing the art of engraving illustrations in copper.

But a cloud soon over shadowed this magnificient movement: the Humanist printers, compromised in both Paris and Lyon by their more or less secret inclinations toward Protestantism, were to be scattered by severe repression. Many of them left the country, going especially to Switzerland; the Estienne and de Tournes families were among them.

By the seventeenth century the little world of the book had lost a great deal of its audience. Economic crisis was rampant; the innumerable editions published in the preceding century had to a certain extent saturated the book market; the wars of religion, and in Germany wars in general, had taken their toll. Printers were not as much in demand as in earlier times; they were looking for work, and accepted numerous types of jobs. They still welcomed the learned men and the scientists, but protected themselves from the severity of the censure of both the Parliament and the court that threatened them. Moreover, centralization had done its work, and the master booksellers and printers were now concentrated in Paris, near the literary salons, the great orders, and the court. Great names were still being written in the history of publishing: Elzévier, Cramoisy, Léonard, Camusat, Desprez, and, at Lyon, Anisson.

The first half of the eighteenth century differed little from the seventeenth, but during this period the booksellers were increasingly influenced by the appearance of the Encyclopedists' movement, and became "philosophers." Special note should be taken of Le Breton, who appears as a hero since he was at the center of the battle. An editor, and perhaps the inspiration, of the *Encyclopédie*, he sustained the courageous band who through their shrewdness and flexibility in confronting a vigilant censorship successfully completed this undertaking, as tremendous as it was courageous, which several years later was to be the spiritual treasure of an often faltering revolution in thought.

But the closing years of the eighteenth century witnessed the appearance of the first signs of the radical transformation that was to change this aging technique, a technique that was still to distinguish itself, before the great onslaught of mechanization, in the search for a new spirit in lettering. This search was to be led by such remarkable technicians as Baskerville, Bodoni, Caslon, and especially the Didot family, whose name marked the period with the indelible stamp of its energetic activity.

Esthetics of the printed book: format and title page

Before we end this chapter on craft printing, we should establish several points concerning the esthetic evolution that was a determining factor in this technical activity.

We have seen that at the time when the industry of the printed book established itself in a new and particularly active world, esthetically speaking it was only a poor relation of the copying and illuminating "industry" of the Middle Ages. It therefore had to "make a name for itself," particularly since the printers – the "weapons manufacturers of civilization," as Lope de Vega later called them – claimed to be the basic instruments of the new knowledge. They could fill this role only by creating *ab nihilo* an esthetics peculiar to their craft and their art.

The printed book, taking up where the manuscript left off, had in the beginning obliged its pioneers to imitate, insofar as it was possible, the works of the copyists and the illuminators. The manuscript book, in its major compositions, was prepared on parchment, with perfect handwriting, and was enriched with illuminations, ornate letters, and rubrics that made it a rare and precious object. The printed book could not attempt to equal these sumptuous forms, but if it hoped to retain the attention of the high-class clientele it at first sought to satisfy, it had to draw its inspiration from, and imitate, these forms.

For some time the printed book thus remained imprisoned within this obligation to copy both the important and the lesser achievements of the manuscript. The early printers, as we have seen, had received from the copyists a perfectly designed lettering, which they utilized slavishly; they illuminated some of their editions to the point of sometimes attempting to produce counterfeits of the sumptuous manuscript.

In truth, however, the goal of the printed page was not the accumulation on vellum of the riches poured out by the brush of the illuminators, but something very different. The printed book was coming to proclaim to the world the preeminently social nature of a culture that until then had been reserved to the clergy. Thus it was no longer possible to preserve the tradition of the "book as a work of art." The book industry had to deliberately choose the path of the "practical book"; above all else it had to make itself understood, to facilitate reading, which was difficult for some people, in a word to place a new spiritual nourishment within reach of a great number of people. This could not be done without a reduction in all the component elements of the book. The spiritual elements were diminished by the Establishment, on a lower level, of a culture that was tending to become more general. Then the material elements lessened in quality. Less precious, much more common, elements permitted rapid multiplication. Paper made with rag wastes replaced the rich parchment and sumptuous vellum skin. Wood engravings were of very poor quality by comparison with the magnificent miniatures. Easily obtainable printer's ink replaced mysterious applications of gold leaf and the rubrication of titles and large letters. It was necessary to be satisfied with a single color, austere engraving, and a meager support.

This is perhaps the reason why the printed book had a kind of inferiority complex vis-à-vis the beautiful medieval manuscripts.

However, the printed book was already giving even the clergy a by no means negligible advantage: the possibility of making numerous and exact copies. Antiquity and the Middle Ages had suffered greatly from incorrect copies that became increasingly worse as additional copies were made. The Massoretes worked for a long time to correct the errors that had gradually crept into the text of the Bible; the Caliph Othman ordered the destruction of all but one of the manuscript copies of the Koran. For the basic texts the University of Paris ordered the establishment of one copy, carefully verified, on which all later copies were based. After the Council of Trent, Pope Clement VIII ordered the printing of an edition of the Vulgate that was thenceforth to be the only text of the Bible to which reference could be made in questions of faith.

This first advantage was complemented by others that were esthetic in nature. The art of the manuscript was essentially an art of goldsmith's work and the miniature. The artist mounted magnificent enamels in a casket that was itself a precious object. Thus he was not concerned with the ornate borders of his pages, which sometimes invaded the entire surface of the parchment page, leaving no space empty, exactly as would have been done with a piece of goldwork.

But what was needed now was the appearance of a "typographic art," and it was precisely the meagerness of its material that was to be the source of this art. Since it was unable to benefit from intrinsic richness of material, it enveloped itself in more subtle and also more intellectual values: the new esthetics was to be established through the secret, reciprocal grouping of texts and illustrations, the harmonious relationship between the sizes of the letters, spacing, margins, and the size of the titles and subtitles. The book would no longer be a massive *objet d'art*: it was to appear as an architectural structure, an ensemble constructed like the façade of a building.

The title page

This ensemble was complemented, and even intensified, by the appearance of a capital and decisive element: the title page, which assembled in several lines of varied and carefully calculated sizes of letters all the information concerning the book: the name of the author, the title, a summary, the name of the printer or publisher. This was, as we shall see, an extremely important event.

Printing seems in fact to have been the first instance of *systematization* of a craft. In all the other professions, production depended on the needs of the buyer, and to a certain extent on his wishes. Printing required the systematic preparation in advance of a great number of "exactly similar" objects, for which it was then necessary to find numerous buyers. This obliged the sellers to keep the book continuously before the eyes of possible purchasers, and therefore to inform and attract them. What better means of information and propaganda than to place information about the component elements of the book, concentrated in a few lines, on its first page? So the title page appeared — on the one hand an appeal, on the other a brief presentation of the subject matter. It was thus one of the first and earliest manifestations of that form of publicity later called "advertising," which gradually invaded the nineteenth century and overflowed, like a tidal wave, in the twentieth.

It is the title page that enables us to grasp the evolution of the book through the centuries. The first printed title pages present an impersonal face; they are generally reduced to an *incipit,* that is, a simple opening of the text, often decorated with a large allegorical engraving, which seeks prominence only through the larger letters of its opening lines. But by the end of the fifteenth century the genuine title page began to appear, sometimes with a large opening line, or with an enormous grotesque letter from which dragons and fabulous animals surged up, direct descendants of the illuminated initial letters of the Carolingian manuscripts.

For a long time the title page remained imprisoned in this heavily rectangular form; its text was often concentrated within an illustrated frame. Gradually, however, the compositor's hand began to free itself from these restrictions; the title page opened up, discovered a form, began to be constructed around personal ideas. At the same time the complete adoption of Roman type and the Latin letters opened the most enriching perspectives of lightness and grace to the typographical art. However, the Books of Hours, which remained stubbornly attached to the tradition of the religious books, and which were influenced by the Gothic lettering and its derivatives, continued for a long time to utilize the heavy decoration of the ornate frames. This tradition remained alive, and the present-day Missal can be considered its modern successor, with its pages framed by heavy pictorial motifs that date back, via the Books of Hours, to the medieval manuscripts.

The sixteenth century, which was an age of transition, utilized two forms of title pages: one that followed in the medieval tradition (Figure 311), the other free of all encumbering ornamentation, and appearing in the simplicity of an extremely light text whose harmonious construction was dominated by the triangular form. This latter title page can be said to be the embodiment of the triumph of the Roman lettering, to whose service all its elements were adapted: margins, distribution of spaces and size of types, union of capital and small letters, straight and slanted letters, and so on. With a single variety of types, whose casting was still mediocre and whose beards and edges were very ragged, the age of Ronsard succeeded in creating minor masterpieces that have never been equaled (Figure 312).

In the seventeenth century, the age of Corneille, the title page was transformed until it achieved the conception of the monumental format, heavy composition, and a "vastitude" that was somewhat monotonous and at times exasperating. It must be remembered that this was the age of the development of science: books were very often in quarto or in folio. The illustrations, in which copperplate engraving practically eliminated black-line engraving, were not without magnificence; they too were monumental and conscientiously worked with the graver, but lacked inspiration and especially imagination. The title page was burdened with an extremely large text, so large, in fact, that sometimes thirty closely spaced lines were needed to complete it, without obstructing the inevitable, pompous allegorical mark of the printer or bookseller, also engraved by the copperplate method, and which often invaded more than two-thirds of the page (Figure 313). Since the requirements of "advertising" were becoming increasingly imperative, it was necessary to develop the elements of information and advertising

CONSILIA
D. BENEDICTI CAPRÆ
& LVDOVICI DE BOLOGNINIS,

IVRECONSVLTORVM CLARISSI-
morum, in quibus de Teſtamentis, materia Feu-
dali, & Subſtitutionibus ampliſſimè tractatur.

Omnia ſedulò recognita, Summariis, & Indice illuſtrata.

LVGDVNI,
Apud Hæredes Iacobi Iunctæ.

1 5 5 6

FIG. 311 Sixteenth-century title page (early style).

C· PLINII SECVNDI
NATVRALIS HISTORIAE
LIBRI TRIGINTASEPTEM,

A Paulo Manutio multis in locis
emendati.

CASTIGATIONES SIGISMVNDI
GELENII.

INDEX PLENISSIMVS.

VENETIIS,
Apud Paulum Manutium, Aldi F.
M D L IX.

FIG. 312 Sixteenth-century title page (later style).

R. P. JOAN. STEPHANI

MENOCHIJ,

DOCTORIS THEOLOGI,

E SOCIETATE JESU,

COMMENTARIJ

TOTIUS

S. SCRIPTURÆ,

EX OPTIMIS QUIBUSQUE

Authoribus collecti

EDITIO NOVISSIMA,

AB AUTHORE POSTREMUM PERLECTA, EMENDATA,

& variis interpretationibus nondùm in lucem editis aucta:

DUOBUS TOMIS COMPREHENSA.

OPUS SANE THEOLOGIS ET CONCIONATORIBUS

maximè utile & pernecessarium.

TOMUS PRIMUS.

LUGDUNI,

Sumptibus FRANCISCI COMBA, viâ Mercatoriâ,

sub signo trium Virtutum.

M. DC. XCVII.

CUM PRIVILEGIO REGIS.

Fig. 313 Seventeenth-century title page.

to the maximum. In order to lighten the whole ensemble, which had become a genuine title-page-description, color was introduced, reminiscent of the rubrication of the old Bibles; red lines alternated with black lines, but did not succeed in eliminating the sensation of a disagreeable superabundance. Of course, not all the seventeenth-century title pages are constructed along this heavy pattern; there are some that preserved the grace of their sixteenth-century predecessors, but it can be said that the first type strongly influenced this monumental century.

With the eighteenth century, the age of Voltaire, we find a certain preciosity and elegance reminiscent of that of the sixteenth century. In the intense technical movement that characterizes the century of the Enlightenment, and that will be studied in the next volume, there can be seen an increasingly pressing need for new basic materials: lettering, ornamental borders, fillets. In his *Gazette de France,* Theophraste Renaudot had already introduced the innovation of imaginative lettering constructed on a herringbone pattern, and Didot was soon to create new shapes for letters. The entire century was keenly alive to these hidden needs. The illustrator now sought "the episode," and enjoyed enlivening his pages with gracious scenes: the age of Voltaire, smiling with the spirit of dilettantism, delicately peeped from behind its fan at the pleasures of existence, and thanks to the ingenuity of its illustrators was able to unite a delightful imagery with the text. It achieved this goal through the subterfuge of vignettes, that is, decorations, fillets, small floral ornaments, and geometrical figures that when cast at the end of a type could be combined to form light frames, ornamental groupings, and decorative, sometimes geometrical constructions (Figure 314). Fournier the younger was a precursor in this field; he engraved vignettes and (remembering the "fantasies" of the *Gazette de France*) created ornate letters in the style of his vignettes. These letters contributed the play of their inexhaustible variety to the title page, and in this way very homogeneous ensembles could be formed. Under Cochin's influence, entire title pages were engraved by the copperplate method, thereby acquiring greater unity of layout and line.

By abandoning the classic stiffness, the eighteenth-century book appears to have attempted to replace the text in a harmoniously created and adapted décor. These daring experiments, need for variety, and always creative curiosity carried typographical art toward a radical transformation of style, and encouraged the engravers of prints and the type founders to seek new methods that would permit this art to liberate itself from old restrictions. We are approaching the great social movements that were to open a new road to the human spirit and permit all phases of technology to work in a climate of absolute liberty, which was propitious for discovery, to be sure, but fraught with danger. While technology, under the pressure of events, won enormous advantages, the esthetics of the book was, by its discovery of the road to absolute liberty in graphic forms and constructions, to offer vast numbers of craftsmen with varying degrees of skill everything its riotous imagination could conceive. It thus formed the foundation of the great esthetic storehouse from which the nineteenth century was to borrow with so much delectation and so little circumspection.

Here, again, "advertising" found its place: it was during this period that "the printed cover of the book" became obligatory. The appearance of the popular

ARMINIUS,

OU

LES CHÉRUSQUES,

TRAGÉDIE;

TIRÉE

DU THÉATRE ALLEMAND,

Par M. BAUVIN, de la Société Littéraire d'Arras.

Le prix eſt de 30 ſols.

A PARIS,

Chez la Veuve DUCHESNE, Libraire, rue Saint-Jacques, au-deſſous de la Fontaine S.-Benoît, au Temple du Goût.

M. DCC. LXXII.

FIG. 314 Eighteenth-century title page with ornamented letters, vignettes, and fillets.

edition, the obligation constantly to increase speed and lower the price obliged publishers to give up the burdensome operation of binding and offer their books under a printed cover that aroused the passerby's attention without requiring him to open the volume.

The printing press

There exist numerous depictions of printing shops of the sixteenth, seventeenth, and eighteenth centuries; almost all of them feature the printing press, the basic piece of equipment of these book-printing shops, in a prominent place. We are thus able to follow the transformations of this press, which displayed a tendency to become lighter so that the work of printing became easier and faster.

The fifteenth-century press, as we have seen, was mounted in wood. Each of its parts was gradually replaced by metal parts that, because of their strength, lightness, and inertia, made possible an increasingly effective improvement in the operation of the press.

The screw appears to have been the first part to be made of metal (Figure 315); this transformation is attributed to one Danner of Nuremberg, around 1550. But despite this reinforcement of the pressure, the double-action printing method continued to be utilized; surviving pictures of such presses clearly show this feature, which we find in the sixteenth-century presses preserved in the Musée Plantin of Antwerp, and even in the press described in Diderot's *Encyclopédie.* This proves that the venerable double-action press was still in use during the third quarter of the eighteenth century. In the next volume we shall learn to which technicians the invention of the single-action press is to be attributed.

It was undoubtedly around 1572 that someone conceived the idea of adapting to the tympan that brought the sheet in contact with the form a second tympan, a frisket formed by a sheet of parchment stretched in a frame and opened at the emplacement of the pages. This frisket, which was folded over the tympan before the latter was in turn folded over the form, fixed the sheet of paper in place and protected it from dirt that might otherwise be left on it by the inking pads touching the edges of the form as well as the types of the pages. In this way the ink marks that sometimes appeared in the margins were eliminated.

Around 1620 an Amsterdam printer named W. Jansw. Blaeu, a former pupil of Tycho Brahe, maker of precision instruments, made a very important perfection in the old handpress: he adapted a counterweight to the press bar by means of which the platen was automatically lifted, so that the pressman did not have to lift the press bar manually. With this instrument, which was christened the "Hollander" press, he obtained a great flexibility of movement and an excellent impression on the paper, which was still further improved by a mobile device in the center of the platen that when pressed by the end of the screw lessened the pressure without any additional effort (Figure 316).

It was a press of this type that in 1638 was transported from England across the sea by one Stephen Daye to become the first printing press introduced into America.

Cabinet presses

A "cabinet" meant a great deal to the important men of the seventeenth and eighteenth centuries: they were very proud of their "cabinets," which served to house their artistic

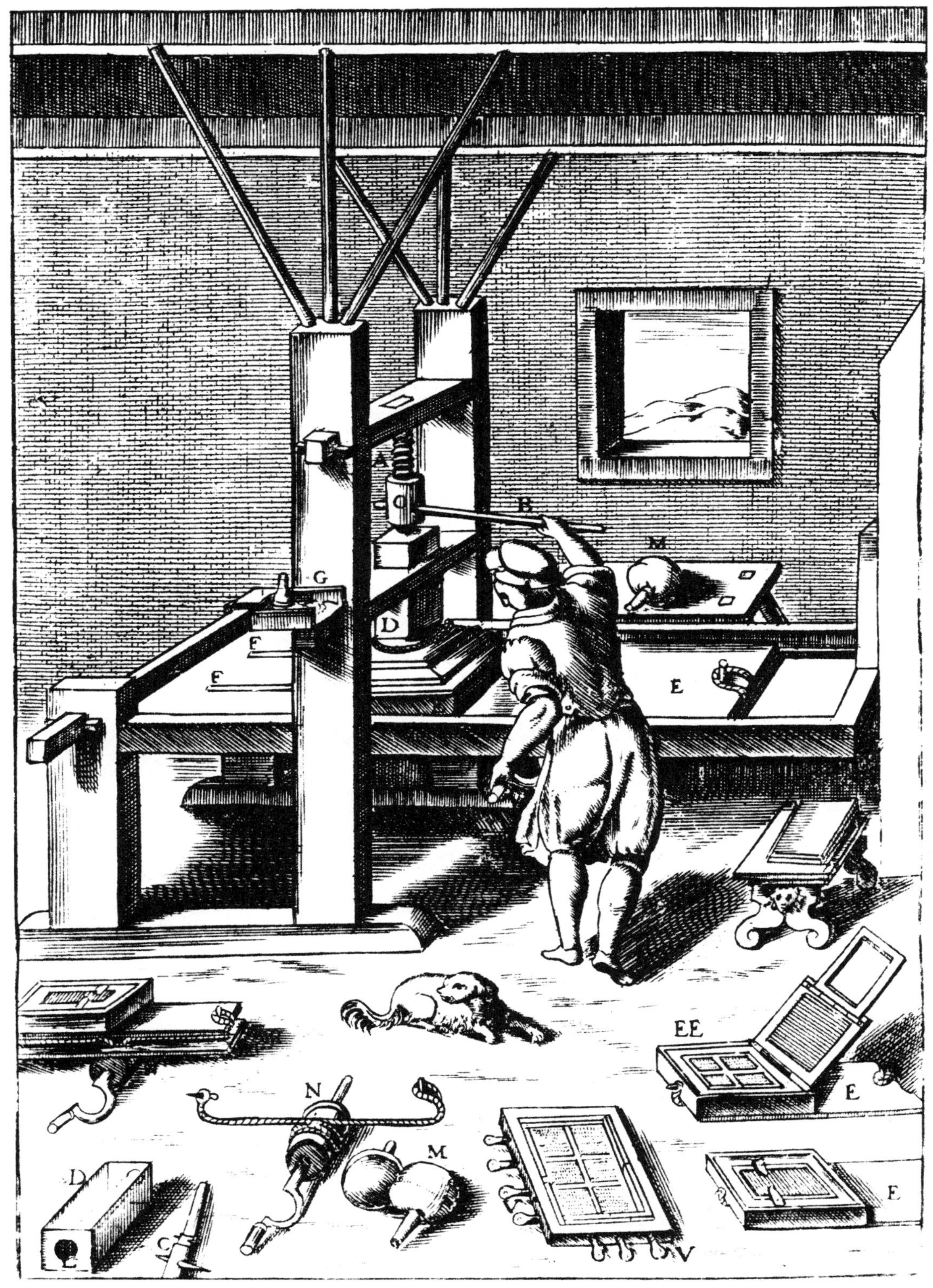

Plate 52. Seventeenth-century printing press. Engraving taken from Zonca, *Novo teatro di machine et edificii. Photo Conservatoire des Arts et Métiers.*

Fig. 315
A sixteenth-century press
(from an anonymous woodcut).

Fig. 316 A printing shop in the eighteenth century.

Plate 53. Press for printing woodcuts. Engraving taken from Zonca, *Novo teatro di machine et edificii. Photo Conservatoire des Arts et Métiers.*

treasures — paintings by the masters, medals, engraved gems, rings, jewels, cartoons for designs, series of prints in their successive stages, rare books, bronzes, marble vases, porcelains, lacquered objects, precious articles of furniture and curiosities. Many of these "cabinets" have been described and studied, especially in the eighteenth century, in very interesting bibliographic manuals.

Some of these "enlightened amateurs," owners of sometimes extremely varied collections, made their cabinets still richer through the installation of small printing presses constructed exactly like those of the printers but on a small scale, on which these men were able to do their own printing without special knowledge of the science of printing. These presses, which are now very rare, were called "cabinet presses" (Figure 317).

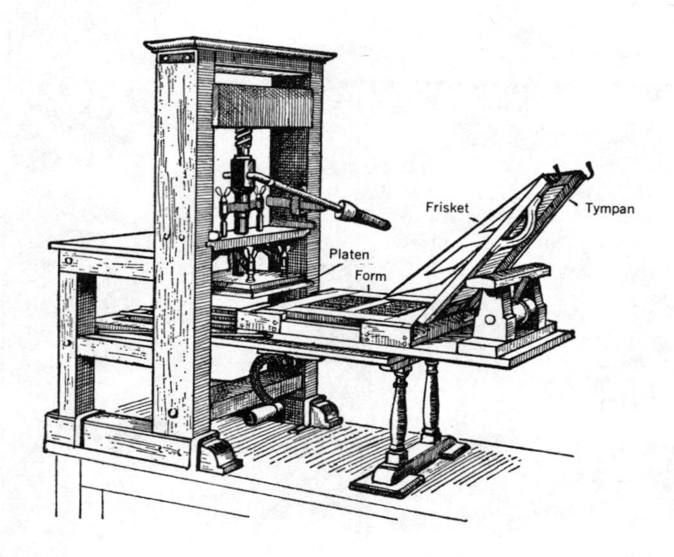

FIG. 317 A "cabinet press," early eighteenth century (Musée lyonnais de l'Imprimerie et de la Banque).

The public's enthusiasm for these machines was very great, and we find an echo of their interest in small installations made available to "enlightened amateurs." An eighteenth-century *Manuel bibliographique* makes an enigmatic offer of a "typographical machine" that was able to print "large books in verse and prose"; it was housed in a secretary-shaped piece of furniture approximately two feet six inches square, and which could be faced with satinwood, rosewood, purple wood, and mahogany. Despite its prolixity, the text does not describe the exact mechanism; the secret was not to be unveiled until after the sale of at least twenty-five, at most one hundred, machines, at the considerable price of 6,000 livres each. The machine appears to have utilized types, but in the manner of intaglio printing — unless we are here in the presence of a small-scale swindle, which is not impossible.

The heralding of a new age

In this chapter we have described the problem of printing as it existed from its beginnings until around the middle of the eighteenth century. This is because the period of three centuries is a complete entity that it would be very difficult to divide into separate parts; this evolution appears firmly unified as well as perfectly logical.

As soon as we reach the middle of the eighteenth century, however, a kind of feverish anxiety appears among the technicians. The age of the mechanical arts was about to be born, and they glimpsed the possibility of utilizing independent motor power; they realized that the new system was to dominate the world of industry, and particularly that of the graphic crafts. This was still only a premonition; the time of the great modern achievements had not yet arrived. In the meantime technicians occupied themselves with minor problems: ease in handling the finished pages, possibility of preserving the elements used, presses with revolutionary shapes better adapted to their functions, the search for more abundant raw materials and more rapid production of paper, and so on.

We shall undoubtedly be astonished by the technical poverty of some of the inventions that were born in this period, but we shall also witness the appearance of several discoveries which though at the time did not appear to be decisive nevertheless paved the way for the great mechanical discoveries the nineteenth century was to promote and the twentieth century to see fully developed.

BIBLIOGRAPHY

ALDIS, H. G., *The Printed Book* (2nd ed., Cambridge, 1941).

BERRY, W. TURNER, and POOLE, H. EDMUND, *Annals of Printing: A Chronological Encyclopaedia from the Earliest Times to 1950* (London, 1966).

BLAND, DAVID, *The Illustration of Books* (3rd ed., London, 1962).

BLUM, ANDRÉ, *On the Origin of Paper* (New York, 1934).

——, *The Origins of Printing and Engraving* (New York, 1940).

BUTLER, P., *The Origin of Printing in Europe* (Chicago, 1940).

CARTER, THOMAS F., *The Invention of Printing and Its Spread Westward,* revised by L. Carrington Goodrich (2nd ed., New York, 1955).

GOLDSCHMIDT, ERNEST P., *The Printed Books of the Renaissance* (Cambridge, 1950).

GOODRICH, L. CARRINGTON, "Printing: Preliminary Report on a New Discovery." *Technology and Culture,* 8 (1967) 376–378.

HUNTER, DARD, *Papermaking: The History and Technique of an Ancient Craft* (2nd ed., London, 1957).

JENNETT, SEAN, *Pioneers in Printing* (Fair Lawn, N.J., 1959).

LEHMANN-HAUPT, HELLMUT, *Gutenberg and the Master of the Playing Cards* (New Haven, Conn., 1966).

McMURTRIE, DOUGLAS C., *The Book: The Story of Printing and Bookmaking* (3rd ed., New York, 1943).

UPDIKE, D. B., *Printing Types: Their History, Forms and Use* (2nd ed., Cambridge, Mass., 1937).

WROTH, LAWRENCE C. (ed.), *A History of the Printed Book* (New York, 1938).

INDEX

A

B

C

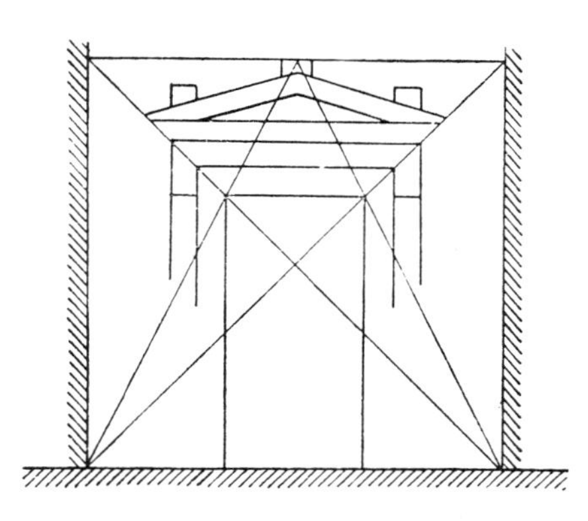

D

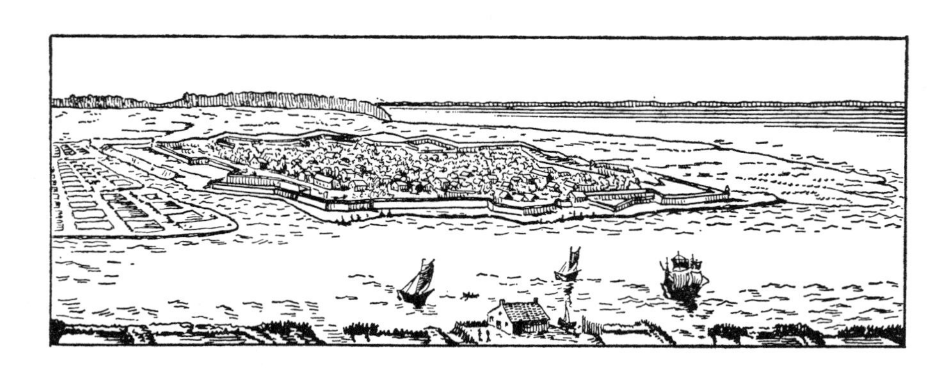

E

F

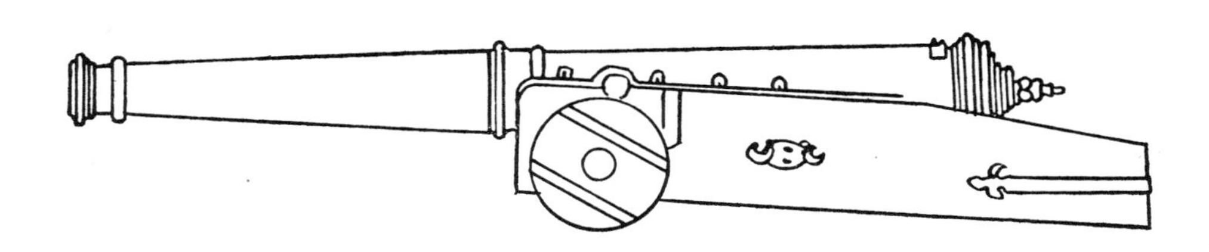

G

H

I

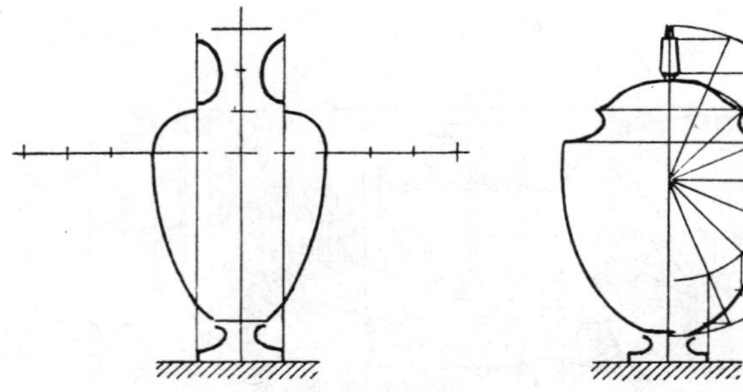

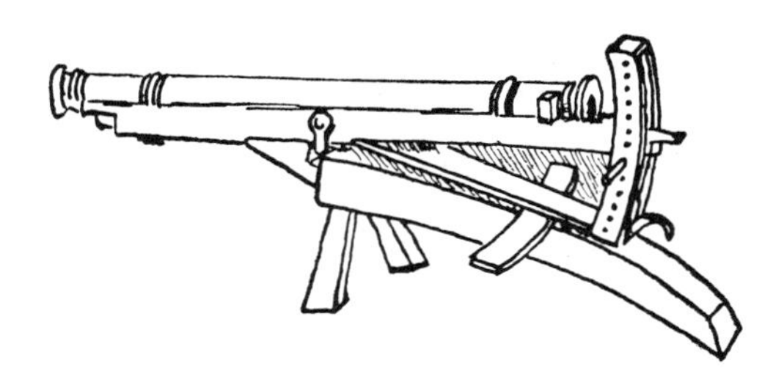

M

N

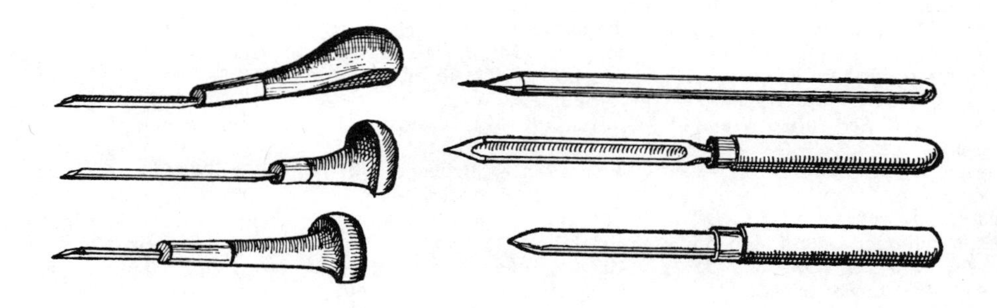

Q

R

S

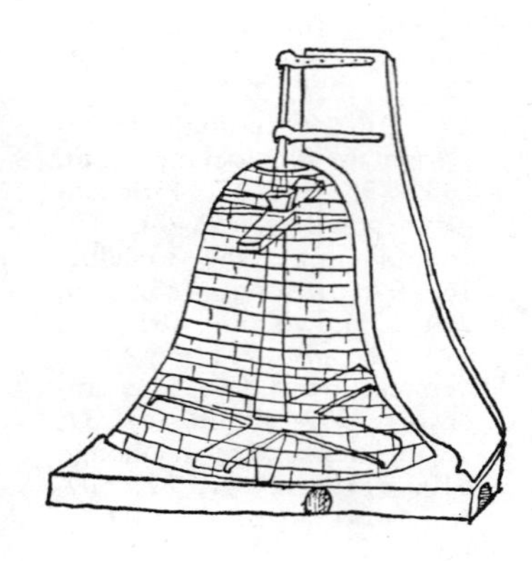

T

U

V

W